# WILTSHIRE BEETLES

Michael Darby

# WILTSHIRE BEETLES

## History, Status, Distribution and Use in Site Assessment

Malthouse Books
in association with the
THE WILTSHIRE NATURAL HISTORY
PUBLICATIONS TRUST

First published in 2009 by
Malthouse Books
Sutton Mandeville, Salisbury SP35LZ
admin@malthousebooks.co.uk



British Library Cataloguing in Publication Data
A catalogue record for this book is available from the British Library

ISBN 978-0-9558506-1-5

Printed in Great Britain by the Choir Press, Gloucester

Cover ilustrations: Notman Wildlife Archive

Front cover: *Rutpela maculata* Poda

FOR LIS

Who has put up with the beetles for more than 30 years without complaint

**Plate 1. Specimen of *Callidium violaceum* from the collection of Marlborough College , probably one of those which destroyed the Headmaster's summer house in 1896 by 'burrowing through the wood in all directions'. It has not been recorded in the county again and only twice in Great Britain since 1970.**

# CONTENTS

# LIST OF PLATES



# LIST OF TABLES



# LIST OF MAPS

# *AUTHOR'S NOTE*

This book is not aimed at the amateur naturalist wanting to study beetles for the first time as the profusion of Latin names and complex data will quickly reveal. Instead, it targets professionals, serious students, and those wanting to follow up information already gleaned from field guides and similar sources.

Beetles are not easy for the amateur wishing to determine specimens to the species level. Field guides lack sufficient information, there are few English names and, since the genitalia are one of the main means of distinguishing species, dissection is often a necessity. This will mean the use of a stereoscopic microscope and, if slides are required, a compound microscope too. In addition, access to up-to-date text books and runs of journals, together with a good reference collection, are important.

A study of Coleopterists' biographies over the last 150 years suggests that the majority acquired the interest as children and that was certainly my experience. My mother used to say that I ate one as a toddler and 'that was what did it'!

That said, the study of entomology remains one of the few scientific disciplines in which the amateur can make an important contribution. The larval and pupal stages of many of the British beetles are still unknown and large parts of Wiltshire are still without any records, even of the commonest species.

My advice to anyone brave enough to want to try is to begin by specialising in a single family. If the large ladybirds were chosen, for example, identification is easiy with a hand lens and there is an authoritative and cheap guide book, *A Year in the Lives of British Ladybirds* by Michael Majerus, Remy Ware and Christina Majerus, recently published by the Royal Entomological Society (www.royensoc.co.uk), and a much more detailed but fascinating book in the Harper Collins, New Naturalist series, by Michael Majerus, *Ladybirds* (1994). These, used in conjunction with the *Coleopterist's Handbook*, (email aespublications@btconnect.com), as a guide to collecting methods, storage, etc., will be perfectly adequate to get started. Who knows, perhaps you might discover the still unknown habitat of the striped ladybird in winter...? Other families and more species will quickly follow, but beware - once hooked, beetles can easily take over your life!

**Plate 2. View on the Hamptworth Estate where David Nash did much of his collecting.**

# ***ACKNOWLEDGEMENTS***

My greatest debt of gratitude must be to David Nash. He has not only allowed me to include all of his almost 5,000 Wiltshire records and provided generous support throughout my time as County Recorder, but also willingly undertook the onerous task of reading the manuscript. The many comments he made have added greatly to the accuracy and integrity of the book. I also owe thanks to Mike Morris and Keith Alexander for their helpful comments on different sections of the manuscript, to Andrew Duff for general advice, and to Miles Bailey at the Choir Press for his unfailing patience and professionalism in dealing with my many queries and problems.

To Peter Hodge I owe enormous gratitude for providing me with copies of the several reports he has compiled while working on Salisbury Plain, Stonehenge and the A303 widening at Stockton.

Ashley Kirk Spriggs kindly typed up all the late Alan Easton's Meligethes' records, Garth Foster provided a print-out of all the water beetle records on the national database, as did Mark Telfer for the ground beetles and Scotty Dodd for the Tenebrionidae and related families. Mark also allowed me to include all his personal ground beetle records as did Darryl Rush. David Boyce kindly rushed me a copy of all the records he and Steve Wright gathered during their recent survey of Salisbury Plain.

Several Coleopterists from outside the county have sent me lists of their Wiltshire records including Keith Alexander, Tony Allen, Max Barclay, Andrew Duff, Adrian Fowles, Martin Harvey, Brian Levey, Michael Salmon, and Malcolm Storey. Within the county I am grateful to several enthusiasts who have regularly forwarded me notices of their findings. These include Ian Gray, Dorothy Graiff, Humphrey Kay and Barbara Last.

Numerous others whose initials appear against the species' entries have provided shorter lists, records of individual species, and specimens for me to identify.

Several individuals have sent me material which they have kindly put aside when working on other groups; these include John Grierson, Steve Covey, Mike Fuller, Martin Askins, Godfrey and Michael Smith, Dave Brotheridge, and Jack Oliver.

To the members of the Wiltshire Archaeological and Natural History Society and the Bath Naturalists I owe thanks for loyally accompanying me on several field trips and providing records. To the Wiltshire staff of Natural England I also owe thanks for their help with contracts, site access, transport and providing access to historical material in their archives.

Amongst my many friends at the Wiltshire Wildlife Trust I owe a special debt of gratitude to Piers Mobsby who has generously provided me with numerous lists of beetle records made during his work on the Trust reserves, and to Michael New who has kept me up to date with his Cerambycid recording. I must also thank my former colleagues, in the Wiltshire and Swindon Biological Records Centre for their help and support over the 16 years during which I have been county recorder. Also my family who have willingly put up with my 'odd' interest, accompanied me on many site visits, and helped with setting up pitfall traps and subsequent monitoring.

All the illustrations are my own with the exception of those of Savernake by Joan Davies and some of the species photographs, particularly those of Staphylinidae, for which I have to thank Marc Tronquet, author and photographer of the excellent *Catalogues Iconographique des Coléoptères des Pyrénées-Orientale* (2006) and his publishers. I am also grateful to Prof. Lech Borowiec of the University of Wroclaw, Poland, for permission to use his equally excellent photographs of Chrysomelidae (www.biol.uni.wroc.pl/cassidae/European Chrysomelidae/Iconography/)

Finally, I owe a huge debt of gratitude to the many landowners in Wiltshire who have allowed me to record on their properties.

**Plate 3. Salisbury Plain looking west from Breach Hill.**

# INTRODUCTION

Although several detailed lists of beetles in the Marlborough area were published in the period 1874-1938, this is the first book devoted to Wiltshire's beetles as a whole. Because it does not discuss their biology or ecology it cannot be said to be definitive. Rather, it sets out to remedy the county's oft-quoted reputation as being one of the most poorly recorded in Britain. The bulk of the text takes the form of a list of species compiled from more than 20,000 records, covering the period 1828 to the present, gathered since I was appointed county recorder in 1993. Because it is the first comprehensive account of these fascinating insects I have felt justified in including more historical material than most contemporary entomologists would expect. This has meant listing not only species which have become extinct in modern times, but also what fossil material is known. The book is better regarded, therefore, as a compilation, rather than a catalogue or checklist, of the county's species.

In the introductory remarks to the systematic list I have outlined the history of collecting in the county, listed the principal sites on which records have been made, and identified species considered as nationally rare; but it is to the accounts of the so-called Indicator species that I wish to draw particular attention.

Looking through the reasons for the notification of many of the 100 or so sites graded as SSSIs in the county for their ecological, as opposed to geological, importance, one is immediately struck by the dependence placed firstly on botanical criteria and secondly on butterflies and birds. These are very important, of course, but have limitations which are not always appreciated. Plants, for example, are particularly vulnerable to pollution both atmospheric and through the application of chemical sprays, and many have airborn seeds enabling dispersion over wide areas. Butterflies and birds are very mobile too, so that they are frequently recorded on sites where they are not breeding. Beetles on the other hand are surprisingly resilient to pollutants, partly because of their tough external skeletons but also because of their secretive habits. Although most possess wings, the majority are more static than butterflies or birds. Many wingless ground beetles are incapable of dispersion over wide areas and often remain on the same site for generations.

It is hardly surprising, therefore, that beetles are being used more and more in site assessment and management planning. For both woodland and grassland lists of so-called 'indicator' species have been compiled, the presence or absence of which are guides to site importance. This book attempts to move this process forward by 'scoring' the grassland list as has already been done for the saproxylic species.

### Recording area

Wiltshire is defined as comprising the two vice counties 7 and 8 (north and south Wiltshire respectively) in the Watsonian vice-county system (Dandy, 1969). This differs in several respects from the modern county. In the south the boundary is extended into Hampshire to include Martin Down and parts of the New Forest around Bramshaw and Nomansland, and, in the north, it encompasses parts of Gloucestershire almost up to Tetbury and Cirencester. In the east small parts of the county to the north of Highworth and Ham Hill are lost to Oxfordshire and Berkshire respectively, and in the west there are two small losses to Dorset immediately to the west of Whitesheet Hill. The boundary between north Wilts (VC7) and south Wilts (VC8) follows the line of the Kennet and Avon canal.

### Information sources

The entomological literature and related scientific publications have been searched, as well as that of local organisations and societies.

Individual amateur and professional entomologists have been approached for records and several local collections checked for determinations and data. Web-based support sites have been widely used and the national databases of certain beetle families searched for Wiltshire references. Reports made by professional entomologists for public and private bodies, including the Department of Transport, Salisbury District Council, Ministry of Defence, Wiltshire Wildlife Trust, Nature Conservancy Council, English Nature, Natural England, and the Institute of Terrestrial Ecology have also been used extensively.

The Wiltshire and Swindon Biological Records Office at their offices in Devizes maintains a computerised database of all Wiltshire Coleoptera compiled by myself as County Recorder, which includes much information for which there was not space in the inevitably abbreviated entries in this volume.

I am aware that I may be criticised for omitting material on the National Biodiversity Network, but at the time of completing my research several important data sets now included were not present, and some that were present lacked much of the detailed information which I required.

**Plate 4. View from the top of Wylye Down showing typical rolling downland of southern Wiltshire.**

**Plate 5. Old oaks and cattle in Savernake Forest (Joan Davies).**

# THE HABITATS OF WILTSHIRE

Wiltshire covers some 348,000 hectares making it the largest county in southern England. The major part comprises chalk downland in the form of Cranborne Chase and Salisbury Plain, which occupy most of VC8 in the south, and Pewsey and Marlborough Downs, in the central and eastern parts of VC7 in the north. Where the ground permits, much has been turned over to the growing of arable crops, formerly cereals, but now oil seed rape, linseed and maize are common. On slopes too steep for cultivation, grassland still predominates, some unimproved, but others much improved by fertilisation in the 1960s and 70s; here cattle and sheep graze.

'Chalk and cheese' is often said to describe the difference between these rolling downlands with their river valleys and large vales where older rocks and greensand are exposed, and the flatter pasture lands, most evident in the north.

In the far north west of the county is a small area of oolitic limestone forming the southernmost extension of the Cotswolds. Here some unimproved grassland also survives. And in the far south west, where the county boundary is extended by the VC8 perimeter, is an area of New Forest acid heathland with a rich beetle fauna including many species not found elsewhere.

The county is drained by three major river systems. In the south virtually all the streams and rivers run into the Avon. The main tributaries are the Ebble, Nadder, Wylye and Bourne which join it near Salisbury. In the west, another Avon is fed by the Mardon, the Gauze Brook and the By Brook, and in the north east more streams run into the Kennet which in turn flows into the Thames. The watershed between these systems is on the Marlborough Downs where Tan Hill and Milk Hill at 295m above sea level are Wiltshire's highest points.

Our largest area of standing water is the Cotswold Water Park in the far north, much of which is in VC7 but not in Wiltshire. Like Langford Lakes (the Wiltshire Wildlife Trust Reserve in the south), the lakes and ponds were fomed as a result of mineral extraction in the last century

Other wetland areas occur in the form of the extensive water meadows which lie alongside many of the rivers. The best known are the Wiltshire Wildlife Trust's reserves at Smallbrook Meadows and Jones's Mill near Pewsey. Two others, traditionally managed meadows on the River Avon near Salisbury at Britford and Lower Woodford, have recently been the subject of an invertebrate study (Drake, 2008). Knowledge of how the water meadows were managed to provide early grazing, especially for ewes prior to lambing, is now largely lost, as are most of ditches, gutters, carriers and hatches. Further examples of other rare wetland survivals in the county are a few reedbeds such as that at the Avon Valley Reserve in Salisbury, and the unique lowland bog at Landford.

Research into Roman and Mediaeval forests in the county makes clear that much larger areas than now were once wooded. The Braydon Forest ran from Zeals, (the name of which derives from Selwood Forest) up the western boundary to embrace Melksham and Chippenham, and then across the north to Purton, but is now extensively cultivated. A second great tract of forest ran up the eastern side of the county from Bentley and Langley woods in the south, through Clarendon and Chute to Savernake, and a third belt, of which Grovely is a survivor, ran from east to west, south of the Plain.

Several of these woodlands including Savernake were once Royal Forests and the county could also boast others at Cranborne Chase and Buckholt. Many such woodlands were managed to allow grazing by animals and the branches cut to a height of 2-3m above the ground to protect the trees. These pasture woodlands, the presence of which is often revealed by the use of 'ley' and 'leigh' (deriving from the anglo saxon 'lēah') in their names, allowed in light and were thus warm which attracted rich invertebrate faunas including beetles. Now, various trials are taking place with a view to returning to this form of management in parts of Savernake, Langley and elsewhere.

Coppice with standards, another form of woodland management, was in decline throughout much of the last century, but has seen a revival recently on various reserves. Like pasture woodland, coppicing also lets in light and warmth which encourages a richer flora and greater biodiversity, before regrowth of the understorey reintroduces shade.

In the last half of the 20th century the Forestry Commission acquired large areas for the planting of commercial timber crops. Many of these suffered badly in the great storms of October 1987 and January 1990. The latter, in particular, damaged trees on the Plain and Savernake, but many older conifers do still survive. Now the Commission's Woodland Grant Scheme offers good incentives to plant broadleaved trees.

The 1987 storm brought down many old deciduous trees in the south of the county, particularly beeches growing on shallow chalk soils. Elms, too, suffered badly at this time, not so much from the storms, as from Dutch Elm disease carried by the beetles *Scolytus scolytus* and *Scolytus multistriatus*. It is estimated that the county lost 683,000 trees through the disease at this time many of which were prominent features in the countryside, as avenues and in hedgerows.

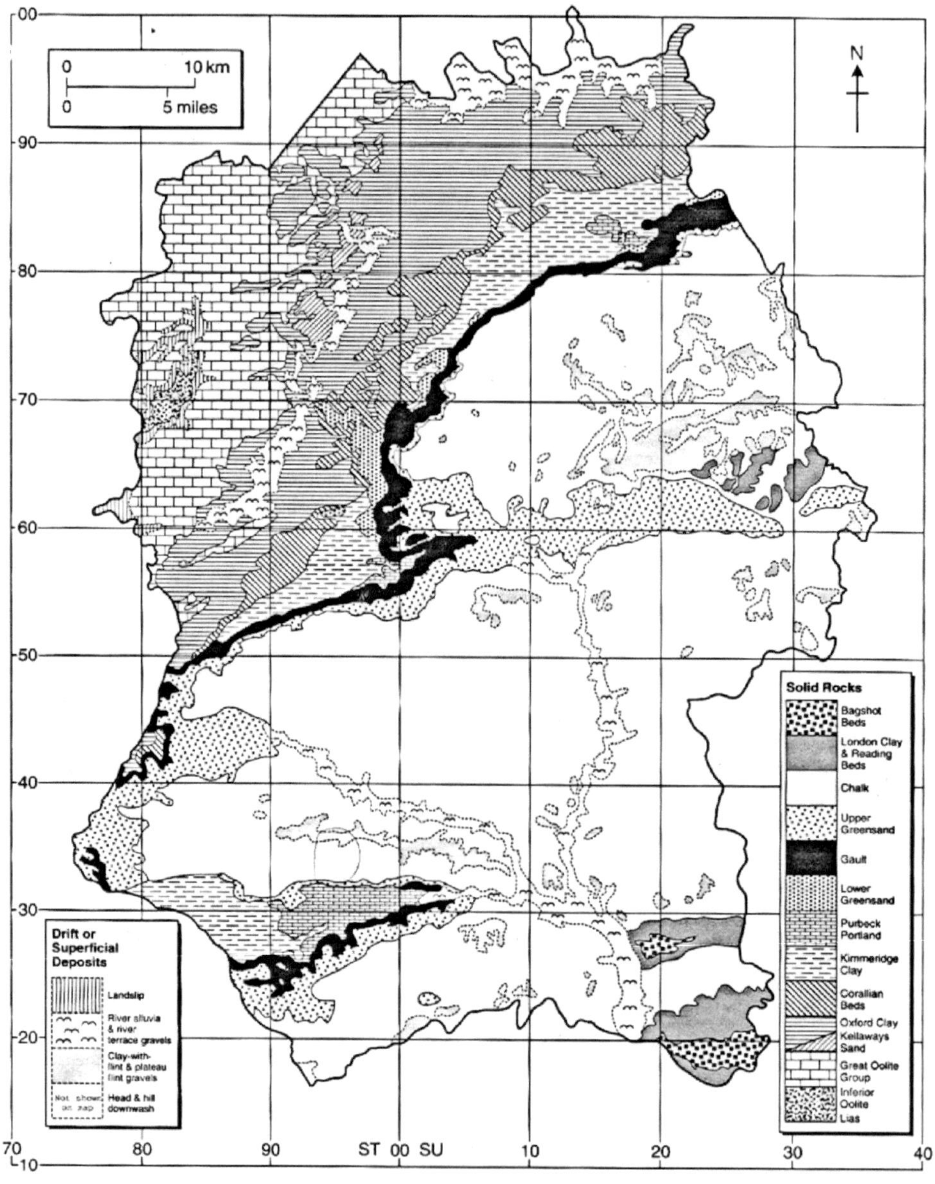


**Map 1. Geological map of Wiltshire (modified from British Geological Survey 1:250.000 Series forSolid rocks and 1:50,000 and 1:360 Series for Drift deposits.**

But the landscape which impresses itself most firmly on the mind of anyone travelling through the centre and south of the county is the chalk downland. McSweeney (1991) calculated that there were 16,165 hectares of calcareous grassland in Wiltshire, including the Ministry of Defence areas on Salisbury Plain and Porton Down, and that this represented 55% of all that remained in England. It is hardly surprising that all 3 Areas of Outstanding Natural Beauty (AONB), 6 of the National Nature Reserves (NNR) and more than 50 Sites of Special Scientific Interest (SSSI) in the county have all been designated on the basis of their calcareous grassland quality.

Chalk grassland develops very slowly and floras such as those present on the best sites may take several hundred years of suitable management to form. Therefore, damage such as that created by ploughing and the drift, or run off, of high levels of fertiliser, pesticides and herbicides from adjacent areas, cannot be reversed in the short-term.

Undoubtedly the most crucial factor in effective management is the control of grazing. Without the short springy turf which can include as many as 30 or more plant species in a square metre, calcareous grassland would convert firstly to scrub and then to woodland. Such grazing may be done by rabbits as well as livestock, although overgrazing by rabbits can lead to the development of different grassland types with longer swards. Overgrazing by cattle can also cause problems, particularly resulting from erosion of the steeper slopes.

Anthills are a characteristic feature of many of the unimproved calcareous grassland sites and provide a warm microclimate encouraging the growth of plants such as wild thyme and common rock rose.

Neutral lowland meadows in the county were surveyed by Wells and King in 1975. As a direct result several which were unimproved were designated as SSSIs, three were purchased by the Wiltshire Wildlife Trust as reserves, and one, North Meadow at Cricklade, was made a NNR. At North Meadow the hay is cut with mechanical cutters rather than scythes, and another change from the past is the use of a vacuum harvester to collect the seeds before the hay is mown. Interestingly, the traditional haystack, was an important habitat for beetles as the early literature shows.

**Plate 6. Martin Down NNR. Some of English Nature's pitfall traps were placed in the centre of this picture.**

Plate 7. Tony Allen's flight intercept trap in Knighton Wood, February 2009.

Plate 8. Pitfall trap covered with chicken wire to keep out small mammals. The earth around has been mined by dung beetles.

Plate 9. Line of pitfall traps in the Outwood area of Langley Wood NNR.

# *HISTORY OF BEETLE RECORDING IN WILTSHIRE*

## Before 1945

The first beetles recorded from Wiltshire are those of James F. Stephens in his two publications *Illustrations of British Entomology* (1828-1832 Coleoptera parts only) and *A Manual of British Coleoptera* (1839). These volumes list thirty one species, the majority found by the **Rev. G.T.Rudd** from Winterbourne Stoke in the south of the county. Because of name changes and the absence of specimens to confirm identifications, the status of some of these records is uncertain. The Rev. Rudd remains a shadowy figure but was clearly serious about his Coleoptera collecting and is recorded to have been both a friend and field companion of Stephens. *Medon ruddii* (now *M. castaneus*), found by Rudd in the New Forest, was named after him by Stephens.

The next records are those known only as fossils from the Purbeck and Wealden Groups which were laid down between 120 and 20 million years ago. These were studied in detail by **Rev. Peter Bellinger Brodie** (1815-1870), the author of *A History of the Fossil Insects in the Secondary Rocks of England* (1845), the first book on the subject. Brodie's Wiltshire finds, made when he was curate at Wylye between 1838-1840, have recently been reviewed by Ross and Jarzembowski (1996) who list 19 species of beetle (plus 3 as *incertae sedis*) all found in deposits in the Vale of Wardour mostly at Dinton and Teffont (see list on p.45). Many of these specimens, which are now in the Natural History Museum in London, are unique types (ie no other examples of the species are known either as fossils or as non-fossil material) and all are now extinct, presumably wiped out by the encroachment of the sea at the onset of the late Lower Cretaceous. The Dinton and Teffont quarries have been designated as SSSIs .

The establishment of periodicals such as the *Entomologist's Annual* in 1856 and the *Entomologist's Monthly Magazine* in 1864, heralded the publication of several articles on the county fauna by well known Coleopterists living elsewhere. In 1870 **Edward Wesley Janson** (1822-1891) published 'Doings in Wiltshire' in the *Entomologists' Weekly Compendium*, for example, and in the following year **Robert Hislop** (1815-1880) sent to the *Entomologist's Monthly Magazine* some 'Notes on Wiltshire Coleoptera'. But, much the most interesting of these early records were those by the Manchester collector **Joseph Sidebotham** (1823-1885). At a meeting of the [Royal] Entomological Society on 7 November 1864 Janson showed (on Sidebotham's behalf) two weevils, *Sibinia pellucens* and *Lixus elongatus* which he had taken near Devizes in June of that year. Both were new to Britain and neither has been recorded since. A ground beetle, *Lebia marginata*, found by Sidebotham in 'a wood near Devizes' was also published at this time (as *L. haemorrhoidalis*) by John Power, a well-known Coleopterist to whom he had sent it. Five years later another weevil, *Baris chlorizans*, also found near Devizes, was published by William Sharp, without a date of capture. Like *S. pellucens* and *L. elongatus*, neither of these insects has been taken in Britain since. All four are, however, common on the continent which it is known that Sidebotham visited frequently at this time.

The three weevil records, along with other doubtful weevil records by Sidebotham, are discussed in detail by Morris and Johnson (2005). who came to the conclusion, reluctantly since Sidebotham was a respected civil personality in Manchester, that the records are probably fraudulent, but whether deliberately or not we may never know.

By far the largest number of records to appear at this time were not in the national press but in a local periodical, the *Report of the Marlborough College Natural History Society*, which began appearing, at first every six months and later annually, in 1865.

The Marlborough College Natural History Society was founded on 9 April 1864 by three pupils with the help of the Rev.T.A.Preston, a master, who had earlier been involved in setting up the Wiltshire Archaeological and Natural History Society. Quickly nicknamed the 'Bug and Beetle', the new society soon attracted an enthusiastic following. One of its members, **Edward Meyrick** (1854-1938), would eventually become one of Wiltshire's greatest and yet, today, almost unknown, naturalists.

Born on the 25 November 1854, the son of the Rev. Edward Meyrick, Vicar of Ramsbury, he became a pupil at the College in 1869 and almost immediately joined the Society. Already his interest was firmly fixed on the Microlepidoptera, and by starting his entomological activities at 4.00 am and working until 7.00, before the school day, he quickly built up substantial collections. So much so that in 1874 he was able to publish lists in the society's annual report which included no fewer than 800 species of moths and butterflies found in the neighbourhood of the school. Not content with this he also applied himself to other groups including Coleoptera, and in the same *Report* published a list of beetles 'by far the largest number of [which]...are due to Mr Janson, who spent a week down here collecting in June, and kindly consented to make up a list of all the species he came across, for our benefit'. Of the 540 species named, 154 were collected by Meyrick himself. This was the county's first known beetle list.

Whether Janson's coming down to Marlborough was at Meyrick's behest is not clear, but it set a trend for involvement with the London collectors which was to considerably benefit the College. In the same *Report* Meyrick recorded that 'Mr **E[dward] C[aldwell] Rye** [1832-1885] kindly offered to name any species that we might send him, and has named various ones for me accordingly'. In 1880 **Rev. William Weekes Fowler** (1849-1923), author of the six volume *Coleoptera of the British Islands* (1887-1891) and one of the editors of the *Entomologist's Monthly Magazine*, recorded *Hydroporus obsoletus* 'among some beetles sent to me for determination from Marlborough'. Thirteen years later Fowler noted another *Hydroporus, marginatus*, 'Among some beetles sent to me to name by **Mr T.N.Hart-Smith**, of Marlborough College.' Later still **James H. Keys** (1855-1941); **W.A.F (Frank) Balfour-Browne** (1874-1967), who lectured to the boys on entomology, and **Kenneth Gloyne Blair** (1882-1952) are also recorded to have determined specimens.

**Plate 10. Portrait of Edward Meyrick (1854-1938)**

After leaving Marlborough Meyrick went up to Trinity College, Cambridge before, in 1878, travelling to Australia and New Zealand where he worked as a teacher of classics. He returned to England in 1887. Throughout his nine years abroad Meyrick continued to work on the Lepidoptera, specialising in the micros, on which he eventually published substantial lists including many new species. Later he would take up South Africa and other countries too, such that by the time of his death he had built up a collection of more than 100,000 specimens (now in the Natural History Museum, London) and described 20,000 new species and more than 1,500 new genera.

Back in England in 1887 Meyrick rejoined Marlborough as a master and quickly took an active role in the Natural History Society. During his absence no beetle recording had been carried out but he quickly encouraged interest in a number of boys and 'over the next four years a good deal of collecting took place'. This eventually led in 1895 to **Arthur Gamul Jebb** (1880-1896) publishing the 'Coleoptera of the Marlborough District', which increased Meyrick's first list to 775 species, although Jebb admitted that this number was 'far short of what it ought to reach, as is clearly shown, by **G[eorge] C[harles] Champion**, [1851-1927] Esq. having added, during a short visit in the spring, no less than 21 new species'.

Tragically Jebb, who also acted as the Society's librarian, died of pneumonia on 11 March 1896, aged 16. In writing his obituary Meyrick made clear that he had lost not just a pupil but a friend, and he complimented Jebb's 'habit of energetic work, ...singular independence of thought', and 'wide and intelligent reading'.

Further collecting at Marlborough led to the publication by Meyrick of a third list in 1905 taking the total to 819 species. Interestingly, this was being compiled at the same time that he received Fellowship of the Royal Society (1904) for his work on Lepidoptera, at that time a unique distinction for a school master.

It should also be mentioned that Meyrick's work for the Society, of which he was now President, extended to much else besides entomology. He acted as leader of many of the sections, and compiled and wrote the bulk of many of the Society's reports, his contributions ranging from annually recording and publishing the height, weight and chest measurements of every boy in the College, to notes on a wide range of subjects from wood carvings to the weather, and to the compilation of lists of plants, birds, mammals and other insects in the Marlborough area. In much of this he was successful not just by dint of persuading the boys, masters and their families to join the Society, but also by contributing a large number of records himself, many observations being made, no doubt, whilst on the 3 to 4 mile daily walks which he continued throughout life

Meyrick retired from teaching in 1915 to pursue his political interests – he was President of the East Wiltshire Unionist society for 25 years – and his work on his favourite microlepidoptera. But he also found time to continue his involvement with the Natural History Society. 'His visits', one of the members later recalled, 'were almost always made on one of his evening walks. We would hear his stick on the stairs and wonder what he was bringing us this time, information or insects, for he never came empty handed.' And another wrote that 'a boy would present himself on a Sunday afternoon [at *Thornhanger*, Meyrick's home] with a box of specimens and might remain for hours; after which, sure of his welcome, he would soon become a regular visitor'. It was quite likely he would meet a distinguished entomologist from London or abroad, for Meyrick rarely left Marlborough, even when younger, always preferring other specialists to come to him.

HANDLIST OF

THE COLEOPTERA OF THE MARLBOROUGH DISTRICT

The entry for each species consists as far as possible of:

Its NUMBER in the *Catalogue of the recorded Coleoptera of the British Isles* (1930 edition) by Sir Thomas Hudson Beare.

Its SCIENTIFIC NAME from the same catalogue. Where the name in Fowler's *British Coleoptera* is different, it is given in brackets.

LETTERS to show in which of the previous lists of Coleoptera it was recorded.

X = 1873 by Mr E. Meyrick.
Y = 1895 by A. G. Jebb.
Z = 1905 by Mr E. Meyrick, F.R.S.

YEARS in which information other than the mere record of capture was given in the *Report*.

The words COMMON or ABUNDANT for those few insects where the evidence seems to justify it. LOCALLY has been added where there is sufficient evidence.

LOCALITIES where the insects which do not appear common have been found. If an insect has seemed common only in certain localities but there is insufficient evidence to put LOCALLY COMMON, localities as well as the word COMMON are inserted.

Localities given usually refer to more recent captures: earlier collectors did not record their data.

Brief notes on its FOOD, FAVOURITE RESORT, etc. taken from the standard books on Coleoptera. We hope that future collectors will find these notes useful. They are strongly recommended to search many types of locality (see the chapter on this subject in *British Beetles* by N. H. Joy) and to consult our *Handlist of Flowering Plants* for the places where local food plants of weevils, etc. are to be found.

HANDLIST OF COLEOPTERA OF MARLBOROUGH DISTRICT 55

The first handlist of local Coleoptera was published in 1873 by Mr E. Meyrick (*M.C.N.H.S. Report for the half-year ending Christmas* 1873, p. 41). Very little collecting had been done before that and the records were mostly those of Mr Janson made during a visit in 1873. Very few localities were inserted but Mr Janson did the bulk of his collecting in the Forest, in the water meadows and at Savernake; species marked in this list with x were therefore probably found in one of these places. The number of species recorded in this list was about 540.

No Coleoptera were recorded between 1873 and 1879. Between 1879 and 1883 there was a good deal of collecting and many species new to our district were recorded.

In 1895 a revised list was published by A. G. Jebb (*M.C.N.H.S. Report for* 1895, p. 55). The total had risen to 775 and more localities were added.

Further collecting between 1900 and 1905 ended in the production of a new handlist by Mr E. Meyrick, F.R.S. (*M.C.N.H.S. Report for* 1905, p. 76). The number of species in this list was 819. The 1905 list was still little more than a list of species, but a great many localities had been added and also a few notes on type of locality and distribution. It is on this 1905 list that the present handlist is based.

Since the 1905 list there has been hardly any collecting apart from a little between 1922 and 1935 and in 1938. The present list contains 919 species, a total still far below what it ought to be.

CICINDELIDAE

2 -Cicindela campestris L.; XYZ; locally common; Forest gravel pits Folly Farm, Calstone; in sandy places.

CARABIDAE

6 -Cychrus rostratus L.; YZ; Savernake; under refuse in woods.
8 -Carabus catenulatus Sp.; YZ; common; under stones and bark, etc.
9 -C. nemoralis Ml.; YZ; under stones and bark, etc.
11 -C. violaceus L.; XYZ; common; under stones and bark, etc.
14 -C. granulatus L.; XYZ; common; in rotten willows or under pond refuse in marshy places.
15 -C. monilis F.; XYZ; the Common, Sun Lane; under stones and bark, etc.
21 -Notiophilus biguttatus F.; XYZ; common; running in the open.
27 -N. palustris Df.; XYZ; Poulton; running in the open.
32 -Leistus fulvibarbis Dj.; Z; Yatesbury; under stones and bark and in moss.
33 -L. ferrugineus L.; XYZ; Rabley; under stones and bark and in moss.

**Plate 11. The final Coleoptera list of the Marlborough College Natural History Society published in the 1938 *Report* which has stood as the county list until now.**

Although there had been a break in beetle recording by the members, apart from a little collecting in the period between 1922-1935, Meyrick himself had not stopped. Right up to the time of his death on 31 March 1938 he worked on a new list, containing 919 species which was published anonymously in the following year as 'A Handlist of the Coleoptera of the Marlborough District' (at the same time as one on the Lepidoptera). This Handlist, although confined to the Marlborough area, has stood by default as the county list until now.

The situation in regard to the College beetle collection is far from clear. The Natural History Society had started a museum as early as 1866, and in 1883 it had grown to such an extent that it moved into new premises in what became known as the Museum Block. Meyrick updated the original *Handbook* in 1894, explaining in the preface that the purpose was to display 'only a series of types of animal life from man down to the lowest animal and in general, therefore, only a few representatives of each family... are exhibited.' These included 'British' beetles and butterflies (no other invertebrates are mentioned) which were shown in the gallery with fossils, minerals, coins, birds eggs and butterflies in locked cabinets 'opened on application'. Whether this British collection was the same as the College collection is not clear. It seems likely, however, that the basis for the collection or collections was one of 500 beetles made by Meyrick whilst a pupil which the 1880 *Report* states was being 'moved into the old moth cabinet' at that time.

The annual reports list many further additions, the most notable being: a collection made by **W.F.H.Blandford** 'in admirable order' in 1883 (added to in 1886), a 'large collection' of British and foreign specimens by **F.C.Pawle**, in 1884, and, in 1903, an anonymous collection of 1,500 specimens. This last was stated to 'have been incorporated with our previous collection and the whole completely re-arranged in [a]... new cabinet'.

Many of the Museum exhibits were not related to natural history and by 1933 occupied so much space that the Natural History Society moved into new headquarters in Mount House. There the members are recorded 'to have begun to make the collections tidy. The lesser known branches of insects [ie other than Lepidoptera] were having their admirers' (Earp, 1965).

In 2000 I made some tentative enquiries at the College about the fate of the beetles and was told that the cabinet containing them had been destroyed when a lorry demolished the corner of Mount House where it was stored. However, after further enquiries with David West, the College Archivist, and Sean Dempster, the Head of Biology, a single cabinet was located in the Biology Department containing drawers of beetles and other insects in very poor condition. The early history of this collection was not known. The College had been to some effort to eradicate *Anthrenus,* which was seriously infesting the drawers, without success, and Alan McKnight, who took over from Sean, kindly allowed me to take the drawers containing the beetles away for further study. Later, when it had become clear that they almost certainly included some of the specimens referred to in the lists, I was told that I could keep them on permanent loan

**Plate 12. Marlborough College Museum in 1895. The beetles were displayed in cases on the balcony.**

Unfortunately very few of the specimens bear any data. Perhaps this was the fault of their having been 'on display' in the museum, although at least one earlier commentator had remarked that many collectors failed to record where their specimens had come from (and possibly the date too?). The lists were presumably compiled from paper records but a search through the College archives has failed to produce any catalogues or other explanatory material.

Attempts to tie up specimens with the published lists has not proved easy. Many published species are not present and many species not published are. It is also clear that many of the latter were not found in the Marlborough area. Boys and masters undoubtedly added specimens found on visits to other parts of the country, and other specimens were contributed by at least one well-known authority. G.C. Champion has already been mentioned, and in 1874 Meyrick recorded that Rye, in addition to naming specimens, had sent him 'type specimens of divers others likely to occur [in the district].'

It is also clear that at an early date the collection was deficient in at least some of the published specimens. **Rev. Edmund James Pearce** (1903-1982), author of the Royal Entomological Society's Pselaphidae *Handbook*, and a water beetle specialist, wrote in March 1934: 'I asked Mr Meyrick if he knew what had become of the specimens on which these lists were based, and he told me that he was under the impression that the specimens, amongst others, were in the Marlborough College Museum. Through the kindness of Mr L.G. Pierson, who has charge of the collections there, I have been able to go through them, and was unable to find any of the local specimens on which the earlier records must have been based; and Mr Pierson does not know what became of them, if they were at any time the property of the College.' Pearce lived locally at this time, having been ordained Priest in 1933 and had his first living at St. Marks, Swindon before entering the Anglican community and moving out of county two years later.

Pearce may have drawn a blank because the Museum had retained some specimens for display when the Natural History Society moved out. On the other hand, perhaps he referred to the water beetles alone of which very few are present. Fortunately, those which formed the basis for his 'A contribution towards a list of the Aquatic Coleoptera of North Wiltshire' (Pearce, 1934) did find their way into the College collection and most are well labelled.

Numerous specimens of early date in groups other than the water beetles are present. These include many singleton 19thC specimens, distinctively mounted on small low cards, which could be Rye's specimens. There are certainly a number of rare species in the groups in which Rye specialised, which are not listed in the *Reports*, ie. were not found in the Marlborough area. Another indicator that at least some of these early specimens may be the ones referred to is the presence of a single example of the very rare 5 spot ladybird, *Coccinella quinquepunctum*. Apart from the specimen published in the 1938 Handlist the species was only known from Scotland at the time. Rye's collection is housed at Bolton Museum but unfortunately long term work there has prevented comparison of mounts and specimens.

As interest declined at Marlborough so it was picked up at Dauntsey's School in West Lavington where the Natural History Society, which had been founded in the early 1930s (Darby, 2002), set out to record everything living within a five mile radius of the school. Dauntsey's Society was a very different one from that at Marlborough. Athough it had its own premises and collections, these were not on the scale of the Marlborough Museum. Furthermore, fewer professional staff and local enthusiasts were available to provide support, and it did not attract the same number of well-known personalities to lecture. Most importantly it could not afford to produce printed and bound reports for national distribution. Instead, what the Society described as 'publications' were typed sheets copied using stencils in small numbers for local distribution only. In 1948 these were brought together in the *Dauntsey Fauna List 1931-1948* which included a list of 150 beetles.

Unlike the Marlborough students and masters, those at Dauntsey's do not seem to have had contact with any of the London specialists (or, surprisingly, with the staff and boys at Marlborough). They relied instead on the use of books to make their determinations. But they appear to have recognised their limitations and produced some high quality research and records nevertheless, particularly in relation to Dew Ponds. Some 26 were being regularly monitored in 1938 and not unsurprisingly produced a good number of water-beetle records.

Central figures in the Dew Pond survey were the pupils **David Le Cren**, whose name appears on many of the records, and **Amyan Macfadyen**. Both later had distinquished careers in natural history, Le Cren as Director of the Freshwater Biological Association and Macfadyen, as Professor of Biology at the University of Ulster and President of the British Ecological Society. Other Dauntsey students who were active members of the natural history society and later became well-known were Anthony Huxley and Desmond Morris.

One might have expected that the work going on at the schools would have been supported and complemented by similar efforts at the Wiltshire Archaeological and Natural History Society, but this was not the case. At its foundation in Devizes in 1853 all the talk was of archaeology and antiquities. Natural History, it turned out, was added to the title almost as an afterthought, and entomology played no recognisable part in the early work of the members.

## After 1945

In 1945 several members of the Society, frustrated by the archaeologists' failure to recognise natural history as other than geology, formed an independent Natural History Section. It was as a direct result of the enthusiasm and perseverance of the members that the Wiltshire Natural History Forum was established, the Museum built its natural history gallery, a natural history curator was appointed (with responsibility for the the Salisbury Museum too), the Wiltshire Trust for Nature Conservation, (now the Wiltshire Wildlife Trust) was formed, and the Biological Records Centre was established (Darby, 2003b). But, although a separate entomological section had by now been set up, its activities were almost entirely confined to butterflies. What work was done on beetles continued to be by enthusiasts, almost all of whom were professionals from outside the county.

In 1965 **Michael George Morris**, then working for the Nature Conservancy Council at Monks Wood, began recording weevils as part of his extensive research into the management of grassland and its effect on invertebrates. This led to his 'Preliminary Account of the Weevils of Wiltshire' published in two parts in 1973 and 1974. with a supplement in 1979. Later Morris became Head of the Institute of Terrestrial Ecology's Furzebrook Research Station at Wareham, from where he would continue to make occasional forays into the county.

At this time too **David Ridley Nash**, a schoolmaster and now the County Recorder for Suffolk, began a series of annual visits to Salisbury which lasted until 1988 and led to the compilation of some 5,000 detailed records. Nash worked mostly in the Salisbury area in the south of the county and, by studying woodland in particular, made some outstanding discoveries. These included most notably the carrion beetle, *Silpha carinata*, thought to be extinct in GB, and the Histerid *Epierus comptus*, new to the GB fauna.

In 1991-92 **Ronald Carr** surveyed water beetles on 20 sites on Salisbury Plain, and in 1995 **Peter J. Hodge** was contracted to carry out an entomological survey of the A303 trunk road between Wylye and Stockton Wood.This turned out to be the first of a series of contracts which saw Hodge return to the county several times. In July 1999, he was at Tilshead Park and Haxton Down on the Plain, and in May and June 2002 back on the Plain for more extensive recording at 7 different sites. Finally he was asked to survey Stonehenge. His efforts resulted in the inclusion of 958 records of almost 300 species in the county list, including many rare weevils.

The Plain was also the subject of other contracted surveys by **Mark Telfer** in 1997, and, in 2008, by **Dave Boyce** and **Steve Wright**. More details of all of these are given on pp.38-39.

Whilst this work was going on **Andy Foster**, another professional entomologist and specialist in beetles, moved into the far north of the county near Malmesbury. He was employed by the Wiltshire Wildlife Trust to survey 4 of their reserves at Landford Bog, Little Durnford, Blackmoor Copse and Jones's Mill.

**Michael Salmon**, the well-known Lepidopterist, lives just outside the county near Fordingbridge, but became Wiltshire's Hemiptera recorder in the mid 1990s and has also taken up beetles. He now has an extensive collection which includes material from Wiltshire, in particular from Cranborne Chase which he surveyed for English Nature in 1997, and the Avon Valley Reserve in Salisbury which we surveyed jointly in 2000 for Chalkhill Environmental Consultants and Salisbury District Council.

Several resident general naturalists and entomologists specialising in other groups should also be mentioned because they have formed small collections. **Piers Mobsby**, on the staff of the Wiltshire Wildlife Trust, who now lives in Salisbury, has made detailed records of many of the beetles he has seen during his work on the Trust reserves over a period of more than 20 years. Piers is principally a botanist but also acts as the county Orthoptera recorder. An earlier Salisbury resident, **H.G. Stokes**, who published a note on *Agrilus sinuatus* in the *Entomologist's Monthly Magazine* in 1952, remains something of an enigma. He certainly formed a collection but by time David Nash met him in the early 1970s he had sold it to E.W.Classey, and its present location is now unknown. There are, however, Cerambycicdae collected by him in the Kaufmann collection in Manchester Museum.

The brothers **Michael and Godfrey Smith,** the county Microlepidoptera recorders, have formed a collection which mainly covers the area around their farm near Trowbridge, and **David Brotheridge**, another Lepidopterist, has a collection from his Avebury garden. The latter is also interested in the Cerambycidae generally.

Two well known Coleopterists should be mentioned who live in the south of the county but have done little or no recording here. **Charles Mackechnie-Jarvis**, now aged 101 , did lead a party from the British Entomological and Natural History Society to Savernake in 1976 but gave up entomology for coins and medals as a result of failing eyesight. He presented his collection to Liverpool Museum and his library was sold by Classey. **Richard Thompson**, the well-known weevil expert on the staff of the Natural History Museum, recently moved into Salisbury, and has been very helpful to me in arranging for the determination of many small Staphylinids.

Finally to myself, **Michael Douglas Darby**. I moved into Sutton Mandeville, near Salisbury, full time in 1992 and was appointed the first county Coleoptera recorder shortly after. Much of my time since has been spent trying to track down records in archives and the literature, as well as those in the possession of both visiting and resident naturalists. I have also concentrated on building up my own reference collection of Wiltshire material and was very fortunate to be able to add to this the collection of Marlborough College as already mentioned. Because of the sad lack of a suitable repository in the county, the Natural History Museum, London has agreed to take these collections on my death.

I have received several contracts which have involved surveys including Langley Wood and Wylye Down NNRs; Bemerton Folly, Salisbury; the Avon Valley Reserve, Salisbury; and Grovely, Great Ridge and Bedwyn Brail woods. Currently I am carrying out a much needed survey of Parsonage Down. General collecting has been confined in particular to woodland but is now switching to calcareous grassland.

# *CHANGES IN WILTSHIRE'S COLEOPTEROUS FAUNA*

Changes in the county fauna have undoubtedly occurred for many reasons over a long period of time. In the pre-historical period inundations of ice from the north and oceans from the south wiped out the entire population on several occasions. Since then weather patterns and the activities of man have been the main causes of most changes, together with, of course, the processes of evolution and adaptation in the organisms themselves.

In the case of weather, temperature is probably the most significant factor because of its effect on all metabolic processes (although beetles being 'cold blooded', are able to withstand a far greater range than mammals). Examples of such changes include: the number of generations produced in a year, reproduction and growth rates, colour patterns, locomotory activity – insects slowed down by the cold, for example,are at greater risk from predators - and even death. Species which have been unable to survive the British winter in the past may be able to do so in the future as a result of global warming.

The response of beetles to winds remains far from clear. Strong winds undoubtedly have the ability to affect transport, particularly in those species that fly (most beetles) as well as the potential, (as with rain, snow and ice), to destroy and change habitats (eg. by felling trees, flooding).

There are also many indirect consequences of climate change. Numerous beetles, for example, are dependent on particular plants which are themselves exposed to climatic influences. (The recent *Atlas of the British and Irish Fauna* by Preston, Pearman and Dines, makes clear the full extent of botanical changes in the last 40 years.) Another example concerns the glow worm and several Silphidae. These species feed on snails which thrive in wet conditions but not in hot, dry weather.

The impact of man's activities on a wide range of habitats in the county from chalk grassland to rivers has been much written about and the consequences for several groups of creatures (eg. butterflies and birds) well recorded. No such documentation exists for beetles, as far as I am aware, although there is some information available about the national picture eg. Jones and Hodge (1995), Hammond (1974) and in early issues of *The Coleopterist*.

Other creatures which have undoubtedly had a considerable impact, are deer and pheasants. The numbers of both have considerably increased in recent years and this is clearly visible in the simplification of the vertical structure of some woodlands both by physical damage and browsing on ground flora, shrubs and branches. A study of the beetles in Langley Wood NNR, in both fenced and unfenced areas, clearly showed more families and species present in the former (Darby, 2005). In the case of pheasants, the impact is mainly on the fringes of smaller blocks of woodland.

Some answers, particularly to questions concerning short term changes, are relatively easily understood. Examples include: the annual fluctuations in ladybird numbers as a result of the death of over-wintering adults in very cold winters; the reduction in dung beetle numbers caused by the use of inappropriate worming treatments; and the occurrence of rare jewel beetles in Trowbridge as a result of the importation of foreign woods in kitchen units. But there are many other questions, particularly those concerning longer term changes, to which answers are not so obvious.

Central to an understanding of the degree of change has to be the ability to compare past and present faunas. Unfortunately, the data to facilitate detailed studies on populations at the local level in the county, their size, constitution, physiological factors and so forth, is almost non-existent.

Putting together the results of collecting and recording to date (c. 20,000 records), it is possible to tabulate the number of new species added to the county list in each decade over the last 150 years.

**Table 1. Number of new species recorded in each decade since 1869.**

| Date | Total |
|---|---|
| - 1869 | 27 |
| 1870-1879 | 524 |
| 1880-1889 | 52 |
| 1890-1899 | 171 |
| 1900-1909 | 52 |
| 1910-1919 | 15 |
| 1920-1929 | 6 |
| 1930-1939 | 107 |
| 1940-1949 | 34 |
| 1950-1959 | 15 |
| 1960-1969 | 1 |
| 1970-1979 | 151 |
| 1980-1989 | 54 |
| 1990-1999 | 248 |
| 2000-2008 | 382 |

The total of 1839 species (excluding the certain sub-fossil records), will undoubtedly rise in the future. It compares with 2,222 species for Somerset, a little over 3,000 for Sussex and 2,049 for Gloucestershire (in 1982). However, because Wiltshire has no coastline, our list will always be smaller than theirs. A recently produced catalogue for the inland county of Warwickshire lists 1,800 species.

In terms of record numbers, however, we are well below these counties. The authors of the Warwickshire list, for example, were able to compile their catalogue from 35,000 records after deleting all those before 1904. Therefore, assuming the proportion of new species to records remains reasonably constant (with an exponentially decreasing total), we can expect our species complement to further exceed theirs at some point in the future.

It would be easy to read into these figures that there has been a considerable increase in the number of new species entering the county in recent years -a finding that appears to fly in the face of contemporary concern about habitat loss and climate change - but that, of course, is not the case. Most will have been here for centuries but no one has looked for them. Different entomologists have different specialisations, and the times of year when individual species appear varies, as do the techniques used to find them.

It is also possible to read into an extrapolation of these figures when related to the records received for all species, that there has been a considerable loss of species in the county over the years. Of the 989 listed before 1950, for example, 282 have not been recorded since. It is tempting to suppose that habitat loss and climate change (the period between 1920 and 1940, like the present, was one of unusual warming in the UK), etc., are again to blame, but the more likely reasons are, as before, the limitations of recorders.

Whilst species-based research may provide grounds for optimism about the present situation, concern about the status of populations is probably still justified. Systematic random sampling to set protocols might help to solve the problem but is unlikely to prove appealing to most entomologists. Small numbers of rare species are inevitably of greater interest to collectors than large numbers of common ones.

There is also a case to be made for putting more effort into recording in areas which are not established reserves, SSSIs, etc. on the basis that many of these are 'insulated', if not from climate change, then from the activities of man. In other words they may not give a true picture of what is happening in most of the county. In the case of Wiltshire, however, so much of the land mass is taken up by downland that this is perhaps not so important as elsewhere.

Just as work specifically focussed on providing data to help answer some of questions around faunal change in the county willl help in future assessments, so study of historically important habitats is also worth more consideration. Published research exists, for example, on the history and distribution of old deer parks in the county which could have a profound influence on understanding changes to the saproxylic beetle fauna in particular.

Similarly, information from mediaeval archives regarding walking the bounds of early forests has enabled historians to construct accurate maps of forests existing in the county, both in the Roman period and at the time of the Domesday book. I am not aware that either of these sources, nor the work of English Nature in the ancient woodland inventory and digital boundary project, has been correlated with contemporary woodlands to direct the recording effort. The geological history of the county in the context of recording beetle changes also requires greater study.

Trying to use the existing information to make statements about specific changes is difficult. The table above covers species but takes no account of distributions. It goes without saying that knowing about the distribution of individual species over time is an essential element in understanding change. However, in one area of the county, Savernake Forest, which was well worked by the Marlborough staff and students, and has recently been the subject of extensive recording by English Nature and others, sufficient data exists to enable some judgements to be made, particularly about species losses and gains.

This is not the place for the detailed discussion which these records demand, but one or two examples which illustrate the importance of systematic recording may be mentioned for interest. Two of the commonest beetles recorded in the Forest in the 1990s - *Longitarsus parvulus*, and *Apthona euphorbiae* (sometimes 100+ specimens in a single trap) do not appear in the Marlborough 1938 list at all. At the other end of the spectrum, no recent records exist for the green tiger beetle, *Cicindela campestris* in the Forest, which was described as frequent in 1938.

The reason for the lack of some beetles may well be the loss of 'heathland' as a result of increasing afforestation. Interestingly, the Marlborough records reveal a surprisingly large number of species not found since and now found almost exclusively in coastal situations, and perhaps these were also dependent on this habitat.

The huge growth in numbers of *L. parvulus* and *A. euphorbiae* is undoubtedly a direct result of larger amounts of linseed and flax, on which they feed, being grown in the county. Their presence in such large numbers at Savernake, however, is evidence of the fact that they are both very mobile and highly polyphagous as adults. What affect they may be having on the indigenous fauna of the Forest is unknown.

Of the list of 282 species not recorded in the county since 1950, it is possible to identify several which are almost certainly extinct. Three RDB taxa listed by Stephens: *Elater sanguineus*, from Salisbury, *Cardiophorus graminius*, from Collingbourne, and *Strangalia attenuata* from Salisbury, are considered to be extinct nationally (though there is an unconfirmed record of the last from West Sussex of 1982). And three further RDB species are known only from a handful of other localities in GB: *Panagaeus cruxmajor*, recorded by Rudd from Winterbourne Stoke in c.1830, *Aphodius subterraneus* recorded by Janson from the Marlborough area in 1873, and *Staphylinus caesareus*, recorded in the same area in 1905.

In respect of species which now appear to be much rarer than previously, I am reluctant to be specific having named species in the past which have since turned up in numbers. *Brachinus crepitans*, the 'bombardier beetle' is one such. In sixteen years I have only come across one specimen, on Wylye Down, and David Nash has not recorded it at all, but recent work on the Plain by Telfer, and Boyce and Wright, appears to show that it is alive and well, certainly in that location.

Another example is that of the glow-worm, *Lampyris noctiluca*. For some time it was thought that the species was declining but a recent appeal for records has revealed that the insect is common throughout large parts of the county. True, the populations appear to be much smaller than many of those nearer London, but in the absence of historical data it is not certain that these low levels have not always existed.

Some new arrivals in the county can be said with certainty to have come from strongholds elsewhere along well documented paths. One might have expected, given publicity about global warming, that these would be from south to north. But that appears not to be the case however, most species appearing to favour an east to west passage thus reflecting ambient temperatures in the UK. Examples include *Leistus rufomarginatus*; *Agrilus pannonicus, Hylis olexai* and most recently *Mordellistona acuticollis* and *Harmonia axyridis*. *Axinotarsus marginellus*, however, was first recorded from South Hampshire in 1966, so presumably arrived in Wiltshire from the south, although it was not recorded here until after being found in most of the eastern counties first.

Other recent arrivals are introductions to the UK which have entered the county in less well known ways. These include *Ptinella simsoni*, the tiny Ptiliid, which arrived from Australia in the 1930s, and *Clambus simsoni*, another Austrlian immigrant, first recorded in south Wales in 1966 and now widespread. Some in this category, *Carcinops pumilo, Euophryum confine, Atomaria lewisi* and *Aridius bifasciatus*, for example, have steadily increased their numbers in recent years and may now be said to be frequent here.

Other species, not recent introductions, which seem to be becoming more frequent include *Pterostichus oblongopunctatus*, *Bitoma crenata, Cerylon ferrugineum, Dirhagus pygmaeus*, and the longhorns *Agapanthia villosvirescens, Paracorymbia fulva, and Phytoecia cylindrica*. Another longhorn for which there are an increasing number of records is *Strangalia quadrifasciata*. .

It is also important to mention that there are several species which may be considered to be 'at risk' in that they are still frequent in Wiltshire but have disappeared in recent years from counties to the north. Two of the most worrying, which are well known 'friends' to all walkers, are the bloody nose beetles *Timarcha tenebricosa* and *Timarcha goettingensis*.

## Conclusions

If we accept that global warming and climate change are priorities, there is a case to be made for putting less effort into recording new species on sites already registered as reserves, SSSI, etc. and more into recording populations of common species in other areas. Given that grassland is so prevalent in the county, more work is required to identify potentially interesting sites, and those locations which are already recognised to be important but which have not been surveyed, or only partially surveyed, should be the subject of proper investigation.

Map 3 shows the total number of species which have been recorded within each 10km square in the county. This is, of course, not so much an accurate map of species distribution as a measure of recording effort. It is immediately obvious, for example, that the most apparently species rich parts of the county happen to be around Marlborough, my home and Salisbury, used as a base by David Nash.The map is, however, a helpful guide to where recording effort might usefully be directed in future.

The far western part of the county stands out as an obvious candidate, particularly in its southern part where Longleat, Stourhead and White Sheet Hill are clearly under-recorded. Similarly the far north east is impoverished though this is probably the least entomologically interesting part of the county. In the.central area, Spye Park and the downs around Morgan's Hill are thin on records, as is Collingbourne

Wood further to the east. But, even in those areas where species numbers are high, there is still a great need for more recording. Many of the county's SSSI's, for example, are still without a single beetle record.

As far as coverage is concerned, it hardly needs stating that the aim should be to extend the 10 km square recording to tetrads, almost half of which are still without a single beetle record. Such a task, however, is almost beyond the capabilities of a single recorder. Given the difficulties of species determination, the establishment of a more formal pattern of recording by out-of-county experts would be invaluable. The example of the botanists is an excellent model to which to aspire, as is that of the Lepidopterists.

Much of the historical data is not as useful as it might be because it is inadequate in one or more respects. Nowadays, we like to think that we record more carefully but it is surprising, nevertheless, to note how many records are submitted in which one or more pieces of vital information are missing. The aim in future must be to try to ensure that records include as much information as possible - certainly the minimum of date, place, map reference, recorder and determiner.

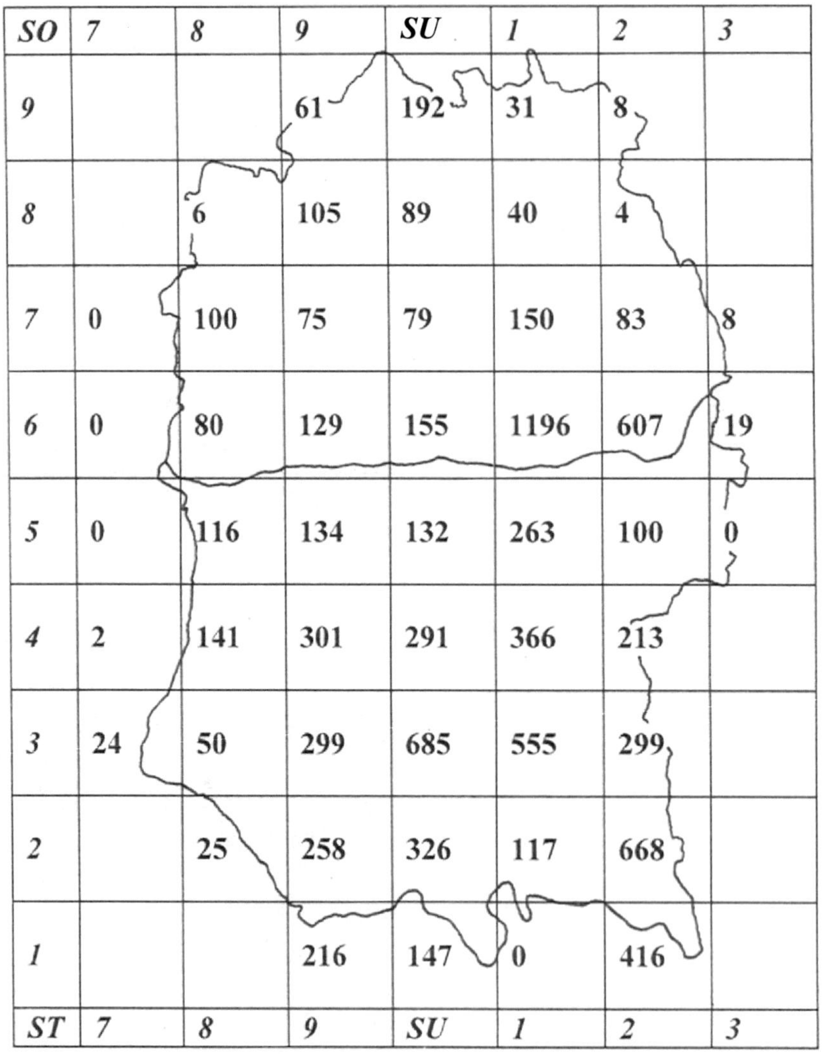


**Map 2. Number of species recorded in each 10 km square in the county.**

# NATIONALLY SCARCE SPECIES IN WILTSHIRE

Wiltshire is rich in species regarded as scarce nationally. No less than 464 of those graded as Red Data Book or notable (Hyman and Parsons, 1992, 1994) occur, or have been recorded as occurring, in the county which is more than a quarter of all the species listed (although a handful of determinations are debatable, marked in the table below with a '?'). Our strength lies not so much in the saproxlic, as the phytophagous species and particularly the weevils. In this last group we can boast 99 representatives which is more than one third of all the notables in the country.

All the scarce species are listed below. The national gradings in the second column are as set out on p.46, and the third column lists the number of sites in which each species has been recorded. Figures in brackets refer to species recorded before 1970, other numbers 1970 to present.

It will be noted that some of the species regarded as nationally scarce have been recorded from more than ten sites.

**Table 2. List of nationally scarce beetles found in Wiltshire**

| **GYRINIDAE** | | |
|---|---|---|
| *?Gyrinus natator* | RDB1 | (1) |
| *Gyrinus urinator* | Nb | 1 |
| **HALIPLIDAE** | | |
| *Haliplus heydoni* | Nb | 1 (1) |
| *Haliplus laminatus* | Nb | (1) |
| **DYTISCIDAE** | | |
| *Agabus biguttatus* | Nb | 6 |
| *Agabus brunneus* | RDB2 | 1 |
| *Agabus labiatus* | Nb | 2 |
| *Agabus uliginosus* | Nb | 4 (1) |
| *Agabus unguicularis* | Nb | 1 |
| *Ilybius chalconatus* | Nb | 4 (2) |
| *Ilybius fenestratus* | Nb | 1 |
| *?Rhantus frontalis* | Nb | 1 |
| *Rhantus suturalis* | Nb | (2) |
| *Hydroglyphus geminus* | Nb | 3 |
| *Deronectes latus* | Nb | 1 |
| *Hydroporus ferrugineus* | Nb | 1 (1) |
| *Hydroporus marginatus* | Nb | 4 (3) |
| *Hydroporus obsoletus* | Nb | 1 (1) |
| *?Hydroporus rufifrons* | RDB2 | 1 |
| *Stictonectes lepidus* | Nb | 1 (1) |
| *?Laccophilus poecilus* | RDB2 | (1) |
| **CARABIDAE** | | |
| *Cicindela sylvatica* | Na | 1 |
| *Brachinus crepitans* | Nb | 10 |
| *Carabus monilis* | Nb | 7 (1) |
| *Blethisa multipunctata* | Nb | (1) |
| *Elaphrus uliginosus* | Nb | 3 (1) |
| *Bembidium clarkii* | Nb | 2 |
| *Bembidion fumigatum* | Nb | 1 |
| *Bembidion quadripustulatum* | Nb | 1 |
| *Poecilus lepidus* | Nb | 1 |
| *Pterostichus oblongopunctatus* | Nb | 9 (4) |
| *Pterostichus quadrifoveatus* | Nb | 2 |
| *Pterostichus anthracinus* | Nb | 10 |
| *Platyderus depressus* | Nb | 3 |
| *Calathus ambiguus* | Nb | 2 |
| *Agonum scitulum* | Na | 1 |
| *Zabrus tenebrioides* | Na | 3 |
| *Amara curta* | Nb | 2 |
| *?Amara lucida* | Nb | (1) |
| *Amara consularis* | Nb | (1) |
| *Amara fulva* | Nb | 1 |
| *Amara equestris* | Nb | 4 |
| *Amara praetermissa* | Nb | 1 |
| *Ophonus azureus* | Nb | 4 |
| *Ophonus schaubergerianus* | Nb | 8 |
| *Anisodactylus nemorivagus* | Na | 1 |
| *?Bradycellus distinctus* | Na | 1 |
| *Acupalpus flavicollis* | Na | 2 |
| *Anthracus consputus* | Nb | 1 |
| *Chlaenius nigricornis* | Nb | 4 (1) |
| *Oodes helopioides* | Nb | 4 (1) |
| *Licinus depressus* | Nb | 3 |
| *Licinus punctatulus* | Na | 2 (1) |
| *Badister unipustulatus* | Nb | 2 (1) |
| *Panagaeus bipustulatus* | Nb | 4 (1) |
| *Panagaeus cruxmajor* | RDB1 | (1) |
| *Lebia chlorocephala* | Nb | 7 |
| **HELOPHORIDAE** | | |
| *Helophorus griseus* | Nb | 2 (2) |
| *Helophorus longitarsis* | RDB3 | 3 |
| *Helophorus nanus* | Nb | (1) |
| *HYDROCHIDAE* | | |
| *Hydrochus elongatus* | RDB3 | (1) |
| **HYDROPHILIDAE** | | |
| *Anacaena bipustulata* | Nb | (1) |
| *Paracymus scutellaris* | Nb | 1 (1) |
| *Berosus affinis* | Nb | 2 |
| *Berosus luridus* | Nb | 1 |
| *Berosus signaticollis* | Nb | 2 |
| *Chaetarthria seminulum* | Nb | 3 |
| *?Enochrus halophilus* | Na | (1) |
| *Laccobius atratus* | Nb | 2 |
| *Laccobius sinuatus* | Nb | (1) |
| *Helochares lividus* | Nb | 3 (2) |
| *Helochares punctatus* | Nb | 1 (2) |

| | | |
|---|---|---|
| *Cercyon granarius* | RDB3 | 3 |
| *Cercyon obsoletus* | Notable | 1 (1) |
| *Cercyon sternalis* | Nb | 1 |
| *Cercyon tristis* | Nb | 1 |
| *Cercyon ustulatus* | Nb | (1) |
| *Cryptopleurum crenatum* | Nb | 1 |
| **HISTERIDAE** | | |
| *Abraeus granulum* | Na | 1 |
| *Plegaderus dissectus* | Nb | 5 |
| *Aeletes atomarius* | RDB3 | 1 |
| *Gnathoncus buyssoni* | Na | 1 |
| *Onthophilus punctatus* | RDBK | 1 |
| *Epierus comptus* | RDBK | 1 |
| *Hister bissexstriatus* | Nb | (1) |
| **HYDRAENIDAE** | | |
| *Hydraena nigrita* | Nb | 1 |
| *Hydraena testacea* | Nb | (1) |
| *Limnebius nitidus* | Nb | 5 (1) |
| *Limnebius papposus* | Nb | (1) |
| **PTILIIDAE** | | |
| *Nossidium pilosellum* | Notable | 3 |
| *Ptenidium turgidum* | RDBK | 2 |
| *Oligella foveolata* | Notable | 1 |
| *Ptilium exaratum* | Notable | 1 |
| *Ptiliolum schwarzi* | RDBK | 1 |
| *Ptinella denticollis* | Notable | (1) |
| *Acrotrichis arnoldi* | RDBK | 4 |
| *Acrotrichis lucidula* | RDBK | 1 |
| *Acrotrichis pumila* | RDBK | 1 |
| *Acrotrichis strandi* | Notable | 1 |
| **LEIODIDAE** | | |
| *Leiodes cinnamomea* | Notable | 2 (1) |
| *Leiodes lucens* | RDBK | 3 |
| *Leiodes oblonga* | Notable | 1 |
| *Leiodes strigipennis* | RDBK | 1 |
| *Agathidium confusum* | RDBi | 1 |
| *Agathidium marginatum* | Notable | 1 |
| *Colon dentipes* | RDBK | 1 |
| *Ptomaphagus varicornis* | RDBK | 2 |
| *Catops longulus* | Notable | 1 |
| *Catopidius depressus* | Notable | 1 |
| **SCYDMAENIDAE** | | |
| *Eutheia scydmaenoides* | Notable | 1 |
| *Neuraphes praeteritus* | Notable | 2 |
| **SILPHIDAE** | | |
| *Aclypea opaca* | Na | 1 (1) |
| *Dendroxena quadrimaculata* | Nb | 1 (1) |
| *Silpha carinata* | RDB1 | 1 |
| *Silpha obscura* | RDB2 | 3 |
| *Nicrophorus interruptus* | Nb | 7 (1) |
| *Nicrophorus vestigator* | Na | 1 |
| **STAPHYLINIDAE** | | |
| *Acidota cruentata* | Nb | 1 (1) |
| *Phyllodrepa salicis* | RDBK | 1 (1) |
| *Omalium rugatum* | Notable | 1 |
| *Mycetoporus punctus* | Notable | (1) |
| *Bibloporus minutus* | Nb | 1 |
| *Pselaphaulax dresdensis* | Notable | (2) |

| | | |
|---|---|---|
| *Claviger testaceus* | Notable | 4 |
| *Mycetoporus piceolus* | Notable | 1 |
| *Sepedophilus bipunctatus* | Nb | 1 |
| *?Tachinus bipustulatus* | RDB1 | (1) |
| *Tachinus rufipennis* | RDB3 | 2 |
| *Aloconota longicollis* | Notable | 1 |
| *Alaobia scapularis* | Notable | 2 |
| *Microdota aegra* | Notable | 1 |
| *Thamiaraea hospita* | Notable | 1 |
| *Gyrophaena manca* | Notable | 2 |
| *Bolitochara pulchra* | Notable | 2 |
| *Lomechusa emarginata* | Notable | (1) |
| *Oliogta apicata* | Notable | 1 |
| *Myllaena kraatzi* | Notable | 1 |
| *Homoeusa acuminata* | RDBK | 1 |
| *Haploglossa marginalis* | Notable | 1 |
| *Ilyobates subopacus* | Notable | 1 |
| *Ilyobates propinquus* | Notable | 1 |
| *Myllaena kraatzi* | Notable | 1 |
| *Placusa tachyporoides* | Notable | 2 |
| *Scaphisoma boleti* | Nb | 2 |
| *Deleaster dichrous* | Nb | 1 |
| *Carpelimus subtilis* | Notable | 1 |
| *Stenus circularis* | Nb | (2) |
| *Stenus canescens* | Nb | 2 |
| *Stenus niveus* | Nb | 1 |
| *Lathrobium zetterstedti* | Nb | (2) |
| *Lathrobium pallidipenne* | Notable | 2 |
| *Lathrobium pallidum* | RDBK | 1 |
| *Paederus fuscipes* | Nb | 3 |
| *Rugilus angustatus* | Notable | (3) |
| *Rugilus similis* | Notable | 1 |
| *Sunius melanocephalus* | Notable | 2 |
| *Philonthus ebeninus* | Notable | 1 (1) |
| *?Philonthus fumarius* | Nb | (2) |
| *?Heterothops dissimilis* | RDBK | (1) |
| *Quedius aetolicus* | Na | 2 |
| *Quedius brevicornis* | Nb | 1 |
| *Quedius invreae* | Nb | 2 |
| *Quedius nigrocaeruleus* | Nb | 1 |
| *Quedius puncticollis* | Nb | 1 |
| *Quedius xanthopus* | Nb | 6 |
| *Platydracus fulvipes* | Nb | 2 |
| *Platydracus latebricola* | Nb | 3 (1) |
| *Staphylinus caesareus* | RDBi | (1) |
| *Ocypus fortunatarum* | Nb | 1 |
| *Ocypus fuscatus* | Nb | (1) |
| *Tasgius pedator* | Na | (1) |
| *Hypnogyra angularis* | Na | 1 |
| **LUCANIDAE** | | |
| *Lucanus cervus* | Nb | 4 (2) |
| *BALBOCERATIDAE* | | |
| *Odonteus armiger* | Na | 2 |
| **GEOTRUPIDAE** | | |
| *Trypocopris pyrenaeus* | Na | 1 |
| **SCARABAEIDAE** | | |
| *?Aphodius coenosus* | Nb | (1) |
| *Aphodius sordidus* | Na | (1) |

| | | |
|---|---|---|
| *Aphodius distinctus* | Nb | |
| *Aphodius paykulli* | Nb | 1 |
| *?Aphodius coenosus* | Nb | (1) |
| *Aphodius subterraneus* | RDB1 | (1) |
| *Aphodius zenkeri* | Nb | 2 |
| *Aphodius arenarius* | Nb | (1) |
| *Omaloplia ruricola* | Nb | 9 (1) |
| *?Onthophagus nuchicornis* | Na | (1) |
| *Euheptaulacus villosus* | Na | (1) |
| **SCIRTIDAE** | | |
| *Elodes elongata* | RDBi | 1 |
| *Prionocyphon serricornis* | Nb | 3 |
| **BUPRESTIDAE** | | |
| *Agrilus viridis* | Na | 2 |
| *Agrilus angustulus* | Nb | 3 |
| *Agrilus biguttatus* | Na | 3 |
| *Agrilus laticornis* | Nb | 8 |
| *Agrilus sinuatus* | Na | 11 |
| *Aphanisticus pusillus* | Nb | 3 |
| *Trachys scrobiculatus* | Na | 5 |
| **BYRRHIDAE** | | |
| *Porcinolus murinus* | Nb | (1) |
| **DRYOPIDAE** | | |
| *Dryops auriculatus* | Nb | 1 |
| *?Dryops griseus* | RDB3 | (1) |
| *Dryops nitidulus* | RDB3 | 1 |
| **PSEPHENIDAE** | | |
| *Eubria palustris* | RDB3 | 1 |
| **EUCNEMIDAE** | | |
| *Melasis buprestoides* | Nb | 2 |
| *Hylis olexai* | RDB3 | 2 |
| *Microrhagus pygmaeus* | RDB3 | 4 |
| **ELATERIDAE** | | |
| *Ctenicera pectinicornis* | Na | (1) |
| *Calambus bipustulatus* | Nb | 2 (1) |
| *Paraphotistus impressus* | Nb | (1) |
| *Ampedus cinnabarinus* | RDB3 | 1 |
| *Ampedus elongantulus* | Na | 3 (1) |
| *Ampedus quercicola* | Nb | 9 |
| *Ampedus sanguineus* | Extinct | (1) |
| *Ampedus sanguinolentus* | Na | 1 (1) |
| *Procraerus tibialis* | RDB3 | 1 (1) |
| *Oedostethus quadripustulatus* | Na | (2) |
| **DRILIDAE** | | |
| *Drilus flavescens* | Na | 2 |
| **LYCIDAE** | | |
| *Platycis minutus* | Nb | 5 |
| **CANTHARIDAE** | | |
| *Cantharis fusca* | RDB3 | 3 (2) |
| *Rhagonycha lutea* | Nb | 3 |
| *Rhagonycha translucida* | Nb | 3 |
| *Silis ruficollis* | Nb | 4 |
| *Malthinus balteatus* | Nb | 2 (1) |
| *Malthinus frontalis* | Nb | 1 |
| *Malthodes fibulatus* | Nb | 2 (1) |
| *Malthodes guttifer* | Nb | 1 |
| *Malthodes maurus* | Nb | 1 |

| | | |
|---|---|---|
| **DERMESTIDAE** | | |
| *Ctesias serra* | Nb | 7 |
| **BOSTRICHIDAE** | | |
| *Lyctus linearis* | Nb | (1) |
| **ANOBIIDAE** | | |
| *Hedobia imperialis* | Nb | 3 (1) |
| *Ptinus subpilosus* | Nb | 3 |
| *Anobium inexspectatum* | Nb | 1 |
| **TROGOSSITIDAE** | | |
| *Thymalus limbatus* | Nb | 3 |
| **CLERIDAE** | | |
| *Tillus elongatus* | Nb | 3 (1) |
| *Korynetes caeruleus* | Nb | (2) |
| **DASYTIDAE** | | |
| *Dasytes niger* | Na | 2 (1) |
| *Dasytes plumbeus* | Nb | 1 |
| **MALACHIIDAE** | | |
| *Malachius aeneus* | RDB3 | 1 (1) |
| **SPHINDIDAE** | | |
| *Sphindus dubius* | Nb | 1 |
| **NITIDULIDAE** | | |
| *Epuraea distincta* | Na | 1 |
| *Epuraea fuscicollis* | Nb | 1 |
| *Epuraea longula* | Nb | 1 |
| *Meligethes atramentarius* | Notable | 2 |
| *Meligethes bidentatus* | RDBK | 2 |
| *Meligethes erichsoni* | Notable | 3 |
| *Meligethes exilis* | Notable | 1 |
| *Meligethes gagathinus* | Notable | 6 |
| *Meligethes incanus* | Notable | 1 |
| *Meligethes lugubris* | Notable | 2 (1) |
| *Meligethes ochropus* | Notable | 1 |
| *Meligethes serripes* | RDBK | 1 |
| *Meligethes solidus* | Notable | 13 |
| *Thalycra fervida* | Nb | 1 |
| *Cryptarcha strigata* | Nb | 1 (1) |
| **MONOTOMIDAE** | | |
| *Rhizophagus nitidulus* | Nb | 3 |
| **SILVANIDAE** | | |
| *Silvanus bidentatus* | Nb | 7 |
| **CUCUJIDAE** | | |
| *Pediacus depressus* | Na | 1 (1) |
| **LAEMOPHLOEIDAE** | | |
| *Cryptolestes spartii* | Na | 1 |
| **PHALACRIDAE** | | |
| *Phalacrus championi* | Na | 2 |
| *Olibrus flavicornis* | RDBK | 5 |
| **CRYPTOPHAGIDAE** | | |
| *Cryptophagus ruficornis* | Notable | 2 |
| *Atomaria nigriventris* | Notable | 1 |
| *Atomaria puncticthorax* | Notable | 2 |
| *Atomaria rubricollis* | RDB1 | 1 |
| **EROTYLIDAE** | | |
| *Tritoma bipustulata* | Na | 1 (1) |
| **BIPHYLLIDAE** | | |
| *Diplocoelus fagi* | Nb | 6 |
| **BOTHRIDERIDAE** | | |

| *Anommatus duodecimstriatus* | Na | 1 |
|---|---|---|
| **CERYLONIDAE** | | |
| *Cerylon fagi* | Nb | 7 |
| **ENDOMYCHIDAE** | | |
| *Symbiotes latus* | Nb | 1 |
| *Lycoperdina bovistae* | RDB3 | 3 |
| **COCCINELLIDAE** | | |
| *Hyperaspis pseudopustulata* | Nb | 1 |
| *Coccinella magnifica* | Na | 1 (1) |
| *Coccinella quinquepunctata* | RDB3 | (1) |
| **CORYLOPHIDAE** | | |
| *Orthoperus nigrescens* | Nb | 8 |
| **LATRIDIIDAE** | | |
| *Enicmus brevicornis* | Notable | 5 |
| **MYCETOPHAGIDAE** | | |
| *Mycetophagus quadriguttatus* | Na | 1 |
| CIIDAE | | |
| *Strigocis bicornis* | Nb | 1 |
| *Cis festivus* | Nb | 2 |
| **TETRATOMIDAE** | | |
| *Hallomenus binotatus* | Nb | 2 |
| *Tetratoma ancora* | Nb | 1 |
| **MELANDRYIDAE** | | |
| *Orchesia micans* | Nb | 4 |
| *Orchesia minor* | Nb | 5 (1) |
| *Abdera biflexuosa* | Nb | 2 (2) |
| *Phloiotrya vaudoueri* | Nb | 1 |
| *Melandrya caraboides* | Nb | 1 (1) |
| *Conopalpus testaceus* | Nb | 3 |
| *Osphya bipunctata* | RDB3 | 1 |
| **MORDELLIDAE** | | |
| *Tomoxia bucephala* | Na | 3 (1) |
| *Variimorda villosa* | Nb | 13 |
| *Mordella holomelaena* | RDBK | 1 |
| *Mordella leucaspis* | RDBK | 1 |
| *Mordellistena brevicauda* | RDBK | 1 (1) |
| *Mordellistena humeralis* | RDBK | 2 |
| *Mordellistena neuwaldeggiana* | RDBK | 1 |
| *Mordellistena parvula* | RDBK | 1 (2) |
| **COLYDIIDAE** | | |
| *Synchita humeralis* | Nb | 1 |
| *Colydium elongatum* | RDB3 | 6 |
| **TENEBRIONIDAE** | | |
| *Eledona agricola* | Nb | 5 |
| *Prionychus ater* | Nb | 3 |
| *Pseudocistela ceramboides* | Nb | (2) |
| *Mycetochara humeralis* | Na | 2 |
| *Scaphidema metallicum* | Nb | 3 (1) |
| **OEDEMERIDAE** | | |
| *Ischnomera caerulea* | RDB3 | 3 (1) |
| *Ischnomera cyanea* | Nb | 8 |
| *Ischnomera sanguinicollis* | Nb | 2 |
| *Oedemera femoralis* | Nb | 1 |
| **MELOIDAE** | | |
| *Meloe violaceus* | Nb | (1) |
| *Meloe rugosus* | RDB3 | 1 |
| **PYROCHROIDAE** | | |
| *Pyrochroa coccinea* | Nb | 10 (1) |

| **SALPINGIDAE** | | |
|---|---|---|
| *Lissodema denticolle* | Nb | 2 (1) |
| **ANTHICIDAE** | | |
| *Omonadus bifasciatus* | Nb | 1 |
| **ADERIDAE** | | |
| *Aderus oculatus* | Nb | 1 |
| **CERAMBYCIDAE** | | |
| *Prionus coriarius* | Na | (2) |
| *?Rhagium inquisitor* | Nb | (1) |
| *Paracorymbia fulva* | RDB3 | 9 |
| *Gracilia minuta* | RDB2 | (1) |
| *Aromia moschata* | Nb | 3 (3) |
| *Poecilium alni* | Nb | (2) |
| *Anaglyptus mysticus* | Nb | 8 |
| *Mesosa nebulosa* | RDB3 | 1 |
| *Stenostola dubia* | Nb | 1 |
| *Phytoecia cylindrica* | Nb | 13 |
| *?Oberea oculata* | RDB1 | 1 |
| **ORSODACNIDAE** | | |
| *Orsodacne humeralis* | Nb | 1 |
| **CHRYSOMELIDAE** | | |
| *Bruchus atomarius* | Nb | 6 (1) |
| *Donacia aquatica* | RDB3 | (1) |
| *?Donacia bicolora* | RDB2 | (1) |
| *Donacia cinerea* | Nb | 1 |
| *Domacia dentata* | Na | (1) |
| *Donacia thalassina* | Nb | 1 (1) |
| *Plateumaris affinis* | Nb | 5 (2) |
| *Plateumaris braccata* | Na | 1 |
| *Cryptocephalus aureolus* | Nb | 8 (1) |
| *Cryptocephalus bipunctatus* | Nb | 2 |
| *Cryptocephalus parvulus* | Nb | (1) |
| *Cryptocephalus punctiger* | Na | (1) |
| *Cryptocephalus sexpunctatus* | RDB2 | (1) |
| *Chrysolina oricalcia* | Nb | 4 (1) |
| *Chrysolina sanguinolenta* | Na | 3 |
| *Chrysolina violacea* | Nb | 5 (2) |
| *Luperus flavipes* | Nb | (1) |
| *Phyllotreta cruciferae* | Nb | 1 |
| *Aphthona nigriceps* | Na | 2 |
| *Longitarsus agilis* | Na | 1 |
| *Longitarsus anchusae* | Nb | 5 |
| *Longitarsus brunneus* | Nb | 1 |
| *Longitarsus curtus* | Na | 1 |
| *Longitarsus dorsalis* | Nb | 13 |
| *Longitarsus fowleri* | Na | 1 |
| *Longitarsus ganglbaueri* | Na | (1) |
| *Longitarsus lycopi* | Nb | 1 |
| *Longitarsus nasturtii* | Nb | (1) |
| *Longitarsus nigrofasciatus* | Na | 1 |
| *Longitarsus parvulus* | Na | 12 |
| *Longitarsus tabidus* | Nb | 2 |
| *Altica brevicollis* | Na | 1 |
| *Altica longicollis* | Nb | 2 |
| *Ochrosis ventralis* | RDB3 | 1 |
| *Crepidodera nitidula* | Nb | 1 (1) |
| *Epitrix atropae* | Nb | 1 |
| *Podagrica fuscicornis* | Nb | (1) |

| | | |
|---|---|---|
| *Mantura rustica* | Nb | 2 (1) |
| *Apteropeda globosa* | Nb | (1) |
| *Psylliodes chalcomera* | Nb | 5 |
| *Psylliodes luteola* | RDBK | 9 |
| *Pilemostoma fastuosa* | Na | 1 |
| *Cassida nebulosa* | RDBi | 2 |
| *Cassida nobilis* | Nb | 1 |
| *Cassida prasina* | Nb | 2 |
| **ANTHRIBIDAE** | | |
| *Anthribus nebulosus* | Nb | 1 |
| *Platyrhinus resinosus* | Nb | 1 |
| *Platystomos albinus* | Nb | 5 |
| *Choragus sheppardi* | Na | 1 |
| **RHYNCHITIDAE** | | |
| *Lasiorhynchites cavifrons* | Nb | 3 (2) |
| *Neocoenorrhinus interpunctatus* | Nb | 2 (1) |
| *Temnocerus longiceps* | Nb | 4 |
| *Temnocerus tomentosus* | Nb | 2 (1) |
| *Byctiscus betulae* | Nb | 2 |
| **APIONIDAE** | | |
| *Aspidapion soror* | Na | 2 |
| *Exapion difficile* | Na | 2 |
| *Melanapion minimum* | RDB3 | (1) |
| *Kalcapion semivittatum* | Na | 1 |
| *Hemitrichapion reflexum* | Na | 4 |
| *Oxystoma cerdo* | Nb | 12 |
| *Catapion pubescens* | Nb | 2 |
| *Protapion difforme* | Nb | 8 (1) |
| *Protapion filirostre* | Nb | 2 |
| *Protapion varipes* | Nb | 3 (1) |
| **ERIRHINIDAE** | | |
| *Grypus equiseti* | Nb | 3 |
| *Notaris scirpi* | Nb | 2 |
| *Thryogenes scirrhosus* | Nb | (1) |
| **CURCULIONIDAE** | | |
| *Curculio betulae* | Nb | 1 |
| *Curculio villosus* | Nb | 2 (1) |
| *Anthonomus ulmi* | Nb | (1) |
| *Ellescus bipunctatus* | Nb | 1 |
| *Dorytomus salicinus* | Nb | (1) |
| *Cleopomiarus graminis* | Nb | 14 |
| *Cleopomiarus plantarum* | RDBK | (1) |
| *Gymnetron melanarium* | Nb | 7 (1) |
| *Gymnetron rostellum* | Na | 1 |
| *Gymnetron veronicae* | Nb | 2 |
| *Gymnetron villosulum* | Nb | 8 (1) |
| *Rhinusa collinum* | Na | 1 |
| *Rhinusa linariae* | Na | 1 |
| *Tachyerges pseudostigma* | RDBK | 5 |
| *Pachytychius haematocephalus* | RDB1 | (1) |
| *Orthochaetes setiger* | Nb | 14 |
| *Tychius lineatulus* | Na | 3 (1) |
| *Tychius schneideri* | Nb | 1 |
| *Tychius tibialis* | Na | 1 |
| *Tychius squamulatus* | Nb | 1 |
| *Bagous tempestivus* | Nb | 1 |
| *Aulacobaris lepidii* | Na | 1 |
| *Aulacobaris picicornis* | Nb | 5 (1) |

| | | |
|---|---|---|
| *Calosirus terminatus* | Nb | 5 |
| *Ceutorhynchus constrictus* | Nb | 2 |
| *Ceutorhynchus hirtulus* | Nb | (1) |
| *Ceutorhynchus unguicularis* | RDB3 | 1 |
| *Coeliodes ruber* | Nb | 2 |
| *C.transversealbofasciatus* | Nb | 4 |
| *Drupenatus nasturtii* | Nb | 1 (1) |
| *Glocianus moelleri* | RDBK | 1 |
| *Glocianus pilosellus* | RDB2 | 1 |
| *Glocianus punctiger* | Nb | 3 |
| *Hadroplontus trimaculatus* | Nb | 3 |
| *Microplontus campestris* | Nb | (1) |
| *Mogulones geographicus* | Nb | (1) |
| *Sirocalodes mixtus* | Nb | 1 |
| *Stenocarus ruficornis* | Nb | 6 (1) |
| *Thamiocolus viduatus* | Nb | 2 |
| *Trichosirocalus horridus* | Na | 1 |
| *Zacladus exiguus* | Nb | (1) |
| *Mononychus punctumalbum* | Na | 1 |
| *Neophytobius muricatus* | Na | 1 |
| *Neophytobius quadrinodosus* | Na | 1 |
| *Pelenomus waltoni* | Nb | (1) |
| *?Rhopalomesites tardyi* | Nb | (1) |
| *Acalles ptinoides* | Nb | 6 |
| *Kyklioacalles roboris* | Nb | 4 |
| *Strophosoma faber* | Nb | 6 |
| *Attactagenus plumbeus* | Nb | 1 |
| *Caenopsis fissirostris* | Nb | 1 |
| *Polydrusus mollis* | Nb | 2 |
| *Polydrusus formosus* | Na | 1 (1) |
| *Polydrusus flavipes* | Nb | 3 |
| *Brachysomus echinatus* | Nb | 2 |
| *Brachysomus hirtus* | RDB3 | 2 |
| *Sitona macularius* | Nb | 3 (1) |
| *Sitona waterhousei* | Nb | 1 |
| *Tanymecus palliatus* | Nb | 1 |
| *Trachyphloeus alternans* | Nb | 4 |
| *Trachyphloeus aristatus* | Nb | 3 |
| *Trachyphloeus asperatus* | Nb | 6 |
| *Trachyphloeus spinimanus* | Nb | 2 |
| *Tropiphorus elevatus* | Nb | 1 (1) |
| *Tropiphorus terricola* | Nb | 1 |
| *Hypera meles* | Na | 3 |
| *Limobius borealis* | Na | 5 |
| *Larinus planus* | Nb | 11 (2) |
| *Rhinocyllus conicus* | Na | 2 |
| *Magdalis barbicornis* | Na | 1 |
| *Magdalis cerasi* | Nb | 3 |
| *Liparus coronatus* | Nb | 1 (1) |
| *?Liparus germanus* | RDB2 | (1) |
| *Leiosoma oblongulum* | Nb | 1 |
| *Anoplus roboris* | Nb | 4 |
| *Ernoporicus fagi* | Na | 2 |
| *Taphrorychus bicolor* | Na | 2 |
| *Trypodendron signatum* | Nb | 3 |
| *Hylesinus orni* | Nb | 1 |
| **PLATYPODIDAE** | | |
| *Platypus cylindrus* | Nb | 5 |

**Plates 13 & 14. Ancient oak and beech in Savernake Forest (Joan Davies).**

**Plate 15. Fallen tree in Cranborne Chase.**

# WOODLAND INDICATOR SPECIES

Harding and Rose (1986) established decaying wood as an important habitat for invertebrates, and suggested a list of 'Indicator' beetle species to assist in the identification and evaluation of sites for conservation purposes. Since then attention has focussed on refining the list and of introducing 'scoring' to help determine priorities. As a result two systems have evolved: the Saproxylic Quality Index (SQI) (Fowles, Alexander, Key, 1999), and the Index of Ecological Continuity (IEC) (Alexander, 2004).

### Saproxylic Quality Index

The SQI upgrades the list of 155 Coleoptera in Harding and Rose to 598 species chosen because of their 'dependence upon microhabitats associated largely with the process of damage and decay in the bark and wood of trees and larger woody shrubs and climbers. This includes sap runs, fungal hyphae or fruiting bodies, rot holes, etc.'. Scores are assigned according to the categories established in Hyman and Parsons (1992, 1994) (see p.47) and the SQI value obtained by adding these, dividing the total by the number of species, and multiplying by 100.
Fowles *et al* (1999) listed 56 sites in Britain arranged in order of importance, but none in Wiltshire were included (although Harding and Rose listed two sites Savernake Forest and Stockton's [sic] Wood). The reason for the omissions was not simply lack of information about the Wiltshire woods, but the fact that 1980 was set as an ideal cut off point for records, and the minimum number of species required to achieve a sensible value was set at 40, which, with the exception of Savernake, none reached. The closest were Great Ridge and Cranborne Chase with 37 species each, none of the remainder achieving more than 25.

Since then considerable efforts have been made both to acquire more historic data and to do more on-site recording so as to achieve the minimum requirement. In some instances this has meant more visits than the 1 to 3 anticipated by Fowles *et al.*, which was possibly the result of the large size of some of the sites, eg. Great Ridge Wood, and consequent difficulty of locating suitable habitats.

The following woods (with the exception of Bentley Wood which has not yet reached the qualifying number of species) are now officially included in the Index which lists 156 sites nationally and can be seen at http://thasos.users.btopenworld.com/sqi.htm: Other wooded areas such as Wardour and Fonthill, and when further surveying has taken place. Parklands such as Stourhead, will surely join this list

**Table 3. Saproxylic Quality Index: scores achieved by Wiltshire Woods.**

| Site | No. Spp | SQS | SQI | GB Status |
|---|---|---|---|---|
| Wiltshire County | 286 | 1450 | 507 | |
| Savernake Forest | 111 | 538 | 484.7 | 36 |
| Langley Wood | 68 | 292 | 429.4 | 50 |
| Great Ridge Wood | 47 | 182 | 387.2 | 68 |
| Grovely Wood | 108 | 406 | 375.9 | 80 |
| Blackmoor Copse | 60 | 223 | 371.7 | 81 |
| Cranborne Chase | 45 | 163 | 362.2 | 86 |
| Devenish Reserve | 75 | 256 | 341.3 | 102 |
| Longleat | 46 | 153 | 332.6 | 107 |
| Bentley Wood* | 36 | 96 | 263.8 | 142 |
| Bedwyn Brail | 40 | 98 | 245 | 148 |

* Officially excluded because based on 36 spp. only

Sites with an SQI of 590 or more are considered to be of international importance, and those of 500-590 of national importance. Thus, Savernake falls just short of achieving a nationally important rating. Interestingly, if Wiltshire is treated as a single site and all the qualifying species taken into account (286 scoring 1450) it achieves an SQI of 507 which would just move us into the nationally important bracket.

### Index of Ecological Continuity

The first IEC was developed (Harding and Rose, 1986) as a means of producing a simple statistic which could be used for grading a site for its significance to nature conservation. The species selection was informed partly by knowledge of the saproxylic fauna of the post glacial wildwood and partly by the degree of association with those sites believed to be the least disturbed by human activities and therefore most closely approximating to the structures and composition of the wildwood. This work was subsequently refined and expanded by Hammond and Harding (1991), Harding and Alexander (1994), and most recently by Alexander

(2004) whose publication forms the basis for the calculations below.

Thus, whereas the SQI operates with a large sub-set of species based on rarity, the IEC works with a much smaller number chosen for their association with, and fidelity to, apparently old forest areas. A cut-off date of 1950 is suggested for records, and simple numeric values of 1-3 apply (see p.47).

The scores for Wiltshire produce a different order of woods to that of the SQI, and suggest that Grovely shows a greater degree of ecological continuity than all but Savernake (see. Table 4. Below).

**Table 4. Index of Ecological Continuity: scores achieved by Wiltshire woods.**

| Site | IEC | GB Status |
|---|---|---|
| | | |
| Savernake Forest | 59 | 37 |
| Grovely Wood | 41 | 49 |
| Langley Wood | 30 | 69 |
| Blackmoor Copse | 20 | 91 |
| Devenish Reserve | 18 | 106 |
| Longleat | 17 | 113 |
| Great Ridge Wood | 16 | 123 |
| Cranborne Chase | 10 | 143 |
| Bentley Wood | - | - |
| Bedwyn Brail | 9 | 145 |

Originally, Harding and Alexander (1994) assigned values of 20 or more to the IEC as indicating national importance, but as new records for sites have been incorporated and previously unrecognised sites have been surveyed, this threshold has been adjusted, above 80 indicates a site of international importance, above 25 one of national importance, and above 15 one of regional importance. Thus three of the Wiltshire woods are shown to be nationally important and a further four regionally so.

Both the IEC and SQI scores put the Wiltshire woods well down the national table, but if one looks at the picture in the South West alone, it tells a very different story. In the case of the SQI, we take the top six positions above woods in Cornwall, Devon, Somerset and Dorset, and in that of the IEC, the top three. Thus the importance of these woodlands in the regional context is clear, as is the need for appropriate conservation priorities.

**Table 5. Saproxylic Quality Index and Index of Ecological Continuity for SW England: positions achieved by Wiltshire Woods.**

| Site | SQI Status in South West | IEC Status in South West |
|---|---|---|
| | | |
| Savernake | 1 | 1 |
| Langley Wood | 2 | 3 |
| Great Ridge Wood | 3 | 9 |
| Grovely Wood | 4 | 2 |
| Blackmoor Copse | 5 | 6= |
| Cranborne Chase | 6 | 11= |
| Devenish Reserve | 8 | 7 |
| Longleat | 10 | 8 |
| Bentley Wood | - | - |
| Bedwyn Brail | 17 | 13 |

It is also worth noting that care needs to be taken before dismissing a site's importance on the basis of one or two visits alone (the figures for Grovely were compiled from records made after more than 30 separate visits). The dates of records need to be taken into account. too. Although it is right that many older records should be excluded, knowledge of the national picture may suggest that others should not.

In regard to future recording at these sites it is worth remembering that expert entomologists making single visits tend to concentrate on rarer species thus producing anomalous results. For sites where a wide range of species have been recorded, however, such visits, and indeed any further increase in recording, is likely to result in a higher index as proportionally fewer new common species will be found. Applying this to Savernake we may expect its rating to increase rather than to decrease, whereas the reverse will probably be true of Grovely where more specialist collecting has taken place. Given the position achieved by the county as a whole in the SQI it seems unlikely that the national status of our woods will be improved unless more new species are added to the county list.

# CALCAREOUS GRASSLAND INDICATOR SPECIES

Calcareous Grassland is defined as being 'developed on shallow lime-rich soils generally overlaying limestone rocks, including chalk' (Habitat Action Plan). Wiltshire is particularly rich in such habitats and no less than 53 SSSIs which include calcareous grassland have been designated in the county.

To assist in the process of evaluating these sites nationally, Alexander (2003) listed 240 beetles (amongst other invertebrates) as indicator species. Each was given a rating (high, medium or low) according to their Fidelity and Continuity within the habitat. These terms he explained as follows:

'Fidelity helps to identify which species should be priorities for conservation. High fidelity species occur primarily in calcareous grassland and are therefore strongly dependent on conservation of their habitat within them. Low fidelity species can be conserved in other habitat types and may not be a priority in calcareous grasslands. Sites with a large number of high fidelity species or with a large proportion of them may be considered to have a high quality for nature conservation.'

'Continuity is a particularly important concept as it identifies sites which have species ... dependent on conservation at site level, and which once lost, may be difficult to recover other than by re-establishment projects. It therefore is especially important in the identification of site quality for nature conservation and site management priorities. As with fidelity, sites with a large number of high continuity species or with a large proportion of them may be considered to have high quality for nature conservation.'

Alexander also pointed out that the 'high', 'moderate' and 'low' categories (see p.46 for more detailed information about these categories) were easily adaptable to a scoring system like that used for the IEC, but, as far as I am aware, that has not been attempted to date. Given the importance of this habitat in Wiltshire, which is greater than in other counties, it does seem to offer a useful tool for assisting in the evaluation of sites which should not be ignored.

Until the recent work of Boyce, Hodge, Telfer and others on Salisbury Plain, and myself on Wylye Down, there were almost no beetle records for any Wiltshire sites to which scoring could be applied. Now, however, lists do exist which are sufficiently comprehesive to permit such a system to be trialled. The table below has been compiled by allocating scores of 1 to 'low', 2 to 'moderate' and 3 to 'high', and the site score has been calculated by adding these together, multiplying by a hundred and dividing by the number of species. I am suggesting that it be called the **Calcareous Grassland Quality Index.**

**Table 6. Calcareous Grassland Quality Index: Scores for Wiltshire sites.**

| Site | Sp. No. | F | C | F+C | Score |
|---|---|---|---|---|---|
| Beacon Hill (SPTAE) | 8 | 21 | 16 | 37 | 462 |
| Stonehenge | 13 | 31 | 18 | 49 | 377 |
| Wylye Down | 24 | 50 | 34 | 84 | 350 |
| Porton Down | 35 | 69 | 52 | 121 | 348 |
| Sidbury Hill (SPTAE) | 20 | 41 | 28 | 69 | 345 |
| Martin Down | 25 | 51 | 34 | 85 | 340 |
| Cockey Down | 8 | 16 | 11 | 27 | 337 |
| Coombe Bissett Down | 6 | 9 | 11 | 20 | 333 |
| Wadmans Coppice (SPTAW) | 10 | 19 | 13 | 32 | 320 |
| Middleton Down | 12 | 22 | 16 | 38 | 317 |
| Fox Covert (SPTAW) | 14 | 26 | 18 | 44 | 314 |
| Black Heath (SPTAC) | 14 | 27 | 17 | 44 | 314 |
| Bulford Ranges (SPTAE) | 9 | 18 | 10 | 28 | 311 |
| Orcheston Down (SPTAC) | 9 | 16 | 11 | 27 | 300 |
| Warminster Ranges (SPTAW) | 9 | 16 | 11 | 27 | 300 |
| Haxton (SPTAE) | 24 | 41 | 30 | 71 | 296 |
| Drivers Training Area (SPTAE) | 21 | 35 | 27 | 62 | 295 |
| Norton Down (SPTAW) | 17 | 28 | 22 | 50 | 294 |
| Cherhill Down | 11 | 19 | 13 | 32 | 291 |
| Beachs | 10 | 18 | 11 | 29 | 290 |

| | | | | | |
|---|---|---|---|---|---|
| Plantation (SPTAE) | | | | | |
| Wexland Hanging (SPTAC) | 21 | 39 | 23 | 61 | 290 |
| Whitesheet Hill | 9 | 15 | 11 | 26 | 289 |
| Bowls Barrow (SPTAW) | 15 | 27 | 16 | 43 | 287 |
| Devil's Ditch | 6 | 12 | 5 | 17 | 283 |
| Little Durnford Down | 6 | 11 | 6 | 17 | 283 |
| Ladywell (SPTAW) | 23 | 41 | 24 | 65 | 283 |
| Ell Barrow (SPTAC) | 9 | 14 | 11 | 25 | 278 |
| Everleigh (SPTAE) | 17 | 27 | 19 | 46 | 271 |
| Dunch Hill (SPTAE) | 19 | 30 | 19 | 49 | 258 |
| Tinker's Track (SPTAW) | 11 | 14 | 14 | 28 | 254 |

It is immediately clear that all of the Wiltshire sites are very far from achieving the 240 species in Alexander's list. The explanation appears to be that it includes species from a variety of different grassland habitat types, eg. limestone and chalk, short sward as well as long, bare earth and rocky places, sparsely vegetated as opposed to thickly, etc.; and also species dependent on a wide range of particular host plants. No single site is likely to fulfil all of these habitat and food plant criteria.

Consequently, if a site is to be considered for inclusion in the list, it seems reasonable to set the target number of species at a fairly low level. I have omitted all sites with less than 5 indicator species (which is why Parsonage Down, which has not been adequately surveyed for beetles, is absent) but I suspect that may be too low. The danger of working with too few species is that a small number of high scores are likely to produce anomalous results, but, on the other hand, a low number of high value species, as recorded at Beacon Hill for example, may be a useful indicator of a site's potential importance within the county.

**Table 7. Calcareous Grassland Quality Index arranged by number of species on a site.**

| Site | Sp No. | F | C | F+C | Site Score |
|---|---|---|---|---|---|
| Porton Down | 35 | 69 | 52 | 121 | 348 |
| Martin Down | 25 | 51 | 34 | 85 | 340 |
| Wylye Down | 24 | 50 | 34 | 84 | 350 |
| Haxton (SPTAE) | 24 | 41 | 30 | 71 | 296 |
| Ladywell (SPTAW) | 23 | 41 | 24 | 65 | 283 |
| Wexland Hanging (SPTAC) | 21 | 39 | 23 | 61 | 290 |
| Drivers Training Area (SPTAE) | 21 | 35 | 27 | 62 | 295 |
| Sidbury Hill (SPTAE) | 20 | 41 | 28 | 69 | 345 |
| Dunch Hill (SPTAE) | 19 | 30 | 19 | 49 | 258 |
| Everleigh (SPTAE) | 17 | 27 | 19 | 46 | 271 |
| Norton Down (SPTAW) | 17 | 28 | 22 | 50 | 294 |
| Bowls Barrow (SPTAW) | 15 | 27 | 16 | 43 | 287 |
| Fox Covert (SPTAW) | 14 | 26 | 18 | 44 | 314 |
| Black Heath (SPTAC) | 14 | 27 | 17 | 44 | 314 |
| Stonehenge | 13 | 31 | 18 | 49 | 377 |
| Middleton Down | 12 | 22 | 16 | 38 | 317 |
| Cherhill Down | 11 | 19 | 13 | 32 | 291 |
| Tinker's Track (SPTAW) | 11 | 14 | 14 | 28 | 254 |
| Beachs Plantation (SPTAE) | 10 | 18 | 11 | 29 | 290 |
| Wadmans Coppice (SPTAW) | 10 | 19 | 13 | 32 | 320 |
| Bulford Ranges (SPTAE) | 9 | 18 | 10 | 28 | 311 |
| Orcheston Down (SPTAC) | 9 | 16 | 11 | 27 | 300 |
| Warminster Ranges (SPTAW) | 9 | 16 | 11 | 27 | 300 |
| Ell Barrow (SPTAC) | 9 | 14 | 11 | 25 | 278 |
| Beacon Hill (SPTAE) | 8 | 21 | 16 | 37 | 462 |
| Cockey Down | 8 | 16 | 11 | 27 | 337 |
| Coombe Bissett Down | 6 | 9 | 11 | 20 | 333 |
| Devil's Ditch | 6 | 12 | 5 | 17 | 283 |
| Little Durnford Down | 6 | 11 | 6 | 17 | 283 |

An alternative system to grading calcareous grassland sites in this way would be to work on the basis of the presence of nationally rare species as does the SQI. But the risk then is, that because of the size and nature of grassland compared with dead wood, very mobile high-scoring species such as dung or carrion beetles, or beetles living in the nests of mammals or birds (eg. *Onthophilus punctatus* on Wyle Down), are more likely to give a false impression of a site's importance.

If a system of scoring beetles as a measure of the importance of calcareous grassland is to be used in the county it will be crucial to work out a protocol for sampling such that each site is treated in a similar way. The method chosen should be relatively simple to operate and not too time-consuming to complete. Year-long sampling, such as took place at Wylye, although highly effective, is not practicable given the funding and manpower available to tackle the large number of sites involved. And it goes without saying, of course, that it should also work in conjunction with other criteria: botanical, entomological, etc..

**Plate 16. The author using a petrol driven vacuum sampler.**

**Plate 17. Pitfall traps on Parsonage Down NNR.**

**Map 3. Showing many of the sites from which Beetles have been recorded in Wiltshire.**

Based on the 1984 & 1988 Ordnance Survey 1:250 000 maps, sheets 8 & 9, with permission of the Controller of Her Majesty's Stationery Office. © Crown Copyright.

# PRINCIPAL SITES

Unfortunately space does not permit the inclusion of information about all of the sites from which beetles have been recorded in the county. The list that follows is of some of the more important. Site descriptions have been taken in many instances from the SSSI designations. The lists of beetles at the end of each entry includes only those species regarded as nationally scarce .

Grid references are in most cases those most commonly used for the site, otherwise they indicate a central pont.

It is worth noting that apart from the sites listed here, and several others on which the low number of beetles recorded does not justify their inclusion, Wiltshire can boast a further 26 calcareous grassland SSSIs for which there are no records at all.

**The mention of a site in this list does not indicate a right of access.**

**BENTLEY WOOD**
SSSI. SU250295. 665 ha.
Owned and managed by a Charitable Trust.
Large complex of woodland to the south of Salisbury, on the site of ancient woodland, which was almost entirely replanted with oak, beech, Norway spruce, Scots pine and Corsican pine after 1945. This treatment eliminated much of the former scientific interest although butterflies and moths flourish. The Trust publishes an annual magazine, *Nature Notes*, to which I contribute occasional articles about the beetles. 215 spp. have been recorded.

*Agabus uliginosus, Ampedus cinnabarinus, Agrilus laticornis, Orthoperus nigrescens, Phalacrus brunnipes, Mordella holomelaena, Variimorda villosus, Diplocoelus fagi, Orsodacne lineola Pyrochroa coccinea, Cassida nobilis, Longitarsus dorsalis, Temnocerus longiceps.*

**BLACMOOR COPSE (Vaughan-Pickard Reserve)**
WWT reserve. SSSI. SU234292. 31.3 ha.
Situated immediately to the west of Bentley Wood on tertiary deposits of the London clay, and in small part the Reading beds. Includes stands of wet ash-maple, pedunculate oak-hazel-ash and hazel-pedunculate oak with a range of soils from light, freely drained, to heavier, more water-logged conditions. Standards of oak and ash, including some large mature trees, occur over a predominantly hazel understorey. Birches are widespread, and field maples local.

Blackmoor has one of the richest floras recorded from a Wiltshire Wood on tertiary strata and is a well-known butterfly site.

The wood was the subject of an invertebrate survey concentrating on saproxylic fauna by Andy Foster in 2004 (Foster, 2004). 157 spp. have been recorded.

*Ilybius chalconatus, Rhantus frontalis, Hydroporus marginatus, Ampedus elongantulus, A. quercicola , Agrilus biguttatus, A. laticornis, Pyrochroa coccinea, Ischnomera cyanea, Anobium inexspectatum, Synchita humeralis, Platycis minutes, Lissodema quadripustulata, Orchesia minor, Silvanus bidentatus, Stenostola dubia, Anaglyptus mysticus, Mesosa nebulosus, Psylliodes luteola, Platystomos albinos, Lasiorhynchites cavifrons, Byctiscus betulae, Brachysomus hirtus, Ellescus bipunctatus, Polydrusus flavipes, P. sericeus.*

**BRIMSDOWN HILL**
SSSI. ST831385, ST838387, ST853377. 193.7 ha
Extensive area of botanically rich calcareous grassland, scrub and woodland, with adjacent acidic types, on the western edge of Salisbury Plain. Occupies an Upper Chalk plateau, surrounded by Middle Chalk slopes of many aspects. Typical thin chalk soils occur on the slopes, while the plateau supports soils of a more acidic nature.

*Protapion filirostre, Orthochaetes setiger*

**CALSTONE and CHERHILL DOWNS**
SSSI. SU047692. 128.6 ha.
Extensive area of herb-rich chalk grassland of exceptional botanical quality on the western edge of the Marlborough Downs. The site occupies a narrow plateau of the Middle and Upper Chalk from which the land descends abruptly to the north and more gently to the south. The southern slopes are dissected by a complex system of narrow, steep sided, dry valleys and coombes, offering sheltered conditions.

*Bruchus atomarius, Longitarsus dorsalis, Microplontus campestris, Cleopomiarus graminis, Glocianus punctiger, Larinus planus*

**COCKEY DOWN**
WWT reserve. SU173320. 15.2ha
An area of botanically rich chalk grassland on a scarp of the Upper Chalk about three kms to the north east of Salisbury, once part of the Clarendon Estate. The steep slopes generally face north west, although narrow depressions have created small scale variations in aspect. Strip lynchets are a prominent feature of some parts of the site. 51 spp. have been recorded here.

**Plate 18. Cockey Down.**

*Nicrophorus interruptus, Omaloplia ruricola, Drilus flavescens, Ampedus elongantulus, Agrilus sinuatus, Rhagonycha fulva, Chrysolina violacea.*

**COOMBE BISSETT DOWN**

WWT reserve, SSSI. SU104245, 110247, 111250. c.18 ha.

Occupies three blocks of botanically rich chalk grassland situated in a series of adjacent dry valleys on the Upper Chalk. There are gentle and steep slopes with lynchet banks which are mostly west facing, and some mixed scrub including hawthorn. 28 spp. have been recorded.

*Silpha obscura, Omaloplia ruricola, Paracorymbia fulva, Agrilus sinuatus, Trachys troglodytes, Longitarsus dorsalis.*

**COTSWOLD WATER PARK**

SSSI. SU0093 to SU2099. 135 ha.

With over 100 lakes formed by mineral extraction from the Upper Thames floodplain in south- east Gloucester and north-west Wiltshire this is the most extensive marl lake system in Britain. The older lakes tend to have been dug wet and often have a varied profile and shallow margins, whereas more recent mineral working has produced steep sided profiles. Several botanical indicators of high water quality are present. 15 spp. have been recorded.

*Carpelimus subtilis.*

**CRANBORNE CHASE**

Part SSSI. ST970180. 451.4 ha.

Situated on the county border and large parts of Chase and Farnham Woods are in Dorset. Derives from an ancient hunting forest and includes remnants of enclosed mediaeval coppice, commonland, wood pasture and 19th century hazel plantation on an undulating plateau of Upper and Middle Chalk with local deposits of Clay with Flints.

From the 11th century until 1829 managed as a chase, sometimes owned privately and sometimes by royalty though never a Royal Forest. Throughout this time the chase accomodated several interests. Large numbers of deer were kept for hunting and during mediaeval times a complex matrix of small areas of wood pasture and extensive coppices (mostly to the south and east) developed managed by commoners.

When the Pitt and later the Pitt-Rivers families acquired the Chase rights after 1714 large beech and lime avenues were planted. In 1829 deer were destroyed in all areas except Rushmore Park.

Four woodland stand-types occur dominated by oak, ash and maple. The structural diversity and complex history have contributed to a rich botanical fauna.

With the exception of ten visits by Michael Salmon between 24 April and 30 July 1997, a single visit by the England Field Unit of NCC, 16 to 25 June 1980, and occasional visits by myself, once in company with WANHS members, the beetles of Cranborne Chase have been very little studied. The present list of 225 spp. includes only ten Staphylinidae.

*Silpha obscura, Microrhagus pygmaeus, Tomoxia bucephala, Mordellistena neuwaldeggiana, Long-*

*itarsus nigrofasciatus, Longitarsus parvulus, Altica brevicollis, Anthribus fasciatus, Rhinocyllus conicus.*

**GREAT RIDGE WOOD**

Fonthill Estate. ST9436.

Largest privately owned woodland in southern Britain. Oak dominant with some downland and scrub, much of which is situated on chalk slopes overlaid with clay-with flint and plateau flint gravels. The site is of considerable archaeological interest with a Roman road running its length, and incorporates areas of chalk grassland.

There is no public access except on public footpaths.

Great Ridge was visited many times by David Nash between 1974 and 1985 who compiled a list of 100 spp. including the remarkable rediscovery in GB of *Silpha carinata*. This was subsequently increased to 176 spp. by myself as a result of a contract from the WWT in 2000.

*Pterostichus oblongopunctatus, Plegaderus dissectus, Abraeus granulum, Acrotrichis arnoldi, Nicrophorus interruptus, Silpha carinata, Quedius xanthopus, Sepedophilus bipunctatus, Hylis olexai, Dasytes niger, Rhizophagus nitidulus, Cerylon fagi, Enicmus brevicornis, Psylliodes luteola, Platystomos albinus, Curculio betulae, Xyloterus signatus.*

**GREEN LANE WOOD**

WWT Reserve. ST8857. 39 ha.

Ancient woodland, dating back to before the time of Domesday, known as Slogrove until the end of the 19th century. The boundary remained unchanged from the 17thc until the 1960s when nearly half was felled for agricultural use leaving the present reserve. Most of the largest trees on this area were felled in the 1930s and many of the present trees have grown from the stools that remained. More recently there has been further clearance of trees for pylons which has resulted in the creation of a long sunny glade. The amount of oak in the wood is surprising and may be the result of a tannery being sited nearby. The wood has long been managed by coppicing.

*Carabus monilis, Hydroporus marginatus, Agrilus angustulus, Agrilus biguttatus, Agrilus laticornis, Agrilus sinuatus, Ischnomera cyanea, Anaglyptus mysticus.*

**GROVELY WOOD**

Forestry Commission and Wilton Estate. SU0534.

Very extensive tract of mixed broadleaved and coniferous woodland situated on chalk overlaid with clay with flint and plateau flint gravels. The site is of considerable archaeological interest with a Roman road running its length, and encloses the WWT's **Little Langford Down** reserve which is an extensive area of botanically rich chalk grassland. Many old trees, and standing and fallen dead wood are features of the western end, whilst the eastern has been replanted by the Forestry Commission.

Grovely was visited on many occasions by David Nash between 1970 and 1988, and formed part of a contract I received from the WWT in 2000. 501 spp. have been recorded.

*Plegaderus dissectus, Ptenidium turgidum, Nicrophorus interruptus, Prionocyphon serricornis, Agrilus angustulus, A. laticornis, Trachys scrobiculatus, Sphindus dubius, Tritoma bipustulata, Colydium elongatum, Abdera biflexuosa, Conopalpus testaceus, Pyrochroa coccinea, Platystomos albinus, Platypus cylindrus*

**Plate 19. The entrance to Great Ridge Wood from Sherrington.**

**Plate 20. Entrance to Grovely Wood from Great Wishford with Ebsbury Copse, where David Nash found *Epierus comptus*, in the distance. The field in the foreground produced hundreds of *Zabrus tenebrioides* in September 1998 (on the cultivated area, not the margins).**

**JONES'S MILL (Vera Jean's Reserve)**
WWT Reserve. SSSI. SU168613. 11.6 ha.
Best example of a calcareous valley mire in Wiltshire. Lies along the headwaters of the Salisbury Avon north east of Pewsey in a shallow valley and was formerly managed as water meadow with water-cress beds. The pattern of vegetation is complex with an intimate mixture of fen, willow-scrub and willow and alder carr. In nutrient terms the site covers a range of conditions from mesotrophic to eutrophic. A detailed description of the site is provided by Heath (2004).

The site was the subject of a survey by Andy Foster in 2002. (Foster, 2002). 197 spp have been recorded.

*Eubria palustris, Scaphisoma boleti, Gyrophaena angustata, Actenicerus sjaelandicus, Eledona agricola, Pyrochroa coccinea, Variimorda villosa, Orchesia micans, Rhagonycha translucida, Plateumaris affinis.*

**KNAPP AND BARNETT'S DOWNS**
SSSI. SU030266, 019273, 024268. 71.4 ha
Large area of botanically rich chalk grassland occupying a complex of dry valleys on Middle Chalk with a wide range of aspects. Parts of the site are heavily grazed and others, dominated by coarse grasses, less so. Hawthorn is common as both dense blocks and scatterered bushes.

All records (18 weevils) are from Mike Morris.
*Orthochaetes setiger*

**KNIGHTON WOOD**
SSSI. SU057223. c.40 ha.
Situated on a plateau of Upper Chalk and composed largely of ash and pedunculate oak with a scrub layer of hazel and scattered field maple. It is partly ancient in origin and much has been previously coppiced. A 3ha glade containing large quantities of dwarf sedge, *Carex humilis*, in species rich chalk grassland, is of particular interest.

All records are from A.J. Allen who has operated a flight intercept trap in the wood, 2007-present, and recorded more than 200 spp.

*Leiodes cinnamomea, Leiodes strigipenne, Colon dentipes, Ptomaphagus varicornis, Neuraphes praeteritus, Omalium rugatum, Tachinus rifipennis, Haploglossa marginalis, Alaobia scapularis, Microdota aegra, Thamiaraea hospita, Holobus apicata, Quedius aetolicus, Q. brevicornis, Q. nigrocaeruleus, Aphodius zenkeri, Ptinus subpilosus, Malachius aeneus, Epuraea fuscicollis, Cryptarcha strigata, Cryptophagus ruficornis, Oncomera femorata, Longitarsus parvulus, Psylliodes luteola, Brachysomus hirtus, Sitona macularius, Acalles roboris.*

**LANDFORD BOG**
WWT Reserve, SSSI. SU259185. 9.1ha.
Valley bog with peat occupying the floor and lower sides of a linear depression, the underlying geology being Bagshot Sands except for a small area of Plateau Gravel. To the west is an area woodland composed primarily of birch. The site resembles the fragile acid mire systems of the New Forest and as such is almost unique in Wiltshire.

The bog was the subject of a survey by Andy Foster in 2001 and 2003. (Foster, 2001, 2003). 123 spp. have been recorded here to date.

*Agabus unguicularis, Laccobius atratus, Helochares punctatus, Platydracus latebricola, Myllaena kraatzi, Ampedus quercicola, Chaetarthria seminulum, Varrimorda villosa, Tomoxia bucephala, Cryptocephalus bipunctatus, Rhynchites tomentosus.*

**Plate 21. Landford Bog.**

**LANGLEY WOOD**

NNR. Hamptworth Estate. SU230206. 219 ha

Extensive tract of unbroken oak high forest with many other woodland stand types, on acid clays, locally overlain by sands and gravels. The site is immediately to the north of the New Forest and contiguous with it. There appears to be no history of grazing or coppicing. Deer are common and five species have been recorded.

An invertebrate survey took place in 1987 which listed 131 spp. This number has since been increased to 367 largely as a result of two contracts which I received from EN for pitfall surveys from 22 April 2004 to 24 March 2005 (Bishops Wood) and 5 August 2006 to 2 July 2007 (Outwood).

*Carabus monilis, Pterostichus oblongopunctatus, Amara curta, Hydroporus obsoletus, Leiodes lucens, Gnathoncus buyssoni, Nicrophorus interruptus, Ilyobates subopacus, Platydracus fulvipes, P. latebricola, Stenagostus villosus, Agrilus laticornis, Hylis olexai, Microrhagus pygmaeus, Platycis minutus, Ptinus subpilosus, Malthodes fibulatus, Lycoperdina bovistae, Phloiophilus edwardsi, Thymals limbatus, Diplocoelus fagi, Triplax aenea, Enicmus brevicornis, Colydium elongatum, Tetratoma fungorum, Pyrochroa coccinea, Orchesia undulata, Longitarsus dorsalis, Altica ericeti, Ochrosis ventralis, Caenopsis fissirostris, Tropiphorus elevatus, Hypera meles, Leiosoma oblongulum, Acalles ptinoides, A. roboris.*

**LITTLE DURNFORD DOWN (DEVENISH RESERVE)**

WWT reserve. SU128349.

Comprises two distinct areas: an impressive beech wood with standing and fallen dead wood, and an area of calcareous grassland grazed by Hebridian sheep. Both are situated on the side and top of a steep, west facing chalk slope.

The reserve was the subject of an invertebrate survey concentrating on saproxylic species by Andy Foster in 2000. 127 spp. have been recorded.

*Nossidium pilosellum, Agrilus sinuatus, Prionocyphon serricornis, Phloiophilus edwardsi, Sphindus dubius, Diplocoelus fagi, Scaphidema metallicum, Lissodema quadripustulata, Pyrochroa coccinea, Orchesia minor, Ischnomera cyanea, Pseudovadonia livida, Strangalia quadrifasciata, Platystomos albinus, Taphrorychus bicolour, Xyleborus dryographus.*

**LONGLEAT WOODS**

Part SSSI of 222ha, the major portion of which is in Somerset.

The SSSI is, ancient, semi-natural, broadleaved woodland with a predominantly high forest structure.

**Plate 22. Langley Wood NNR.**

A wide range of woodland types are present including.some which are more characteristic of central England and rare in the West. Outside the SSSI there is a large area of parkland, CentreParcs occupy a part of the woodland, and another part is given over to a game reserve for wild animals.

Very little invertebrate recording appears to have been carried out. What beetle records exist are largely the work of Max Barclay, 1997-2002. 146 species have been recorded including 35 SQI species.

*Pterostichus oblongopunctatus, Scaphisoma boleti, Thymalus limbatus, Sulcacis bicornis, Prionychus ater, Hallomenus binotatus, Malthodes guttifer, Chrysolina oricalcia, C. staphylaea, Platyrhinus resinosus, Larinus planus, Platypus cylindrus.*

**MARTIN DOWN**

NNR. 342ha. The site is in Hampshire but in VC8, S. Wilts., for recording purposes.

Extensive area of chalk grassland and scrub with some woodland in the far north and south. Common grazing rights exist on the reserve. To the south the grassland has been unploughed for centuries, but in the north ploughing took place for a few years in the 1940s and 50s. The presence of clay, silt and flint deposits has resulted in patches of more acid soils.

The southern end of the site was the subject of a pitfall trapping survey by English Nature in 1999.

*Zabrus tenebriodes, Licinus depressus, Panagaeus bipustulatus, Lebia chlorocephala, Lamprinodes saginatus, Nicrophorus interruptus, Meligethes solidus, Lycoperdina bovistae, Atomaria rubricollis, Psylliodes chalcomera, Longitarsus parvulus, Brachysorus echinatus, Hemitrichapion reflexum, Cleopomiarus graminis, Orthochaetes setiger*

**PARSONAGE DOWN**

NNR. SU050412. 188.6 ha.

Outstanding calcareous downland site situated on the Upper Chalk of the southern edge of Salisbury Plain, with an extensive area of flat and gently sloping land, much of which has escaped ploughing and other agricultural improvement during the last 130 years. Most of the site is botanically rich grassland, but gorse and hawthorn occur locally as clumps particularly in the west.

Some recording was carried out at the time of preparing the Invertebrate Site Register but I have not seen these records. Since then no serious beetle recording appears to have taken place until very recently when I began a pitfall trapping programme that is still in progress.

*Agrilus sinuatus, Paracorymbia fulva, Agapanthia villosoviridescens, Anisoxya fuscula, Variimorda villosa, Malthinus balteatus, Temnocerus longiceps, Protapion varipes, Rhinocyllus conicus, Magdalis barbicornis, Tychius lineatulus.*

**PORTON DOWN**

Defence Estates. Part SSSI.

A large part of Porton Down falls in Hampshire which has caused problems in assigning records to Wiltshire when recorders have not been specific about locations. BothTower Hill and Roche Court Down, for example, fall in both counties. Although not included in the SSSI Idmiston Down and Boscombe Down East in Wiltshire, both of which abut the site, have caused further confusion.

Porton Down is an extensive area of grassland, scrub and woodland on a gently undulating plateau of Upper Chalk. Several shallow dry valleys are incorporated together with many features of archaeological significance. Documentray evidence exists showing that the grasslands, of which four main types have been identified, range from 40 to 200 years old.

Beetle recording appears not to have been as thorough as the site deserves, partly, no doubt, because of the difficulties of access. I was presented with a list of 111 species from the Site Dossier (cited in the Systematic List as Anon, 2001) but this did not include names of recorders, dates, or any habitat information. Furthermore, almost all the species were common and from Hampshire. I am also aware that surveys have taken place by eg. Monks Wood Lowlands Grassland Group in the 1970s, but have not had sight of the records many of which have been destroyed. The species are listed here are primarily the result of casual visits by individuals.

*Licinus depressus, Panagaeus bipustulatus, Pterostichus oblongopunctatus, Quedius xanthopus, Odonteus armiger, Omaloplia ruricola, Epitrix atropae, Diplocoelus fagi, Variimorda villosa, Dasytes niger, Malthinus balteatus, Cryptocephalus bipunctatus, Longitarsus parvulus, L. anchusae, Lasiorhynchites cavifrons, Neocoenorrhinus interpunctatus, Protapion filirostre, Gymnetron melanarium, Sirocalodes mixtus, Coeliodes rubber, C.transverse-albofasciatus, Sitona mcularius, Trachyphloeus spinimanus, Larinus planus*

**ROUNDWAY DOWN & COVERT**

SSSI. SU000646. 86 ha.

Extensive tract of chalk downland to the north of Devizes with an area of woodland and scrub. Occupies a series of steep, generally south-west facing slopes on the Lower Chalk. There are small outcrops of Upper Greensand on the lower slopes, while Middle Chalk occurs at the highest point. The majority of the grassland is ungrazed and tussocky, but shorter turf appears on lynchets, etc. where the soil is thinner.

The site is famous (infamous?) as being the locality where Joseph Sidebotham recorded several very rare species not since found in GB (see p.9). No serious recording has taken place since and no nationally notable species are known.

**Plate 23. Avon Valley reedbed, Salisbury. Area of transition to alder carr.**

**SALISBURY: AVON VALLEY RESERVE**

LNR. SU3. 27 ha.

The site is bounded on the east by the River Avon, on the west by allotments and a public park, and on the north by a species-rich meadow. It divides into two distinct parts: a wet reed spinney and a reedbed, both rare habitats in the county.

Because of concern about the impact of falling water levels, the author and Michel Salmon were commissioned by Chalkhill Consultants to carry out a survey on behalf of Salisbury District Council and English Nature, 7 April to 19 June 2000. 218, mostly nationally common spp. were recorded (Darby, 2001)

*Cercyon tristis, Epuraea distincta, Apion varipes, Baris lepidii, Tychius tibialis.*

**SALISBURY: BEMERTON FOLLY**

LNR. SU119316

Small area of mixed woodland with large beech and larch trees, and two open glades, once part of the Wilton Estate park and now surrounded by a modern housing estate. A public footpath runs the length of the wood. I carried out a survey for Salisbury District Council 7 March - 15 June 2002 and recorded 52 species (Darby, 2002d). Only six SQI species were found. The presence of a considerable amount of garden waste deposited along the east perimeter has undoubtedly had an impact on the fauna, not least by the introduction of a number of adventive and synanthropic species.

*Pterostichus oblongopunctatus, Platyderis ruficollis, Oligota apicata, Orthoperus nigrescens, Scaphidema metallicum.*

**SALISBURY PLAIN**

SSSI. 19,689 ha

Largest expanse (size of the Isle of Wight) of unimproved chalk downland in north west Europe and represents 41% of Britain's remaining area of this habitat. Large parts were acquired by the Ministry of Defence early in the 20th century (Salisbury Plain Training Area divided into three blocks, East, Central and West by the A342 and A360 roads) and these have not been subject to intensive farming. In addition to chalk downland, the site supports scrub and woodland habitats, temporary and permanent pools, and the Nine Mile River winterbourne. Geologically it occupies a plateau of Upper and Middle Chalk.

A diversity of soil types, slope, aspect, and past and present land use, has given rise to various grassland communities some of which are of great antiquity. Parts of the East block and the periphery of the Central and West blocks comprise areas of grassland managed for grazing pasture and hay cutting, whilst the middle of Centre and West are ungrazed.

The Coleoptera of the Plain received surprisingly little attention until comparatively recently. At the time of its designation as an SSSI in 1993 only one beetle, *Cantharis fusca*, was thought worthy of mention. Ron Carr had surveyed water beetles on 20 sites in 1991,1992 but his list of the 49 species he recorded was not published until 1997. Mark Telfer was involved in survey work after Defence Estates issued a contract to the Centre for Ecology and Hydrology, Monks Wood, in June 1997, and the Ministry of Defence commissioned Peter Hodge to do a general insect survey of Tilshead Park and Haxton Down in July 1999. Three years later Hodge was back when English Nature commissioned him to survey 8 Sites in May and June 2002. The most recent work has been by Dave Boyce who (with Steve Wright) received a contract from Natural England to survey 12 sites (some duplicating the earlier sites) in June and August 2008.

These surveys have resulted in the listing of more than 200 species including many rare weevils and other phytophagous species. The sites involved are:

| | | |
|---|---|---|
| SPTAW | Bowls Barrow | ST9247 |
| | Fox Covert | SU0244 |
| | Ladywell | ST9248, 9348 |
| | Norton Down | ST9245 |
| | Tilshead Park | SU0247 |
| | Tinker's Track | ST9846 |
| | Wadman's Coppice | ST9448 |
| | Warminster Ranges | ST9148 |
| SPTAC | Black Heath | SU0650 |
| | Ell Barrow | SU0851 |
| | Orcheston Down | SU0748 |
| | Wexland Hanging | SU1147, 1148 |
| SPTAE | Beach's Plantation | SU1850 |
| | Beacon Hill | SU2144 |
| | Bulford Ranges | SU2046 |
| | Drivers Training Area | SU2149 |
| | Dunch Hill | SU2049 |
| | Everleigh | SU1754 |
| | Haxton | SU1949, 1950 |
| | Nine Mile River | |

Some of the more interesting species are: *Agabus labiatus, Hydroglyphus geminus, Brachinus crepitns, Carabus monilis, Bembidion clarkii, Amara praetermissa, Ophonus azureus, O. Schaubergerianus, Chlaenius nigricornis, Licinus depressus, Lebia chlorocephala, Panagaeus bipustulatus, Helophorus griseus, H. longitarsus, Berosus affinis, B. signaticollis, Aloconota longicollis, Aphanisticus pusillus, Trachys scrobiculatus, Malthinus balteatus, Mordellistena brevicauda, Meligethes bidentatus, M. exilis, M. solidus, Olibrus flavicornis, Phalacrus brunnipes, Scymnus femoralis, Pyrochroa coccinea, Phytoecia cylindrica, Bruchus atomarius, Chrysolina sanquinolenta, Cryptocephalus aureolus, Longitarsus anchusae, L. curtus, L. dorsalis, L. parvulus,*

*Crepidodera nitidula, Psylliodes chalcomera, Exapion difficile, Protapion filirostre, Hemitrichapion reflexum, Oxystoma cerdo, Notaris scirpi, Cleopomiarus graminis, Gymnetron collinum, G. melanarium, Tychius lineatulus, T. squamulatus, Aulacobaris picicornis, Glocianus moelleri, G. pilosellus, Limobius borealis, Trachyphloeus alternans, T. aristatus, T. asperatus, T. spinimanus, Hadroplontus trimaculatus, Trichosirocalus horridus, Calosirus terminatus, Larinus planus.*

**SAVERNAKE FOREST**

SSSI. 904.7ha

Extensive area of ancient woodland with over one thousand years of documented history. Lies on a plateau of Upper Chalk, covered with Clay with Flints and dissected by dry valleys. The Chalk is exposed along the valley sides, but small deposits of Valley Gravels, Reading Beds and Bagshot Sands, together with the clay, give rise to particularly acid conditions in places.

Relics of the ancient wood pasture remain, represented by the distinctive, open-crowned specimens of sessile and pedunculate oak which are scattered throughout the site. 18th and early 19th century beech and oak plantations have now assumed a semi-natural structure, but the forest consists mainly of 20th century beech or oak plantation.

The forest grassland shows both acidic and calcareous characteristics with a rich flora, and there is an exceptional diversity of fungi numbering well over 500 spp. There are a few ponds and a small area of chalk heath.

At first sight Savernake looks as good for Coleoptera as Windsor or the New Forest, but this is deceptive. As long ago as 1873 Meyrick noted that Janson 'expressed it as his opinion, that he had never seen fewer Coleoptera in any district than ours, and he was especially disappointed in the Longhorns, for which he considered the Forest eminently suited, yet very few occurred' (Meyrick, 1874). And almost exactly a hundred years later A.A.Allen wrote: 'I may say in passing that despite the abundance of old beech timber in this potentially rich locality, the lignicolous Coleoptera appeared on the whole to be few and disappointing – so far as several days work there permitted me to judge' (Allen, 1971). Like Meyrick, Allen thought a more prolonged acquaintance with the Forest would produce better results, but as Meyrick would later discover, and as we now know, it has not.

The forest was visited by the members of the British Entomological and Natural History Society in May 1976 and by the England Field Unit of the NCC in 1999, and since then it has been the subject of extensive trapping exercises by EN as follows:

1993. Intercept and Malaise trapping at White Road South (SU205676) and White Road North (SU20986819) between 20 April -11 October 1993; and Red Vein Bottom West (SU21746770) and Red Bottom East (SU22456782) between 16 June - 11 October 1993. Determinations by S.A.Williams.

1994. Intercept and Malaise trapping at Long Harry (SU20826768) between 28 April – 18 May, 9 June – 19 July, 30 August – 11 October 1994; Church Walk (SU20746746) between 18 May - 11 October 1994; and Grey Road (SU20536738) between 28 April - 11 October 1994. Determinations by S.A.Williams

1995. Intercept and Malaise trapping at Marie Louise Ride (SU21376582); Sawpit Drive (SU21536567); Charcoal Burners (SU21756542); and Column Ride (SU21746535) at various dates between 31 May - 16 November 1995. Determinations by M.D.Darby.

In spite of this programme and numerous individual visits by a large number of Coleopterists, Savernake does not boast any RDB1 or RDB2 species. (Darby & Alexander, 2003). The number of species recorded for the Forest is 436.

*Ptenidium turgidum, Quedius xanthopus, Placusa tachyporoides, Calambus bipustulatus, Procraerus tibialis, Malthodes fibulatus, M. maurus, Platycis minuta, Ctesias serra, Ptinomorphus imperialis, Dorcatoma flavicornis, Ptinus subpilosus, Epuraea longula, Pediacus depressus, Silvanus bidentatus, Diplocoelus fagi, Cerylon fagi, Lathridius consimilis, Mycetophagus piceus, Colydium elongatum, Eledona agricola, Pseudocistela ceramboides, Mycetochara humeralis, Pyrochroa coccinea, Phloiotrya vaudoueri, Melandrya caraboides, Conopalpus testaceus, Tomoxia bucephala, Ischnomera cyanea, I. caerula, I. sanguinicollis, Prionus corarius, Leptura sexguttata, Anaglyptus mysticus, Platypus cylindrus*

**SPYE PARK**

Part SSSI occupying 90.3 ha.

Complex habitat mosaic including large expanses of some of the finest undisturbed alderwoods in the county. Many spp. of lichens, bryophytes and vascular plants in the wet soil conditions. The geology comprises Oxford Clay, Corallian Beds and Lower Greensand within a valley system of gentle to moderate slopes. There are many wet flushes and streams. Almost no Coleoptera recording has taken place.

*Trypodendron signatum*

**STARVEALL & STONY DOWN**

Part SSSI occupying 22.5 ha. ST991400.

Large block of botanically rich chalk grassland on the steep sides of a dry valley north of the River Wylye. Most of the slopes are west, south west, or east facing on exposures of Middle Chalk though there are small areas of Upper Chalk at the highest levels.

*Orthochaetes setiger*

**WYLYE DOWN**
NNR. SU002361. c.60 ha.
Extensive block of floristically rich calcareous grassland designated as a NNR in 1951. Occupies a shallow, dry north-south valley showing evidence of an extensive Celtic field system. The uniformly rich, short turf has been maintained by many years of grazing by sheep and cattle.

7 spp. of weevils were recorded from the site by Mike Morris before 2006 when I received a contract from English Nature to carry out a survey. Using pitfall traps over a twelve month period, supplemented by some experimental flight intercept trapping and sweeping of the taller vegetation, 203 further species were recorded, including two RDB species.

*Brachinus crepitans, Carabus monilis, Licinus depressus, Panageaus bipustulatus, Onthophilus punctatus, Agathidium marginatum, Nicrophorus interruptus, Choleva fagniezi, C. oblonga, Ptomaphagus varicornis, Acidota cruentata, Aleochara cuniculorum, Ischnosoma splendidum, Lathrobium multipunctatum Megalinus glabratus, Mycetoporus nigricollis, Ocypus fortunatorum, Philonthus carbonarius, Platydracus fulvipes, Quedius cinctus, Q. schatzmayri, Q. scintillans, Sepedophilus lusitanicus, Tasgius winkleri, Aphanisticus pusillus, Mordellistena variegata, Rhagonycha lutea, Cassida prasina, Atomaria rubricollis, Longitarsus dorsalis, L. atricillus, Psylliodes luteola, Catapion pubescens, Larinus planus, Cleopomiarus graminis, Orthochaetes setiger, Stenocarus umbrinus, Strophosoma faber, Trachyphloeus alternans.*

**Plate 24. The managed water meadows at Britford, site of a survey by C. Martin Drake in 2002 (Drake, 2009).**

# ABBREVIATIONS
## Collectors, Recorders and Determiners

| | |
|---|---|
| AAA | A.A. Allen |
| AC | A. Coultiss |
| ACW | A.C. Warne |
| AES | A.E. Stubbs |
| AGC | A.G. Chichester |
| AGD | A.G. Duff |
| AGJ | A.G. Jebb |
| AH | A. Halstead |
| AJM | A.J. Maddock |
| AJWA | A.J.W. Allen |
| AKS | A. Kirk-Spriggs |
| ALJ | A.L. Jackson |
| AM | A. McKee |
| AMF | A.M. Fairbairn |
| AOC | A.O. Chase |
| AP | A. Percival |
| APF | A.P. Foster |
| APFe | A.P. Fowles |
| AS | A. Skinner |
| AWM | A.W. Mahaffy |
| BB | B. Blumsom |
| BG | B. Gillam |
| BH | B. Heath |
| BJR | B.J. Reynolds |
| BL | B. Last |
| Ble | B. Levey |
| BM | B. Maxfield |
| BNP | B.N. Parker |
| BPM | B.P. Moore |
| BT | B. Taylor |
| BWHC | B.W.H. Coulson |
| CA | Sir C. Andrewes |
| CD | ? |
| CET | C.E. Tottenham |
| CH | C. Hosie |
| CHH | C.H. Harding |
| CJ | C. Johnson |
| CM | Mrs C. Mardon |
| CMD | C.M. Drake |
| CMJ | C. MacKechnie-Jarvis |
| ColH | Colonel Havergal |
| CPH | C.P. Hurst |
| CT | C. Tilbury |
| CTS | C.T.Sidgwick |
| DAC | D.A. Cooling |
| DB | D. Bretheridge |
| DBS | D.B. Shirt |
| DC | D. Clements |
| DCL | D.C. Lloyd |
| DG | D. Graiff |
| DH | D. Halstead |
| DGB | D.G. Brown |
| DHM | D.H. Murray |
| DJG | D.J. Gibbs |
| DJL | D.J. Leeming |
| DK | D. Keen |
| DL | D. Lambert |
| DLC | D. LeCren |
| DMR | D.M. Rush |
| DP | D. Procter |
| DRN | D.R. Nash |
| DT | D. Turner |
| DTB | D.T. Bilton |
| DS | D. Simpson |
| DSu | D. Sumner |
| EBB | E.B. Badcock |
| ECD | E.C. Darby |
| EEdeS | E.E. de Satgé |
| EJP | E.J. Pearce |
| EKC | E.K. Chambers |
| ERBS | E.R.B. Stiddard |
| ET | E. Tigwell |
| EWJ | E.W. Janson |
| FBB | F. Balfour-Browne |
| FEDH | F.E. Drummund-Hay |
| FSA | F.S. Alston |
| GAW | G.A. Wood |
| GW | G. Walton |
| GCC | G.C. Champion |
| GEW | G.E. Wooroffe |
| GH | G. Heaton |
| Gha | G. Harris |
| GHW | G.H. Wiltshire |
| GJK | G.J. Kerrich |
| GLR | G.L. Rolleston |
| GNF | G.N. Foster |
| GO | G. Orledge |
| GOD | G. O'Donovan |
| GSB | G.S. Banwell |
| GT | G. Terry |
| GTR | G.T. Rudd |
| GTKM | G.T.K. Maurice |
| GW | G. Watson |
| GWa | G. Walton |
| GWG | G.W. Green |
| GWGH | G.W.G. Head |
| GY | Mrs G. Yerrington |
| HCB | H.C. Berwick |
| HFC | H.F. Cory |
| HGM | H.G. Morgan |
| HGS | H.G. Stokes |
| HM | H. Mendel |
| HMP | H.M. Prescott |
| HT | H. Tierney |
| HW | H. West |
| HS | H. Senior |
| IJG | I.J. Gray |
| IR | I. Rendell |
| ISC | I.S. Carter |
| ITH | I.T. Hamilton |
| KJA | K.J. Atkins |
| JA | J. Acton |
| JAO | J.A. Owen |
| JAW | J.A. Wanklyn |
| JC | J. Cooter |
| JCK | J.C. Krüger |
| JED | J.E. Druett |
| JEO | J.E. Oliver |
| JES | J.E. Smith |
| JET | J.E. Thorpe |
| JF | J. Fishwick |
| JHB | J.H. Bratton |
| JHK | J.H. Keys |
| JKG | J.K. Grearson |
| JL | J. Langworthy |
| JLE | J.L. Evans |
| JM | J. Moon |
| JMW | J.M. Walters |
| JN | J. Notman |
| JP | J. Paul |
| JS | J. Sidebotham |
| JSD | J.S. Denton |
| JSm | J. Smith |
| JVB | J.V. Blackford |
| JW | J. Walker |
| JWe | J. Webb |
| JWo | J. Woolger |
| JeW | J. Wall |
| KC | K. Crawford |
| KGA | K.G. Allenby |
| KGB | K.G. Blair |
| KF | K. French |
| KNAA | K.N.A. Alexander |
| LJW | L. Jones-Walters |
| LN | L. Newcombe |
| MA | M. Ausden |
| MAL | Mrs A. Lange |
| MAS | M.A. Salmon |
| MC | Mr Coleman |
| MCH | M.C. Harvey |
| MDD | M.D. Darby |
| MG | M. Gurney |
| MGM | M.G. Morris |
| MGT | M.G. Telfer |
| MH | Miss Haddock |
| MHo | M.Howe |
| MIR | M.I. Russell |
| ML | M. Lang |
| MLC | M.L. Cox |
| Mlu | M. Lush |

| | |
|---|---|
| MNS | M.N. Smith |
| MP | Miss Preston |
| MPS | Mrs P. Simpson |
| MPT | Mrs P. Thomas |
| MrT | Mr Twisk |
| MS | M. Sammes |
| MSt | M. Stone |
| MW | M. Waters |
| NH | N. Hutchins |
| NHJ | N.H. Joy |
| NNM | N.N. Marsh |
| NR | N. Rea |
| OFW | O.F. Waterfield |
| PAH | P.A. Harwood |
| PCT | P.C. Tinning |
| PCP | P.C. Phillips |
| PD | P. Darby |
| PH | P. Harwood |
| PHM | P.H. Maddock |
| PJC | P.J. Chandler |
| PJH | P.J. Hodge |
| PJW | P.J. Watson |
| PLB | P.L. Brown |
| PLP | P.L. Philpot |
| PM | P. Mobsby |
| PMH | P.M. Hammond |
| PMe | P. Merritt |
| PMP | P.M. Pavett |
| PP | Mrs P. Parker |
| PS | P. Skidmore |
| PVS | P.V. Strong |
| RAJ | R.A. Jones |
| RAW | R.A. Waddilove |
| RB | R. Beckett |
| RCS | R.C. Stern |
| RD | R. Davies |
| RL | R. Large |
| RM | R. Montague |
| RMC | R.M. Charles |
| RRUK | R.R. Uththoff-Kaufmann |
| RS | R. Scagell |
| RSK | R.S. Key |
| RWL | R.W. Lloyd |
| RWR | R.W. Rickards |
| SAC | S.A. Clarke |
| SB | S. Bottom |
| SC | S. Covey |
| SD | S. Day |
| SE | S. Edwards |
| SG | S. Gaskell |
| SGB | S.G. Ball |
| SW | S. Warrington |
| SWa | S. Walker |
| TAP | Rev T.A. Preston |
| TGW | T.G. Winter |
| TJ | T. James |
| TM | T. Marsham |
| TNHS | Rev.T.N. Hart-Smith |
| TWED | T.W.E. Darby |
| VH | V. Headland |
| WC | W. Crowther |
| WES | W.E. Sharpe |
| WFB | W.F. BLandford |
| WBP | W.B. Price |
| WN | W. Nicholas |
| WRQ | W. & R. Quantrill |

# Other abbreviations

| | |
|---|---|
| AONB | Area of Outstanding Natural Beauty |
| BAP | Biodiversity Action Plan |
| BBI | Beetles (British Isles) web site |
| c. | circa |
| Conf. | Confirmed by |
| DFL | Dauntsey Fauna List (see Bibliography) |
| EN | English Nature |
| GB | Great Britain |
| HAP | Habitat Action Plan |
| IEC | Index of Ecological Continuity |
| ISR | Invertebrate Site Register |
| ITE | Institute of Terrestrial Ecology |
| MCC | Marlborough College Collection |
| LNR | Local Nature Reserve |
| NCC | Nature Conservancy Council |
| NE | Natural England |
| n.d. | no date |
| NHM | Natural History Museum, London |
| NNR | National Nature Reserve |
| nr. | near |
| R. | River |
| SQI | Saproxylic Quality Index |
| SPTA | Salisbury Plain Training Area |
| SSSI | Site of Special Scientific Interest |
| UK | United Kingdom |
| WANHS | Wiltshire Archaeological and Natural History Society |
| WSBRC | Wiltshire and Swindon Biological Records Centre |
| WWT | Wiltshire Wildlife Trust |

**Plate 25. Flight intercept traps on the author's compost bins. The traps are made by cutting holes in the sides of plastic drinks bottles and the preservative used is Ethylene glycol (antifreeze).**

# SYSTEMATIC LIST

Because this is the first full account of Wiltshire beetles all known records have been listed, even though this has meant including historical records not always incorporated in such works. Fossil records are listed separately because none of the species has survived in the county into modern times.

The Marlborough College records have presented me with a problem in so far as duplication is concerned. In his 1938 publication, Meyrick makes specific reference to whether a species was included in the earlier lists of Jebb and himself, but does not mention whether the same species was also found later. Because I felt that all records specifically attributable to individuals, particularly well-known coleopterists such as Rye, Champion, Janson and Blair, should be mentioned, together with references to Meyrick's 1938 publication, duplication almost certainly occurs in many instances.

Nomenclature of taxa follows the latest *British Checklist* (Duff, 2008). Space could probably have been saved by not including all authors' names and dates, but, as a result of the laws of priority, history is an integral and often fascinating part of entomological science. Furthermore, it does no harm to be reminded of the considerable achievements, often at a very early date, of our predecessors. At a practical level, recent designations are often the result of the splitting of an earlier species into two or more separate species, and hence act as a warning to be particularly careful when making determinations.

Because of the paucity of records compared with many other counties, a depth of information on which to draw for distributional data is lacking. What records we have seem to me to reflect collecting effort better than species distribution. For this reason, and because of the format of the entries which makes clear the vice county and 10 km square distribution, I felt that the provision of individual species maps was not only unnecessary, but likely to be misleading.

It should be noted that some of the records listed have not been authenticated or checked by specialists.

## Guidance notes for users

The format of the entries is set out in the following example (partially fabricated for explanatory purposes):

**Superfamily CURCULIONOIDEA**
**Family CURCULIONIDAE**
Subfamily COSSONINAE
Tribe COSSONINI [(1)]

**RHOPALOMESITES Wollaston, 1873** [(2)]
[SUBGENERIC NAME] [(3)]
Holly Weevil [(4)]
*?**Rhopalomesites tardyi*** (Curtis, 1825) [(5)]
**Nb,** [(6)] **IEC-3,** [(7)] **SQS-8** [(8)] **[LCGR-FI CI]** [(9)]
7 [(10)] SU05 [(11)]: Devizes, Roundway,[(12)] 2-Jun-1864 [(13)] (JHB) [(14)] (Sidebotham, 1968) [(15)]
Resident.[(16)] Very local; [(17)] in pile of cut vegetation [under bark of oak] in wet area. [(18)] Two examples in MCC.[(19)]

**(1) Family and other names**
Family, superfamily, and other supra-generic names are included without authorities and citations.

**(2) Generic name**
The name of the genus into which the species is placed together with the authority and date of citation.

**(3) Subgeneric name**
When applicable, sub-generic names are given, together with authorities and dates of citations. These are sometimes a helpful guide to earlier, well known genera.

**(4) English name**
English names are given when they exist (rarely) These names have been extracted from a range of sources both old and new.

**(5) Latin species name**
The Latin name, authority and date of citation are as Duff (2008). Brackets around the name of the authority indicate that the species was originally cited in a different genus. A square bracket preceding the entry indicates that the species has never existed in the county but is included in the systematic list because it has been previously published as such. A ? before the entry indicates that there is doubt about the species' inclusion in the county list.

**(6) Status in GB**
See chart on p.47. The classifications are taken from Hyman and Parsons, (1992, 1994). All species not fitting into these categories are considered to be either common or local.

**(7) Index of Ecological Continuity (IEC)**
See p.26 and table on p.47. Species were selected for inclusion in the Index according to the extent to which they have been consistently recorded from areas of ancient woodland with continuity of dead-wood habitats, particularly in pasture woodlands.

The threshold figures for calculating sites of international, national and regional importance are:
International importance.................................over 80
National importance.......................................25-80
Regional importance.......................................15-80

**(8) Saproxylic Quality Index (SQI)**
See p.25 and table on p.47. All saproxylic species are scored according to their status in GB as set out by Hyman and Parsons (1992, 1994).

The threshold figures for calculating sites of national and regional importance are tentatively set as follows:
International importance................................over 590
National importance.........................................500- 590

**(9) Lowland Calcareous Grassland Review**
See pp.27-29 and table on p.47. Alexander (2003) lists 250 species which, in his view, are strongly associated with, and dependent upon lowland calcareous grassland, and graded the species according to their fidelity and continuity in the habitat.

**(10) Vice County.**
Recording is based on the Watsonian Vice County system (Dandy, 1969). See p.3: Recording area.
7-North Wiltshire,
8-South Wiltshire.

**(11) 10km square.**
As far as possible each record has been assigned to a 10km square of the national grid. This system, rather than the listing of more detailed map references follows practice elsewhere, and it is hoped that it will assist the production of distrubutional maps in future. It also assists in relating distributional information to that for other Wiltshire animals and plants already treated in this way (eg. Gillam, 1993).

In some instances where locations fall into two or more 10 km squares - Devizes and Salisbury, for example, - I have opted for a single square as denoting the location of a record when no specific information to enable more precise documentation has been provided. Where information has been provided it goes without saying that the same location will be listed in two or more squares. Similarly, in the case of squares bridging the vice county boundary the same square number may appear in both vice counties

**(12) Location**
Within a 10km square records are sorted alphabetically by location. Further information about many of the more important sites is provided above.

**The mention of a location does not indicate a right of access.**

**(13) Date of record**
Year dates before 1920 are written in full. After 1920 they are abbreviated to decades except when the year is mentioned alone when it is written in full. When continuous recording has taken place over time, eg. when intercept, malaise or pitfall traps have been used, the beginning and end dates are separated by 'to'. When recording has not been continuous, but involved more than 8 records from the same site, a hyphen is used between the dates.

**(14) Initials of recorders**
Full surnames and initials are given in the table of Abbreviations above. Occasionally names of determiners and confirmers, when separate from the recorder, are added, in which case they are preceded by det[ermined by] or conf[irmed by].

**(15) Citations**
All published references to support individual records have been included, where known (see Bibliography). Similarly, many more recent records taken from unpublished surveys commissioned by private and public bodies are referenced in the same way.

**(16) Historical status of the species in Wiltshire**
I have adopted the terms used by Duff (1993), ie.

Resident: a species with a self-sustaining population over a long period.

Casual: a species with a self-sustaining population over a few seasons, then dying out.

Vagrant: a species with no self-sustaining populations established at any time.

Species are explicitly considered to be native, i.e. they arrived in Wiltshire in ways not directly dependent upon Man's activities; otherwise they are explicitly considered to be introductions.

**(17) Dispersion**
Again, I have adopted the terms used by Duff (1993). Widespread, local and very local are used to indicate dispersion and refer directly to the number and location of the 10km squares in which a species has been recorded. VC7 and VC8 fall into 53 squares of which 8 may be ignored because the areas concerned are so small or their entomological significance negligible (see maps, pp.18, 30). Like Duff I have adopted subjective dispersion categories for the remaining 45 as follows: under 1-3 10km squares: very local; 3 – 7 10 km squares: local; 7 – 20 10 km squares: widespread (but only when there is a spread of squares in both vice counties); above 20 10km squares: ubiquitous.

**(18) Habitat**
Habitat and other ecological information is included at the end of each species entry. Where information from other sources, both national and regional, provides insights not revealed by local data, it has been given in square brackets. Where specific local habitat data is given (ie not in square brackets) it may relate to one or more records only in a longer series when no other information is available from that series.

**(19) Notes**

Notes on recent name changes, conservation significance, national status, location of specimens, etc.. Comments on the existence of species or not in the Marlborough College Collection (MCC) referred to by Meyrick (1938), are confined only to those species of which Meyrick's account is the only indication of their presence in the county, or because they are considered nationally or locally rare.

## Species illustrations

The illustrations of species fall into two distinct categories. 130 are confined to nationally rare and scarce species, and to one or two which are not nationally rare but are scarce or unusual in Wiltshire. Very small and similar species such as Cryptophagidae, Nitidulidae, Ptiliidae and, many of the Curculionidae, etc. have not been included. The remaining 25 illustrations are confined to some of those species which I am most frequently asked to identify as the county recorder.

The majority of illustrations are of Wiltshire specimens, the few exceptions being mainly in the Staphylinidae and Chrysomelidae. I have not introduced scale bars assuming that most readers will already be familiar with relative sizes or have access to literature which provides specific measurements.

## Status designations

Table 8 which follows is placed on a separate page so that it may be more easily read in conjunction with the systematic list.

## List of fossil beetles

The following fossils are listed by Ross and Jarzembowski, 1996. (see p.9). I have not followed the example of Duff (1993) and included them in the main systematic list because none have survived into the present time and the genera cannot be associated with even the wider European fauna.

***Anapiptus brodiei*** Handlirsch
Dinton (Brodie, 1845) NHM I3515

***Apistotes purbeccensis*** (Giebel)
Vale of Wardour (Brodie, 1845) NHM 11963

***Cerylonopsis striata*** (Brodie)
Dinton (Brodie, 1845) NHM 3514

***Coleopteron rugostriatus*** (Giebel)
Teffont (Brodie, 1845) NHM 11959

***Coleopteron vetustus*** Giebel
Dinton (Brodie, 1845) NHM 3516

***Diaperidium mithrax*** Westwood
Vale of Wardour

***Helophoropsis brodiei*** (Giebel)
Dinton (Brodie, 1845) NHM 3524

***?Helopidium brodiei*** (Giebel)
Vale of Wardour (Brodie, 1845) NHM 11962

***Helopidium westwoodi*** (Giebel)Dinton, Vale of Wardour (Brodie, 1845) HNM 3500, 3946

***Hydrobiites purbeccensis*** (Giebel)
Dinton (Brodie, 1845) NHM 11966

***Hyperomima antiqua*** (Giebel)
Dinton (Brodie, 1845) NHM 3504

***Kakoselia angliae (***Giebel)
Dinton (Brodie, 1845) NHM 3528

***Kamaroma breve*** Handlirsch
Dinton (Brodie, 1845) NHM 11968

***Katapontisus brodiei*** (Giebel)
Vale of Wardour (Brodie, 1845) NHM 11964

CUPEDIDAE
***Omma elongata*** (Brodie)
Dinton (Brodie, 1845) NHM 3527, 12149

***Pseudocymindis antiqua*** (Giebel)
Vale of Wardour (Brodie, 1845) NHM 11965

***Stictulus brodiei*** Handlirsch
Dinton (Brodie, 1845) NHM 3519

***Tychon antiquum*** (Giebel)
Vale of Wardour (Brodie, 1845) NHM 11967

**Table 8: Status Designations**

| | |
|---|---|
| **RDB1(Endangered)** | Taxa in danger of extinction in Britain and whose survival is unlikely if causal factors that have contributed to their rarity continue operating. |
| **RDB 2 (Vulnerable)** | Taxa believed likely to move into the Endangererd category in the near future if the causal factors that have contributed to their rarity continue operating. |
| **RDB 3 (Rare)** | Taxa with small populations that are not at present Endangered or Vulnerable, but are at risk of becoming so. |
| **RDB I (Indeterminate)** | Taxa considered to be Endangered, Vulnerable or Rare, but where there is not enough information to say which of the three categories (RDB 1 to 3) is appropriate. |
| **RDB App (Extinct)** | Taxa that were formerly native to Britain but have not been recorded since 1900. |
| **RDB K (Insufficien-tly known)** | Taxa that are suspected but not definitely known to belong to any of the above categories, because of lack of information |
| **Na (Notable A)** | Taxa that do not fall within RDB categories but which are none-the-less uncommon in Great Britain and thought to occur in 30 or fewer 10km squares of the National Grid or, for less well-recorded groups, within 7 or fewer vice-counties. |
| **Nb (Notable B)** | Taxa that do not fall within RDB categories but which are none-the-less uncommon in GB and thought to occur in between 31 and 100 10km squares of the National Grid or, for less well-recorded groups between 8 and 20 vice-counties |
| **Notable:** | Species that are estimated to occur within the range of 16 to 100 km squares, but the subdividing of which into Notable A and Notable B was not attempted by Hyman and Parsons (1992, 1994). |
| **IEC-1** | Species which are known to have occurred in recent times only in areas believed to be ancient woodland, mainly pasture woodland. |
| **IEC-2** | Species which occur mainly in areas believed to be ancient woodland with abundant dead-wood habitats, but which also appear to have been recorded from areas that may not be ancient woodland or for which the locality data is imprecise. |
| **IEC-3** | Species which occur widely in wooded land, but which are collectively characteristic of ancient woodland with dead-wood habitats. |
| **SQS-1** | Common species occurring in dead wood habitats |
| **SQS-2** | Local species occurring in dead wood habitats |
| **SQS-4** | Very local/uncertain species occurring in dead wood habitats |
| **SQS-8** | Nationally notable B species occurring in dead wood habitats |
| **SQS-16** | Nationally notable A / Red Data Book RDBK, Insufficiently Known, species occurring in dead wood habitats. |
| **SQS-24** | Red Data Book RDBI, Indeterminate /RDB3, Rare species occurring in dead wood habitats. |
| **SQS-32** | Red Data Book RDB1, Endangered/ RDB2, Vulnerable and RDBApp, Extinct species occurring in dead wood habitats. |
| **LCGR-Fh** | High fidelity. Species which occur primarily in lowland calcareous grassland and are therefore strongly dependent on their habitat. |
| **LCGR-Fl** | Low fidelity. Species which can be conserved in other habitat types and as a result may not be a priority. |
| **LCGR-Fm** | Medium Fidelity. Species which fall between the above two categories |
| **LCGR-Ch** | High continuity. Species which are dependent on conservation at the site level and which once lost may be difficult to recover other than by re-establishment projects. |
| **LCGR-Cl** | Low continuity. Species which are not dependent on conservation at the site level and which once lost could probably be re-covered without re-estblishment projects. |
| **LCGR-Cm** | Medium continuity. Species which fall between the above two categories |

# Order COLEOPTERA
## Suborder MYXOPHAGA
### Superfamily SPHAERIUSOIDEA
### Family SPHAERIUSIDAE

**SPHAERIUS Waltl, 1838**
**[*Sphaerius acaroides* Waltl, 1838**
**RDBK**
7 SU16: Marlborough area, 1900-1905 (Meyrick, 1906)**]**
Previously known as *Microsporus*. The specimen probably referred to by Meyrick is in the MCC labelled *Chilton 5/7/02*, and is not a beetle but a small seed pod.

### Family GYRINIDAE
Subfamily GYRININAE
Tribe GYRININI

**GYRINUS Geoffroy, 1762**
**Mariner Whirligig Beetle**
***Gyrinus marinus* Gyllenhal, 1808**
7 SU16: Lockeridge, 1994 (JEO)
8 SU95: Great Cheverell, 15-May-41 (LeC) (DFL, 1948)
Resident. Very local [in open water on lakes and large ponds].

**Shady Whirligig Beetle**
**?*Gyrinus natator* (Linnaeus, 1758)**
**RDB1**
7 SU16: Lockeridge, 1994 (JEO); Marlborough area, 1870-1938 (Meyrick, 1938); 14-Mar-1879 (WFB) (Meyrick, 1880)
These records almost certainly refer to *G. substriatus*. *G. natator* is known in Britain from two tens in Westmorland and Cumberland only.

**Common Whirligig Beetle**
***Gyrinus substriatus* Stephens, 1828**
7 ST98: Dauntsey, silt pond, 20-May, 15-Jun-33 (EJP) (Pearce, 1934, as *G.natator,* but the specimens are in the MCC and are this species)
SU09: Ashton Keynes, 15-Mar-00 (JKG det MDD); Cotswold Water Park, 28-Apr-82 (GNF)
8 ST94: Ladywell (SPTAW), pond, 3-Jun - 12-Aug-08 (Boyce & Wright, 2009); Wadman's Coppice (SPTAW), pond, 3-Jun - 12-Aug-08 (Boyce & Wright, 2009)
ST95: West Lavington, Jun to Aug-41 (LeC) (DFL, 1948)
SU14: Nine Mile R., Bulford (SPTAE), 23-Jun-93 (CMD)
SU21: Hamptworth Estate, pond, 3-Aug-73 (DRN); Pound Bottom, 9-Apr-74 (DRN)
SU22: East Grimstead, bomb crater, 1-Aug-78 (DRN)
SU25: Limmer Ponds, 28-Jul-92 (RC) (Carr, 1997)
Resident. Local; on smaller man-made ponds and ditches in shaded or exposed conditions.

Gyrinus urinator

***Gyrinus urinator* Illiger, 1807**
**Nb**
7 ST96: Semington Brook, 1-Jul-97 (MNS)
Resident. Very local; [flotillas may be found in slower stretches of partly shaded streams and rivers amongst overhanging vegetation].

Tribe ORECTOCHILINI

**ORECTOCHILUS Régimbart, 1882**
**Hairy Whirligig Beetle**
***Orectochilus villosus* (Müller, O.F, 1776)**
7 ST98: Dauntsey, silt pond, 1932-1933 (EJP) (Pearce, 1934)
8 ST96: Semington Brook, 1-Jul-97 (MNS)
SU03: R. Wylye, Chilhampton, 14-Aug-86 (DRN)
Resident. Very local; [in rivers and wave-washed shores of exposed lakes]. Forms large flotillas at night but is usually seen singly during the day. Flightless.

### Family HALIPLIDAE

**BRYCHIUS Thomson, C.G, 1859**
***Brychius elevatus* (Panzer, 1793)**
7 ST87: Castle Combe, 1928 (EJP); Nettleton Brook, 1-May-79 (ISC)
ST88: Sherston, 1933 (EJP) (Pearce, 1934)
ST98: Dauntsey, R. Avon, 1929-1932 (EJP) (Pearce, 1934)
SU09: Brinkworth, 1932 (EJP)
SU16: R. Kennet, Manton, 1975 (DAC) (Cooling, 1981)
SU28: R. Cole, nr Bourton, 10-Aug-29 (EJP)
SU37: R. Kennet, Littlecote, 1971-1980 (DAC) (Cooling, 1981)
Resident. Local; in rivers and streams.

**HALIPLUS Latreille, 1802**
HALIPLUS Latreille, 1802
***Haliplus confinis* Stephens, 1828**

7 ST98: Dauntsey, silt pond, 27-Jun-32 (EJP) (Pearce, 1934)
SU16: Marlborough area, 1905-1938 (Meyrick, 1938)
SU06: Kennet & Avon Canal, 1933 (EJP)
SU08: Wootton Bassett, 1930 (EJP)
SU09: Cricklade, 2-Jun-99 (KGA); Minety, 1930 (EJP)
8 SU22: Blackmoor Copse, Jun-1995 (SAC)
Resident. Local; [in rivers, canals, ditches and lakes].

***Haliplus obliquus*** (Fabricius, 1787)
7 ST88: Sherston, 1933 (EJP)
ST98: Dauntsey, silt pond, 4-Sep-32 (EJP) (Pearce, 1934); Little Hinton, 24-Sep-32 (EJP)
SU09: Cotswold Water Park, 28-Apr-82 (GNF); Emmett Hill, 2-Oct-90 (JL); Minety, 1930 (EJP)
SU16: Marlborough area, 1869-1938 (Meyrick, 1938)
SU17: Barbury Hill, 1928 (EJP)
8 ST94: Ladywell (SPTAW), pond, 3-Jun - 12-Aug-08 (Boyce & Wright, 2009)
SU04: Winterbourne Stoke, c. 1828 (GTR) (Stephens, 1828)
SU14: Nine Mile R., Bulford (SPTAE), 23-Jun-93 (CMD)
Resident. Local; [in rivers, streams, canals, ditches and lakes].

NEOHALIPLUS Netolitzky, 1911

***Haliplus lineatocollis*** (Marsham, 1802)
7 ST87: Castle Combe, 1928 (EJP)
ST98: Dauntsey, silt pond, 27-Jun-32 (EJP) (Pearce, 1934)
ST99: R. Thames, 29-Oct-00 (APF); R. Thames, side stream, 9-May-99 (APF)
SU06: Kennet & Avon Canal, 1933 (EJP)
SU08: Minety, 5-Jun-1930 (EJP)
SU09: Ashton Keynes, 31-May-96 (KGA); Cotswold Water Park, 28-Apr-82 (GNF)
SU16: Marlborough area, 1905-1938 (Meyrick, 1938); R. Kennet, Pan Bridge, 1975 (DAC)
SU17: Barbury Hill, 1928 (EJP)
SU18: Coate Reservoir, 1928 (EJP)
SU26: Savernake Canal, 1930-1932 (EJP)
8 ST85: Westbury, 1929 (EJP)
ST94: Codford St. Peter, chalk stream and pool, 1991-1992 (RC) (Carr, 1997); Ladywell (SPTAW), pond, 3-Jun - 12-Aug-08 (Boyce & Wright, 2009)
SU03: R. Wylye, 5-Aug-91 (RC) (Carr, 1997); Wylye, Apr-81 (APF)
SU04: R.Till, Rollestone, 17-May-91 (RC)
SU05: West Lavington, dew ponds, 29-Mar-41(LeC); 6-Oct-45 (DCL) (DFL, 1948)
SU14: Bulford (SPTAE), muddy pond, 19-Jul-91 (RC) (Carr, 1997); Bulford, R. Avon, 29-Mar, 9-Apr-91 (RC); Bulford Ranges (SPTAE), pond, 3-Jun - 12-Aug-08 (Boyce & Wright, 2009); Figheldean, R. Avon, 20-Jul-91 (RC)
SU15: R. Avon, Upavon, 19-Jul-91 (RC) (Carr, 1997); R. Avon, Rushall, 1931 (EJP)
SU16: R. Kennet, Pan Bridge, 1971-1980 (DAC) (Cooling, 1981)
Resident. Widespread and frequent in most water courses, lakes and ponds.

HALIPLINUS Guignot, 1939

***Haliplus fluviatilis*** Aubé, 1836
7 ST96: Kennet & Avon Canal, 2-Jul-97 (MNS); Semington Brook, 1-Jul-97 (MNS)
ST98: Dauntsey, silt pond, 1932-1933 (EJP) (Pearce, 1934)
SU06: Kennet & Avon Canal, 1933 (EJP)
SU08: Wootton Bassett, 1930 (EJP)
SU09: Brinkworth, 1932 (EJP); Cotswold Water Park, 28-Apr-82 (GNF)
SU18: South Marston, 1928 (EJP)
SU19: Stanton Fitzwarren, 1932 (EJP)
SU26: Savernake Canal, 1930-1932 (EJP)
SU28: Little Hinton, 1932 (EJP)
8 SU03: R. Wylye, 5-Aug-91 (RC) (Carr, 1997)
SU14: R. Avon, Amesbury, 19-Jul-91 (RC) (Carr, 1997); R. Avon, Figheldean (SPTAE), 20-Jul-91 (RC)
SU15: R. Avon, Rushall, 1931 (EJP); R. Avon, Upavon, 19-Jul-91 (RC) (Carr, 1997)
Resident. Local; in rivers and ponds.

***Haliplus heydeni*** Wehncke, 1875
**Nb**
7 ST18 Swindon, 1932-1933 (EJP) (Pearce, 1934: 'in ponds')
SU16: Marlborough area, 1905-1938 (Meyrick, 1938: 'Pewsey Road pond and pond east of Wernham. In ponds')
SU18: South Marston, 1928 (EJP)
SU19: Hannington, Swindon, 11-May-29 (AF)
Resident. Very local; [in well-vegetated stagnant water]. None of these specimens in the MCC.

***Haliplus immaculatus*** Gerhardt, 1877
7 ST98: Dauntsey, silt pond, 20-May-33 (EJP) (Pearce, 1934)
SU16: Martinsell, 1905-1938 (Meyrick, 1938)
SU17: Barbury Hill, 1928 (EJP)
8 SU12: Petersfinger, gravel pit lake, 25-Jul-91 (DRN)
Resident. Local; in moving and standing water.

***Haliplus lineolatus*** Mannerheim, 1844
7 ST98: Dauntsey, silt pond, 4-Oct-32 (EJP) (Pearce, 1934)
8 SU14: R. Avon. Figheldean, 1991-1992 (RC) (Carr, 1997)

Resident. Very local; in moving and standing water.

***Haliplus ruficollis*** (De Geer, 1774)
7 ST87: Stowell Farm, 7-Nov-90 (ITE)
ST98: Dauntsey, silt pond, 1932-1933 (EJP) (Pearce, 1934); Nails, Box Road pond, 9-May-05 (SW)
SU16: Marlborough area, 30 May 1879 (RAW) (Meyrick, 1880); 1869-1938 (Meyrick, 1938)
SU17: Barbury Hill, 1928 (EJP)
SU18: South Marston, 1928 (EJP)
8 SU03: Wylye, Apr-81 (APF)
SU05: West Lavington, dew ponds, 29-Mar-41, 30-Mar-41 (LeC) (DFL, 1948)
SU12: Petersfinger, gravel pit lake, 25-Jul-91 (DRN)
SU22: Blackmoor Copse, 1-Jun-95 (SAC)
Resident. Local; in moving and standing water. Duff (1993) points out that Balfour- Browne considered records of this species before about 1911 to be very uncertain.

***Haliplus sibiricus*** Motschulsky, 1860
7 ST87: Castle Combe, 1928 (EJP)
ST88: R. Avon, 19-Oct-90 (ITE)
ST98: Dauntsey, silt pond, 1932-1933 (EJP) (Pearce, 1934)
SU09: Brinkworth, 1932 (EJP)
SU15: R. Avon, Rushall, 1931 (EJP)
SU16: Wernham, nr Marlborough, 1905-1938 (Meyrick, 1938)
SU26: Savernake Canal, 1930 –1932 (EJP)
SU28: Little Hinton, 1932 (EJP)
Resident. Local. Both records from ponds, but known from wide range of moving and standing water in GB. No specimens in MCC.

LIAPHLUS Guignot, 1928
***Haliplus flavicollis*** Sturm, 1834
7 ST98: Dauntsey, silt pond, 1, 27-Jun-32 (EJP) (Pearce, 1934)
SU09: Ashton Keynes, 27-May-96 (KGA); Cotswold Water Park, 28-Apr-82 (GNF); Cricklade, 2-Jun-99 (KGA)
SU18: South Marston, 1928 (EJP)
Su26: Savernake Canal, 1930-1932 (EJP)
8 ST94: Ladywell (SPTAW), pond, 3-Jun - 12-Aug-08 (Boyce & Wright, 2009)
Resident. Very local; [in a wide range of standing and moving water].

***Haliplus fulvus*** (Fabricius, 1801)
7 ST98: Dauntsey, silt pond, 1932-1933 (EJP) (Pearce, 1934);
SU08: Wootton Bassett, 1930 (EJP)
SU16: Marlborough area, 1869-1938 (Meyrick, 1938: 'not uncommon... in ponds where there is gravel'.)
SU25: Dod's Down, 1869-1938 (Meyrick, 1938)
Resident. Local. [In Somerset greatly decreased; in ditches, reservoirs and ponds].

***Haliplus laminatus*** (Schaller, 1783)
**Nb**
7 ST87: Castle Combe, 1928 (EJP)
ST98: Dauntsey, silt pond, (EJP) (Pearce, 1934); R. Avon, Dauntsey, 1-Jun-32 (EJP)
SU06: Kennet & Avon Canal, 1933 (EJP)
SU15: R. Avon, Rushall, 1931 (EJP)
SU27: R. Kennet, Ramsbury, 16-Aug-00 (BMF)
Resident. Very local; [in slow-moving parts of rivers and base-rich pools and canals with restricted vegetation]. Pearce recorded that he found two specimens in the Dauntsey silt pond but the only specimens bearing his name in the MCC are from the R. Avon.

## Family NOTERIDAE

**NOTERUS Clairville, 1806**
***Noterus clavicornis*** (De Geer, 1774)
7 ST98: Dauntsey, silt pond, 27-Jun-32 (EJP) (Pearce, 1934)
ST96: Kennet & Avon Canal, 2-Jul-97 (MNS)
SU09: Cotswold Water Park, 28-Apr-82 (GNF)
SU16: Thornhill Pond, 1939 (det KGB) (Knight, 1940)
8 ST94: Wadman's Coppice (SPTAW), pond, 3-Jun - 12-Aug-08 (Boyce & Wright, 2009)
SU14: Bulford Ranges (SPTAE), pond, 3-Jun - 12-Aug-08 (Boyce & Wright, 2009)
SU22: Blackmoor Copse, 1-Jun-95 (SAC)
Resident. Local; [not uncommon in slow-moving water, lakes and ponds].

## Family PAELOBIIDAE

**HYGROBIA Latreille, 1804**
**Screech Beetle**
***Hygrobia hermanni*** (Fabricius, 1775)
7 ST98: Dauntsey, silt pond, 1932-1933 (EJP) Pearce, 1934
SU09: Ashton Keynes, 16-Mar-02 (JKG det MDD)
SU16: Marlborough area, 1882, 1923-1924 (Meyrick, 1938)
8 SU05: West Lavington, dew ponds, Jul-32, May, Sep-41, Jun- 44 (LeC, ITH, DCL) (DFL, 1948)
SU14: Bulford Ranges (SPTAE), 3-Jun - 12-Aug-08 (Boyce & Wright, 2009)
SU22: East Grimstead, bomb crater, 1-Aug-78 (DRN)
SU25: Limmer Ponds, 28-Jul-92 (RC) (Carr, 1997)

Resident. Likely to be widespread in mainly standing water.

## Family DYTISCIDAE
### Subfamily AGABINAE

**AGABUS Leach, 1817**

ACATODES Thomson, C.G, 1859

***Agabus sturmii*** (Gyllenhal, 1808)
7 ST86: Monkton Farleigh, 27-Jun-79 (ISC)
ST98: Dauntsey, silt pond, 1932-1933 (EJP) (Pearce, 1934)
SU16: Marlborough area, 1905-1938 (Meyrick, 1938)
8 ST84: Longleat, 7-Mar-99 (MB)
Resident: Local; [in slow-flowing and standing water].

AGABUS Leach, 1817

***Agabus labiatus*** (Brahm, 1791)
**Nb**
8 SU05: Ell Barrow (SPTA C), 3-Jun - 12-Aug-08 (Boyce & Wright, 2009)
SU14: Bulford, Nine Mile R. (SPTAE), 23-Jun-93 (CMD)
Resident. Very local; associated with temporary pools.

***Agabus uliginosus*** (Linnaeus, 1761)
**Nb**
7 SU16 Marlborough area, 28-May-1879 (WFB) (Meyrick, 1880); 1869-1938 (Meyrick, 1938: in ponds)
SU27: Ramsbury, 17-Mar-30 (FBB)
8 ST94: Codford St. Peter, intermittent chalk stream and adjacent pool, 1991-1992 (RC) (Carr, 1997)
SU04: Winterbourne Stoke, Oct-1828 (GTR) (Stephens, 1829)
SU13: R. Bourne, Idmiston, 19-Jul-91(RC) (Carr, 1997)
SU14: Bulford, Nine Mile R. (SPTAE), 23-Jun-93 (CMD)
SU22: Bentley Wood, 27-May-98 (MDD)
Resident. Likely to be widespread in temporary still water.

GAURODYTES Thomson, C.G, 1859

***Agabus affinis*** (Paykull, 1798)
8 SU21: Landford Bog, 9-Jun-01 (APF) (Foster, 2001)
Resident. Very local; in peaty pools and ponds.

***Agabus biguttatus*** (Olivier, 1795)
**Nb**
7 SU16: R. Kennet, Clatford Farm, 1971-1980 (DAC) (Cooling, 1981); R. Kennet, Pan Bridge, 1971-1980 (DAC) (Cooling, 1981); R. Kennet, West Kennet, 1971-1980 (DAC) (Cooling, 1981)
8 ST94: Codford St. Peter, intermittent chalk stream and adjacent pool, 1991-1992 (RC) (Carr, 1997)
SU04: R. Till, Elston, 10-Apr-91 (RC); Winterbourne Stoke, 21-Jul-91(RC) (Carr, 1997)
SU13: R. Bourne, Idmiston, 5-Aug-82 (DRN)
SU14: Bulford, 19-Jul-91 (RC); Nine Mile R., Bulford (SPTAE), 23-Jun-93 (CMD)
Resident. Very local. Likes underground situations [and most frequently found after heavy rain when adults have been flushed out].

Agabus biguttatus

***Agabus bipustulatus*** (Linnaeus, 1767)
7 ST96: Semington, 1-Jul-97 (MNS)
ST98: Dauntsey, silt pond, 1932-1933 (EJP) (Pearce, 1934); Nails, Box Road pond, 9-May-05 (SW)
ST99: R. Thames, 29-Oct-00 (APF)
SU08: Purton, Swindon, 14-Oct-99 (KGA); Wootton Bassett, 3-Jun-99 (KGA)
SU09: Ashton Keynes, 16-Mar-02 (JKG det MDD); 2 -Oct-90 (JL); Cotswold Water Park, 28-Apr-82 (GNF); Cricklade, 2-Jun-99 (KGA)
SU16: Marlborough area, Jun-1873 (EWJ) (Meyrick, 1874); 6-Feb-1879 (JA) (Meyrick, 1880); 1869-1938 (Meyrick, 1938)
8 ST85: Green Lane Wood, 7-Jun-95 (WANHS det MDD)
ST92: Sutton Mandeville, 16-Aug-93 (MDD)
ST94: Ladywell (SPTAW), pond, 3-Jun - 12-Aug-08 (Boyce & Wright, 2009); Wadman's Coppice (SPTAW), pond, 3-Jun - 12-Aug-08 (Boyce & Wright, 2009)
ST95: West Lavington, 10-Aug-41 (LeC)
SU03: Berwick St. James, 1-Aug-94 (BL); R. Wylye, Apr-81 (APF); 5-Aug-91(RC) (Carr, 1997)
SU04: Chitterne Down (SPTAW), dew pond, 17-Aug-91 (RC) (Carr, 1997); R. Till, Winterbourne Stoke, 21-Jul-91 (RC) (Carr, 1997)
SU05: The Warren, chalk stream and fish pond 10-Apr-91 (RC) (Carr, 1997)
SU13: R. Bourne, Idmiston, 19-Jul-91 (RC) (Carr, 1997)
SU14: Bulford (SPTAE), 1991-1992 (RC) (Carr,

1997); 3-Jun - 12-Aug-08 (Boyce & Wright, 2009)
SU21: Hamptworth, 26-Jan-97 (DTB); Landford Bog, 24-Jun, 24-Aug-01 (APF) (Foster, 2001)
SU22: Blackmoor Copse, 1-Jun-95 (SAC); East Grimstead, 8-Jun-05 (JSD); Glazier's Copse, 22-Jul-73 (DRN); Langley Wood, 19-Jun-87 (RSK)
SU23: West Winterslow, 30-Aug-01 (NB)
SU25: Limmer Ponds, 28-Jul-92 (RC) (Carr, 1997)

Resident. Widespread and abundant in streams and standing water.

***Agabus brunneus*** (Fabricius, 1798)
**RDB2, UKBAP, WiltsBAP**
8 SU21: Hamptworth Estate, 17-Aug-76 (DRN) (Nash, 1979)

Resident. A single specimen was taken in damp flood refuse from a dried out pool in a gravel-bottomed, acid flush. Such intermittent water sources are the insect's usual habitat [and it is sometimes found deep in the gravel]. Only other known locations are in Cornwall, Dorset and the New Forest.

***Agabus didymus*** (Olivier, 1795)
7 ST98: Baydon, dew pond, 1932-1933 (EJP) (Pearce, 1934)
ST99. R. Thames, 29-Oct-00 (APF)
SU09: Cotswold Water Park, 28-Apr-82 (GNF)
SU16: Marlborough area, 1905-1938 (Meyrick, 1938); 1907 (WES)
SU37: R. Kennet, Littlecote, 1971-1980 (DAC) (Cooling, 1981)
8 ST85: Biss Farm, 20-Oct-90 (ITE)
SU04: Chitterne Down (SPTAW), dew pond, 17-Aug-91, May-92 (RC) (Carr, 1997); R. Till, Winterbourne Stoke, 21-Jul, 5-Aug-91 (RC) (Carr, 1997)
SU05: The Warren, chalk stream and fish pond, 1o-Apr-91 (RC) (Carr, 1997)
SU13: R. Bourne, Idmiston, 19-Jul-91 (RC) (Carr, 1997)
SU14: Bulford (SPTAE), muddy pond, 19-Jul-91 (RC) (Carr, 1997); pond, 3-Jun - 12-Aug-08 (Boyce & Wright, 2009), Nine Mile R. (SPTAE), 23-Jun-93 (CMD)

Resident. Local. Carr notes that its occurrence in a remote muddy pond at Chitterne Down raises questions about the species' supposed flightlessness.

***Agabus guttatus*** (Paykull, 1798)
7 ST87: West Yatton Down, 24-Apr-87 (JHB)
8 SU21: Burnt Ground Wood, 19-Apr-74; Pound Bottom, 9-Apr-74 (DRN)

Resident. Very local; in streams and springs often in woodland.

***Agabus nebulosus*** (Forster, 1771)
7 ST98: Dauntsey, silt pond, 1932-1933 (EJP) (Pearce, 1934)
ST87: Stowell Farm, 7-Nov-90 (ITE)
SU09: Cricklade, 2-Jun-99 (KGA)
SU16: Marlborough area, 14-Mar-1879 (WFB) (Meyrick, 1880); Manton, nr. Marlborough, 1869-1938 (Meyrick, 1938)
SU17: Barbury Hill, 1928, (EJP)
SU26: Leigh Hill, pond, 1932 (Meyrick, 1938)
8 ST94: Ladywell (SPTAW), pond, 3-Jun - 12-Aug-08 (Boyce & Wright, 2009); Wadman's Coppice SPTAW), 3-Jun - 12-Aug-08 (Boyce & Wright, 2009)
ST95: West Lavingtom, 1938 (PLP) (DFL, 1948); 29-Mar-41 (LeC) (DFL, 1948)
SU04: PTAW), dew pond, 17-Aug-91 (RC) (Carr, 1997); R. Till, Winterbourne Stoke, 21-Jul, 5-Aug-91 (RC) (Carr, 1997)
SU13: R. Bourne, Idmiston, 19-Jul-91 (RC) (Carr, 1997)
SU14: Bulford (SPTAE), muddy pond, 19-Jul-91 (RC) (Carr, 1997); Nine Mile River, 23-Jun-93 (CMD)
SU22: East Grimstead, bomb crater, Hawk's Grove, 1-Aug-78 (DRN)
SU25: Limmer Ponds, 28-Jul-92 (RC) (Carr, 1997)

Resident. Widespread in ditches, lakes and ponds.

***Agabus paludosus*** (Fabricius, 1801)
7 ST96: Semington, 1-Jul-97 (MNS)
SU16: Marlborough, 1907 (WES)
SU17: Ogbourne Maizey, 1869-1938 (Meyrick, 1938)
SU37: R. Kennet, Littlecote, 1971-1980 (DAC) (Cooling, 1981)
8 ST85: Biss Farm, 20-Oct-90 (ITE)
SU03: R. Wylye, 5-Aug-91 (RC) (Carr, 1997)
SU04: R.Till, Winterbourne Stoke, 1991-1992 (RC) (Carr, 1997)
SU05: The Warren, chalk stream and fish pond, 10-Apr-91 (RC) (Carr, 1997)
SU13: Idmiston, 25-Aug-72 (DRN); R. Bourne, Idmiston, 1991-1992 (RC) (Carr, 1997)
SU14: Bulford, Nine Mile R. (SPTAE), 29-Mar-91 (RC) (Carr, 1997); Bulford Ranges (SPTAE), pond, 3-Jun - 12-Aug-08 (Boyce & Wright, 2009)
SU15: R. Avon, Upavon, 19-Jul-91 (RC) (Carr, 1997)

Resident. Widespread in rivers, streams and ponds.

***Agabus unguicularis*** Thomson, C.G, 1867
**Nb**
8 SU21: Landford Bog, 2001-2003 (APF) (Foster, 2001, 2003)

Resident. Very local. [Associated with areas that dry out, either with moss in mesotrophic mires or in grassy edges of base-rich ponds].

**ILYBIUS Erichson, 1832**

***Ilybius ater*** (De Geer, 1774)

7 ST98: Dauntsey, silt pond, 1932-1933 (EJP) (Pearce, 1934)
ST96: Melksham, 1-Jul-97 (MNS)
SU16: Marlborough area, 1905-1938 (Meyrick, 1938)
8 ST95: West Lavington, dew pond 10-Aug-41 (LeC) (DFL, 1948)
SU04: Chitterne Down (SPTAW), dew pond, 17-Aug-91 (RC) (Carr, 1997)
SU22: Blackmoor Copse, 1-Jun-95 (SAC)
Resident. Local; in ditches and ponds.

***Ilybius chalconatus*** (Panzer, 1796)
**Nb**
7 ST98: Dauntsey, silt pond, 1932-1933 (EJP) (Pearce, 1934)
ST99: R. Thames side stream, 9-May-99 (APF)
SU26: Folly Farm, 1939 (det KGB) (Knight, 1940)
8 ST94: Codford St. Peter, intermittent chalk stream and adjacent pool, 1991-1992 (RC) (Carr, 1997)
SU14: Nine Mile R., Bulford (SPTAE), 9-Apr-91 (RC) (Carr, 1997)
SU22: Blackmoor Copse, 3-Aug-03 (APF) (Foster, 2004); East Grimstead, 8-Jun-05 (JSD); Homan's Copse, washed shingle, 20-Jul-73 (DRN)
Resident. Local; in shaded pools and ditches often among dead leaves. Formerly included in *Agabus.*

***Ilybius fenestratus*** (Fabricius, 1781
**Nb**
7 SU09: Cotswold Water Park, 30-May-97 (JSD)
8 SU12: Petersfinger, 25-Jul-91 (DRN)
Resident. Very local. [Confined to lowland ponds, lakes and canals with patches of marginal vegetation].

***Ilybius fuliginosus*** (Fabricius, 1792)
7 ST98: Dauntsey, silt pond, 27-Jun-32 (EJP) (Pearce, 1934)
SU08: Wootton Bassett, 28-Oct-94 (KGA)
SU16: Marlborough area, 22-Sep-1879 (WFB) (Meyrick, 1880); 1869-1938 (Meyrick, 1938)
SU26: Savernake Canal, 1930-1932 (EJP)
8 ST85: Bullen Hill Farm, 1982-2002 (EGS det MDD)
ST92: Donhead St. Mary, 18, 19-Jul-95 (HK det MDD)
ST94: Ladywell (SPTAW), pond, 3-Jun - 12-Aug-08 (Boyce & Wright, 2009)
ST95: West Lavington, dew ponds, farm pond and stream, 5-Oct-39 (KJA), 13, 15-Jun-44 (PJW, GSB, WN) (DFL, 1948)
SU03: R. Wylye, 5-Aug-91(RC) (Carr, 1997)
SU04: R.Till, Winterbourne Stoke, 5-Aug-91 (RC)
SU14: Bulford (SPTAE), muddy pond, 19-Jul-91 (RC) (Carr, 1997); pond, 3-Jun - 12-Aug-08 (Boyce & Wright, 2009)
SU15: R. Avon, Upavon, 19-Jul-91 (RC) (Carr, 1997)
SU21: Landford Bog, 24-Aug-01 (APF) (Foster, 2001)
SU22: Blackmoor Copse, 1-Jun-95 (SAC)
SU25: Limmer Ponds, 28-Jul-92 (RC) (Carr, 1997)
Resident. Widespread in standing and moving water. One record at light.

***Ilybius montanus*** (Stephens, 1828)
7 SU08: Wootton Bassett, 3-Jun-99 (KGA)
8 SU21: Hamptworth Estate, 20-Aug-76 (DRN); Pound Bottom, 9-Apr-74 (DRN)
Resident. Very local; under stones and in roots in dry bed. Transferred from *Agabus.*

***Ilybius quadriguttatus*** (Lacordaire in Boisduval & Lacordaire, 1835)
7 ST98: Dauntsey, silt pond, 1932-1933 (EJP) (Pearce, 1934)
Resident. Very local. Known in GB to inhabit other still water habitats.

**PLATAMBUS Thomson, C.G, 1859**

***Platambus maculatus*** (Linnaeus, 1758)
7 ST86: Castle Combe, Nettleton Brook, 1-May-79 (ISC)
ST88: R. Avon, 10-Oct-90 (ITE)
SU16: Marlborough area, 30-May-1879 (RAW) (Meyrick, 1880); 1869-1938 (Meyrick, 1938); R. Kennet, Clatford Farm, 1971-1980 (DAC) (Cooling, 1981)
SU27: Ramsbury, 17-Mar-30 (FBB)
8 SU13: Idmiston, edge of R. Bourne, 25-Aug-72 (DRN); Lower Woodford, R. Avon flood refuse, 2-Aug-82 (DRN)
SU14: R. Avon Figheldean, 20-Jul-91 (RC) Carr, 1997)
SU15: R. Avon, Rushall, 1931 (EJP); R. Avon, Upavon, 19-Jul-91 (RC) (Carr, 1997)
Resident. Probably widespread in moving water.

## Subfamily COLYMBETINAE

**COLYMBETES Clairville, 1806**

***Colymbetes fuscus*** (Linnaeus, 1758)
7 ST98: Dauntsey, silt pond, 1932-1933 (EJP) (Pearce, 1934)
SU09: Cotswold Water Park, 28-Apr-82 (GNF); Emmett Hill, 2-Oct-90 (JL)
SU16: Marlborough area, 25-Sep-1879 (WFB) (Meyrick, 1880); Wernham, nr Marlborough,

1869-1938 (Meyrick, 1938)
SU26: Savernake Forest, 1870-1873 (Meyrick, 1874)
8 ST85: Biss Wood, 11-Jun-84 (EGS det MDD)
ST95: West Lavington, 1938 (PLP) (DFL, 1948)
SU03: R. Wylye, 5-Aug-91 (RC) (Carr, 1997)
SU04: Chitterne Down (SPTAW), dew pond, 17-Aug-91 (RC) (Carr, 1997); R. Till, Rollestone, 17-May-91 (RC) (Carr, 1997); R. Till, Winterbourne Stoke, 21-Jul-91 (RC) (Carr, 1997)
SU13: R. Bourne, Idmiston, 19-Jul-91(RC) (Carr, 1997)
SU14: Bulford Ranges (SPTAE), pond, 3-Jun - 12-Aug-08 (Boyce & Wright, 2009); Nine Mile R., 23-Jun-93 (CMD)
SU21: Landford Bog, 24-Aug-01 (APF) (Foster, 2001)
Resident. Widespread in rivers, streams, ditches and ponds.

**RHANTUS Dejean, 1833**
RHANTUS Dejean, 1833
?***Rhantus frontalis*** (Marsham, 1802)
**Nb**
8 SU22: Blackmoor Copse, Jun-95 (SAC)
Resident. Very local [in stagnant water usually on exposed sites].There is doubt about this determination and it was omitted by Foster, 2001, 2003.

Rhantus frontalis

***Rhantus suturalis*** (MacLeay, 1825)
**Nb**
7 ST98: Dauntsey, silt pond, 27-Jun-1932 (EJP) (Pearce, 1934)
SU16: Marlborough area, 30-May-1879 (GTKM) (Meyrick, 1880); Marlborough area, 1873-1938 (Meyrick, 1938)
Resident. Very local; [in exposed ponds and ditches amongst vegetation].

Subfamily COPELATINAE
**LIOPTERUS Dejean, 1833**
***Liopterus haemorrhoidalis*** (Fabricius, 1787)
7 SU08: Wooton Bassett, pond, 1932-1933 (EJP) (Pearce, 1934)
Resident. Very local, but likely to be widespread in streams, ditches and ponds. Formerly assigned to *Copelatus*.

Subfamily DYTISCINAE
Tribe ACILIINI

**ACILIUS Leach, 1817**
***Acilius sulcatus*** (Linnaeus, 1758)
7 SU09: Emmett Hill, 2-Oct-90 (JL)
SU16: Marlborough area, 1869-1938 (Meyrick, 1938)
8 SU21: Hamptworth Estate, in water-filled skip, 18-Apr-84 (DRN)
SU25: Limmer Ponds, 28-Jul-92 (RC) (Carr, 1997)
Resident. Local, but likely to be widespread in ditches and ponds.

Tribe DYTISCINI

**DYTISCUS Linnaeus, 1758**
**Great Diving Beetle**
***Dytiscus marginalis*** Linnaeus, 1758
7 ST96: Melksham, 1-Jul-97 (MNS)
ST98: Dauntsey, silt pond, 1932-1933 (EJP) (Pearce, 1934)
SU09: Ashton Keynes, 15-Mar-00 (JKG det MDD)
SU16: Marlborough area, 14-Mar-1879 (WFB) (Meyrick, 1880); 1869-1938 (Meyrick, 1938)
SU27: Ramsbury, 1870-1873 (Meyrick, 1874)
8 ST85: Bullen Hill Farm, 6-Feb-82 (EGS det MDD)
SU21: Burnt Ground Wood, 24-Jul-73 (DRN); Hamptworth Estate, fish pond, 3-Aug-73 (DRN)
Resident. Widespread in ponds. The species is known to be attracted to light in spring and early summer.

***Dytiscus semisulcatus*** Müller, O.F, 1776
'*Plentiful everywhere in Hants and Wilts*' c.1828 (GTR) (Stephens, 1829)
7 ST18: Swindon, pond, 1932-1933 (EJP) (Pearce, 1934)
SU16: Marlborough area, 1905-1938 (Meyrick, 1938)
8 ST85: Bullen Hill Farm, 6-Feb-82 (EGS det MDD)
SU21: Hamptworth Estate, fish pond, 3-Aug-73 (DRN)
Resident. Local; in ponds and known to fly to light in summer.

Subfamily HYDROPORINAE
Tribe BIDESSINI

**HYDROGLYPHUS Motschulsky, 1853**

***Hydroglyphus geminus*** (Fabricius, 1792)

8 ST94: Ladywell (SPTAW), pond, 3-Jun - 12-Aug-08 (Boyce & Wright, 2009); Wadman's Coppice (PTAW), pond, 3-Jun - 12-Aug-08 (Boyce & Wright, 2009)
SU04: Chitterne Down (SPTAW), 17-Aug-91 (RC) (Carr, 1997)
SU14: Nine Mile R., Bulford (SPTAE), 23-Jun-93 (CMD)
Resident. Very local, in ponds.

Tribe HYDROPORINI

**DERONECTES Sharp, 1882**

***Deronectes latus*** (Stephens, 1829)
**Nb**
7 SU16: Marlborough area, 1869-1938 (Meyrick, 1938)
Resident. Very local. [Confined to running water and usually found amongst the moss *Fontinalis* in partial shade]. No specimen under this name in the MCC.

**GRAPTODYTES Seidlitz, 1887**

***Graptodytes pictus*** (Fabricius, 1787)
7 ST98: Dauntsey, silt pond, 27-Jun-1932 (EJP) (Pearce, 1934)
SU16: Marlborough area, 1905-1938 (Meyrick, 1938)
8 SU04: Winterbourne Stoke, 1828 (GTR)
SU22: East Grimstead, 1-Aug-78 (DRN)
Resident. Local; in moving and still water.

**HYDROPORUS Clairville, 1806**

***Hydroporus angustatus*** Sturm, 1835
7 ST98: Dauntsey, silt pond, 1-Aug-32 (EJP) (Pearce, 1934)
SU08: Wootton Bassett, 3-Jun-99 (KGA)
8 SU14: Bulford Ranges (SPTAE), pond, 3-Jun - 12-Aug-08 (Boyce & Wright, 2009)
Resident. Very local; [in streams, ditches and ponds].

***Hydroporus discretus*** Fairmaire & Brisout, 1859
7 SU09: Cotswold Water Park, 28-Apr-82 (GNF)
8 ST94: Codford St. Peter, chalk stream and adjacent pool, 1991-1992 (RC) (Carr, 1997)
ST96: Semington, 1-Jul-97 (MNS)
SU14: Nine Mile River, Bulford (SPTAE), 9-Apr-91 (RC) (Carr, 1997); 23-Jun-93 (CMD)
SU15: R. Avon, Upavon, 19-Jul-91 (RC) (Carr, 1997)
Resident. Local; in streams, rivers and ponds.

***Hydroporus erythrocephalus*** (Linnaeus, 1758)
*'common everywhere in Hants and Wilts-abundantly'* c.1828 (GTR) (Stephens, 1829)
7 ST98: Dauntsey, silt pond, 1932-1933 (EJP) (Pearce, 1934)
SU16: Marlborough area, 1869-1938 (Meyrick, 1938)
8 SU21: Landford Bog, 24-Aug-01 (APF) (Foster, 2001)
Resident. Local; in still water.

***Hydroporus ferrugineus*** Stephens, 1829
**Nb**
8 ST94: Codford St. Peter, intermittent chalk stream and adjacent pool, 1991-1992 (RC) (Carr, 1997)
SU25: Collingbourne Wood, 1820s (GTR) These specimens were the basis for Stephens' description of the species (Fowler, 1887).
Resident. Very local; [ and frequently found near to and in springs after heavy rain]. A subterranean, flight-less species.

***Hydroporus gyllenhalii*** Schiödte, 1841
8 SU21: Landford Bog, 24-Aug-01 (APF) (Foster, 2001)
Resident. Very local; in peaty ponds and pools.

***Hydroporus marginatus*** (Duftschmid, 1805)
**Nb**
7 ST99: R. Thames side stream, 9-May-99 (APF)
SU09: Cotswold Water Park, 28-Apr-82 (GNF)
SU16: Marlborough, (TNH-S det WWF) (Fowler, 1880, 1887); 1907 (WES)
SU27: Ramsbury, spring-1894 (Jebb, 1895); 1873-1938 (Meyrick, 1938); 1895 (RWL); 1949 CET)
8 ST85: Green Lane Wood, 7-Jun-95 (WANHS det MDD)
SU04: R.Till, Elston, 10-Apr-91 (RC) (Carr, 1997)
SU14: R. Avon, Figheldean (SPTAE), 20-Jul-91(RC) (Carr, 1997); Nine Mile R., Bulford (SPTAE), 23-Jun-93 (CMD)
SU22: Blackmoor Copse, 1-Jun-95 (SAC)
SU25: Collingbourne Wood, Apr - Jul-1915 (PAH) (Harwood, 1915)
Resident. Widespread [and often found in association with temporary, base-rich water, including intermittent parts of chalk streams].

***Hydroporus memnonius*** Nicolai, 1822
7 SU08: Minety, 28-May-96 (KGA)
8 ST94: Codford St. Peter, intermittent chalk stream and adjacent pool, 1991-1992 (RC) (Carr, 1997)
SU21: Hamptworth Estate, grubbed out of dry bed, 20-Aug-76 (DRN)
Resident. Very local; in ponds and pools.

***Hydroporus nigrita*** (Fabricius, 1792)
7 ST99: R. Thames side stream, 9-May-99 (APF)
SU26: Savernake Forest, 1869-1938 (Meyrick,

1938)
8 SU03: Wylye, Apr-81 (APF)
SU21: Hamptworth Estate, grubbed from dry bed, 20-Aug-76 (DRN); Landford Bog, 24-Aug-01 (APF) (Foster, 2001)
Resident. Very local; [in streams, ditches, ponds and pools often overlying peat].

***Hydroporus obsoletus*** Aubé, 1838
**Nb**
7 SU27: Ramsbury, 1869-1938 (Meyrick, 1938) This is presumably the specimen noticed by Fowler as found at Marlborough (Fowler, 1893)
8 SU22: Langley Wood NNR, pitfall trap, 6 to 26-Oct-06 (MDD) (Darby, 2008)
Resident. Very local. [Occurs in springs and seepages, and sometimes drainage pipes. Many accounts in flood refuse and at mouths of springs in heavy rain].

***Hydroporus palustris*** (Linnaeus, 1761)
7 ST98: Dauntsey, silt pond, 1932-1933 (EJP) (Pearce, 1934)
ST99: R. Thames, 29-Oct-00 (APF)
SU09: Cotswold Water Park, 28-Apr-82 (GNF)
SU16: Marlborough area, 1869-1938 (Meyrick, 1938)
SU26: R. Kennet, Savernake, 1974-1976, (DAC) (Cooling, 1981); Savernake, 1905 (Meyrick, 1938)
SU27: Ramsbury, 17-Mar-30 (FBB); 1949 (CET)
8 ST95: West Lavington, dew ponds, 14-Nov-40 (KJA); 29, 30- Mar-41 (LeC) (DFL, 1948)
ST96: Semington, 1-Jul-97 (MNS)
SU03: R. Wylye, 5-Aug-91 (RC) (Carr, 1997)
SU04: R.Till, Winterbourne Stoke, 21-Jul, 5-Aug-91(RC) (Carr, 1997); R. Till, Rollestone, 17-May 91(RC) (Carr, 1997)
SU05: The Warren, chalk stream and fish pond, 10-Apr-91 (RC) (Carr, 1997)
SU14: R. Avon, Amesbury, 19-Jul-91 (RC) (Carr, 1997); Bulford (SPTAE), muddy pond, 1991-1992 (RC) (Carr, 1997); pond, 3-Jun – 12-Aug-08 (Boyce & Wright, 2009) R. Avon, Figheldean, 20-Jul-91 (RC) (Carr, 1997); Nine Mile R., Bulford (SPTAE), 29-Mar-91 (RC) (Carr, 1997); 23-Jun-93 (CMD)
SU15: R. Avon, Upavon, 19-Jul-91 (RC) (Carr, 1997)
SU17: Bulford, 19-Jul-91 (RC)
SU21: Landford Bog, 24-Aug-01 (APF) (Foster, 2001)
SU22: East Grimstead, 1-Aug-78 (DRN)
Resident. Widespread and frequent in both standing and moving water.

***Hydroporus planus*** (Fabricius, 1781)
*'Common everywhere in Hants and Wilts'* [as *melanocephalus*] c.1828 (GTR) (Stephens, 1829)
7 ST87: Stowell Farm, 7-Nov-90 (ITE)
ST98: Dauntsey, silt pond, 1932-1933 (EJP) (Pearce, 1934)
SU08: Purton, Swindon, 14-Oct-99 (KGA)
SU09: Ashton Keynes, 31-May-96 (KGA); Cricklade, 2-Jun-99 (KGA); Emmett Hill, 2-Oct-90 (JL); Cotswold Water Park, 28-Apr-82 (GNF); 30-May-97 (JSD)
SU16: Marlborough area, 6-Feb-1879 (JA) (Meyrick, 1880); 1869-1938 (Meyrick, 1938)
8 ST94: Ladywell (SPTAW), pond, 3-Jun - 12-Aug-08 (Boyce & Wright, 2009); Summer Down (SPTAE), 7-Aug-92 (JHB); Wadman's Coppice (SPTAW), pond, 3-Jun - 12-Aug-08 (Boyce & Wright, 2009)
ST95: West Lavington, dew ponds, 29-Mar-41 (LeC) (DFL, 1948)
SU03: R. Wylye, 5-Aug-91(RC) (Carr, 1997)
SU04: Chitterne Down (SPTAW), dew pond, 17-Aug-91 (RC) (Carr, 1997); R. Till, Winterbourne Stoke, 5-Aug-91(RC) (Carr, 1997); Salisbury Plain, 25-Jun-02 (PJH) (Hodge, 2002)
SU14: Bulford Ranges (SPTAE), pond, 3-Jun - 12-Aug-08 (Boyce & Wright, 2009); Nine Mile R., Bulford, 23-Jun-93 (CMD)
SU21: Landford Bog, 24-Aug-01 (APF) Foster, 2001)
SU25: R. Bourne, 9-Apr-91 (RC) (Carr, 1997)
Resident. Widespread and frequent in still and running water.

***Hydroporus pubescens*** (Gyllenhal, 1808)
7 ST98: Dauntsey, silt pond, 1932-1933 (EJP) (Pearce, 1934)
SU09: Cotswold Water Park, 28-Apr-82 (GNF)
SU26: Folly Farm, 1939 (det KGB) (Knight, 1940)
8 ST94: Codford St. Peter, 1991-1992 (RC) (Carr, 1997)
SU03: Wylye, Apr-81 (APF)
SU14: R. Avon, Figheldean (SPTAE), 20-Jul-91(RC) (Carr, 1997)
SU21: Hamptworth Estate, 9-Apr-74 (DRN); Landford Bog, 24-Aug-01 (APF) (Foster, 2001)
Resident. Widespread in still and moving water.

?***Hydroporus rufifrons*** (Müller, O.F, 1776)
**RDB2**
7 SU26: Savernake Forest, 1979 (AAA)
This is a remarkable record of a species known since 1950 only from the north of England and Scotland, but its status is unclear. The record occurs in an anonymous printout of Coleoptera dated 1970-1993 in NE's archives. It is not mentioned in Foster, et al, 2008.

***Hydroporus tessellatus*** (Drapiez, 1819)
7 ST98: Dauntsey, silt pond, 27-Jun-1932 (EJP) (Pearce, 1934)

ST99: R. Thames side stream, 9-May-99 (APF)
SU08: Purton, Swindon, 14-Oct-99 (KGA); Wootton Bassett, 3-Jun-99 (KGA)
SU09: Ashton Keynes, 31-May-96 (KGA); 15-Mar-00 (JKG det MDD); Cotswold Water Park, 28-Apr-82 (GNF)
SU16: Marlborough area, 30 Mar 1879 (WFB) (Meyrick, 1880); 1869-1938 (Meyrick, 1938)
8 ST85: Biss Farm, 20-Oct-90 (ITE)
SU14: Bulford Ranges (SPTAE), pond, 3-Jun - 12-Aug-08 (Boyce & Wright, 2009)
SU21: Landford Bog, 9-Jun-01 (APF) (Foster, 2001)
Resident. Local; in ponds and pools.

**NEBRIOPORUS Régimbart, 1906**
NEBRIOPORUS Régimbart, 1906
***Nebrioporus elegans*** (Panzer, 1794)
7 ST87: Castle Combe, 1-May-79 (ISC)
ST88: R. Avon, 19-Oct-90 (ITE)
ST89: Brinkworth, 1932-1933 (EJP) (Pearce, 1934)
ST96: Semington Brook, 1-Jul-97 (MNS)
SU16: Marlborough area, 30 May 1879 (RAW) (Meyrick, 1880); 1869-1938 (Meyrick, 1938)
SU17: Milk Hill, 6-Oct-45 (DCL) (DFL, 1948)
SU26: R. Kennet, Savernake, 1974-1976 (DAC) (Cooling, 1981 as *depressus /elegans*)
SU27: Ramsbury, 1949 (CET)
SU37: Chitterne Forest, 1934, (EJP)
8 ST95: West Lavington, ?1940 (GHW) (DFL, 1948)
SU02: Knighton Wood, Sep-07 (AJA)
SU03: R. Wylye, 5-Aug-91(RC) (Carr, 1997)
SU04: Winterbourne Stoke, c.1828 (GTR) Stephens, 1828)
SU14: R. Avon, Figheldean (SPTAE), 20-Jul-91(RC) (Carr, 1997)
SU15: R. Avon, Rushall, 1931 (EJP); R. Avon, Upavon, 19-Jul-91 (RC) (Carr, 1997)
Resident. Widespread. All the above were recorded as *N. depressus* (Fabricius, 1775) but this species is now known to inhabit northern lakes and lochs, usually with deep water, and to be replaced by *elegans* in running water and man-made, still water habitats elsewhere.

**OREODYTES Seidlitz, 1887**
***Oreodytes sanmarkii*** (Sahlberg, C.R, 1826)
7 ST88: Sherston, 1935 (GAW)
SU06, Marden, 27-Mar-07 (DJL)
SU07: Cherhill, Rivers Brook, 27-Mar-07 (DJL)
SU26: R. Kennet, Savernake, 1974-1976 (DAC) (Cooling, 1981); R. Kennet, Littlecote, 1971-1980 (DAC) (Cooling, 1981)
8 ST94: R. Wylye, Norton Bavant, 1971-1980 (DAC) (Cooling, 1981); Tytherington, 1991-1992 (RC)
Resident. Local; in gravelly rivers and streams.

**PORHYDRUS Guignot, 1945**
***Porhydrus lineatus*** (Fabricius, 1775)
7 ST98: Dauntsey, silt pond, 27-Jun-32 (EJP) (Pearce, 1934)
SU08: Wootton Bassett, 1930 (EJP)
Resident. Very local; [in running and still water]. Likely to be more widespread.

**STICTONECTES Brinck, 1943**
***Stictonectes lepidus*** (Olivier, 1795)
**Nb**
8 ST95: West Lavington, dew pond, 29-Mar-41 (LeC) (DFL, 1948)
SU21: Hamptworth Estate, 18-Apr-84 (DRN)
Resident. Very local. [Characteristic habitats are 'stream pools' ie dam ponds, natural backwaters or deeper pockets of water left in otherwise dried out stream-beds].

**STICTOTARSUS Zimmermann, 1919**
***Stictotarsus duodecimpustulatus*** (Fabricius, 1792)
7 ST89: Brinkworth, Woodbridge Brook, 1932-1933 (EJP) (Pearce, 1934)
8 ST96: Semington Brook, 1-Jul-97 (MNS)
SU21: Hamptworth, 26-Jan-97 (DTB)
Resident. Very local; [in both still and running water]. Likely to be more widespread.

Tribe HYGROTINI

**HYGROTUS Stephens, 1828**
COELAMBUS Thomson, C.G, 1860
***Hygrotus confluens*** (Fabricius, 1787)
7 SU17: Barbury Hill, 1928 (EJP)
8 ST94: Ladywell (SPTAW), pond, 3-Jun - 12-Aug-08 (Boyce & Wright, 2009); Summer Down (SPTAE), 7-Aug-92 (JHB)
SU04: Chitterne Down (SPTAW), 17-Aug-91 (RC) (Carr, 1997)
Resident. Very local; in ponds and pools.

***Hygrotus impressopunctatus*** (Schaller, 1783)
7 SU16: Marlborough area, 1869-1938 (Meyrick, 1938)
8 SU14: Bulford Ranges (SPTAE), pond, 3-Jun - 12-Aug-08 (Boyce & Wright, 2009)
Resident. Local. Two specimens under this name in the MCC without data.

HYGROTUS Stephens, 1828
***Hygrotus inaequalis*** (Fabricius, 1777)
7 ST96: Melksham, 1-Jul-97 (MNS)
ST98: Box Road pond, nr Brinkworth, 9-May-05 (SW); Dauntsey, silt pond, 27-Jun-32 (EJP) (Pearce, 1934)
SU09: Cotswold Water Park, 28-Apr-82 (GNF)
SU16: Wernham, nr Marlborough, 1905-1938 (Meyrick, 1938)

SU26: Savernake Canal. 1930-1932 (EJP)
8 ST95: West Lavington, dew ponds, May, Aug, Sep-41 (LeC)
SU03: Wylye, Apr-81 (APF)
SU04: R.Till, Winterbourne Stoke, 21-Jul-91(RC) Carr, 1997)
SU24: Bulford Ranges (SPTAE), pond, 3-Jun - 12-Aug-08 (Boyce & Wright, 2009)
SU22: Blackmoor Copse, 1-Jun-95 (SAC)
Resident. Local; in still and running water.

***Hygrotus versicolor*** (Schaller, 1783)
7 ST98: Dauntsey, silt pond, 1932-1933 (EJP) (Pearce, 1934)
SU09: Cricklade, 31-May-99 (KGA)
SU16: Marlborough area, 1869-1938 (Meyrick, 1938)
Resident. Local; [in still and running water]. Likely to be more widespread.

### Tribe HYPHYDRINI

**HYPHYDRUS Illiger, 1802**
***Hyphydrus ovatus*** (Linnaeus, 1761)
7 ST98: Dauntsey, silt pond, 1932-1933 (EJP) (Pearce, 1934)
SU09: Cotswold Water Park, 30-May-97 (JSD)
SU16: Manton, nr. Marlborough, 1869-1938 (Meyrick, 1938)
SU26: Savernake Canal, 1930-1932 (EJP)
8 ST95: West Lavington, dew ponds and a farm pond, Aug-Sep-41 (LeC); May-Jun-44 (JES, PJW, GSB) (DFL, 1948)
SU22: Blackmoor Copse, 1-Jun-95 (SAC); East Grimstead, bomb crater, Hawk's Grove, 1-Aug-78 (DRN)
Resident. Local; in running water and ponds. Lilkely to be more widespread.

## Subfamily LACCOPHILINAE

**LACCOPHILUS Leach, 1815**
***Laccophilus hyalinus*** (De Geer, 1774)
7 ST96: Kennet & Avon Canal, 2-Jul-97 (MNS)
SU16: Marlborough area, 1869-1938 (Meyrick, 1938)
SU26: Savernake Forest, 1869-1938 (Meyrick 1938)
SU27: R.Kennet, Rabley, 1869-1938 (Meyrick, 1938)
8 ST96: Semington Brook, 1-Jul-97 (MNS)
SU03: R. Wylye, 5-Aug-91 (RC) (Carr, 1997)
SU12: Petersfinger, gravel pit lake, 25-Jul-91 (DRN)
Resident. Local; in still and running water. Likely to be more widespread.

***Laccophilus minutus*** (Linnaeus, 1758)
7 ST98: Dauntsey, silt pond, 1932-1933 (EJP) (Pearce, 1934)
SU09: Cotswold Water Park, 28-Apr-82 (GNF)
SU16: Marlborough area, 1869-1938 (Meyrick, 1938); R. Kennet, Overton, 1975 (DAC)
8 ST94: Ladywell (SPTAW), pond, 3-Jun - 12-Aug-08 (Boyce & Wright, 2009); Wadman's Coppice (SPTAW), pond, 3-Jun - 12-Aug-08 (Boyce & Wright, 2009)
ST95: West Lavington, dew ponds, 12-Oct-40 (GHW det LeC) (DFL, 1948); 30-Mar, 10-Aug-41 (LeC) (DFL, 1948)
SU04: Chitterne Down (SPTAW), 17-Aug-91 (RC) (Carr, 1997)
SU25: Limmer ponds, 28-Jul-92 (RC) (Carr, 1997)
Resident. Local; in still and running water. Likely to be more widespread

**?*Laccophilus poecilus*** Klug, 1834
**RDB2**
8 ST95: West Lavington, Black Dog Farm pond, 29-Jun-44 (DCL, GAW) (DFL, 1948 as *L. variegatus*)
In the absence of the specimen(s) there must be some doubt about this determination. *L.poecilus* is a fen inhabiting species usually found near the coast and only ever recorded reliably in this country on eleven occasions. At the present time it is only known from the Lewes levels.

# Family CARABIDAE
## Subfamily CICINDELINAE

Cicindela campestris

**CICINDELA Linnaeus, 1758**
**Green Tiger Beetle**
***Cicindela campestris*** Linnaeus, 1758
**LCGR-FI CI**
7 SU16: Marlborough area, 1869-1938 (Meyrick, 1938); Lockeridge, 1994 (JEO)
SU26: Savernake Forest, Jun-1873 (EWJ) (Meyrick, 1874)
8 ST84: Longleat, 22-Jun-95, 29-May, 1-Aug-96, 28-Jul-97 (PM); 25-May-05 (DJG)

ST91: Cranborne Chase, 7-May-97 (MAS)
ST94: Bowls Barrow (SPTAW), 16-May-02 (PJH) (Hodge, 2002); Ladywell (SPTAW), 15-May-02 (PJH) (Hodge, 2002); 3-Jun - 12-Aug-08 (Boyce & Wright, 2009); Salisbury Plain, 31-May-92 (SP); Tinker's Track SPTAW), 3-Jun - 12-Aug-08 (Boyce & Wright, 2009); Wadman's Coppice (SPTAW) 3-Jun - 12-Aug-08 (Boyce & Wright, 2009); Warminster Ranges (SPTAW), 3-Jun - 12-Aug-08 (Boyce & Wright, 2009)
ST95: Salisbury Plain, 18-May-40 (PLP)
SU04: Shrewton Folly, 20-Jun-93 (SP)
SU05: Black Heath (SPTAC), 3-Jun - 12-Aug-08 (Boyce & Wright, 2009); Ell Barrow (SPTAC), 3-Jun - 12-Aug-08 (Boyce & Wright, 2009)
SU14: Haxton (SPTAE), 17-May-02 (PJH) (Hodge, 2002); Wexland Hanging (SPTAC), 16-May-02 (PJH) (Hodge, 2002)
SU15: Casterley Camp, 21-Apr-02 (BG); Salisbury Plain, 24-Apr-99 (BG); Water Dean (SPTA), 24-Apr-99 (BG)
SU21: Hamptworth Estate, 25-Apr-84 (DRN); Landford Bog, 9-Jun-01 (APF) (Foster, 2001); 22-Apr-97, 17-Aug-00 (PM)
SU22: Bentley Wood, 2-Jul-89, 22-May-90 (PM); 23-May-04 (AGD); 8-May-08 (PM); Blackmoor Copse, 4-May-92 - 7-May-04, 3-Jun-06, 27-Apr-07 (PM); 17-May-02 (APF) (Foster, 2004); Landford Bog, 14-Apr-06 (PM); Langley Wood NNR, 22-Apr to 3-Jun-04 (MDD) (Darby, 2005)
SU23: Porton Down, 1970-2001 (Anon, 2001)
SU24: Beacon Hill (SPTAE), 3-Jun – 12-Aug-08 (Boyce & Wright, 2009)
SU25: Sidbury Hill (SPTAE), 21-May-02 (AS det MDD)
SU26: Cobham Frith, 26-May-05 (IJG)

Resident. Widespread; both in open grassland and woodland particularly on sandy and peaty ground. The Bentley Wood 8-May-08 record was of more than 50 beetles in a single 10 metre length of track.

Cicindela sylvatica

**Wood Tiger Beetle**
***Cicindela sylvatica*** Linnaeus, 1758
**Na**

8 ST92: Tisbury, early Jun-04 (JH)

This record is based on a single sighting and would be suspect if the determiner were not known to be reliable. The insect has been recorded in recent times only from dry sandy heaths in a narrow strip from Surrey to Dorset. The Tisbury location is an ungrazed meadow adjacent to the river Nadder

## Subfamily BRACHININAE

**BRACHINUS Weber, 1801**
**Bombardier Beetle**
***Brachinus crepitans*** (Linnaeus, 1758)
**Nb, LCGR-Fm Cl**

8 ST94: Wadman's Coppice (SPTAW) 3-Jun - 12-Aug-08 (Boyce & Wright, 2009); Warminster Ranges (SPTAW), 3-Jun - 12-Aug-08 (Boyce & Wright, 2009)
SU03: Wylye Down NNR, 10-Apr to 4-May-07 (MDD) (Darby, 2008)
SU04: Tilshead Down, 1, 4-Jun-97 (MGT)
SU05: Black Heath (SPTAC), 3-Jun - 12-Aug-08 (Boyce & Wright, 2009); Ell Barrow (SPTAC), 3-Jun - 12-Aug-08 (Boyce & Wright, 2009)
SU15: Beach's Plantation (SPTAE), 13, 19-Jun-97, 1-May-05 (MGT),
SU24: Dunch Hill (SPTAE), 16-Jun-97 (MGT); Sidbury Hill (SPTAE), 29-May-97 (MGT);
SU25: Sidbury Hill (SPTAE), 28, 29-May-97 (MGT)

Resident. Very local. Probably much declined in the county and most recent GB records are coastal.

Brachinus crepitans

## Subfamily CARABINAE
### Tribe CARABINI

**CARABUS Linnaeus, 1758**
CARABUS Linnaeus, 1758
***Carabus granulatus*** Linnaeus, 1758

7 SU16: Marlborough area, Jun-1873 (EWJ) (Meyrick, 1874); 20-May-1879 (RAW) (Meyrick, 1880); 1869-1938 (Meyrick, 1938)

8 ST74: Longleat, 19-Sep-82 (EGS det MDD)
ST85: Bullen Hill Farm, 26-Nov-83 (EGS det MDD)
SU13: Great Durnford, 11-Apr-71 (DRN); Lower Woodford, 8-Apr-70 (DRN); Stratford Sub Castle, 11-Apr-71 (DRN)

Resident. Local. Hygrophilus species occurring in marshes and areas subject to winter flooding.

MORPHOCARABUS Géhin, 1885

***Carabus monilis*** Fabricius, 1792

**Nb**

7 SU16: Marlborough area, Jun-1873 (EWJ) (Meyrick, 1874), 1869-1938 (Meyrick, 1938)
8 ST85: Green Lane Wood, 7-Jun-95 (WANHS det MDD)
ST92: Sutton Mandeville, 7-Jun-92, 20-Aug-92 (MDD)
SU01: Martin Down NNR, 1999 (EN, 1999)
SU03: Wylye Down NNR, 10-Apr to 4-Sep-2007 (MDD) (Darby, 2009)
SU15: Beach's Plantation (SPTAE), 10 to 17-Jun-97 (MGT)
SU22: Langley Wood NNR, 6-Jun to 2-Jul-07 (MDD) (Darby, 2008)
SU24: Nr Sidbury Hill (SPTAE), 12,13-Jun-97 (MGT)

Resident. Local; in a variety of habitats including cultivated land. Declined nationally in recent years.

Carabus monilis

ARCHICARABUS Seidlitz, 1887

***Carabus nemoralis*** Müller, O.F, 1764

7 SU07: Calne, 6-Aug-00; Lyneham, 19-Jul-00 (DMR); Yatesbury, 1991-1992 (BG)
SU08: Ravensroost Wood, 17-Jun-05 (MA det MDD)
SU16: Marlborough area, 1869-1938 (Meyrick, 1938); Lockeridge, 1994 (JEO)
SU17: Avebury, 13-Apr, 7-May-03 (DB)
SU26: Savernake Forest, 15-Jun-73 (SAW); 28-Jul to 21-Sep-93 (EN det SAW)
8 ST85: Clanger Wood, 1995-1995 (MA det MDD)
ST92: Sutton Mandeville, 10-Mar-07 (MDD)
ST93: Great Ridge Wood, 18-Apr-74 (DRN) (Nash, 1979); Stockton Wood, 1995-1995 (MA det MDD)
SU01: Martin Down NNR, 1979-1992 (RGB); 1999 (EN, 1999)
SU02: Gardiner Forest, 1-Sep-72 (DRN) (Nash, 1979)
SU03: Grovely Wood, 30-May-74, 16-Apr-84 (DRN) (Nash, 1979); Wylye Down NNR, 27-Jan-07 to 10-Oct-07 (MDD) (Darby, 2009)
SU05: West Lavington, 9-May-34 (ITH) (DFL, 1948)
SU15: Beach's Plantation (SPTAE), 1-May-05 (MGT); 4-Sep-05 (JMW)
SU17: Barbury Hill, 1995-1995 (MA det MDD)
SU22: Langley Wood NNR, 1-Apr-04 to 24-Mar-05, 25-Aug-06 to 6-Jun-07 (MDD) (Darby, 2005, 2008)
SU23: West Winterslow, 11-Aug-01 (NB)

Resident. Widespread; mainly in winter/spring in a range of habitats including woodland, calcareous grassland and gardens

MESOCARABUS Thomson, C.G, 1875

***Carabus problematicus*** Herbst, 1786

7 SU16: Marlborough area, 30 Jun 1879 (AGC) (Meyrick, 1880); 1869-1938 (Meyrick, 1938)
SU26: Savernake Forest, 2-Apr-70 (DRN); 27-May to 16-Jun-93, 1 to 19-Jul, 30-Aug to 9-Sep-94 (EN det SAW), 11-May-95 (EN det MDD)
8 ST84: Longleat, 25-Mar-99 - 28-Aug-00 (MB)
ST91: Cranborne Chase, 16 to 25-Jun-80 (NCC)
SU00: Barbury Castle, 8-Apr-07 (SH)
SU01: Martin Down NNR, 1999 (EN, 1999)
SU02: Gardiner Forest, 1-Sep-72 (DRN); Knighton Wood, 1-Jan-05, Apr-07 (AJWA)
SU03: Grovely Wood, 10-Apr-1970 - 11-Aug-87 (DRN); Wylye Down NNR, 16-Oct-06 to 10-Oct-07 (MDD) (Darby, 2009)
SU04: Tilshead Down (SPTAW), 4-Jun-97 (MGT)
SU15: Beach's Plantation (SPTAE), 13 to 17-Jun-97 (MGT)
SU21: Hamptworth Estate, 3-Aug-73, 27-Jul-78 (DRN)
SU22: Langley Wood NNR, 5-Aug-85 (DRN); 14-May to10-Dec-04, 5-Aug-06 to 2-Jul-07 (MDD) (Darby, 2008)
SU23: Porton Down, 1970-2001 (Anon, 2001)
SU24: Nr Sidbury Hill (SPTAE), 12 to 17-Jun-97, 1-May-05 (MGT); 4-Sep-05 (JMW)
SU25: Collingbourne Forest, 25-Aug-72 (DRN)

Resident. Widespread; in a range of habitats especially long gassland, woodland and heaths. Often mistaken for the Violet Ground Beetle, *C. violaceus*.

MEGODONTUS Solier, 1848

**Violet Ground Beetle**

***Carabus violaceus*** Linnaeus, 1758

7 SU06: Morgan's Hill, 13-May, 24-Sep-84 (DMR)
SU07: Yatesbury, 1991-1992 (BG)
SU16: Lockeridge, 1994 (JEO); Marlborough area, 1-Jun-1873 (EWJ) (Meyrick, 1874); 1-Apr-1879 (JL) (Meyrick, 1880); 1869-1938 (Meyrick, 1938); 1994 (JEO); 2-Feb-84 (DMR)
SU17: Avebury, 11-Jun, 5-Aug-03 (DB); Fyfield Down, 21-Jul-80 (DMR)
SU26: Marlborough, 2-Feb-84 (DMRSavernake Forest, 3-Apr-70 (DRN); 25-Apr to 16-Jun-93, 9-Aug to 20-Sep-94 (EN det SAW); 15-Jul-95 (EN det MDD)
SU27: High Clear Down, 8-Jun-05 (PM)
8 ST84: Longleat, 1-Jul, 29-Aug-99, 28-Aug-00 (MB); 7-Jul-99 (MDD)
ST91: Cranborne Chase, 8-Jul-97 (MAS); 22-Jun-99 (MDD)
ST92: Sutton Mandeville, 1-Aug-92, 15-Oct-92 (MDD)
ST93: Pits Wood, Chilmark, 18-Feb-98 (MDD)
SU01: Martin Down NNR, 1999 (EN, 1999)
SU02: Middleton Down, 1988 -1989 (BL)
SU03: Grovely Wood, 29-Oct-70, 11-Aug-87 (DRN); 1-Jul-98, 17-Feb-98 (MDD); Wylye Down NNR, 29-Mar to 10-Oct-07 (MDD) (Darby, 2009)
SU06: Devizes, 10-Mar-02 (BG)
SU14: Bourne Bottom (SPTAE), 13-Jun-97 (MGT)
SU15: Upavon, 26-Aug-01 (DMR)
SU17: Barbury Hill, 1-Jan-95 (MA det MDD)
SU21: Hamptworth Estate, 3-Aug-73, 4-Apr-75 (DRN)
SU22: Langley Wood NNR, 14-May to 8-Oct-04, 5-Aug-06 to 2-Jul-07 (MDD) (Darby, 2005, 2008)
SU23: Porton Down, 11-Jul-93 (PTW det MDD); West Winterslow, 9-Jul-01 (NB)
SU24: Newton Tony, 25-Aug to 20-Sep-06, (MA det MGT)

Resident. Widespread; in a wide range of habitats particularly gardens, woodland and on calcareous soils. Some records may be mis-identified as a result of confusion with *C. problematicus*.

Tribe CYCHRINI

**CYCHRUS Fabricius, 1794**

***Cychrus caraboides*** (Linnaeus, 1758)

ssp. ***rostratus*** (Linnaeus, 1761)

7 SU16: Marlborough area, 6 Feb 1879 (WFB) (Meyrick, 1880)
SU26: Savernake Forest, 1869-1938 (Meyrick, 1938); 28-Jul to 28-Aug-93 (EN det SAW); 11-May to 3-Aug-95 (EN det MDD); 9-Nov-01 (IJG)
8 ST92: Sutton Mandeville, 28-Jun-95 (MDD)
SU03: Grovely Wood, 6-Apr-72, 8-May-75, 17-Apr-84 (DRN)
SU05: West Lavington, 1931-48 (DFL, 1948)
SU13: Cockey Down, 1-Aug-88 (PM)
SU22: Langley Wood NNR, 14-May to 8-Oct-04, 5-Aug-06 to 6-Jun-07 (MDD) (Darby, 2005, 2008)

Resident. Widespread ; mainly in woodland where it feeds on snails.

Tribe NEBRIINI

**LEISTUS Frölich, 1799**

POGONOPHORUS Latreille, 1802

***Leistus rufomarginatus*** (Duftschmid, 1812)

7 SU16: Gopher Wood, 13-Jul-96 (MA det MDD)
SU26: Savernake Forest, 31-May-73, 22-Jun-78 (KNAA) (Alexander, 1979)
8 ST84: Longleat, 8-May-99 (MB)
ST85: Green Lane Wood, 7-Jun-95 (WANHS det MDD)
ST91: Cranborne Chase, 16 to 25-Jun-80 (NCC)
SU15: Beach's Plantation (SPTAE), 4-Sep-05 (JMW)
SU21: Hamptworth Estate, 27-Jul-78 (DRN)
SU22: Langley Wood NNR, 22-Apr-04 to 17-Nov-06 (MDD) (Darby, 2005, 2008)

Resident, recent introduction (1942). Widespread; feeds on Collembola in woodland.

***Leistus spinibarbis*** (Fabricius, 1775)

**LCGR-FI CI**

7 ST96: Chittoe, 9-May-00 (EGS det MDD)
SU06: Beckhampton, 3-Jun-00 (DMR)
SU07: Calne, 30-Sep, 13-Oct-84 (DMR)
SU17: Avebury, 3-May, 14-Jun-03, 7-May-07 (DB det MDD)
8 ST92: Sutton Mandeville, 1-Apr-92 (MDD)
ST93: Great Ridge Wood, 2-May-00 (MDD)
SU02: Bishopstone, 7-May-89 (MDD)
SU03: Grovely Wood, 8-Apr-72 (DRN); 14-Jun-01 (WANHS det MDD)
SU13: Little Durnford Down, 13-Sep-99 (MDD)
SU15: Beach's Plantation (SPTAE), 1-May-05 (MGT)
SU21: Franchises Wood, 3-May-95 (MAS); Hamptworth Estate, 31-Oct-73, 15-Apr-74 (DRN); Pound Bottom, 9-Apr-74 (DRN)
SU22: Harestock, 26-Oct-71 (DRN); Langley Wood, 14-May to 8-Oct-04 (MDD) (Darby, 2005)
SU23: Porton Down, 20-Jul-97 (MAS)
SU24: Driver Training Area (SPTAE), 1-May-05 (MGT)
SU26: Bedwyn Brail, 5-Sep, 5-Nov-03 (MDD)

Resident. Widespread; particularly in woodland.

LEISTOPHORUS Reitter, 1905

***Leistus fulvibarbis*** Dejean, 1826

7 ST97: Christian Malford, 8-May-80 (DMR); Foxham, 24-Apr-80 (DMR)
SU07: Catcomb Wood, 1-May-96 (MA det MDD); Goatacre, 23-Apr-80 (DMR)
SU09: Emmett Hill, 1-Jan-95 (MA det MDD)
SU16: Marlborough area, 1869-1938 (Meyrick, 1938)
8 SU02: Knighton Wood, Jun-08 (AJWA)
SU03: Wylye Down NNR, 4-Sep to 10-Oct-07 (MDD) (Darby, 2009)
SU13: Salisbury, 10-Apr-72 (DRN); Great Durnford, 6-Apr-70 (DRN)

Resident. Widespread; mainly in litter in damp woodland.

LEISTUS Frölich, 1799

***Leistus ferrugineus*** (Linnaeus, 1758)

7 ST97: Christian Malford, 8-May-80 (DMR)
ST99: Flisteridge Wood, 1-Jan-80 (KNAA)
SU07: Calne, 21-Oct-01 (DMR); Yatesbury, 13-Jun-73 (BG)
SU08: Lyneham, 23-Jun-80 (DMR)
SU16: Marlborough area, Jun-1873 (EWJ) (Meyrick, 1874); 18 Jun 1879 (FED-H) (Meyrick, 1880); 1869-1938 (Meyrick, 1938)
SU26: Savernake Forest, 28-Jul to 28-Aug-93 (EN det SAW); 31-May-95 (EN det MDD)
8 ST82: Oysters Coppice, 6-Oct-02 (MDD)
ST92: Sutton Mandeville, 7-Jun-96 (MDD)
ST93: Great Ridge Wood, 11-Aug-80 (DRN)
ST94: Ladywell (SPTAW), 23-Jun-02 (PJH) (Hodge, 2002)
SU01: Martin Down NNR, 1999 (EN, 1999)
SU03: Grovely Wood, 29-Oct-70, 11-Aug-87 (DRN)
SU15: Beach's Plantation (SPTAE), 1-May-05 (MGT); Upavon, 30-Sep-02 (IJG); Manningford, 4-Aug-88 (DRN)
SU25: Sidbury Hill (SPTAE), Jun-97 (MGT)

Resident. Widespread and frequent in dryish open spaces such as fields and gardens, but also in woodland.

***Leistus terminatus*** (Hellwig in Panzer, 1793)

7 ST87: Castle Combe, 22-May-61 (PJMG) (Greenslade, 1963)
SU07: Catcomb Wood, 1-May-96 (MA det MDD)
SU26: Savernake Forest, 28-Aug to 20-Sep-93 (EN det SAW)
SU27: Knighton, 11-Jun-85 (DMR)
8 ST93: Great Ridge Wood, 18-Mar-00 (MDD)
SU16: Jones's Mill, 21-Jun-97 (MDD)
SU21: Hamptworth Common, 20-Aug-76 (DRN)

Resident. Widespread; in damp grasslands, woodlands and gardens.

**NEBRIA Latreille, 1802**

NEBRIA Latreille, 1802

***Nebria brevicollis*** (Fabricius, 1792)

7 ST86: Whaddon, 13-May-00 (DMR)
ST87: Rack Hill, 1-Jun-96 (MA det MDD); Slaughterford, 10-Jun-00 (DMR)
ST96: Chittoe, 16-Jul-00 (DMR)
ST97: Kellaways, 29-Apr-80 (DMR)
ST98: Malmesbury, 13-Jul-80 (DMR)
SU06: Beckhampton, 3-Jun-00 (DMR); Morgan's Hill, 13-May-84, 24-Sep-84 (DMR)
SU07: Calne, 5-May-84 (DMR); Catcomb Wood, 1-May-96 (MA det MDD); Clyffe Pypard, 9-Jun-80 (DMR); Lyneham, 10, 17-Apr-80 (DMR); Yatesbury, 13-Jun-73 (BG)
SU09: Upper Waterhay, 5-Jun-07, 5-Jun-05 (JKG det MDD)
SU16: Gopher Wood, 1-May, 13-Jul-96 (MA det MDD); Marlborough area, 6-Feb-1879 (TNH-S) (Meyrick, 1880); 1869-1938 (Meyrick, 1938); East Kennet, 3-Jun-01 (DMR); Wilcot, 1-May-00 (DMR)
SU17: Avebury, 4-May-03 (DB); Fyfield Down, 21-Jul-80 (DMR); Ogbourne Maizey, 20-May-01 (DMR)
SU26: Crofton, 26-May-01 (DMR)
SU27: Knighton, 11-Jun-85 (DMR)
8 ST82: Oysters Coppice, 6-Oct-02 (MDD)
ST84: Longleat, 4-May-98, 29-Aug-99, 28-Aug-00 (MB)
ST85: Clanger Wood, 1-Jan-95 (MA det MDD); Green Lane Wood, 7-Jun-95 (MDD); Bullen Hill Farm, 1-Jan-82 (EGS det MDD)
ST92: Sutton Mandeville, 1-Apr-92 (MDD); Sutton Row, 14-Sep-95 (MDD)
ST93: Stockton Wood, 1-Jan-95 (MA det MDD); Stockton Down, 1-Jan-95 (MA det MDD); Great Ridge Wood, 7-May-00 (MDD)
ST95: Coombe Hill, 1-Jan-95 (MA det MDD)
SU02: Middleton Down, 14-May-1989, 4-Apr-90 (BL)
SU03: Wylye Down NNR, 29-Mar to 10-Oct-07 (MDD) (Darby, 2009); Grovely Wood, 11-Jun-00, 21-Aug-01 (MDD)
SU04: Tilshead Down, 1 - 4-Jun-97 (MGT)
SU13: Salisbury, Folly, 7-Mar - 15-Jun-02 (MDD) (Darby, 2002d)
SU14: Boscombe Down, 1-Jan-94 (MA det MDD); Bourne Bottom (SPTAE), 13 to 17-Jun-97 (MGT)
SU15: Beach's Plantation (SPTAE), 1-May-05 (MGT); 4-Sep-05 (JMW)
SU16: Jones's Mill, 17-Oct-01 (APF) (Foster, 2002); Wilcot, 1-May-00 (DMR)
SU21: Landford Bog, 1-Jan-03 (APF) (Foster, 2003); SU24: Newton Tony, 25-Aug to 20-Sep-06, (MA det MGT)
SU22: Langley Wood NNR, 25-May-94 (WANHS det MDD); 22-Apr-04 to 11-Feb-05, 5-Aug-

06 to 2-Jul-07 (MDD) (Darby, 2005, 2008); roadside, 14-Sep-55 (BPM, GEW) (Moore, 1956)
SU22: Blackmoor Copse, 4-May-03 (APF) (Foster, 2004); Mapleway, Dean Copse, 31- Aug-03 (EG)
SU23: Porton Down, 1-Aug-91 (PTW det MDD); West Winterslow, 1-Sep-01 (NB)
SU24: Driver Training Area (SPTAE), 1-May-05 (MGT); Sidbury Hill (SPTAE), 19-May, 13 to 17-Jun-97 (MGT)
SU26: Bedwyn Brail, 2-Jun-03 (MDD)
Resident. Ubiquitous; in a wide range of situations but prefers woodland to grassland.

***Nebria salina*** Fairmaire & Laboulbène, 1854
8 ST94: Warminster Ranges (SPTAW), 3-Jun - 12-Aug-08 (Boyce & Wright, 2009)
SU03: Wylye Down NNR, 31-Dec-06 to 27-Jan-07 (MDD) (Darby, 2009)
SU21: Burnt Ground Wood, 31-Oct-75 (DRN); Hamptworth Estate, 31-Oct-75 (DRN);
SU22: roadside, 14-Sep-55 (BPM, GEW) (Moore, 1956)
SU24: Beacon Hill (SPTAE), 3-Jun - 12-Aug-08 (Boyce & Wright, 2009); Newton Tony, 25-Aug to 20-Sep-06, (MA det MGT)
Resident. Local. Amongst roots, under bark, under cut weeds. Not as frequent as *N.brevicollis,* which it closely resembles, and prefers drier situations.

Tribe NOTIOPHILINI

**NOTIOPHILUS Duméril, 1806**
***Notiophilus aquaticus*** (Linnaeus, 1758)
8 SU21: Cloven Hill, 19-Aug-88 (DRN)
SU16: Jones's Mill, 1-Jan-97 (MAS)
Resident. Very local; under stones, etc. on open heaths and short grassland.

***Notiophilus biguttatus*** (Fabricius, 1779)
'*Not very frequent in Hants and Wilts*' pre.1828 (GTR) (Stephens, 1828)
7 ST86: Whaddon, 13-May-00 (DMR)
ST98: Malmesbury, 13-Jul-80 (DMR)
SU06: Beckhampton, 3-Jun-00 (DMR); Cherhill, 27-Aug-01 (DMR); Cherhill Down, 1-Jun-04 (BLe); Morgan's Hill, 13-May, 24-Sep-84 (DMR)
SU07: Bewick Basset, 12-Aug-00 (DMR); Calne, 5-May-84 (DMR); Catcomb Wood, 1-May-96 (MA det MDD); Clyffe Pypard, 9-Jun-80 (DMR); Lyneham, 29-Dec-79, 4-Apr-80, 27-May-80 (DMR); Yatesbury, 13-Jun-73 (BG)
SU08: Tockenham Wick, 4-Apr-80 (DMR); Webbs Wood, 7-Mar-02 (JKG det MDD)
SU09: Somerford Keynes, 4-Jun-04 (BLe)
SU16: East Kennet, 3-Jun-01 (DMR); Gopher Wood, 1-May-96, 13-Jul-96 (MA det MDD); Marlborough area, 16 Feb 1879 (OFW) (Meyrick, 1880); Marlborough area, 1873-1938 (Meyrick, 1938); West Stowell, 3-Sep-00 (DMR); West Woods, 1-May-02 (MDD)
SU17: Fyfield Down, 21-Jul-80 (DMR); Ogbourne Maizey, 20-May-01 (DMR); Sells Green, 20-Aug-00 (DMR)
SU26: Froxfield, 9-Apr-85 (DMR); Savernake Forest, 28-Jul to 28-Aug-93, 28-Apr to 18-May, 19-Jul to 9-Aug-94 (EN det SAW); 25-Apr to 26-May-95 (EN det MDD); 22-Jun-98 (GS det MDD)
8 ST84: Longleat, 7-Jul-99 (MDD)
ST85: Bullen Hill Farm, 1-Jan-82, 8-May-03 (EGS det MDD)
ST91: Cranborne Chase, 16 to 25-Jun-80 (NCC); 24-May-97 (MAS)
ST92: Sutton Mandeville, 1-Jul-92 (MDD); Wardour, 13-Nov-95 (MDD)
ST93: Great Ridge Wood, 18-Apr-74 (DRN); Stockton Wood, 1-Jan-95 (MA det MDD); Great Ridge Wood, 2-May-00 (MDD)
ST94: Wadman's Coppice (SPTAW) 3-Jun - 12-Aug-08 (Boyce & Wright, 2009)
SU02: Bishopstone, 7-May-89 (MDD); Knighton Wood, 1-Jan-05, Mar-Apr-07 (AJWA)
SU03: Grovely Wood, 11-Aug-70, 13-Aug-71, 11-Apr-72, 18-Jul-73 (DRN); 12-Jun-00 (MDD); Wylye Down NNR, 31-Dec-06 to 27-Jan-07 (MDD) (Darby, 2009)
SU05: West Lavington, 4-Jul-01 (DMR)
SU13: Salisbury, Oct-73 (DRN); Salisbury, Avon Valley Reserve, 1-Apr-00 (MAS); 13-Apr-00 (MDD) (Darby, 2001); Old Sarum, 31-May-97 (MAS)
SU14: Bourne Bottom (SPTAE), 19-Jun-97 (MGT)
SU15: Beach's Plantation (SPTAE), 19-Jun-97, 1-May-05 (MGT); Manningford, 4-Aug-88 (DRN); North Newton, 23-Jun-01 (DMR)
SU16: Jones's Mill, 1-Jan-97 (MAS); Wilcot, 1-May-00 (DMR);
SU21: Burnt Ground Wood, 21-Jul-73, 31-May-74 (DRN); Hamptworth Estate, 9-Apr-74, 5-Aug-85 (DRN); Pound Bottom, 9-Apr-74 (DRN)
SU22: Bentley Wood, 3-Sep-97 (WANHS det MDD); Langley Wood NNR, 6-Jun-02, 22-Apr-04 to 24-Mar-05, 5-Aug-06 to 2-Jul-07 (MDD) (Darby, 2008); roadside, 14-Sep-55 (BPM, GEW) (Moore, 1956); White-parish, 22-Jul-73 (DRN)
SU23: Porton Down, 3-Jul-99 (MAS)
SU24: Newton Tony, 25-Aug to 20-Sep-06, (MA det MGT); Sidbury Hill (SPTAE), 18, 19, 28-May, 4, 6, 16-Jun-97 (MGT)
SU25: Blackmoor Down, 25-Aug-72 (DRN)
Resident. Widespread; in almost all habitats, particularly woodland.

***Notiophilus germinyi*** Fauvel, 1863
8 SU03: Grovely Wood, 4-Aug-78 (DRN)
SU05: Ell Barrow (SPTAC), 3-Jun - 12-Aug-08 (Boyce & Wright, 2009)
SU22: Langley Wood NNR, 14-May-04 (MDD) (Darby, 2005)
SU24: Beacon Hill (SPTAE), 3-Jun - 12-Aug-08 (Boyce & Wright, 2009)
Resident. Very local; [usually on heaths and in dry grassland].

***Notiophilus palustris*** (Duftschmid, 1812)
7 ST97: Stanley, 2-Jul-80 (DMR)
SU16: Marlborough area, 1869-1938 (Meyrick, 1938)
8 ST91: Cranborne Chase, 16 to 25-Jun-80 (NCC)
ST94: Norton Down, 23-Jun-02 (PJH) (Hodge, 2002)
ST95: Coombe Hill, 1-Jan-95 (MA det MDD)
SU03: Wylye Down NNR, 22-Feb to 29-Mar, 10-Apr to 4-May-07 (MDD) (Darby, 2009)
SU04: Shrewton, 2-Aug-83 (DRN)
SU14: Wexland Hanging (SPTAC), 22-Jun-02 (PJH) (Hodge, 2002)
SU15: Manningford, 4-Aug-88 (DRN); Upavon, 26-Aug-01 (DMR)
SU21: Burnt Ground Wood, 24-Jul-73 (DRN); Hamptworth Estate, 20-Aug-76 (DRN); Pound Bottom, 19-Apr-74, 4-Aug-82 (DRN)
SU22: Langley Wood NNR, 22-Apr to 17-Sep-04 (MDD) (Darby, 2005); Bentley Wood, 31-May-97 (MDD)
SU23: Porton Down, 1970-2001 (Anon, 2001)
SU24: Dunch Hill (SPTAE), 16-Jun-97 (MGT)
Resident. Widespread In damp grasslands and woodland in shady habitats.

***Notiophilus rufipes*** Curtis, 1829
7 ST86: Whaddon, 13-May-00 (DMR)
SU26: Savernake Forest, 30-May-73 (DS)
8 ST85: Green Lane Wood, 7-Jun-95 (WANHS det MDD)
ST93: Great Ridge Wood, 7-May-00 (MDD)
SU22: Langley Wood NNR, 25-Jun-04, 5-Aug-06 to 2-Jul-07 (MDD) (Darby, 2005, 2008)
Resident. Local; in litter etc. generally in shaded habitats.

***Notiophilus substriatus*** Waterhouse, G.R, 1833
7 ST87: Castle Combe, 22-May-61 (PJMG) (Greenslade, 1963)
ST94: Ladywell (SPTAW), 3-Jun - 12-Aug-08 (Boyce & Wright, 2009); Warminster Ranges (SPTAW), 3-Jun - 12-Aug-08 (Boyce & Wright, 2009)
SU06: Beckhampton, 3-Jun-00 (DMR); Cherhill, 27- Aug-01 (DMR); Morgan's Hill, 24-Sep-84 (DMR)
SU07: Bewick Basset, 12-Aug-00 (DMR);Calne, 6-Aug-00 (DMR); Goatacre, 6-Sep-80 (DMR); Yatesbury, 13-Jun-73 (BG)
SU09: Somerford Keynes, 4-Jun-04 (BLe)
SU16: Savernake Forest, 30-Jun-46 (Holmes & Tennant, 1947)
SU17: Avebury, 7-Oct-07 (DB det MDD); Ogbourne Maizey, 20-May-01 (DMR)
SU26: Crofton, 26-May-01 (DMR)
8 SU03: Grovely Wood, 29-Oct-70 (DRN); Wylye Down NNR, 22-Feb to 29-Mar-07 (MDD) (Darby, 2009)
SU04: Shrewton, 2-Aug-83 (DRN)
SU05: West Lavington, 4-Jul-01 (DMR)
SU13: Salisbury, Avon Valley Reserve, 1-Apr-00 (MAS) (Darby, 2001)
SU14: Larkhill west (SPTAC), 2-Jun-97 (MGT)
SU15: Upavon, 4-Aug-88 (DRN)
SU21: Hamptworth Estate, 6-Aug-79 (DRN)
SU22: Langley Wood NNR, 17-Sep to 8-Oct-04, 26-Oct to 17-Nov-06 (MDD) (Darby, 2005, 2008)
SU24: Bulford Ranges (SPTAE), 3-Jun - 12-Aug-08 (Boyce & Wright, 2009)
SU25: Sidbury Hill (SPTAE), 29-May-97 (MGT)
Resident. Widespread; in open, dry, often sandy situations, less frequently in woodland.

Tribe ELAPHRINI

Blethisa multipunctata

**BLETHISA Bonelli, 1810**
***Blethisa multipunctata*** (Linnaeus, 1758)
**Nb**
7 SU16: Marlborough area, 28 May 1879 (TNH-S) (Meyrick, 1880); 1869-1938 (Meyrick, 1938)
Resident. Very local; [on vegetated marshy lake shores and in fens]. Four specimens in the MCC without data.

**ELAPHRUS Fabricius, 1775**
ELAPHRUS Fabricius, 1775
***Elaphrus cupreus*** Duftschmid, 1812
7 ST87: Slaughterford, 10-Jun-00 (DMR)

ST96: Calne, 14-Apr-80 (DMR)
SU08: Lyneham, 3-Jul-80 (DMR)
SU16: Marlborough area, 14 May 1879 (WBP) (Meyrick, 1880)
SU26: Mildenhall, 12-Feb-05 (Meyrick, 1874)
8 ST92: Wardour, 5-Jun-96 (MDD, MAS)
SU05: West Lavington, 20-May-44 (DCL) (DFL, 1948)
SU12: Charlton-All-Saints, 2-Jun-95, 9-May-96 (MAS); Salisbury, 30-Jul-73 (DRN)
SU13: Great Durnford, 3-Aug-88 (DRN); Lower Woodford, 1-Aug-83, 8-Aug-86 DRN); Middle Woodford, 27-Jul-77 (DRN); Salisbury, Avon Valley Reserve, 1-Apr-00 (MAS) (Darby, 2001)
SU14: Figheldean, 28-Jul-82 (DRN); Nine Mile R. (SPTAE) pond, 5-Jun-97 (MGT)
SU16: Jones's Mill, 19-Jun-01(APF) (Foster, 2002)
SU21: Hamptworth Estate, 24-Apr-84 (DRN)
SU22: Langley Wood NNR, 22-Apr to 3-Jun-04 (MDD) (Darby, 2005); roadside, 14-Sep-55 (BPM, GEW) (Moore, 1956)
SU26: Bedwyn Brail, 19-Jun-03 (MDD)

Resident. Widespread; in wet areas of grassland and woodland.

Elaphrus uliginosus

***Elaphrus uliginosus*** Fabricius, 1792
**Nb**

7 SU16: Marlborough area, 1-Jan-33 (Meyrick, 1938)
8 ST95: Seend Cleve, 31-May-85 (DMR)
SU12: Britford, Jun – 15-Sep-02 (CMD) (Drake, 2008)
Lower Woodford, Jun – 15-Sep-02 (CMD) (Drake, 2008)

Resident. Very local; [in oligotrophic fens and bogs in lowland areas]. There is a single specimen in the MCC without data.

TRICHELAPHRUS Semenov, 1926
***Elaphrus riparius*** (Linnaeus, 1758)

7 ST96: Calne, 14-Apr-80 (DMR)
ST97: Studley, 18-May-80 (DMR)
ST98: Malmesbury, 13-Jul-80 (DMR)
SU06: Beckhampton, 3-Jun-00 (DMR)
SU16: East Kennet, 3-Jun-01 (DMR); Marlborough area, 1869-1938 (Meyrick, 1938)
SU17: Ogbourne Maizey, 20-May-01 (DMR); Sells Green, 20-Aug-00 (DMR)
8 ST92: Wardour, 5-Jun-96 (MDD)
SU13: Lower Woodford, 27-Jul-82, 1-Aug-83, 8-Aug-86 (DRN); Middle Woodford, 27-Jul-76 (DRN); Salisbury, 25-Jul-78 (DRN); Stratford Sub Castle, 26-May-74 (DRN)
SU14: Nine Mile R., pond (SPTAE), 5-Jun-97 (MGT)
SU21: Landford Bog, 25-May-98 (MDD)
SU22: Langley Wood NNR, 14-May to 6-Aug-04 (MDD) (Darby, 2005); Redlynch, 13-Apr-74 (DRN)
SU24: Driver Training Area (SPTAE), 1-May-05 (MGT)

Resident. Widespread; on margins of still water where there is damp bare ground.

Tribe LORICERINI

**LORICERA Latreille, 1802**
***Loricera pilicornis*** (Fabricius, 1775)

7 ST86: Whaddon, 13-May-00 (DMR)
ST96: Calne, 14-Apr-80 (DMR); Chittoe, 16-Jul-00 (DMR)
ST97: Kellaways, 13-Sep-80 (DMR)
ST98: Malmesbury, 13-Jul-80 (DMR)
SU07: Bewick Basset, 12-Aug-00 (DMR); Clyffe Pypard, 4-Sep-80 (DMR); Lyneham, 6-Apr, 3-May-80 (DMR)
SU08: Tockenham Wick, 25-May-80 (DMR); Yatesbury, 13-Jun-73 (BG)
SU09: Ashton Keynes, 25-Mar-01 (JKG det MDD)
SU16: East Kennet, 3-Jun-01 (DMR); Marlborough area, 6-Feb-1879 (GTKM) (Meyrick, 1880); 1869-1938 (Meyrick, 1938); West Stowell, 3-Sep-00 (DMR)
SU26: Savernake Forest, 3-Apr-70 (DRN)
8 ST82: Oysters Coppice, 6-Oct-02 (MDD)
ST84: Longleat, 15-Jun-73, 31-May-96 (PM); 25-Mar, 29-Aug-99 (MB)
ST91: Cranborne Chase, 16 to 25-Jun-80 (NCC)
ST92: Sutton Mandeville, 4-Apr-92, 1-Feb-93 (MDD); Wardour, 13-Nov-95 (MDD)
ST95: Whorton Common, 28-Aug-00 (DMR)
SU01: Martin Down NNR, 1999 (EN, 1999)
SU02: Knighton Wood, Mar-Apr-07 (AJWA); Middleton Down, 14-May-89, 1-Jul-89 (BL)
SU03: Barford Down, 10-Apr-72 (DRN); Grovely Wood, 17-Apr-71, 8-Apr, 21-Aug-72 (DRN)
SU12: Salisbury, 30-Jul-73 (DRN)
SU13: Great Durnford, 18-Aug-72, 3-Aug-88 (DRN); Lower Woodford, 28-Jul-78, 2-Aug-82, 8-Aug-86 (DRN); Middle Woodford, 5-Aug-82 (DRN); Salisbury, Avon Valley Reserve, 1-May-00 (MDD) (Darby, 2001); Stratford sub Castle, 6-Aug-87 (DRN)

SU14: Figheldean, 28-Jul-82 (DRN)
SU15: Beach's Plantation (SPTAE), 1-May-05 (MGT)
SU16: Jones's Mill, 17-Oct-01 (Foster, 2002)
SU15: North Newton, 23-Jun-01 (DMR)
SU21: Hamptworth Estate, 3-Aug-73, 15-Apr-74 (DRN)
SU22: Blackmoor Copse, 7-May-02, 3-Aug-03 (APF) (Foster, 2004); Langley Wood NNR, 22-Apr-04 to 24-Mar-05, 5 to 25-Aug-06 (MDD) (Darby, 2005, 2008)
SU23: Porton Down, 1970-2001 (Anon, 2001)
SU25: Sidbury Hill (SPTAE), 6-Jun-97 (MGT)
Resident. Widespread; in a variety of habitats.

Tribe SCARITINI

**CLIVINA Latreille, 1802**

***Clivina collaris*** (Herbst, 1784)

7 SU16: Marlborough area, 1869-1938 (Meyrick, 1938)
Resident. Very local; under refuse in damp places. [On clay, sand and silt near fresh water].

***Clivina fossor*** (Linnaeus, 1758)

7 ST86: Whaddon, 13-May-00 (DMR)
ST98: Malmesbury, 13-Jul-80 (DMR)
SU07: Yatesbury, 13-Jun-73 (BG)
SU09: Emmett Hill, 1-Jan-95 (MA det MDD)
SU16: Marlborough area, 1869-1938 (Meyrick, 1938); Oare, 23-May-02 (IJG); Poulton, Marlborough district, Jun-1873 (EWJ) (Meyrick, 1874)
SU17: Ogbourne Maizey, 20-May-01 (DMR)
8 ST84: Longleat, 25-Mar-99 (MB)
ST92: Fovant, 1-May-95 (MDD)
SU02: Fovant, 31-May-95 (MDD)
SU13: Lower Woodford, 2-Aug-82 (DRN); Salisbury, Avon Valley Reserve, 12-Apr-00 (MDD) (Darby, 2001)
SU16: Wilcot, 1-May-00 (DMR)
SU21: Hamptworth Estate, 1-Apr-75, 6-Apr-77 (DRN)
SU22: Langley Wood NNR, 5 to 22-Apr-04 (MDD) (Darby, 2005)
SU23: West Winterslow, 11-Aug-01 (NB)
Resident. Widespread; in most open habitats especially arable land, pasture and gardens.

**DYSCHIRIUS Bonelli, 1810**

DYSCHIRIODES Jeannel, 1941

***Dyschirius globosus*** Herbst, 1784

7 SU09: Somerford Keynes, 4-Jun-04 (BLe)
SU16: Marlborough area, 1869-1938 (Meyrick, 1938); Poulton, Marlborough district, Jun 1873 (EWJ) (Meyrick, 1874)
SU17: Ogbourne Maizey, 20-May-01 (DMR)
8 SU13: Lower Woodford, 1-Aug-83 (DRN); Salisbury, Avon Valley Reserve, 1-Apr-00 (MAS); 2-May-00 (MDD) (Darby, 2001)
SU16: Jones's Mill, 1-Jan-97 (MAS)
Resident. Local; on damp, bare, peaty and sandy ground near water.

***Dyschirius luedersi*** Wagner, 1915

8 SU14: Nine Mile R., pond (SPTAE), 5-Jun-97 (MGT)
Resident. Very local; [on bare mud near water both coastal and inland].

Tribe TRECHINI

**TRECHUS Clairville, 1806**

EPAPHIUS Leach in Samouelle, 1819

***Trechus secalis*** (Paykull, 1790)

7 SU07: Yatesbury, 13-Jun-73 (BG); Bewick Basset, 12-Aug-00 (DMR)
SU16: Marlborough area, Jun-1873 (EWJ) (Meyrick, 1874); 1869-1938 (Meyrick, 1938)
8 SU21: Hamptworth Estate, 20-Aug-76 (DRN)
SU22: Langley Wood NNR, 27-Aug to 17-Sep-04 (MDD) (Darby, 2005)
Resident. Local; in damp grassland, arable fields and woodland.

TRECHUS Clairville, 1806

***Trechus obtusus*** Erichson, 1837

7 SU06: Roundway Camp, Jul-40 (FDB) (Buck, 1956)
SU07: Yatesbury, 13-Jun-73 (BG)
SU16: Marlborough area, 22-Mar-1879 (FED-H) (Meyrick, 1880); 1869-1938 (Meyrick, 1938)
SU26: Savernake Forest, 3-Apr-70 (DRN)
8 ST82: Oysters Coppice, 1-Jan-02 (MDD)
ST91: Cranborne Chase, 30-Jun-99 (MDD)
ST91: Ladywell (SPTAW), 23-Jun-02 (PJH) (Hodge, 2002)
SU02: Gardiner Forest, 1-Sep-72 (DRN)
SU03: Grovely Wood, 21-Aug-72, 1-Aug-80 (DRN)
SU21: Hamptworth Estate, 15-Apr-74, 28-Oct-75, 20-Aug-76 (DRN)
SU22: Langley Wood NNR, 27-Aug-04 to 24-Mar-05; 25-Aug-06 to 20-Feb-07 (MDD) (Darby, 2005, 2008)
SU23: Thorny Down woods, 16-Oct-00 (MDD)
Resident. Widespread; in fields, gardens, and woodland.

***Trechus quadristriatus*** (Schrank, 1781)

7 ST96: Chittoe, 16-Jul-00 (DMR)
ST99: Crudwell, 19-Jun-73 (APF)
SU06: Beckhampton, 3-Jun-00 (DMR)
SU07: Bewick Basset, 12-Aug-00 (DMR); Calne, 28-May, 30-Sep-84 (DMR); Clyffe Pypard, 4-Sep-80 (DMR); Yatesbury, 13-Jun-73 (BG)

SU09: Somerford Keynes, 4-Jun-04 (BLe)
SU16: Marlborough area, 10 Mar 1879 (WFB) Meyrick, 1880), 1869-1938 (Meyrick, 1938); West Stowell, 3-Sep-00 (DMR)
SU26: Savernake Forest, 4-Sep-90 (TGW); 30-Aug to 20-Sep-94 (EN det SAW)
8 ST94: Ladywell, 1-Sep-97 (EGS det MDD)
SU03: Grovely Wood, 11-Aug-71, 11-Aug-87 (DRN); Wylye Down NNR, 4-Sep-07 (MDD) (Darby, 2009)
SU04: Shrewton, 6-Aug-82 (DRN); SU04: Tilshead Down, 1-Jun-97 (MGT)
SU13: Great Durnford, 3-Aug-83, 3-Aug-88 (DRN); Salisbury, 10-Apr-72 (DRN)
SU14: Boscombe Down, 1-Jan-94 (MA det MDD); Bourne Bottom (SPTAE), 13-Jun-97 (MGT)
SU22: Langley Wood NNR, 6-Jun to 2-Jul-07 (MDD) (Darby, 2008); Redlynch, 4-Aug-82 (DRN); roadside, 14-Sep-55 (BPM, GEW) (Moore, 1956)
SU23: Porton Down, 1970-2001 (Anon, 2001)
SU24: Dunch Hill (SPTAE), 16-Jun-97 (MGT); Newton Tony, 25-Aug to 20-Sep-06, (MA det MGT); Sidbury Hill (SPTAE), 28-May-97 (MGT)

Resident. Widespread; particularly in open and dry disturbed situations such as agricultural land and gardens.

**TRECHOBLEMUS Ganglbauer, 1891**
***Trechoblemus micros*** (Herbst, 1784)
8 SU02: Knighton Wood, Sep-07 (AJWA)
Resident. Very local; [in damp grasslands and various habitats near water, probably in association with small mammal burrows].

Tribe BEMBIDIINI

**ASAPHIDION des Gozis, 1886**
*A.stierlini* and *A.curtum* were only recently (1986) separated from *A. flavipes* so that determinations before then may be erroneus.

***Asaphidion curtum*** (Heyden, 1870)
8 ST84: Longleat, 7-Mar, 8-May-99 (MB)
ST92: Sutton Mandeville, 20-Apr-92 (MDD)
SU22: Langley Wood NNR, 22-Apr-04 to 24-Mar-05, 5-Jan-07 to 6-Jun-08 (MDD) (Darby, 2005, 2008)
Resident. Local; on open ground on heavy soils. Probably more widespread.

***Asaphidion flavipes*** (Linnaeus, 1761)
7 SU07: Yatesbury, 13-Jun-73 (BG)
SU16: Marlborough area, Jun-1873 (EWJ) (Meyrick, 1874)
8 ST91: Cranborne Chase, 16 to 25-Jun-80 (NCC)
SU21: Hamptworth Estate, 15-Apr-74, 24-Apr-84 (DRN)
SU22: Langley Wood NNR, 25-May-94 (WANHS det MDD)
Resident. Local; on open soils near water. Probably more widespread

***Asaphidion stierlini*** (Heyden, 1880)
7 SU16: East Kennet, 3-Jun-01 (DMR)
SU26: Crofton, 26-May-01 (DMR)
8 SU13: Great Durnford, 18-Aug-72 (DRN)
Resident. Very local; in open situations on light soils. These appear to be the most westerly records in GB.

**OCYS Stephens, 1828**
***Ocys harpaloides*** (Audinet-Serville, 1821)
8 ST82: Oysters Coppice, 6-Oct-02 (MDD)
ST92: Wardour, 3-Apr-96 (EN); 18-Jun-02 (MDD) Sutton Mandeville, 13-Jan-99 (MDD)
ST93: Great Ridge Wood, 18-Apr-74, 29-Oct-75 (DRN)
SU02: Knighton Wood, 7-Oct-06 (AJWA); Wilton, 8-Apr-70 (DRN)
SU03: Grovely Wood, 11-May-00 (MDD det RT); 29-Oct-73, 11-Aug-71, 3, 8-Apr-75 (DRN)
SU13: Salisbury, Bemerton Folly, 7-Mar - 15-Jun-02 (MDD) (Darby, 2002d)
SU22: Blackmoor Copse, 31-Aug-01, 17-Mar, 3-Aug-03 (APF) (Foster, 2004); Glazier's Copse, 16-Aug-88 (DRN)
Resident. Local; most specimens under bark in woodland. Probably widespread.

**BEMBIDION Latreille, 1802**
NEJA Motschulsky, 1864
***Bembidion nigricorne*** Gyllenhal, 1827
**Nb**
8 SU21: Pound Bottom, 24-Jul-73 (DRN)
Resident. Very local; running near heather. Probably confined to the far south of the county.

METALLINA Motschulsky, 1850
***Bembidion lampros*** (Herbst, 1784)
7 ST86: Whaddon, 13-May-00 (DMR)
ST87: Slaughterford, 10-Jun-00 (DMR)
ST97: Kellaways, 13-Sep-80 (DMR)
ST98: Malmesbury, 13-Jul-80 (DMR)
SU06: Beckhampton, 3-Jun-00 (DMR); Morgan's Hill, 24-Sep-84 (DMR); Roundway Camp, Jul-40 (FDB) (Buck, 1956)
SU07: Yatesbury, 13-Jun-73 (BG)
SU08: Lyneham, 3-Jul-80 (DMR)
SU16: East Kennet, 3-Jun-01 (DMR); Marlborough area, Jun-1873 (EWJ) (Meyrick, 1874); 30-Mar-1879 (WFB) (Meyrick, 1880); 1873-1938 (Meyrick, 1938); West Stowell, 3-Sep-00 (DMR)
SU17: Ogbourne Maizey, 20-May-01 (DMR)
SU26: Crofton, 26-May-01 (DMR); Froxfield, 9-Apr-85 (DMR)
8 ST85: Bullen Hill Farm, 1-Jan-82 (EGS det MDD)

ST91: Cranborne Chase, 16 to 25-Jun-80 (NCC)
ST92: Sutton Mandeville, 3-Apr-92, 1-May, 1-Jul-93 (MDD); Wardour, 5-Jun-96 (MAS)
ST93: Great Ridge Wood, 29-Jul-82 (DRN)
ST94: Ladywell (SPTAW), 3-Jun - 12-Aug-08 (Boyce & Wright, 2009)
ST95: Whorton Common, 28-Aug-00 (DMR)
SU01: Martin Down NNR, 1-Jan-79 (RGB)
SU03: Grovely Wood, 6, 13-Apr-71, 6, 12-Apr-72 (DRN); 21-Aug-01 (MDD)
SU05: West Lavington, 4-Jul-01 (DMR)
SU08: Grovely Wood, 1-Aug-73 (DRN)
SU14: Stonehenge, 22-Apr-03 (PJH) (Hodge, 2003)
SU15: Beach's Plantation (SPTAE), 10, 16-Jun-97, 1-May-05 (MGT); North Newton, 23-Jun-01 (DMR)
SU16: Jones's Mill, 1-Jan-97 (MAS); West Kennett, 4-May-03 (MGT)
SU21: Franchises Wood, 3-May-95 (MAS); Hamptworth Estate, 3-Aug-73, 15-Apr-74, 28-Jul-77 (DRN); Landford Bog, 9-Jun-01 (APF) (Foster, 2001); Pound Bottom, 24-Jul, 4-Aug-73, 9-Apr-74 (DRN)
SU22: Glazier's Copse, 16-Aug-88 (DRN); Langley Wood, 22-Apr-04 to 11-Feb-05, (MDD) (Darby, 2005); roadside, 14-Sep-55 (BPM, GEW) (Moore, 1956); Whiteparish Common, 22-Jul-73 (DRN)
SU23: Porton Down, 1970-2001 (Anon, 2001)
SU24: Dunch Hill (SPTAE), 16-Jun-97 (MGT); Newton Tony, 25-Aug to 20-Sep-06, (MA det MGT); Sidbury Hill (SPTAE), 6, 13-Jun-97 (MGT)

Resident. Ubiquitous; in all dry sunny situations especially gardens and agricultural land.

***Bembidion properans*** (Stephens, 1828)

7 ST97: Stanley, 2-Jul-80 (DMR)
ST98: Malmesbury, 13-Jul-80 (DMR)
SU07: Calne, 30-Sep-84, 6-Aug-00 (DMR); Yatesbury, 13-Jun-73 (BG)
SU16: East Kennet, 3-Jun-01 (DMR)
SU16/17: Avebury, 7-Jun-06 (DB det MDD); 4-May-03 (MGT)

8 ST91: Cranborne Chase, 16 to 25-Jun-80 (NCC)
ST92: Wardour, 13-Nov-95 (MDD)
ST94: Wadman's Coppice (SPTAW) 3-Jun - 12-Aug-08 (Boyce & Wright, 2009)
SU04: Shrewton, 6-Aug-82 (DRN)
SU13: Great Durnford, 18-Aug-72 (DRN); Lower Woodford, 22-Jul-78, 1-Aug-83 (DRN); Salisbury, Avon Valley Reserve, 1-Apr-00 (MAS) (Darby, 2001)
SU15: Beach's Plantation (SPTAE), 1-May-05 (MGT)
SU16: Jones's Mill, 19-Jun-01 (APF) (Foster 2002); Wilcot, 1- May-00 (DMR)
SU21: Franchises Wood, 3-May-95 (MAS); Pound Bottom, 4-Aug-82 (DRN)

Resident. Widespread; on dry, open clay soils.

PRINCIDIUM Motschulsky, 1864

***Bembidion punctulatum*** Drapiez, 1821

7 SU16: Marlborough area, Jun-1873 (EWJ) (Meyrick, 1874) 1869-1938 (Meyrick, 1938)

Resident. Very local; [in gravel and shingle by fresh water]. There is a single specimen without data in the MCC.

EUPETEDROMUS Netolitzky, 1911

***Bembidion dentellum*** (Thunberg, 1787)

7 ST86: Whaddon, 13-May-00 (DMR)
ST87: Castle Combe, 22-May-61 (PJMG)
ST97: Kellaways, 29-Apr-80 (DMR); Sutton Benger, 8-May-80 (DMR)
ST98: Malmesbury, 13-Jul-80 (DMR)
SU06: Beckhampton, 3-Jun-00 (DMR)
SU07: Lyneham, 4-Oct-80 (DMR)
SU09: Somerford Keynes, 4-Jun-04 (BLe)
SU16: East Kennet, 1-May-00, 3-Jun-01 (DMR); Marlborough area, 1869-1938 (Meyrick, 1938)
SU17: Ogbourne Maizey, 20-May-01 (DMR)
SU27: Ramsbury, 12-Feb-05 (Meyrick, 1874)

8 ST94: Ladywell (SPTAW), 3-Jun - 12-Aug-08 (Boyce & Wright, 2009)
SU12: Salisbury, 30-Jul-73 (DRN)
SU13: Great Durnford, 16-Aug-69 (DRN)
SU14: Nine Mile R., pond (SPTAE), 5-Jun-97 (MGT)
SU21: Hamptworth Estate, 20-Aug-76, 24-Apr-84 (DRN)
SU25: Sidbury Hill (SPTAE), 19-May-97 (MGT)
SU26: Bedwyn Brail, 19-Jun-03 (MDD)

Resident. Widespread in marshy sites near water.

NOTAPHUS Dejean, 1821

***Bembidion varium*** (Olivier, 1795)

8 SU13: Middle Woodford, 5-Aug-82 (DRN)

Resident. Very local; on partly bare ground near water.

OCYDROMUS Clairville, 1806

***Bembidion bruxellense*** Wesmael, 1835

8 SU12: Charlton-All-Saints, 9-May-96 (MAS)
SU13: Lower Woodford, 8-Aug-86 (DRN)
SU21: Hamptworth Estate, 3-Aug-73 (DRN)

Resident. Local; on mud, sand and gravel at edges of water, and sometimes in damp grassland.

***Bembidion deletum*** Audinet-Serville, 1821

7 ST97: Studley, 3-Oct-80 (DMR)
SU07: Lyneham, 2-Sep-84 (DMR)
SU16: Marlborough area, 1869-1938 (Meyrick, 1938)
SU26: Savernake Forest, 3-Apr-70 (DRN)
SU27: Ramsbury (Meyrick, 1874)

8 ST91: Cranborne Chase, 16 to 25-Jun-80 (NCC)
SU21: Burnt Ground Wood, 21-Jul-73 (DRN); Hamptworth Estate, 3-Apr-78, 25-Apr-84 (DRN); Pound Bottom, 24-Jul-73, 9, 19-Apr-74 (DRN)
SU22: Glazier's Copse, 31-Jul-85 (DRN); Homan's Copse, 20-Jul-73 (DRN); Whiteparish Common, 22-Jul-73 (DRN); Redlynch, 13-Apr-74 (DRN)
Resident. Widespread; in damp, shaded situations, often in woodland, usually near trickling water.

***Bembidion femoratum*** Sturm, 1825
7 SU16: Marlborough area, 1869-1938 (Meyrick, 1938)
8 SU13: Salisbury, 28-Oct-73 (DRN)
SU21: Hamptworth Estate, 24-Jul-73 (DRN)
SU22: Harestock, 26-Oct-71 (DRN); Redlynch, 13-Apr-74 (DRN); roadside, 14-Sep-55 (BPM, GEW) (Moore, 1956)
Resident. Local; in open, sometimes damp, coarse sand and gravel not always near water.

***Bembidion tetracolum*** Say, 1825
7 ST86: Whaddon, 13-May-00 (DMR)
ST87: Slaughterford, 10-Jun-00 (DMR)
ST96: Calne, 14-Apr-80 (DMR); Chittoe, 16-Jul-00 (DMR)
ST97: Kellaways, 29-Apr-80 (DMR); Stanley, 2-Jul-80 (DMR); Studley, 18-May-80 (DMR); Sutton Benger, 8-May-80 (DMR)
ST98: Malmesbury, 13-Jul-80 (DMR)
SU07: Calne, 5-May, 30-Sep-84, 30-Dec-89, 6-Aug-00 (DMR); Goatacre, 23-Apr, 28-Jun-80 (DMR); Yatesbury, 13-Jun-73 (BG)
SU08: Lyneham, 11-Apr, 23-Jun-80 (DMR)
SU09: Cotswold Water Park, 21-Jun-90 (APFw); Somerford Keynes, 4-Jun-04 (BLe)
SU16: Marlborough area, 1 Jan 1870 (Meyrick, 1874)
SU17: Ogbourne Maizey, 20-May-01 (DMR); Sells Green, 20-Aug-00 (DMR)
SU27: Knighton, 11-Jun-85 (DMR)
8 ST84: Longleat, 25-Mar-99 (MB)
ST92: Wardour, 13-Nov-95 (MDD)
ST95: Whorton Common, 28-Aug-00 (DMR)
SU03: Wilton, 14-Aug-76 (DRN)
SU13: Lower Woodford, 30-Jul-82 (DRN); Salisbury, 15-Aug-88 (DRN): Stratford sub Castle, 5-Apr-72 (DRN)
SU16: Wilcot, 1-May-00 (DMR)
SU21: Hamptworth Estate, 29-Mar-78 (DRN)
SU22: Redlynch, 13-Apr-74 (DRN); Harestock, 26-Oct-71 (DRN)
SU23: Porton Down, 1970-2001 (Anon, 2001)
Resident. Widespread; on open, not too dry soil, especially near water.

NEPHA Motschulsky, 1864

***Bembidion illigeri*** Netolitzky, 1914
7 ST97: Christian Malford, 9-Aug-80 (DMR); Studley, 18-May-80 (DMR)
SU07: Calne, 14-Apr-80, 5-May, 30-Sep-84 (DMR)
SU16: Marlborough area, 1869-1938 (Meyrick, 1938)
SU17: Sells Green, 20-Aug-00 (DMR)
8 ST94: Ladywell (SPTAW), 3-Jun - 12-Aug-08 (Boyce & Wright, 2009)
SU22: Redlynch, 13-Apr-74 (DRN)
SU23: West Winterslow, 8-Jul-01 (NB)
Resident. Local; [on open, sunny sites near water].

SINECHOSTICTUS Motschulsky, 1864

***Bembidion stomoides*** Dejean, 1831
**Nb**
7 ST99: Flisteridge Wood, 1-Jan-80 (KNAA)
Resident. Very local; [in shingle by rivers].

Bembidion stomoides

SEMICAMPA Netolitzky, 1910

***Bembidion gilvipes*** Sturm, 1825
**Nb**
8 SU13: Stratford sub Castle, 5-Apr-72 (DRN)
Resident. Very local; [on moist, clay soils in shaded marshy habitats including woods].

DIPLOCAMPA Bedel, 1896

***Bembidion assimile*** Gyllenhal, 1810
7 SU09: Somerford Keynes, 4-Jun-04 (BLe)
8 SU12: Petersfinger, 25-Jul-91 (DRN)
Resident. Very local; [in marshy and swampy areas].

***Bembidion clarkii*** (Dawson, 1849)
**Nb**
8 ST94: Ladywell (SPTAW), 3-Jun - 12-Aug-08 (Boyce & Wright, 2009)
SU25: Sidbury Hill (SPTAE), 19-May-97 (MGT)
Resident. Very local; [in shaded wet sites near water].

***Bembidion fumigatum*** (Duftschmid, 1812)
**Nb**
8 SU14: Nine Mile R., pond (SPTAE), 5-Jun-97 (MGT)

Resident. Very local; [in lowland fens on clay soils].

BEMBIDION Latreille, 1802

***Bembidion quadrimaculatum*** (Linnaeus, 1761)

7 ST86: Whaddon, 13-May-00 (DMR)
ST97: Kellaways, 29-Apr, 13-Sep-80 (DMR); Stanley, 2-Jul-80 (DMR)
ST98: Malmesbury, 13-Jul-80 (DMR); Dauntsey Green, 5-Apr-80 (DMR)
SU07: Bewick Basset, 12-Aug-00 (DMR); Calne, 30-Dec-1899, 6-Aug-00 (DMR); Clyffe Pypard, 4-Sep-80 (DMR); Goatacre, 23-Apr-80 (DMR); Lyneham, 13-May-80 (DMR); Yatesbury, 13-Jun-73 (BG)
SU08: Lyneham, 19-Apr-80 (DMR)
SU09: Somerford Keynes, 4-Jun-04 (BLe)
SU16: Marlborough area, 25 Mar 1879 (TNH-S); 30-Mar-1879 (WFB) (Meyrick, 1880); 1873-1938 (Meyrick, 1938); West Stowell, 3-Sep-00 (DMR)
8 ST84: Longleat, 4-May-98 (MB)
ST92: Fovant, 25-Apr-95 (MDD); Sutton Mandeville, 25-Apr-92 (MDD); Wardour, 5-Jun-96 (MAS)
ST93: Great Ridge Wood, 2-May-00 (MDD)
ST95: Whorton Common, 28-Aug-00 (DMR)
SU13: Salisbury, 7-Apr-74 (DRN); Lower Woodford, 28-Jul-78 (DRN)
SU15: East Sharcott, 31-Aug-00 (IJG); North Newton, 23-Jun-01 (DMR)
SU16: Wilcot, 1-May-00 (DMR)
SU21: Burnt Ground Wood, 21, 24-Jul-73 (DRN)
SU22: Redlynch, 13-Apr-74 (DRN); roadside, 14-Sep-55 (BPM, GEW) (Moore, 1956)
Resident. Widespread; on open, dry soils including arable fields.

***Bembidion quadripustulatum*** Audinet-Serville, 1821

**Nb**

8 SU25: Sidbury Hill (SPTAE), 19-May, 6-Jun-97 (MGT)
Resident. Very local; [on damp, sometimes shaded bare mud or clay]. Nationally very scarce.

TREPANEDORIS Netolitzky, 1918

***Bembidion doris*** (Panzer, 1796)

8 SU21: Hamptworth Estate, 24-Apr-84 (DRN)
SU26: Bedwyn Brail, 19-Jun-03 (MDD)
Resident. Very local; in wet situations [such as marshes].

TREPANES Motschulsky, 1864

***Bembidion articulatum*** (Panzer, 1795)

7 ST86: Whaddon, 13-May-00 (DMR)
ST96: Calne, 14-Apr-80 (DMR); Chittoe, 16-Jul-00 (DMR)
ST98: Malmesbury, 13-Jul-80 (DMR)
SU06: Beckhampton, 3-Jun-00 (DMR)
SU07: Calne, 5-May-84 (DMR)
SU08: Lyneham, 3-Jul-80 (DMR); Tockenham Wick, 25-May-80 (DMR)
SU09: Somerford Keynes, 4-Jun-04 (BLe)
SU16: East Kennet, 3-Jun-01 (DMR)
SU17: Sells Green, 20-Aug-00 (DMR)
8 ST94: Ladywell (SPTAW), 3-Jun - 12-Aug-08 (Boyce & Wright, 2009)
ST95: Whorton Common, 28-Aug-00 (DMR)
SU13: Stratford sub Castle, 26-May-74 (DRN); Salisbury, 25-Jul-78 (DRN); Lower Woodford, 28-Jul-78, 1-Aug-83, 8-Aug-86, 1-Aug-89 (DRN)
8 SU14: Nine Mile R., pond (SPTAE), 5-Jun-97 (MGT)
SU23: West Winterslow, 8-Jul-01 (NB)
Resident. Widespread in cracks in bare sand or mud near water.

PHYLA Motschulsky, 1844

***Bembidion obtusum*** Audinet-Serville, 1821

7 SU06: Morgan's Hill, 24-Sep-84 (DMR); Cherhill, 27- Aug-01 (DMR)
SU07: Clyffe Pypard, 4-Sep-80 (DMR); Yatesbury, 13-Jun-73 (BG); Lyneham, 27-May-80 (DMR)
SU16: East Kennet, 3-Jun-01 (DMR); Marlborough area, Jun-1873 (EWJ) (Meyrick, 1874); 4-Mar-1879 (FED-H) (Meyrick, 1880); 1873-1938 (Meyrick, 1938); West Stowell, 3-Sep-00 (DMR)
8 ST93: Great Ridge Wood, 29-Jul-82 (DRN)
SU02: Knighton Wood, Sep-07 (AJWA)
SU03: Grovely Wood, 26-Jul-82 (DRN); Wylye Down NNR, 21-Jan-07 (MDD) (Darby, 2008a)
SU14: Bourne Bottom (SPTAE), 13-Jun-97 (MGT)
SU15: Beach's Plantation (SPTAE), 13 - 17-Jun-97, 1-May-05 (MGT); 4-Sep-05 (JMW)
SU16: West Kennett, 4-May-03 (MGT)
SU21: Hamptworth Estate, 28-Oct-75 (DRN)
SU22: Redlynch, 4-Aug-82 (DRN)
SU26: Cobham Frith, 24-Jul-82 (DRN)
Resident. Widespreadl; on open, often cultivated ground.

PHILOCHTHUS Stephens, 1828

***Bembidion aeneum*** Germar, 1824

7 ST86: Whaddon, 13-May-00 (DMR)
ST87: Slaughterford, 10-Jun-00 (DMR)
ST96: Calne, 14-Apr-80 (DMR); Lacock, 16-Jun-82 (KNAA)
SU06: Beckhampton, 3-Jun-00 (DMR)
SU07: Clyffe Pypard, 4-Sep-80 (DMR); Yatesbury, 13-Jun-73 (BG); Calne, 5-May, 30-Jun-84 (DMR)
SU08: Lyneham, 9-Mar, 11-Apr-80 (DMR)
SU16: East Kennet, 3-Jun-01 (DMR)
SU17: Ogbourne Maizey, 20-May-01 (DMR)

8 ST92: Sutton Mandeville, 5-Apr-92 (MDD)
Resident. Widespread on damp, poorly draining soils often with vegetation.

***Bembidion biguttatum*** (Fabricius, 1779)
7 ST86: Whaddon, 13-May-00 (DMR)
ST96: Calne, 14-Apr-80 (DMR); Chittoe, 16-Jul-00 (DMR)
ST97: Christian Malford, 8-May-80 (DMR); Studley, 18-May-80 (DMR); Foxham, 24-Apr-80 (DMR); Kellaways, 29-Apr, 13-Sep 80 (DMR); Sutton Benger, 8-May-80 (DMR)
ST98: Dauntsey, silt pond, 5-Apr-80 (DMR); Malmesbury, 13-Jul-80 (DMR)
SU06: Theobalds Green, 28-Jun-80 (DMR); Beckhampton, 3-Jun-00 (DMR)
SU07: Compton Bassett, 28-Jun-80 (DMR); Yatesbury, 13-Jun-73 (BG)
SU08: Lyneham, 21-Mar, 11-Apr-80 (DMR); Tockenham Wick, 25-May-80 (DMR)
SU09: Emmett Hill, 1-Jan-95 (MA det MDD); Somerford Keynes, 4-Jun-04 (BLe)
SU16: East Kennet, 1-May-00, 3-Jun-01 (DMR); Marlborough area, 1869-1938 (Meyrick, 1938)
SU17: Ogbourne Maizey, 20-May-01 (DMR); Sells Green, 20-Aug-00 (DMR)
SU26: Crofton, 26-May-01 (DMR)
8 ST84: Longleat, 28-Aug-00 (MB)
ST94: Ladywell (SPTAW), 3-Jun - 12-Aug-08 (Boyce & Wright, 2009); Wadman's Coppice (SPTAW), 3-Jun - 12-Aug-08 (Boyce & Wright, 2009)
SU07: New Zealand, 5-Jan-80 (DMR)
SU13: Great Durnford, 3-Aug-88 (DRN); Salisbury, Avon Valley Reserve, 1-Apr-00 (MAS) (Darby, 2001)
SU15: Fittleton, 29-Jul-83 (DRN)
SU21: Landford Bog, 25-May-98 (MDD)
SU22: East Grimstead, 5-Apr-77 (DRN)
SU25: Sidbury Hill (SPTAE), 19-May, 5, 6-Jun-97 (MGT)
Resident. Widespread on open mud and silty ground often near fresh water.

***Bembidion guttula*** (Fabricius, 1792)
7 ST86: Whaddon, 13-May-00 (DMR)
ST87: Slaughterford, 10-Jun-00 (DMR)
ST97: Stanley, 2-Jul-80 (DMR); Studley, 18-May-80 (DMR); Sutton Benger, 8-May-80 (DMR)
SU07: Goatacre, 30-Dec-79, 23-Apr-80 (DMR); Lyneham, 17-Apr, 13-May-80 (DMR); Yatesbury, 13-Jun-73 (BG)
SU08: Lyneham, 9-Mar-80 (DMR); Somerford Common, 23-Apr-01 (JKG det MDD)
SU16: Marlborough area, 1869-1938 (Meyrick, 1938)
SU17: Ogbourne Maizey, 20-May-01 (DMR)
SU18: Sevenfields LNR, Swindon, 30-Mar-01 (JKG det MDD)
SU27: Ramsbury, 1870-1873 (Meyrick, 1874)
8 SU03: Grovely Wood, 10-Apr-70 (DRN)
SU12: Petersfinger, 25-Jul-91 (DRN); Salisbury, 30-Jul-73 (DRN)
SU13: Great Durnford, 3-Aug-83 (DRN); Lower Woodford, 2-Aug-82 (DRN); Salisbury, Avon Valley Reserve, 7, 13-Apr, 1, 12-May-00 (MDD) (Darby, 2001)
SU21: Hamptworth Estate, 28-Oct-75, 20-Aug-76, 6-Aug-79, 24-Apr-84 (DRN); Landford Bog, 1-Jan-03 (APF) (Foster, 2003)
SU22: Bentley Wood, 31-Mar-78 (DRN); East Grimstead, 5-Apr-77 (DRN); Redlynch, 13-Apr-74 (DRN)
Resident. Widespread and frequent in a wide range of habitats, especially near water.

***Bembidion lunulatum*** (Geoffroy in Fourcroy, 1785) 7
ST86: Whaddon, 13-May-00 (DMR)
ST96: Calne, 14-Apr-80 (DMR); Chittoe, 16-Jul-00 (DMR)
ST97: Kellaways, 13-Sep-80 (DMR); Studley, 18-May-80 (DMR)
SU07: Calne, 5-May-84 (DMR); Lyneham, 13-May-80 (DMR); Yatesbury, 13-Jun-73 (BG)
SU08: Lyneham, 23-Jun-80 (DMR)
8 ST91: Cranborne Chase, 16 to 25-Jun-80 (NCC)
ST94: Ladywell (SPTAW), 3-Jun - 12-Aug-08 (Boyce & Wright, 2009)
SU15: Fittleton, 29-Jul-83 (DRN); North Newnton, 23-Jun-01 (DMR)
SU21: Franchises Wood, 4-Aug-77 (DRN); Hamptworth Estate, 3-Aug-73, 20-Aug-76 (DRN)
SU22: Bentley Wood, 31-Mar-78 (DRN); Homan's Copse, 20-Jul-73 (DRN); Redlynch, 13-Apr-74 (DRN)
SU25: Sidbury Hill (SPTAE), 6-Jun-97 (MGT)
Resident. Widespread; on damp bare ground near water.

***Bembidion mannerheimii*** Sahlberg, C.R, 1827
8 ST94: Ladywell (SPTAW), 3-Jun - 12-Aug-08 (Boyce & Wright, 2009)
SU13: Salisbury, 25-May-00 (MAS) (Darby, 2001)
SU21: Hamptworth Estate, 15-Apr-74, 24-25-Apr-84, 19-Aug-88 (DRN)
SU22: East Grimstead, 5-Apr-77 (DRN); Langley Wood, 22-Apr-04 to 24-Mar-05, 15-Sep-06 to 25-Apr-07 (MDD) (Darby, 2005, 2008)
Resident. Local; in damp grassland and shaded habitats especially woodland.

Tribe PATROBINI

**PATROBUS Dejean, 1821**
***Patrobus atrorufus*** (Ström, 1768)
7 ST97: Stanley, 2-Jul-80 (DMR)
ST98: Malmesbury, 13-Jul-80 (DMR)
ST99: Flisteridge Wood, 1980-1982 (KNAA)
SU07: Yatesbury, 13-Jun-73 (BG)
SU16: Marlborough area, 24-Feb-1879 (WFB) (Meyrick, 1880); 1869-1938 (Meyrick, 1938)
8 SU21: Franchises Wood, 25-Jul-73 (DRN)
SU22: roadside, 14-Sep-55 (BPM, GEW) (Moore, 1956)
Resident. Local; in woodlands and damp grasslands.

Tribe PTEROSTICHINI

**STOMIS Clairville, 1806**
***Stomis pumicatus*** (Panzer, 1795)
7 ST86: Whaddon, 13-May-00 (DMR)
SU06: Beckhampton, 3-Jun-00 (DMR)
SU07: Yatesbury, 13-Jun-73 (BG)
SU16: West Stowell, 3-Sep-00 (DMR)
SU26: Savernake Forest, 22-Mar-46 (Holmes & Tennant, 1947)
8 ST93: Great Ridge Wood, 18-Apr-74 (DRN)
SU01: Martin Down NNR, 1999 (EN, 1999)
SU03: Grovely Wood, 8-Apr-74, 17-Apr-84 (DRN)
SU13: Middle Woodford, 27-Jul-77 (DRN); Stratford sub Castle, 12-Apr-71 (DRN)
SU15: Milton (Lilbourne) Hill Quarry, 30-Aug-02 (IJG)
SU22: Blackmoor Copse, 4-18-May-02 (APF) (Foster, 2004); roadside, 14-Sep-55 (BPM, GEW) (Moore, 1956)
SU26: Bedwyn Brail, 2-Jun-03 (MDD)
Resident. Widespread; in woodlands, grasslands and on disturbed ground.

**POECILUS Bonelli, 1810**
***Poecilus cupreus*** (Linnaeus, 1758)
7 ST87: Castle Combe, 22-May-61 (PJMG) Greenslade, 1963)
SU08: Minety (Stonehill Wood), 13-Apr-02 (JKG det MDD)
SU16: Marlborough area, Jun-1873 (EWJ) (Meyrick, 1874)
8 ST85: Clanger Wood, 1995 (MA det MDD)
ST92: Wardour, 5-Jun-96 (MDD); Sutton Mandeville, 27-Apr-08 (MDD)
ST94: Tinker's Track (SPTAW), 3-Jun - 12-Aug-08 (Boyce & Wright, 2009)
ST95: Cheverell, 30-May-97 (MGT)
SU04: Tilshead Down, 1 - 4-Jun-97 (MGT)
SU03: Grovely Wood, 16-Apr-74 (DRN); Wylye Down NNR, 29-Mar-07 (MDD) (Darby, 2008a)
SU14: Bourne Bottom (SPTAE), 13 to 19-Jun-97 (MGT)
SU15: Beach's Plantation (SPTAE), 10, 13 - 17-Jun-97, 1-May-05 (MGT); 4-Sep-05 (JMW)
SU21: Hamptworth Estate, 15-Apr-74, 1-Apr-75, 3-Apr-78 (DRN); Pound Bottom, 19-Apr-74 (DRN)
SU22: Langley Wood NNR, 29-May-03, 1 to 22-Apr-04, 15-Sep-06 to 6-Jun-07 (MDD) (Darby, 2005, 2008); Redlynch, 13-Apr-74 (DRN)
SU23: Bentley Wood, 5-Apr-77 (DRN)
SU24: Driver Training Area (SPTAE), 1-May-05 (MGT); Dunch Hill (SPTAE), 3-Jun - 12-Aug-08 (Boyce & Wright, 2009); Sidbury Hill (SPTAE), 29-May-97 (MGT)
Resident. Widespread; in dry habitats, including fields and woodland, often running in the sun. Easily confused with *P. versicolor.*

***Poecilus lepidus*** (Leske, 1785)
**Nb**
8 SU21: Pound Bottom, 4-Aug-73 (DRN)
Resident. Very local; under stone [on dry open heathland]. Probably confined to the far south of the county.

***Poecilus versicolor*** (Sturm, 1824)
8 ST93: Great Ridge Wood, 18-Apr-74 (DRN)
SU04: Shrewton, 6-Aug-82 (DRN)
SU21: Burnt Ground Wood, 31-May-74 (DRN); Hamptworth Estate, 31-May-74 (DRN); Pound Bottom, 4-Aug-82 (DRN)
SU22: Langley Wood NNR, 22-Apr to 19-Nov-04, 25- Aug to 15-Sep-06 (MDD) (Darby, 2005, 2008); roadside, 14-Sep-55 (BPM, GEW) (Moore, 1956)
Resident. Local; in grassland and arable land, and occasionally woods, often running in the sun. Easily confused with *P. cupreus.*

**PTEROSTICHUS Bonelli, 1810**
STEROPUS Stephens, 1828
**Black Clock**
***Pterostichus madidus*** (Fabricius, 1775)
7 ST87: Rack Hill, 1995 (MA det MDD); Jun-96 (MA det MDD)
ST96: Chittoe, 9-May-00 (EGS det MDD); 16-Jul-00 (DMR)
ST98: Bradfield Wood, 7-Apr-96 (MA det MDD); Malmesbury, 13-Jul-80 (DMR); Startley, 30-Apr-85 (DMR)
SU06: Beckhampton, 3-Jun-00 (DMR)
SU07: Catcomb Wood, May-96 (MA det MDD); Lyneham, 3-Mar-80, 7-Apr-80 (DMR); Calne, 6-Aug-00 (DMR)
SU08: Ravensroost Wood, 1995 (MA det MDD)
SU16: Gopher Wood, May-96 (MA det MDD); 13-Jul-96 (MA det MDD); Marlborough area, Jun-1873 (EWJ) (Meyrick, 1874); 3-Jun-1879 (ALJ) (Meyrick, 1880); 1869-1938 (Meyrick, 1938)

SU26: Savernake Forest, Jul to Aug-93 (EN det SAW)
8 ST84: Longleat, 29-Aug-99 (MB)
ST85: Clanger Wood, 1-Jan-95 (MA det MDD); Green Lane Wood, 26-Jun-98 (JN); 7-Jun-95 (MDD)
ST91: Cranborne Chase, 7-May-97 (MAS); 16-Jun-80 (NCC)
ST92: Wardour, 3-Apr-96 (EN)
ST93: Stockton Down, 1995 (MA det MDD); Great Ridge Wood, 13-Aug-81 (DRN); 10-May-00 (MDD)
ST94: Tinker's Track (SPTAW), 3-Jun - 12-Aug-08 (Boyce & Wright, 2009); Warminster Ranges (SPTAW), 3-Jun - 12-Aug-08 (Boyce & Wright, 2009)
ST95: Coombe Hill, 1995 (MA det MDD)
SU01: Martin Down NNR, 1999 (EN, 1999)
SU02: Broad Chalke, 2000 (BL); Gardiner Forest, 1-Sep-72 (DRN); Middleton Down, 3-Apr-97 (MDD); 6-Nov-73, May-98 (BL)
SU03: Grovely Wood, 10 records 6-Aug-70 – 1-Aug-1988 (DRN); 1998 - 2001 (MDD); 14-Jun-01 (WANHS det MDD); Wylye Down NNR, 13-Nov-06 to 10-Oct-07 (MDD) (Darby, 2009)
SU04: Orcheston Down (SPTA), 18-May-02 (PJH) (Hodge, 2002); Shrewton, 28-Jul-82 (DRN)
SU05: Black Heath (SPTAC), 3-Jun - 12-Aug-08 (Boyce & Wright, 2009); West Lavington, 1938 (PLP) (DFL, 1948)
SU12: Coombe Bissett Down, May, Jun-00 (MDD); Petersfinger, 7-Aug-88 (DRN)
SU13: Salisbury, 10-Apr-72 (DRN); Salisbury, Bemerton Folly, 7-Mar - 15-Jun-02 (MDD) (Darby, 2002d); Old Sarum, 31-May-97 (MAS)
SU14: Boscombe Down, 1994 (MA det MDD)
SU15: Beach's Plantation (SPTAE), 13 - 19-Jun-97, 1-May-05 (MGT); 4-Sep-05 (JMW); Manningford, 13-Aug-87, 4-Aug-88 (DRN); Upavon, 26-Aug-01 (DMR); Widdington, 5-Aug-83 (DRN)
SU17: Barbury Hill, 1995 (MA det MDD)
SU21: Franchises Wood, 3-May-95 (MAS); Hamptworth Estate, 3-Aug-73 (DRN); Landford Bog, 24-Aug-01 (APF) (Foster, 2001)
SU22: Glazier's Copse, 16-Aug-88 (DRN); Langley Wood, 12-Apr-74, 5-Aug-85 (DRN); 25-May-94 (WANHS det MDD); 22-Apr-05 to 2-Jul-07 (MDD) (Darby, 2005, 2008); Whiteparish Common, 22-Jul-73 (DRN)
SU23: Bentley Wood, 18-Jun-05 (JN); 27-May-97 (MDD); Porton Down, 12-Jun-94 (PTW det MDD); 1970-2001 (Anon); West Winterslow, 9-Jul-01 (NB)
SU24: Beacon Hill (SPTAE), 3-Jun - 12-Aug-08 (Boyce & Wright, 2009); Bulford Ranges (SPTAE), 3-Jun - 12-Aug-08 (Boyce & Wright, 2009); Driver Training Area (SPTAE), 1-May-05 (MGT); Dunch Hill (SPTAE), 16, 19-Jun-97 (MGT); SU24: Newton Tony, 25-Aug to 20-Sep-06, (MA det MGT); Sidbury Hill (SPTAE), 18, 28, 29-May, 5, 13 to 17-Jun-97 (MGT)

Resident. Ubiquitous; in woodlands, grasslands and gardens.

ADELOSIA Stephens, 1835

***Pterostichus macer*** (Marsham, 1802)

7 SU16: Marlborough area, 1869-1938 (Meyrick, 1938)
SU18: Sevenfields LNR, Swindon, 19-Apr-01 (JKG det MDD)
8 SU05: West Lavington (PLP) (DFL, 1948)

Resident. Local; [often in cracks in clay soil and sometimes under bark].

PLATYSMA Bonelli, 1810

***Pterostichus niger*** (Schaller, 1783)

7 SU07: Yatesbury, 13-Jun-73 (BG)
SU16: Marlborough area, 30 Jun 1879 (GLR) (Meyrick, 1880); spring-1894 (GCC) (Jebb, 1895); 1869-1938 (Meyrick, 1938)
SU26: Savernake Forest, 2, 3-Apr-70 (DRN)
8 ST84: Longleat, 5-May-86 (AGD); 7-Jul-99 (MDD); 29-Aug-99 (MB)
ST91: Cranborne Chase, 16 to 25-Jun-80 (NCC); 7-May-97 (MAS)
ST92: Sutton Mandeville, May-93 (MDD)
ST93: Great Ridge Wood, 18-Apr-74; 12-Apr-77; 2-Aug-82 (DRN)
SU01: Martin Down NNR, 1999 (EN, 1999)
SU03: Grovely Wood, 8-Apr-75 (DRN); 17-Feb-98 (MDD); Wylye Down NNR, 4-Sep-07 (MDD) (Darby, 2009)
SU04: Tilshead Down, 4-Jun-97 (MGT)
SU21: Hamptworth Estate, 3-Aug-73, 15-Apr-74 (DRN) Harestock, 26-Oct-71 (DRN); Redlynch, 9-Apr-74 (DRN)
SU22: Bentley Wood, 23-May-04 (AGD); Langley Wood NNR, 12-Apr-74, 5-Aug-85 (DRN); 22-Apr to 8-Oct-04, 5-Aug-06 to 2-Jul-07 (MDD) (Darby, 2005, 2008)
SU26: Bedwyn Brail, 2-Jun-03 (MDD)

Resident. Widespread; in woodland and damp grassland.

BOTHRIOPTERUS Chaudoir, 1838

?***Pterostichus adstrictus*** Eschscholtz, 1823

7 SU26: Savernake Forest, 1939 (det KGB) (Knight, 1940)

There must be some doubt about this determination. This is a northern species of upland moors with no records in England south of the Wash. It can fly, however, and Luff, 1998 records that it has turned up in 'more lowland localities'. There is a single specimen without data in the MCC.

***Pterostichus oblongopunctatus*** (Fabricius, 1787)
**Nb**
7 SU16: Marlborough area, 6-Feb-1879 (GTKM) Meyrick, 1880); spring-1894 (GCC) (Jebb, 1895); 1869-1938 (Meyrick, 1938)
SU26: Savernake Forest, 2-Apr-70 (DRN); 6-Jun-73 (APF); 6-Dec-73 (APFw, PS); 22-Jun-90 (APFw); 26-May to 16-Jun-93 (EN det SAW); 31-May to 23-Jun-95 (EN de3 t MDD); 7-Nov-01, 17-Oct-02 (IJG)
SU36: Stype Wood, Jun-1870 (RH) (Hislop, 1871)

Pterostichus oblongopunctatus

8 ST84: Longleat, 25-Mar-99 (MB)
ST92: Sutton Mandeville, 27-Apr-08 (MDD)
ST93: Stockton Wood, 1995 (MA det MDD); Great Ridge Wood, 11-Aug-80 (DRN)
SU03: Grovely Wood, 15 records 10-Apr-70 to 26-Jul-82 (DRN); 17-Feb-98 to 12-Jun-00 (MDD)
SU13: Salisbury, Bemerton Folly, 7-Mar - 15-Jun-02 (MDD) (Darby, 2002d)
SU22: Langley Wood NNR, 1-Apr-04 to 2-Jul-07 (MDD) (Darby, 2005, 2008)
SU23. Porton Down, 25-Apr-93 (PTW det MDD); 1970-2001 (Anon)
Resident. Widespread and frequent in woodland.

***Pterostichus quadrifoveolatus*** Letzner, 1852
**Nb**
8 SU21: Peaked Wood, 3-Aug-73, 15-Apr-74 (DRN) (Nash,1978)
Resident, arrived in GB (c.1900). Very local; under stones, etc. in a dry drainage ditch in a recently clear-felled area with burnt logs, and under burnt logs. [Recent records are often from unburnt heaths and woodland on sandy and peaty soils]. Formerly known as *P. angustatus*.

OMASEUS Stephens, 1828

***Pterostichus melanarius*** (Illiger, 1798)

7 ST87: Slaughterford, 10-Jun-00 (DMR)
ST97: Stanley, 2-Jul-80 (DMR); Kellaways, 13-Sep-80 (DMR)
ST98: Malmesbury, 13-Jul-80 (DMR)
SU06: Roundway Camp, Jul-40 (FDB) (Buck, 1956)
SU07: Calne, 25-Aug-85 (DMR); Bewick Basset, 12-Aug-00 (DMR); Yatesbury, 13-Jun-73 (BG)
SU08: Lyneham, 3-May-80 (DMR)
SU09: Emmett Hill, 1995 (MA det MDD); 1873-1938 (Meyrick, 1938)
SU17: Fyfield Down, 21-Jul-80 (DMR)
8 ST91: Cranborne Chase, 16 to 25-Jun-80 (NCC)
ST93: Stockton Down, 1995 (MA det MDD); Great Ridge Wood, 12-Apr-77, 19-Apr-84 (DRN)
SU01: Martin Down NNR, 1999 (EN, 1999)
SU02: Middleton Down, 6-Nov-73 (BL)
SU03: Barford Down, 29-Oct-73 (DRN); Grovely Wood, 13-Aug-70, 13, 14-Apr-71, 8-Apr-72, 29-Oct-73, 1-Aug-88 (DRN); 11-Jun-00 (MDD); Wylye Down NNR, 16-Oct-06 to 10-Oct-07 (MDD)
SU05: West Lavington, 4-Jul-01 (DMR)
SU12: Salisbury, 30-Jul-73 (DRN)
SU15: Beach's Plantation (SPTAE), 4-Sep-05 (JMW); Manningford, 4-Aug-88 (DRN); Upavon, 26-Aug-01 (DMR); North Newton, 23-Jun-01 (DMR)
SU21: Burnt Ground Wood, 19-Apr-74 (DRN)
SU22: roadside, 14-Sep-55 (BPM, GEW) (Moore, 1956); Langley Wood NNR, 25-May-94 (WANHS det MDD); 14-May to 17-Sep-04, 5-Aug-06 to 2-Jul-07 (MDD) (Darby, 2005, 2008)
SU24: Newton Tony, 25-Aug to 20-Sep-06, (MA det MGT)
SU25: Sidbury Hill (SPTAE), 5-Jun-97 (MGT)
Resident. Widespread; in open habitats including grassland and cultivated fields.

PSEUDOMASEUS Chaudoir, 1838

***Pterostichus anthracinus*** (Panzer, 1795)
**Nb**
7 ST86: Whaddon, 13-May-00 (DMR)
ST97: Kellaways, 29-Apr-80 (DMR)
SU07: Calne, 28-May-84 (DMR)
SU08: Lyneham, 3-Jul-80 (DMR)
SU16: East Kennet, 3-Jun-01 (DMR); West Kennet, 4-May-03 (MGT)
8 SU12: Britford, Jun - 15-Sep-02 (CMD) (Drake, 2008)
SU13: Salisbury, 7-Apr-74 (DRN); Lower Woodford, 8-Aug-86 (DRN); Jun - 15-Sep-02 (CMD) (Drake, 2008)
Resident. Local; [in shaded habitats where there is a good litter layer often close to water].

Pterostichus anthracinus

SU21: Hamptworth Estate, 3-Aug-73, 15-Apr-74, 20-Aug-76, 3-Apr-78 (DRN)
SU22: roadside, 14-Sep-55 (BPM, GEW) (Moore, 1956); Redlynch, 4-Aug-82 (DRN); Langley Wood, 22-Apr-04 to 24-Mar-05 (MDD) (Darby, 2005)
Resident. These records may include the sibling species *P.rhaeticus* Heer 1837 (Luff 1990). *P. nigrita sensu stricto* inhabits wet, well vegetated habitats whereas *rhaeticus* is usually found in more open areas, but the two do overlap. Both are widespread and frequent in GB.

***Pterostichus rhaeticus*** Heer, 1837
8 SU21: Redlynch, 31-Oct-73 (DRN)
Resident. Under wood in a gravel pit. See *P. nigrita*.

***Pterostichus minor*** (Gyllenhal, 1827)
7 ST97: Christian Malford, 8-May-80 (DMR)
SU08: Lyneham, 3-May-80 (DMR)
SU16: Marlborough area, 6-Feb-1879 (GTKM) (Meyrick, 1880); 1869-1938 (Meyrick, 1938)
8 SU13: Salisbury, Avon Valley Reserve, 13-Apr to 19-Jun-00 (MDD) (Darby, 2001); Great Durnford, 3-Aug-88 (DRN)
SU14: Nine Mile R., pond (SPTAE), 5-Jun-97 (MGT)
SU21: Burnt Ground Wood, 19-Apr-74 (DRN); Hamptworth Estate, 4-Aug-73, 15-Apr-74, 20-Aug-76, 6-Apr-77 (DRN)
SU22: East Grimstead, Hawks Grove, 5-Apr-77 (DRN)
Resident. Widespread; on vegetated sites near water.

***Pterostichus nigrita*** (Paykull, 1790)
7 ST87: Slaughterford, 10-Jun-00 (DMR)
ST96: Calne, 14-Apr-80 (DMR)
ST97: Kellaways, 29-Apr, 13-Sep, 3-Oct-80 (DMR); Christian Malford, 8-May-80 (DMR)
SU07: Goatacre, 23-Apr-80 (DMR)
SU08: Lyneham, 21-Mar-80 (DMR); Tockenham Wick, 25-May-80 (DMR)
SU16: Marlborough area, 6-Feb-1879 (GTKM) (Meyrick, 1880); 1869-1938 (Meyrick, 1938);
SU17: Sells Green, 20-Aug-00 (DMR)
8 ST94: Ladywell (SPTAW), 3-Jun - 12-Aug-08 (Boyce & Wright, 2009); Wadman's Coppice (SPTAW), 3-Jun - 12-Aug-08 (Boyce & Wright, 2009)
SU02: Fovant, 31-May-95 (MDD)
SU03: Wilton, 14-Aug-76 (DRN)
SU12: Salisbury, 30-Jul-73 (DRN); Charlton-All-Saints, 9-May-96 (MAS)
SU13: Lower Woodford, 1-Aug-89 (DRN); Middle Woodford, 27-Jul-77 (DRN); Great Durnford, 3-Aug-88 (DRN)
SU14: Figheldean, 28-Jul-82 (DRN)
SU16: Jones's Mill, 16-Jun-73 (PM)

LAGARUS Chaudoir, 1838

***Pterostichus vernalis*** (Panzer, 1795)
7 ST87: Slaughterford, 10-Jun-00 (DMR)
ST97: Studley, 9-Apr-85 (DMR)
SU06: Morgan's Hill, 24-Sep-84 (DMR)
SU07: Goatacre, 6-Sep-80 (DMR); Calne, 5-May-84 (DMR)
SU09: Upper Waterhay, 17-May to 4-Jun-05, 11-Jun-07 (JKG det MDD)
SU16: East Kennet, 3-Jun-01 (DMR); Marlborough area, 1869-1938 (Meyrick, 1938)
8 ST84: Longleat, 7-Mar-99 (MB)
ST92: Wardour, 5-Jun-96 (MAS)
SU13: Stratford sub Castle, 5-Apr-72 (DRN)
SU21: Hamptworth Estate, 15-Apr-74 (DRN)
Resident. Widespread; in damp grassland with some litter often near water.

ARGUTOR Stephens, 1828

***Pterostichus diligens*** (Sturm, 1824)
7 SU07: Goatacre, 30-Dec-79 (DMR); Calne, 28-May-84 (DMR)
SU16: Marlborough area, 1869-1938 (Meyrick, 1938);
8 SU13: Salisbury, Avon Valley Reserve, 7-Apr to 19-Jun-00 (MDD); Apr-00 (MAS) (Darby, 2001)
SU16: Jones's Mill, 19-Jun-01 (APF) (Foster, 2002)
SU21: Hamptworth Estate, 15-Apr-74, 6-Apr-77 (DRN); Landford Bog, 2003 (APF) (Foster 2003); Redlynch, 31-Oct-73 (DRN)
SU23: Porton Down, 1970-2001 (Anon, 2001)
SU26: Bedwyn Brail, 19-Jun-03 (MDD)
Resident. Widespread; in marshy areas and shaded and damp grassland.

***Pterostichus strenuus*** (Panzer, 1796)
7 ST86: Whaddon, 13-May-00 (DMR)
ST87: Castle Combe, 22-May-61 (PJMG) Greenslade, 1963); Slaughterford, 10-Jun-00 (DMR)

ST96: Chittoe, 16-Jul-00 (DMR)
ST97: Studley, 3-Oct-80 (DMR); Kellaways, 29-Apr-80 (DMR); Stanley, 2-Jul-80 (DMR); Studley, 9-Apr-85 (DMR)
SU06: Morgan's Hill, 13-May-84, 24-Sep-84 (DMR) Beckhampton, 3-Jun-00 (DMR)
SU07: Calne, 27-Apr-85 (DMR); Goatacre, 23-Apr, 28-Jun-80 (DMR); Yatesbury, 13-Jun-73 (BG); Clyffe Pypard, 9-Jun-80 (DMR)
SU08: Lyneham, 21-Mar-80 (DMR)
SU09: Emmett Hill, 1-Jan-95 (MA det MDD)
SU16: Avebury, 4-May-03 (MGT); Marlborough area, 10-Mar-1879 (WFB) (Meyrick, 1880); 1869-1938 (Meyrick, 1938); West Stowell, 3-Sep-00 (DMR); East Kennet, 3-Jun-01 (DMR)
SU17: Ogbourne Maizey, 20-May-01 (DMR)
SU98: Braydon Pond, 5-Apr-80 (DMR)
8 ST82: Oysters Coppice, 6-Oct-02 (MDD)
ST85: Green Lane Wood, 26-Jun-98 (JN)
ST91: Cranborne Chase, 16 to 25-Jun-80 (NCC)
ST93: Stockton Wood, 1995 (MA det MDD); Stockton Down, 1995 (MA det MDD); Great Ridge Wood, 18-Apr-74 (DRN)
ST94: Ladywell (SPTAW), 3-Jun - 12-Aug-08 (Boyce & Wright, 2009)
SU02: Fovant, 31-May-95 (MDD)
SU03: Grovely Wood, 6-Apr-71, 8-Apr-72 (DRN); 1-Aug-80 (MDD)
SU04: Tilshead Down, 4-Jun-97 (MGT)
SU06: All Cannings, 19-Nov-01 (IJG)
SU13: Salisbury, Avon Valley Reserve, 12, 13-Apr-00, 12-May-00 (MDD); Apr-00 (MAS) (Darby, 2001)
SU15: Beach's Plantation, 1-May-05 (MGT)
SU16: Jones's Mill, 21-Jun-97 (MDD); Wilcot, 1-May-00 (DMR)
SU17: Hallam, 6-Aug-83 (DRN)
SU21: Hamptworth Estate, 4-Aug-73, 9 to 15-Apr-74, 29-Mar-78 (DRN); Landford Bog, 4-Apr-75 (DRN); 25-May-98 (MDD)
SU22: Langley Wood NNR, 6-Jun-02 (MDD); Bentley Wood, 31-Mar-78 (DRN); 22-Apr-04 to 24-Mar-05 (MDD) (Darby, 2005)
SU23: Redlynch, 31-Oct-73 (DRN); Porton Down, 1970-2001 (Anon); 12-Nov-96 (MAS)
SU25: Sidbury Hill (SPTAE), 19-May, 6-Jun-97 (MGT)
SU26: Bedwyn Brail, 2-Jun-03 (MDD)

Resident. Ubiquitous; in almost all habitats.

**ABAX Bonelli, 1810**

***Abax parallelepipedus*** (Piller & Mitterpacher, 1783)

7 ST87: Rack Hill, 1995, Jun-96 (MA det MDD); West Yatton Down, 1995 (MA det MDD); Slaughterford, 10-Jun-00 (DMR)
SU06: Morgan's Hill, 13-May-84 (DMR)
SU07: Catcomb Wood, May-96 (MA det MDD); Yatesbury, 13-Jun-73 (BG); Clyffe Pypard, 4-Sep-80 (DMR); Goatacre, 28-Jun-80 (DMR)
SU08: Minety (Stonehill Wood), 13-Apr-02 (JKG det MDD); 8-Feb-02 (MS); Ravensroost Wood, 1995 (MA det MDD); Lyneham, 19-Apr-80 (DMR)
SU16: Marlborough area, Jun-1873 (EWJ) Meyrick, 1874); 20-Jun-1879 (JA) (Meyrick, 1880); 1869-1938 (Meyrick, 1938); Gopher Wood, 13-Jul-96 (MA det MDD)
SU17: Avebury, 19-May-03 (DB)
SU26: Hen's Wood, 3-Jul-02 (JKG det MDD); Savernake Forest, Jun-96 (MA det MDD); Jul-Jun-93 (EN det SAW); 1-May-95 (EN det MDD)
8 ST84: Longleat, May to Jul-99, 28-Aug-00 (MB)
ST85: Bullen Hill Farm, 1982-2002 (EGS det MDD); Clanger Wood, 1995 (MA det MDD); 7-Jun-95 (MDD); 26-Jun-98 (JN)
ST91: Cranborne Chase, 16 to 25-Jun-80 (NCC); 30-Jun-99 (MDD)
ST93: Stockton Down, 1995 (MA det MDD); Stockton Wood, 1995 (MA det MDD); Great Ridge Wood, 18-Apr-74, 13-Aug-81, 29-Jul-85 (DRN)
SU01: Martin Down NNR, 1999 (EN, 1999)
SU02: Middleton Down, 8-May-89 (BL); Gardiner Forest, 1-Sep-72 (DRN)
SU03: Grovely Wood, 11-Aug-71, 8,10, 12-Apr-72, 4-Apr-77, 11-Aug-81, 1-Aug-88 (DRN); 21-Aug-01 (MDD)
SU04: Salisbury Plain, 25-Jun-02 (PJH) (Hodge, 2002)
SU05: West Lavington, 1938 (PLP) (DFL, 1948)
SU13: Salisbury, Bemerton Folly, 7-Mar - 15-Jun-02 (MDD) (Darby, 2002d)
SU15: Beach's Plantation (SPTAE), 1-May-05 (MGT); 4-Sep-05 (JMW); Manningford, 4-Aug-88 (DRN)
SU21: Hamptworth Estate, 15, 19-Apr-1974, 4-Aug- 77 (DRN); Burnt Ground Wood, 24-Jul-73 (DRN); Landford Bog, 2003 (APF) (Foster, 2003)
SU22: Bentley Wood, 11- Aug-01 (MDD); 5-Jun-04 (AGD); Glazier's Copse, 16-Aug-88 (DRN); Homan's Copse, 24-Jul-73 (DRN); Langley Wood NNR, 5-Aug-85, 6-Jun-02 (DRN); 19-Jun-87 (RSK); 22-Apr to 10-Dec-04, 5-Aug-06 to 2-Jul-07 (MDD)(Darby, 2006, 2008); Whiteparish Common, 2-Jul-95 (MAS)
SU23: Porton Down, 1990-1993 (PTW det MDD); 1970-2001 (Anon)
SU24: Beacon Hill (SPTAE), 3-Jun - 12-Aug-08 (Boyce & Wright, 2009); Driver Training Area (SPTAE), 13-Jun-97, 1-May-05 (MGT)

Resident. Ubiquitous; in woodland particularly. (On the continent regarded as an indicator of ancient forests).

Tribe SPHODRINI

**PLATYDERUS Stephens, 1828**
***Platyderus depressus*** (Audinet-Serville, 1821)
**Nb**
8 SU12: Petersfinger, 7-Aug-88 (DRN)
SU13: Salisbury, 10-Apr-72, 28-Oct-73, 13-Aug-81 (DRN); Salisbury, Bemerton Folly, 7-Mar - 15-Jun-02 (MDD) (Darby, 2002d)
Resident. Very local; in fields, gardens and waste ground in open situations on dry soil.

Platyderus depressus

**SYNUCHUS Gyllenhal, 1810**
***Synuchus vivalis*** (Illiger, 1798)
7 SU07: Yatesbury, 13-Jun-73 (BG)
8 SU02: Grovely, 11-Aug-87 (DRN)
SU15: Beach's Plantation (SPTAE), 12-Jun-97 (MGT)
SU22: Langley Wood NNR, 27-Aug to17-Sep-04 (MDD) (Darby, 2005)
SU24: Newton Tony, 25-Aug to 20-Sep-06, (MA det MGT)
Resident. Very local; in gardens, grasslands and woodland.

**CALATHUS Bonelli, 1810**
AMPHIGYNUS Haliday, 1841
***Calathus rotundicollis*** Dejean, 1828
7 SU16: Marlborough area, 1869-1938 (Meyrick, 1938)
SU17: Fyfield Down, 21-Jul-80 (DMR)
SU26: Savernake Forest, spring-1894 (GCC) (Jebb, 1895)
8 ST84: Longleat, 8-May-99 (MB); 7-Jul-99 (MDD)
ST92: Sutton Mandeville, 21-Sep-95 (MDD)
ST93: Stockton Down, 1995 (MA det MDD); Great Ridge Wood, 29-Oct-75 (DRN)
SU01: Martin Down NNR, 1999 (EN, 1999)
SU02: Gardiner Forest, 1-Sep-72 (DRN)
SU03: Grovely Wood, 11-Aug-87, 13-Aug-71 (DRN)
SU13: Salisbury, Bemerton Folly, 7-Mar - 15-Jun-02 (MDD) (Darby, 2002d)
SU15: Beach's Plantation (SPTAE), 4-Sep-05 (JMW)
SU21: Hamptworth Estate, 26-Jul-77, 22-Jul-78 (DRN)
SU22: Langley Wood NNR, 3-Jun-17-Sep-04, 5-Aug-06 to 2-Jul-07 (MDD)
Resident. Widespread; in woodland. Formerly *C. piceus*.

CALATHUS Bonelli, 1810
***Calathus ambiguus*** (Paykull, 1790)
**Nb**
8 SU24: Dunch Hill (SPTAE), 16, 19-Jun-97 (MGT); Sidbury Hill (SPTAE), 28-May-97 (MGT)
Resident. Very local; [in open, dry sandy sites both inland and on the coast].

***Calathus cinctus*** Motschulsky, 1850
8 SU14: Bourne Bottom (SPTAE), 19-Jun-97 (MGT)
SU24: Dunch Hill (SPTAE), 16, 19-Jun-97 (MGT); Sidbury Hill (SPTAE), 5-Jun-97 (MGT)
Resident. Very local; [on coastal dunes and dry, lowland heaths].

***Calathus fuscipes*** (Goeze, 1777)
7 ST87: Slaughterford, 10-Jun-00 (DMR); Castle Combe, 13-Nov-84 (DMR)
ST97: Kellaways, 13-Sep-80 (DMR)
SU07: Yatesbury, 13-Jun-73 (BG); Goatacre, 6-Sep-80 (DMR); Lyneham, 10-Apr-80 (DMR)
SU16: Marlborough area, 1869-1938 (Meyrick, 1938); Pewsey Downs, 22-Jun-90 (APFw)
SU17: Fyfield Down, 9-Apr-85 (DMR)
8 ST83: Chicklade, 29-Jul-82 (DRN)
ST93: Great Ridge Wood, 18-Apr-74, 13-Aug-81 (DRN)
ST94: Warminster Ranges (SPTAW), 3-Jun - 12-Aug-08 (Boyce & Wright, 2009)
SU01: Martin Down NNR, 1999 (EN, 1999)
SU02: Middleton Down, 6-Nov-73 (BL)
SU03: Wylye Down, 14-Jun-97 (MGT); Wylye Down NNR, 1 to 16-Oct-06, 10-Jul to 4-Sep- 07 (MDD) (Darby, 2009); Grovely Wood, 14-Apr-71, 9-Aug-71, 8-Apr-72 (DRN)
SU04: Fox Covert (SPTAW), 3-Jun - 12-Aug-08 (Boyce & Wright, 2009); Tilshead Down, 1-Jun-97 (MGT)
SU05: Ell Barrow (SPTAC), 3-Jun - 12-Aug-08 (Boyce & Wright, 2009)
SU14: Larkhill west (SPTAC), 2-Jun-97 (MGT)
SU15: Beach's Plantation (SPTAE), 13 - 19-Jun-97, 1-May-05 (MGT); 4-May-05 (JMW)
SU17: Barbury Hill, 1995 (MA det MDD)
SU21: Pound Bottom, 28-Jul-77 (DRN)
SU23: Porton Down, 31-May-92, 15-Aug-93 (PTW det MDD); 1970-2001 (Anon); 3-Jul-99

(MAS)
SU24: Bulford Ranges (SPTAE), 3-Jun - 12-Aug-08 (Boyce & Wright, 2009)Driver Training Area (SPTAE), 1-May-05 (MGT); Dunch Hill (SPTAE), 3-Jun - 12-Aug-08 (Boyce & Wright, 2009); Newton Tony, 25-Aug to 20-Sep-06, (MA det MGT); Sidbury Hill (SPTAE), 18, 28, 29-May, 5, 16-Jun-97 (MGT)
Resident. Widespread; in open grassland situations, gardens and cultivated land.

***Calathus melanocephalus*** (Linnaeus, 1758)
7 SU07: Lyneham, 13-May-80 (DMR)
SU16: Marlborough area, 30-Mar-1879 (WFB) (Meyrick, 1880); 1869-1938 (Meyrick, 1938)
SU17: Fyfield Down, 21-Jul-80 (DMR); 9-Apr-85 (DMR)
8 SU01: Martin Down NNR, 1999 (EN, 1999)
SU03: Wylye Down NNR, 16-Oct-06 (MDD) (Darby, 2009)
SU05: West Lavington, Jun-39, (PLP) (DFL, 1948)
SU14: Bourne Bottom (SPTAE), 19-Jun-97 (MGT); Haxton (SPTAE), 3-Jun - 12-Aug-08 (Boyce & Wright, 2009)
SU15: Beach's Plantation (SPTAE), 16-Jun-97 (MGT); 4-Sep-05 (JMW)
SU22: roadside, 14-Sep-55 (BPM, GEW) (Moore, 1956)
SU23: Porton Down, 1970-2001 (Anon, 2001)
SU24: Dunch Hill (SPTAE), 3-Jun - 12-Aug-08 (Boyce & Wright, 2009); Sidbury Hill (SPTAE), 16-Jun-97 (MGT)
SU25: Haxton Down, 17-Aug-85 (DMR)
Resident. Widespread; in open grassland, gardens and arable land.

**LAEMOSTENUS Bonelli, 1810**
PRISTONYCHUS Dejean, 1828
***Laemostenus terricola*** (Herbst, 1784)
8 SU23: Porton Down, 1970-2001 (Anon, 2001)
SU03: Wylye Down NNR, 5-Jun to 4-Sep-07 (MDD) (Darby, 2009); Grovely Wood, 11-Aug-87 (DRN)
SU05: West Lavington 5-Oct-40 (KJA) (DFL, 1948)
SU22: Langley Wood NNR, 17-Sep-04 (MDD) (Darby, 2005)
Resident. Local; in woods, gardens and buildings.

Tribe PLATYNINI

**OLISTHOPUS Dejean, 1828**
***Olisthopus rotundatus*** (Paykull, 1790)
**LCGR- Fl Cl**
7 SU07: Calne, 6-Aug-00 (DMR)
SU16: Marlborough area, Jun-1873 (EWJ) Meyrick, 1874); 1869-1938 (Meyrick, 1938)
8 ST94: Ladywell (SPTAW), 3-Jun - 12-Aug-08 (Boyce & Wright, 2009)
SU15: Manningford, 4-Aug-88 (DRN)
SU21: Redlynch, 4-Aug-73 (DRN); Pound Bottom, 24-Jul-73, 28-Jul-77, 4-Aug-82 (DRN)
SU24: Driver Training Area (SPTAE), 13-Jun-97 (MGT); Dunch Hill (SPTAE), 16-Jun-97 (MGT); 3-Jun - 12-Aug-08 (Boyce & Wright, 2009)
Resident. Local; on dry heathland.

**OXYPSELAPHUS Chaudoir, 1843**
***Oxypselaphus obscurus*** (Herbst, 1784)
7 SU09: Emmett Hill, 1995 (MA det MDD)
SU16: Avebury, 4-May-03 (MGT)
8 SU05: Patney, 23-Nov-01 (IJG)
SU13: Great Durnford, 3-Aug-88 (DRN)
SU21: Hamptworth Estate, 4-Aug-73, 6-Apr-77, 26-Jul-77, 18-Apr-84 (DRN); Landford Bog, 25-May-98 (MDD); East Grimstead, 5-Apr-77 (DRN)
SU22: Langley Wood NNR, 29-May-03 (MDD); Bentley Wood, 31-Mar-78 (DRN)
SU24: Sidbury Hill (SPTAE), 19-May, 6-Jun-97 (MGT)
Resident. Local; in marshy areas and damp shaded situations including woodland.

**PARANCHUS Lindroth, 1974**
***Paranchus albipes*** (Fabricius, 1796)
7 SU16: Marlborough area, Jun-1873 (EWJ) (Meyrick, 1874)
8 ST82: Oysters Coppice, 6-Oct-02 (MDD)
SU13: Salisbury, 8-Aug-70 (DRN); Middle Woodford, 27-Jul-77 (DRN)
SU21: Pound Bottom, 20-Jul-73 (DRN)
Resident. Local; at margins of water. Probably widespread.

**ANCHOMENUS Bonelli, 1810**
***Anchomenus dorsalis*** (Pontoppidan, 1763)
7 ST87: Castle Combe, 22-May-61 (PJMG) Greenslade, 1963)
SU16: Marlborough area, Jun-1873 (EWJ) Meyrick, 1874); 4-Mar-1879 (FED-H) (Meyrick, 1880); 1869-1938 (Meyrick, 1938)
SU17: Avebury, 26-May, 15-Jun-03; 7-Feb-04; 14-Apr, 3-Jun-07 (DB det MDD)
8 ST93: Stockton Down, 1995 (MA det MDD)
ST94: Warminster Ranges (SPTAW), 3-Jun - 12-Aug-08 (Boyce & Wright, 2009)
SU03: Barford Down, 6-Apr-72 (DRN); Grovely Wood, 6,8,10-Apr-72 (DRN); Wylye Down NNR, 10-Jul to 3-Aug-07 (MDD) (Darby, 2008a)
SU04: Tilshead Down, 1-Jun-97 (MGT)
SU05: Ell Barrow (SPTAC), 3-Jun - 12-Aug-08 (Boyce & Wright, 2009)
SU13: Stratford sub Castle, 5-Apr-72 (DRN)
SU15: Beach's Plantation (SPTAE), 13-Jun-97, 1-

May-05 (MGT); 4-May05 (JMW)
SU23: Pitton, 6-Jun-61 (PJMG) (Greenslade, 1963)
SU24: Driver Training Area (SPTAE), 1-May-05 (MGT); Sidbury Hill (SPTAE), 28-May, 6-Jun-97 (MGT)
Resident. Widespread; in arable fields, gardens and waste land.

**PLATYNUS Bonelli, 1810**
***Platynus assimilis*** (Paykull, 1790)
7 SU26: Savernake Forest, 9-Apr-02 (IJG)
8 SU21: Hamptworth Estate, 15-Apr-74, 24-Apr-84 (DRN)
SU22: roadside, 14-Sep-55 (BPM, GEW) (Moore, 1956); Langley Wood NNR, 12-Apr-74 (DRN); 25-Jun-04 to 14-May-05 (MDD) (Darby, 2005); Bagfield Copse, 7-Aug-86 (DRN)
Resident. Local; in woods especially near water.

**AGONUM Bonelli, 1810**
EUROPHILUS Chaudoir, 1859
***Agonum fuliginosum*** (Panzer, 1809)
7 ST86: Whaddon, 13-May-00 (DMR)
ST96: Chittoe, 16-Jul-00 (DMR)
ST98: Malmesbury, 13-Jul-80 (DMR)
SU07: Goatacre, 23-Apr-80, 13, 28-Jun-80 (DMR)
SU08: Lyneham, 21-Mar-80, 19-Apr-80 (DMR)
SU16: E.Kennet, 1-May-00, 3-Jun-01 (DMR); Marlborough area, 1869-1938 (Meyrick, 1938)
SU17: Ogbourne Maizey, 20-May-01 (DMR)
SU27: Ramsbury, 1870-1873 (Meyrick, 1874)
SU98: Braydon Pond, 5-Apr-80 (DMR)
8 ST82: Oysters Coppice, 6-Oct-02 (MDD)
SU13: Salisbury, Avon Valley Reserve, Apr-Jun-00 (MDD) (Darby, 2001); Great Durnford, 3-Aug-88 (DRN)
SU14: Bulford Ranges (SPTAE), 3-Jun - 12-Aug-08 (Boyce & Wright, 2009)
SU15: Fittleton, 29-Jul-83 (DRN)
SU21: Hamptworth Estate, 9-Apr, 29-May-74, 1, 4-Apr-75, 20-Aug-76, 26-Jul-77, 24-Apr-84 (DRN)
SU22: Langley Wood NNR, 14-May to 19-Nov-04 (MDD) (Darby, 2005); Bentley Wood, 5-Apr-77 (DRN)
SU26: Bedwyn Brail, 19-Jun-03 (MDD)
Resident. Widespread and frequent in damp situations in woods and grassland.

***Agonum micans*** Nicolai, 1822
7 ST97: Kellaways, 29-Apr-80 (DMR)
Resident. Very local; in damp shaded situations. Overwinters under bark.

***Agonum piceum*** (Linnaeus, 1758)
7 SU18: Swindon, 15-May-01 (JKG det MDD)
8 SU13: Salisbury, Avon Valley Reserve, 13-Apr-00 (MDD) (Darby, 2001)
Resident. Very local; in marshy situations [on heavy soils].

***Agonum scitulum*** Dejean, 1828
**Na**
7 ST97: Kellaways, 29-Apr, 13-Sep-80 (DMR)
Resident. Very local; [in shaded marshes and carrs].

***Agonum thoreyi*** Dejean, 1828
7 ST96: Calne, 14-Apr-80 (DMR)
ST97: Kellaways, 29-Apr-80 (DMR)
SU08: Lyneham, 21-Mar to 3-Jul-80 (DMR)
8 SU13: Salisbury, 26-May-74 (DRN); Salisbury, Avon Valley Reserve,13-Apr-00 (MDD) (Darby, 2001)
Resident. Local; in marshy areas and reedbeds.

AGONUM Bonelli, 1810
***Agonum emarginatum*** (Gyllenhal, 1827)
7 SU07: Calne, 6-Aug-00 (DMR)
SU16: E.Kennet, 1-May-00 (DMR)
8 ST93: Great Ridge Wood, 19-Apr-84 (DRN)
ST94: Wadman's Coppice (SPTAW), 3-Jun - 12-Aug-08 (Boyce & Wright, 2009)
SU13: Stratford sub Castle, 5-Apr-72 (DRN); Salisbury, Avon Valley Reserve, 12-May-00 (MDD); Great Durnford, 16-Aug-69, 3-Aug-88 (DRN)
Resident. Local; in marshy areas.

***Agonum marginatum*** (Linnaeus, 1758)
7 ST86: Whaddon, 13-May-00 (DMR)
ST96: Calne, 14-Apr-80 (DMR)
ST97: Christian Malford, 9-Aug-80 (DMR)
SU08: Lyneham, 11-Apr-80 (DMR)
SU16: Marlborough area, Jun-1873 (EWJ) Meyrick, 1874); 1869-1938 (Meyrick, 1938)
8 ST92: Wardour, 5-Jun-96 (MDD, MAS)
ST94: Ladywell (SPTAW), 3-Jun - 12-Aug-08 (Boyce & Wright, 2009)
SU12: Charlton-All-Saints, 9-May-96 (MAS)
SU13: Lower Woodford, 1-Aug-83, 30-Jul-82 (DRN); Salisbury, 25-Jul-78 (DRN)
Resident. Widespread; in wet, vegetated areas usually near water.

***Agonum muelleri*** (Herbst, 1784)
7 ST86: Whaddon, 13-May-00 (DMR)
ST96: Chittoe, 16-Jul-00 (DMR)
ST97: Kellaways, 13-Sep-80 (DMR)
SU07: Calne, 28-May-84 (DMR); Yatesbury, 13-Jun-73 (BG)
SU15: Manningford Bruce, 17-Feb-1879 (GTKM) (Meyrick, 1880); 17-Jul-00 (IJG); West Stowell, 3-Sep-00 (DMR); Marlborough area, 1869-1938 (Meyrick, 1938)
8 ST84: Longleat, 29-Aug-99 (MB)
SU04: Salisbury Plain, 2-Aug-83 (DRN)

SU12: Salisbury, 30-Jul-73 (DRN)
SU13: Lower Woodford, 2-Aug-82, 1-Aug-83, 8-Aug-86, 1-Aug-89 (DRN); Middle Woodford, 8-Aug-79 (DRN)
SU16: Wilcot, 1-May-00 (DMR)

Resident. Widespread; in damp situations in grassland, gardens, fields and open woodland.

***Agonum viduum*** (Panzer, 1796)

7 ST97: Kellaways, 29-Apr-80 (DMR); Studley, 9-Apr-85 (DMR)
SU08: Lyneham, 3-Jul-80 (DMR); 3-May-80 (DMR)
SU16: Marlborough area, 1869-1938 (Meyrick, 1938)
8 ST94: Ladywell (SPTAW), 3-Jun - 12-Aug-08 (Boyce & Wright, 2009); Wadman's Coppice (SPTAW), 3-Jun - 12-Aug-08 (Boyce & Wright, 2009)
ST95: Seend Cleve, 28-May-85 (DMR)
SU13: Middle Woodford, 27-Jul-77 (DRN)
SU22: roadside, 14-Sep-55 (BPM, GEW) (Moore, 1956)
SU23: Porton Down, 1970-2001 (Anon, 2001)

Resident. Local; on well vegetated ground near water.

Zabrus tenebrioides

Tribe ZABRINI

**ZABRUS Clairville, 1806**

***Zabrus tenebrioides*** (Goeze, 1777)

**Na**

8 SU03: Grovely Wood, 1-Sep-98 (MDD)
SU23: West Winterslow, 1-Sep-01 (NB)
SU24: Newton Tony, 25-Aug to 20-Sep-06, (MA det MGT)

Resident. Very local; in cultivated fields. At Grovely Wood in hundreds walking around edges of a ploughed field (plate 19).

**AMARA Bonelli, 1810**

ZEZEA Csiki, 1929

***Amara plebeja*** (Gyllenhal, 1810)

7 ST96: Chittoe, 16-Jul-00 (DMR)
ST97: Stanley, 2-Jul-80 (DMR)
ST98: Startley, 30-Apr-85 (DMR)
SU06: Morgan's Hill, 13-May, 24-Sep-84 (DMR)
SU07: Clyffe Pypard, 4-Sep-80, 5-May-84, 30-Dec-89 (DMR); Yatesbury, 13-Jun-73 (BG)
SU09: Somerford Keynes, 4-Jun-04 (BLe)
SU16: Marlborough area, 1869-1938 (Meyrick, 1938); West Kennet, 4-May-03 (MGT); West Stowell, 3-Sep-00 (DMR)
SU26: Savernake Forest, 4-Aug-95 (EN)
SU27: Ramsbury, 1870-1873 (Meyrick, 1874)
SU98: Braydon Pond, 5-Apr-80 (DMR)
8 ST91: Cranborne Chase, 16 to 25-Jun-80 (NCC)
ST93: Stockton Down, 1995 (MA det MDD); Great Ridge Wood, 18-Apr-74, 29-Oct-75, 19-Apr-84 (DRN)
ST94: Norton Down (SPTAW), 23-Jun-02 (PJH) (Hodge, 2002)
SU03: Grovely Wood, 6-Apr-71, 16-Apr-84 (DRN)
SU13: Lower Woodford, 28-Jul-78 (DRN)
SU14: Bourne Bottom (SPTAE), 13-Jun-97 (MGT)
SU15: Beach's Plantation (SPTAE), 10 - 19-Jun-97, 1-May-05 (MGT); North Newton, 23-Jun-01 (DMR)
SU21: Hamptworth Estate, 15-Apr-74, 3-Apr-78, 24, 25-Apr-84 (DRN)
SU22: Homan's Copse, 31-Jul-85 (DRN); Redlynch, 4-Aug-82 (DRN); Langley Wood NNR, 25-May-94 (WANHS det MDD)
SU24: Driver Training Area (SPTAE), 13-Jun-97 (MGT)

Resident. Local and frequent in damp grasslands and well vegetated moist habitats including arable fields, often near water.

AMARA Bonelli, 1810

**Sun Beetle**

***Amara aenea*** (De Geer, 1774)

7 ST96: Conigre Meadow, 20-May-01 (JKG det MDD); Chittoe, 16-Jul-00 (DMR)
ST97: Kellaways, 13-Sep-80 (DMR)
SU06: Morgan's Hill, 24-Sep-84 (DMR)
SU07: Calne, 5-May-30-Sep-84 (DMR); Goatacre, 28-Jun - 6-Sep-80 (DMR); Yatesbury, 13-Jun-73 (BG)
SU08: Minety, Stonehill Wood, 13-May-02 (JKG det MDD); Lyneham, 3-Jul-80 (DMR)
SU09: Upper Waterhay, 30-Apr-00 (JKG det MDD)
SU16: Marlborough area, Jun-1873 (EWJ) (Meyrick, 1874); 8-May-1879 (JAW) (Meyrick, 1880); 1869-1938 (Meyrick, 1938)
SU17: Ogbourne Maizey, 20-May-01 (DMR)
SU18: Rushy Platt, 20-May-01 (JKG det MDD); Swindon, 15-May-01 (JKG det MDD)
SU98: Braydon Pond, 5-Apr-80 (DMR)
8 ST91: Cranborne Chase, 16 to 25-Jun-80 (NCC)
ST92: Sutton Mandeville, Apr-92 (MDD)
ST94: Wadman's Coppice (SPTAW) 3-Jun - 12-Aug-08 (Boyce & Wright, 2009)
SU02: Knighton Wood, 07 (AJWA); Middleton

Down, May-89 (BL)
SU03: Wylye Down NNR, 29-Mar to 10-Apr-07 (MDD) (Darby, 2009); Grovely Wood, 12-Jun-00 (MDD)
SU04: Shrewton, 28-Jul-82 (DRN); Tilshead Down, 4-Jun-97 (MGT)
SU13: Old Sarum, 31-May-97 (MAS); Salisbury, Avon Valley Reserve, 1-Apr-00 (MAS) (Darby, 2001); Boscombe Down, 9-Oct-94 (MDD)
SU14: Bourne Bottom (SPTAE), 13, 19-Jun-97 (MGT); Haxton (SPTAE), 3-Jun - 12-Aug-08 (Boyce & Wright, 2009); Larkhill west, 2-Jun- 97 (MGT)
SU15: Beach's Plantation (SPTAE), 10 - 19-Jun-97, 1-May-05 (MGT)
SU21: Hamptworth Estate, 15-Apr-74, 6-Apr-77, 3-Apr-78 (DRN); Pound Bottom, 9-Apr-74 (DRN)
SU22: Langley Wood NNR, 14-May to 3-Jun-04 (MDD) (Darby, 2005); Blackmoor Copse, 4 to 18-May-02 (APF) (Foster, 2004)
SU23: Porton Down, 1970-2001 (Anon, 2001); West Winterslow, 25-Aug-01 (NB)
SU24: Bulford Ranges (SPTAE), 3-Jun - 12-Aug-08 (Boyce & Wright, 2009); Driver Training Area (SPTAE), 13-Jun-97, 1-May-05 (MGT); Dunch Hill (SPTAE), 16-Jun-97 (MGT); 3-Jun - 12-Aug-08 (Boyce & Wright, 2009); Sidbury Hill (SPTAE), 18-May-97 (MGT)

Resident. Ubiquitous; in dry fields, grasslands waste ground and gardens.

***Amara communis*** (Panzer, 1797)

7 SU07: Goatacre, 30-Dec-79 (DMR)
SU16: Marlborough area, 1869-1938 (Meyrick, 1938)
8 SU03: Wylye Down NNR, 4-May to 5-Jun-07 (MDD) (Darby, 2008); Barford Down, 6-Apr-72 (DRN)
SU05: Ell Barrow (SPTAC), 3-Jun - 12-Aug-08 (Boyce & Wright, 2009); West Lavington, 4-Jul-01 (DMR)
SU15: Beach's Plantation (SPTAE), 13-Jun-97 (MGT)Upavon, 4-Aug-88 (DRN)
SU21: Pound Bottom, 9-Apr-74 (DRN); Hamptworth Estate, 15-Apr-74 (DRN)
SU24: Bulford Ranges (SPTAE), 3-Jun - 12-Aug-08 (Boyce & Wright, 2009)

Resident. Local; in open habitats in grassland, gardens and woodland.

***Amara convexior*** Stephens, 1828

7 SU16: Wrest Kennet, 4-May-03 (MGT)
8 ST94: Ladywell (SPTAW), 3-Jun - 12-Aug-08 (Boyce & Wright, 2009); Warminster Ranges (SPTAW), 3-Jun - 12-Aug-08 (Boyce & Wright, 2009)
SU01: Martin Down NNR, 1979-1992 (RGB); 1999 (Anon, 1999)
SU03: Barford Down, 10-Apr-72 (DRN); Wylye Down NNR, 29-Mar to 10-Jul-07 (MDD) (Darby, 2009); Grovely Wood, 13-Apr-71, 10-11-Apr-72, 16-Apr-74 (DRN)
SU04: Orcheston Down, 22-Jun-02 (PJH) (Hodge, 2002)
SU05: Black Heath (SPTAC), 3-Jun - 12-Aug-08 (Boyce & Wright, 2009)
SU14: Haxton (SPTAE), 3-Jun - 12-Aug-08 (Boyce & Wright, 2009)
SU15: Beach's Plantation (SPTAE), 13 - 17-Jun-97 (MGT); Everleigh (SPTAE), 24-Jun-02 (PJH) (Hodge, 2002)
SU21: Hamptworth Estate, 29-Mar-78 (DRN)
SU23: Porton Down, 1970-2001 (Anon, 2001)
SU24: Driver Training Area (SPTAE), 1-May-05 (MGT); Sidbury Hill (SPTAE), 12 to 17-Jun-97 (MGT)

Resident. Local; in open sunny sites including gravel pits.

***Amara curta*** Dejean, 1828

**Nb, LCGR-Fm Cm**

8 SU04: Tilshead Down, 1 - 4-Jun-97 (MGT)
SU15: Beach's Plantation (SPTAE), 10-Jun-97, 1-May-05 (MGT)

Resident. Very local; [in dry grasslands, heath and dunes].

***Amara eurynota*** (Panzer, 1796)

7 ST96: Chittoe, 16-Jul-00 (DMR)
SU07: Bewick Basset, 12-Aug-00 (DMR); Yatesbury, 13-Jun-73 (BG)
SU16: Marlborough area, 16 Mar 1879 (CTS) (Meyrick, 1880); 1869-1938 (Meyrick, 1938); West Stowell, 3-Sep-00 (DMR)
SU27: Ramsbury, 1995 (Anon) (Jebb, 1895)
8 SU02: Knighton Wood, Apr-07, May-08 (AJWA)
SU03: Wylye Down NNR, 3-Aug to 4-Sep-07 (MDD) (Darby, 2009)
SU04: Tilshead Down, 4-Jun-97 (MGT)
SU15: Beach's Plantation (SPTAE), 13-Jun-97, 1-May-05 (MGT)
SU24: Dunch Hill (SPTAE), 16-Jun-97 (MGT)
SU25: Haxton Down, 17-Aug-85 (DMR)

Resident. Local; in rather dry open situations.

***Amara familiaris*** (Duftschmid, 1812)

7 ST87: Rudloe, 10-Apr-85 (DMR)
ST96: Calne, 14-Apr-80 (DMR)
SU06: Morgan's Hill, 24-Sep-84 (DMR)
SU07: Calne, 5-May, 27-Aug, 30-Sep-84 (DMR); Yatesbury, 13-Jun-73 (BG); Goatacre, 6-Sep-80 (DMR)
SU09: Somerford Keynes, 4-Jun-04 (BLe); Blakehill Farm, 11-May-01 (JKG det MDD)
SU16: Marlborough area, Jun-1873 (EWJ)

(Meyrick, 1874); 8-May-1879 (FED-H) (Meyrick, 1880); 1869-1938 (Meyrick, 1938);
SU26: Savernake Forest, 16-Jun to 7-Jul-93, 28-Apr to 18-May-94 (EN det SAW)
8 ST91: Cranborne Chase, 24-May-97 (MAS)
ST93: Great Ridge Wood, 18-Apr-74, 19-Apr-84 (DRN)
ST94: Bowls Barrow (SPTAW), 22-Jun-02 (PJH) (Hodge, 2002)
SU02: Fovant, 31-May-95 (MDD); Knighton Wood, May-08 (AJWA); Middleton Down, May-89 (BL)
SU03: Barford Down, 10-Apr-72 (DRN); Grovely Wood, 11-Apr, 21-Aug-72, 8-Apr, 30-May-74, 3-Apr-75, 16-Apr-84 (DRN); Wylye Down NNR, 4-May to 5-Jun-07 (MDD) (Darby, 2009)
SU14: Bourne Bottom (SPTAE), 13-Jun-97 (MGT)
SU15: Beach's Plantation (SPTAE), 10, 16, 19-Jun-97 MGT); Everleigh (SPTAE), 18-May-02 (PJH) (Hodge, 2002)
SU16: Jones's Mill, 1997 (MAS)
SU21: Hamptworth Estate, 15-Apr-74, 28-Oct-75, 29-Mar-78, 24-Apr-84 (DRN); Pound Bottom, 4-Aug-72, 4-Aug-73 (DRN)
SU22: Langley Wood NNR, 22-Apr to 14-May-04 (MDD) (Darby, 2005)
SU04: Tilshead Down, 4-Jun-97 (MGT)
Resident. Widespread in open grasslands and on heaths. Less frequent in woodland.

?**_Amara lucida_** (Duftschmid, 1812)
**Nb**
7 SU16: Marlborough area, Jun-1873 (EWJ) (Meyrick, 1874); 1869-1938 (Meyrick, 1938)
This record must be considered doubtful.Two specimens under this name in the MCC are not this species and the insect is normally found in dunes and dry grassland around the coast.

**_Amara lunicollis_** Schiödte, 1837
7 SU07: Calne, 6-Aug-00 (DMR)
SU16: Marlborough area, 12-Feb-84 (DMR) (Meyrick, 1880); 1869-1938 (Meyrick, 1938)
SU26: Savernake Forest, 15-Jun-73 (SAW); 27-May to 16-Jun-93, 28-Apr to 18-May-94 (EN det SAW)
8 ST91: Cranborne Chase, 16 to 25-Jun-80 (NCC)
ST94: Warminster Ranges (SPTAW), 3-Jun - 12-Aug-08 (Boyce & Wright, 2009)
SU01: Martin Down NNR, 1999 (EN, 1999)
SU03: Wylye Down NNR, 5-Jun to 10-Jul-07 (MDD) (Darby, 2009)
SU05: Ell Barrow (SPTAC), 3-Jun - 12-Aug-08 (Boyce & Wright, 2009)
SU13: Salisbury, 28-Oct-73 (DRN)
SU14: Bourne Bottom (SPTAE), 13 to 17-Jun-97 (MGT)
SU15: Beach's Plantation (SPTAE), 13-Jun-97 (MGT)
SU21: Hamptworth Estate, 15-Apr-74, 26-Jul-76, 25-Apr-84 (DRN); Pound Bottom, 24-Jul-73, 9-Apr-74 (DRN)
Resident. Widespread; in open or semi-open habitats on well drained and not too dry soils.

**_Amara montivaga_** Sturm, 1825
8 SU03: Little Langford, 3-Jul-96 (MAS)
Resident. Very local; in open sandy or chalky sites with ruderal vegetation

**_Amara ovata_** (Fabricius, 1792)
7 SU06: Morgan's Hill, 24-Sep-84 (DMR)
SU16: Marlborough area, 24-Jun-1879 (FSA) (Meyrick, 1880); 1869-1938 (Meyrick, 1938)
8 SU03: Wylye Down NNR, 29-Mar to 5-Jun-07 (MDD) (Darby, 2009); Grovely Wood, 12-Jun-00 (MDD)
SU15: Beach's Plantation (SPTAE), 1-May-05 (MGT); Upavon, 26-Aug-01 (DMR)
Resident. Local; in open, dry fields and gardens.

**_Amara similata_** (Gyllenhal, 1810)
7 SU06: Morgan's Hill, 24-Sep-84 (DMR)
SU07: Calne, 30-Sep-84 (DMR); Yatesbury, 13-Jun-73 (BG)
8 ST92: Wardour, 13-Nov-95 (MDD)
SU02: Knighton Wood, Jul-08 (AJWA)
SU03: Grovely Wood, 8-Apr-72 (DRN)
SU14: Bourne Bottom (SPTAE), 13-Jun-97 (MGT)
SU15: Beach's Plantation (SPTAE), 10, 16-Jun-97 (MGT)
SU21: Hamptworth Estate, 15-Apr-74, 1-Apr-75, 24-Apr-84 (DRN)
SU22: Langley Wood NNR, 22-Apr to 14-May-04, 15-Jul to 5-Aug-06 (MDD) (Darby, 2005, 2008); Redlynch, 4-Aug-82 (DRN)
SU24: Sidbury Hill (SPTAE), 5-Jun-97 (MGT)
Resident. Local; usually in open fields and gardens often near water.

**_Amara tibialis_** (Paykull, 1798)
8 SU21: Pound Bottom, 4-Aug-82 (DRN)
Resident. Very local; [on heaths and well drained open ground]. May be confined to the far south of the county.

CELIA Zimmermann, 1832
**_Amara bifrons_** (Gyllenhal, 1810)
7 ST97: Kellaways, 13-Sep-80 (DMR)
SU07: Yatesbury, 13-Jun-73 (BG)
SU08: Lyneham, 3-Jul-80 (DMR)
SU09: Cotswold Water Park, 21-Jun-90 (APFw)
8 SU04: Shrewton, 6-Aug-82 (DRN)
SU22: roadside, 14-Sep-55 (BPM, GEW) (Moore, 1956)
SU04: Tilshead Down, 1-Jun-97 (MGT)

SU14: Bourne Bottom (SPTAE), 19-Jun-97 (MGT); Haxton (SPTAE), 3-Jun - 12-Aug-08 (Boyce & Wright, 2009)
SU15: Beach's Plantation (SPTAE), 16-Jun-97 (MGT)
SU25: Haxton Down (SPTAE), 17-Aug-85 (DMR)
Resident. Widespread; on open sites on well drained soils.

***Amara praetermissa*** (Sahlberg, C.R., 1827)
**Nb, LCGR-FI CI**
8 ST94: Tinker's Track (SPTAW), 3-Jun - 12-Aug-08 (Boyce & Wright, 2009)
Resident. Very local; [in open, dry, well drained and often isturbed habitats, including grassland, dunes and 'brownfield' sites].

BRADYTUS Stephens, 1827
***Amara apricaria*** (Paykull, 1790)
7 ST96: Chittoe, 16-Jul-00 (DMR)
ST97: Kellaways, 13-Sep-80 (DMR)
SU07: Calne, 28-May-84 (DMR); Goatacre, 6-Sep-80 (DMR)
SU16: Marlborough area, 1869-1938 (Meyrick, 1938)
8 SU03: Grovely Wood, 11-Aug-71, 29-Oct-73, 3-Apr-75 (DRN)
SU15: Beach's Plantation (SPTAE), 16 - 19-Jun-97, 1-May-05 (MGT)
SU24: Sidbury Hill (SPTAE), 29-May-97 (MGT)
Resident. Local; on open, often cultivated sites where there is ruderal vegetation.

***Amara consularis*** (Duftschmid, 1812)
**Nb**
8 SU22: roadside, 14-Sep-55 (BPM, GEW) (Moore, 1956)
Resident. Very local; [in dry sandy areas and gravel pits with vegetation, also sandy heaths].

Amara fulva

***Amara fulva*** (Müller, O.F, 1776)
**Nb**
7 SU07: Calne, 30-Sep-84 (DMR)
Resident. Very local; [in sandy and gravelly sites often near water].

PERCOSIA Zimmermann, 1832
***Amara equestris*** (Duftschmid, 1812)
**Nb, LCGR- FI CI**
8 SU21: Hamptworth Common, 4-Aug-77 (DRN) (Nash, 1978); Pound Bottom, 24-Jul-73 (DRN) (Nash, 1978); Franchises Wood, 4-Aug-77 (DRN) (Nash, 1978)
SU36: Ham Hill, 14-Jun-97 (MGT)
Resident. Very local; under stones besides a gravel path, dead in a water-filled trench and under stone in a heathy clearing. Possibly confined to the far south of the county.

**CURTONOTUS Stephens, 1827**
***Curtonotus aulicus*** (Panzer, 1796)
8 ST94: Warminster Ranges (SPTAW), 3-Jun -12-Aug-08 (Boyce & Wright, 2009)
ST95: Imber Ranges (SPTAW), 7-Aug-97 (MCH)
SU01: Martin Down NNR, 1999 (EN, 1999)
SU03: Wylye Down NNR, 5-Jun to 10-Oct-07 (MDD) (Darby, 2009); Grovely Wood, 30-Jul, 5-Aug-79 (DRN)
SU04: Orcheston Down (SPTA), 22-Jun-02 (PJH) (Hodge, 2002); Shrewton, 28-Jul-82 (DRN); Tilshead Down, 4-Jun-97 (MGT)
SU05: West Lavington, 1938 (PLP) (DFL, 1948)
SU14: Bourne Bottom (SPTAE), 13, 19-Jun-97 (MGT)
SU15: Beach's Plantation (SPTAE), 13-Jun-97 (MGT); Everleigh (SPTAE), 24-Jun-02 (PJH) (Hodge, 2002)
SU22: Bagfield Copse, 7-Aug-86 (DRN)
SU24: Sidbury Hill (SPTAE), 18, 28, 29-May, 4, 5, 16-Jun-97 (MGT)
Resident. Local; in open, dry habitats where there is herbaceous vegetation in seed.

Tribe HARPALINI

**HARPALUS Latreille, 1802**
PSEUDOOPHONUS Motschulsky, 1844
**Strawberry Seed Beetle**
***Harpalus rufipes*** (De Geer, 1774)
7 ST96: Chittoe, 16-Jul-00 (DMR)
SU07: Yatesbury, 13-Jun-73 (BG); Lyneham, 13-May-80 (DMR)
SU09: Upper Waterhay, 17-May, 5-Jun-05, 5-Jun-07 (JKG det MDD)
SU16: Avebury, 4-May-03 (MGT); East Kennet, 3-Jun-01 (DMR); Marlborough area, Jun-1873 (EWJ) (Meyrick, 1874); 1869-1938 (Meyrick, 1938); West Kennet, 4-May-03 (MGT); West Stowell, 3-Sep-00 (DMR)
SU17: Ogbourne Maizey, 20-May-01 (DMR)
SU26: Crofton, 26-May-01 (DMR)
8 ST85: Bullen Hill Farm, 31-Jul-02 (EGS det MDD)
SU01: Martin Down NNR, 1999 (EN, 1999)
SU02: Middleton Down, 1-May-89 (BL)

SU03: Barford Down, 10-Apr-72 (DRN); Grovely Wood, 14-Apr-71, 6 to 10-Apr-72, 18-Jul-73, 8-Apr-74, 3-Apr-75 (DRN); 1-Jun- 00 (MDD); Wylye Down NNR, 5-Jun-07 (MDD) (Darby, 2009)
SU04: Fox Covert (SPTAW), 3-Jun - 12-Aug-08 (Boyce & Wright, 2009); Orcheston Down (SPTAC), 22-Jun-02 (PJH) (Hodge, 2002); Shrewton, 6-Aug-82 (DRN); Tilshead Down, 1-Jun-97 (MGT)
SU13: Salisbury, 15-Aug-88 (DRN)
SU14: Bourne Bottom (SPTAE), 13, 19-Jun-97 (MGT)
SU15: Beach's Plantation (SPTAE), 13, 16, 19-Jun-97, 1-May-05 (MGT)
SU16: Wilcot, 1-May-00 (DMR)
SU21: Hamptworth Estate, 3-Aug-73, 15-Apr-74 (DRN); Landford Bog, 24-Jun-01 (APF) (Foster, 2001)
SU22: roadside, 14-Sep-55 (BPM, GEW) (Moore, 1956); Langley Wood NNR, 16-Jul to 6-Aug-04, 25-Apr to 6-Jun-07 (MDD) (Darby, 2005, 2008)
SU23: West Winterslow, 1-Oct-01 (NB)
SU24: Bulford Ranges (SPTAE), 3-Jun - 12-Aug-08 (Boyce & Wright, 2009); Dunch Hill (SPTAE), 16-Jun-97 (MGT); Newton Tony, 25-Aug to 20-Sep-06, (MA det MGT)
SU25: Blackmore Down, 25-Aug-82 (DRN)

Resident. Ubiquitous; in open, dry situations on light soils, especially arable fields.

HARPALUS Latreille, 1802

***Harpalus affinis*** (Schrank, 1781)

7 ST86: Whaddon, 13-May-00 (DMR)
ST96: Chittoe, 16-Jul-00 (DMR)
ST97: Studley, 18-May-80, 3-Oct-80 (DMR)
SU07: Yatesbury, 13-Jun-73 (BG); Calne, 28-May-84 (DMR); Bewick Basset, 12-Aug-00 (DMR); Goatacre, 6-Sep-80 (DMR)
SU08: Lyneham, 11-Apr-80 (DMR)
SU16: West Stowell, 3-Sep-00 (DMR); Marlborough area, 1869-1938 (Meyrick, 1938)
SU00: Cotswold Water Park, 10-Apr-97 (APF)
8 ST94: Ladywell (SPTAW), 3-Jun - 12-Aug-08 (Boyce & Wright, 2009)
ST95: Great Cheverill Hill, 11-Sep-06 (JP)
SU03: Grovely Wood, 8-Apr-74, 4-Aug-78 (DRN)
SU04: Fox Covert, 3-Jun - 12-Aug-08 (Boyce & Wright, 2009); Salisbury Plain, 25-Jun-02 (PJH) (Hodge, 2002)
SU13: Salisbury, 10-Apr-72 (DRN); 20-My-97 (KNAA)
SU14: Tockenham Wick, 4-Apr-80 (DR); Bourne Bottom (SPTAE), 13-Jun-97 (MGT); Bulford House, 2-Jun-97 (MGT)
SU15: Beach's Plantation (SPTAE), 10, 13, 16-Jun-97, 1-May-05 (MGT); 4-Sep-05 (JMW)
SU21: Pound Bottom, 24-Jul-73, 19-Apr-74 (DRN)
SU22: roadside 14-Sep-55 (BPM, GEW) (Moore, 1956); Redlynch, 13-Apr-74 (DRN)
SU24: Driver Training Area (SPTAE), 1-May-05 (MGT); Dunch Hill (SPTAE), 16-Jun-97 (MGT); Sidbury Hill (SPTAE), 18, 28, 29-May, 6-Jun-97 (MGT); Tilshead Down, 1, 4-Jun-97 (MGT)

Resident. Widespread; in most dry open situations including fields, gardens, waste ground and grassland.

***Harpalus latus*** (Linnaeus, 1758)

7 SU16: Marlborough area, 1869-1938 (Meyrick, 1938)
8 ST91: Cranborne Chase, 16 to 25-Jun-80 (NCC)
ST94: Warminster Ranges (SPTAW), 3-Jun -12-Aug-08 (Boyce & Wright, 2009)
SU01: Martin Down NNR, 1999 (EN, 1999)
SU03: Wylye Down NNR, 10-Jul to 4-Sep-07 (MDD) (Darby, 2009); Grovely Wood, 10-Apr-72 (DRN)
SU13: Salisbury, 10-Apr-72 (DRN)
SU21: Franchises Wood, 3-May-95 (MAS); Pound Bottom, 9-Apr-74 (DRN
SU22: Langley Wood NNR, 14-May to 29-Oct-04, 25-Apr to 6-Jun-07 (MDD) (Darby, 2005, 2008)
SU23: Porton Down, 1970-2001 (Anon, 2001)
SU24: Bulford Ranges (SPTAE), 3-Jun - 12-Aug-08 (Boyce & Wright, 2009)

Resident. Widespread in the south of the county in dry grassland.

***Harpalus rubripes*** (Duftschmid, 1812)

7 SU07: Calne, 30-Sep-84 (DMR); Yatesbury, 13-Jun-73 (BG); Calne, 6-Aug-00 (DMR)
8 ST94: Ladywell (SPTAW), 3-Jun - 12-Aug-08 (Boyce & Wright, 2009)
SU13: Salisbury, 10-Apr-72 (DRN); Salisbury, Avon Valley Reserve,Apr-00 (MAS) (Darby, 2001)
SU15: Manningford, 4-Aug-88 (DRN)
SU16: Jones's Mill, 16-May-02 (IJG)

Resident. Local; in open, dry sandy and chalky habitats.

***Harpalus tardus*** (Panzer, 1796)

8 SU21: Landford Bog, 9-Jun-01 (APF) (Foster, 2001)

Resident. Very local; in open, usually dry habitats.

**OPHONUS Dojean, 1821**

OPHONUS Dejean, 1821

***Ophonus azureus*** (Fabricius, 1775)

**Nb, LCGR-Fm Cm**

8 ST94: Warminster Ranges (SPTAW), 3-Jun -12-Aug-08 (Boyce & Wright, 2009)
SU14: Haxton (SPTAE), 3-Jun - 12-Aug-08 (Boyce & Wright, 2009)

SU24: Dunch Hill (SPTAE), 3-Jun - 12-Aug-08 (Boyce & Wright, 2009); Sidbury Hill (SPTAE), 29-May-97 (MGT)
Resident. VerylLocal; [in open coastal sites, also inland on warm chalk or limestone slopes].

METOPHONUS Bedel, 1895

***Ophonus puncticeps*** Stephens, 1828

8 ST15: Manningford, 13-Aug-87 (DRN)
ST94: Ladywell (SPTAW), 19-Aug-05 (KNAA); 3-Jun - 12-Aug-08 (Boyce & Wright, 2009); Warminster Ranges (SPTAW), 3-Jun -12-Aug- 08 (Boyce & Wright, 2009)
SU01: Martin Down NNR, 1999 (EN, 1999)
SU03: Wylye Down NNR, 3-Aug to 4-Sep-07 (MDD) (Darby, 2009)
SU05: Black Heath (SPTAC), 3-Jun - 12-Aug-08 (Boyce & Wright, 2009); Ell Barrow (SPTAC), 3-Jun - 12-Aug-08 (Boyce & Wright, 2009)
SU15: Beach's Plantation (SPTAE), 16-Jun-97 (MGT); West Everleigh Down, 25-Sep-01 (IJG)
SU24: Bulford Ranges (SPTAE), 3-Jun - 12-Aug-08 (Boyce & Wright, 2009); Dunch Hill (SPTAE), 16-Jun-97 (MGT)
Resident. Local; on open, dry, usually sandy soils.

***Ophonus rufibarbis*** (Fabricius, 1792)

7 SU16: Marlborough area, 1870-1938 (Meyrick, 1938); West Kennet, 4-May-03 (MGT)
8 ST94: Ladywell (SPTAW), 3-Jun - 12-Aug-08 (Boyce & Wright, 2009)
SU15: Beach's Plantation (SPTAE), 12, 19-Jun-97, 1-May-05 (MGT); 4-Sep-05 (JMW)
Resident. Very local; [in partly vegetated dry habitats on almost all soils]. Three examples in MCC without data.

***Ophonus schaubergerianus*** (Puel, 1937)
**Nb, LCGR-Fm Cl**

8 ST94: Ladywell (SPTAW), 3-Jun - 12-Aug-08 (Boyce & Wright, 2009)
SU01: Martin Down NNR, 1999 (EN, 1999)
SU03: Grovely Wood, 8 to 12-Apr-72, 23-Jul-73 (DRN)
SU04: Tilshead Down, 1, 4-Jun-97 (MGT)
SU15: Beach's Plantation (SPTAE), 1-May-05 (MGT)
SU24: Driver Training Area (SPTAE), 1-May-05 (MGT); Dunch Hill (SPTAE), 16-Jun-97 (MGT); Sidbury Hill (SPTAE), 19-May, 5-Jun-97 (MGT)
Resident. Very local; in dry grassland and open scrub, more rarely in woodland.

**ANISODACTYLUS Dejean, 1829**

***Anisodactylus binotatus*** (Fabricius, 1787)

7 ST96: Chittoe, 16-Jul-00 (DMR)
ST97: Studley, 3-Oct-80 (DMR)
SU07: Calne, 5 to 28-May-84 (DMR)
8 SU05: Market Lavington, 2-Aug-83 (DRN); West Lavington, 1938 (PLP) (DFL, 1948)
SU13: Salisbury, 10-Apr-72 (DRN)
SU21: Hamptworth Estate, 26-Jul-77 (DRN)
Resident. Local; in damp habitats on poorly draining soils.

***Anisodactylus nemorivagus*** (Duftschmid, 1812)
(var. ***atricornis***, with black legs)
**Na**

8 SU21: Pound Bottom, 19-Apr-74, 31-May-74 (DRN) (Nash, 1978)
Resident. Very local; under stones besides a gravel track [on dry sandy heaths]. Probably confined to the far south of the county.

**DICHEIROTRICHIS Jacquelin du Val, 1855**

[***Dicheirotrichis gustavii*** Crotch, 1871
8 SU13: Salisbury, 1-Apr-00 (MAS) (Darby, 2001)]
This record was in error. *D gustavii* is a coastal species only recorded from saline habitats.

**TRICHOCELLUS Ganglbauer, 1892**

***Trichocellus placidus*** (Gyllenhal, 1827)

7 SU16: East Kennet, 1-May-00 (DMR)
SU17: Ogbourne Maizey, 20-May-01 (DMR)
8 SU13: Salisbury, 1-Apr-00 (MAS); 7-Apr-00 (MDD) (Darby, 2001)
SU15: Fittleton, 29-Jul-83 (DRN)
Resident. Local; in damp grassland and woodland litter.

**BRADYCELLUS Erichson, 1837**

?***Bradycellus distinctus*** (Dejean, 1829)
**Na**

7 SU16: Marlborough area, Jun-1873 (EWJ) (Meyrick, 1874); 1869-1938 (Meyrick, 1938)
Resident. Very local; in moss. This record almost certainly refers to *B. sharpi* which had not been separated from *distinctus* in 1873. *B. distinctus* was only found on dry and gravelly sites around the south coast at this time and is now known only from Dungeness in Kent. The single specimen under this name in the MCC is *sharpi* (but it does show some puncturation at the centre front of the prothorax).

***Bradycellus harpalinus*** (Audinet-Serville, 1821)

7 SU07: Yatesbury, 13-Jun-73 (BG)
SU16: Marlborough area, 1869-1938 (Meyrick, 1938)
8 ST92: Donhead St. Mary, 19-Jul-95 (HK det MDD)
SU01: Martin Down NNR, 1999 (EN, 1999)
SU03: Grovely Wood, 16-Aug-71, 24-Aug-72 (DRN)
SU21: Hamptworth Estate, 3-Aug-73, 15-Apr-74, 20-Aug-76, 6-Apr-77, 19-Aug-88 (DRN);

Pound Bottom, 9-Apr-74 (DRN)
SU22: Langley Wood NNR, 22-Apr to 14-May-04, 20-Feb to 24-Mar-07 (MDD) (Darby, 2005, 2008)
Resident. Local; in a wide range of dry habitats.

***Bradycellus ruficollis*** (Stephens, 1828)
8 ST93: Great Ridge Wood, 29-Jul-82 (DRN)
SU21: Pound Bottom, 9-Apr-74 (DRN)
Resident. Very local; usually on well-drained heathland.

***Bradycellus sharpi*** Joy, 1912
8 SU04: Tilshead Down, 4-Jun-97 (MGT)
SU15: Beach's Plantation (SPTAE), 16-Jun-97 (MGT)
SU21: Hamptworth Estate, 15-Apr-74, 6-Apr-77, 5-Aug-85 (DRN)
SU22: Bagfield Copse, 7-Aug-86 (DRN); Homan's Copse, 20-Jul-73 (DRN); Langley Wood NNR, 9-Nov to 10-Dec-04 (MDD) (Darby, 2005); East Grimstead, 5-Apr-77 (DRN)
SU24: Sidbury Hill (SPTAE), 18-May-97 (MGT)
Resident. Local; in damp, shaded sites in woodland and grassland.

***Bradycellus verbasci*** (Duftschmid, 1812)
7 SU07: Calne, 28-May, 30-Sep-84, 25-Aug-85 (DMR); Lyneham, 29-Dec-79 (DMR)
SU09: Cotswold Water Park, 21-Jun-90 (APFw) Marlborough area, 1869-1938 (Meyrick, 1938)
8 SU13: Salisbury, 28-Oct-73 (DRN)
SU21: Hamptworth Estate, 29-Mar-78 (DRN)
Resident. Local; in open situations such as arable fields and waste ground.

**STENOLOPHUS Dejean, 1821**
***Stenolophus mixtus*** (Herbst, 1784)
7 SU07: Calne, 28-May-84, 6-Aug-00 (DMR)
SU16: East Kennet, 3-Jun-01 (DMR)
8 ST92: Wardour, 5-Jun-96 (MAS)
SU12: Charlton-All-Saints, 9-May-96 (MAS)
SU14: Stonehenge, 22-Apr-03 (PJH) (Hodge, 2003)
SU21: Hamptworth Estate, 20-Aug-76 (DRN)
SU26: Bedwyn Brail, 19-Jun-03 (MDD)
Resident. Local; in marshland and at edge of water [especially on clay soils].

**ACUPALPUS Latreille, 1829**
***Acupalpus dubius*** Schilsky, 1888
7 SU07: Goatacre, 25-Jan-80 (DMR)
SU09: Somerford Keynes, 4-Jun-04 (BLe)
SU16: Marlborough area, 1869-1938 (Meyrick, 1938)
8 ST92: Wardour, 5-Jun-96 (MAS)
SU21: Hamptworth Estate, 4-Apr-75 - 12-Aug-86 (DRN); Hamptworth Common, 5-Aug-80, 24-Apr-84 (DRN)
SU22: East Grimstead, Hawks Grove, 5-Apr-77 (DRN); Langley Wood NNR, 29-May-03, 14-May to 3-Jun-04 (MDD) (Darby, 2005)
SU26: Bedwyn Brail, 19-Jun-03 (MDD)
Resident. Local; in litter, moss and tussocks near water.

***Acupalpus flavicollis*** (Sturm, 1825)
**Na**
8 SU21: Windyeats Farm, 31-Oct-73 (DRN det AAA) (Nash, 1978); Hamptworth Estate, 25-Apr-84 (DRN)
Resident. Very local; in damp, sandy and gravelly situations near water. Both these records in moss.

***Acupalpus meridianus*** (Linnaeus, 1761)
7 ST98: Malmesbury, 13-Jul-80 (DMR)
SU06: Beckhampton, 3-Jun-00 (DMR)
SU07: Calne, 27-Apr, 16-Jun-85 (DMR); Lyneham, 9-May-80 (DMR)
8 SU13: Salisbury, Avon Valley Reserve, 7 to 13-Apr-00 (MDD) (Darby, 2001)
SU22: Redlynch, 13-Apr-74 (DRN); Langley Wood NNR, 14-May to 3-Jun-04 (MDD) (Darby, 2005)
Resident. Local; in gardens and fields on open clay or peat soils.

***Acupalpus parvulus*** (Sturm, 1825)
7 SU07: Calne, 28-May-84 (DMR)
8 SU21: Hamptworth Estate, 12-Aug-86 (DRN)
Resident. Very local; [in damp habitats near vegetation].

**ANTHRACUS Motschulsky, 1850**
***Anthracus consputus*** (Duftschmid, 1812)
**Nb**
7 SU09: Somerford Keynes, 4-Jun-04 (BLe)
Resident. Very local; [under stones and in dense vegetation and litter near standing water].

Tribe CHLAENIINI

Chlaenius nigricornis

**CHLAENIUS Bonelli, 1810**
***Chlaenius nigricornis*** (Fabricius, 1787)

**Nb**
7 SU07: Yatesbury, 13-Jun-73 (BG)
SU16: Marlborough area, Jun-1873 (EWJ); 6-Feb-1879 (TNH-S) (Meyrick, 1880); spring-1894 (GCC) (Jebb, 1895); 1869-1938 (Meyrick, 1938)
8 SU12: Britford, Jun – 15-Sep-02 (CMD) (Drake, 2008)
SU13: Lower Woodford, 27-Jul-82, 1-Aug-83, 8-Aug-86 (DRN); Jun – 15-Sep-02 (CMD) (Drake, 2008)
Resident. Very local; in damp grass- and marshland.

***Chlaenius vestitus*** (Paykull, 1790)
7 SU07: Calne, 16-Oct-00 (DMR)
SU09: Somerford Keynes, 4-Jun-04 (BLe); Cotswold Water Park, 21-Jun-90 (APFw)
SU16: Wooton Rivers, 18-Jun-46 (Holmes & Tennant, 1947)
8 ST94: Wadman's Coppice (SPTAW), 3-Jun - 12-Aug-08 (Boyce & Wright, 2009)
Resident. Very local; [in mud and clay cracks near water].

Tribe OODINI

**OODES Bonelli, 1810**
***Oodes helopioides*** (Fabricius, 1792)
**Nb**
7 SU16: Marlborough area, 1869-1938 (Meyrick, 1938)
8 ST95: Seend Cleve, 31-May-85 (DMR)
SU12: Britford, Jun - 15-Sep-02 (CMD) (Drake, 2008)
SU13: Great Durnford, 19-Aug-69 (DRN); Lower Woodford, Jun - 15-Sep-02 (CMD) (Drake, 2008)
Resident. Very local; in litter and vegetation at edge of water and in reedbeds. Able to survive being submerged.

Tribe LICININI

**LICINUS Latreille, 1802**
***Licinus depressus*** (Paykull, 1790)
**Nb, LCGR-Fh Ch**
8 SU01: Martin Down NNR, 1999 (EN, 1999)
SU03: Wylye Down NNR, 3-Aug to 4-Sep-07 (MDD) (Darby, 2009)
SU23: Porton Down, 1970-2001 (Anon, 2001)
Resident. Very local; at roots, etc. in calcareous grassland.

***Licinus punctatulus*** (Fabricius, 1792)
**Na, LCGR-Fh Ch**
7 SU26: Little Frith, 1905-1938 (Meyrick, 1938)
8 SU06: Cherhill, 27-Aug-01 (DMR)
SU07: Lyneham, 17-Apr-80 (DMR)
Resident. Very local; on dry, sandy or chalky soils. Three specimens in the MCC without data.

Licinus punctulatus

**BADISTER Clairville, 1806**
BADISTER Clairville, 1806
***Badister bullatus*** (Schrank, 1798)
7 SU16: Marlborough area, 22 Mar 1879 (PCP) (Meyrick, 1880); 1869-1938 (Meyrick, 1938); East Kennet, 1-May-00, 3-Jun-01, (DMR); Gopher Wood, 7-Jun-00 (IJG); West Kennet, 4-May-03 (MGT)
SU17: Ogbourne Maizey, 20-May-01 (DMR)
ST87: Long Dean, 24-Apr-87 (JB det RSK)
ST97: Studley, 9-Apr-85 (DMR)
8 ST92: Wardour, 5-Jun-96 (MAS)
ST93: Great Ridge Wood, 19-Apr-84 (DRN)
ST94: Wadman's Coppice (SPTAW) 3-Jun - 12-Aug-08 (Boyce & Wright, 2009)
SU01: Martin Down NNR, 1999 (EN, 1999)
SU02: Bishopstone, 1 to 7-May-98, Jul-1991 (MDD)
SU03: Berwick St. James, Jun-96 (MDD); Grovely Wood, 4-Apr-78 (DRN); Wylye Down, 27-Jan to 10-Jul-07 (MDD) (Darby, 2009)
SU04: Salisbury Plain, 25-Jun-02 (PJH) (Hodge, 2002); Shrewton, 6-Aug-82 (DRN)
SU06: Morgan's Hill, 13-May-84 (DMR)
SU13: Salisbury, 12-Apr-71 (DRN); Salisbury, Avon Valley Reserve,Apr-00 (MAS) (Darby, 2001); Salisbury, Bemerton Folly, 7-Mar - 15-Jun-02 (MDD) (Darby, 2002d)
SU14: Bulford House, 18-May-97 (MGT); Haxton, 24-Jun-02 (PJH) (Hodge, 2002)
SU15: SU15: Beach's Plantation (SPTAE), 1-May-05 (MGT); 4-Sep-05 (JMW); Everleigh (SPTAE), 18-May-02 (PJH) (Hodge, 2002)
SU22: Langley Wood NNR, 3-Jun to 16-Jul-04 (MDD) (Darby, 2005)
SU24: Sidbury Hill (SPTAE), 18-May, 12, 13-Jun-97 (MGT); Tidworth, 30-May-04 (BLe)
Resident. Widespread. All examples that I have seen are of the larger form with more extensive elytral

markings usually found in open situations such as heaths and grasslands. The other form is on average smaller and found at the edge of rivers.

***Badister unipustulatus*** Bonelli, 1813
**Nb**
*Wiltshire* (Fowler, 1887)
7 SU16: East Kennet, 3-Jun-01 (DMR)
8 SU17: Barbury Hill, 1995 (MA det MDD)
Resident. Very local; in marshes and damp litter on well vegetated sites.

TRIMORPHUS Stephens, 1828
***Badister sodalis*** (Duftschmid, 1812)
7 SU26: Cobham Frith, 30-Jun-46 (Holmes & Tennant, 1947)
8 ST93: Great Ridge Wood, 19-Apr-84 (DRN)
Resident. Very local; in moss [and litter in damp woodlands].

Tribe PANAGAEINI

**PANAGAEUS Latreille, 1802**
***Panagaeus bipustulatus*** (Fabricius, 1775)
**Nb, LCGR-Fm Cl**
8 SU01: Martin Down NNR, 1979-1992 (RGB); 1999 (EN, 1999)
SU03: Wylye Down, 4-May to 5-Jun-07 (MDD) (Darby, 2009)
SU05: West Lavington, 1938 (PLP) (DFL, 1948)
SU14: Wexland Hanging (SPTAC), 16-May-02 (PJH) (Hodge, 2002)
SU23: Porton Down, 20-Jul-97 (MAS); 1970-2001 (Anon); 3-Jul-99 (PMP)
Resident. Local; on calcareous grassland.

Panagaeus bipustulatus

***Panagaeus cruxmajor*** Linnaeus, 1758)
**RDB1**
8 SU04: Winterbourne Stoke, c.1828 (GTR) (Stephens, 1828)
Resident, probably extinct. [Amongst lush vegetation on soft soil or mud at margins of standing or slowly moving water, in fens]. Formerly widespread through central and southern England but now only known from the coast of Carmarthenshire and Wicken Fen.

Tribe PERIGONINI

**PERIGONA Laporte, 1835**
***Perigona nigriceps*** (Dejean, 1831)
8 ST91: Rushmore Park, 23-Sep-01 (AJWA)
Resident. Very local; in a dung heap.

Tribe LEBIINI

**LEBIA Latreille, 1802**
LAMPRIAS Bonelli, 1810
***Lebia chlorocephala*** (Hoffmann, J, 1803)
**Nb, LCGR-Fm Cl**
7 ST86: Box Hill, 1952 (BNP)
8 SU03: Grovely Wood, 8-Apr-72 (DRN)
SU01: Martin Down NNR, 3-May-08 (R.Thomas email on BBI web site, 2008)
SU14: Haxton (SPTAE), 3-Jun - 12-Aug-08 (Boyce & Wright, 2009); nr Sidbury Hill (SPTAE), 9, 13 to 17-Jun-97 (MGT)
SU25: Ludgershall, 29-Mar-15 (PAH); Sidbury Hill (SPTAE), 5-Jun-97 (MGT)
Resident. Local; in open grassland especially where there are tussocks, usually associated with *Chrysolina* spp. The Ludgershall specimen 'running over an oak stump'.

Lebia chlorocephala

[***Lebia marginata*** (Geoffroy in Fourcroy, 1785
**Extinct**
7 SUO6: Devizes, Jun-1864 (JS)]
Power (1865) was sent this insect by Sidebotham and described it as: 'a fine specimen, beaten out by Mr Sidebotham himself from a wood near Devizes'. There is one other 19th century record from Shropshire, but the insect has not been recorded in GB since (though included in Duff, 2008). (See p.9)

**DEMETRIAS Bonelli, 1810**
DEMETRIAS Bonelli, 1810
***Demetrias atricapillus*** (Linnaeus, 1758)
7 ST98: Bradfield Wood, 7-Apr-96 (MA det MDD)
SU06: Roundway Camp, Jul-40 (FDB) (Buck,

1956); Beckhampton, 3-Jun-00 (DMR)
SU07: Yatesbury, 13-Jun-73 (BG)
SU08: Somerford Common, 23-Apr-01 (JKG det MDD)
SU09: Ashton Keynes, 26-Apr-01 (JKG det MDD)
SU16: East Kennet, 3-Jun-01 (DMR); Marlborough area, Jun-1873 (EWJ) (Meyrick, 1874); 1869-1938 (Meyrick, 1938); Oare, 2, 4-Oct-00 (IJG); West Kennet, 4-May-03 (MGT); Wooton Rivers, 10-Aug-81 (DRN)
SU17: Ogbourne Maizey, 20-May-01 (DMR) (Meyrick, 1938)
SU26: Savernake Forest, 27-Oct-95 (EN) (Meyrick, 1938)
8 ST92: Sutton Down, 10-Apr-77 (DRN); Sutton Mandeville, 22-Nov-92 (MDD)
ST95: Seend Cleve, 31-May-85 (DMR)
SU02: Knighton Wood, Sep-06 (AJWA)
SU03: Barford Down, 6-Apr-72 (DRN); Grovely Wood, 11-Aug-71 (DRN); 19-Aug-95 (MAS); Wylye Down, 31-Dec-06 to 27-Jan-07 (MDD) (Darby, 2009)
SU13: Stratford sub Castle, 5-Apr-72, 26-May-74 (DRN); Salisbury, Avon Valley Reserve, 1-May-00 (MDD); Apr-00 (MAS) (Darby, 2001)
SU14: Bulford House, 18-May-97 (MGT); Stonehenge, 22-Apr-03 (PJH) (Hodge, 2003)
SU15: Beach's Plantation (SPTAE), 16-Jun-97, 1-May-05 (MGT); Everleigh (SPTAE), 18-May-02 (PJH) (Hodge, 2002)
SU16: Wilcot, 1-May-00 (DMR); Woodborough, 4-Aug-83 (DRN)
SU21: Hamptworth Estate, 15-Apr-74 (DRN)
SU23: Porton Down, 1870-2001 (Anon); 15-Oct-95, 12-Nov-96 (MAS)
SU24: Dunch Hill (SPTAE), 16-Jun-97 (MGT); Sidbury Hill (SPTAE), 19-May-97 (MGT)
SU25: Sidbury Hill (SPTAE), 29-May-97 (MGT)

Resident. Ubiquitous; in tussocky grassland and agricultural fields.

**PARADROMIUS Fowler, 1887**

***Paradromius linearis*** (Olivier, 1795)

7 ST86: Whaddon, 13-May-00 (DMR)
ST97: Studley, 9-Apr-85 (DMR)
SU06: Beckhampton, 3-Jun-00 (DMR)
SU07: Calne, 16-Jun-85 (DMR)
SU16: West Kennet, 4-May-03 (MGT); West Woods, 30-Sep-45 (Holmes & Tennant, 1947)
SU17: Avebury, 3-Jun-06 (DB det MDD)
8 ST85: Green Lane Wood, 7-Jun-95 (WANHS det MDD) (Meyrick, 1938); 17-Aug-04 (MDD)
ST91: Cranborne Chase, 16 to 25-Jun-80 (NCC)
ST93: Great Ridge Wood, 18-Apr-74 (DRN)
ST94: Norton Down, 23-Jun-02 (PJH) (Hodge, 2002)
SU01: Martin Down NNR, 1999 (EN, 1999)
SU02: Bishopstone, 19-May-91 (MDD)
SU03: Grovely Wood, 13-Aug-70, 11-Aug-71, 4-Apr-77 (DRN); 3-Aug-04 (MDD); Wylye Down NNR, 22-Feb to 29-Mar, 5-Jun to 10-Jul-07 (MDD) (Darby, 2009)
SU04: Shrewton, 28-Jul-82 (DRN); Orcheston Down (SPTA), 18-May-02 (PJH) (Hodge, 2002)
SU07: New Zealand, 25-Jan-80 (DMR)
SU13: Boscombe Down, 9-Oct-94 (MDD); Salisbury, Avon Valley Reserve, Apr to 12-May-00 (MAS, MDD) (Darby, 2001)
SU15: Beach's Plantation (SPTAE), 1-May-05 (MGT); Fittleton, 29-Jul-83 (DRN); West Everleigh Down, 4-Oct-01 (IJG);
SU16: Woodborough, 4-Aug-83 (DRN)
SU22: Glazier's Copse, 16-Aug-88 (DRN); Langley Wood NNR, 29-May-03 (MDD)
SU24: Dunch Hill (SPTAE), 16-Jun-97 (MGT); 3-Jun - 12-Aug-08 (Boyce & Wright, 2009); Sidbury Hill (SPTAE), 16-Jun-97 (MGT)
SU25: Haxton Down, 17-Aug-85 (DMR)

Resident. Ubiquitous; in dry, tussocky grassland.

**DROMIUS Bonelli, 1810**

***Dromius agilis*** (Fabricius, 1787)

7 ST87: Corsham Park, 3-Jun-04 (BLe)
SU16: Marlborough area, Jun-1873 (EWJ) Meyrick, 1874); 17-Feb-1879 (WFB) (Meyrick, 1880); 1869-1938 (Meyrick, 1938)
SU26: Savernake Forest, 2-Apr-70 (DRN)
8 SU03: Grovely Wood, 18-Jul-73, 4-Aug-78 (DRN)

Resident. Local; in both deciduous and coniferous woodland. Arboreal.

***Dromius meridionalis*** Dejean, 1825

7 ST98: Lacock, 29-May-04 (BLe)

Resident. Very local; mostly on deciduous trees. Arboreal.

***Dromius quadrimaculatus*** (Linnaeus, 1758)

7 ST86: Neston Park, 3-Jun-04 (BLe, KNAA)
ST96: Spye Park, 31-May-04 (KNAA)
ST99: Flisteridge Wood, 23-Mar-82 (KNAA)
SU16: Marlborough area, Jun-1873 (EWJ) (Meyrick, 1874); 17-Feb-1879 (WFB) (Meyrick, 1880); 1869-1938 (Meyrick, 1938)
SU26: Savernake Forest, 2-Apr-70 (DRN); 4-Aug-95 (EN det MDD)
8 ST85: Bullen Hill Farm, 1982-2002 (EGS det MDD)
ST91: Cranborne Chase, 30-Jun-99 (MDD)
ST92: Sutton Mandeville, 6-Sep-97, 2-Oct-95 (MDD); 1-Sep-96 (JAO)
ST93: Great Ridge Wood, 18-Apr-74 (DRN); 4-Oct-04 (MDD)
SU02: Knighton Wood, 7-Oct-06, Aug-07 (AJWA);

Wilton, 16-Aug-88 (DRN)
SU03: Grovely Wood, 29-Oct-70 - 4-Apr-78 (DRN); 3-Aug-04 (MDD)
SU13: Little Durnford Down, 26-Nov-00 (APF) (Foster, 2000)
SU16: Jones's Mill, 17-Oct-01 (APF) (Foster, 2002)
SU22: Blackmoor Copse, 31-Aug-01, 17-Mar-02 (APF) (Foster, 2004); Hamptworth Estate, 20-Jul-73 (DRN); Langley Wood NNR, 6-Aug-86 (DRN); 29-May-03, 27-Aug to 8-Oct-04 (MDD)
SU23: Bentley Wood, 21-Oct-90 (PM)
SU26: Bedwyn Brail, 20-Sep-04 (MDD)
Resident. Widespread; arboreal species found mainly on deciduous trees.

**CALODROMIUS Reitter, 1905**
***Calodromius spilotus*** (Illiger, 1798)
7 ST86: Neston Park, 3-Jun-04 (BLe, KNAA)
ST87: Corsham Park, 3-Jun-04 (BLe)
ST96: Sandy Lane, 19-Aug-84 (DMR)
SU16: Marlborough area, Jun-1873 (EWJ) (Meyrick, 1874); 8-Mar-1879 (WFB) (Meyrick, 1880); Marlborough area, 1873-1938 (Meyrick, 1938)
8 SU02: Knighton Wood, Oct, Dec-06, Oct-07 (AJWA)
SU03: Grovely Wood, 11-Aug-71, 6-Apr-72, 4-Apr-77, 1-Aug-88 (DRN)
SU04: Parsonage Down NNR, 2, 22-Jul-08 (MDD)
SU16: Jones's Mill, 17-Oct-01 (APF) (Foster, 2002)
SU22: Blackmoor Copse, 31-Aug-01, 3-Aug-03 (APF) Foster, 2004)
SU26: Bedwyn Brail, 20-Sep-04 (MDD)
Resident. Widespread arboreal species found mainly, but not always, on coniferous trees. Formerly *D.quadrinotatus*.

**PHILORHIZUS Hope, 1838**
***Philorhizus melanocephalus*** (Dejean, 1825)
7 ST97: Studley, 3-Oct-80, 9-Apr-85 (DMR)
SU06: Roundway Camp, Jul-40 (FDB) (Buck, 1956)
SU07: Calne, 6-Aug-00 (DMR); Bewick Basset, 12- Aug-00 (DMR)
SU16: Marlborough area, Jun-1873 (EWJ) (Meyrick, 1874); 1869-1938 (Meyrick, 1938); West Stowell, 3-Sep-00 (DMR)
8 ST93: Great Ridge Wood, 4-Oct-04 (MDD)
SU02: Knighton Wood, Dec 06 (AJWA)
SU22: Blackmoor Copse, 31-Aug-01 (APF) (Foster, 2004)
Resident. Local; in dry grassland.

**MICROLESTES Schmidt-Göbel, 1846**
***Microlestes maurus*** (Sturm, 1827)
7 ST86: Whaddon, 13-May-00 (DMR)
8 SU24: Driver Training Area (SPTAE), 1-May-05 (MGT)
Resident. Very local; on dry soils with some litter and often with shade from vegetation

**SYNTOMUS Hope, 1838**
***Syntomus foveatus*** (Geoffroy in Fourcroy, 1785)
8 SU03: Barford Down, 6-Apr-72 (DRN)
SU13: Salisbury, Avon Valley Reserve, Apr-00 (MAS) (Darby, 2001)
SU14: Haxton (SPTAE), 3-Jun - 12-Aug-08 (Boyce Wright, 2009); Stonehenge, 22-Apr-03 (PJH) (Hodge, 2003)
SU15: Beach's Plantation (SPTAE), 10-Jun-97 (MGT)
SU21: Hamptworth Estate, 15-Apr-74 (DRN)
SU22: roadside, 14-Sep-55 (BPM, GEW) (Moore, 1956)
SU24: Beacon Hill (SPTAE), 3-Jun - 12-Aug-08 (Boyce & Wright, 2009); Driver Training Area (SPTAE), 1-May-05 (MGT); 24-Jun-02 (PJH) (Hodge, 2002); Sidbury Hill (SPTAE), 18-May-97 (MGT)
SU25: Sidbury Hill (SPTAE), 28-May, 5, 13 to 17-Jun-97 (MGT)
Resident. Local; on dry sandy heaths and grassland

***Syntomus obscuroguttatus*** (Duftschmid, 1812)
7 SU09: Somerford Keynes, 4-Jun-04 (BLe)
8 SU21: Hamptworth Estate, 15-Apr-74, 29-Mar-78 (DRN)
SU22: Redlynch, 4-Aug-82 (DRN)
Resident. Very local; in moss and litter on heavy soils and in damp conditions.

# Suborder POLYPHAGA
## Superfamily HYDROPHILOIDEA
## Family HELOPHORIDAE

**HELOPHORUS Fabricius, 1775**
EMPLEURUS Hope, 1838
**Wheat Mud Beetle**
***Helophorus nubilus*** Fabricius, 1777
7 SU16: Marlborough area, 1869-1938 (Meyrick, 1938)
8 ST94: Wadman's Coppice (SPTAW) 3-Jun - 12-Aug-08 (Boyce & Wright, 2009)
SU04: Shrewton, 6-Aug-82 (DRN)
Resident. Very local; in rough grassland at roots.

***Helophorus rufipes*** (Bosc d'Antic, 1791)
7 SU16: Marlborough area, Jun-1873 (EWJ) (Meyrick, 1874); 1869-1938 (Meyrick, 1938)
8 SU15: Beach's Plantation (SPTAE), 10, 13 to 17-Jun-97 (MGT)
Resident. Very local; at roots of plants. No specimen under this name in the MCC.

MEGAHELOPHORUS Kuwert, 1886

**Helophorus aequalis** Thomson, C.G. 1868

7 SU08: Wootton Bassett, 3-Jun-99 (KGA)
SU09: Ashton Keynes, 31-May-96 (KGA)
SU16: Marlborough area, 1869-1938 (Meyrick, 1938)
8 ST94: Ladywell (SPTAW), pond, 3-Jun - 12-Aug-08 (Boyce & Wright, 2009); Wadman's Coppice (SPTAW), pond, 3-Jun - 12-Aug-08 (Boyce & Wright, 2009)
SU21: Hamptworth Estate, grubbed in dry bed, 20-Aug-76 (DRN); Landford Bog, 9-Jun, 24-Jun-01 (APF) (Foster, 2001)
SU22: Blackmoor Copse, 1-Jun-95 (SAC); East Grimstead, 11-Aug-78 (DRN); 8-Jun-05 (JSD)
SU26: Bedwyn Brail, 19-Jun-03 (MDD); Wilton Brail, 19-Jun-03 (MDD)

Resident. Local; in shallow streams, ditches, ponds and pools.

**Helophorus grandis** Illiger, 1798

7 ST95: Semington Brook, 10-Jun-91 (EGS det MDD)
ST96: Melksham, 1-Jul-97 (MNS)
ST98: Box Road pond, nr Brinkworth, 1-Jul-97 (MNS); Conygre Mead, Malmesbury, 5-Apr-02 (JKG det MDD); Dauntsey, silt pond, 1932-1933 (EJP) (Pearce, 1934)
ST99: R. Thames, side stream, 9-May-99 (APF)
SU08: Minety, 28-May-96 (KGA); Purton, Swindon, 14-Oct-99 (KGA); Wootton Bassett, 28-Oct-94, 26-May-96 (KGA)
SU09: Ashton Keynes, 16-Mar-02 (JKG det MDD); Cotswold Water Park, 28-Apr-82 (GNF); 30-May-97 (JSD); Upper Waterhay, 5 to 11-Jun-07 (JKG det MDD); Emmett Hill, 2-Oct-90 (JL)
SU16: Marlborough area, Jun-1873 (EWJ) (Meyrick, 1874); 24-Mar-1879 (WFB) (Meyrick, 1880)
SU26: Savernake Forest, 1 to 31-May-95 (EN det MDD)
8 ST94: Codford St. Peter, intermittent chalk stream and adjacent pool, 1991-1992 (RC) (Carr, 1997); Wadman's Coppice (SPTAW), pond, 3-Jun - 12-Aug-08 (Boyce & Wright, 2009)
SU95: West Lavington, dew ponds, 29, 30-Mar-41 (LeC) (DFL, 1948)
SU01: Martin Down NNR, 1-Jan-79, 11-Apr-92 (RGB)
SU03: Grovely Wood, 17-Apr-71 (DRN)
SU04: Chitterne Down (SPTAW), dew pond, 13-Jun-73 (RC) (Carr, 1997); R. Till, Elston, 10-Apr-91 (RC) (Carr, 1997)
SU05: The Warren, 10-Apr-91 (RC) Carr, 1997)
SU14: Nine Mile River, Bulford (SPTAE), 9-Apr-91 (RC) (Carr, 1997)
SU21: Hamptworth Estate, stream, 27-Jul-78 (DRN)
SU22: Langley Wood NNR, 1 to 22-Apr-04, 26-Oct to 17-Nov-06, 5 to 22-Jan-07 (MDD) (Darby, 2005, 2008); East Grimstead, Hawks Grove, 1-Aug-78 (DRN)

Resident. Widespread in shallow streams, ditches, ponds and pools.

ATTRACTHELOPHORUS Kuwert, 1886

**Helophorus brevipalpis** Bedel, 1881

7 ST96: Melksham, 1-Jul-97 (MNS); Semington Brook, 1-Jul-97 (MNS)
ST98: Box Road pond, nr. Brinkworth, 9-May-05 (SW); Dauntsey, silt pond, 1932-1933 (EJP) (Pearce, 1934)
SU08: Wootton Bassett, 1930 (EJP); 5-Aug-95 (JHB); 3-Jun, Oct-99 (KGA)
SU09: Emmett Hill, 2-Oct-90 (JL); Cotswold Water Park, 28-Apr-82 (GNF); 21-Jun-90 (APFw); 30-May-97 (JSD)
SU16: Marlborough area, 1869-1938 (Meyrick, 1869-1938)
SU26: Savernake Canal, 1930 – 1932 (EJP)
SU27: Ramsbury, 17-Mar-30 (FBB)
8 ST85: Biss Farm, 20-Oct-90 (ITE)
ST91: Cranborne Chase, 22-Jun-99 (MDD)
ST92: Sutton Mandeville, Jul-92 (MDD)
ST94: Codford St. Peter, 1991-1992 (RC) (Carr, 1997); Ladywell (SPTAW), pond, 3-Jun - 12-Aug-08 (Boyce & Wright, 2009); Wadman's Coppice (SPTAW), pond, 3-Jun - 12-Aug-08 (Boyce & Wright, 2009)
ST95: Semington Brook, 20-Apr-06 (DJL)
SU04: Salisbury Plain, 25-Jun-02 (PJH) (Hodge, 2002)
SU05: The Warren, chalk stream and fish pond, 10-Apr-91 (RC) (Carr, 1997)
SU13: Idmiston, R. Bourne, 5-Aug-82 (DRN); Lower Woodford, 28-Jul-78 (DRN); Great Durnford, 16- Aug-69, 7-Aug-70 (DRN)
SU14: Bulford Ranges (SPTAE), 3-Jun - 12-Aug-08 (Boyce & Wright, 2009); Nine Mile R., Bulford (SPTAE), 23-Jun-93 (CMD)
SU21: Hamptworth Estate, 6-Aug-79 (DRN); Landford Bog, 24-Aug-01 (APF) (Foster, 2001)
SU22: Blackmoor Copse, Jun-95 (SAC); East Grimstead, 8-Jun-05 (JSD)
SU26: Bedwyn Brail, 19-Jun-03 (MDD); Wilton Brail, 19-Jun-03 (MDD)

Resident. Widespread; in still water and often in flight and coming to light in summer. Many specimens taken from vegetation next to water.

HELOPHORUS Fabricius, 1775

**Helophorus flavipes** Fabricius, 1792

7 ST98: Dauntsey, silt pond, 1932-1933 (EJP) (Pearce, 1934)
8 ST95: West Lavington, 6-May, 10-Aug-41 (LeC) (DFL, 1948)

SU13: Great Durnford, 7-Aug-70 (DRN)
SU21: Franchises Wood, 4-Aug-77 (DRN); Hamptworth Estate, 20-Aug-76 (DRN); Landford Bog, 24-Jun-01 (APF) (Foster, 2001)
Resident. Local; in streams, springs and ponds, where it is sometimes abundant

***Helophorus granularis*** (Linnaeus, 1760)
7 SU08: Wootton Bassett, 1930 (EJP)
SU09: Cotswold Water Park, 28-Apr-82 (GNF)
SU16: Marlborough area, 24-Mar-1879 (WFB) (Meyrick, 1880)
Resident. Very local; [rare, in exposed, temporary marshes and on the fluctuating margins of hard-bottomed ponds]. No specimen under this name in the MCC.

***Helophorus griseus*** Herbst, 1793
**Nb**
7 SU09: Ashton Keynes, 30-May-97 (JSD); Cotswold Water Park, 30-May-97 (JSD)
SU16: Marlborough area, 30-May-1879 (RAW) (Meyrick, 1880); Marlborough area, 1873-1938 (Meyrick, 1938)
SU26: Savernake Forest, 1870-1873 (Meyrick, 1874)
8 ST94: Wadman's Coppice (SPTAW), pond, 3-Jun - 12-Aug-08 (Boyce & Wright, 2009)
Resident. Very local; in exposed, temporary, clayey pools. Two specimens in MCC without data.

***Helophorus longitarsis*** Wollaston, 1864
**RDB3**
8 ST94: Ladywell (SPTAW), pond, 3-Jun - 12-Aug-08 (Boyce & Wright, 2009); Wadman's Coppice (SPTAW), pond, 3-Jun - 12-Aug-08 (Boyce & Wright, 2009)
Resident. Very local; by pond netting and ground search around pond edge.

***Helophorus minutus*** Fabricius, 1775
7 SU08: Purton, Swindon, 14-Oct-99 (KGA); Wootton Bassett, 5-Aug-95 (JHB)
SU09: Cotswold Water Park, 28-Apr-82 (GNF)
8 ST95: West Lavington, 29-Mar-41 (LeC) (DFL, 1948)
SU03: R. Wylye, 5-Aug-91 (RC) (Carr, 1997)
SU04: Chitterne Down (SPTAW), dew pond, 17-Aug-91 (RC) (Carr, 1997); R.Till, Winterbourne Stoke, 21-Jul-91 (RC)
SU13: Middle Woodford, 27-Jul-77 (DRN); R. Bourne, Idmiston, 19-Jul-91 (RC) (Carr, 1997)
SU14: R. Avon, Amesbury, 19-Jul-91 (RC) (Carr, 1997); Bulford (SPTAE), muddy pond, 19-Jul-91 (RC) (Carr, 1997); R. Avon, Figheldean, 20-Jul-91 (RC) (Carr, 1997); Nine Mile R., Bulford (SPTAE), 23-Jun- 93 (CMD)
SU15: R. Avon, Upavon, 19-Jul-91 (RC) (Carr, 1997)
SU25: Limmer Ponds, 28-Jul-92(RC) (Carr, 1997)
Resident. Widespread in ditches, streams, lakes and ponds.

***Helophorus nanus*** Sturm, 1836
**Nb**
7 ST99: Thames Head, 7-Apr-1871 (WM)
Resident. Very local; [in marshy edges of temporary ponds and fens, particularly where there is growth of mossess].

***Helophorus obscurus*** Mulsant, 1844
7 SU16: Marlborough area, 1869-1938 (Meyrick, 1938)
ST92: Wardour, 5-Jun-96 (MAS)
ST94: Codford St. Peter, intermittent chalk stream and adjacent pool, 1991-1992 (RC) (Carr, 1997); Ladywell (SPTAW), pond, 3-Jun- 12-Aug-08 (Boyce & Wright, 2009); Wadman's Coppice (SPTAW), pond, 3-Jun - 12-Aug-08 (Boyce & Wright, 2009)
SU04: Chitterne Down (SPTAW), dew pond, 1991-1992 (RC) (Carr, 1997)
SU09: Ashton Keynes, 31-May-96 (KGA); Cotswold Water Park, 28-Apr-82 (GNF)
SU13: Lower Woodford, 27-Jul-82 (DRN) Salisbury, Avon Valley Reserve, Apr-00 (MAS) (Darby, 2001)
8 ST96: Semington, 1-Jul-97 (MNS)
Resident. Local; in still and running water.

## Family HYDROCHIDAE

**HYDROCHUS Leach, 1817**
***Hydrochus elongatus*** (Schaller, 1783)
**RDB3**
7 ST98: Dauntsey, silt pond, 1932-1933 (EJP) (Pearce, 1934)
Resident. Very local; [In shallow, well-vegetated, still water, often in reedbeds and other areas with rich emergent vegetation over clay, in low lying areas]. *Hydrochus* spp. are aquatic but do not swim. No specimens of this species are located with Pearce's other water beetles in the MCC.

## Family HYDROPHILIDAE
### Subfamily HYDROPHILINAE

**ANACAENA Thomson, C.G, 1859**
***Anacaena bipustulata*** (Marsham, 1802)
**Nb**
7 ST98: Dauntsey, silt pond and R. Avon 1932-33 (EJP) (Pearce, 1934)
SU09: Cotswold Water Park, 28-Apr-82 (GNF)

Resident. Very local. Usually found in ditches and ponds in former fenland. No specimens of this species are located with Pearce's other water beetles in the MCC.

***Anacaena globulus*** (Paykull, 1798)

7 ST98: Dauntsey, silt pond and R. Avon, 1932-1933 (EJP) (Pearce, 1934)
SU09: Cotswold Water Park, 28-Apr-82 (GNF)
SU16: Marlborough area, 1869-1938 (Meyrick, 1938)
8 ST85: Biss Farm, 20-Oct-90 (ITE)
ST94: Codford St. Peter, intermittent chalk stream and adjacent pool, 1991-1992 (RC) (Carr, 1997)
ST95: West Lavington, dew pond, 19-Jun-41 (LeC) (DFL, 1948)
SU02: Fovant, water-cress beds, 25-Apr, 31-May-95 (MDD)
SU03: R. Wylye, 5-Aug-91 (RC) (Carr, 1997)
SU13: Salisbury, Avon Valley Reserve, Apr to Jun-00 (MDD) (Darby, 2001)
SU14: Bulford, Nine Mile R. (SPTAE), 29-Mar-91 (RC) (Carr, 1997)
SU15: R. Avon., Upavon, 19-Jul-91 (RC) (carr, 1997)
SU16: Jones's Mill, 21-Jun-97 (MDD)
SU21: Hamptworth, 26-Jan-97 (DTB); Landford Bog, 2003 (PM); 9-Jun, 24-Aug-01 (APF) (Foster, 2001)
SU22: Langley Wood NNR, 22-Apr-04 to 3-Mar-05, 15-Sep to 26-Oct-06 (MDD) (Darby, 2005, 2008); East Grimstead, 11-Aug-86 (DRN)

Resident. Widespread and frequent both in water, and moss and litter near water.

***Anacaena limbata*** (Fabricius, 1792)

7 ST96: Kennet & Avon Canal, 2-Jul-97 (MNS)
ST98: Dauntsey, silt pond, 1932-1933 (EJP) (Pearce, 1934)
ST99: R. Thames, side stream, 9-May-99 (APF)
SU09: Ashton Keynes, 31-May-96 (KGA); Emmett Hill, 2-Oct-90 (JL)
SU16: Marlborough area, 1869-1938 (Meyrick, 1938)
8 ST94: Codford St. Peter, intermittent chalks stream and adjacent pool, 1991-1992 (RC) (Carr, 1997); Wadman's Coppice (SPTAW), pond, 3-Jun - 12-Aug-08 (Boyce & Wright, 2009)
ST95: West Lavington, dew pond, 29-Mar-41 (LeC) (DFL, 1948)
SU02: Fovant, water cress beds, 31-May-95 (MDD)
SU03: R. Wylye, 5-Aug-91 (RC) Carr, 1997)
SU04: R.Till, Rollestone, 17-May-91 (RC) (Carr, 1997)
SU12: Petersfinger, 25-Jul-91 (DRN)
SU13: Great Durnford, 7, 17-Aug-70 (DRN)
SU14: R. Avon, Amesbury, 19-Jul-91 (RC) (Carr, 1997); R. Avon, Figheldean (SPTAE), 20-Jul-91 (RC) (Carr, 1997); Nine Mile River, Bulford (SPTAW), 13-Jun-73 (RC) (Carr, 1997);1-23-Jun-93 (CMD); 3-Jun – 12-Aug-08 (Boyce & Wright, 2009)
SU22: Bentley Wood, 11-Aug-86 (DRN)
SU25: Limmer Ponds, 28-Jul-92 (RC) (Carr, 1997)

Resident. Widespread; in ditches, marshes, silt ponds and peaty pools. *A limbata* was confused with *A. lutescens* until 1986 so that records before then should be considered doubtful.

***Anacaena lutescens*** (Stephens, 1829)

8 SU03: R. Wylye, 5-Aug-91 (RC) (Carr, 1997)
SU13: Salisbury, Avon Valley Reserve, 13-Apr, 12-May, 19-Jun-00 (MDD) (Darby, 2001)
SU21: Landford Bog, 9-Jun, 24-Aug-01 (APF) (Foster, 2001)
SU22: Langley Wood NNR, 22-Apr to 14-May-04 (MDD) (Darby, 2005)
SU26: Bedwyn Brail, 2-Jun-03 (MDD)

Resident. Local; in ditches and peaty pools.

**PARACYMUS Thomson, C.G, 1867**

***Paracymus scutellaris*** (Rosenhauer, 1856)

**Nb**

7 SU16: Marlborough area, 1869-1938 (Meyrick, 1938)
8 SU16: Jones's Mill, 21-Jun-97 (MDD)
SU21: Hamptworth Estate, 10-Aug-84 (DRN)

Resident. Very local; [typically in base-rich flushes in otherwise acidic areas where there is high rainfall]. There are 7 specimens under this name in the MCC.

**BEROSUS Leach, 1817**

BEROSUS Leach 1817

***Berosus affinis*** Brullé, 1835

**Nb**

8 ST94: Ladywell (SPTAW), pond, 3-Jun - 12-Aug-08 (Boyce & Wright, 2009); Wadman's Coppice (SPTAW), pond, 3-Jun - 12-Aug-08 (Boyce & Wright, 2009)

Resident. Very local, by pond-netting. [Mainly associated with slow-flowing drains and ponds on coastal levels, with some vegetation over mud or silt].

***Berosus luridus*** (Linaaeus, 1761)

**Nb**

7 ST99: R. Thames, side stream, 9-May-99 (APF)

Resident. Very local; [in ponds and drains with a peaty substratum].

***Berosus signaticollis*** (Charpentier, 1825)

**Nb**

8 ST94: Ladywell (SPTAW), pond, 3-Jun - 12-Aug-08 (Boyce & Wright, 2009); Wadman's

Coppice (SPTAW), 3-Jun - 12-Aug-08 (Boyce & Wright, 2009)
Resident. Very local; [associated with shallow, rain-water filled, temporary pools].

**CHAETARTHRIA Stephens, 1833**
***Chaetarthria seminulum*** sens. lat.
**Nb**
8 SU21: Hamptworth Estate, 10-Aug-84 (DRN); Landford Bog, 24-Aug-01, 2003 (APF) (Foster 2001, 2003)
SU22: East Grimstead, bomb crater and Hawks Grove, 11-Aug-86 (DRN)
Resident. Very local; in [shallow water with sand, mud or moss [often associated with exposed conditions]. Confused until 2003 with *C. simillima* so that these records require confirmation.

**CYMBIODYTA Bedel, 1881**
***Cymbiodyta marginellus*** (Fabricius, 1792)
7 SU08: Wootton Bassett, 1930 (EJP); 3-Jun-99 (KGA)
Resident. Very local; [in ditches and ponds].

**ENOCHRUS Thomson, C.G, 1859**
***Enochrus coarctatus*** (Gredler, 1863)

7 ST98: Dauntsey, silt pond, 27-Jun-32 (EJP) (Pearce, 1934); R. Avon, 1929-1932 (EJP)
8 SU22: Blackmoor Copse, Jun-95 (SAC)
Resident. Very local; in ditches and small ponds.

?***Enochrus halophilus*** (Bedel, 1878)
**Na**
7 SU16: Manton pond, 20-Mar-46 (Holmes & Tennant, 1947)
There must be some doubt about the determination of this species which is normally associated with brackish water or inland areas receiving saline seepage. Two specimens under this name in the MCC are both *melanocephalus* (Olivier, 1792).

***Enochrus testaceus*** (Fabricius, 1801)
7 ST98: Dauntsey, silt pond, 20-May-33 (EJP) (Pearce, 1934)
Resident. Very local; in ditches and ponds. Two specimens in the MCC.

**HELOCHARES Mulsant, 1844**
***Helochares lividus*** (Forster, 1771)
**Nb**
7 SU09: Ashton Keynes, 30-May-97 (JSD); 24-Mar-1879 (WFB) (Meyrick, 1880); 1869-1938 (Meyrick, 1938); Cotswold Water Park, 30-May-97 (JSD)
8 ST94: Ladywell (SPTAW), pond, 3-Jun - 12-Aug-08 (Boyce & Wright, 2009); Wadman's Coppice (SPTAW), pond, 3-Jun - 12-Aug-08 (Boyce & Wright, 2009)
ST95: West Lavington, dew ponds, 10-Aug, 16-Sep-41 (LeC) (DFL, 1948)
SU04: Chitterne Down (SPTAW), dew pond, 17-Aug-91 (RC) (Carr, 1997)
SU14: Bulford Ranges (SPTAE), pond, 3-Jun - 12-Aug-08 (Boyce & Wright, 2009)
SU25: Limmer Ponds, 28-Jul-92 (RC) (Carr, 1997)
Resident. Local; in still and running water. Recomm-ended to lose its Notable status.

***Helochares punctatus*** Sharp, 1869
**Nb**
7 ST98: Dauntsey, silt pond, 1932-1933 (EJP) (Pearce, 1934)
SU16: Marlborough area, 1869-1938 (Meyrick, 1938)
8 SU21: Landford Bog, 24-Aug-01 (APF) (Foster, 2001)
Resident. Very local; in low lying peaty areas, part-icularly wet heaths. 9 specimens in MCC.

**HYDROBIUS Leach, 1815**
***Hydrobius fuscipes*** (Linnaeus, 1758)
7 ST98: Box Road pond, nr. Brinkworth, 9-May-05 (SW); Dauntsey, silt pond, 1932-1933 (EJP) (Pearce, 1934)
ST99: R. Thames, 29-Oct-00 (APF); R. Thames, side stream, 9-May-99 (APF)
SU08: Wootton Bassett, 3-Jun-99 (KGA)
SU09: Ashton Keynes, 2-Mar-02 (JKG det MDD); Cotswold Water Park, 28-Apr-82 (GNF); 21-Jun-90 (APFw); 30-May-97 (JSD)
SU16: Marlborough area, 30-Mar-1879 (GTKM) (Meyrick, 1880); 1869-1938 (Meyrick, 1938)
8 ST94: Codford St. Peter, 1991-1992 (RC) (Carr, 1997); Ladywell (SPTAW), pond,3-Jun – 12-Aug-08 (Boyce & Wright, 2009); Wadman's Coppice (SPTAW), pond, 3-Jun - 12-Aug-08 (Boyce & Wright, 2009)
SU95: West Lavington, 7-Oct-40 (KJA) (DFL, 1948)
SU03: R. Wylye, 5-Aug-91 (RC) (Carr, 1997)
SU04: Chitterne Down (SPTAW), dew pond, 17-Aug-91(RC) (Carr, 1997); R. Till, Rollestone, 17-May-91(RC) (Carr, 1997); R. Till, Winterbourne Stoke, 5-Aug-91 (RC) (Carr, 1997)
SU05: The Warren, 10-Apr-91 (RC) (Carr, 1997)
SU13: Lower Woodford, 2-Aug-82 (DRN)
SU14: Bulford (SPTAE), muddy pond, 19-Jul-91 (RC) (Carr, 1997); pond, 3-Jun - 12-Aug-08 (Boyce & Wright, 2009); Nine Mile River, Bulford, 9-Apr-91 (RC) (Carr, 1997); Figheldean, 28-Jul-82 (DRN)
SU21: Landford Bog, 24-Aug-01 (APF) (Foster, 2001)
SU22: Langley Wood NNR, 22-Apr to 14-May-04 (MDD)

Resident. Widespread; in mainly stagnant water.

**LACCOBIUS Erichson, 1837**

***Laccobius atratus*** Rottenburg, 1874

**Nb**

8 SU21: Pound Bottom, 9-Apr-74 (DRN) (Nash, 1977a); Landford Bog, 9-Jun-01 (APF) (Foster, 2001)

Resident. Very local; [associated with hillside flushes, watershed and valley mires].

***Laccobius bipunctatus*** (Fabricius, 1775)

7 ST96: Kennet & Avon Canal, 2-Jul-97 (MNS)
ST98: Dauntsey, silt pond, 1932-1933 (EJP) (Pearce, 1934)
SU09: Cotswold Water Park, 28-Apr-82 (GNF)
SU26: Folly Farm, 1939 (Det KGB) (Knight, 1940)

8 SU05: The Warren, chalk stream and fish pond, 10-Apr-91 (RC) (Carr, 1997)
SU14: Bulford (SPTAE), muddy pond, 19-Jul-91 (RC) (Carr, 1997)

Resident. Local; in still and running water.

***Laccobius colon*** (Stephens, 1829)

7 ST98: Dauntsey, silt pond, 1932-1933 (EJP) (Pearce, 1934)

8 SU04: Chitterne Down (SPTAW), dew pond,, 1991-1992 (RC) (Carr, 1997)
SU25: Limmer Ponds, 1991-1992 (RC) (Carr, 1997)

Resident. Very local; in ditches, lakes and ponds.

***Laccobius minutus*** (Linnaeus, 1758)

7 ST98: Box Road pond, nr Brinkworth, 9-May-05 (SW)
SU09: Cotswold Water Park, 28-Apr-82 (GNF); 30-May-97 (JSD)

Resident. Very local; in lakes and ponds.

***Laccobius sinuatus*** Motschulsky, 1849

**Nb**

7 ST99: R. Thames side stream, 9-May-99 (APF)
SU09: Cotswold Water Park, 28-Apr-82 (GNF)
SU16: Marlborough area, 1906 (Meyrick, 1938)

Resident. Very local; on plants in water, and in damp muddy places.

***Laccobius striatulus*** (Fabricius, 1801)

7 SU09: Cotswold Water Park, 28-Apr-82 (GNF)
SU16: Marlborough area, 1869-1938 (Meyrick, 1938)

8 ST94: Codford St. Peter, intermittent chalk stream and pool, 1991-1992 (RC) (Carr, 1997)
SU14: R. Avon, Amesbury, 19-Jul-91 (RC) (Carr, 1997)

Resident. Very local; in streams, ditches and ponds.

Subfamily SPHAERIDIINAE

**COELOSTOMA Brullé, 1835**

***Coelostoma orbiculare*** (Fabricius, 1775)

7 ST98: Dauntsey, silt pond, 1932-1933 (EJP) (Pearce, 1934); R. Avon, 1929-1932 (EJP) (Pearce, 1934)

8 SU13: Salisbury, Apr-00 (MAS) (Darby, 2001)

Resident. Very local; in well-vegetated ponds.

**CERCYON Leach, 1817**

CERCYON Leach, 1817

***Cercyon granarius*** Erichson, 1837

**RDB3**

8 SU12: Britford, Jun - 15-Sep-02 (CMD) (Drake, 2008)
SU13: Lower Woodford, 18-Jun, 17-Jul, 26-Aug-02 (CMD) (Drake, 2008)

Resident. Very local; [in flood debris, floating weed and flooded grass tussocks in marshy ground].

***Cercyon haemorrhoidalis*** (Fabricius, 1775)

7 SU16: Marlborough area, 20 Mar 1879 (WFB) (Meyrick, 1880); 1869-1938 (Meyrick, 1938)
SU26: Savernake Forest, 21-Sep to 11-Oct-93, 1 to 19-Jul-94 (EN det SAW)

8 SU02: Knighton Wood, 6-Jul-07 (AJWA)
SU03: Wylye Down, 10-Jul to 3-Aug-07 (MDD) (Darby, 2009); Grovely Wood, 10-Apr-70 (DRN)
SU13: Salisbury, Avon Valley Reserve, Apr-00 (MAS) (Darby, 2001);
Great Durnford, 27-Jul-77 (DRN)
SU22: Bentley Wood, 31-May-97 (MDD); Langley Wood, 25-Apr to 6-Jun-07 (MDD) (Darby, 2008)

Resident. Widespread; in dung and rotting vegetation.

***Cercyon impressus*** (Sturm, 1807)

7 SU16: Marlborough area, Jun-1873 (EWJ) (Meyrick, 1874); 1869-1938 (Meyrick, 1938)
SU26: Savernake Forest, 2-Jul-93 (EN det SAW)

8 SU02: Middleton Down, 4-May-95 (MDD)
SU03: Wylye Down, 10-Jul to 4-Sep-07 (MDD) (Darby, 2009)

Resident. Local; in dung and damp litter in grassland.

***Cercyon laminatus*** Sharp, 1873

7 SU16: Marlborough area, 17-Feb-1879 (WFB) (Meyrick, 1880); 1869-1938 (Meyrick, 1938)

8 SU16: Jones's Mill, Sep-97 (HK det MDD);
SU22: Langley Wood NNR, 15-Jul to 5-Aug-06 (MDD) (Darby, 2008)
SU23: Porton Down, 3-Jul-99 (BL)

Resident. Local; in litter in woodland and grassland.

***Cercyon lateralis***(Marsham, 1802)

8 SU02: Knighton Wood, Sep-07 (AJWA)

Resident. Very local; in litter.

***Cercyon marinus*** Thomson, C.G, 1853
8 SU13: Salisbury, 25-Jul-78 (DRN)
Resident. Very local; in damp litter in marshy areas.

***Cercyon melanocephalus*** (Linnaeus, 1758)
7 ST97: North Draycot Park, 22-May-00 (MDD)
SU16: Marlborough area, Jun-1873 (EWJ) (Meyrick, 1874); 1869-1938 (Meyrick, 1938)
8 SU02: Knighton Wood, Oct-06 (AJWA)
SU03: Grovely Wood, 25-Oct-71 (DRN)
SU16: Jones's Mill, 19-Jun-01 (APF) (Foster, 2002)
Resident. Local and frequent in dung. Probably widespread.

***Cercyon obsoletus*** (Gyllenhal, 1808)
**Notable**
7 SU16: Marlborough area, 1869-1938 (Meyrick, 1938)
SU26: Savernake Forest, 8 to 28-Jul-93 (EN det SAW)
8 SU14: Stonehenge, summer-03 (PJH) (Hodge, 2003)
Resident. Very local; in dung, carcasses and rotting vegetation in woodland, parkland and grassland. Formerly *C. lugubris.*

***Cercyon pygmaeus*** (Illiger, 1801)
7 SU16: Marlborough area, Jun-1873 (EWJ) (Meyrick, 1874); 1869-1938 (Meyrick, 1938)
8 SU02: Middleton Down, 4-May-95 (MDD)
Resident. Local; in dung.

***Cercyon quisquilius*** (Linnaeus, 1761)
7 SU16: Marlborough area, 1869-1938 (Meyrick, 1938)
8 SU03: Grovely Wood, 3-Aug-79 (DRN)
SU22: Harestock, 13-Apr-72 (DRN)
Resident. Local; in dung and sometimes at light.

***Cercyon sternalis*** Sharp, 1918
**Nb**
7 SU09: Somerford Keynes, 4-Jun-04 (BLe)
Resident. Very local; in litter in ponds and ditches, also comes to light.

***Cercyon tristis*** (Illiger, 1801)
**Nb**
8 SU13: Salisbury, Avon Valley Reserve, 13-Apr-00 (MDD) (Darby, 2001)
Resident. Very local; in muddy litter in wetland.

***Cercyon unipunctatus*** (Linnaeus, 1758)
7 SU16: Marlborough area, Jun-1873 (EWJ) (Meyrick, 1874); 1869-1938 (Meyrick, 1938)
8 ST92: Sutton Mandeville, 6-Jul-93 (MDD)
Resident. Local. Synanthropic species found in compost heaps and garden refuse.

PARACERCYON Seidlitz, 1888

***Cercyon analis*** (Paykull, 1798)
7 SU16: Marlborough area, 1869-1938 (Meyrick, 1938)
8 SU21: Hamptworth Estate, 15-Apr-74, 1-Aug-79 (DRN)
Resident. Very local; in rotting vegetation.

DICYRTOCERCYON Ganglbauer, 1904

***Cercyon ustulatus*** (Preyssler, 1790)
**Nb**
7 SU16: Marlborough area, 1869-1938 (Meyrick, 1938)
Resident. Very local; in wet litter and mud at the edges of ponds. One specimen only under this name in the MCC correctly identified.

**MEGASTERNUM Mulsant, 1844**

***Megasternum concinnum*** (Marsham, 1802)
7 SU16: Marlborough area, 17-Feb-1879 (WFB) (Meyrick, 1880); 1869-1938 (Meyrick, 1938)
SU26: Savernake Forest, 8 to 28-Aug-93, 18-May to 19-Jul, 20-Sep to 11-Oct-94 (EN det SAW)
8 ST91: Rushmore Park, 23-Sep-01 (AJWA)
ST92: Sutton Mandeville, 30-Sep, 15-Oct-95, 6-Jul-07 (MDD)
SU01: Martin Down NNR, 1999 (Anon 1999)
SU02: Knighton Wood, Sep-06, Sep-07 (AJWA)
SU03: Grovely Wood, 6-Aug-70, 4-Aug-78 (DRN); 21-Aug-01 (MDD); Wylye Down NNR, 16-Oct-06 to 4-Sep-07) (MDD) (Darby, 2009)
SU13: Salisbury, Avon Valley Reserve, 19-Jun-00 (MDD) (Darby, 2001); Salisbury, Bemerton Folly, 7-Mar - 15-Jun-02 (MDD) (Darby, 2002d)
SU14: Haxton, 24-Jun-02 (PJH) (Hodge, 2002)
SU16: Jones's Mill, Sep-97 (HK det MDD)
SU21: Hamptworth Estate, 15-Apr-74 (DRN)
SU22: Bentley Wood, 3-Sep-97 (WANHS det MDD); 31-May-97 (MDD); Landford, 12- Aug-87 (DRN); Langley Wood NNR, 15-Sep-06 to 2-Jul-07 (MDD) (Darby, 2008)
SU26: Bedwyn Brail, 19-Jun-03 (MDD)
Resident. Widespread; in dung and rotting vegetation. Formerly *M.obscurum.*

**CRYPTOPLEURUM Mulsant, 1844**

***Cryptopleurum crenatum*** (Kugelann, 1794)
**Nb**
8 SU12: Britford, Jun - 15-Sep-02 (CMD) (Drake, 2008)

***Cryptopleurum minutum*** (Fabricius, 1775)
7 SU16: Marlborough area, 5-May-1879 (WFB) (Meyrick, 1880); 1869-1938 (Meyrick, 1938)
SU26: Savernake Forest, 1-Jul to 9-Aug-94 (EN det SAW)

8 ST84: Longleat, 24-Sep-89 (AGD)
ST91: Rushmore Park, 23-Sep-01 (AJWA)
SU02: Knighton Wood, May, Sep, Oct-07, May-08 (AJWA); Middleton Down, 4-May-95 (MDD)
SU03: Grovely Wood, 12-Jun-00 (MDD); Wylye Down, 4-May to 5-Jun-07 (MDD) (Darby, 2008a)
SU22: Bentley Wood, 11-Aug-01 (MDD)

Resident. Local; in rotting vegetation and dung. *C. subtile* was not separated from *minutum* until 1967 so that some earlier records may be incorrect.

***Cryptopleurum subtile*** Sharp, 1884

8 ST91: Rushmore Park, 23-Sep-01 (AJWA)
SU02: Knighton Wood, May-07 (AJWA)

Resident. Very local. Synanthropic species found in dung and rotting vegetation. Recently (1967) shown to be separate from *C. minutum.*

**SPHAERIDIUM Fabricius, 1775**

***Sphaeridium bipustulatum*** Fabricius, 1781

7 SU16: Marlborough area, 1869-1938 (Meyrick, 1938)
8 SU03: Grovely Wood, 1-Aug-88 (DRN)

Resident. Very local; in dung.

***Sphaeridium lunatum*** Fabricius, 1792

7 SU26: Savernake Forest, 4-Aug-95 (EN)
8 ST91: Rushmore Park, 23-Sep-01 (AJWA)
SU02: Middleton Down, 4-May-95 (MDD)
SU03: Grovely Wood, 11-Aug-87 (DRN); Wylye Down, 3-Aug to 4-Sep-07 (MDD) (Darby, 2008a)
SU22: Bentley Wood, 11-Aug-01 (MDD); Langley Wood, 27-Aug to 17-Sep-04 (MDD) (Darby, 2005)

Resident. Local; in dung.

***Sphaeridium scarabaeoides*** (Linnaeus, 1758)

7 ST86: Neston Park, 3-Jun-04 (BLe)
SU09: Ashton Keynes, 26-Apr-01 (JKG det MDD); Upper Waterhay, 6-Jul-07 (JKG det MDD)
SU16: Marlborough area, Jun-1873 (EWJ) (Meyrick, 1874); 21-May-1879 (WFB) (Meyrick, 1880)
8 SU03: Grovely Wood, 11-Aug-87 (DRN); Wylye Down, 5-Jun to 10-Jul-07 (MDD) (Darby, 2008a)
SU21: Landford Bog, 24-Aug-01 (APF) (Foster, 2001)

Resident. Local; in dung. *S. marginatum* was not separated from *scarabaeoides* until 1989 so that some records may refer to that species.

## Family HISTERIDAE

Subfamily ABRAEINAE

Tribe ABRAEINI

**ABRAEUS Leach, 1817**

***Abraeus granulum*** Erichson, 1839

**Na, IEC-1, SQS-8**

8 ST93: Great Ridge Wood, 10-May-00 (MDD)

Resident. Very local; a single specimen in rotten beech. [Also found in rotting wood of other deciduous trees].

***Abraeus perpusillus*** (Marsham, 1802)

**IEC-likely addition, SQS-4**

7 SU26: Savernake Forest, 7-Apr-70 (DRN); 23-Oct-01 (MDD); 29, 30-May-76 (MacKechnie-Jarvis, 1976)
8 ST91: Cranborne Chase, 21-May-02 (MDD)
ST93: Great Ridge Wood, 10-May-00 (MDD)
SU02: Knighton Wood, May-07 (AJWA)
SU03: Grovely Wood, 25-Oct-71, 6, 8-Apr, 23-Aug-72, 16-Apr-74, 16-Apr-84 (DRN); 27-May-02 (MDD)
SU13: Little Durnford Down, 27-May-97, 20-May-03 (MDD); Salisbury, Avon Valley Reserve, 19-Jun-00 (MDD) (Darby, 2001)
SU23: Thorny Down Woods, 16-Oct-00 (MDD); West Winterslow, 1-Aug-01 (NB)
SU26: Bedwyn Brail, 5-Nov-03 (MDD)

Resident. Widespread; in damp rotting wood. Formerly *A. globosus.*

Tribe PLEGADERINI

**PLEGADERUS Erichson, 1834**

***Plegaderus dissectus*** Erichson, 1839

**Nb, IEC-2, SQS-8**

7 ST96: Spye Park, 24-Apr-87 (KNAA)
8 ST93: Great Ridge Wood, 10-May-00 (MDD)
SU03: Grovely Wood, 3-5-Aug-71 (DRN)
SU21: Hamptworth Estate, 6-Aug-79 (DRN)
SU22: Langley Wood NNR, 6-Jun-02 (MDD)

Resident. Local; in old decaying tree stumps and logs, mainly oak and beech.

***Plegaderus vulneratus*** (Panzer, 1797)

8 SU21: Hamptworth Estate, 26-Jul-77 (DRN); Pound Bottom, 4-Aug-82, 5-Aug-85 (DRN)
SU26: Bedwyn Common, 24-Jul-82 (DRN)

Resident. Very local; under bark of conifers. Recent introduction to GB first notified in 1968 and now naturalised.

Tribe ACRITINI

**AELETES Horn, 1873**

***Aeletes atomarius*** (Aubé, 1842)

**RDB3, IEC-1, SQS-16**

8 SU26: Wilton Brail, 19-Jun-03 (MDD)

Resident. Very local; in a rotten beech log. [Under bark and in rotten wood of other trees].

## Subfamily SAPRININAE

**SAPRINUS Erichson, 1834**
***Saprinus aeneus*** (Fabricius, 1775)
7 SU26: Savernake Forest, 25-Apr to 26-May-93 (EN det SAW)
Resident. Very local; in dung.

***Saprinus semistriatus*** (Scriba, 1790)
7 SU16: Marlborough area, Jun-1873 (EWJ) (Meyrick, 1874); 1869-1938 (Meyrick, 1938)
8 ST92: Sutton Mandeville, 1-Oct-96 (MDD)
ST93: Stockton Wood, 7-Jul-96 (MAS)
8 SU03: Grovely Wood, 23-Jul-73 (DRN); Wylye Down, 4-May to 5-Jun-07 (MDD) (Darby, 2008a)
Resident. Local and frequent in carrion and, more rarely, dung.

**GNATHONCUS Jacquelin du Val, 1857**
***Gnathoncus buyssoni*** Auzat, 1917
**Na**
8 SU22: Langley Wood NNR, 15-Jul to 5-Aug-06 (MDD) (Darby, 2007)
Resident. Very locall; ia single example in a pitfall trap. [In nests of birds and small mammals and also in fungi].

***Gnathoncus nannetensis*** (Marseul, 1862)
8 SU02: Knighton Wood, 7-Oct-06, Apr, May-07, May-08 (AJWA)
Resident. Very local; in a flight intercept trap. [In bird's nests, carrion and fungi].

***Gnathoncus rotundatus*** (Kugelann, 1792)
8 SU02: Knighton Wood, 7-Oct-06, Apr-07 (AJWA)
Resident. Very local; in aerial litter, bird nests, etc..

## Subfamily DENDROPHILINAE
### Tribe DENDROPHILINI

**DENDROPHILUS Leach 1817**
***Dendrophilus punctatus*** (Herbst, 1792)
8 SU02: Knighton Wood, 7-Oct-06 (AJWA)
Resident. Very local; in aerial litter, bird nests, etc.. [Also found with the ant *Formica rufa* and in carrion].

**KISSISTER Marseul, 1862**
***Kissister minimus*** (Aubé, 1850)
8 SU13: Salisbury, under stone, Old Sarum, 11-Apr-71 (DRN)
SU14: Haxton (SPTAE), 24-Jun-02 (PJH) (Hodge, 2002); Stonehenge, 22-Apr-03 (PJH) (Hodge, 2003)
Resident. Very local; at roots especially in sandy places.

### Tribe PAROMALINI

**CARCINOPS Marseul, 1855**
***Carcinops pumilio*** (Erichson, 1834)
8 SU02: Grovely Wood, 4-Aug-70 (DRN); Knighton Wood, 7-Oct-06 (AJWA)
SU05: Market Lavington, 2-Aug-83 (DRN)
SU13: Salisbury, 1-Aug-73 (DRN)
Resident. Very local; in carrion, rotting vegetable matter, bat dung, etc..

**PAROMALUS Erichson, 1834**
***Paromalus flavicornis*** (Herbst, 1792)
**IEC- likely addition, SQS-2**
7 ST87: Corsham Court Park, 3-Jun-04 (KNAA)
ST96: Spye Park, 24-Apr-87 (KNAA)
SU26: Savernake Forest, 2-Apr-70 (DRN); 29-May-76 (MacKechnie-Jarvis, 1976); 6-Jun-73 (APF); 15-Jun-73, 29-May-83, 5-May-90, 30-May-03 (KNAA); 6-Dec-73, 22-Jun-90 (APFw); 26-May to 16-Jun-93 (EN det SAW)
8 ST73: Stourhead, 11-Feb-84 (KNAA) (Alexander, 1996); Tucking Mill Hanging, 11-Feb-84 (KNAA)
ST84: Longleat, 28-Aug-00 (MB)
ST91: Cranborne Chase, 21-May-02 (MDD)
ST93: Great Ridge Wood, 8-May-00 (MDD)
SU03: Grovely Wood, 10-Apr-70, 8-Apr-75, 4-Aug-78, 16-Apr-84, 4-Aug-87 (DRN); 11-May-00 (MDD); 14-Jun-01(WANHS det MDD)
SU13: Little Durnford Down, 13-Sep-99 (MDD)
SU22: Blackmoor Copse, 3-Aug-03 (APF) (Foster, 2004); Langley Wood NNR, 6-Jun-02 (MDD)
Resident. Widespread under bark of beech and oak.

## Subfamily ONTHOPHILINAE

Onthophilus punctatus

**ONTHOPHILUS Leach, 1817**
***Onthophilus punctatus*** (Müller, O.F, 1776)
**RDBK**
8 SU03: Wylye Down, 31-Dec-06 to 27-Jan-07, 22-Feb to 29-Mar-07 (MDD) (Darby, 2009)
Resident. Very local; mainly recorded from mole nests. One specimen in a pitfall trap in each period. Only three records apart from these in GB since 1970.

***Onthophilus striatus*** (Forster, 1771)
8 ST93: Great Ridge Wood, 18-Apr-74 (DRN)
SU02: Knighton Wood, 6-Jul-07, May-08 (AJWA)
SU15: Beach's Plantation (SPTAE), 10-Jun-97 (MGT)
SU21: Hamptworth Estate, 28-Oct-75 (DRN)
SU22: Langley Wood NNR, 17-Sep to 8-Oct-04 (MDD) (Darby, 2005)
Resident. Local; in dung and rotting vegetation.

## Subfamily TRIBALINAE

**EPIERUS Erichson, 1834**
***Epierus comptus*** Erichson, 1834
**RDBK, IEC-possible addition, SQS-16**
8 SU03: Grovely Wood, Aug-80, 4-11-Aug-87, Aug-88 (DRN) (Nash, 1999)
The status of this species in the county is unclear. Found by DRN in numbers under bark on several occasions but more recent searches by MDD have failed to find it again. Apart from a single record from the New Forest, these remain the only ones in GB.

Epierus comptus

## Subfamily HISTERINAE

**MARGARINOTUS Marseul, 1854**
PTOMISTER Houlbert & Monnot, 1923
***Margarinotus brunneus*** (Fabricius, 1775)
7 SU16: Marlborough area, Jun-1873 (EWJ) (Meyrick, 1874); 3-Jun-1879 (RWR) (Meyrick, 1880); 1869-1938 (Meyrick, 1938)
SU26: Savernake Forest, 25-Apr to 11-May-95 (EN det MDD))
8 SU02: Knighton Wood, 7-Oct-06 (AJWA)
SU22: Langley Wood NNR, 15-Jul to 5-Aug-06, 6-Jun to 2-Jul-07 (MDD) (Darby, 2008)
Resident. Local; in dung, carrion, bird nests and rotting vegetation.

***Margarinotus merdarius*** (Hoffmann, J. 1803)
8 SU22: Langley Wood NNR, 16-Jul to 6-Aug-04, 15-Jul to 5-Aug-06 (MDD) (Darby, 2005, 2008)

Resident. Very local; in a pitfall trap. [In rotting vegetation].

***Margarinotus striola*** (Sahlberg, C.R. 1819)
ssp. ***succicola*** (Thomson, C.G., 1862)
7 SU16: Marlborough area, Jun-1873 (EWJ) (Meyrick, 1874)
SU26: Savernake Forest, 28-Apr to 30-Aug-94 (EN det SAW); 25-Apr to 30-May, 5-Jul to 3-Aug-95 (EN det MDD)
8 SU03: Grovely Wood, 15-Jun-69 (DGB); 4-Aug-70, 16-Aug-71 (DRN)
SU22: Langley Wood NNR, 15-Jul to 5-Aug-06 (MDD) (Darby, 2008)
Resident. Local; in carrion, dung, putrid fungi, etc.. Probably widespread.

PARALISTER Bickhardt, 1917
***Margarinotus neglectus*** (Germar, 1813)
7 SU16: Marlborough area, Jun-1873 (EWJ) (Meyrick, 1874)
8 SU22: Langley Wood NNR, 16-Jul to 6-Aug-04 (MDD) (Darby, 2005)
Resident. Very local; in carrion, flood refuse, moss, etc..

***Margarinotus purpurascens*** (Herbst, 1792)
7 ST86: Neston Park, 3-Jun-04 (KNAA)
8 SU14: Stonehenge, 22-Apr-03 (PJH) (Hodge, 2003)
SU15: Beach's Plantation (SPTAE), 12-Jun-97 (MGT)
SU24: Newton Tony, 1-Jun-03 (DG det MAS)
Resident. Very local; [in rotting vegetation, compost heaps, etc.].

***Margarinotus ventralis*** (Marseul, 1854)
7 SU16: Marlborough area, 1869-1938 (Meyrick, 1938)
SU26: Savernake Forest, 1-Jul to 9-Aug-94 (EN det SAW); 11 to 31-May, 15-Jul to 3-Aug-95 (EN det MDD)
8 SU03: Grovely Wood, 1-Aug-88 (DRN det DH)
SU22: Langley Wood NNR, 6 to 27-Aug-04, 2-Jul to 5-Aug-06 (MDD) (Darby, 2005, 2008)
SU23: West Winterslow, 1-Aug-01 (NB)
Resident. Local; in carrion, dung, rotting vegetation, etc.. There is some confusion about this species and *carbonarius* sensu auct Britt which it replaced.

**HISTER Linnaeus, 1758**
***Hister bissexstriatus*** Fabricius, 1801
**Nb**
7 SU16: Martinsell, 18-Jun-46 (Holmes & Tennant, 1947)
Resident. Local; [in dung on grassland]. There is a single specimen without data in the MCC.

Hister bissexstriatuis

***Hister unicolor*** Linnaeus, 1758
7 SU16: Marlborough area, 1869-1938 (Meyrick, 1938)
8 ST91: Cranborne Chase, 16 to 25-Jun-80 (NCC)
SU03: Grovely Wood, 14-Apr-71, 30-Jul-79, 1-Aug-88 (DRN); Wylye Down NNR, 4-May to 5-Jun-07 (MDD) Darby, 2009)
SU16: Jones's Mill, 21-Jun-97 (MDD)
SU22: Langley Wood NNR, 6 to 27-Aug-04 (MDD) (Darby, 2005)
Resident. Local; in compost heaps, dung, rotting fungi, etc..

**ATHOLUS Thomson, G.C, 1859**
***Atholus duodecimstriatus*** (Schrank, 1781)
7 ST87: Castle Combe, 3-Jun-04 (KNAA)
SU16: Marlborough area, 1869-1938 (Meyrick, 1938)
8 ST73: Stourhead, 4-Jul-95 (KNAA)
ST91: Rushmore Park, 23-Sep-01 (AJWA)
SU05: Market Lavington, 2-Aug-83 (DRN)
Resident. Very local; in dung and compost heaps.

## Family HYDRAENIDAE
### Subfamily HYDRAENINAE

**HYDRAENA Kugelann, 1794**
***Hydraena nigrita*** Germar, 1824
**Nb**
8 SU22: East Grimstead, 8-Jun-05 (JSD)
Resident. Very local; [mainly amongst gravel and sand in partial shade at the edge of streams. It also occurs in pools in stream channels].

***Hydraena riparia*** Kugelann, 1794
7 ST98: R. Avon, Dauntsey, 1929-1932 (EJP)
SU16: Marlborough area, 1869-1938 (Meyrick, 1938); Poulton, spring-1894 (GCC) (Jebb, 1895); 17-Mar-30 (FBB)
SU27: Ramsbury, spring-1894 (GCC) (Jebb, 1895)
8 ST94: Codford St. Peter, intermittent chalk stream and adjacent pool, 1991-1992 (RC) (Carr, 1997)
SU04: R.Till, Rollestone, 1991-1992 (RC) (Carr, 1997)
SU13: Lower Woodford, R. Avon tributary, 30-Jul-82 (DRN); R. Bourne, St Thomas' Bridge, Apr-81 (APF)
SU14: R. Avon, Amesbury, 19-Jul-91 (RC) (Carr, 1997)
SU22: East Grimstead, bomb crater, Hawks Grove, 1-Aug-78 (DRN)
Resident. Widespread; in thickly vegetated water courses.

***Hydraena testacea*** Curtis, 1830
**Nb**
7 ST98: Dauntsey, silt pond, 1932-1933 (EJP) (Pearce, 1934)
SU09: Ashton Keynes, 31-May-96 (KGA)
Resident. Very local; [In stagnant water with well-developed marginal vegetation, also in slow moving water in canals and streams]. Two specimens in MCC without data.

**LIMNEBIUS Leach, 1815**
***Limnebius nitidus*** (Marsham, 1802)
**Nb**
7 ST98: Dauntsey, silt pond, 1932-1933 (EJP) (Pearce, 1934)
8 ST95: Erlestoke, 1991-1992 (RC) (Carr, 1997)
SU14: R. Avon, Amesbury, 19-Jul-91 (Carr, 1997); Bulford (SPTAE), muddy pond, 19-Jul-91 (RC) (Carr, 1997); R. Avon, Figheldean (SPTAE), 20-Jul-91 (RC) (Carr, 1997)
SU15: R. Avon, Upavon, 19-Jul-91 (RC) (Carr, 1997)
Resident. Local; in mud and silt besides both still and slow-moving water.

***Limnebius papposus*** Mulsant, 1844
**Nb**
7 SU27: Ramsbury, 17-Mar-30 (FBB)
Resident. Very local; [largely confined to fen areas in drains and ponds, usually with rich vegetation and detritus].

***Limnebius truncatellus*** (Thunberg, 1794)
7 SU09: Ashton Keynes, 31-May-96 (KGA); Cotswold Water Park, 28-Apr-82 (GNF)
SU16: Marlborough area, 1869-1938 (Meyrick, 1938)
SU26: Savernake Canal, 1930-1932 (EJP)
8 ST94: Codford St. Peter, intermittent chalk stream and adjacent pool, 1991-1992 (RC) (Carr, 1997)
ST95: West Lavington, dew pond, 19-Sep-41 (LeC) (DFL, 1948)
SU03: R. Wylye, 5-Aug-91 (RC) (Carr, 1997)
SU04: R.Till, Winterbourne Stoke, 21-Jul-91 (RC) (Carr, 1997)
SU21: Hamptworth Estate, stream near fish pond, 27-Jul-78 (DRN)
SU22: East Grimstead, 1-Aug-78 (DRN); 8-Jun-05

(JSD)
Resident. Widespread; in mud besides both still and moving water.

### Subfamily OCHTHEBIINAE

**OCHTHEBIUS Leach, 1815**
ASIOBATES Thomson, G.C, 1859
***Ochthebius bicolon*** Germar, 1824
**Nb**
[7 ST98: Dauntsey, silt pond, 1932-1933 (EJP) (Pearce, 1934) GF tells me that these specimens were later det. by CET as *dilatatus*.]
8 SU12: Britford, Jun - 15-Sep-02 (CMD) (Drake, 2008)

***Ochthebius dilatatus*** Stephens, 1829
7 ST98: R. Avon, Dauntsey, 1932 (EJP)
8 SU22: East Grimstead, 11-Aug-86 (DRN)
Resident. Very local. DRN: three specimens in moss by a boggy area, [more usually found in mud besides water]. See *O. bicolon* above.

HOMALOCHTHEBIUS Kuwert, 1887
***Ochthebius minimus*** (Fabricius, 1792)
7 SU08: Wootton Bassett, 1930 (EJP)
SU09:Cotswold Water Park, 28-Apr-82 (GNF); 30-May97 (JSD); Emmett Hill, 2-Oct-90 (JL)
SU16: Marlborough area, 1869-1938 (Meyrick, 1938)
8 SU03: R. Wylye, 5-Aug-91 (RC) (Carr, 1997)
SU04: Chitterne Down (SPTAW), dew pond, 13-Jun-73 (RC) (Carr, 1997); R. Till, Rollestone, 17-May-91 (RC) (Carr, 1997); R. Till, Winterbourne Stoke, 21-Jul-91 (RC) (Carr, 1997)
SU05: The Warren, chalk stream and fish pond, 10-Apr-91 (RC) (Carr, 1997)
SU13: Great Durnford, 7-Aug-70, 3-Aug-88 (DRN) Middle Woodford, 5-Aug-82 (DRN); R. Bourne at St. Thomas' bridge, Apr-81 (APF)
SU14: R. Avon, Amesbury, 19-Jul-91 (RC) (Carr, 1997); Bulford (SPTAE), muddy pond, 29-Mar-91 (RC) (Carr, 1997); pond, 3-Jun – 12-Aug-08 (Boyce & Wright, 2009); Nine Mile River, Bulford (SPTAE), 9-Apr-91 (RC) (Carr, 1997); R. Avon, Figheldean (SPTAE), 20-Jul-91 (RC) (Carr, 1997)
Resident. Widespread; mainly in still water.

## Family PTILIIDAE

### Subfamily PTILIINAE

#### Tribe PTENIDIINI

**NOSSIDIUM Erichson, 1845**
***Nossidium pilosellum*** (Marsham, 1802)
**Notable, IEC- possible addition, SQS-8**
8 SU13: Great Durnford, 18-Aug-72 (DRN) (Nash, 1976); Little Durnford Down, 13-Sep-99 MDD)
SU14: Netheravon, 29-Jul-83 (DRN)
Resident. Very local; in rotten wood, particularly elm, [but also of other trees including poplar and beech].

**PTENIDIUM Erichson, 1845**
MATTHEWSIUM Flach, 1889
***Ptenidium laevigatum*** Erichson, 1845
7 SU26: Savernake Forest, 23-Oct-01 (MDD)
8 ST92: Sutton Mandeville, 30-Sep-95 (MDD)
SU26: Bedwyn Brail, 2-Jun-03 (MDD)
Resident. Local; in damp or marshy habitats amongst plant litter and also in woodland.

***Ptenidium turgidum*** Thomson, C.G, 1855
**RDBK, IEC-2, SQS-16**
7 SU26: Savernake Forest, 23-Oct-01 (MDD)
8 SU03: Grovely Wood, 27-May-02 (MDD)
Resident. Very local; in rotten wood, particularly beech and elm, but also oak.

WANKOWIZIUM Flach, 1889
***Ptenidium intermedium*** Wankowicz, 1869
8 SU13: Salisbury, Avon Valley Reserve, Apr to Jun-00 (MDD) (Darby, 2001)
Resident. Very local; in marsh and other wetland litter, and flood refuse.

PTENIDIUM Erichson, 1845
***Ptenidium formicetorum*** Kraatz, 1851
8 SU03: Grovely Wood, 27-May-02 (MDD)
Resident. Very local; in litter, etc. in woodland, not confined to ant nests as its name suggests.

***Ptenidium fuscicorne*** Erichson, 1845
8 ST93: Fonthill, 27-May-95 (MDD)
SU03: Burcombe, 23-Jun-95 (MDD)
SU13: Great Durnford, 3-Aug-88 (DRN)
Resident. Very local; wetland species found in flood refuse, marsh litter and at roots of grass tufts. Nationally scarce

***Ptenidium longicorne*** Fuss, 1868
8 SU13: Lower Woodford, 5-Aug-82, 1-Aug-83 (DRN); Middle Woodford, 8-Aug-79 (DRN)
SU14: Figheldean, 7-Aug-81 (DRN)
SU15: Fittleton, 5-Aug-82, 29-Jul-83 (DRN)
Resident. Very local; crawls on mud at edge of rivers.

***Ptenidium pusillum*** (Gyllenhal, 1808)
7 ST96: Rowde, 5-Aug-83 (DRN)
SU16: Marlborough area, Jun-1873 (EWJ) (Meyrick, 1874); 1869-1938 (Meyrick, 1938); Wootton Rivers, 10-Aug-81 (DRN)
SU17: Hallam, 6-Aug-83 (DRN)
SU26: Savernake Forest, 31-May-00 (WANHS det

MDD)
8 ST83: Chicklade, 30-Jul-82 (DRN)
ST91: Cranborne Chase, 30-Jun-99 (MDD)
ST92: Sutton Mandeville, 1-Jun-93, 25-Jun-07 (MDD); Swallowcliffe, 28-Sep-95 (MDD)
ST93: Great Ridge, 29-Jul-82 (DRN)
SU03: Grovely Wood, 2-Aug-79, 8-Aug-80 (DRN)
SU05: Market Lavington, 2-Aug-83 (DRN)
SU13: Boscombe Down, 8, 9-Oct-94 (MDD); Lower Woodford, 27-Jul, 5-Aug-82 (DRN); Salisbury, 13-Aug-81 (DRN); Salisbury, Avon Valley Reserve, 1, 19-Jun-00 (Darby, 2001); Salisbury, Bemerton Folly, 7-Mar – 15-Jun-02 (MDD) (Darby, 2002d)
SU15: North Newnton, 5-Aug-83 (DRN)
SU16: Jones's Mill, 21-Jun-97 (MDD); Woodborough, 4-Aug-83 (DRN)
SU21: Hamptworth Estate, 6-Aug-79 (DRN)
SU22: East Grimstead, 12-Aug-87 (DRN); Redlynch, 4-Aug-82 (DRN)
SU23: Porton Down, 1970-2001 (Anon, 2001); Thorny Down woods, 16-Oct-00 (MDD)
Resident. Widespread and frequent in all kinds of plant refuse especially in open situations.

GILLMEISTERIUM Flach, 1889

***Ptenidium nitidum*** (Heer, 1841)
7 SU16: Wootton Rivers, 10-Aug-81 (DRN)
SU26: Savernake Forest, 31-May-00 (WANHS det MDD)
8 ST92: Sutton Mandeville, 1-Jun-93, 9-May-95, 25-Jun-07 (MDD); Sutton Row, 9-May-95 (MDD)
ST93: Great Ridge Wood, 11-Aug-81, 29-Jul-82 (DRN)
SU03: Burcombe, 23-Jun-95 (MDD); Grovely Wood, 2-Aug-79 (DRN)
SU13: Lower Woodford, 5-Aug-82 (DRN); Middle Woodford, 8-Aug-79 (DRN)
SU15: Fittleton, 29-Jul-83 (DRN)
SU16: Woodborough, 4-Aug-83 (DRN)
SU21: Hamptworth Estate, 6-Aug-79 (DRN)
SU22: Bentley Wood, 12-Aug-87 (DRN); 3-Sep-97 (WANHS det MDD); East Grimstead, 12-Aug-87 (DRN)
SU26: Bedwyn Brail, 19-Jun-03 (MDD); Wilton Brail, 19-Jun-03 (MDD)
Resident. Widespread and frequent in similar situations to *pusillum* with which it is often found.

Tribe PTILIINI

**PTILIOLA Haldeman, 1848**

***Ptiliola kunzei*** (Heer, 1841)
8 ST92: Sutton Mandeville, 6-Jul-93, 27-Apr-95, 14-Aug-07 (MDD)
ST93: Great Ridge Wood, 27, 29-Jul-82 (DRN)
SU03: Burcombe, 23-Jun-95 (MDD)
SU04: Shrewton, 28-Jul-82 (DRN)
SU22: East Grimstead, 12-Aug-87 (DRN)
Resident. Local; in putrefying substances in a range of more open woodland habitats.

**PTILIOLUM Flach, 1888**

PTILIOLUM Flach, 1888

***Ptiliolum fuscum*** (Erichson, 1845)
8 ST93: Great Ridge Wood, 2-May-00 (MDD); Pits Wood, Chilmark, 2-Jun-95 (MDD);
SU03: Burcombe, 23-Jun-95 (MDD)
Resident. Very local; in grassland, but also in woodlands and gardens, in animal dung.

***Ptiliolum spencei*** (Allibert, 1844)
8 ST92: Sutton Mandeville, 25-Jul-07, 25-Sep-07 (MDD); Swallowcliffe, 28-Sep-95 (MDD)
SU03: Burcombe, 23-Jun-95 (MDD)
SU21: Hamptworth Estate, 1-Aug-78 (DRN)
Resident. Local; grassland species mainly in dungheaps and grasspiles.

EUPTILIUM Flach, 1889

***Ptiliolum schwarzi*** (Flach, 1887)
**RDBK**
8 SU21: Hamptworth Common, 24-Mar-03 (MDD)
Resident. Very local; on an old rabbit carcass in woodland. Very few records nationally.

**OLIGELLA Motschulsky, 1869**

***Oligella foveolata*** (Allibert, 1844)
**Notable**
8 ST92: Sutton Mandeville, 30-Sep-95, 1-Apr-98 (MDD det CJ)
Resident. Very local; these specimens from litter in woodland and the base of a rotten stump two feet below ground level. [More usually in old dungheaps and compost heaps in gardens].

**PTILIUM Gyllenhal, 1827**

***Ptilium exaratum*** (Allibert, 1844)
**Notable**
8 ST92: Sutton Mandeville, 25-Jun-07 (MDD)
Resident. Very local; in a garden compost heap.

***Ptilium horioni*** Rosskothen, 1934
8 ST91: Cranborne Chase, 30-Jun-99 (MDD)
ST92: Sutton Mandeville, 30-Sep-95 (MDD)
SU03: Burcombe, 23-Jun-95 (MDD)
Resident. Very local; in grass piles and compost heaps in gardens and woodland.

Tribe PTINELLINI

**PTINELLA Motschulsky, 1844**

***Ptinella aptera*** (Guérin-Méneville, 1839)
**IEC-possible addition, SQS-2**
8 ST91: Cranborne Chase, 21-May-02 (MDD)

SU03: Grovely Wood, 1-Aug-80, 26-Jul-82 (DRN)
SU22: Bentley Wood, 5-May-03 (MDD)
Resident. Very local; under bark and in rotten wood of dead broadleaved trees.

***Ptinella denticollis*** (Fairmaire, 1858)
**Notable, IEC-possible addition, SQS-8**
7 SU16: Marlborough area, Jun-1873 (EWJ) (Meyrick, 1874); 1869-1938 (Meyrick, 1938)
Resident. Very local; under bark of dead, broad-leaved trees. No specimens in MCC.

***Ptinella errabunda*** Johnson, 1975
8 ST92: Sutton Mandeville, 30-Sep-95, 29-Jan-98, 14-Aug-07 (MDD)
SU03: Burcombe, 23-Jun-95 (MDD)
SU21: Hamptworth Estate, 6-Aug-79 (DRN)
Resident, recent introduction (1975) thought to come from New Zealand . Local; under tight bark of dead broadleaved and coniferous trees.

***Ptinella simsoni*** (Matthews, A. 1878)
8 ST92: Sutton Mandeville, 1-Apr-98 (MDD)
SU03: Burcombe, 23-Jun-95 (MDD)
Resident, recent introduction (1930s) thought to have arrived from New Zealand. Very local; in woodland and grass piles in gardens. Previously *Plitium subvariolosum.*

**PTERYX Matthews, A, 1858**
***Pteryx suturalis*** (Heer, 1841)
**IEC- possible addition, SQS-2**
7 SU16: Marlborough area, 1869-1938 (Meyrick, 1938)
SU26: Savernake Forest, spring-1894 (GCC) (Jebb, 1895); 23-Oct-01 (MDD)
8 ST84: Longleat, 6-Jul-99 (MDD)
ST92: Fovant, 25-Apr-95 (MDD)
ST93: Great Ridge Wood, 29-Jul-85 (DRN)
Resident. Local; under bark and in rotten wood of dead broadleaved trees, rarely coniferous.

## Subfamily ACROTRICHINAE
### Tribe NEPHANINI

**NEPHANES Thomson, C.G, 1859**
***Nephanes titan*** (Newman, 1834)
7 ST96: Rowde, 5-Aug-83 (DRN)
8 ST83: Chicklade, 29, 30-Jul-82 (DRN)
ST92: Sutton Mandeville, 30-Sep, 19-Oct-95; 30-Jul-07; 14-Aug-07 (MDD); Swallowcliffe, 28-Sep-95 (MDD)
SU03: Grovely Wood, 28-Jul-83 (DRN)
SU05: Market Lavington, 2-Aug-83 (DRN)
SU15: Fittleton, 29-Jul-83 (DRN); North Newnton, 5- Aug-83 (DRN)
Resident. Local; in old dungheaps and garden compost.

### Tribe ACROTRICHINI

**ACROTRICHIS Motschulsky, 1848**
***Acrotrichis arnoldi*** Rosskothen, 1935
**RDBK**
8 ST92: Sutton Mandeville, 30-Sep-95 (MDD)
ST93: Great Ridge Wood, 16-May-00 (MDD)
SU26: Bedwyn Brail, 19-Jun-03 (MDD); Wilton Brail, 19-Jun-03 (MDD)
Resident. Very local; in grass piles in grassland and woodland. Probably not deserving its RDB status

***Acrotrichis atomaria*** (De Geer, 1774)
7 SU16: Marlborough area, Jun-1873 (EWJ) (Meyrick, 1874); 1869-1938 (Meyrick, 1938)
8 SU03: Berwick St. James, 20-Jun-95 (MDD); Burcombe, 23-Jun-95 (MDD); Steeple Langford, 19-Jun-95 (MDD)
SU13: Great Durnford, 3-Aug- 88 (DRN); Lower Woodford, 5-Aug-82 (DRN); Salisbury, 12-Apr-71 (DRN); Salisbury, Avon Valley Reserve, Apr - Jun-00, (MDD) (Darby, 2001); Salisbury, Bemerton Folly, 7-Mar – 15-Jun-02 (MDD) (Darby, 2002d)
SU14: Figheldean, 7-Aug-81 (DRN)
SU15: Fittleton, 29-Jul-83 (DRN)
SU21: Hamptworth Estate, 18, 24-Apr-84, 5-Aug-85 (DRN)
SU22: East Grimstead, 11-Aug-86 (DRN); 3-Sep-97 (WANHS det MDD)
SU23: Porton Down, 1970-2001 (Anon, 2001); Thorny Down Woods, 16-Oct-00 (MDD)
Resident. Widespread; in litter and moss in damp situations.

***Acrotrichis cognata*** (Matthews, A, 1877)
7 SU16: Wootton Rivers, 10-Aug-81 (DRN)
SU26: Savernake Forest, 31-May-00 (WANHS det MDD)
8 ST91: Cranborne Chase, 30-Jun-99 (MDD)
ST92: Sutton Mandeville, 25-Jun, 25-Jul-07 (MDD)
ST93: Great Ridge Wood, 27, 29-Jul-82 (DRN); 2 to 16-May-00 (MDD)
SU03: Burcombe, 23-Jun-95 (MDD); Grovely Wood, 3-Aug-79 (DRN); 21-Aug-01 (MDD)
SU14: Figheldean, 7-Aug-81 (DRN)
SU22: East Grimstead, 12-Aug-87 (DRN)
SU26: Bedwyn Brail, 19-Jun-03 (MDD); Wilton Brail, 19-Jun-03 (MDD)
Resident, recent introduction (1967) from N. America. Widespread; in all kinds of putrefying substances, often in shaded localities.

***Acrotrichis fascicularis*** (Herbst, 1793)
7 SU17: Hallam, 6-Aug-83 (DRN)
8 ST92: Sutton Mandeville, 27-Apr-95, 25-Jun-07 (MDD)
ST93: Great Ridge Wood, 29-Jul-82 (DRN)
SU02: Knighton Wood, Sep-06 (AJWA)

SU03: Grovely Wood, 3-Aug-71, 2-Aug-79, 8-Aug-80 (DRN)
SU13: Boscombe Down, 8, 9-Oct-94 (MDD); Lower Woodford, 5-Aug-82 (DRN); Salisbury, 13-Aug-81 (DRN); Salisbury, Avon Valley Reserve, 13-Apr-00 (Darby, 2001); Salisbury, Bemerton Folly, 7-Mar - 15-Jun-02 (MDD) (Darby, 2002d)
SU14: Figheldean, 7-Aug-81 (DRN)
SU15: Fittleton, 29-Jul-83 (DRN)
SU16: Jones's Mill, 21-Jun-97 (MDD); Woodborough, 4-Aug-83 (DRN)
SU21: Hamptworth Estate, 29-Mar, 6-Aug-78 (DRN)
SU22: East Grimstead, 12-Aug-87 (DRN)
SU26: Wilton Brail, 19-Jun-03 (MDD)
Resident. Widespread; in heaps of cut vegetation, mostly in open country and around farms and houses.

***Acrotrichis insularis*** (Mäklin, 1852)
8 ST91: Cranborne Chase, 30-Jun-99 (MDD)
SU03: Burcombe, 23-Jun-95 (MDD)
SU13: Salisbury, Bemerton Folly, 7-Mar - 15-Jun-02 (MDD) (Darby, 2002d)
SU16: Jones's Mill, 21-Jun-97 (MDD)
SU21: Hamptworth Estate, 3-Apr-78 (DRN)
Resident. Local; in a range of putrefying substances usually in woodland. North American species first recorded in GB in 1966.

***Acrotrichis intermedia*** (Gillmeister, 1845)
7 SU26: Savernake Forest, 31-May-00 (WANHS det MDD)
8 ST84: Longleat, 6-Jul-99 (MDD)
ST91: Cranborne Chase, 30-Jun-99 (MDD)
ST92: Sutton Mandeville, 11-May-95 (MDD)
ST93: Great Ridge Wood, 19-Apr-84 (DRN)
SU02: Fovant, 30-Apr-95 (MDD)
SU03: Grovely Wood, 8-Aug-80 (DRN)
SU06: Horton, 5-Aug-83 (MDD)
SU13: Salisbury, Bemerton Folly, 7-Mar - 15-Jun-02 (MDD) (Darby, 2002d)
SU21: Hamptworth Estate, 24-Apr-84, 5-Aug-85 (DRN)
SU22: Langley Wood NNR, 25-May-94 (WANHS det MDD); 22-Apr to 14-May-04 (MDD)
SU23: Porton Down, 1970-2001 (Anon, 2001); Thorny Down, 16-Oct-00 (MDD)
SU26: Bedwyn Brail, 2-Jun-03, 5-Nov-03 (MDD)
Resident. Widespread; the commonest *Acrotrichis* in woodland litter.

***Acrotrichis lucidula*** Rosskothen, 1935
**RDBK**
8 SU15: Fittleton, 29-Jul-83 (DRN det CJ)
Resident. Very local; [in wet moss in damp locations].

***Acrotrichis montandonii*** (Allibert, 1844)
7 SU16: Wootton Rivers, 10-Aug-81 (DRN)
SU17: Hallam, 6-Aug-83 (DRN)
8 ST91: Cranborne Chase, 30-Jun-99 (MDD)
ST92: Sutton Mandeville, 27-Apr-95 (MDD)
ST93: Great Ridge Wood, 29-Jul-82 (DRN)
SU03: Burcombe, 23-Jun-95 (MDD)
SU04: Shrewton, 28-Jul-82 (DRN)
SU05: Market Lavington, 2-Aug-83 (DRN)
SU13: Boscombe Down, 8-Oct-94 (MDD)
SU26: Bedwyn Brail, 2-Jun-03 (MDD)
Resident. Local; in woodlands, grasslands and gardens in plant refuse, old dungheaps, etc..

***Acrotrichis pumila*** (Erichson, 1845)
**RDBK**
8 SU03: Burcombe, 23-Jun-95 (MDD)
Resident. Very local; these specimens were in a grass pile in a damp, shaded ditch. Nationally very few records.

***Acrotrichis rosskotheni*** Sundt, 1971
8 SU03: Grovely Wood, 8-Aug-80 (DRN)
SU22: Langley Wood NNR, 25-May-94 (WANHS det MDD)
Resident. Very local; primarily in leaf litter in woodland.

***Acrotrichis rugulosa*** Rosskothen, 1935
8 SU03: Burcombe, 23-Jun-95 (MDD)
SU13: Salisbury, Bemerton Folly, 7-Mar - 15-Jun-02 (MDD) (Darby, 2002d)
Resident. Very local; in dung and decaying vegetation, often in wooded habitats.

***Acrotrichis sericans*** (Heer, 1841)
7 ST96: Rowde, 5-Aug-83 (DRN)
SU16: Wootton Rivers, 10-Aug-81 (DRN)
8 ST83: Chicklade, 30-Jul-82 (DRN)
ST92: Sutton Mandeville, 1-Jun, 6-Jul-93, 25-Aug-07 (MDD)
ST93: Great Ridge Wood, 29-Jul-82 (DRN)
SU02: Middleton Down, 4, 21-May-95 (MDD)
SU03: Grovely Wood, 2-Aug-79, 5-Aug-81 (DRN)
SU04: Shrewton, 28-Jul-82 (DRN)
SU13: Lower Woodford, 27-Jul-82
SU15: North Newnton, 5-Aug-83 (DRN)
SU16: Woodborough, 4-Aug-83 (DRN)
SU22: Redlynch, 4-Aug-82 (DRN)
Resident. Widesoread; in decaying vegetable matter in a variety of situations. Probably widespread.

***Acrotrichis sitkaensis*** (Motschulsky, 1845)
8 SU03: Burcombe, 23-Jun-95 (MDD)
SU13: Middle Woodford, 8-Aug-79 (DRN)
Salisbury, Avon Valley Reserve, 13-Apr-00 (MDD) (Darby, 2001)
SU14: Figheldean, 7-Aug-81 (DRN)
SU15: Fittleton, 29-Jul-83 (DRN)
Resident. Local; in litter in damp situations. Probably widespread.

***Acrotrichis strandi*** Sundt, 1958
**Notable**
8 SU14: Netheravon, 30-Jul-85 (DRN det CJ)
Resident. Very local; [in all kinds of damp litter]. This specimen from a cut *Carex* pile at the edge of the R. Avon.

***Acrotrichis thoracica*** (Waltl, 1838)
7 ST96: Rowde, 5-Aug-83
SU16: Wootton Rivers, 10-Aug-81 (DRN)
SU17: Hallam, 6-Aug-83 (DRN)
8 SU03: Grovely Wood, 5-Aug-81 (DRN); Little Langford, 3-Jul-96 (MAS)
SU04: Shrewton., 28-Jul-82 (DRN)
SU05: Market Lavington, 2-Aug-83 (DRN)
SU13: Lower Woodford, 27-Jul-82 (DRN)
SU15: Fittleton, 29-Jul-83 (DRN); North Newnton, 5-Aug-83 (DRN)
SU22: East Grimstead, 12-Aug-87 (DRN); Redlynch, 4-Aug-82 (DRN)
Resident. Local; in dung and grass heaps.

CTENOPTERYX Flach, 1889
***Acrotrichis grandicollis*** (Mannerheim, 1844)
8 ST93: Great Ridge Wood, 11-Aug-81, 29-Jul-82 (DRN); 2-May-00 (MDD)
SU02: Wilton, 6-Sep-95 (MDD); Middleton Down, 4-May-95 (MDD)
SU03: Burcombe, 23-Jun-95 (MDD); Grovely Wood, 5-Aug-81 (DRN)
SU15: Fittleton, 29-Jul-83 (DRN)
SU16: Wootton Rivers, 10-Aug-81 (DRN)
SU22: East Grimstead, 12-Aug-87 (DRN); Redlynch, 4-Aug-82 (DRN)
Resident. Local; in grass piles and dung.

## Family LEIODIDAE
### Subfamily LEIODINAE
#### Tribe LEIODINI

**LEIODES Latreille, 1796**
***Leiodes badia*** (Sturm, 1807)
**SQS-16**
7 SU16: Marlborough area, Jun-1873 (EWJ) (Meyrick, 1874); 1869-1938 (Meyrick, 1938)
Resident. Very local; by evening sweeping [in woodland] possibly associated with subterranean fungi. No specimen under this name in MCC.

***Leiodes calcarata*** (Erichson, 1845)
7 SU16: Marlborough area, Jun-1873 (EWJ) (Meyrick, 1874); 1869-1938 (Meyrick, 1938)
SU26: Savernake Forest, 15-Jun-73 (SAW); 20-Apr, 28-Aug-93 (EN det SAW)
8 SU02: Knighton Wood, Sep-06, 6-Jul, Sep-07, May, Jul-08 (AJWA)
SU03: Grovely Wood, 24-Aug-72, 5-Aug-79 (DRN); 17-Feb-98 (MDD)
SU13: Great Durnford, 6-Aug-71 (DRN)
SU22: Langley Wood NNR, 3 to 25-Jun-04 (MDD det JC); 15-Jul to 5-Aug-06 (MDD) (Darby, 2005, 2008)
Resident. Local. Most of these specimens taken in flight intercept traps or by evening sweeping. Associated with subterranean fungi, possibly truffles, *Tuber* spp.. This is the most frequently encountered member of the genus.

***Leiodes cinnamomea*** (Panzer, 1793)
**Notable**
7 SU16: Marlborough area, 1869-1938 (Meyrick, 1938)
8 SU02: Knighton Wood, Oct-06 (AJWA) (Allen, 2008)
SU12: Salisbury, Oct-1907 (NHJ)
Resident. Very local; chiefly under beech trees [in beech woodland and parkland on chalky soils]. Associated with subterranean fungi including truffles*Tuber* spp..

***Leiodes litura*** Stephens, 1832
7 SU16: Marlborough area, 1869-1938 (Meyrick, 1938)
SU26: Savernake Forest, 26-May to 8-Jul-93 (EN det SAW)
8 SU02: Knighton Wood, Oct, Nov-06, Nov-07, Jul-08 (AJWA)
SU03: Wylye Down NNR, 4-May to 5-Jun-07 (MDD) (Darby, 2008); Grovely Wood, 10-Apr-70, 25-Oct-71 (DRN)
Resident. Local; by evening sweeping in Autumn; on coat in sun; sweeping under trees. Possibly associated with subterranean fungi.

***Leiodes lucens*** (Fairmaire, 1855)
**RDBK**
7 SU26: Savernake Forest, 20-Sep to 11-Oct-94 (EN det SAW); 25-Aug -14-Sep-95 (EN det MDD)
8 SU22: Langley Wood NNR, 6 to 27-Aug-04 (MDD det JC) (Darby, 2005)
Resident. Very local; in woodland, possibly associated with subterranean fungi.

***Leiodes oblonga*** (Erichson, 1845)
**Notable**
8 SU03: Grovely Wood, 11-Aug-71 (DRN)
Resident. Very local; swept on a ride at dusk. Possibly associated with subterranean fungi.

***Leiodes strigipennis* Daffner, 1983**
**RDBK**
8 SU02: Knighton Wood, May-07 (AJWA) (Allen, 2008)
Resident. Local; in a flight intercept trap. In woodland [and disturbed ground] possibly associated with subterranean fungi.

**LIOCYRTUSA Daffner, 1982**
***Liocyrtusa vittata*** (Curtis, 1840)
8 SU14: Wexland Hanging (SPTAC), 22-Jun-02 (PJH) (Hodge, 2002)
SU24: Driver Training Area (SPTAE), 24-Jun-02 (PJH) (Hodge, 2002)
Resident. Very local; on vegetation in calcareous grassland.

Tribe PSEUDOLIODINI

**COLENSIS Erichson, 1845**
***Colenis immunda*** (Sturm, 1807)
7 SU16: Marlborough area, Jun-1873 (EWJ) (Meyrick, 1874); 1869-1938 (Meyrick, 1938)
8 SU02: Knighton Wood, 6-Jul-07 (AJWA)
Resident. Very local; in woodland possibly associated with subterranean fungi.

Tribe AGATHIDIINI

**ANISOTOMA Panzer, 1797**
***Anisotoma humeralis*** (Fabricius, 1792)
**SQS-2**
7 SU26: Savernake Forest, 25-Apr to 20-Sep-93, 18-May to 20-Sep-94 (EN det SAW); 25-Apr to 23-Jun, 4 to 24-Aug-95 (EN det MDD)
8 SU02: Knighton Wood, Sep-06, Apr, May, 6-Jul, Aug-07, (AJWA)
SU03: Grovely Wood, 4, 6-Aug-70, 3,13-Aug-71, 23-Aug-72 (DRN)
SU22: Blackmoor Copse, 17-May-02 (APF) (Foster, 2004); Langley Wood, 28-May-74 (DRN)
Resident. Local; on fungi in woodlands. This is the commonest member of the genus.

***Anisotoma orbicularis*** (Herbst, 1792)
**SQS-2**
7 SU16: Marlborough area, Jun-1873 (EWJ) (Meyrick, 1874); 1869-1938 (Meyrick, 1938)
SU26: Savernake Forest, 9-Jun to 1-Jul-94 (EN det SAW); 11 to 30-May-95 (EN det MDD)
8 ST92: Wardour, 18-Jun-02 (MDD)
SU02: Knighton Wood, 6-Jul-07, May-08 (AJWA)
SU03: Grovely Wood, 5-Aug-71, 24-Aug-72 (DRN); 14-Jun-01 (WANHS det MDD)
Resident. Local; in woodlands on fungi on rotten stumps and logs.

**AMPHICYLLIS Erichson, 1845**
***Amphicyllis globus*** (Fabricius, 1792)
7 SU16: Marlborough area, Jun-1873 (EWJ) Meyrick, 1874);1869-1938 (Meyrick, 1938)
Resident. Very local; associated with fungi on rotten wood in woodland.

**AGATHIDIUM Panzer, 1797**
CYPHOCEBLE Thomson, C.G, 1859
***Agathidium nigrinum*** Sturm, 1807
**SQS-2**
7 SU16: Marlborough area, Jun-1873 (EWJ) (Meyrick, 1874); 1869-1938 (Meyrick, 1938)
SU26: Savernake Forest, 28-Aug to 20-Sep-93, 1 to 19-Jul-94 (EN det SAW)
8 SU03: Grovely Wood, 29-Oct-70, 25-Oct-71 (DRN)
SU22: Blackmoor Copse, 4 to18-May-02 (APF) (Foster, 2004)
Resident. Local; in litter in woodlands.

NEOCEBLE des Gozis, 1886
***Agathidium confusum*** Brisout, 1863
**RDBi, SQS-24**
7 SU26: Savernake Forest, 21-Sep to 9 Nov-93, 18-May to 9 Jun-94 (EN det SAW)
Resident. Very local; in flight intercept traps. [Fungi in woodland]. Only been a handful of records in GB since 1900.

***Agathidium convexum*** Sharp, 1866
8 SU22: Langley Wood NNR, 17-Sep to 8-Oct-04 (MDD det JC) (Darby, 2005)
Resident. Very local; in [moss, tussocks and] woodland litter.

***Agathidium marginatum*** Sturm, 1807
**Notable**
8 SU03: Wylye Down NNR, 3-Aug to 4-Sep-07 (MDD det JC) (Darby, 2009)
Resident. Local; at the roots of plants and grasses on chalk.

***Agathidium nigripenne*** (Fabricius, 1792)
**SQS-2**
7 SU16: West Woods, May-02 (MDD det JC)
SU26: Savernake Forest, 3 to 11-May-95 (EN det MDD)
8 SU02: Knighton Wood, Sep, Oct-06, Apr, Oct-07 (AJWA)
SU03: Grovely Wood, 16-Apr-74, 4-Apr-78 (DRN)
SU21: Hamptworth Estate, 3-Aug-73 (DRN)
Resident. Local; under bark of coniferous and deciduous trees.

***Agathidium rotundatum*** (Gyllenhal, 1827)
**SQS-2**
8 SU03: Grovely Wood, 11-Aug-71 (DRN)
SU22: Langley Wood NNR, 6, 8-Jun-02 (MDD det JC)

SU25: Ludgershall, Apr - Jul-1915 (PAH) (Harwood, 1915)
Resident. Very local; under bark of coniferous and deciduous trees.

***Agathidium varians*** Beck, 1817

**SQS-2**
7 SU16: Marlborough area, Jun-1873 (EWJ) (Meyrick, 1874); 1869-1938 (Meyrick, 1938)
8 SU03: Grovely Wood, 1-Aug-80 (DRN), 21-Aug-01 (MDD det JC)
Resident. Very local; in damp woodland litter.

AGATHIDIUM Panzer, 1797

***Agathidium atrum*** (Paykull, 1798)
7 SU16: Marlborough area, Jun-1873 (EWJ) (Meyrick, 1874); 1869-1938 (Meyrick, 1938)
8 SU03: Grovely Wood, 18-Jul-73 (DRN)
SU21: Hamptworth Estate, 24-Apr-84 (DRN)
SU25: Ludgershall, Apr - Jul-1915 (PAH) (Harwood, 1915)
Resident. Local; in moss, tussocks and woodland litter.

***Agathidium laevigatum*** Erichson, 1845
8 SU03: Grovely Wood, 23-Jul-73 (DRN)
SU04: Shrewton, 1-Aug-85 (DRN)
Resident. Very local; in old straw.

***Agathidium seminulum*** (Linnaeus, 1758)
**SQS-2**
7 SU26: Savernake Forest, 30-Aug to 20-Sep-94 (EN det SAW)
8 SU03: Grovely Wood, 12-Jun-00 (MDD)
SU22: Langley Wood NNR, 3 to 25-Jun-04 (MDD det JC) (Darby, 2005)
Resident. Very local; under bark of coniferous and deciduous trees.

## Subfamily COLONINAE

**COLON Herbst, 1797**

MYLOECHUS Latreille, 1807

***Colon brunneum*** (Latreille, 1807)
7 SU16: Marlborough area, Jun-1873 (EWJ) (Meyrick, 1874); 1869-1938 (Meyrick, 1938)
8 SU02: Knighton Wood, May, Jul-08 (AJWA)
Resident. Very local. This is the most frequently encountered member of this very poorly understood family, the larvae of which have never been identified. No specimen under this name in the MCC.

***Colon dentipes*** (Sahlberg, C.R, 1822)
**RDBK**
8 SU02: Knighton Wood, Nov-06, Jul-08 (AJWA) (Allen, 2008)
Resident. Very local; in a flight intercept trap. [In open areas in woodland and woodland edges, as well as in grassland]. Very few recent records nationally.

## Subfamily CHOLEVINAE

Tribe CHOLEVINI

**NARGUS Thomson, C.G, 1867**

NARGUS Thomson, C.G, 1867

***Nargus velox*** (Spence, 1813)
8 SU01: Martin Down NNR, 1999 (EN, 1999)
SU13: Great Durnford, 6-Apr-70 (DRN)
SU21: Hamptworth Estate, 15-Apr-74 (DRN)
SU22: Langley Wood NNR, 8-Oct-04 to 24-Mar-05, 15-Sep to 17-Nov-06 (Darby, 2005, 2008)
SU23: Porton Down, 1970-2001 (Anon, 2001); Thorny Down Woods, 16-Oct-00 (MDD)
SU26: Bedwyn Brail, 20-Sep-04 (MDD)
Resident. Local; in leaf litter and other decaying organic matter.

***Nargus wilkinii*** (Spence, 1813)
8 SU22: Langley Wood NNR, 8-Oct-04 to 24-Mar-05, 5-Aug to 26-Oct-06 (MDD) (Darby, 2005, 2008)
Resident. Very local; in leaf litter and other decaying organic matter.

DEMOCHRUS Thomson, C.G, 1867

***Nargus anisotomoides*** (Spence, 1813)
7 SU16: Marlborough area, 1869-1938 (Meyrick, 1938)
Resident. Very local; in vegetable refuse.

**CHOLEVA Latreille, 1796**

CHOLEVOPSIS Jeannel, 1922

***Choleva spadicea*** (Sturm, 1839)
8 SU21: Franchises Wood, 3-May-95 (MAS)
Resident. Very local; in leaf litter and other decaying organic matter.

CHOLEVA Latreille, 1796

***Choleva agilis*** (Illiger, 1798)
7 SU16: Marlborough area, 3-Apr-1879 (RAW) (Meyrick, 1880); 1869-1938 (Meyrick, 1938)
Resident. Very local; in vegetable refuse, especially mole nests.

***Choleva angustata*** (Fabricius, 1781)
7 SU16: Marlborough area,05 May 1879 (WFB) (Meyrick, 1880); 1869-1938 (Meyrick, 1938)
SU26: Savernake Forest, 20 to 25-Apr-93 (EN det SAW)
8 SU02: Knighton Wood, Jul-08 (AJWA)
SU03: Wylye Down NNR, 16-Oct to 13-Nov-06 (MDD) (Darby, 2009)
Resident. Very local; in leaf litter and other decaying organic matter.

***Choleva fagniezi*** Jeannel, 1922
8 SU03: Wylye Down NNR, 31-Dec-06 to 27-Jan-07 (MDD) (Darby, 2009)
Resident. Very local; at roots of plants.

***Choleva jeanneli*** Britten, 1922

8 SU22: Langley Wood NNR, 31-Dec-04 to 3-Mar-05 (MDD) (Darby, 2005)
Resident. Very local; in leaf litter and other decaying organic matter.

***Choleva oblonga*** Latreille, 1807
8 SU03: Wylye Down NNR, 31-Dec-06 to 22-Feb-07 (MDD) (Darby, 2009)
Resident. Very local; in leaf litter and other decaying organic matter.

**SCIODREPOIDES Hatch, 1933**
***Sciodrepoides fumatus*** (Spence, 1813)
7 SU16: Marlborough area, 1869-1938 (Meyrick, 1938)
SU26: Savernake Forest, 8-Jul to 28-Aug-93 (EN det SAW)
8 SU02: Knighton Wood, Apr-07 (AJWA)
SU03: Grovely Wood, 4 -13-Aug-70 (DRN)
SU22: Langley Wood NNR, 5 to 25-Aug-05, 15-Jul to 5-Aug-06 (MDD) (Darby, 2005, 2008)
Resident. Local; in leaf litter and other decaying organic matter.

***Sciodrepoides watsonl*** (Spence, 1813)
7 SU16: Marlborough area, Jun-1873 (EWJ) (Meyrick, 1874); 1869-1938 (Meyrick, 1938)
SU26: Savernake Forest, 16-Jun to 28-Jul-93; 9-Jun to 19-Jul-94 (EN det SAW); 11 to 30-May, 4 to 24-Aug-95 (EN det MDD) MDD, SAW)
8 SU02: Knighton Wood, Apr, Sep, Oct-07 (AJWA)
SU03: Wylye Down NNR, 10-Jul to 3-Aug-07 (MDD) (Darby, 2009)
SU13: Great Durnford, 17-Jul-73 (DRN)
SU22: Langley Wood NNR, 25-May-94 (WANHS det MDD); 16-Jul to 17-Sep-04, 5 to 25-Aug-06 (MDD) (Darby, 2005, 2008)
Resident. Local; in leaf litter and other decaying organic matter.

**CATOPS Paykull, 1798**
***Catops chrysomeloides*** (Panzer, 1798)
7 SU16: Marlborough area, 1869-1938 (Meyrick, 1938)
SU26: Savernake Forest, 1869-1938 (Meyrick, 1938); 11 to 30-May-95 (EN det MDD); West Woods, 1869-1938 (Meyrick, 1938)
8 SU23: Porton Down, 15-Oct-95 (MAS); 1970-2001 (Anon)
Resident. Very local; in leaf litter and other decaying organic matter.

***Catops coracinus*** Kellner, 1846
7 SU16: Marlborough area, 1869-1938 (Meyrick, 1938)
SU26: Savernake Forest, 11 to 30-May-95 (EN)
8 SU02: Knighton Wood, Sep-06, Apr, Sep-07 (AJWA)
SU22: Langley Wood NNR, 25-May-94 (WANHS det MDD); 5 to 25-Aug-06 (MDD) (Darby, 2008)
Resident. Local; in leaf litter and other decaying organic matter.

***Catops fuliginosus*** Erichson, 1837
7 SU26: Savernake, 1 to 19-Jul, 9 to 30-Aug-94 (EN det SAW)
8 SU02: Knighton Wood, Nov-07 (AJWA)
SU03: Wylye Down, 1 to 16-Oct-06 (MDD) (Darby, 2008a)
SU13: Boscombe Down, 9-Oct-94 (MDD)
SU22: Langley Wood NNR, 22-Apr to 14-May-04 (MDD) (Darby, 2005)
Resident. Local; in decaying organic matter and leaf litter.

***Catops fuscus*** (panzer, 1794)
8 SU02: Knighton Wood, Aug, Oct-07 (AJWA)
Resident. Very local; in a flight intercept trap.

***Catops grandicollis*** Erichson, 1837
7 SU16: Marlborough area, 1869-1938 (Meyrick, 1938)
SU26: Savernake Forest, 26-May to 16-Jun, 8 to 28-Aug-93 (EN det SAW)
8 SU02: Knighton Wood, Sep, Oct-06 (AJWA)
SU03: Wylye Down NNR, 16-Oct-06 to 10-Oct-07 (MDD) (Darby, 2009)
Resident. Local; in leaf litter and other decaying organic matter.

***Catops kirbii*** (Spence, 1813)
7 SU26: Savernake Forest, 8 to 28-Aug-93 (EN det SAW)
8 SU02: Knighton Wood, Sep-06 (AJWA)
SU22: Langley Wood NNR, 25-May-94 (WANHS det MDD), 6 to 26-Oct-06 (MDD) (Darby, 2008)
Resident. Very local; in leaf litter and other decaying organic matter.

Catops longulus

***Catops longulus*** Kellner, 1846
**Notable**
7 SU26: Savernake Forest, 11 to 30-May-95, 4 to

24-Aug-95 (EN det MDD)
Resident. Very local.; in flight intercept traps. [In fungi and carrion in broadleaved woodland].

***Catops morio*** (Fabricius, 1787)
7 SU16: Marlborough area, Jun-1873 (EWJ) (Meyrick, 1874); 1869-1938 (Meyrick, 1938)
Resident. Very local; [in leaf litter and other decaying organic matter].

***Catops nigricans*** (Spence, 1813)
7 SU16: Marlborough area, Jun-1873 (EWJ) (Meyrick, 1874); 24-Mar-1879 (WFB) (Meyrick, 1880); 1869-1938 (Meyrick, 1938)
8 SU01: Martin Down NNR, 1999 (EN, 1999)
ST92: Wardour, 5-Jun-96 (MAS)
SU16: Jones's Mill, 1997 (MAS)
Resident. Local; in leaf litter and other decaying organic matter.

***Catops nigrita*** Erichson, 1837
7 SU16: Marlborough area, Jun-1873 (EWJ) (Meyrick, 1874); 1869-1938 (Meyrick, 1938)
SU26: Savernake Forest, 25-Apr to 16-Jun-93, 9-Jun to 20-Sep-94 (EN det SAW); 11 to 30-May-95 (EN det MDD)
8 ST93: Great Ridge Wood, 6-Aug-85 (DRN)
SU02: Knighton Wood, Sep-06, Sep-07 (AJWA)
SU22: Langley Wood NNR, 25-Aug-06 to 2-Jul-07 (MDD) (Darby, 2008)
Resident. Local; in leaf litter and other decaying organic matter.

***Catops tristis*** (Panzer, 1793)
7 SU16: Marlborough area, 24-Mar-1879 (WFB) (Meyrick, 1880); 1869-1938 (Meyrick, 1938)
SU26: Savernake Forest, 21-Sep to 11-Oct-93, 20-Sep to 11-Oct-94 (EN det SAW); 6 to 27-Oct-95 (EN det MDD)
8 SU02: Knighton Wood, Sep, Oct-06, Oct, Nov-07 (AJWA)
SU03: Grovely Wood, 6-Apr-71 (DRN)
SU13: Great Durnford, 6-Apr-70 (DRN)
SU22: Langley Wood NNR, 26-Oct to 8-Dec-06 (MDD) (Darby, 2008)
Resident. Local; in leaf litter and other decaying organic matter.

**CATOPIDIUS Jeannel, 1908**
***Catopidius depressus*** (Murray, 1856)
**Notable**
8 ST92: Sutton Mandeville, 21-Apr-92 (MDD)
Resident. Very local; in pitfll traps in rabbit burrows.

**PARABATHYSCIA Jeannel, 1922**
***Parabathyscia wollastoni*** (Janson, 1857)
7 ST99: Crudwell, 19-Jun-73 (APF)
8 SU12: Salisbury, The Close, n.d. (CMJ)
Resident. Very local; in leaf litter and other decaying organic matter.

Tribe PTOMAPHAGINI

**PTOMAPHAGUS Illiger, 1798**
***Ptomaphagus medius*** Rey, 1889
8 SU02: Knighton Wood, 7-Oct-06 (AJWA)
SU03: Wylye Down NNR, 10-Jul to 3-Aug-07 (MDD) (Darby, 2008)
SU04: Shrewton, 1-Aug-85 (DRN)
SU22: Langley Wood NNR, 6 to 26-Oct-06 (MDD) (Darby, 2008)
SU24: Newton Tony, 25-Aug to 20-Sep-06, (MA det MGT)
Resident. Local; in leaf litter and other decaying organic matter.

***Ptomaphagus subvillosus*** (Goeze, 1777)
7 SU16: Marlborough area, Jun-1873 (EWJ) (Meyrick, 1874); 1869-1938 (Meyrick, 1938)
8 ST92: Sutton Mandeville, 5-Apr-92 (MDD)
SU03: Wylye Down NNR, 27-Jan to 4-Sep-07 (MDD) (Darby, 2009)
SU22: Langley Wood NNR, 14-May-04 to 24-Mar-05 (MDD) (Darby, 2005)
SU24: Newton Tony, 25-Aug to 20-Sep-06, (MA det MGT)
Resident. Local; in leaf litter and other decaying organic matter.

***Ptomaphagus varicornis*** (Rosenhauer, 1847)
**RDBK, LCGR-Fm C?**
8 SU02: Knighton Wood, Sep-06 (AJWA)
SU03: Wylye Down NNR, 1 to 16-Oct-06 (MDD) (Darby, 2009)
Resident. Very local; in flight intercept and pitfall traps. [In leaf litter and other decaying organic matter].

Subfamily PLATYPSYLLINAE

**LEPTINUS Müller, P.W.J, 1817**
***Leptinus testaceus*** Müller, P.W.J, 1817
8 SU03: Grovely Wood, 29-Oct-73 (DRN)
Resident. Very local; in the nest of a ?vole.

## Family SCYDMAENIDAE
Subfamily SCYDMAENINAE
Tribe EUTHEIINI

**EUTHEIA Stephens, 1830**
***Eutheia scydmaenoides*** Stephens, 1830
**Notable**
8 SU22: Hamptworth Estate, 4-Aug-82 (DRN)
Resident. Very local; in old straw near a fishpond.

Tribe CEPHENNIINI

**CEPHENNIUM Müller, P.W.J. & KUNZE, 1822**
***Cephennium gallicum* Ganglbauer, 1899**
8 ST91: Cranborne Chase, 22-Jun-99 (MDD)
SU22: Langley Wood NNR, 25-May-94 (WANHS det MDD); 29-May-03 (MDD)
SU26: Bedwyn Brail, 2-Jun-03 (MDD)
Resident. Very local; in woodland litter.

Tribe CYRTOSCYDMINI

**EUCONNUS Thomson, C.G, 1859**
EUCONNUS Thomson, C.G., 1859
***?Euconnus fimetarius* (Chaudoir, 1845)**
7 SU26/27: Axford, 1869-1938 (Meyrick, 1938)
SU27: Ramsbury, 1869-1938 (Meyrick, 1938)
Resident. Very local; in vegetable refuse.
No specimens under this name in the MCC. Meyrick noted that the species may have been confused with *E. hirticollis.*

***Euconnus hirticollis* (Illiger, 1798)**
7 SU16: Marlborough area, 1869-1938 (Meyrick, 1938)
SU26/27: Axford, Spring 1894 (GCC) (Jebb, 1895)
SU27: Ramsbury, Spring 1894 (GCC) (Jebb, 1895)
8 SU21: Hamptworth Common, 24-Mar-03 (MDD)
Resident. Local; in sedge refuse and woodland litter. See *E. fimetarius.*

**NEURAPHES Thomson, C.G, 1859**
NEURAPHES Thomson C.G., 1859
***Neuraphes angulatus* (Müller, P.W.J. & Kunze, 1822)**
8 SU25: Ludgershall, Apr - Jul-1915 (PAH) (Harwood, 1915)
Resident. Very local; in moss, under bark and in fungi.

***Neuraphes elongatulus* (Müller, P.W.J. & Kunze, 1822)**
7 SU16: Marlborough area, Jun-1873 (EWJ) (Meyrick, 1874); 1869-1938 (Meyrick, 1938)
SU26: Savernake Forest, 19-Jul to 9-Aug-94 (EN det SAW)
8 ST93: Great Ridge Wood, 29-Jul-85 (DRN)
SU03: Grovely Wood, 4-Aug-80 (DRN)
SU22: Langley Wood NNR, 25-Jun to 16-Jul-04 (MDD) (Darby, 2005)
Resident. Local; in litter and rotten wood in woodland.

PARARAPHES Reitter, 1891
***Neuraphes praeteritus* Rye, 1872**
**Notable**
8 SU02: Knighton Wood, 6-Jul-07, May, Jul-08 (AJWA)
SU22: Langley Wood NNR, 24-May-03 (MDD)
Resident. Very local; in rotten wood in woodland.

**STENICHNUS Thomson, C.G, 1859**
***Stenichnus bicolor* (Denny, 1825)**
**SQS-4**
8 SU21: Hamptworth Estate, 6-Aug-79 (DRN)
Resident. Very local; under bark of a dead elm.

***Stenichnus collaris* (Müller, P.W.J. & Kunze, 1822)**
7 SU16: Marlborough area, Jun-1873 (EWJ) (Meyrick, 1874); 1869-1938 (Meyrick, 1938)
SU26: Savernake Forest, 23-Oct-01 (MDD)
8 ST92: Sutton Row, 21-Sep-95 (MDD); Fovant, 25-Apr-95 (MDD)
SU03: Grovely Wood, 4-Aug-80 (DRN); Wylye Down, 10-Jul-07 (MDD) (Darby, 2009)
SU21: Hamptworth Estate, 6-Apr-77, 5-Aug-85 (DRN); Landford Bog, 2003 (APF) (Foster, 2003)
SU22: Langley Wood NNR, 25-Jun to 16-Jul-04 (MDD) (Darby, 2005)
SU26: Bedwyn Brail, 5-Nov-03 (MDD)
Resident. Widespread; in moss and litter in woodland.

***Stenichnus scutellaris* (Müller, P.W.J. & Kunze, 1822)**
7 SU16: Marlborough area, Jun-1873 (EWJ) (Meyrick, 1874); 1869-1938 (Meyrick, 1938)
8 SU02: Knighton Wood, Apr-07 (AJWA)
Resident. Very local; in litter in woodland.

Tribe SCYDMAENINI

**SCYDMAENUS Latreille, 1802**
SCYDMAENUS Latreille, 1802
***Scydmaenus tarsatus* Müller, P.W.J. & Kunze, 1822**
8 ST91: Rushmore Park, 23-Sep-01 (AJWA)
ST92: Sutton Mandeville, 1-Jun-93, 27-Apr-95; 25-Jun, 14-Aug-07 (MDD); Swallowcliffe, 28-Sep-95 (MDD)
SU03: Grovely Wood, 2-Aug-79 (DRN)
SU13: Lower Woodford, 27-Jul, 2-Aug-82 (DRN)
SU21: Hamptworth Estate, 6-Aug-79 (DRN)
SU22: Langley Wood NNR, 16-Jul to 6-Aug-04 (MDD) (Darby, 2005)
Resident. Local; in grass piles and rotting vegetation.

## Family SILPHIDAE
### Subfamily SILPHINAE

**NECRODES Leach, 1815**
***Necrodes littoralis* (Linnaeus, 1758)**
7 SU16: Marlborough area, 12 Feb 1879 (GTKM) (Meyrick, 1880); 1869-1938 (Meyrick, 1938)
SU17: Avebury, 30-May, 31-Jul-03, 17-Oct-06 (DB det MDD)

8 ST85: Bullen Hill Farm, 1-Jan-82 (EGS det MDD)
SU03: Grovely Wood, 3-Aug-71, 4-Aug-78 (DRN)
Resident. Local; on carrion and often at light. Probably widespread.

**THANATOPHILUS Leach, 1815**
***Thanatophilus rugosus*** (Linnaeus, 1758)
7 SU16: Marlborough area, 18 May 1879 (TNH-S) (Meyrick, 1880); 1869-1938 (Meyrick, 1938)
SU27: Ramsbury, 1870-1873 (Meyrick, 1874)
8 ST93: Stockton Wood, 7-Jul-96 (MAS)
ST95: West Lavington, 1938 (PLP) (DFL, 1948)
SU02: Middleton Down, 3-Apr-97 (MDD)
SU03: Grovely Wood, 8, 9-Apr-71, 10, 11-Apr-72, 8-Apr-74, 4-Aug-78 (DRN); Wylye Down NNR, 3-Aug to 10-Oct-07 (MDD) (Darby, 2009);
SU22: Langley Wood NNR, 6-Jun to 2-Jul-07 (MDD) (Darby, 2008)
Resident. Widespread; on carrion.

***Thanatophilus sinuatus*** (Fabricius, 1775)
7 SU16: Marlborough area, 1869-1938 (Meyrick, 1938)
8 ST93: Great Ridge Wood, 16-Aug-76, 2-Aug-82 (DRN); 7-May-00 (MDD)
SU02: Chickengrove Bottom, 23-May-01 (BL det MDD)
SU03: Grovely Wood, 11-Aug-70, 9, 17-Apr-71, 10-Apr-72, 8-Apr-74 (DRN); Wylye Down NNR, 10-Jul to 10-Oct-07 (MDD) (Darby, 2009)
SU13: Lower Woodford, 27-Jul-82 (DRN)
Resident. Local; on carrion including dead fish

**OICEOPTOMA Leach, 1815**
***Oiceoptoma thoracicum*** (Linnaeus, 1758)
7 SU16: Marlborough area, 1869-1938 (Meyrick, 1938)
SU26: Savernake Forest, 30-May-73 (DS), 15-Jul-95 (EN det MDD)
8 ST91: Cranborne Chase, 16 to 25-Jun-80 (NCC)
ST93: Great Ridge Wood, 2-Aug-82 (DRN); 7-May-00 (MDD); Stockton, 7-Jul-96 (MAS)
SU02: Knighton Wood, un, Jul-07 (AJWA)
SU03: Little Langford Down, 6-Apr-99 (PM); Grovely Wood, 10 records 11-Aug-70 – 3-Aug-79 (DRN);
SU22: Blackmoor Copse, 22-May-04 (PM); Langley Wood NNR, 16-Jul to 27Aug-04, 5-Aug-06 to 2-Jul-07 (MDD) (Darby, 2005, 2008)
Resident. Widespread; on carrion.

**ACLYPEA Reitter, 1884**
***Aclypea opaca*** (Linnaeus, 1758)
**Na**
7 SU16: Marlborough area, 1869-1938 (Meyrick, 1938)
SU26: Savernake Forest, 2-Aug-87 (DRN)
Resident. Very local; climbing stem after rain [at roots of plants, particularly Chenopodiaceae] Appears to be declining in GB and very few recent records.

Dendroxina quadrimaculata

**DENDROXENA Motschulsky, 1858**
***Dendroxena quadrimaculata*** (Scopoli, 1772)
**Nb**
7 SU16: Marlborough area, Jun-1873 (EWJ) (Meyrick, 1874); 7-Jun-1879 (PHM) (Meyrick, 1880); 1869-1938 (Meyrick, 1938)
8 SU21: Burnt Ground Wood, 31-May-74 (DRN)
Resident. Very local; on oaks. Predatory on Lepidopterous larvae.

**SILPHA Linnaeus, 1758**
***Silpha atrata*** Linnaeus, 1758
7 ST87: Rack Hill, Jun-96 (MA det MDD)
SU16: Marlborough area, Jun-1873 (EWJ) (Meyrick, 1874); 6-Feb-1879 (GTKM) (Meyrick, 1880); 1869-1938 (Meyrick, 1938); Oare, 10-Sep-00 (IJG)
SU26: Savernake Forest, 2-Apr-70 (DRN); 3-Nov-99, 7-Jul-00 (IJG); 21-Sep-93 (EN det SAW)
8 ST82: Oysters Coppice, 6-Oct-02 (MDD)
ST93: Great Ridge Wood, 18-Apr-74, 6-Aug-85 (DRN)
ST84: Longleat, 7-Mar-99 (MB); 7-Jul-99 (MDD)
ST91: Cranborne Chase, 24-Apr-97 (MAS)
ST92: Wardour, 13-Nov-95 (MDD)
ST93: Great Ridge Wood, 4-May-82 (EGS det MDD); 18-Mar, 7-May-00 (MDD)
ST94: Ladywell (SPTAW), 3-Jun - 12-Aug-08 (Boyce & Wright, 2009)
ST95: Coombe Hill, 1995 (MA det MDD); West Lavington, 1938 (PLP) (DFL, 1948)
SU00: Barbury Castle, 2-Apr-07 (SH det MDD)
SU01: Martin Down NNR, 1999 (EN, 1999)
SU02: Middleton Down, 2-Jul-89 (BL); Wilton, 16-Aug-88 (DRN)
SU03: Grovely Wood, 10-Apr-70, 8, 10-Apr-72, 31-Jul-79, 1-Aug-88 (DRN); Wylye Down NNR, 10-Apr to 4-May-07 (MDD) (Darby, 2009)
SU04: Salisbury Plain, 25-Jun-02 (PJH) (Hodge, 2002); Shrewton, 30-Jul, 6-Aug-82 (DRN)

SU05: Redholm, 2000 (BL)
SU13: Cockey Down South, 13-Jun-86 (PM); Great Durnford, 6-Apr-70 (DRN); Salisbury, 28-Oct-73 (DRN); 12-Apr-00 (MDD) (Darby, 2001)
SU15: Beach's Plantation (SPTAE), 1-May-05 (MGT)
SU22: Blackmoor Copse, 4-18-May-02 (APF) (Foster, 2004); Langley Wood NNR, 22-Apr to 14-May-04, 15-Sep-06 to 2-Jul-07 (MDD) (Darby, 2005, 2008)
SU23: Porton Down, 12-Jun-94 (PTW det MDD)
SU24: Driver Training Area (SPTAE), 1-May-05 (MGT)

Resident. Frequent; in woodland and grassland under stones, in litter and rotten wood. Feeds on snails.

***Silpha carinata*** Herbst, 1783
**RDB1**
8 ST93: Great Ridge Wood, 18-Apr-74, 12, 16-Aug-76, 12-Apr-77, 13-Aug-80, 2-Aug-82, 6-Aug-5 (DRN) (Nash, 1975, 1977b)

Resident. Discovered by Nash after more than 150 years in GB without any records. After the initial finding of a single specimen and placing of baited pitfall traps, subsequent visits to the site produced further examples which confirmed the insect's status as a British breeding species.

Silpha carinata

***Silpha laevigata*** Fabricius, 1775
**LCGR-Fm C?**
7 SU16: Marlborough area, 1869-1938 (Meyrick, 1938)
SU27: Ramsbury, 1870-1873 (Meyrick, 1874)
8 ST93: Great Ridge Wood, 7-May-00 (MDD)
ST95: Bratton Castle, 20-May-05 (SC det MDD)
SU01: Martin Down NNR, 1999 (EN, 1999)
SU02: Middleton Down, May-97 (BL det MDD)
SU03: Wylye Down NNR, 4 to 10-Oct-07 (MDD) (Darby, 2009)
SU13: Cockey Down, 13-Jun-86, 1-Aug-88 (PM); Little Durnford Down, 28- Apr-93 (PM); Salisbury, 1-Apr-75 (DRN)
SU14: Stonehenge, 22-Apr-03 (PJH) (Hodge, 2003)
SU15: Beach's Plantation (SPTAE), 13, 16-Jun-97 (MGT)
SU17: Barbury Hill, 1995 (MA det MDD)
SU23: Porton Down, 19-Apr-92 (PTW det MDD); West Winterslow, 12-Aug-01 (NB)
SU24: Dunch Hill (SPTAE), 16-Jun-97 (MGT); Sidbury Hill (SPTAE), 18-May-97 (MGT)
SU24: Sidbury Hill (SPTAE), 28-May-97 (MGT)

Resident. Widespread; usually in open situations particularly on chalky soils.

***Silpha obscura*** Linnaeus, 1758
**RDB2**, **LCGR-Fl Cl**
8 ST91: Cranborne Chase, 16 to 25-Jun-80 (NCC)
SU12: Coombe Bissett Down, 15-May-97 (PM)
SU13: Little Durnford Down, 18-May-93 (PM)

Resident. Very local; under stones and at roots of plants on calcareous and sandy ground. Very few recent records in GB and species appears to be declining.

***Silpha tristis*** Illiger, 1798
**LCGR-Fm Cl**
7 SU06: Cherhill Down, 31-May-04 (BLe)
8 ST84: Longleat, 15-Jun-73 (PM)
ST93: Great Ridge Wood, 12-Apr-77, 6-Aug-85 (DRN); 7-May-00 (MDD)
ST95: Coombe Hill, 1995 (MA det MDD)
SU01: Martin Down NNR, 1999 (EN, 1999)
SU02: Middleton Down, 6-Nov-73, 1-May-89, May-97 (BL)
SU03: Grovely Wood, 14-Apr-71, 8-Apr-72, 23-Jul-73, 30-May-74, 4-Apr-77 (DRN); Wylye Down NNR, 4- May to 10-Oct-07 (MDD) (Darby, 2009)
SU13: Great Durnford, 17-Jul-73 (DRN); Little Durnford Down, 20-Apr-92 (PM); Salisbury, 20-Mar-94 (PM)
SU15: Beach's Plantation (SPTAE), 1-May-05 (MGT)
SU17: Barbury Hill, 1995 (MA det MDD)
SU22: Homan's Copse, 20-Jul-73 (DRN); Langley Wood, 15-Jul to 5-Aug-06 (MDD) (Darby, 2008)
SU24: Dunch Hill (SPTAE), 16-Jun-97 (MGT); Sidbury Hill (SPTAE), 13-Jun-97 (MGT)

Resident. Widespread; in dry carrion, under stones, in moss and wandering on ground.

## Subfamily NICROPHORINAE

**NICROPHORUS Fabricius, 1775**
***Nicrophorus humator*** (Gleditsch, 1767)
7 ST96: Chittoe, 9-May-00 (EGS det MDD)

SU16: Marlborough area, 13 Jun 1879 (WFB) (Meyrick, 1880); 1869-1938 (Meyrick, 1938)
SU17: Avebury, 30-Mar-03 (DB)
SU26: Savernake Forest, 15-Jul to 4-Aug-95 (EN det MDD)
8 ST85: Bullen Hill Farm, 1982-2002 (EGS det MDD)
ST93: Stockton Wood, 7-Jul-96 (MAS); Great Ridge Wood, 2-Aug-82 (DRN)
ST95: West Lavington, 16-Sep-39 (ITH) (DFL, 1948); 16-Jul-45 (FA) (DFL, 1948)
SU02: Middleton Down, 14-Aug-89 (BL)
SU03: Dinton, 1983-1993 (SP); Grovely Wood, 9,17-Apr, 3-Aug-71, 8-Apr-74, 4-Aug-78 (DRN); Wylye Down NNR, 3-Aug to 10-Oct-07 (MDD) (Darby, 2009)
SU13: Cockey Down, 10-Aug-89 (PM)
SU15: Manningford, 13-Aug-87 (DRN)
SU22: Langley Wood NNR, 6-Aug to 8-Oct-04, 15-Jul to 5-Aug-06 (MDD) (Darby, 2005, 2008)
SU23: Porton Down, 8-Aug-93 (PTW det MDD)
Resident. Widespread; in carrion and at light.

Nicrophorus interruptus

**_Nicrophorus interruptus_** Stephens, 1830
**Nb**
7 SU16: Marlborough area, 1869-1938 (Meyrick, 1938)
8 ST93: Great Ridge Wood, 2-Aug-82 (DRN)
SU01: Martin Down NNR, 1999 (Anopn, 1999)
SU03: Grovely Wood, 3-Aug-71, 4-Aug-78 (DRN); Wylye Down NNR, 10-Jul to 4-Sep -07 (MDD) (Darby, 2009)
SU13: Cockey Down, 10-Aug-89 (PM)
SU15: Manningford, 13-Aug-87 (DRN)
SU22: Langley Wood NNR, 16-Jul to 6-Aug-04 (MDD) (Darby, 2005)
Resident. Local; in carrion.

**_Nicrophorus investigator_** Zetterstedt, 1824
7 SU16: Marlborough area, 1869-1938 (Meyrick, 1938)
SU26: Savernake Forest, 15-Jul to 4-Aug-95 (EN det MDD)
8 ST93: Great Ridge Wood, 16-Aug-76, 2-Aug-82 (DRN)
ST95: West Lavington, 1938 (PLP) (DFL, 1948); 30-Oct-43 (ITH) (DFL, 1948)
SU02: Middleton Down, 6-Nov-73 (BL)
SU03: Grovely Wood, 3-Aug-71 (DRN); Wylye Down NNR, 4-Sep to 10-Oct-07 (MDD) (Darby, 2009)
SU13: Cockey Down, 10-Aug-89 (PM)
SU15: Manningford, 13-Aug-87 (DRN)
Resident. Widespread; in carrion and at light.

**_Nicrophorus vespillo_** (Linnaeus, 1758)
7 SU16: Marlborough area, 18 Sep 1879 (GTKM) (Meyrick, 1880); 1869-1938 (Meyrick, 1938)
SU17: Avebury, 9-Aug-2003 (DB)
8 ST84: Berkley Woods, 21-May-00 (AGD)
ST85: Bullen Hill Farm, 4-Jun-82, 17-Jul-86 (EGS det MDD)
ST93: Great Ridge Wood, 16-Aug-76, 2-Aug-82 (DRN)
SU01: Martin Down NNR, 1999 (EN, 1999)
SU03: Wylye Down NNR, 3-Aug to 10-Oct-07 (MDD) (Darby, 2009)
SU12: Coombe Bissett Down, May-Jun-00 (MDD)
SU13: Cockey Down, 10-Aug-89 (PM)
SU15: West Everleigh Down, 19-Jun-01 (IJG)
SU16: Jones's Mill, 16-May-02 (IJG)
SU22: Langley Wood NNR, 16-Jul to 6-Aug-04, 15-Jul to 5-Aug-06 (MDD) (Darby, 2005, 2008)
SU95: West Lavington, 1938 (PLP) (DFL, 1948)
Resident. Widespread; in carrion and at light.

**_Nicrophorus vespilloides_** Herbst, 1783
7 SU16: Marlborough area, 5-May-1879 (WFB) (Meyrick, 1880); 1869-1938 (Meyrick, 1938)
SU26: Savernake Forest, 30-May-73 (DS), 15-Jun-73 (SAW); 25-Apr to 9-Nov-93, 9-Jun to 20-Sep-94 (EN det SAW); 31-May to 21-Oct-95 (EN det MDD)
8 ST92: Sutton Row, 21-Oct-95 (MDD)
ST93: Great Ridge Wood, 16-Aug-76, 13-Aug-81, 2-Aug-82, 6-Aug-85 (DRN)
ST95: West Lavington, 29-May-34 (ITH) (DFL, 1948)
SU02: Knighton Wood, Sep-06, May, Jul, Aug-07 (AJWA)
SU03: Grovely Wood, 13-Apr, 3-Aug, 25, 27- Oct-71, 8-Apr-74 (DRN); Wylye Down NNR, 10-Jul to 3-Aug-07 (MDD) (Darby, 2009)
SU16: Jones's Mill, 16-Jun-73 (PM)
SU22: Langley Wood NNR, 14-May to 18-Oct-04, 5- Aug to 2-Jul-07 (MDD) (Darby, 2005,

2008)
Resident. Widespread; in carrion, fungi and at light. The commonest species of the genus in woodland.

***Nicrophorus vestigator*** Herschel, 1807
**Na**
8 SU16: Pewsey area, 25-Aug-99 (JW det MDD)
Resident. Very local; in carrion. Very few recent records in GB.

# Family STAPHYLINIDAE
## Subfamily OMALINAE
### Tribe ANTHOPHAGINI

Acidota cruentata

**ACIDOTA Stephens, 1829**
***Acidota cruentata*** Mannerheim, 1830
**Nb**
7 SU27: Ramsbury, 1870-1873 (Meyrick, 1874)
8 SU03: Wylye Down NNR, 27-Jan to 22-Feb-07 (MDD) (Darby, 2009)
Resident. Very local; on calcareous grassland at roots of plants and grasses. Possibly overlooked because of time of appearance.

**ANTHOBIUM Leach, 1819**
***Anthobium atrocephalum*** (Gyllenhal, 1827)
7 ST99: Flisteridge Wood, 21-Mar-82 (KNAA)
SU08: Ravensroost Wood, 1995 (MA det MDD)
SU16: Marlborough area, 1869-1938 (Meyrick, 1938)
8 ST92: Sutton Mandeville, 5-Apr-92 (MDD)
SU02: Knighton Wood, Oct-06, Nov-07 (AJWA)
SU13: Old Sarum, 5-Apr-72 (DRN); Great Durnford, 6-Apr-70 (DRN)
SU22: Langley Wood NNR, 19-Nov-04 to 24-Mar-05, 17 Nov 06 to 6 Jun-07 (MDD) (Darby, 2005, 2008)
SU23: Porton Down, 1970-2001 (Anon, 2001); Thorny Down woods, 16-Oct-00 (MDD)
Resident. Widespread; in carrion and litter in both grassland and woodland.

***Anthobium unicolor*** (Marsham, 1802)
7 ST99: Flisteridge Wood, 21-Mar-82 (KNAA)
SU16: Marlborough area, 1869-1938 (Meyrick, 1938)
SU26: Savernake Forest, 21-Sep to 11-Oct-93 (EN det SAW); 6 to 27-Oct-95 (EN det MDD)
8 ST92: Sutton Mandeville, 15-Oct-95, 1-May-97 (MDD)
ST93: Stockton Wood, 19958 (MA det MDD)
SU02: Knighton Wood, Apr-07 (AJWA)
SU03: Grovely Wood, 27-May-02 (MDD)
SU13: Great Durnford, 6-Apr-70 (DRN)
SU22: Langley Wood NNR, 8-Dec-06 to 5-Jan-07 (MDD) (Darby, 2008)
Resident. Widespread; in carrion and litter in both grassland and woodland.

**LESTEVA Latreille, 1796**
***Lesteva longoelytrata*** (Goeze, 1777)
7 SU16: Marlborough area, 28 May 1879 (WFB) (Meyrick, 1880); 1869-1938 (Meyrick, 1938)
SU18: Haydon Meadow, 10-Apr-02 (JKG det MDD)
SU26: Savernake Forest, 30-Aug to 20-Sep-94 (EN det SAW); 25-Apr to 30-May-95 (EN det MDD)
Resident. Local; in damp litter and moss in woods and grassland.

***Lesteva pubescens*** Mannerheim, 1830
7 SU16: Marlborough area, 1869-1938 (Meyrick, 1938)
8 SU21: Peaked Wood, 29-May-74 (DRN)
Resident. Very local; in damp litter and moss.

***Lesteva sicula*** Erichson, 1840
ssp. ***heeri*** Fauvel, 1871
7 SU26: Savernake Forest, 31-May-95 (EN det MDD)
8 ST92: Sutton Row, 9-May-95 (MDD)
SU02: Fovant, old water cress beds, 1-May-95 (MDD)
SU13: Salisbury, Avon Valley Reserve, Apr-May-00 (MDD); 1-Apr-00 (MAS) (Darby, 2001)
SU16: Jones's Mill, 21-Jun-97 (MDD); 1997 (MAS)
SU21: Redlynch, 3-Aug-73 (DRN)
SU22: Langley Wood NNR, 3-Jun to 31-Dec-04 (MDD) (Darby, 2005)
Resident. Local; in damp litter and moss in wet, shaded areas.

**OLOPHRUM Erichson, 1838**
***Olophrum piceum*** (Gyllenhal, 1810)
7 SU07: Catcomb Wood, May-96 (MA det MDD)
SU08: Ravensroost Wood, 1995 (MA det MDD)
SU16: Marlborough area, 1869-1938 (Meyrick, 1938)
SU26: Savernake Forest, 19958 (MA det MDD)
8 SU21: Hamptworth Estate, 6-Apr-77 (DRN); Redlynch, 31-Oct-73 (DRN)

SU22: Langley Wood NNR, 29-May-03, 22-Apr-04 to 24-Mar-05, 22-Jan to 20-Feb-07 (MDD) (Darby, 2005, 2008)
Resident. Local; in moss and litter in woods.

Tribe CORYPHIINI

***Coryphium angusticolle*** Stephens, 1834
8 SU03: Grovely Wood, 29-Oct-73 (DRN)
Resident. Local; under beech bark.

Tribe EUSPHALERINI

**EUSPHALERUM Kraatz, 1858**
***Eusphalerum luteum*** (Marsham, 1802)
7 SU16: Alton Priors, 30-Jun-00 (IJG); West Woods, 19-Jul-00 (IJG); Marlborough area, 1873-1938 (Meyrick, 1938)
SU26: Savernake Forest, 28-Jul-93 (EN det SAW)
8 ST91: Cranborne Chase, 22-30-Jun-99 (MDD)
ST93: Fonthill, 19-Jul-95, 26-Jun-02 (MDD)
SU02: Knighton Wood, May-07 (AJWA)
SU03: Grovely Wood, 9-Aug-71, 4-Aug-87 (DRN)
SU23: West Winterslow, 1-Aug-01 (NB det MDD)
Resident. Widespread on umbelliferae in or on the edges of woodland.

***Eusphalerum minutum*** (Fabricius, 1792)
7 SU16: Chilton, 1870-1905 (Meyrick, 1906); Marlborough area, 1869-1938 (Meyrick, 1938)
SU26/27: Axford, 1895-1938 (Meyrick, 1906)
8 SU12: Charlton-All-Saints, 2-Jun-95 (MS); 9-May-96 (MAS)
SU13: Salisbury, 26-May-74 (DRN); Great Durnford, 17-Jul-73 (DRN)
Resident. Local; on Umbelliferae near water.

***Eusphalerum primulae*** (Stephens, 1834)
7 SU16: Marlborough area, 1869-1938 (Meyrick, 1938)
SU26: Savernake Forest, 6-Jan-73 (TCT) (Meyrick, 1874)
8 SU03: Grovely Wood, 27-May-74, 16-Apr-84 (DRN)
Resident. Local; on primroses, *Primula vulgaris,* and hawthorn.

***Eusphalerum sorbi*** (Gyllenhal, 1810)
7 SU16: Marlborough area, 1869-1938 (Meyrick, 1938)
Resident. Very local; [on Umbelliferae in woods]. There is one specimen without data in the MCC.

***Eusphalerum torquatum*** (Marsham, 1802)
7 SU16: Marlborough area, Jun-1873 (EWJ) (Meyrick, 1874); 1869-1938 (Meyrick, 1938)
Resident. Very local; [on flowering plants in woods].There are four specimens without data in the MCC.

Tribe OMALIINI

**ACRULIA Thomson, C.G, 1858**
***Acrulia inflata*** (Gyllenhal, 1813)
**SQS-2**
7 SU16: West Woods, May-02 (MDD)
8 SU03: Grovely Wood, 27-Oct-71, 23-Aug-72, 16-Apr-74, 4-Aug-80 (DRN)
Resident. Very local; under bark of beech and oak in woodland.

**DROPEPHYLLA Mulsant and Rey, 1880**
***Dropephylla ioptera*** (Stephens, 1832)
**SQS-1**
7 SU16: Marlborough area, Jun-1873 (EWJ) Meyrick, 1874)
8 ST93: Great Ridge Wood, 28-Apr-00 (MDD conf. RT)
SU22: Langley Wood NNR, 29-May-03 (MDD)
Resident. Very local; under bark in woodland.

***Dropephylla vilis*** (Erichson, 1840)
**SQS-1**
7 SU16: Marlborough area, Jun-1873 (EWJ) (Meyrick, 1874)
8 ST82: Oysters Coppice, 6-Oct-02 (MDD)
Resident. Very local; under bark.

**HAPALARAEA Thomson, C.G, 1858**
***Hapalaraea pygmaea*** (Paykull, 1800)
**SQS-2**
7 SU16: Marlborough area, 1869-1938 (Meyrick, 1938)
8 SU02: Knighton Wood, 7-Oct-06, Oct-07 (AJWA)
Resident. Very local; the Knighton specimen in a flight intercept trap [under bark, on fungi and in bird nests]. One specimen in the MCC with label insc. *Marlborough 16/6/02.*

**OMALIUM Gravenhorst, 1802**
***Omalium caesum*** Gravenhorst, 1806
7 SU16: Tidcombe, 1895-1905 (Meyrick, 1905)
8 ST92: Sutton Mandeville, 19-Oct-95 (MDD)
SU02: Knighton Wood, Jun, Oct-07 (AJWA)
SU16: Jones's Mill, 1-Sep-97 (HK det MDD)
SU21: Hamptworth Estate, 29-Mar to 3-Apr-78 (DRN)
SU22: Langley Wood NNR, 15-Jul to 5-Aug-06 (MDD) (Darby, 2008)
Resident. Local; in straw and leaf litter.

***Omalium excavatum*** Stephens, 1834
8 ST91: Cranborne Chase, 22-Jun-99 (MDD)
Resident. Very local; in litter.

***Omalium italicum*** Bernhauer, 1902

8 SU02: Knighton Wood, Mar, Jun-07 (AJWA)
SU16: Jones's Mill, 21-Jun-97 (MDD)
Resident. Very local; in decaying vegetation.

***Omalium rivulare*** (Paykull, 1789)
7 SU16: Marlborough area, Jun-1873 (EWJ) (Meyrick, 1874); 24 May 1879 (WFB) (Meyrick, 1880); 1869-1938 (Meyrick, 1938)
SU26: Savernake Forest, 1870-1873 (Meyrick, 1874), 11 to 30-May-95 (EN det MDD)
8 ST92: Sutton Mandeville, 15-Oct-95 (MDD)
SU02: Knighton Wood, Oct-06, Mar, Apr, Jun, Nov-07 (AJWA)
SU03: Grovely Wood, 29-Oct-73, 8-Apr, 30-Oct-75 (DRN)
SU21: Hamptworth Estate, 15-Apr-74 (DRN)
SU22: Langley Wood NNR, 25-May-94 (WANHS det MDD)
Resident. Widespread; in straw, decomposing vegetation and litter, under bark, often in damp situations.

***Omalium rugatum*** Mulsant & Rey, 1880
**Notable**
8 SU02: Knighton Wood, Mar, Nov-07 (AJWA)
Resident. Very local; in a flight intercept trap. [Wide range of situations involving decomposing vegetation and carrion].

**PHLOEONOMUS Heer, 1839**
***Phloeonomus punctipennis*** Thomson, C.G, 1867
**SQS-2**
7 SU16: West Woods, May-02 (MDD)
8 ST84: Longleat, 25-Mar-99 (MB)
ST92: Wardour, 3-Apr-96 (EN); 13-Nov-95 (MDD)
ST93: Fonthill, 26-Jun-02 (MDD); Great Ridge Wood, 2-May-00 (MDD)
SU03: Grovely Wood, 8-Apr-75 (DRN)

SU21: Burnt Ground Wood, 21-Jul-73 (DRN)
SU26: Bedwyn Brail, 2-Jun-03 (MDD)
Resident. Widespread; under beech and oak bark.

***Phloeonomus pusillus*** (Gravenhorst, 1806)
**SQS-2**
7 SU16: Marlborough area, Jun-1873 (EWJ) (Meyrick, 1874); 1869-1938 (Meyrick, 1938)
8 ST92: Wardour, 13-Nov-95 (MDD)
Resident. Very local; under bark of conifers. One specimen in MCC without data.

**PHLOEOSTIBA Thomson, C.G, 1858**
***Phloeostiba plana*** (Paykull, 1792)
**SQS-2**
7 SU26: Savernake Forest, 29, 30-May-76 (MacKechnie-Jarvis, 1976)
8 ST84: Longleat, 25-Mar-99 (MB)
ST92: Sutton Mandeville, 24-Oct-95 (MDD)
SU02: Knighton Wood, Sep-06, Mar-07 (AJWA)
Resident. Local; under bark.

***Phyllodrepa floralis*** (Paykull, 1789)
7 SU16: Marlborough area, Jun-1873 (EWJ) (Meyrick, 1874); 1869-1938 (Meyrick, 1938)
8 SU02: Knighton Wood, Mar-07 (AJWA)
Resident. Very local; [in woodland].There is a single specimen in the MCC.

Phyllodrepa salicis

***Phyllodrepa salicis*** (Gyllenhal, 1810)
**RDBK**
7 SU16: Marlborough area, Jun-1873 (EWJ) (Meyrick, 1874); 1869-1938 (Meyrick, 1938)
Resident. Very local; [in squirrel dreys, carrion, and found at rowan blossom and fermenting fruit residues. It has possibly also been found in haystack litter]. No specimens in MCC.

**XYLODROMUS Heer, 1839**
***Xylodromus concinnus*** (Marsham, 1802)
7 SU16: Marlborough area, 1869-1938 (Meyrick, 1938)
8 SU21: Hamptworth Estate, 1-Aug-79 (DRN)
SU26: Bedwyn Brail, 2-Jun-03 (MDD)
Resident. Very local; in straw and a grass pile.

Subfamily PROTEININAE

**MEGARTHRUS Stephens, 1829**
***Megarthrus bollevoyei*** (Saulcy, 1862)
7 SU16: Marlborough area, 1869-1938 (Meyrick, 1938)
8 ST92: Sutton Mandeville, 25-Jul-07 (MDD)
SU03: Grovely Wood, 8-Aug-80 (DRN)
Resident. Very local; in decaying vegetation including cut conifer strippings.

***Megarthrus denticollis*** (Beck, 1817)
8 ST92: Sutton Mandeville, 1-Jun, 6-Jul-93 (MDD)
ST93: Great Ridge Wood, 16-May-00 (MDD)
SU02: Knighton Wood, May-08 (AJWA)
SU03: Grovely Wood, 2-Aug-79 (DRN)
SU13: Great Durnford, 17-Jul-73 (DRN)
SU21: Hamptworth Estate, 6-Aug-79 (DRN)

Resident. Local; in decaying vegetation.

***Megarthrus depressus*** (Paykull, 1789)
7 SU16: Marlborough area, Jun-1873 (EWJ) (Meyrick, 1874); 1869-1938 (Meyrick, 1938)
SU26: Savernake Forest, 25-Apr to 16-Jun-93 (EN det SAW), 11 to 30-May-95 (EN det MDD)
8 SU03: Grovely Wood, 21-Aug-01 (MDD)
SU16: Jones's Mill, 1-Sep-97 (HK det MDD)
SU21: Hamptworth Estate, 6-Aug-79 (DRN)
Resident. Local; in decaying vegetation. Probably widespread.

**METOPSIA Wollaston, 1854**
***Metopsia clypeata*** (Müller, P.W.J, 1821)
8 ST94: Ladywell (SPTAW), 23-Jun-02 (PJH) (Hodge, 2002)
SU13: Old Sarum, 5-Apr-72 (DRN)
SU24: Driver Training Area (SPTAE), 17-May-02 (PJH) (Hodge, 2002)
Resident. Very local; in decaying vegetation.

**PROTEINUS Latreille, 1796**
***Proteinus atomarius*** Erichson, 1840
7 SU16: Marlborough area, Jun-1873 (EWJ) (Meyrick, 1874); 1869-1938 (Meyrick, 1938)
8 SU02: Knighton Wood, May-08 (AJWA)
SU13: Salisbury, Bemerton Folly, 7-Mar - 15-Jun-02 (MDD) (Darby, 2002d)
SU22: Langley Wood NNR, 16-Jul to 6-Aug-04 (MDD) (Darby, 2005)
Resident. Very local; in fungi and litter.

***Proteinus brachypterus*** (Fabricius, 1792)
7 SU26: Savernake Forest, 9-Oct-91 (AGD), 11 to 30-May-95 (EN det SAW)
8 ST92: Wardour, 29-Oct-95 (MDD)
SU02: Knighton Wood, Sep, Oct-06, Oct, Nov-07 (AJWA)
SU03: Grovely Wood, 30-Oct-75 (DRN)
SU13: Salisbury, Avon Valley Reserve, Apr-00 (MAS) (Darby, 2001)
SU22: Bentley Wood, 10-Oct-01 (MDD); Langley Wood NNR, 19-Nov-04 to 24-Mar-05, 6-Oct-06 to 20-Feb-07 (MDD) (Darby, 2005, 2008)
SU23: Porton Down, 1970-2001 (Anon, 2001); 15-Oct-95 (MAS); Thorny Down woods, 16-Oct-00 (MDD)
SU26: Bedwyn Brail, 5-Nov-03 (MDD)
Resident. Widespread; abundant in fungi in the autumn.

***Proteinus leavigatus*** Hochhuth, 1871
8 SU13: Great Durnford, 6-Aug-71, 17-Jul-73 (DRN)
Resident. Very local; swept from vegetation by the R. Avon [in fungi and rotting vegetation].

***Proteinus ovalis*** Stephens, 1834
7 SU06: Roundway Camp, Jul-40 (FDB) (Buck, 1956)
SU16: Marlborough area, Jun-1873 (EWJ) (Meyrick, 1874); 1869-1938 (Meyrick, 1938)
8 SU02: Knighton Wood, Nov-07 (AJWA)
ST92: Sutton Mandeville, 15, 19-Oct-95 (MDD)
SU23: Porton Down, 1970-2001 (Anon, 2001); Thorny Down Woods, 16-Oct-00 (MDD)
Resident. Local; in fungi and rotting vegetation.

Subfamily MICROPEPLINAE

**ARRHENOPEPLUS Koch, 1937**
***Arrhenopeplus tesserula*** (Curtis, 1828)
8 ST92: Sutton Mandeville, 25-Jul-07 (MDD)
Resident. Very local; in rotting vegetation.

**MICROPEPLUS** Latreille, 1809
***Micropeplus fulvus*** Erichson, 1840
7 SU26: Savernake Forest, 15-Jun-73 (SAW); 20-Apr-93 (EN det SAW)
8 ST91: Rushmore Park, 23-Sep-01 (AJWA)
SU02: Knighton Wood, Oct-06 (AJWA)
SU03: Burcombe, 23-Jun-95 (MDD)
SU14: Stonehenge, 22-Apr-03 (PJH) (Hodge, 2003)
SU21: Hamptworth Estate, 15-Apr-74, 28-Oct-76 (DRN)
Resident. Local; in rotting vegetation including straw bales.

***Micropeplus porcatus*** (Paykull, 1789)
7 SU16: Marlborough area, Jun-1873 (EWJ) Meyrick, 1874); 1869-1938 (Meyrick, 1938)
Resident. Very local; [in decaying vegetation and on mud and soil]

***Micropeplus staphylinoides*** (Marsham, 1802)
7 SU16: West Woods, 29-May-01 (IJG)
8 SU01: Martin Down NNR, 1999 (EN, 1999)
SU02: Knighton Wood, Oct-07 (AJWA)
SU03: Grovely Wood, 13-Apr-71(DRN); Wylye Down NNR, 27-Jan to 10-Oct-07 (MDD) (Darby, 2009)
SU04: Salisbury Plain, 6-Aug-82 (DRN)
SU13: Old Sarum, 2-Apr-71 (DRN)
SU16: Jones's Mill, 21-Jun-97 (MDD)
SU21: Hamptworth Estate, 29-Mar-78 (DRN)
SU22: Langley Wood NNR, 27-Aug to 17-Sep-04 (MDD) (Darby, 2005)
SU24: Sidbury Hill (SPTAE), 13 to 17-Jun-97 (MGT)
Resident. Widespread; in rotting vegetation and under stones.

Subfamily PSELAPHINAE
Supertribe CLAVIGERITAE

Tribe CLAVIGERINI

**CLAVIGER Preyssler, 1790**
**Yellow Ants'-nest Beetle**
***Claviger testaceus*** Preyssler, 1790
**Notable, LCGR-Fm Cu**
7 ST87: West Kington, 5-Jan, 17-Mar-07 (MLu); Yatton Down, 5-Jan-07 (MLu); Yatton Keynell, 5-Jan-07 (MLu)
8 SU03: Grovely Wood, 8 to 10-Apr-72 (DRN)
Resident. Very local; in nests of the ant *Lasius flavus* under stones.

Supertribe EUPLECTITAE
Tribe EUPLECTINI

**EUPLECTUS Leach, 1817**
***Euplectus karstenii*** (Reichenbach, 1816)
7 SU16: Marlborough area, Jun-1873 (EWJ) (Meyrick, 1874)
Resident. Very local; [under bark and in rotten wood]. A single specimen under this name in MCC.

***Euplectus piceus*** Motschulsky, 1835
8 SU03: Grovely Wood, 2-Aug-79 (DRN)
SU22: Langley Wood NNR, 6-Jun-02 (MDD)
Resident. Very local; in mouldy grass and rotten wood.

***Euplectus sanguineus*** Denny, 1825
8 ST92: Swallowcliffe, 28-Sep-95 (MDD)
SU21: Hamptworth Estate, 1-Apr-75, 1-Aug-79 (DRN)
SU22: Redlynch, 4-Aug-82 (DRN)
Resident. Very local; in grass pile and straw.

Tribe TRICHONYCHINI
Subtribe BIBLOPORINA

**BIBLOPORUS Thomson, C.G, 1859**
***Bibloporus bicolor*** (Denny, 1825)
8 SU02: Knighton Wood, Oct-07 (AJWA)
SU03: Grovely Wood, Aug-71 to Apr-75 (DRN)
SU21: Hamptworth Estate, 31-Oct-73 (DRN)
SU22: Langley Wood NNR, 6-Jun-02 (MDD); Bentley Wood, 5-May-03 (MDD)
Resident. Very local; under beech bark.

***Bibloporus minutus*** Raffray, 1914
**Nb**
8 SU03: Grovely Wood, 16-Apr-84 (DRN)
Resident. Very local; under bark of oak branch.

Subtribe PANAPHANTINA

**BIBLOPLECTUS Reitter, 1882**
***Bibloplectus ambiguus*** (Reichenbach, 1816)
8 SU21: Hamptworth Estate, 24-Apr-84 (DRN)
Resident. Very local; in moss in sallow carr.

Tribe BRACHYGLUTINI

**BRACHYGLUTA Thomson, C.G, 1859**
***Brachygluta fossulata*** (Reichenbach, 1816)
7 SU16: Marlborough area, Jun-1873 (EWJ) (Meyrick, 1874); 10 Mar 1879 (WFB) (Meyrick, 1880); 1869-1938 (Meyrick, 1938)
8 SU21: Hamptworth Estate, 6-Apr-77 (DRN)
SU22: Langley Wood NNR, 22-Apr to 14-May-04 (MDD) (Darby, 2005); Bentley Wood, 11-Aug-86 (DRN)
Resident. Very local; in moss in damp situations.

**FAGNIEZIA Jeannel, 1950**
***Fagniezia impressa*** (Panzer, 1803)
7 SU16: Marlborough area, 1869-1938 (Meyrick, 1938)
Resident. Very local; [in moss and grass tussocks, especially on the banks of rivers and ponds]. There is a single specimen in the MCC.

**REICHENBACHIA Leach, 1826**
***Reichenbachia juncorum*** (Leach, 1817)
8 SU13: Salisbury, Avon Valley Reserve, Apr-00 (MAS) (Darby, 2001)
SU21: Hamptworth Estate, 18, 25-Apr-84 (DRN)
SU22: Bentley Wood, 11-Aug-86 (DRN)
Resident. Very local; in same situations as the preceeding but much more common.

Reichenbachia juncorum

**RYBAXIS Saulcy, 1876**
***Rybaxis longicornis*** (Leach, 1817)
8 SU22: East Grimstead, Hawks Grove, crater pond 11-Aug-86 (DRN)
Resident. Very local; in moss.

Tribe BYTHININI
**BRYAXIS Kugelann, 1794**
***Bryaxis bulbifer*** (Reichenbach, 1816)
7 SU16: Marlborough area, 1869-1938 (Meyrick, 1938)
SU27: Ramsbury, Spring 1894 (GCC) (Jebb, 1895)

8 SU13: Great Durnford, 3-Aug-88 (DRN); Salisbury, Avon Valley Reserve, Apr to Jun-00 (MDD); Apr-00 (MAS) (Darby, 2001)
SU15: Fittleton, 29-Jul-83 (DRN)
SU16: Jones's Mill, 21-Jun-97 (MDD)
SU21: Hamptworth Estate, 6-Apr-77, 24-Apr-84, 5-Aug-85 (DRN)
SU22: East Grimstead, 5-Apr-77, 11-Aug-86 (DRN)
ST93: Fonthill, 27-May-95 (MDD)
Resident. Widespread; in moss, weed refuse, grass tussocks, etc..

***Bryaxis curtisii*** (Leach, 1817)
8 ST93: Great Ridge Wood, 19-Apr-84 (DRN)
SU02: Knighton Wood, Mar, May, Oct-07, Jul-08 (AJWA)
SU03: Grovely Wood, 4-Aug-70, 3-Aug-71, 16-Apr-74 (DRN)
SU13: Salisbury, Avon Valley Reserve, Apr-00 (MDD) (Darby, 2001)
SU22: Blackmoor Copse, 4-May-02 (APF) (Foster, 2004; Langley Wood NNR, 6 to 27-Aug-04 (MDD) (Darby, 2005)
Resident. Local; in moss and litter in woodland.

***Bryaxis puncticollis*** (Denny, 1825)
8 SU21: Hamptworth Estate, 6-Apr-77, 18-Apr-84 (DRN)
Resident. Very local; in moss in damp situations.

**BYTHINUS Leach, 1817**
***Bythinus macropalpus*** Aubé, 1833
8 SU22: Langley Wood NNR, 25-Jun to16-Jul-04 (MDD) (Darby, 2005)
Resident. Very local; in a pitfall trap [in moss and rotten wood, sometimes in nests of small mammals].

Tribe TYCHINI

**TYCHUS Leach, 1817**
***Tychus niger*** (Paykull, 1800)
8 SU03: Wylye Down NNR, 10-Apr to 4-May-07 (MDD) (Darby, 2009)
SU13: Great Durnford, 17-Jul-73 (DRN)
Resident. Very local; one swept and one in a pitfall trap [in a wide variety of habitats]. Formerly one of the commonest Pselaphids in GB but possibly now declining.

Supertribe PSELAPHITAE
Tribe PSELAPHINI

**PSELAPHAULAX Reitter, 1909**
***Pselaphaulax dresdensis*** (Herbst, 1792)
**Notable**
7 SU16: Froxfield, c.1873 (Meyrick, 1873); Marlborough area, 1869-1938 (Meyrick, 1938)
SU36: Stype Wood, Jun-1870 (RH) (Hislop, 1871)
Resident. Very local; at roots of grass in damp places. No specimens in MCC.

**PSELAPHUS Herbst, 1792**
***Pselaphus heisei*** Herbst, 1792
8 SU21: Landford Bog, 2003 (APF) (Foster, 2003)
SU22: Langley Wood NNR, 22-Apr to 14-May-04 (MDD) (Darby, 2005)
Resident. Very local; in moss and tussocks in boggy ground.

## Subfamily PHLOEOCHARINAE

**PHLOEOCHARIS Mannerheim, 1830**
***Phloeocharis subtilissima*** Mannerheim, 1830
8 SU02: Knighton Wood, Apr, Nov-07 (AJWA)
Resident. Very local; in a flight intercept trap [under bark, in moss on trees, etc.].

## Subfamily TACHYPORINAE
Tribe MYCETOPORINI

**BOLITOBIUS Leach, 1819**
***Bolitobius castaneus*** (Stephens, 1832)
7 SU09: Upper Waterhay, 17-May to 4-Jun-05 (JKG det MDD)
8 SU01: Martin Down NNR, 1999 (EN, 1999)
SU02: Knighton Wood, May-08(AJWA)
SU03: Wylye Down NNR, 5-Jun to 10-Jul-07 (MDD) (Darby, 2009)
SU13: Salisbury, 7-Apr-74 (DRN)
Resident. Local; in grass tussocks, under stones, etc. in open situations.

***Bolitobius cingulatus*** Mannerheim, 1830
7 SU26: Savernake Forest, 15-Jun-73 (SAW); 27-May-93 (EN det SAW);
8 SU03: Wylye Down NNR, 10-Jul-07 (MDD) (Darby, 2008a)
SU22: Langley Wood NNR, 5-Aug-06 to 6-Jun-07 (MDD) (Darby, 2008)
Resident. Very local; in damp moss in woodland, so that the Wylye Down record is unusual.

**ISCHNOSOMA Stephens, 1829**
***Ischnosoma splendidum*** (Gravenhorst, 1806)
7 SU16: Marlborough area, 1869-1938 (Meyrick, 1938)
8 SU03: Wylye Down NNR, 27-Jan to 22-Feb-07 (MDD) (Darby, 2009)
Resident. Very local; amongst roots, etc. in grassland. Apparently a winter insect so possibly under recorded.

**LORDITHON Thomson, C.G, 1858**
***Lordithon exoletus*** (Erichson, 1839)
7 SU26: Savernake Forest, 16-Jun to 8-Jul-93 (EN det SAW)
8 SU02: Knighton Wood, Aug-07 (AJWA)
Resident. Very local; on fungi in woodland.

***Lordithon lunulatus*** (Linnaeus, 1761)
7 SU16: Marlborough area, 1869-1938 (Meyrick, 1938)
SU26: Savernake Forest, 25-Apr to 20-Sep-93, 1 to 19-Jul, 20-Sep to 11-Oct-94 (EN det SAW); 15-Jul to 3-Aug-95 (EN det MDD)
8 SU02: Knighton Wood, Sep-06, Apr, 6-Jul, Aug-07, May, Jul-08 (AJWA)
SU16: Jones's Mill, 19-Jun-01 (APF) (Foster, 2002)
Resident. Local; on fungi in wooded situations.

***Lordithon thoracicus*** (Fabricius, 1777)
7 SU16: Marlborough area, 1869-1938 (Meyrick, 1938)
SU26: Savernake Forest, 9-Oct-91 (AGD); 1 to 19-Jul-94 (EN det SAW)
8 SU23: Porton Down, 1970-2001 (Anon, 2001); 15-Oct-95 (MAS)
Resident. Very local; on fungi in woodland.

***Lordithon trinotatus*** (Erichson, 1839)
8 SU02: Knighton Wood, Nov-06 (AJWA)
SU03: Grovely Wood, 29-Oct-73 (DRN)
SU22: Langley Wood NNR, 6-Jun-02 (MDD)
Resident. Very local; on fungi on beech.

**MYCETOPORUS Mannerheim, 1830**
***Mycetoporus clavicornis*** (Stephens, 1832)
8 SU22: Langley Wood NNR, 3-Mar-05, 20-Feb to 24-Mar-07 (MDD) (Darby, 2005, 2008)
Resident. Very local; in litter in woodland.

***Mycetoporus lepidus*** (Gravenhorst, 1802)
7 SU16: Marlborough area, 1869-1938 (Meyrick, 1938)
SU26: Savernake Forest, 1 to 19-Jul-94 (EN det SAW)
Resident. Very local; in litter in woodland. One specimen without data in MCC.

***Mycetoporus longulus*** Mannerheim, 1830
7 SU16: Marlborough area, Jun-1873 (EWJ) (Meyrick, 1874); 1869-1938 (Meyrick, 1938)
Resident. Very local; [in woodland]. One specimen without data in MCC.

***Mycetoporus nigricollis*** Stephens, 1835
8 SU02: Knighton Wood, May-08 (AJWA)
SU03: Wylye Down, 27-Jan to 29-Mar-07 (MDD) (Darby, 2009)
Resident. Very local; in tussocks, etc. in grassland.

***Mycetoporus piceolus*** Rey,1883
**Notable, LCGR Fm Cu**
8 SU24: Dunch Hill (SPTAE), 3-Jun - 12-Aug-08 (Boyce & Wright, 2009)
Resident. Very local; [associated with dry, chalky and sandy places].

Mycetoporus punctus

***Mycetoporus punctus*** (Gravenhorst, 1806)
**Notable**
7 SU16: Marlborough area, Jun-1873 (EWJ) (Meyrick, 1874); 1869-1938 (Meyrick, 1938)
Resident. Very local; [in rotten stumps, dead leaves, reed litter and fungi in woodland and fens, also on cattle-grazed grassland]. One specimen without data in the MCC.

***Mycetoporus rufescens*** (Stephens, 1832)
8 SU22: Langley Wood NNR, 17-Nov-06 to 22-Jan-07 (MDD) (Darby, 2008)
Resident. Very local; in a pitfall trap [in moss in woodlands].

**PARABOLITOBIUS Li, Zhao & Sakai, 2000**
***Parabolitobius inclinans***(Gravenhorst, 1806)
8 SU22: Langley Wood NNR, 3-24-Mar-05 (MDD) (Darby, 2005)
Resident. Very local; in a pitfall trap [in damp moss and litter in woodland and grassland].

Tribe TACHYPORINI

**CILEA Jacquelin du Val, 1856**
***Cilea silphoides*** (Linnaeus, 1767)
8 ST91: Rushmore Park, 23-Sep-01 (AJWA)
Resident. Very local; in a dung heap.

**LAMPRINODES Luze, 1901**
***Lamprinodes saginatus*** (Gravenhorst, 1806)
8 SU01: Martin Down NNR, 1999 (EN, 1999)
Resident. Very local; in a pitfall trap. [At roots of grass and in moss in damp places, especially on heaths].

**SEPEDOPHILUS Gistel, 1856**

***Sepedophilus bipunctatus*** (Gravenhorst, 1802)
**Nb, SQS-8**
8 ST93: Great Ridge Wood, 19-Apr-84 (DRN)
Resident. Very local; in a fungoid log.

Sepedophilus bipunctatus

***Sepedophilus immaculatus*** (Stephens, 1832)
**LCGR-Fm Cu**
8 ST93: Great Ridge Wood, 18-Apr-74 (DRN)
Resident. Very local; in straw on heather area.

***Sepedophilus littoreus*** (Linnaeus, 1758)
**SQS-2**
7 SU16: Marlborough area, 1869-1938 (Meyrick, 1938)
SU26: Savernake, spring-1894 (GCC) (Jebb, 1895); 9-Jun to 1-Jul-94 (EN det SAW)
SU21: Hamptworth Common, 30-Mar-03 (MDD); Hamptworth Estate, 15-Apr-74 (DRN)
SU22: Blackmoor Copse, 17-Mar-02 (APF) (Foster, 2004)
Resident. Local; in litter and old straw bales.

***Sepedophilus lusitanicus*** Hammond, 1973
**SQS-2**
8 SU03: Wylye Down NNR, 10-Apr to 4-May-07 (MDD) (Darby, 2009)
Resident. Very local. Grassland appears to be an unusual habitat for this recently described species which is normally associated with coniferous woodland.

***Sepedophilus marshami*** (Stephens, 1832)
8 ST93: Great Ridge Wood, 10-May-00 (MDD det RT)
SU22: Blackmoor Copse, 17-Mar-02 (APF) (Foster, 2004)
Resident. Very local; in rotten wood and litter.

***Sepedophilus nigripennis*** (Stephens, 1832)
8 ST94: Bowls Barrow (SPTAW), 16-May-02 (PJH) (Hodge, 2002); Ladywell (SPTAW), 3-Jun – 12-Aug-08 (Boyce & Wright, 2009); Wadman's Coppice (SPTAW) 3-Jun - 12-Aug-08 (Boyce & Wright, 2009)
SU03: Wylye Down NNR, 2-Dec-06 to 29-Mar-07 (MDD) (Darby, 2008)
SU04: Salisbury Plain, 16-May-02 (PJH) (Hodge, 2002)
SU05: Ell Barrow (SPTAC), 3-Jun - 12-Aug-08 (Boyce & Wright, 2009)
SU14: Stonehenge, 22-Apr-03 (PJH) (Hodge, 2003)
SU24: Dunch Hill (SPTAE), 3-Jun - 12-Aug-08 (Boyce & Wright, 2009)
Resident. Local; in dry situations at roots of various plants in grassland.

**TACHINUS Gravenhorst, 1802**

***?Tachinus bipustulatus*** (Fabricius, 1792)
**RDB1, IEC-possible addition, SQS-32**
7 SU26: Savernake Forest, 1-Oct-45 (Holmes & Tennant, 1947)
Resident. Very local; [at exuding sap of *Cossus* infected trees and also under bark]. There must be some doubt about this determination. The species has not been recorded in GB since the 1930s and no specimen exists in the MCC.

***Tachinus humeralis*** Gravenhorst, 1802
7 SU16: Marlborough area, Jun-1873 (EWJ) (Meyrick, 1874); 1869-1938 (Meyrick, 1938)
SU26: Savernake Forest, 3-Apr-70 (DRN); 9-Oct-91 (AGD); 9-Jun to 19-Jul, 20-Sep to 11-Oct-94 (EN det SAW); 31-May to 23-Jun-95 (EN det MDD)
8 ST92: Sutton Mandeville, 15-Oct-95 (MDD)
SU02: Knighton Wood, Oct-06, Apr, Nov-07, May-08 (AJWA)
SU22: Bentley Wood, 31-May-97 (MDD); Langley Wood, 25-May-94 (WANHS det MDD); 15-Sep to 6-Oct-06 (MDD) (Darby, 2008)
SU26: Wilton Brail, 19-Jun-03 (MDD)
Resident. Widespread; in decaying vegetation and fungi in woodland.

***Tachinus laticollis*** Gravenhorst, 1802
7 SU16: Marlborough area, 22-Mar-1879 (CHH) (Meyrick, 1880); 1869-1938 (Meyrick, 1938)
SU26: Savernake Forest, 15-Jun-73 (SAW); 16-Jun to 28-Aug-93 (EN det SAW); 11 to 30-May-95 (EN det MDD)
8 ST92: Sutton Mandeville, Jun-94 (MDD det RT)
SU22: Blackmoor Copse, 4-18-May-02 (APF) (Foster, 2004)
Resident. Local; in decaying vegetation in woodland and a garden.

***Tachinus marginellus*** (Fabricius, 1781)
7 SU16: Marlborough area, 30 Mar 1879 (WFB) (Meyrick, 1880);1869-1938 (Meyrick, 1938)
8 ST91: Cranborne Chase, 30-Jun-99 (MDD)
SU13: Salisbury, Avon Valley Reserve, 13-Apr-00 (MDD) (Darby, 2001)
SU23: Porton Down, 1970-2001 (Anon, 2001)

Resident. Very local; in decaying vegetation in woodland and grassland.

***Tachinus pallipes*** (Gravenhorst, 1806)
8 ST93: Great Ridge Wood, 2-May-00 (MDD)
Resident. Very local; on fungi in woodland.

Tachinus rufipennis

***Tachinus proximus*** Kraatz, 1855
8 ST92: Sutton Mandeville, 15-Oct-95 (MDD)
SU22: Langley Wood NNR, 25-Apr to 6-Jun-07 (MDD) (Darby, 2008)
Resident. Very local; in a wine trap and pitfall traps in woodland.

***Tachinus rufipennis*** Gyllenhal, 1810
**RDB3**
7 SU26: Savernake Forest, 3-Apr-70 (DRN)
8 SU02: Knighton Wood, Apr-07 (AJWA)
Resident. Very local; in a flight intercept trap and [in dung, vegetable refuse and carrion].

***Tachinus rufipes*** (Linnaeus, 1758)
7 SU09: Ashton Keynes, 7-May-01 (JKG det MDD)
SU16: Marlborough area, 17-Feb-1879 (WFB) (Meyrick, 1880); 1869-1938 (Meyrick, 1938)
SU26: Savernake Forest, 16-Jun to 21-Sep-93 (EN det SAW)
8 ST92: Sutton Mandeville, 1-Apr-92 (MDD)
ST93: Great Ridge Wood, 18-Apr-74 (DRN)
SU03: Wylye Down NNR, 27-Jan to 22-Feb-07 (MDD) (Darby, 2009)
SU13: Salisbury, Avon Valley Reserve, 13-Apr-00 (MDD) (Darby, 2001)
SU16: Jones's Mill, Sep-97 (HK det MDD)
SU21: Hamptworth Estate, 15-Apr-74 (DRN)
SU22: Langley Wood NNR, 14-May to 27-Aug-04, 15-Sep-06 to 6-Jun-07 (MDD) (Darby, 2005, 2008)
SU26: Bedwyn Brail, 2-Jun-03 (MDD)
Resident. Widespread; in rotting vegetation including straw bales, under stone and in dung in woodland and grassland. Formerly *T.signatus.*

***Tachinus subterraneus*** (Linnaeus, 1758)
7 ST99: Crudwell, 19-Jun-73 (APF);
SU16: Marlborough area, Jun-1873 (EWJ) (Meyrick, 1874); 1869-1938 (Meyrick, 1938);
SU26: Savernake Forest, 3-Apr-70 (DRN); 14-Nov-89 (AGD)
8 ST92: Wardour, 13-Nov-95 (MDD); Sutton Mandeville, 21-Apr-92 (MDD)
Resident. Local; in litter, fungi, garden compost, etc.

**TACHYPORUS Gravenhorst, 1802**
***Tachyporus atriceps*** Stephens, 1832
7 SU16: Marlborough area, 1869-1938 (Meyrick, 1938)
8 ST94: Ladywell (SPTAW), 3-Jun - 12-Aug-08 (Boyce & Wright, 2009)
Resident. Very local; [in moss in grassland and woods]

***Tachyporus chrysomelinus*** (Linnaeus, 1758)
7 SU16: Marlborough area, Jun-1873 (EWJ) (Meyrick, 1874); 17-Feb-1879 (WFB) (Meyrick, 1880); 1869-1938 (Meyrick, 1938)
SU26: Savernake Forest, 20-Sep to 11-Oct-94 (EN det SAW)
8 SU02: Knighton Wood, Mar, Apr, Jun-07, May-08 (AJWA)
SU03: Wylye Down NNR, 10-Apr to 4-May-07 (MDD) (Darby, 2009)
SU13: Salisbury, Avon Valley Reserve, 2-May-00 (MDD) (Darby, 2001)
SU26: Bedwyn Brail, 2-Jun-03 (MDD)
Resident. Local; in litter in woodland and grassland. Shown by Booth, 1988, to comprise two sibling species: *dispar* and *chrysomelinus* so that records before then are indeterminate.

***Tachyporus dispar*** (Paykull, 1789)
7 SU26: Savernake Forest, 28-Apr to 18-May-94 (EN det SAW)
8 ST94: Wadman's Coppice (SPTAW) 3-Jun - 12-Aug-08 (Boyce & Wright, 2009)
SU01: Martin Down NNR, 1999 (EN, 1999)
SU02: Knighton Wood, Nov-07 (AJWA)
SU05: Ell Barrow (SPTAC), 3-Jun - 12-Aug-08 (Boyce & Wright, 2009)
SU14: Haxton (SPTAE), 3-Jun - 12-Aug-08 (Boyce & Wright, 2009); Wexland Hanging (SPTAE), 3-Jun - 12-Aug-08 (Boyce & Wright, 2009)
SU24: Beacon Hill (SPTAE), 3-Jun - 12-Aug-08 (Boyce & Wright, 2009); Bulford Ranges (SPTAE), 3-Jun - 12-Aug-08 (Boyce & Wright, 2009)
Resident. Very local; in litter and amongst roots in woodland and grassland. See *T. chrysomelinus.*

***Tachyporus hypnorum*** (Fabricius, 1775)

7 ST26: Flat Holm, 10-May-02 (JKG det MDD)
ST98: Rodbourne, 29-Mar-02 (JKG det MDD)
SU08: Somerford Common, 23-Apr-01 (JKG det MDD)
SU09: Ashton Keynes 26-Apr-01 (JKG det MDD); Blakehill Farm, 11-May-01 (JKG det MDD)
SU16: Marlborough area, Jun-1873 (EWJ) (Meyrick, 1874); 17-Feb-1879 (WFB) (Meyrick, 1880); 1869-1938 (Meyrick, 1938);
SU18: Swindon, 30-Mar-01 (JKG det MDD)
SU26: Savernake Forest, 25-Apr to 26-May-93, 28-Apr to 18-May, 1 to 19-Jul, 20-Sep to 11-Oct-94 (EN det SAW)
8 ST91: Cranborne Chase, 23-May-97 (MAS)
ST93: Great Ridge Wood, 18-Apr-74 (DRN)
ST94: Bowls Barrow (SPTAW), 16-May-02 (PJH) (Hodge, 2002); Norton Down (SPTAW), 23-Jun-02 (PJH) (Hodge, 2002)
SU02: Knighton Wood, Sep, Oct-06, Mar, Nov-07, (AJWA)
SU03: Wylye Down NNR, 2-Dec-06 to 3-Aug-07 (MDD) (Darby, 2009)
SU13: Boscombe Down, 9-Oct-94 (MDD); Old Sarum, 31-May-97 (MAS); Salisbury, Avon Valley Reserve, Apr-00 (MAS) (Darby, 2001); Salisbury, Bemerton Folly, 7-Mar - 15-Jun-02 (MDD) (Darby, 2002d)
SU14: Stonehenge, 22-Apr-03 (PJH) (Hodge, 2003)
SU16: Jones's Mill, 1997 (MAS)
SU22: Langley Wood NNR, 20-Feb to 24-Mar-05, 8-Dec-06 to 5-Jan-07 (MDD) (Darby, 2005, 2008); Bentley Wood, 27-May-97 (MDD)
SU23: Porton Down, 1970-2001 (Anon, 2001)
SU24: Bulford Ranges (SPTAE), 3-Jun - 12-Aug-08 (Boyce & Wright, 2009)

Resident. Widespread; on vegetation and at roots in grassland and woodland.

***Tachyporus nitidulus*** (Fabricius, 1781)

7 ST97: North Draycot Park, 22-May-00 (MDD)
ST99: Crudwell, 19-Jun-73 (APF)
SU16: Marlborough area, Jun-1873 (EWJ) (Meyrick, 1874); 1869-1938 (Meyrick, 1938)
SU26: Savernake Forest, 16-Jun to 7-Jul-93, 30-Aug to 11-Oct-94 (EN det SAW); 4 to 24-Aug, 27-Oct to 16-Nov-95 (EN det MDD)
8 ST91: Cranborne Chase, 30-Jun-99 (MDD)
ST92: Sutton Mandeville, 20-Apr-92 (MDD)
ST93: Great Ridge Wood, 16-May-00 (MDD)
SU01: Martin Down NNR, 1979-1992, 21-Aug-89 (RGB)
SU02: Knighton Wood, Sep-06, Mar, Apr-07 (AJWA)
SU03: Wylye Down NNR, 22-Feb to 3-Aug-07 (MDD) (Darby, 2009); Grovely Wood, 11-Aug-71 (DRN)
SU13: Boscombe Down, 9-Oct-94 (MDD); Salisbury, Bemerton Folly, 7-Mar - 15-Jun-02 (MDD) (Darby, 2002d)
SU14: Stonehenge, 22-Apr-03 (PJH) (Hodge, 2003)
SU22: Langley Wood NNR, 14-May-04 to 24-Mar-05 (MDD) (Darby, 2005)

Resident. Widespread; in litter, amongst roots and in moss in woodland and grassland.

***Tachyporus obtusus*** (Linnaeus, 1767)

7 SU16: Marlborough area, Jun-1873 (EWJ) (Meyrick, 1874); 17-Feb-1879 (WFB) (Meyrick, 1880); 1869-1938 (Meyrick, 1938)
8 ST91: Cranborne Chase, 30-Jun-99 (MDD)
ST92: Wardour, 5-Jun-96 (MAS)
SU13: Salisbury, Avon Valley Reserve, Apr-00 (MAS) (Darby, 2001)
SU14: Stonehenge, 22-Apr-03 (PJH) (Hodge, 2003)
SU15: Beach's Plantation (SPTAE), 1-May-05 (MGT)
SU22: Blackmoor Copse, 17-Mar-02 (APF) (Foster, 2004)

Resident. Local; on vegetation in woodland.

***Tachyporus pallidus*** Sharp, 1871

7 SU26: Savernake Forest, 11 to 30-May-95 (EN det MDD)
8 SU13: Salisbury, Avon Valley Reserve, Apr-00 (MAS) (Darby, 2001)
SU16: Jones's Mill, 21-Jun-97 (MDD); 1997 (MAS)
SU22: Bentley Wood, 31-May-96 (MDD)
SU23: Porton Down, 1970-2001 (Anon, 2001)

Resident. Local; in damp situations amongst litter, etc.

***Tachyporus pusillus*** Gravenhorst, 1806

7 SU16: Marlborough area, 15-Feb-1879 (TNH-S) (Meyrick, 1880); 1869-1938 (Meyrick, 1938)

Resident. Very local; [in woodland and grassland amongst litter, moss, etc.]. Two specimens without data in MCC.

***Tachyporus solutus*** Erichson, 1839

7 SU16: Marlborough area 1869-1938 (Meyrick, 1938)
SU26: Savernake Forest, 16-Jun-93 (EN det SAW)
8 SU13: Salisbury, Avon Valley Reserve, Apr-00 (MAS) (Darby, 2001)
SU22: Whiteparish Common, 2-Jul-95 (MAS)

Resident. Local; in woodland, grassland and wetland.

***Tachyporus tersus*** Erichson, 1839

7 SU26: Mildenhall, 1939 (det KGB) (Knight, 1940)

Resident. Very local; [in moss]. Single specimen in MCC under this name without data.

***Tachyporus transversalis*** Gravenhorst, 1806

7 SU16: Marlborough area, 1869-1938 (Meyrick, 1938)
SU27: Ramsbury, spring-1894 (GCC) (Jebb, 1895)
Resident. Very local; [in marshy places in moss]. Single specimen in MCC without data.

## Subfamily HABROCERINAE

**HABROCERUS Erichson, 1839**

***Habrocerus capillaricornis*** (Gravenhorst, 1806)

7 SU16: Marlborough area, 1869-1938 (Meyrick, 1938)
SU26: Savernake Forest, spring-1894 (GCC) (Jebb, 1895); 9-Oct-91 (AGD); 28-Apr to 18-May-94 (EN det SAW); 23-Oct-01 (MDD)
8 ST93: Great Ridge Wood, 2-May-00 (MDD)
SU02: Knighton Wood, Jul-08 (AJWA)
SU13: Salisbury, Bemerton Folly, 7-Mar - 15-Jun-02 (MDD) (Darby, 2002d)
SU16: Jones's Mill, 21-Jun-97 (MDD)
SU21: Hamptworth Estate, 1-Aug-79 (DRN)
SU22: Langley Wood NNR, 3 to 25-Jun-04 (MDD) (Darby, 2005)
SU26: Bedwyn Brail, 2, 19-Jun-03, 5-Nov-03 (MDD); Wilton Brail, 19-Jun-03 (MDD)
Resident. Local; in litter in woodland.

## Subfamily ALEOCHARINAE
### Tribe ALEOCHARINI

**ALEOCHARA Gravenhorst, 1802**

ALEOCHARA Gravenhorst, 1802

***Aleochara curtula*** (Goeze, 1777)

7 SU16: Marlborough area, Jun-1873 (EWJ) (Meyrick, 1874); 1869-1938 (Meyrick, 1938)
SU26: Savernake Forest, 25-Apr to 11-May-95 (EN)
8 SU03: Wylye Down NNR, 3-Aug-07 (MDD) (Darby, 2008a)
SU22: Langley Wood NNR, 16-Jul to 27-Aug-04, 5-Aug to 15-Sep-06 (MDD) (Darby, 2005, 2008)
SU23: Porton Down, 3-Jul-99 (PJH) (Hodge, 2002)
Resident. Local; in carrion. Probably widespread.

***Aleochara lata*** Gravenhorst, 1802

7 SU16: Marlborough area, 1869-1938 (Meyrick, 1938)
8 ST92: Sutton Mandeville, Jul-02 (MDD)
SU03: Wylye Down NNR, 4-Sep-07 (MDD) (Darby, 2009)
Resident. Very local; in carrion.

BARYODMA Thomson, C.G, 1858

***Aleochara intricata*** Mannerheim, 1830

8 SU23: Porton Down, 3-Jul-99 (BL)
Resident. Very local; [in horse dung particularly]

COPROCHARA Mulsant and Rey, 1874

***Aleochara bilineata*** Gyllenhal, 1810

8 SU13: Salisbury, Avon Valley Reserve, 13-Apr-00 (MDD) (Darby, 2001)
Resident. Very local; in rotting vegetation.

***Aleochara bipustulata*** (Linnaeus, 1761)

8 ST91: Rushmore Park, 23-Sep-01 (AJWA)
Resident. Very local: in a dung heap.

XENOCHARA Mulsant and Rey, 1874

***Aleochara cuniculorum*** Kraatz, 1858

8 SU03: Wylye Down NNR, 4-May to 5-Jun-07 (MDD) (Darby, 2009)
Resident. Very local; in pitfall traps [in rabbit burrows but also from nests and burrows of other mammals and birds] in this case probably mole nests.

***Aleochara lanuginosa*** Gravenhorst, 1802

7 SU16: Marlborough area, 1869-1938 (Meyrick, 1938);
SU18: Swindon: R.mead, 7-May-01 (JKG det MDD)
SU26: Savernake Forest, 1-Jul to 19-Aug-94 (EN det SAW); 15-Jul to 3-Aug-95 (EN det MDD)
8 SU03: Wylye Down NNR, 2-Dec-06 (MDD) (Darby, 2008a)
SU04: Orcheston Down (SPTA), 22-Jun-02 (PJH) (Hodge, 2002)
SU16: Jones's Mill, 21-Jun-97 (MDD)
Resident. Local; in cow dung.

***Aleochara sparsa*** Heer, 1839

8 SU02: Knighton Wood, Nov-06, Mar, May, Sep – Nov-07 (AJWA)
SU13: Salisbury, 13-Apr-00 (MDD) (Darby, 2001)
SU22: Blackmoor Copse, 3-Aug-03 (APF) (Foster, 2004
Resident. Very local; [in decaying vegetable matter and carrion].

**TINOTUS Sharp, 1883**

***Tinotus morion*** (Gravenhorst, 1802)

7 SU26: Savernake Forest, 23-Oct-01 (MDD)
8 SU13: Salisbury, Avon Valley Reserve, 13-Apr-00 (MDD) (Darby, 2001)
Resident. Very local; in dung.

### Tribe ATHETINI
#### Subtribe ATHETINA

**ACROTONA Thomson, C.G, 1859**

***Acrotona parvula*** (Mannerheim, 1830)

8 ST93: Great Ridge Wood, 2-May-00 (MDD det RT)

Resident. Very local; in leaf litter.

**ALAOBIA Thomson, C.G. 1858**
***Alaobia gagatina*** (Baudi, 1848)
8 SU02: Knighton Wood, Apr-07 (AJWA)
SU22: Langley Wood NNR, 6 to 27-Aug-04 (MDD) (Darby, 2005)
Resident. Very local; in litter, etc..

***Alaobia scapularis*** (Sahlberg, C.R, 1831)
**Notable, LCGR-FI Cu**
7 ST87: Rack Hill, 1995 (MA det MDD)
8 SU02: Knighton Wood, 6-Jul-07 (AJWA)
Resident. Very local; [in moss and fungi on chalk downland and in a quarry]

***Alaobia sodalis*** (Erichson, 1837)
8 SU13: Salisbury, Bemerton Folly, 7-Mar - 15-Jun-02 (MDD) (Darby, 2002d)
Resident. Very local; in woodland litter.

***Alaobia taxiceroides*** Munster, 1932
8 SU02: Knighton Wood, Sep, Nov-06 (AJWA)
Resident. Very local; in a flight intercept trap. [In nests of small mammals] A recent (1971) addition to the British list when it was only known from the New Forest.

***Alaobia trinotata*** (Kraaatz, 1856)
8 SU02: Knighton Wood, Mar, Apr-07 (AJWA)
Resident. Very local; in a flight intercept trap.

**ALOCONOTA Thomson, C.G, 1858**
ALOCONOTA Thomson, C.G., 1858
***Aloconota gregaria*** (Erichson, 1839)
7 SU26: Savernake Forest, 8 to 28-Jul-93, 1 to 19-Jul-94 (EN det SAW)
8 ST94: Ladywell (SPTAW), 3-Jun - 12-Aug-08 (Boyce & Wright, 2009)
Resident. Very local; in malaise and flight intercept traps. [In litter and grass tussocks, often near fresh water or in flood litter]

Aloconota longicollis

***Aloconota longicollis*** (Mulsand & Rey, 1852)
**Notable**
8 ST94: Ladywell (SPTAW), 3-Jun - 12-Aug-08 (Boyce & Wright, 2009)
Resident. Very local, ground search around edge of pond.

**AMISCHA Thomson, C.G, 1858**
***Amischa analis*** (Gravenhorst, 1802)
7 SU16: Marlborough area, Jun-1873 (EWJ) (Meyrick, 1874); 1869-1938 (Meyrick, 1938)
SU26: Savernake Forest, 27-May-93 (EN), 23-Oct-01 (MDD)
8 SU01: Martin Down NNR, 21-Aug-89 (RGB)
SU02: Knighton Wood, Oct-06, May-07 (AJWA)
SU14: Haxton (SPTAE), 3-Jun - 12-Aug-08 (Boyce & Wright, 2009); Wexland Hanging (SPTAE), 3-Jun - 12-Aug-08 (Boyce & Wright, 2009)
SU21: Hamptworth Common, 24-Mar-03 (MDD)
Resident. Widespread and frequent, in grassland and woodland.

***Amischa decipiens*** (Sharp, 1869)
8 SU01: Martin Down NNR, 1999 (EN, 1999)
Resident. Very local; in pitfall traps. [In marshy places; in moss; in rotten wood]

**ANOPLETA Mulsant and Rey, 1874**
***Anopleta corvina*** (Thomson, C.G, 1856)
8 ST92: Sutton Mandeville, 15-Oct-95 (MDD det RGB); Wardour, 13-Nov-95 (MDD det RGB)
SU13: Salisbury, 29-Apr to 25-May-00 (MAS det RGB, RT)
Resident. Very local; in decaying fungi and vegetable refuse.

**ATHETA Thomson, C.G, 1858**
***Atheta aeneicollis*** (Sharp, 1869)
8 SU02: Knighton Wood, Apr-07 (AJWA)
Resident., Very local; in a flight intercept trap.

***Atheta aquatica*** (Thomson, C.G, 1852)
8 SU02: Knighton Wood, Sep-06 (AJWA)
Resident. Very local; in a flight intercept trap. [In grass heaps and tussocks; marshy places at roots of grass, in moss, etc.]

***Atheta brunneipennis*** (Thomson, C.G, 1852)
7 SU26: Savernake Forest, 15-Jun-73 (SAW), 20-Apr to 7-Jul-93, 20-Sep to 11-Oct-94 (EN det SAW)
8 SU22: Langley Wood NNR, 15-Sep to 17-Nov-06 (MDD) (Darby, 2008)
Resident. Very local; in litter and vegetable refuse.

***Atheta castanoptera*** (Mannerheim, 1830)
8 ST91: Cranborne Chase, 22-Jun-99 (MDD det RT)
SU02: Knighton Wood, Sep-06, Apr, Sep, Oct-07

(AJWA)
SU13: Salisbury, Avon Valley Reserve, Apr-00 (MAS) (Darby, 2001)
SU16: Jones's Mill, Sep-97 (HK det MDD)
Resident. Local; in deacying vegetable refuse and fungi.

***Atheta graminicola*** (Gravenhorst, 1806)
7 SU16: Marlborough area, Jun-1873 (EWJ) (Meyrick, 1874)
SU26: Savernake Forest, 23-Oct-01 (MDD)
SU27: Ramsbury, spring-1894 (GCC) (Jebb, 1895)
8 ST93: Great Ridge Wood, 28-Apr-00 (MDD det RT)
SU22: Langley Wood NNR, 25-Apr to 6-Jun-07 (MDD)(Darby, 2008)
Resident. Local; in damp litter.

***Atheta hypnorum*** (Kiesenwetter, 1850)
7 SU26: Savernake Forest, 23-Oct-01 (MDD)
Resident. Very local; in leaf litter.

***Atheta xanthopus*** (Thomson, C.G, 1856)
8 SU13: Salisbury, Avon Valley Reserve, Apr to 12-May-00 (MAS); 19-Jun-00 (MDD) (Darby, 2001)
SU23: Porton Down, 1970-2001 (Anon, 2001)
Resident. Very local; in litter in damp situations.

**Species provisionally assigned to *Atheta***

***Atheta britanniae*** Bernhauer & Scheerpeltz, 1926
7 SU26: Savernake Forest, 20 to 25-Apr-93 (EN det SAW)
Resident. Very local; in a flight intecept trap.

***Atheta crassicornis*** (Fabricius, 1792)
7 SU26: Savernake Forest, 21-Sep to 11-Oct-93, 28-Apr to 18-May, 19-Jul to 9-Aug-94 (EN det SAW)
8 SU02: Knighton Wood, Apr, Nov-07 (AJWA)
SU21: Hamptworth Common, 24-Mar-03 (MDD)
SU22: Langley Wood NNR, 25-Apr to 6-Jun-07 (MDD) (Darby, 2008)
Resident. Local; in damp litter.

***Atheta fungicola*** (Thomson, C.G, 1852)
7 SU26: Savernake Forest, 4-Aug-95 (EN det MDD)
8 SU03: Wylye Down, 3-Aug-07 (MDD) (Darby, 2008a)
SU22: Langley Wood NNR, 25-May-94 (WANHS det MDD), 16-Jul to 6-Aug-04, 15-Jul to 5-Aug-06 (MDD) (Darby, 2005, 2008)
Resident. Very local; in woodland litter.

***Atheta harwoodi*** Williams, 1930
8 SU02: Knighton Wood, 7-Oct-06 (AJWA)
Resident. Very local; in a flight intercept trap.

***Atheta liturata*** (Stephens, 1832)
8 SU22: Langley Wood NNR, 5-Aug-85 (DRN det CJ)
SU16: Jones's Mill, Sep-97 (HK det MDD)
Resident. Very local; in fungi.

***Atheta nidicola*** (Johansen, 1914)
8 SU03: Grovely Wood, 12-Apr-72 (DRN det AAA)
Resident. Very local; in a grey squirrel's drey.

***Atheta ravilla*** (Erichson, 1839)
8 ST92: Wardour, 13-Nov-95 (MDD det RGB)
Resident. Very local; in woodland litter.

***Atheta vaga*** (Heer, 1839)
7 SU26: Savernake Forest, 29, 30-May-76 (MacKechnie-Jarvis, 1976)
8 SU02: Knighton Wood, Oct-06, Mar, Apr-07 (AJWA)
Resident. Very local; in woodland.

**COPROTHASSA Thomson, C.G, 1859**
***Coprothassa melanaria*** (Mannerheim, 1830)
8 SU02: Knighton Wood, Sep-07 (AJWA)
Resident. Very local; in a flight intercept trap. [in decaying vegetable matter and dung].

**DALOTIA Casey, 1910**
***Dalotia coriaria*** (Kraatz, 1856)
8 ST92: Sutton Mandeville, 25-Jun-07 (MDD)
Resident. Very local; in garden compost.

**DATOMICRA Mulsant and Rey, 1874**
***Datomicra celata*** (Erichson, 1837)
8 ST92: Sutton Mandeville, 6-Jul-07 (MDD)
SU13: Salisbury, Bemerton Folly, 7-Mar - 15-Jun-02 (MDD) (Darby, 2002d)
Resident. Very local; in garden compost.

***Datomicra sordidula*** (Erichson, 1837)
8 SU23: Bentley Wood, 3-Sep-97 (MDD det RT)
Resident. Very local; in leaf litter.

**DIMETROTA Mulsant & Rey, 1873**
***Dimetrota atramentaria*** (Gyllenhal, 1810)
8 SU02: Knighton Wood, Sep-06 (AJWA)
Resident. Very local; in a flight intercept trap.

***Dimetrota marcida*** (Erichson, 1837)
8 SU02: Knighton Wood, Nov-07 (AJWA)
Resident. Very local: in a flight intercept trap.

**DINARAEA Thomson, C.G, 1858**
***Dinaraea aequata*** (Erichson, 1837)
**SQS-1**
7 SU16: Marlborough area, 1869-1938 (Meyrick,

1938)
SU26: Savernake, spring-1894 (GCC) (Jebb, 1895)
8 ST93: Great Ridge Wood, 4-May-00 (MDD det RT); 29-Oct-75 (DRN)
SU03: Grovely Wood, 16-Apr-74 (DRN)
SU13: Little Durnford Down, 13-Sep-99 (MDD)
SU22: Langley Wood NNR, 12-Apr-74 (DRN)
Resident. Local; under bark.

***Dinaraea linearis*** (Gravenhorst, 1802)
**SQS-2**
7 SU16: Marlborough area, 1869-1938 (Meyrick, 1938)
SU26: Savernake, spring-1894 (GCC) (Jebb, 1895)
8 ST91: Cranborne Chase, 21-May-02 (MDD)
SU22: Langley Wood NNR, 6-Jun-02 (MDD)
Resident. Local; under bark.

**GEOSTIBA Thomson, C.G, 1858**
***Geostiba circellaris*** (Gravenhorst, 1806)
7 SU16: Marlborough area, 1869-1938 (Meyrick, 1938)
8 ST92: Wardour, 29-Oct-95 (MDD det RT)
SU13: Salisbury, Avon Valley Reserve, Apr-00 (MAS); 7, 12-Apr-00 (MDD) (Darby, 2001)
SU22: Langley Wood NNR, 29-May-03 (MDD)
SU26: Bedwyn Brail, 2-Jun-03 (MDD)
Resident. Local; in woodland and grassland in litter and at roots of plants.

**LIOGLUTA Thomson, C.G, 1858**
***Liogluta longiuscula*** (Gravenhorst, 1802)
7 SU16: Marlborough area, 1869-1938 (Meyrick, 1938)
SU27: Ramsbury, spring-1894 (GCC) (Jebb, 1895)
8 SU02: Knighton Wood, Jul-08 (AJWA)
Resident. Very local; in a flight intercept trap and [in damp decaying vegetation]. Two specimens in MCC one labelled *Marlborough 19/10/02.*

Microdota aegra

**MICRODOTA Mulsant and Rey, 1873**
***Microdota aegra*** (Heer, 1841)
**Notable**
8 SU02: Knighton Wood, Nov-06 (AJWA)
Resident. Very local; in a flight intercept trap [in moss, and amongst leaf litter] possibly subterranean.

***Microdota amicula*** (Stephens, 1832)
8 ST92: Sutton Mandeville, 19-Oct-95 (MDD det RGB)
Resident. Very local; in fungus in garden.

**MOCYTA Mulsant and Rey, 1874**
***Mocyta amplicollis*** (Mulsant & Rey, 1873)
7 SU08: Somerford Common, 27-May-01 (JG det RT)
8 SU04: Parsonage Down NNR, 22-Jul-08 (MDD)
Resident. Very local; on vegetation. A recent addition (1965) to the British list.

***Mocyta clientula*** (Erichson, 1839)
8 ST93: Great Ridge Wood, 25-May-00 (MDD det RGB)
Resident. Very local; in damp litter.

***Mocyta fungi*** (Gravenhorst, 1806)
7 SU16: Marlborough area, 17-Feb-1879 (TNH-S) (Meyrick, 1880)
SU26: Savernake Forest, 28-Apr to 18-May, 9-Jun to 9-Aug, 20-Sep to 11-Oct-94 (EN det SAW); 2001 (MDD)
8 ST93: Great Ridge Wood, 7-May-00 (MDD det RT)
ST94: Ladywell (SPTAW), 3-Jun - 12-Aug-08 (Boyce & Wright, 2009)
SU01: Martin Down NNR, 1999 (EN, 1999)
SU02: Knighton Wood, Sep-06 (AJWA)
SU03: Grovely Wood, 12-Jun-00 (MDD det RGB)
SU13: Salisbury, 29-Apr to 25-May-00 (MAS det RT, RGB) (Darby, 2001); Salisbury, Bemerton Folly, 7-Mar - 15-Jun-02 (MDD) (Darby, 2002d)
SU22: Langley Wood NNR, 5-Aug-06 to 6-Jun-07 (MDD) (Darby, 2008)
SU23: Bentley Wood, 3-Sep-97 (MDD det RGB)
Resident. Widespread and abundant in decaying vegetation and litter. *M. fungi* is regarded as a species complex so that these records may refer to more than one species.

**MYCETOTA Ádám, 1987**
***Mycetota laticollis*** (Stephens, 1832)
8 SU13: Salisbury, 28-Apr-00 (MDD det RT) (Darby, 2001)
Resident. Very local; in litter.

**NEHEMITROPIA Lohse, 1971**
***Nehemitropia lividipennis*** Mannerheim, 1830
7 SU16: Marlborough area, 1869-1938 (Meyrick, 1938)

8 ST91: Rushmore Park, 23-Sep-01 (AJWA)
Resident. Very local; in a dung heap and [in grass heaps].

**NOTOTHECTA Thomson, C.G, 1858**
***Notothecta flavipes*** (Gravenhorst, 1806)
7 SU16: Marlborough area, Jun-1873 (EWJ) (Meyrick, 1874); 1869-1938 (Meyrick, 1938)
Resident. Very local; [in nests of the ant *Formica rufa*].

**PHILHYGRA Mulsant and Rey, 1873**
***Philhygra luridipennis*** (Mannerheim, 1830)
7 SU16: Marlborough area, 1873-1895 (Anon) (Jebb, 1895)
Resident. Very local; [in shingle on the banks of rivers].

**PLATARAEA Thomson, C.G, 1858**
***Plataraea brunnea*** (Fabricius, 1798)
7 SU16: Marlborough area, 1869-1938 (Meyrick, 1938)
8 ST94: Ladywell (SPTAW), 15-May-02 (PJH) (Hodge, 2002)
SU03: Grovely Wood, 11-Aug-71 (DRN)
Resident. Very local; on low vegetation [in moss and grass tussocks].

**TRICHIUSA Casey, 1893**
***Trichiusa immigrata*** Lohse, 1984
8 ST91: Rushmore Park, 23-Sep-01 (AJWA)
Resident. Very local; in a dung heap.

Subtribe THAMIARAEINA

**THAMIARAEA Thomson, C.G, 1858**
***Thamiaraea cinnamomea*** (Gravenhorst, 1802)
7 SU16: Marlborough area, Jun-1873 (EWJ) (Meyrick, 1874); 1869-1938 (Meyrick, 1938)
8 SU02: Knighton Wood, Sep-07 (AJWA)
Resident. Very local; in a flight intercept trap and [at sap of trees infected by *Cossus*; and in decaying vegetation].

***Thamiaraea hospita*** (Märkel, 1844)
**Notable**
8 SU02: Knighton Wood, Sep-06, Apr-07 (AJWA)
Resident. Very local; in a flight intercept trap [at sap exuding from old oak also from larval burrows of *Cossus*].

Tribe AUTALIINI

**AUTALIA Leach, 1819**
***Autalia impressa*** (Olivier, 1795)
7 SU26: Savernake Forest, 9-Oct-91 (AGD)
8 ST92: Sutton Mandeville, 15, 19-Oct-95 (MDD); Wardour, 13-Nov-95 (MDD)
SU02: Knighton Wood, Oct-07 (AJWA)
SU23: Porton Down, 1970-2001 (Anon, 2001); Thorny Down Woods, 16-Oct-00 (MDD)
Resident. Local; on decaying fungi.

***Autalia longicornis*** Scheerpeltz, 1947
7 SU26: Savernake Forest, 14-Nov-89 (AGD)
8 SU02: Knighton Wood, Oct, Nov-07 (AJWA)
SU03: Grovely Wood, 29-Oct-73 (DRN)
SU13: Salisbury, Apr-00 (MDD) (Darby, 2001)
Resident. Very local; on fungus.

***Autalia rivularis*** (Gravenhorst, 1802)
8 ST92: Sutton Mandeville, 1-Jun-93 (MDD)
ST93: Great Ridge Wood, 2, 16-May-00 (MDD)
SU02: Middleton Down, 4-May-95 (MDD)
Resident. Very local; in litter and decaying vegetation.

Tribe DEINOPSINI

**DEINOPSIS Matthews, A.H, 1838**
***Deinopsis erosa*** (Stephens, 1832)
8 SU16: Jones's Mill, 11, 21-Jun-97 (MDD)
Resident. Very local; in litter in marshy situations.

Tribe FALAGRIINI

**CORDALIA Jacobs, 1925**
***Cordalia obscura*** (Gravenhorst, 1802)
8 SU12: Coombe Bissett Down, 1-Apr-97 (MDD)
SU13: Salisbury, Avon Valley Reserve, Apr to 12-May-00 (MAS); 13-Apr, 2-May, 1-Jun-00 (MDD) (Darby, 2001)
Resident. Very local; in decaying vegetable matter.

Tribe HOMALOTINI
Subtribe BOLITOCHARINA

**BOLITOCHARA Mannerheim, 1830**
***Bolitochara bella*** Märkel, 1844
7 SU16: Marlborough area, Jun-1873 (EWJ) (Meyrick, 1874); 1869-1938 (Meyrick, 1938)
SU26: Savernake Forest, spring-1894 (GCC) (Jebb, 1895); 6-Jan-73 (AAA)
8 SU02: Knighton Wood, May-08 (AJWA)
SU03: Grovely Wood, 11-Aug-70 (DRN)
Resident. Very local; in fungi.

***Bolitochara lucida*** (Gravenhorst, 1802)
**SQS-2**
8 SU02: Knighton Wood, May, Nov-07, Sep-08 (AJWA)
SU03: Grovely Wood, 4-Aug-70 (DRN)
Resident. Very local; in bracket fungus.

***Bolitochara obliqua*** Erichson, 1837

7 SU16: West Woods, 1-May-02 (MDD)
SU03: Grovely Wood, 13-Aug-70 (DRN)
Resident. Very local; in fungi.

***Bolitochara pulchra*** (Gravenhorst, 1806)
**Notable, SQS-8**
8 ST91: Cranborne Chase, 30-Jun-99 (MDD)
ST92: Wardour, 18-Jun-02 (MDD)
Resident. Very local; in fungi.

Bolitochara pulchra

**LEPTUSA Kraatz, 1856**
***Leptusa fumida*** Kraatz, 1856
**SQS-1**
7 ST99: Flisteridge Wood, 2-May-82 (KNAA det PM)
SU26: Savernake, 1 to 19-Jul-94 (EN det SAW)
8 ST84: Longleat, 7-Jul-99 (MDD)
ST92: Wardour, 18-Jun-02 (MDD); Sutton Mandeville, 1-Apr-98 (MDD det RGB)
ST93: Fonthill, 26-Jun-02 (MDD)
SU02: Knighton Wood, Oct-06 (AJWA)
SU13: Salisbury, Avon Valley Reserve, 24-May-00 (Darby, 2001); Salisbury, Bemerton Folly, 7-Mar - 15-Jun-02 (MDD) (Darby, 2002d)
SU22: Bentley Wood, 10-Oct-01; 5-May-03 (MDD); Langley Wood NNR, 12-Apr-74 (DRN)
SU26: Bedwyn Brail, 5-Nov-03 (MDD)
Resident. Widespread; under bark and on fungi in woodland.

***Leptusa pulchella*** (Mannerheim, 1830)
7 SU26: Savernake Forest, 6-Dec-73, 22-Jun-90 (APFw); 1 to 19-Jul-94 (EN det SAW)
8 ST93: Great Ridge Wood, 18-Apr-74, 29-Jul-85 (DRN); 4-Oct-04 (MDD)
Resident. Very local; under bark in woodland.

***Leptusa ruficollis*** (Erichson, 1839)
7 SU16: Marlborough area, 6-Feb-1879 (WFB) (Meyrick, 1880); 1869-1938 (Meyrick, 1938)
8 SU02: Knighton Wood, Sep-06, Sep, Oct-07 (AJWA)
SU13: Salisbury, Avon Valley Reserve, 24-May-00 (MDD) (Darby, 2001)
Resident. Very local; under bark in woodland.

Subtribe GYROPHAENINA

**AGARICOCHARA Kraatz, 1856**
***Agaricochara latissima*** (Stephens, 1832)
8 SU02: Knighton Wood, Sep-06, Nov-07 (AJWA)
SU03: Steeple Langford, 13-Jan-04 (MDD)
SU26: Bedwyn Brail, 20-Sep-04 (MDD)
Resident. Very local; on fungi in woodland.

**ENCEPHALUS Stephens, 1832**
***Encephalus complicans*** Stephens, 1832
7 SU16: Marlborough area, Jun-1873 (EWJ) Meyrick, 1874); 1869-1938 (Meyrick, 1938);
8 SU03: Grovely Wood, 27-May-74 (DRN)
SU16: Jones's Mill, 21-Jun-97 (MDD)
Resident. Very local; in moss, flood refuse.

**GYROPHAENA Mannerheim, 1830**
***Gyrophaena affinis*** Mannerheim, 1830
8 SU03: Grovely Wood, 21-Aug-01 (MDD)
Resident. Very local; in fungus.

***Gyrophaena fasciata*** (Marsham, 1802)
8 ST82: Oysters Coppice, 6-Oct-02 (MDD)
Resident. Very local; under bark.

Gyrophaena manca

***Gyrophaena manca*** Erichson, 1839
**Notable**
8 SU16: Jones's Mill, 21-Jun-97 (MDD)
Resident. Very local; in fungus. Formerly *G. angustata*.

Subtribe HOMALOTINA

**ANOMOGNATHUS Solier, 1849**
***Anomognathus cuspidatus*** (Erichson, 1839)
**SQS-2**
7 SU26: Savernake Forest, 22-Jun-90 (APFw)
8 ST91: Cranborne Chase, 21-May-02 (MDD)
ST93: Fonthill, 26-Jun-02 (MDD)
SU03: Grovely Wood, 16-Apr-74 (DRN)
SU13: Little Durnford Down, 20-May-03 (MDD)
SU26: Bedwyn Brail, 2-Jun, 5-Nov-03 (MDD)
Resident. Local; under bark, usually beech.

**HOMALOTA Mannerheim, 1830**
***Homalota plana*** (Gyllenhal, 1810)
**SQS-2**
8 SU13: Little Durnford Down, 20-May-03 (MDD)
SU26: Bedwyn Brail, 5-Nov-03 (MDD)
Resident. Very local; under bark.

Tribe HYGRONOMINI

**HYGRONOMA Erichson, 1837**
***Hygronoma dimidiata*** (Gravenhorst, 1806)
7 SU16: Marlborough area, 1869-1938 (Meyrick, 1938)
SU27: Ramsbury, spring-1894 (GCC) (Jebb, 1895)
8 SU13: Great Durnford, 18-Aug-72 (DRN)
SU23: Porton Down, 3-Jul-99 (PJH) (Hodge, 2002)
Resident. Local; at roots and in tussocks on grassland.

Tribe HYPOCYPHTINI

**CYPHA Leach, 1819**
***Cypha longicornis*** (Paykull, 1800)
7 SU26: Savernake Forest, 23-Oct-01 (MDD)
8 ST92: Sutton Mandeville, 1-Jun-93 (MDD)
Fovant, 25-Apr-95 (MDD)
SU02: Knighton Wood, Sep-06, Mar-07 (AJWA)
SU03: Grovely Wood, 21-Aug-01 (MDD); Little Langford, 3-Jul-96 (MAS); Grovely Wood, 11-Aug-71 (DRN)
SU13: Boscombe Down, 8, 9-Oct-94 (MDD); Salisbury, Avon Valley Reserve, 13-Apr-00 (MDD) (Darby, 2001); Salisbury, Bemerton Folly, 7-Mar - 15-Jun-02 (MDD) (Darby, 2002) Stratford sub Castle, 7 Aug 71 (DRN)
SU14: Haxton (SPTAE), 17-May-02 (PJH) (Hodge, 2002); Stonehenge, 22-Apr-03 (PJH) (Hodge, 2003)
SU15: Everleigh, 24-Jun-02 (PJH) (Hodge, 2002)
SU22: Langley Wood NNR, 25-May-94 (WANHS det MDD), 14-May to 3-Jun-04 (MDD) (Darby, 2005)
SU23: Porton Down, 1970-2001 (Anon); Thorny Down woods, 16-Oct-00 (MDD); West Winterslow, 1-Aug-01 (NB)
Resident. Widespread in woodland and grassland.

**OLIGOTA Mannerheim, 1830**
***Oligota apicata*** (Erichson, 1837)
**Notable**
8 SU02: Knighton Wood, Sep, Oct-07 (AJWA)
SU13: Salisbury, Bemerton Folly, 7-Mar - 15-Jun-02 (MDD) (Darby, 2002d)
Resident. Very local; in a flight intercept trap [on fungi] and in litter etc. in woodland.

Oligota apicata

***Oligota inflata*** (Mannerheim, 1830)
8 SU13: Salisbury, Avon Valley Reserve, 24-May-00 (MDD) (Darby, 2001)
Resident. Very local; in a grass pile trap in wetland.

***Oligota parva*** Kraatz, 1862
8 ST92: Swallowcliffe, 28-Sep-95 (MDD)
Resident. Very local; in a grass pile.

***Oligota picipes*** (Stephens, 1832)
7 SU16: Marlborough area, Jun-1873 (EWJ) (Meyrick, 1874); 1869-1938 (Meyrick, 1938)
8 SU13: Salisbury, Bemerton Folly, 7-Mar - 15-Jun-02 (MDD) (Darby, 2002d)
Resident. Very local; in a grass pile in woodland and in vegetable refuse. There is a single specimen under this name (as *O. atomaria*) without data in the MCC.

***Oligota punctulata*** Heer, 1839
8 ST92: Sutton Mandeville, Mar-95 (MDD det RT)
Resident. Very local; in garden compost.

***Oligota pusillima*** (Gravenhorst, 1806)
7 SU16: Marlborough area, Jun-1873 (EWJ) (Meyrick, 1874); 1869-1938 (Meyrick, 1938)
Resident. Very local; n vegetable refuse. There is a single specimen without data in the MCC.

Tribe LOMECHUSINI
Subtribe LOMECHUSINA

**LOMECHUSA Gravenhorst, 1806**
***Lomechusa emarginata*** (Paykull, 1789)
**Notable**
8 ST95: West Lavington, 23-Mar-45 (DCL) (DFL, 1948)
Resident. Very local; [in nests of various species of ant in woodland, heathland and grassland].

Subtribe MYRMEDONIINA

**DRUSILLA Leach, 1819**
***Drusilla canaliculata*** (Fabricius, 1787)
7 ST87: Rack Hill, 1995 (MA det MDD)
SU16: Marlborough area, 1869-1938 (Meyrick, 1938)
8 ST93: Great Ridge Wood, 18-Apr-74 (DRN)
ST94: Ladywell (SPTAW), 3-Jun - 12-Aug-08

(Boyce & Wright, 2009); Norton Down, 23-Jun-02 (PJH) (Hodge, 2002); Wadman's Coppice (SPTAW) 3-Jun - 12-Aug-08 (Boyce & Wright, 2009)
SU01: Martin Down NNR, 1999 (EN, 1999)
SU03: Wylye Down NNR, 16-Oct-06 to 10-Oct-07 (MDD) (Darby, 2009)
SU04: Fox Covert, 3-Jun - 12-Aug-08 (Boyce & Wright, 2009);
SU05: Black Heath (SPTAC), 3-Jun - 12-Aug-08 (Boyce & Wright, 2009); Ell Barrow (SPTAC), 3-Jun - 12-Aug-08 (Boyce & Wright, 2009)
SU13: Salisbury, Avon Valley Reserve, Apr-00 (MAS) (Darby, 2001)
SU14: Boscombe Down, 1994 (MA det MDD); Stonehenge, 22-Apr-03 (PJH) (Hodge, 2003); Wexland Hanging SPTAC) 22-Jun-02 (PJH) (Hodge, 2002)
SU15: Beach's Plantation (SPTAE), 1-May-05 (MGT)
SU21: Landford Bog, 25-May-98 (MDD); 2003 (APF) (Foster, 2003)
SU22: Bentley Wood, 3-Sep-97 (WANHS det MDD)
SU24: Beacon Hill (SPTAE), 3-Jun - 12-Aug-08 (Boyce & Wright, 2009); SU24: Bulford Ranges (SPTAE), 3-Jun - 12-Aug-08 (Boyce & Wright, 2009); Driver Training Area (SPTAE), 1-May-05 (MGT)

Resident. Widespread; particularly on calcareous grassland (at Wylye the dominant staphylinid).

**ZYRAS Stephens, 1835**
***Zyras limbatus*** (Paykull, 1789)
8 SU13: Salisbury, 7-Apr-74 (DRN)
Resident. Very local; near ants' nest in garden.

Tribe MYLLAENINI

**MYLLAENA Erichson, 1837**
***Myllaena dubia*** (Gravenhorst, 1806)
8 SU13: Salisbury, Avon Valley Reserve, Apr-00 (MAS) (Darby, 2001)
SU23: Porton Down, 1970-2001 (Anon, 2001)
Resident. Very local; in moss in reedbed and marshy areas.

***Myllaena infuscata*** Kraatz, 1853
8 SU13: Salisbury, 12-Apr-00 (MDD) (Darby, 2001)
SU16: Jones's Mill, 2000 (MDD)
Resident. Very local; in moss in reedbed and at edge of water.

***Myllaena kraatzi*** Sharp, 1871
**Notable**
8 SU21: Landford Bog, 24-Jun-01 (APF) (Foster, 2001)
Resident. Very local; [in *Sphagnum* moss].

Myllaena kraatzi

Tribe OXYPODINI
Subtribe DINARDINA

**HOMOEUSA Kraatz, 1856**
***Homoeusa acuminata*** (Märkel, 1842)
**RDBK, LCGR-Fl Cu**
8 SU02: Knighton Wood, 2005-2007 (AJWA) (Allen, 2008)
Resident. Very local; in flight intercept trap. [In nests of several species of ant in woodland and chalk downland].

Subtribe OXYPODINA

**HAPLOGLOSSA Kraatz, 1856**
***Haploglossa marginalis*** (Gravenhorst, 1806)
**Notable**
8 SU02: Knighton Wood, 7-Oct-06 (AJWA)
Resident. Very local; in aerial litter, bird nests, etc.. [In rotten wood and nests of birds in tree hollows, in woodland and parkland].

***Haploglossa villosula*** (Stephens, 1832)
8 SU02: Knighton Wood, Apr-07 (AJWA)
Resident. Very local; in a flight intercept trap.

**ILYOBATES Kraatz, 1856**
***Ilyobates bennetti*** Donisthorpe, 1914
**Notable**
8 SU22: Langley Wood NNR, 25-Aug to 15-Sep-06 (MDD) (Darby, 2008)
Resident. Very local; in damp litter.

***Ilyobates propinquus*** (Aubé, 1850)
**Notable**
8 SU14: Stonehenge, summer-03 (PJH) (Hodge, 2003)
Resident. Very local; [in sandy places and flood meadows, under stones and in litter].

Ilyobates propinquus

**OCALEA Erichson, 1837**
***Ocalea picata*** (Stephens, 1832)
8 SU22: Langley Wood NNR, 14-May to 3-Jun-04 (MDD) (Darby, 2005)
Resident. Very local; in moss, leaf litter, etc..

**OCYUSA Kraatz, 1856**
***Ocyusa picina*** (Aubé, 1850)
7 SU16: Marlborough area, 1869-1938 (Meyrick, 1938)
Resident. Very local; [in marshy places in moss and in roots of vegetation].

**OXYPODA Mannerheim, 1830**
***Oxypoda acuminata*** (Stephens, 1832)
7 SU16: 1869-1938 (Meyrick, 1938)
SU26: Savernake Forest, 11-30-May-95 (EN det MDD)
8 ST92: Sutton Mandeville,Apr-00 (MDD)
SU02: Knighton Wood, Oct, Nov-06, Apr, Oct-07 (AJWA)
SU22: Langley Wood NNR, 19-Nov-04 to 24-Mar-05, 26-Oct-06 to 24-Mar-07 (MDD) (Darby, 2005, 2008)
Resident. Local; in litter. Formerly *O. lividipennis.*

***Oxypoda alternans*** (Gravenhorst, 1802)
7 SU26: Savernake Forest, 9-Oct-91 (AGD)
8 SU02: Knighton Wood, Oct, Nov-07 (AJWA)
ST91: Cranborne Chase, 23-May-97 (MAS)
Resident. Very local; in woodland in fungi.

***Oxypoda annularis*** Mannerheim, 1830
8 SU23: Thorny Down Woods, 16-Oct-00 (MDD)
Porton Down, 1970-2001 (Anon, 2001)
Resident. Very local; [in moss, leaf litter, etc.].

***Oxypoda elongatula*** Aubé, 1850
7 SU16: Marlborough area, 1869-1938 (Meyrick, 1938)
Resident. Very local; [in swamps].

***Oxypoda opaca*** (Gravenhorst, 1802)
8 ST91: Cranborne Chase, 30-Jun-99 (MDD)
SU13: Salisbury, Avon Valley Reserve, 7-Apr-00 (MDD) (Darby, 2001); Boscombe Down, 9-Oct-94 (MDD)
Resident. Very local; in vegetable detritus and litter.

**PHLOEOPORA Erichson, 1837**
***Phloeopora testacea*** (Mannerheim, 1830)
**SQS-1**
7 ST99: Flisteridge Wood, 2-May-82 (KNAA)
SU26: Savernake Forest, 8-Jan-90 (AGD)
8 SU22: Hamptworth Estate, 12-Apr-74 (DRN)
Resident. Very local; under bark and in fungi.

Subtribe TACHYUSINA

**ISCHNOPODA Stephens, 1835**
***Ischnopoda leucopus*** (Marsham, 1802)
7 SU16: Marlborough area, Jun-1873 (EWJ) (Meyrick, 1874
Resident. Very local; [on banks of rivers].

Tribe PLACUSINI

**PLACUSA Erichson, 1839**
***Placusa tachyporoides*** (Waltl, 1838)
**Notable**
7 SU26: Savernake Forest, 29-May-73 (CM-J)
8 SU02: Knighton Wood, Mar-07 (AJWA)
Resident. Very local; [under sappy bark and in sappy wood].

## Subfamily SCAPHIDIINAE
Tribe SCAPHIDIINI

**SCAPHIDIUM Kirby, 1837**
***Scaphidium quadrimaculatum*** Olivier, 1790
**SQS-2**
7 SU16: Four Mile Copse, 3-Jun-46 (Holmes and Tennant, 1947); West Woods, 1905-1938 (Meyrick, 1938)
SU26: Savernake Forest, 14-Jun-02 (IJG)
8 ST92: Wardour, 13-Nov-95 (MDD); Sutton Mandeville, Oct-92 (MDD); Wardour, 19-Apr-95 (MDD); Fovant,, 25-Apr-95 (MDD)
ST93: Great vRidge Wood, 18-Apr-74 (DRN)
SU02: Knighton Wood, May, Jun-08 (AJWA)
SU03: Grovely Wood, 10-Apr-70, 13-Aug-71, 16, 18-Apr, 27-May-74, 26-Jul-82 (DRN); 21-Aug-01 (MDD)
SU12: Coombe Bissett Down, 13-May-99 (PM)
SU22: Bentley Wood, 9-May-91 (PM); Blackmoor Copse, 3-Aug-03 (APF) (Foster, 2004); Langley Wood, 28-May-74 (DRN);Mapleway Dean Copse, 31-Aug-03 (EG)
SU23: Porton Down, 1970-2001 (Anon, 2001)
SU25: Ludgershall, Apr - Jul-1915 (PAH)

(Harwood, 1915)
Resident. Widespread; in litter, under logs, etc. in woodland.

Tribe SCAPHISOMATINI

**SCAPHISOMA Leach, 1815**
***Scaphisoma agaricinum*** (Linnaeus, 1758)
**SQS-2**
7 SU16: Marlborough area, Jun-1873 (EWJ) (Meyrick, 1874); 1869-1938 (Meyrick, 1938)
SU26: Savernake Forest, 30-Aug to 20-Sep-94 (EN det SAW)
8 SU02: Knighton Wood, May-08 (AJWA)
SU03: Grovely Wood, 6-Aug-70 (DRN), 5, 6-Apr-71 (DRN)
SU13: Boscombe Down, 9-Oct-94 (MDD)
SU21: Hamptworth Common, 24-Mar-03 (MDD); Landford Bog, 25-May-98 (MDD)
SU23: West Winterslow, 1-Aug-01 (NB)
SU26: Wilton Brail, 19-Jun-03 (MDD)
Resident. Widespread; in rotten wood and on fungi in woodland.

***Scaphisoma boleti*** (Panzer, 1793)
**Nb, SQS-8**
8 ST84: Longleat, 7-Mar-99 (MB)
SU16: Jones's Mill, 21-Jun-97 (MDD)
Resident. Very local; in fungi and rotten wood in woodland.

Scaphisoma boleti

Subfamily PIESTINAE

**SIAGONIUM Kirby, 1815**
***Siagonium quadricorne*** Kirby, 1815
**SQS-2**
7 SU16: Marlborough area, Jun-1873 (EWJ) (Meyrick, 1874); 1869-1938 (Meyrick, 1938)
SU26: Savernake Forest, 3, 7-Apr-70 (DRN)
8 SU02: Knighton Wood, Aug-07 (AJWA)
SU03: Grovely Wood, 10-Apr-70, 25-Oct-71 (DRN)
SU13: Great Durnford, 18-Aug-72 (DRN)
Resident. Very local; under beech bark.

Subfamily OXYTELINAE
Tribe COPROPHILINI

**COPROPHILUS Latreille, 1829**
***Coprophilus striatulus*** (Fabricius, 1792)
8 SU22: Langley Wood NNR, 16-Jul to 6-Aug-04 (MDD) (Darby, 2005)
Resident. Very local; in litter.

Tribe DELEASTERINI

**DELEASTER Erichson, 1839**
***Deleaster dichrous*** (Gravenhorst, 1802)
**Nb**
8 SU13: Lower Woodford, 1-Aug-83 (DRN)
Resident. Very local; edge of tributary of R. Avon.

**SYNTOMIUM Curtis, 1828**
***Syntomium aeneum*** (Müller, P.W.J, 1821)
7 SU26: Savernake Forest, 15-Jun-73 (SAW), 8 to 28-Jul-93, 1 to 19-Jul (EN det SAW); 23-Jun to 15-Jul-95 (EN det MDD)
8 SU13: Salisbury, Bemerton Folly, 7-Mar - 15-Jun-02 (MDD) (Darby, 2002d)
Resident. Very local; in damp moss and litter.

Tribe OXYTELINI

**ANOTYLUS Thomson, C.G, 1859**
***Anotylus complanatus*** (Erichson, 1839)
8 SU02: Knighton Wood, Apr-07 (AJWA)
Resident. Very local: in a flight intercept trap.

***Anotylus inustus*** (Gravenhorst, 1806)
7 ST87: Rack Hill, 3-Jun-96 (MA det MDD)
SU16: Marlborough area, 1869-1938 (Meyrick, 1938);
SU26: Savernake Forest, 16-Jun-93 (EN det SAW); 31-May-95 (EN det MDD)
8 ST93: Great Ridge Wood, 16-May-00 (MDD)
SU01: Martin Down NNR, 1999 (EN, 1999)
SU02: Fovant, 31-May-95 (MDD); Middleton Down, 4-May-95 (MDD)
SU03: Grovely Wood, 30-May-74 (DRN); 12-Jun-00 (MDD); Wylye Down NNR, 29-Mar-07 to 5-Jun-07, 3-Aug to 4-Sep-07 (MDD) (Darby, 2009)
SU14: Haxton (SPTAE), 3-Jun - 12-Aug-08 (Boyce & Wright, 2009)
SU16: Jones's Mill, 2000 (MDD)
SU22: Langley Wood NNR, 14-May-04, 5-Aug-06 to 2-Jul-07 (MDD) (Darby, 2005, 2008)
SU26: Bedwyn Brail, 19-Jun-03 (MDD)
Resident. Widespread; in herbivore dung and decaying vegetation.

***Anotylus nitidulus*** (Gravenhorst, 1802)
7 SU16: Marlborough area, 1869-1938 (Meyrick, 1938)
Resident. Very local; [in carrion].

Anotylus nitidulus

***Anotylus rugosus*** (Fabricius, 1775)
7 SU16: Marlborough area, 17-Feb-1879 (WFB) (Meyrick, 1880); 1869-1938 (Meyrick, 1938)
SU18: Swindon, 7-May, 30-Mar-01 (JKG det MDD)
SU26: Savernake Forest, 8 to 28-Jul-93, 28-Apr to 18-May, 1-Jul to 9-Aug-94 (EN det SAW); 15-Jul to 4-Aug-95 (EN det MDD)
8 ST94: Norton Down (SPTAW), 23-Jun-02 (PJH) (Hodge, 2002)
SU03: Dinton, R. Nadder, 9-Jul-95 (MDD); Grovely Wood, 12-Jun-00 (MDD)
SU13: Salisbury, Avon Valley Reserve, 13-Apr, 2, 12-May-00 (MDD), Apr-00 (MAS) (Darby, 2001)
SU16: Jones's Mill, Sep-97 (HK det MDD)
SU21: Hamptworth Estate, 4-Apr, 15-Apr-74, 28-Oct-75 (DRN); Hamptworth Common, 24-Mar-03 (MDD)
SU22: Langley Wood NNR, 5 to 25-Aug-06 (MDD) (Darby, 2008)
SU26: Bedwyn Brail, 19-Jun-03 (MDD)
Resident. Widespread in grassland, more rarely in woodland in tussocks, etc..

***Anotylus sculpturatus*** (Gravenhorst, 1806)
7 SU16: Marlborough area, Jun-1873 (EWJ) (Meyrick, 1874); 1869-1938 (Meyrick, 1938)
SU26: Savernake Forest, var. dates 20-Apr-93 to 4-Aug-95 (EN det SAW, MDD)
8 ST92: Sutton Mandeville, 21-Apr-92 (MDD)
ST93: Stockton Wood, 7-Jul-96 (MAS)
ST94: Ladywell (SPTAW), 15-May-02 (PJH) (Hodge, 2002)
SU01: Martin Down NNR, 1999 (EN, 1999)
SU13: Boscombe Down, 8, 9-Oct-94 (MDD)
SU22: Bentley Wood, 31-May-97 (MDD); Langley Wood, 14-May-04 to 24-Mar-05, 5-Aug-06 to 17-Nov-06 (MDD) (Darby, 2005, 2008); Whiteparish Common, 2-Jul-95 (MAS)
SU23: Porton Down, 1970-2001 (Anon, 2001)
SU24: Bulford Ranges (SPTAE), 3-Jun - 12-Aug-08 (Boyce & Wright, 2009)
Resident. Widespread and abundant in herbivore dung, decaying vegetation and often in flight.

***Anotylus tetracarinatus*** (Block, 1799)
7 SU16: Marlborough area, 1869-1938 (Meyrick, 1938)
SU26: Savernake Forest, 28-Aug-93, 1-Jul to 9-Aug-94 (EN det SAW)
8 ST91: Cranborne Chase, 24-Apr-97 (MAS); 30-Jun-99 (MDD)
ST92: Sutton Mandeville, 21-Apr-92, 6-Jul-93, 25-Jun-07 (MDD)
ST93: Great Ridge Wood, 16-May-00 (MDD
SU02: Knighton Wood, Jun-07, Jul-08 (AJWA); Middleton Down, 4-May-95 (MDD)
SU03: Grovely Wood, 12-Jun-00 (MDD); Wylye Down NNR, 5-Jun-07 to 3-Aug-07 (MDD) (Darby, 2009);
SU13: Old Sarum, 31-May-97 (MAS); Salisbury, Avon Valley Reserve, Apr-00 (MAS); 13-Apr-00 (Darby, 2001); Salisbury, Bemerton Folly, 7-Mar - 15-Jun-02 (MDD) (Darby, 2002d)
SU21: Hamptworth Estate, 6-Aug-79 (DRN)
SU22: Bentley Wood, 11-Aug-01 (MDD); Black-moor Copse, 4-18-May-02 (Foster, 2004); Langley Wood NNR, 25-May-94 (WANHS det MDD), 3-Jun-04 to 16-Jul-04, 5-Aug-06 (MDD) (Darby, 2005, 2008);
SU23: Porton Down, 3-Jul-99 (MAS)
SU26: Bedwyn Brail, 19-Jun-03 (MDD)
Resident. Widespread in herbivore dung and rotting vegetation.

**OXYTELUS Gravenhorst, 1802**

***Oxytelus laqueatus*** (Marsham, 1802)
7 ST26: Flat Holm, 10-May-02 (JKG det MDD)
SU16: Marlborough area, 1869-1938 (Meyrick, 1938)
SU26: Savernake Forest, 23-Oct-01 (MDD)
8 ST93: Great Ridge Wood, 2-May-00 (MDD)
SU21: Hamptworth Estate, 6-Aug-79 (DRN)
SU22: Bentley Wood, 11-Aug-01 (MDD); Langley Wood, 6 to 26-Oct-06 (MDD) (Darby, 2008)
Resident. Local; in dung and cut grass.

***Oxytelus migrator*** Fauvel, 1904
8 ST91: Rushmore Park, 23-Sep-01 (AJWA)
Resident. Very local; in a dung heap.

***Oxytelus sculptus*** Gravenhorst, 1806
8 SU14: Stonehenge, 22-Apr-03 (PJH) (Hodge, 2003)

**PLATYSTETHUS Mannerheim, 1830**
CRAETOPYCRUS Tottenham, 1939

***Platystethus capito*** Heer, 1839
8 ST94: Ladywell (SPTAW), 3-Jun - 12-Aug-08 (Boyce & Wright, 2009)

SU14: Wexland Hanging (SPTAE), 3-Jun - 12-Aug-08 (Boyce & Wright, 2009)

***Platystethus cornutus*** (Gravenhorst, 1802)
7 SU16: Marlborough area, Jun-1873 (EWJ) (Meyrick, 1874)
8 SU03: Berwick St. James, 20-Jun-95 (MDD); Steeple Langford, 19-Jun-95 (MDD)
Resident. Very local; on mud near water.

***Platystethus nitens*** (Sahlberg, C.R, 1832)
7 SU09: Somerford Keynes, 4-Jun-04 (BLe)
Resident. Very local; [in mud and damp litter near water].

PLATYSTETHUS Mannerheim, 1830
***Platystethus arenarius*** (Fourcroy, 1785)
8 SU01: Martin Down NNR, 1979-1992 (RGB)
SU04: Parsonage Down, 2-Jul-08 (MDD)
SU21: Pound Bottom, 9-Apr-74 (DRN)
Resident. Very local; in decaying vegetation and dung.

### Tribe THINOBIINI

**BLEDIUS Leach, 1819**
HESPEROPHILUS Curtis, 1829
***Bledius longulus*** Erichson, 1839
7 SU16: Marlborough area, Jun-1873 (EWJ) (Meyrick, 1874); 1869-1938 (Meyrick, 1938)
SU26: Savernake Forest, 6-Jan-73 (AAA)
Resident. Very local; [in sandy places near streams].

**CARPELIMUS Leach, 1819**
***Carpelimus bilineatus*** Stephens, 1834
7 SU09: Somerford Keynes, 4-Jun-04 (BLe)
SU16: Marlborough area, 1869-1938 (Meyrick, 1938)
8 ST94: Ladywell (SPTAW), 3-Jun - 12-Aug-08 (Boyce & Wright, 2009)
SU13: Salisbury, Avon Valley Reserve, 1-May-00 (MDD) (Darby, 2001)
Resident. Very local; on mud at edge of water.

***Carpelimus corticinus*** (Gravelnhorst, 1806)
8 ST94: Ladywell (SPTAW), 3-Jun - 12-Aug-08 (Boyce & Wright, 2009); Wadman's Coppice (SPTAW), 3-Jun - 12-Aug-08 (Boyce & Wright, 2009)
Resident. Very local; on ground at edge of pond.

***Carpelimus elongatulus*** (Erichson, 1839)
8 SU02: Fovant, var. dates Apr-Jun-00 (MDD)
SU13: Salisbury, Avon Valley Reserve, 1-May-00 (MDD) (Darby, 2001)
SU21: Hamptworth Estate, 24-Apr-84 (DRN)
Resident. Very local; in moss at edge of Salix carr and vegetation by old watercress beds.

***Carpelimus pusillus*** (Gravenhorst, 1802)
7 SU27: Rabley Wood, 1939 (det KGB) (Knight, 1940)
Resident. Very local; [on damp soil at edge of water]

***Carpelimus subtilis*** (Erichson, 1839)
**Notable**
7 SU09: Cotswold Water Park, 21-Jun-90 (APFw)
Resident. Very local; [on margins of rivers and ponds in damp sand and a shingle bank].

### Subfamily OXYPORINAE

**OXYPORUS Fabricius, 1775**
***Oxyporus rufus*** (Linnaeus, 1758)
8 SU03: Grovely Wood, 4-Aug-70 (DRN)
SU14: Netheravon, 29-Jul-83 (DRN)
Resident. Very local; in fungus.

### Subfamily STENINAE

**DIANOUS Leach, 1819**
***Dianous coerulescens*** (Gyllenhal, 1810)
7 SU16: Marlborough area, 1869-1938 (Meyrick, 1938)
SU27: Ramsbury, spring-1894 (GCC) (Jebb, 1895)
Resident. Very local; [in hilly regions in wet moss by streams].

**STENUS Latreille, 1796**
HEMISTENUS Motschulsky, 1860

Stenus aceris

***Stenus aceris*** Stephens, 1833
8 ST94: Ladywell (SPTAW), 3-Jun - 12-Aug-08 (Boyce & Wright, 2009); Wadman's Coppice (SPTAW) 3-Jun - 12-Aug-08 (Boyce & Wright, 2009)
SU03: Grovely Wood, 12-Jun-00 (MDD)
SU05: Black Heath (SPTAC), 3-Jun - 12-Aug-08 (Boyce & Wright, 2009)
SU13: Salisbury, Apr-00 (MAS) (Darby, 2001)
SU22: Langley Wood NNR, 8-Oct-04 to 24-Mar-05

(MDD) (Darby, 2005)
SU24: Bulford Ranges (SPTAE), 3-Jun - 12-Aug-08 (Boyce & Wright, 2009); Dunch Hill (SPTAE), 3-Jun - 12-Aug-08 (Boyce & Wright, 2009)
Resident. Very local; in moss and roots.

***Stenus impressus*** Germar, 1824
7 SU16: Marlborough area, 17-Feb-1879 (TNH-S) (Meyrick, 1880); 1869-1938 (Meyrick, 1938)
SU26: Savernake Forest, 14-Nov-89 (AGD); 20-Sep to 11-Oct-94 (EN det SAW)
SU27: Ramsbury, 1870-1873 (Meyrick, 1874)
8 SU02: Knighton Wood, Apr-07 (AJWA)
SU13: Salisbury, Bemerton Folly, 7-Mar - 15-Jun-02 (MDD) (Darby, 2002d)
SU14: Wexland Hanging (SPTAC) 22-Jun-02 (PJH) (Hodge, 2002)
SU22: Bentley Wood, 31-May-97 (MDD); Langley Wood NNR, 25-May-94 (WANHS det MDD)
Resident. Widespread; on vegetation in woodland and grassland.

***Stenus ossium*** Stephens, 1833
7 SU16: Marlborough area, 1869-1938 (Meyrick, 1938)
SU17: Preshute, 1870-1873 (Meyrick, 1874)
8 SU02: Knighton Wood, Mar-07 (AJWA)
Resident. Very local; [in tussocks or litter near water].

***Stenus subaeneus*** Erichson, 1840
7 SU16: Marlborough area, 1869-1938 (Meyrick, 1938)
Resident. Very local; [damp places in moss, flood and haystack refuse; often in sandpits].

HYPOSTENUS Rey, 1884

***Stenus cicindeloides*** (Schaller, 1783)
7 SU16: Marlborough, spring-1894 (GCC) (Jebb, 1895); Marlborough area, 1869-1938 (Meyrick, 1938)
SU27: Ramsbury, spring-1894 (GCC) (Jebb, 1895)
8 ST91: Cranborne Chase, 19-Jul-97 (MAS)
SU12: Charlton-All-Saints, 9-May-96 (MAS)
SU13: Salisbury, Apr-00 (MAS) (Darby, 2001)
SU14: Stonehenge, 22-Apr-03 (PJH) (Hodge, 2003)
SU16: Jones's Mill, 19-Jun-01 (APF) (Foster, 2002), 1997 (MAS)
SU21: Landford Bog, 24-Aug-01 (APF) (Foster, 2001)
SU26: Bedwyn Brail, 19-Jun-03 (MDD)
Resident. Widespread; in marshy places.

***Stenus fulvicornis*** Stephens, 1833
7 SU09: Somerford Keynes, 4-Jun-04 (BLe)
SU16: Marlborough area, 1869-1938 (Meyrick, 1938)
8 SU13: Salisbury, Apr-00 (MAS) (Darby, 2001)
SU22: Langley Wood NNR, 29-May-03 (MDD)
Resident. Local; in litter near water.

***Stenus latifrons*** Erichson, 1839
7 SU16: Marlborough area, 1869-1938 (Meyrick, 1938)
8 SU21: Hamptworth Estate, 15-Apr-74 (DRN)
Resident. Very local; in old straw bales, [litter near water].

***Stenus similis*** (Herbst, 1784)
7 SU16: Marlborough area, 1869-1938 (Meyrick, 1938)
8 SU22: Blackmoor Copse, 4 -18-May-02 (APF) (Foster, 2004)
Resident. Very local; [in drier situations].

***Stenus solutus*** Erichson, 1840
8 SU12: Charlton-All-Saints, 9-May-96 (MAS)
SU13: Stratford sub Castle, 7-Aug-71 (DRN); Lower Woodford, 28-Jul-78 (DRN)
Resident. Very local; on vegetation near the R. Avon.

***Stenus tarsalis*** Ljungh, 1810
7 SU09: Somerford Keynes, 4-Jun-04 (BLe)
SU16: Marlborough area, 31-Jan-1879 (GTKM) (Meyrick, 1880); 1869-1938 (Meyrick, 1938)
8 SU12: Lower Woodford, 28-Jul-78 (DRN); Charlton-All-Saints, 9-May-96 (MAS)
SU13: Stratford sub Castle, 7-Aug-71 (DRN)
Resident. Very local; on vegetation near water.

METASTENUS Adám, 2001

***Stenus binotatus*** Ljungh, 1804
7 SU16: Marlborough area, 1869-1938 (Meyrick, 1938)
Resident. Very local; [on vegetation near water].

***Stenus canescens*** Rosenhauer, 1856
**Nb**
8 SU12: Britford, Jun - 15-Sep-02 (CMD) (Drake, 2008)
SU13: Idmiston, 13-Aug-87 (DRN)
Resident. Very local; from watercress.

***Stenus flavipes*** Stephens, 1833
7 SU18: Swindon, 30-Mar-01 (JKG det MDD)
8 ST91: Cranborne Chase, 19-Jul-97 (MAS)
ST94: Bowls Barrow (SPTAW), 16-May-02 (PJH) (Hodge, 2002)
SU16: Jones's Mill, 1997 (MAS)
SU21: Landford Bog, 24-Jun, 24-Aug-01 (Foster, 2001)
SU22: Langley Wood NNR, 29-May-03 (MDD)
Resident. Widespread; on vegetation near water.

***Stenus nitidiusculus*** Stephens, 1833
8 SU21: Hamptworth Estate, 20-Aug-76 (DRN)
SU22: Langley Wood NNR, 29-May-03 (MDD)

Resident. Very local; in a dry stream bed and litter by water.

***Stenus niveus*** Fauvel, 1865
**Nb**
8 SU16: Jones's Mill, 21-Jun-97 (MDD)
Resident. Very local; in detritus at edge of water.

***Stenus pallitarsis*** Stephens, 1833
8 ST84: Longleat, 7-Jul-99 (MDD)
SU22: Bentley Wood, 31-May-97 (MDD)
Resident. Very local; in detritus near water.

***Stenus picipes*** Stephens, 1830
8 ST91: Cranborne Chase, 22-Jun-99 (MDD), 19-Jul-97 (MAS)
ST93: Great Ridge Wood, 4-May-00 (MDD det RT)
SU13: Salisbury, Apr-00 (MAS) (Darby, 2001)
SU16: Jones's Mill, 1-Jan-9731-Dec-97 (MAS)
Resident. Local; in litter and reed tussocks near water.

***Stenus pubescens*** Stephens, 1833
7 SU16: Marlborough area, Jun-1873 (EWJ) (Meyrick, 1874); 1869-1938 (Meyrick, 1938)
SU27: Ramsbury, spring-1894 (GCC) (Jebb, 1895)
8 SU12: Charlton-All-Saints, 9-May-96 (MAS); Salisbury, 30-Jul-73 (DRN)
Resident. Very local; on vegetation by water.

STENUS Latreille, 1796

***Stenus biguttatus*** (Linnaeus, 1758)
8 SU22: Redlynch, 13-Apr-74 (DRN)
Resident. Very local; in wet gravel pit.

***Stenus bimaculatus*** Gyllenhal, 1810
7 SU16: Marlborough area, 1869-1938 (Meyrick, 1938)
SU26: Mildenhall, 1870-1873 (Meyrick, 1874)
8 ST92: Sutton Row, 9-May-95 (MDD); Fovant, 1, 31-May-95 (MDD)
SU13: Salisbury, 13-Apr to 19-Jun-00 (MDD); Apr-00 (MAS) (Darby, 2001)
SU21: Hamptworth Estate, 4-Apr-75 (DRN)
SU22: Bentley Wood, 27-May-97 (MDD)
Resident. Widespread; in litter, etc. near water.

***Stenus boops*** Ljungh, 1810
7 SU16: Marlborough area, 1869-1938 (Meyrick, 1938
SU27: Ramsbury, spring-1894 (GCC) (Jebb, 1895)
8 ST92: Wardour, 5-Jun-96 (MAS)
SU12: Charlton-All-Saints, 9-May-96 (MAS)
SU13: Salisbury, 1-May-00 (MDD) (Darby, 2001)
Resident. Local; in litter at edge of water and in reed bed.

***Stenus canaliculatus*** Gyllenhal, 1827
7 SU09: Somerford Keynes, 4-Jun-04 (BLe)
Resident. Very local; [in grass tussocks near water].

***Stenus circularis*** Gravenhorst, 1802
**Nb**
7 SU16: Marlborough area, 1869-1938 (Meyrick, 1938)
SU27: Ramsbury, 1870-1873 (Meyrick, 1874)
Resident. Very local; [in damp moss by water].

***Stenus clavicornis*** (Scopoli, 1763
7 SU16: Marlborough area, 1869-1938 (Meyrick, 1938)
8 SU03: Grovely Wood, 29-Oct-73 (DRN); Wylye Down NNR, 3-Aug to 10-Oct-07 (MDD) (Darby, 2009)
SU04: Orcheston Down, 22-Jun-02 (PJH) Hodge, 2002)
SU21: Hamptworth Estate, 15-Apr-74, 6-Aug-79 (DRN)
SU22: Bentley Wood, 3-Sep-97 (WANHS det MDD)
SU24: Sidbury Hill (SPTAE), 13 to 17-Jun-97 (MGT)
Resident. Local; in cut grass, old straw bales, litter, etc.

***Stenus guttula*** Müller, P.W.J, 1821
7 SU16: Marlborough area, Jun-1873 (EWJ) (Meyrick, 1874); 1869-1938 (Meyrick, 1938)
Resident. Very local; at margins of ponds and rivers].

***Stenus juno*** (Paykull, 1789)
7 SU16: Marlborough area, 6-Feb-1879 (TNH-S) (Meyrick, 1880); 1869-1938 (Meyrick, 1938)
8 ST84: Longleat, 7-Mar-99 (MB)
SU13: Salisbury, 13-Apr-00 (MDD); Apr-00 (MAS) (Darby, 2001)
SU16: Jones's Mill, 19-Jun, 2-Jul-01 (APF) (Foster, 2002)
SU21: Landford Bog, 24-Jun-01 (APF) (Foster, 2001)
Resident. Local; in litter, etc. near water. Probably widespread.

***Stenus nanus*** Stephens, 1833
7 SU16: Marlborough area, 1869-1938 (Meyrick, 1938)
SU27: Ramsbury, 1870-1873 (Meyrick, 1874)
Resident. Very local; [common, near fresh water].

***Stenus providus*** Erichson, 1839
8 SU21: Hamptworth Estate, 6-Aug-79 (DRN); Landford Bog, 25-May-98 (MDD)
SU22: Langley Wood NNR, 14-May to 27-Aug-04 (MDD) (Darby, 2005)
Resident. Very local; in grass piles and woodland litter.

***Stenus pusillus*** Stephens, 1833
7 SU09: Somerford Keynes, 4-Jun-04 (BLe)
SU16: Marlborough area, Jun-1873 (EWJ) (Meyrick, 1874); 1869-1938 (Meyrick, 1938)
8 SU12: Petersfinger, 25-Jul-91 (DRN)
SU13: Salisbury, 13-Apr, 1-May-00 (MDD) (Darby, 2001)
Resident. Local; in detritus at edge of water.

TESNUS Rey, 1884
***Stenus brunnipes*** Stephens, 1833
7 SU16: Marlborough area, 1869-1938 (Meyrick, 1938); Jones's Mill, 1997 (MAS)
8 SU01: Martin Down NNR, 1999 (EN, 1999)
SU03: Grovely Wood, 24-Aug-72, 3-Aug-79 (DRN); 31-May-97 (MAS); Wylye Down NNR, 16-Oct-06 to 3-Aug-07 (MDD) (Darby, 2008a)
SU05: Black Heath (SPTAC), 3-Jun - 12-Aug-08 (Boyce & Wright, 2009)
SU13: Salisbury, 7-Apr-74 (DRN)
SU14: Wexland Hanging (SPTAC) 22-Jun-02 (PJH) (Hodge, 2002)
SU22: Bentley Wood, 3-Sep-97 (WANHS det MDD); Langley Wood NNR, 3 to 24-Mar-05 (MDD) (Darby, 2005)
SU24: Dunch Hill (SPTAE), 3-Jun - 12-Aug-08 (Boyce & Wright, 2009)
Resident. Widespread; in grassland and woodland not necessarily near water.

***Stenus crassus*** Stephens, 1833
7 SU16: Marlborough area, 1869-1938 (Meyrick, 1938)
SU26: Ramsbury, spring-1894 (GCC) (Jebb, 1895)
Resident. Very local; in vegetation at edge of water.

## Subfamily EUAESTHETINAE

**EUAESTHETUS Gravenhorst, 1806**
***Euaesthetus ruficapillus*** Boisduval & Lacordaire, 1835
8 SU21: Hamptworth Estate, 24-Apr-84 (DRN)
Resident. Very local; in moss in sallow carr.

## Subfamily PAEDERINAE

**ASTENUS Dejean, 1833**
ASTENUS Dejean, 1835
***Astenus lyonessius*** (Joy, 1908)
7 SU06: Roundway Camp, Jul-40 (FDB) (Buck, 1956)
SU16: Marlborough area, 1869-1938 (Meyrick, 1938)
8 ST94: Ladywell (SPTAW), 23-Jun-02 (PJH) (Hodge, 2002)
SU03: Grovely Wood, 23-Jul-73, 29-Oct-73 (DRN)
SU21: Hamptworth Estate, 1-Apr-75 (DRN)
Resident. Local; in detritus near water.

**LATHROBIUM Gravenhorst, 1802**
LATHROBIUM Gravenhorst, 1802
***Lathrobium brunnipes*** (Fabricius, 1792)
7 SU16: Marlborough area, 24 Feb 1879 (WFB) (Meyrick, 1880), 1869-1938 (Meyrick, 1938);
SU26: Mildenhall, 1870-1873 (Meyrick, 1874)
8 ST93: Great Ridge Wood, 18-Apr-74 (DRN)
SU03: Grovely Wood, 30-May-74, 8-Apr-75 (DRN)
SU13: Salisbury, Apr-00 (MAS); 13-Ap, 12-May, 19-Jun-00 (MDD) (Darby, 2001)
SU16: Jones's Mill, 21-Jun-97 (MDD)
SU21: Hamptworth Estate, 3-Aug-73, 1, 4-Apr-75 (DRN)
SU22: Langley Wood NNR, 25-May-94 (WANHS det MDD); 6-Aug to 17-Sep-04 (MDD) (Darby, 2005)
SU26: Bedwyn Brail, 2-Jun-03 (MDD)
Resident. Widespread; in straw and litter.

***Lathrobium elongatum*** (Linnaeus, 1767)
7 SU16: Marlborough area, 6-Feb-1879 (WFB) (Meyrick, 1880); 1869-1938 (Meyrick, 1938)
SU26: Mildenhall, 1870-1873 (Meyrick, 1874)
Resident. Very local; [in flood refuse, litter, etc. in marshy places].

***Lathrobium fovulum*** Stephens, 1833
8 SU13: Salisbury, 7-Apr-00 (MDD) (Darby, 2001)
SU22: Langley Wood NNR, 14-May to 3-Jun-04 (MDD) (Darby, 2005)
Resident. Very local; in detritus in marshy places.

***Lathrobium fulvipenne*** (Gravenhorst, 1806)
7 SU16: Marlborough area, 1869-1938 (Meyrick, 1938)
8 SU13: Salisbury, Apr-00 (MAS) (Darby, 2001)
Resident. Very local; in moss, detritus near water [often under bark].

Lathrobium geminum

***Lathrobium geminum*** Kraatz, 1857
7 SU16: Marlborough area, 1869-1938 (Meyrick, 1938)
8 ST93: Great Ridge Wood, 18-Apr-74 (DRN)
SU13: Salisbury, 13-Apr-00 (MDD) (Darby, 2001)
SU22: Langley Wood NNR, 22-Apr to 25-Jun-04 (MDD) (Darby, 2005)
Resident. Local; in moss, straw, etc..

***Lathrobium pallidipenne*** Hochhuth, 1851
**Notable**
8 ST92: Sutton Mandeville, 21-Jul-93 (MDD)
SU21: Hamptworth Estate, 28-Oct-75 (DRN)
Resident. Very local; in litter near water.

***Lathrobium pallidum*** Nordmann, 1837
**RDBK**
8 SU14: Stonehenge, Summer-03 (PJH) (Hodge, 2003)
Resident. Very local; taken in a pitfall trap. [Probably associated with riversides, marshes, etc. Also under stones on short turf on a mountain].

TETARTOPEUS Czwalina, 1888
***Lathrobium quadratum*** (Paykull, 1789)
7 SU16: Marlborough area, 1869-1938 (Meyrick, 1938)
SU26: Mildenhall, 1870-1873 (Meyrick, 1874)
Resident. Very local; [in marshy places, at roots of grass, in moss, flood refuse, etc].

***Lathrobium terminatum*** Gravenhorst, 1802
8 SU21: Hamptworth Estate, 20-Aug-76 (DRN)
SU26: Bedwyn Brail, 19-Jun-03 (MDD)
SU16: Jones's Mill, 21-Jun-97 (MDD)
Resident. Very local; in marshy places in litter, etc..

***Lathrobium zetterstedtil*** Rye, 1872
**Nb**
7 SU16: Marlborough area, 1869-1938 (Meyrick, 1938)
SU27: Ramsbury, spring-1894 (GCC) (Jebb, 1895 as *L.punctatum*)
Resident. Very local; [marshy places, in dead leaves, moss, flood refuse, etc.].

**LITHOCHARIS Dejean, 1833**
***Lithocharis nigriceps*** Kraatz, 1859
8 ST91: Rushmore Park, 23-Sep-01 (AJWA
SU03: Grovely Wood, 29-Oct-73, 2-Aug-79 (DRN)
SU21: Hamptworth Estate, 1-Apr-75 (DRN)
Resident. Very local; in straw and mouldy grass. A recent introduction (1962) to GB from tropical Asia.

***Lithocharis ochracea*** (Gravenhorst, 1802)
8 SU03: Great Wishford, Hamshill Ditches, 29-Oct-73 (DRN)
Resident. Very local; [in haystack refuse, dung heaps, hot beds, etc.].

**LOBRATHIUM Mulsant and Rey, 1878**
***Lobrathium multipunctum*** Gravenhorst, 1802
7 SU16: Marlborough area, 24-Mar-1879 (WFB) (Meyrick, 1880); 1869-1938 (Meyrick, 1938)
8 SU03: Wylye Down NNR, 27-Jan to 22-Feb-07 (MDD) (Darby, 2009)
Resident. Very local; in roots of vegetation [under stones on sides of hills].

**MEDON Stephens, 1833**
***Medon brunneus*** (Erichson, 1839)
7 SU16: Marlborough area, Jun-1873 (EWJ) (Meyrick, 1874); 1869-1938 (Meyrick, 1938)
8 ST93: Great Ridge Wood, 18-Apr-74 (DRN), 28-Apr-00 (MDD)
SU24: Beacon Hill (SPTAE), 3-Jun - 12-Aug-08 (Boyce & Wright, 2009)
Resident. Very local; in litter, straw and nests of (?) voles.

Paederus fuscipes

**PAEDERUS Fabricius, 1775**
***Paederus fuscipes*** Curtis, 1826
**Nb**
8 SU12: Britford, Jun - 15-Sep-02 (CMD) (Drake, 2008)
SU13: Lower Woodford, Jun - 15-Sep-02 (CMD) (Drake, 2008)
SU22: Landford, 12-Aug-87 (DRN)
Resident. Very local; [on mud in moss, etc. in marshes, bogs and on the margins of standing water]

***Paederus littoralis*** Gravenhorst, 1802
7 SU16: Marlborough area, Jun-1873 (EWJ) (Meyrick, 1874); 1869-1938 (Meyrick, 1938); Pewsey Downs, 22-Jun-90 (APFw)
8 ST85: Bullen Hill Farm, 1-Jan-82 to 31-Dec-02, 14-Aug-03 (EGS det MDD)
ST93: Great Ridge Wood, 18-Apr-74 (DRN)
ST94: Ladywell (SPTAW), 3-Jun - 12-Aug-08 (Boyce & Wright, 2009);Tinker's Track

(SPTAW), 3-Jun - 12-Aug-08 (Boyce & Wright, 2009); Wadman's Coppice (SPTAW) 3-Jun - 12-Aug-08 (Boyce & Wright, 2009)
SU02: Bishopstone, 19-May-91 (MDD)
SU03: Barford Down, 6-Apr-72 (DRN); Wylye Down NNR, 16-Oct-06 to 10-Oct-07 (MDD) (Darby, 2009); Grovely Wood, 6- Apr-71 (DRN)
SU04: Salisbury Plain, 25-Jun-02 (PJH) (Hodge, 2002)
SU12: Coombe Bissett Down, 1-Apr-97 (MDD), 9-Jun-04 (JN)
SU14: Haxton (SPTAE), 17-May, 24-Jun-02 (PJH) (Hodge, 2002); Stonehenge, 22-Apr-03 (PJH) (Hodge, 2003); Wexland Hanging (SPTAC), 16-May, 22-Jun-02 (PJH) (Hodge, 2002)
SU23: West Winterslow, 11-Aug-01 (NB)
SU24: Driver Training Area (SPTAE), 17-May-02 (PJH) (Hodge, 2002); 1-May-05 (MGT)
Resident. Widespread and abundant particularly on chalk grassland at all times of the year.

***Paederus riparius*** (Linnaeus, 1758)
7 SU16: Marlborough area, Jun-1873 (EWJ) (Meyrick, 1874); 1869-1938 (Meyrick, 1938)
8 SU02: Fovant, 31-May-95 (MDD)
SU13: Great Durnford, 6-Aug-71, 3-Aug-88 (DRN); Salisbury, 7 to 13-Apr-00 (MDD), Apr-00 (MAS) (Darby, 2001)
SU14: Bulford Ranges (SPTAE), 3-Jun - 12-Aug-08 (Boyce & Wright, 2009); Stonehenge, 22-Apr-03 (PJH) (Hodge, 2003)
SU16: Jones's Mill, 1997 (MAS)
SU23: West Winterslow, 9-Jul-01 (NB)
Resident. Local; on mud at margins of water.

**RUGILUS Leach, 1819**
***Rugilus angustatus*** (Geoffroy, 1785)
**Notable**
7 SU16: Marlborough area, Jun-1873 (EWJ) (Meyrick, 1874);
SU26: Savernake, spring-1894 (GCC) (Jebb, 1895)
SU27: Ramsbury plantation, spring-1894 (GCC) (Jebb, 1895)
Resident. Very local; [in litter in damp places but also found in drier situations]. Recorded as *R. fragilis*.

***Rugilus erichsoni*** (Fauvel, 1867)
7 SU16: Gopher Wood, 13-Jul-96 (MA det MDD); Marlborough area, 1869-1938 (Meyrick, 1938)
8 SU16: Jones's Mill, 1997 (MAS)
Resident. Very local; in damp litter in marshland.

***Rugilus orbiculatus*** (Paykull, 1789)
7 SU16: Marlborough area, Jun-1873 (EWJ) (Meyrick, 1874); 1869-1938 (Meyrick, 1938)
SU27: Ramsbury, 1870-1873 (Meyrick, 1874)
8 ST91: Rushmore Park, 23-Sep-01 (AJWA)
SU03: Grovely Wood, 10-Apr-72, 29-Oct-73 (DRN)
SU21: Hamptworth Estate, 15-Apr-74, 3-Apr-78 (DRN)
Resident. Local; in damp litter including straw and under flints.

***Rugilus rufipes*** Germar, 1835
7 SU16: Marlborough area, Jun-1873 (EWJ) (Meyrick, 1874); 1869-1938 (Meyrick, 1938);
8 SU13: Salisbury, 13-Apr-00 (MDD) (Darby, 2001)
SU16: Jones's Mill, 21-Jun-97 (MDD)
SU22: Langley Wood NNR, 14-May to 25-Jun-04 (MDD) (Darby, 2005)
Resident. Local; in damp litter.

***Rugilus similis*** (Erichson, 1839)
**Notable, LCGR-Fm Cu**
8 SU14: Stonehenge, 8-Apr-03 (PJH) (Hodge, 2003)
Resident. Very local; on vegetation in grassland.

Sunius melanocephalus

**SUNIUS Stephens, 1829**
***Sunius melanocephalus*** (Fabricius, 1792 )
**Notable**
8 ST94: Tinker's Track (SPTAW), 3-Jun - 12-Aug-08 (Boyce & Wright, 2009)
SU03: Grovely Wood, 14-Apr-71 (DRN)
Resident. Very local; under stone with ants (*Lasius flavus*) and [in litter mostly in open areas on sandy or other light soils].

***Sunius propinquus*** (Brisout, 1867)
8 SU02: Knighton Wood, 7-Oct-06 (AJWA)
SU21: Hamptworth Estate, 28-Oct-75 (DRN)
Resident. Very local; in straw near a pond.

Subfamily STAPHYLININAE

Tribe OTHIINI

**ATRECUS Jacquelin du Val, 1856**

***Atrecus affinis*** (Paykull, 1789)

**SQS-1**

7 ST99: Flisteridge Wood, 6-May-98 (KNAA)
SU16: Marlborough area, Jun-1873 (EWJ) (Meyrick, 1874); 1869-1938 (Meyrick, 1938)
SU26: Savernake Forest, 9-Jun to 1-Jul, 19-Jul to 9-Aug-94 (EN det SAW); 31-May-00 (WANHS det MDD)
8 ST84: Longleat, 29-Aug-99 (MB)
ST93: Great Ridge Wood, 8, 10-May-00 (MDD)
SU03: Grovely Wood, 8-Apr-74 (DRN); 12-Jun-00 (MDD); 14-Jun-01 (WANHS det MDD)
SU12: Burnt Ground Wood, 31-Oct-75 (DRN)
SU13: Little Durnford Down, 20-May-03 (MDD); Salisbury, Bemerton Folly, 7-Mar - 15-Jun-02 (MDD) (Darby, 2002d)
SU22: Bentley Wood, 11-Aug-01 (MDD); Langley Wood, 6-Jun-02 (MDD)
SU26: Bedwyn Brail, 2-Jun-03 (MDD)
Resident. Widespread; under bark and in rotten wood.

**OTHIUS Stephens, 1829**

***Othius angustus*** Stephens, 1833

8 SU22: Bentley Wood, 3-Sep-97 (WANHS det MDD)
Resident. Very local; in litter.

***Othius laeviusculus*** Stephens, 1833

7 SU16: Marlborough area, Jun-1873 (EWJ) (Meyrick, 1874); 1869-1938 (Meyrick, 1938)
SU26: Savernake Forest, 11 to 30-May-95 (EN det MDD)
8 SU03: Wylye Down NNR, 2-Dec-06 to 22-Feb-07 (MDD) (Darby, 2009)
Resident. Very local; at roots of vegetation in grassland.

***Othius punctulatus*** (Goeze, 1777)

7 SU16: Marlborough area, Jun-1873 (EWJ) (Meyrick, 1874); 1869-1938 (Meyrick, 1938)
SU26: Savernake Forest, 25-Apr-95 to 30-May-95 (EN det MDD)
8 ST91: Cranborne Chase, 30-Jun-99 (MDD)
ST93: Great Ridge Wood, 18-Apr-74 (DRN)
SU03: Grovely Wood, 17-Apr-71, 3-Apr-75 (DRN); Wylye Down NNR, 22-Feb to 29-Mar-07 (MDD) (Darby, 2009)
SU13: Boscombe Down, 9-Oct-94 (MDD)
SU21: Burnt Ground Wood, 19-Apr-74 (DRN); Hamptworth Estate, 15-Apr-74, 28-Oct-75 (DRN)
SU22: Langley Wood NNR, 22-Apr-04 to 24-Mar-05, 15-Sep to 6-Oct-06 (MDD) (Darby, 2005, 2008)
Resident. Widespread; in litter and under stones in woodland and grassland.

***Othius subuliformis*** Stephens, 1833

7 SU26: Savernake Forest, 23-Oct-01 (MDD)
8 SU13: Salisbury, Bemerton Folly, 7-Mar - 15-Jun-02 (MDD) (Darby, 2002d)
SU22: Langley Wood NNR, 14-May-04 to 24-Mar-05, 25-Aug-06 to 24-Mar-07 (MDD) (Darby, 2005, 2008)
Resident. Very local; in moss and litter in woodland. Formerly *O. myrmecophilus.*

Tribe STAPHYLININI

Subtribe PHILONTHINA

**BISNIUS Stephens, 1829**

***Bisnius cephalotes*** (Gravenhorst, 1802)

8 SU03: Grovely Wood, 8-Apr-74 (DRN)
Resident. Very local; in a rabbit carcass.

***Bisnius fimetarius*** (Gravenhorst, 1802)

7 SU16: Marlborough area, 4-May-1879 (FED-H) (Meyrick, 1880)
8 ST91: Rushmore Park, 23-Sep-01 (AJWA)
ST92: Sutton Mandeville, Apr-00 (MDD)
ST93: Great Ridge Wood, 2-May-00 (MDD det RT)
SU02: Knighton Wood, Mar, Oct-07, Jul-08 (AJWA)
SU23: Bentley Wood, 31-May-97 (MDD det RT)
Resident. Local; in moss and litter mainly in woodland.

***Bisnius pseudoparcus*** Brunne, 1976

8 ST91: Rushmore Park, 23-Sep-01 (AJWA)
Resident. Very local; in a dung heap.

***Bisnius sordidus*** (Gravenhorst, 1802)

7 SU16: Poulton, 1870-1873 (Meyrick, 1874)
8 SU13: Salisbury, Apr-00 (MDD) (Darby, 2001)
Resident. Very local; [in dung, haystack refuse, etc.].

***Bisnius subuliformis*** (Gravenhorst, 1802)

8 SU03: Grovely Wood, 12-Apr-72 (DRN det AAA)
Resident. Very local; in a grey squirrel's drey.

**GABRIUS Stephens, 1829**

***Gabrius breviventer*** (Sperk, 1835)

8 SU13: Salisbury, 19-Jun-00 (MDD) (Darby, 2001)
Resident. Very local; in moss in sallow carr. Formerly *G.pennatus.*

***Gabrius nigritulus*** (Gravenhorst, 1802)

7 SU16: Marlborough area, 1869-1938 (Meyrick, 1938)
8 SU13: Salisbury, Apr-00 (MAS) (Darby, 2001)
Resident. Very local; in litter in reed bed.

***Gabrius piliger*** Mulsant & Rey, 1876

7 SU26: Savernake Forest, 7 to 28-Aug-93 (EN det SAW)
Resident. Very local; [in damp litter or carrion].

***Gabrius splendidulus*** (Gravenhorst, 1802)
8 ST82: Oysters Coppice, 6-Oct-02 (MDD)
ST84: Longleat, 4-May-92 (AGD); 12-Apr-02 (MB)
ST93: Fonthill, 26-Jun-02 (MDD); Great Ridge Wood, 2, 8-May-00 (MDD)
SU03: Grovely Wood, 8, 16-Apr-74, 8-Apr-75 (DRN); 17-Feb-98, 27-May-02 (MDD); 14-Jun-01 (WANHS det MDD)
SU21: Hamptworth Common, 30-Mar-03 (MDD); Burnt Ground Wood, 19-Apr-74 (DRN)
SU22: Bentley Wood, 5-May-03 (MDD); Blackmoor Copse, 3-Aug-03 (APF) (Foster, 2004); Langley Wood NNR, 12-Apr-74 (DRN); 6-Jun-02, 29-May-03, 24-Mar to 6-Jun-07 (MDD) (Darby, 2008)
SU23: Porton Down, 1970 -2001 (Anon)
SU26: Bedwyn Brail, 2-Jun-03 (MDD)
Resident. Widespread; under bark and in litter in woodland.

***Gabrius trossulus*** (Nordmann, 1837)
7 SU16: Wootton Rivers, 1869-1938 (Meyrick, 1938)
8 SU13: Salisbury, 7, 13-Apr, 2-May-00 (MDD) (Darby, 2001)
Resident. Very local; in damp places and litter in a reedbed.

**NEOBISNIUS Ganglbauer, 1895**
***Neobisnius villosulus*** (Stephens, 1833)
8 SU13: Salisbury, Apr, 3-May-00 (MAS) (Darby, 2001)
Resident. Very local; in litter in reedbed.

**PHILONTHUS Stephens, 1829**
***Philonthus addendus*** Sharp, 1867
8 SU22: Langley Wood NNR, 15-Jul to 5-Aug-06 (MDD) (Darby, 2008)
Resident. Very local; in litter.

***Philonthus carbonarius*** (Gravenhorst, 1802)
7 SU16: Marlborough area, Jun-1873 (EWJ) (Meyrick, 1874); 24 Feb 1879 (WFB) (Meyrick, 1880); 1869-1938 (Meyrick, 1938)
SU26: Savernake Forest, 20-Apr to 21-Sep-93 (EN det SAW)
8 ST82: Oysters Coppice, 6-Oct-02 (MDD)
ST93: Great Ridge Wood, 18-Apr-74, 29-Oct-75 (DRN); 2-May-00 (MDD det RT)
ST94: Warminster Ranges (SPTAW), 3-Jun -12-Aug-08 (Boyce & Wright, 2009)
SU02: Knighton Wood, Sep-06, Mar, Oct-07 (AJWA)
SU03: Grovely Wood, 3- Apr-75 (DRN); 21-Aug-01 (MDD); Wylye Down NNR, 31-Dec to 21-Jan-07, 5-Jun to 10-Jul-07 (MDD) (Darby, 2008a)
SU14: Haxton (SPTAE), 3-Jun - 12-Aug-08 (Boyce & Wright, 2009)
SU21: Hamptworth Estate, 15-Apr-74 (DRN)
SU22: Langley Wood NNR, 17-Sep to 19-Nov-04 (MDD) (Darby, 2005)
Resident. Widespread; under stones and in moss, straw, litter, etc.. Formerly *P. varius*.

***Philonthus cognatus*** Stephens, 1832
7 SU09: Upper Waterhay, 17-May-054-Jun-05 (JKG det MDD)
SU16: Marlborough area, 1869-1938 (Meyrick, 1938)
SU26: Savernake Forest, 20-Apr to 21-Sep-93, 28-Apr to 18-May, 30-Aug to 20-Sep-94 (EN det SAW)
8 ST82: Oysters Coppice, 6-Oct-02 (MDD)
ST85: Clanger Wood, 1-Apr-85 (AGD)
ST93: Great Ridge Wood, 18-Apr-74 (DRN)
SU01: Martin Down NNR, 1999 (EN, 1999)
SU02: Knighton Wood, Apr, Sep-07 (AJWA); Middleton Down, 6-Nov-73 (BL)
SU03: Grovely Wood, 24-Aug-72 (DRN); Wylye Down NNR, 22-Feb to 22-Dec-07 (MDD) (Darby, 2009)
SU22: Blackmoor Copse, 4-18-May-02 (APF) (Foster, 2004); Langley Wood NNR, 15-Sep-06 to 6-Jun-07 (MDD) (Darby, 2008)
Resident. Widespread; in litter, on vegetation, in moss on grassland and in woodland.

***Philonthus debilis*** (Gravenhorst, 1802)
8 ST93: Great Ridge Wood, 18-Apr-74 (DRN)
Resident. Very local; in straw in heather area.

***Philonthus decorus*** (Gravenhorst, 1802)
7 SU16: Marlborough area, 1869-1938 (Meyrick, 1938); 06 Feb 1879 (WFB) (Meyrick, 1880)
8 ST93: Great Ridge Wood, 7-May-00 (MDD)
SU13: Salisbury, 15-Jun-02 (MDD)
SU22: Homan's Copse, 20-Jul-73 (DRN); Langley Wood, 14-May to 17-Sep-04, 5-Aug-06 to 2-Jul-07 (MDD) (Darby, 2005, 2008)
SU23: Porton Down, 1970-2001 (Anon, 2001)
Resident. Local; in litter, plant roots, etc., in woodland.

***Philonthus discoideus*** (Gravenhorst, 1802)
8 ST91: Rushmore Park, 23-Sep-01 (AJWA)
Resident. Very local; in a dung heap.

***Philonthus ebeninus*** (Gravenhorst, 1802)
**Notable**
7 SU16: Marlborough area, Jun-1873 (EWJ) (Meyrick, 1874), 1869-1938 (Meyrick, 1938)
8 SU22: Langley Wood NNR, 15-Jul to 5-Aug-06 (MDD) (Darby, 2007)

Resident. Very local; in litter [in cow dung, moss, and a rotting fence post on sandhills, peat bogs and peat moorland]

Philonthus ebeninus

?***Philonthus fumarius*** (Gravenhorst, 1806)
**Nb**
7 SU16: Marlborough area, 1869-1938 (Meyrick, 1938)
SU27: Ramsbury, spring-1894, (GCC) (Jebb, 1895)
There must be some doubt about these (this?) determinations. *P. fumarius* is primarily a fen species which has also been found in a few other marshes, especially on the coast, in reed or sedge debris and in moss]. There is no specimen under this name in the MCC.

***Philonthus intermedius*** (Boisduval & Lacordaire, 1835)
8 ST85: Clanger Wood, 1-Apr-85 (AGD)
SU23: Porton Down, 3-Jul-99 (MAS)
Resident. Very local; [in rotting grass and herbivore dung].

***Philonthus laminatus*** (Creutzer, 1799)
7 SU16: Marlborough area, 6-Feb-1879 (GTKM) (Meyrick, 1880); 1869-1938 (Meyrick, 1938)
SU26: Savernake Forest, 21-Sep to 11-Oct-93, 19-Jul to 9-Aug-94 (EN det SAW); 15-Jul to 3-Aug-95 (EN det MDD)
8 ST92: Sutton Mandeville, 27-Mar-08 (MDD)
SU03: Wylye Down NNR, 22-Feb to 10-Jul-07 (MDD) (Darby, 2009)
SU13: Great Durnford, 16-Aug-69 (DRN)
SU21: Hamptworth Estate, 28-Oct-75 (DRN)
SU22: Langley Wood NNR, 8-Oct to 10-Dec-04 (MDD) (Darby, 2005)
Resident. Widespread; in damp litter, straw, and dung;

***Philonthus longicornis*** Stephens, 1832
7 SU16: Marlborough area, Jun-1873 (EWJ) (Meyrick, 1874); 1869-1938 (Meyrick, 1938)
Resident. Very local; [in moss, dung, haystack refuse, etc.].

***Philonthus marginatus*** (Ström, 1768)
7 SU16: Marlborough area, 1869-1938 (Meyrick, 1938)
SU26: Savernake Forest, 1-Jul to 9-Aug-94 (EN det SAW); 15-Jul to 3-Aug-95 (EN det MDD)
8 ST85: Clanger Wood, 1-Apr-85 (AGD)
SU02: Knighton Wood, Sep-06, Aug-07, May-08 (AJWA)
Resident. Local; in litter, dung, etc., in woodland.

***Philonthus micans*** (Gravenhorst, 1802)
7 SU16: Overton, 1870-1873 (Meyrick, 1874); Marlborough area, 1869-1938 (Meyrick, 1938)
Resident. Very local; [in marshy places, roots of grass, moss, etc.].

***Philonthus nigrita*** (Gravenhorst, 1806)
7 SU16: Marlborough area, 1869-1938 (Meyrick, 1938)
SU27: Ramsbury, spring-1894 (GCC) (Jebb, 1895)
Resident. Very local; [in marshy places, roots of grass, sphagnum, etc.].

***Philonthus politus*** (Linnaeus, 1758)
7 SU16: Marlborough area, 24 Feb 1879 (WFB) (Meyrick, 1880), 1869-1938 (Meyrick, 1938); Overton, 1870-1873 (Meyrick, 1874)
SU26: Savernake Forest, 9-Jun to 1-Jul-94 (EN det SAW)
8 ST93: Great Ridge Wood, 16-Aug-76 (DRN)
SU22: Langley Wood NNR, 6 to 27-Aug-04, 5-Aug to 26-Oct-06 (MDD) (Darby, 2005, 2008)
Resident. Local; in decaying vegetation. Probably widespread.

***Philonthus quisquiliarius*** (Stephens, 1832)
8 SU13: Lower Woodford, 28-Jul-78 (DRN)
Resident. Very local; on mud near R. Avon.

***Philonthus rectangulus*** Sharp, 1874
7 SU26/27: Axford, 1939 (det KGB) (Knight, 1940)
8 ST91: Rushmore Park, 23-Sep-01 (AJWA)
Resident, recent introduction (1935, from the East). Very local; in a dung heap [garden compost, etc.]. Synanthropic species.

***Philonthus rufipes*** (Stephens, 1832)
7 SU16: Marlborough area, 1869-1938 (Meyrick, 1938)
Resident. Very local; [in wet places, in moss, at roots of grass and occasionally damp haystack refuse]. Formerly known as *P. immundus*.

***Philonthus sanguinolentus*** (Gravenhorst, 1802)
7 SU26: Savernake Forest, 28-Jul-93 (EN)

8 ST85: Clanger Wood, 1-Apr-85 (AGD)
SU03: Grovely Wood, 3-Apr-75 (DRN)
SU21: Hamptworth Estate, 15-Apr-74 (DRN)
Resident. Local; in litter, moss and old straw.

***Phlionthus spinipes*** Sharp, 1874
8 ST91: Rushmore Park, 23-Sep-01 (AJWA)
Resident. Very Local; in a dung heap.

***Philonthus splendens*** (Fabricius, 1792)
7 SU16: Marlborough area, 24 Feb 1879 (WFB) (Meyrick, 1880); 1869-1938 (Meyrick, 1938)
SU26: Savernake Forest, 15-Jul to 3-Aug-95 (EN det MDD)
8 SU22: Langley Wood NNR, 15-Jul to 5-Aug-06 (MDD) (Darby, 2008)
Resident. Very local; in dung [and occasionally carrion].

***Philonthus succicola*** Thomson, C. G, 1860
7 SU16: Poulton, 1870-1873 (Meyrick, 1874) Marlborough area, 1869-1938 (Meyrick, 1938)
SU26: Savernake Forest, 15-Jun-73 (SAW); 19-Jul to 9-Aug-94 (EN det SAW)
8 ST93: Great Ridge Wood, 16-Aug-76 (DRN)
Resident. Very local; in rotting vegetation and fungi.

***Philonthus tenuicornis*** Mulsant & Rey, 1853
7 SU26: Savernake Forest, 1-Jul to 9-Aug-94 (EN det SAW)
8 ST91: Rushmore Park, 23-Sep-01 (AJWA)
SU02: Knighton Wood, 6-Jul-07 (AWJA)
SU22: Langley Wood NNR, 5-Aug-06 to 26-Oct-07 (MDD) (Darby, 2008)
Resident. Very local; in rotting vegetation etc. in woodland.

***Philonthus umbratilis*** (Gravenhorst, 1802)
7 SU16: Marlborough area, 1869-1938 (Meyrick, 1938)
Resident. Very local; in moss.

***Philonthus varians*** (Paykull, 1789)
7 SU16: Marlborough area, Jun-1873 (EWJ) (Meyrick, 1874); 1869-1938 (Meyrick, 1938)
SU26: Savernake Forest, 27-May to 16-Jun-93, 28-Apr to 9-Jun, 20-Sep to 11-Oct-94 (EN det SAW)
8 ST94: Norton Down (SPTAW), 15-May-02 (PJH) (Hodge, 2002)
SU03: Grovely Wood, 5-Apr-71 (DRN); Wylye Down NNR, 27-Jan to 22-Feb-07 (MDD) (Darby, 2009)
SU22: Langley Wood NNR, 3 to 24-Mar-05 (MDD) (Darby, 2005)
Resident. Local; in decaying vegetation, herbivore dung, etc.

Subtribe QUEDIINA

**HETEROTHOPS Stephens, 1829**

?***Heterothops dissimilis*** (Gravenhorst, 1802)
**RDBK**
7 SU16: Marlborough area, 1869-1938 (Meyrick, 1938)
Resident. Very local; [in grass piles]. This species has been confused with *H. minutus* and most early records probably refer to that species. Consequently this record should be considered as doubtful.

***Heterothops niger*** Kraatz, 1868
8 SU01: Martin Down NNR, 1999 (EN, 1999)
Resident. Very local; [in mole nests].

**QUEDIUS Stephens, 1829**

DISTICHALIUS Casey, 1915

***Quedius cinctus*** (Paykull, 1790)
7 SU16: Marlborough area, 1869-1938 (Meyrick, 1938)
8 SU03: Wylye Down NNR, 20-Sep-05, 4-May to 5-Jun-07 (MDD) (Darby, 2009)
SU21: Hamptworth Estate, 28-Oct-75 (DRN)
SU22: Bentley Wood, 31-May-97 (MDD)
Resident. Local; in grass trap in woodland, a cut thistle pile, and at roots of vegetation in grassland.

MICROSAURUS Dejean, 1833

***Quedius aetolicus*** Kraatz, 1858
**Na, IEC-3, SQS-16**
8 ST91: Cranborne Chase, 21-May-02 (MDD) (Meyrick, 1938)
SU02: Knighton Wood, Sep-06 (AJWA)
Resident. Very local; under bark and in flight intercept trap. [Most records are from squirrel dreys].

***Quedius brevicornis*** (Thomson, C. G, 1860)
**Nb, SQS-8**
8 SU02: Knighton Wood, 7-Oct-06 (AJWA) (Allen, 2008)
Resident. Very local; in aerial litter, bird nests, etc.. [Under bark, in rot holes, in bird nests, in hornet nests, in fungi and under stones].

***Quedius cruentus*** (Olivier, 1795)
8 ST85: Green Lane Wood, 31-May-04 (KNAA)
ST93: Fonthill, 26-Jun-02 (MDD); Great Ridge Wood, 29-Oct-75 (DRN)
SU02: Knighton Wood, Sep-06, Apr, May, 6-Jul, Sep-07 (AJWA)
SU03: Grovely Wood, 6-Aug-70 (DRN)
SU13: Salisbury, Apr-00 (MAS) (Darby, 2001)
SU14: Haxton (SPTAE), 17-May-02 (PJH) (Hodge, 2002)
SU22: Blackmoor Copse, 3-Aug-03 (APF) (Foster, 2004)
Resident. Local; in decaying vegetation.

***Quedius invreae*** Gridelli, 1924
**Nb**
7 SU26: Savernake Forest, 20-Sep to 11-Oct-94 (EN det SAW)
8 SU02: Knighton Wood, Sep, Nov-06 (AJWA) (Allen, 2008)
Resident. Very local; in flight intercept traps. [Under bark and at sap from trees particularly those damaged by *Cossus* larvae, also in rotting vegetation and wasps' nests].

***Quedius lateralis*** (Gravenhorst, 1802)
7 SU26: Savernake Forest, 21-Sep to 11-Oct-93, 20-Sep to 11-Oct-94 (EN det SAW)
8 ST84: Longleat, 29-Aug-99 (MB)
ST91: Cranborne Chase, 22-Jun-99 (MDD)
SU03: Grovely Wood, 25-Oct-71, 18-Jul-73 (DRN)
SU13: Salisbury, Bemerton Folly, 7-Mar - 15-Jun-02 (MDD) (Darby, 2002d)
SU22: Langley Wood NNR, 14-May to 19-Nov-04, 15-Sep to 6-Oct-06 (MDD) (Darby, 2005, 2008)
Resident. Local; in litter, etc. in woodland. Mainly an autumn species.

***Quedius maurus*** (Sahlberg, C.R, 1830)
**IEC-3**, **SQS-4**
7 SU26: Savernake Forest, 15-Jun-73 (SAW); 20-Apr to 27-May-93 (EN det SAW)
8 SU22: Langley Wood NNR, 29-May-03 (MDD)
Resident. Very local; under bark in woodland.

***Quedius mesomelinus*** (Marsham, 1802)
7 SU26: Savernake Forest, 20 to 25-Apr-93 (EN det SAW)
8 SU02: Knighton Wood, Sep-06, Apr-07 (AJWA)
SU12: Burnt Ground Wood, 31-Oct-75 (DRN)
Resident. Very local; in fungus on oak [grass heaps nd carrion].

***Quedius nigrocaeruleus*** Fauvel, 1874
**Nb**
8 SU02: Knighton Wood, Sep-06, Sep-07 (AJWA)
Resident. Very local; in a flight intercept trap. [In nests of moles, blue tits and wasps, a rabbit burrow, a sandpit, a wooded bog, etc.].

***Quedius puncticollis*** (Thomson, C.G., 1867)
**Nb**
8 SU02: Knighton Wood, Sep-07, Sep-08 (AJWA)
Resident. Very local; in a flight intercept trap [mainly in mole nests, although it has also been found in rabbit burrows and a hornet's nest].

***Quedius xanthopus*** Erichson, 1839
**Nb**, **IEC-3**, **SQS-4**
7 SU16: Marlborough area, 1869-1938 (Meyrick, 1938)
SU26: Savernake Forest, spring-1894 (GCC) (Jebb, 1895); 6-Dec-73 (RCS); 15-Jun-73 (SAW); 23-Oct-90 (TGW); 11-Oct to 9-Nov-93, 9-Jun to 19-Jul-94 (EN det SAW); 15-Sep to Oct-95 (EN det MDD); 6-Oct-96 (KNAA)
8 ST73: Stourhead, 11-Feb-84, 4-Jul-95 (KNAA) (Alexander, 1996); Tucking Mill Hanging, 11-Feb-84 (KNAA)
ST93: Great Ridge Wood, 29-Oct-75 (DRN)
SU03: Grovely Wood, 8-Apr-75 (DRN)
SU23: Porton Down, 1970-2001 (Anon, 2001); Thorny Down woods, 16-Oct-00 (MDD)
Resident. Widespread; in fungi and under bark in woodland.

*Quedius xanthopus*

QUEDIONUCHUS Sharp, 1884

***Quedius curtipennis*** Bernhauer, 1908
8 ST93: Great Ridge Wood, 18-Apr-74 (DRN)
SU03: Wylye Down NNR, 27-Jan to 22-Feb-07 (MDD) (Darby, 2009); Grovely Wood, 3-Apr-75 (DRN)
SU22: Langley Wood NNR, 3 to 25-Jun-04, 25-Apr to 6-Jun-07 (MDD) (Darby, 2005, 2008)
SU24: Beacon Hill (SPTAE), 3-Jun - 12-Aug-08 (Boyce & Wright, 2009)
Resident. Very local; in litter and at roots of plants.

QUEDIUS Stephens, 1829

***Quedius fuliginosus*** (Gravenhorst, 1802)
7 SU16: Marlborough area, Jun-1873 (EWJ) (Meyrick, 1874); 24 Feb 1879 (WFB) (Meyrick, 1880), 1869-1938 (Meyrick, 1938)
8 ST93: Great Ridge Wood, 28-Apr-00 (MDD)
SU03: Wylye Down NNR, 16-Oct to 13-Nov-06 (MDD) (Darby, 2009)
SU21: Hamptworth Estate, 15-Apr-74 (DRN)
SU22: Bentley Wood, 3-Sep-97 (WANHS det MDD); Langley Wood NNR, 1 to 22-Apr-04, 17-Nov-06 to 6-Jun-07 (MDD) (Darby, 2005, 2008)
Resident. Widespread; in litter. The Marlborough records are indeterminate because *Q. curtipennis* was not separated from *fuliginosus* until 1943.

***Quedius levicollis*** (Brullé, 1832)
7 SU16: Marlborough area, 1869-1938 (Meyrick, 1938)
8 ST91: Rushmore Park, 23-Sep-01 (AJWA)
ST94: Ladywell (SPTAW), 3-Jun - 12-Aug-08 (Boyce & Wright, 2009)
SU01: Martin Down NNR, 1999 (EN, 1999)
SU21: Hamptworth Estate, 28-Oct-75 (DRN); Landford Bog, 2003 (APF) (Foster, 2003)
SU22: Langley Wood NNR, 6 to 26-Oct-06 (MDD) (Darby, 2008)
SU23: Porton Down, 1970-2001 (Anon, 2001)
Resident. Local; in litter, under logs and stones. Formerly known as *Q. tristis.*

***Quedius molochinus*** (Gravenhorst, 1806)
7 SU16: Marlborough area, 17-Feb-1879 (WFB) (Meyrick, 1880); 1869-1938 (Meyrick, 1938)
8 ST93: Great Ridge Wood, 18-Apr-74 (DRN)
SU03: Grovely Wood, 8-Apr-72 (DRN); Wylye Down NNR, 10-Jul to 3-Aug-07 (MDD) (Darby, 2009)
Resident. Local; in straw and litter, under stone and at roots of plants.

RAPHIRUS Stephens, 1829

***Quedius fumatus*** (Stephens, 1833)
7 SU16: Marlborough area, 1869-1938 (Meyrick, 1938)
8 ST82: Cann Wood, n.d. (JHK)
SU13: Salisbury, Apr-00 (MAS) (Darby, 2001)
SU21: Hamptworth Estate, 1-Apr-75 (DRN)
SU23: Porton Down, 1970-2001 (Anon, 2001), 16-Oct-00 (MDD)
Resident. Local; in damp litter, and straw near a pond.

***Quedius humeralis*** Stephens, 1832
8 SU16: Jones's Mill, 21-Jun-97 (MDD)
SU13: Salisbury, Bemerton Folly, 7-Mar - 15-Jun-02 (MDD) (Darby, 2002d)
Resident. Very local; in damp moss and litter.

***Quedius maurorufus*** (Gravenhorst, 1806)
7 SU16: Marlborough area, Jun-1873 (EWJ) (Meyrick, 1874); 1869-1938 (Meyrick, 1938)
SU27: Ramsbury, spring-1894 (GCC) (Jebb, 1895)
8 SU13: Salisbury, 7, 13-Apr-00 (MDD) (Darby, 2001)
SU22: Langley Wood NNR, 22-Apr to 10-Dec-04, 6 to 22 Jan 07 (MDD) (Darby, 2005, 2008)
Resident. Local; in wet vegetation in woodland and a reedbed.

***Quedius nigriceps*** Kraatz, 1857
8 SU13: Salisbury, 29-May-02 (MDD)
SU22: Langley Wood NNR, 6-Oct-06 to 20-Feb-07 (MDD) (Darby, 2008)
Resident. Very local; in litter in woodland.

***Quedius nitipennis*** (Stephens, 1833)
7 SU16: Marlborough area, 1869-1938 (Meyrick, 1938)
8 SU22: Langley Wood NNR, 8-Dec-06 (MDD) (Darby, 2008)
Resident. Very local; in damp litter.

***Quedius picipes*** (Mannerheim, 1830)
7 ST96: Spye Park, 25-Apr-87 (KNAA)
SU16: Marlborough area, Jun-1873 (EWJ) (Meyrick, 1874); 1869-1938 (Meyrick, 1938)
SU26: Savernake Forest, 31-May-04 (BLe)
8 ST94: Norton Down (SPTAW), 15-May-02 (PJH) (Hodge, 2002)
SU01: Martin Down NNR, 1999 (EN, 1999)
SU03: Grovely Wood, 10-Apr-72, 8-Apr-74 (DRN); Wylye Down NNR, 4-May to 5-Jun-07 (MDD) (Darby, 2009)
SU14: Boscombe Down, 1994 (MA det MDD)
SU21: Hamptworth Estate, 15-Apr-74 (DRN)
SU22: Langley Wood NNR, 22-Apr to 19-Nov-04 (MDD) (Darby, 2005)
Resident. Widespread; in litter and at roots of plants in woodland and grassland.

***Quedius schatzmayri*** Gridelli, 1922
8 SU03: Wylye Down NNR, 31-Dec-06 to 27-Jan-07 (MDD) (Darby, 2009)
Resident. Very local; at roots of plants in grassland. May be under-recorded as a winter insect.

***Quedius scintillans*** (Gravenhorst, 1806)
8 SU03: Wylye Down NNR, 31-Dec-06 to 27-Jan-07 (MDD) (Darby, 2009)
Resident. Very local; at roots of plants in grassland. May be under-recorded as a winter insect

***Quedius semiaeneus*** (Stephens, 1833)
7 SU16: Marlborough area, 1-Jan-0031-Dec-05 (Meyrick, 1906); 1869-1938 (Meyrick, 1938)
SU26: Savernake Forest, 20-Sep to 11-Oct-94 (EN det SAW)
Resident. Very local; [in moss, haystack refuse, etc. often under stones, and often by the sea]

***Quedius semiobscurus*** (Marsham, 1802)
7 SU16: Marlborough area, 1869-1938 (Meyrick, 1938)
8 ST94: Ladywell (SPTAW), 3-Jun - 12-Aug-08 (Boyce & Wright, 2009)
SU01: Martin Down NNR, 1999 (EN, 1999)
SU03: Wylye Down NNR, 4-May to 5-Jun, 3-Aug to 4-Sep-07 (MDD) (Darby, 2008)
SU13: Boscombe Down, 9-Oct-94 (MDD)
Resident. Local; at roots of plants in grassland.

Subtribe STAPHYLININA

**CREOPHILUS Leach, 1819**

***Creophilus maxillosus*** (Linnaeus, 1758)
7 SU16: Marlborough area, 30 May 1879 (RAW) (Meyrick, 1880);1869-1938 (Meyrick, 1938)
8 ST93: Great Ridge Wood, 16-Aug-76 (DRN)
ST95: West Lavington, 6-Apr-40 (ITH) (DFL, 1948)
SU03: Grovely Wood, 3-Aug-71 (DRN)
SU04: Winterbourne Stoke, 22-Aug-40 (KJA) (DFL, 1948)
Resident. Local; in carrion.

**OCYPUS Leach, 1819**
OCYPUS Leach, 1819
**Devil's Coach-horse**
***Ocypus olens*** (Müller, O.F, 1764)
7 SU16: Gopher Wood, May-96 (MA det MDD); Knap Hill, 24-Sep-01 (IJG); Lockeridge, 16-Jun-73 (JEO); Marlborough area, Jun-1873 (EWJ) (Meyrick, 1874); 1869-1938 (Meyrick, 1938)
SU26: Savernake Forest, 28-Jul to 28-Aug-93; 19-Jul to 9-Aug-94 (EN det SAW)
8 ST84: Longleat, 28-Aug-00 (MB)
ST93: Stockton Wood, 1995 (MA det MDD)
ST95: Phantom Wood, 8-May-41 (GHW) (DFL, 1948); West Lavington, 1938 (PLP) (DFL, 1948)
SU01: Martin Down NNR, 1979-1992 (RGB); 1999 (EN, 1999)
SU03: Wylye Down NNR, 16-Oct-06 to 10-Oct-07 (MDD) (Darby, 2009)
SU12: Coombe Bissett Down, May-Jun-00 (MDD)
SU13: Cockey Down, 1-Aug-88 (PM); 9-Sep-97 (CH)
SU15: Beach's Plantation (SPTAE), 1-May-05 (MGT); 4-Sep-05 (JMW)
SU22: Bentley Wood, 12-Feb-01 (NB det MDD), 6-Oct-97 (JN); Langley Wood NNR, 22-Apr to 10-Dec-04, 5- Aug-06 to 20-Feb-07 (MDD) (Darby, 2005, 2008)
SU23: Porton Down, 10-Aug-96 (PTW det MDD); Oct-90 (PTW det MDD)
SU24: Driver Training Area (SPTAE), 1-May-05 (MGT)
SU26: Easton Royal, 12-Oct-01 (IJG)
Resident. Widespread and frequent; at roots of plants under stones and in litter, in grassland and woodland.

PSEUDOCYPUS Mulsand and Rey, 1876
***Ocypus aeneocephalus*** (De Geer, 1774)
7 ST87: Rack Hill, 1995 (MA det MDD)
SU16: Marlborough area, Jun-1873 (EWJ) (Meyrick, 1874); 8-May-1879 (WFB) (Meyrick, 1880); 1869-1938 (Meyrick, 1938)
8 SU01: Martin Down NNR, 1999 (EN, 1999)
SU03: Grovely Wood, 8, 14-Apr-71 (DRN); Wylye Down NNR, 16-Oct to 13-Nov-06 (MDD) (Darby, 2009)
SU15: Manningford, 4-Aug-88 (DRN)
SU22: Langley Wood NNR, 10-Dec-04 to 24-Mar-05, 15-Sep-06 to 2-Jul-07 (MDD) (Darby, 2005, 2008)
Resident. Widespread; at roots of plants, under stones and in litter, in woodland and grassland.

***Ocypus fortunatarum*** Wollaston, 1871
**Nb**
8 SU03: Wylye Down NNR, 27-Jan-07 (MDD) (Darby, 2009)
Resident. Very local; at roots of plants in calcareous grassland.

***Ocypus fuscatus*** (Gravenhorst, 1802)
**Nb**
7 SU16: Marlborough area, 1870-1873 (Meyrick, 1874), 1869-1938 (Meyrick, 1938)
Resident. Very local; [in dry dung, at roots of plants, under stones, etc. in grassland, downland, fens and possibly woodland].

Ocypus fortunatorum

**ONTHOLESTES Ganglbauer, 1895**
***Ontholestes murinus*** (Linnaeus, 1758)
7 SU16: Marlborough area, Jun-1873 (EWJ) (Meyrick, 1874); 1869-1938 (Meyrick, 1938)
SU26: Savernake Forest, 25-Apr to 11-May-95, 15-Jul to 3-Aug-95 (EN det MDD)
8 ST84: Longleat, 29-Aug-99 (MB)
ST92: Sutton Mandeville, Jun-96 (MDD)
ST93: Great Ridge Wood, 2-Aug-82 (DRN)
ST94: Bowls Barrow (SPTAW), 16-May-02 (PJH) (Hodge, 2002)
SU02: Middleton Down, Jul-89 (BL) (Meyrick, 1938)
SU03: Grovely Wood, 3-Aug-71 (DRN)
SU14: Stonehenge, 22-Apr-03 (PJH) (Hodge, 2003)
SU22: Bentley Wood, 27-May-97 (MDD); Langley Wood, 15-Jun to 5-Aug-06 (MDD) (Darby, 2008)

***Ontholestes tessellatus*** (Fourcroy, 1758)

7 SU16: Marlborough area, 1869-1938 (Meyrick, 1938)
SU26: Savernake Forest, 15-Jul to 3-Aug-95 (EN det MDD)
8 ST93: Great Ridge Wood, 2-Aug-82 (DRN)
SU03: Grovely Wood, 11-Aug-71 (DRN)
SU15: Manningford, 13-Aug-87 (DRN)
Resident. Local; in carcases.

**PLATYDRACUS Thomson, C.G, 1858**
***Platydracus fulvipes*** (Scopoli, 1763)
**Nb**
8 SU03: Wylye Down NNR, 4-May to 5-Jun-07 (MDD) (Darby, 2009)
SU22: Langley Wood NNR, 3 to 25-Jun-04 (MDD) (Darby, 2005)
Resident. Very local; in litter and at roots of plants in grassland and woodland.

Platydracus fulvipes

***Platydracus latebricola*** (Gravenhorst, 1806)
**Nb, LCGR-Fm Cu**
8 SU03: Wylye Down NNR, 29-Mar to 3-Aug-07 (MDD) (Darby, 2009)
SU21: Landford Bog, 24-Jun-01, 1-Dec-03 (APF) (Foster, 2001, 2003)
SU22: Langley Wood NNR, 14-May to 25-Jun-04 (MDD) (Darby, 2005)
SU25: Ludgershall, Apr - Jul-1915 (PAH) (Harwood, 1915)
Resident. Local; in litter and under stones.

***Platydracus stercorarius*** (Olivier, 1795)
8 SU01: Martin Down NNR, 1999 (EN, 1999)
SU02: Wilton, 13-Sep-95 (MDD)
SU03: Wylye Down NNR, 5-Jun to 10-Jul-07 (MDD) (Darby, 2009)
SU13: Salisbury, 28-Jul-83 (DRN)
SU15: Manningford, 4-Aug-88 (DRN)
Resident. Local; at roots of plants and walking in the open.

Platydracus latebricola

**STAPHYLINUS Linnaeus, 1758**
***Staphylinus caesareus*** Cederhjelm, 1798
**RDBi**
7 SU16: Marlborough area, 1869-1938 (Meyrick, 1938)
Resident. Very local; [under a stone on a sandy roadside verge] There must be some doubt about this determination. *S. caesareus* was confused with *S. dimidiaticornis* until the latter was split from it after 1930. Only recorded from South Devon and South Hampshire before 1970.

***Staphylinus dimidiaticornis*** Gemminger, 1851
8 SU21: Hamptworth Estate, 15-Apr-74 (DRN); Pound Bottom, 9-Apr-74 (DRN)
Resident. Very local; flying near cow dung and in old straw.

**TASGIUS Stephens, 1829**
RAYACHEILA Motschulsky, 1845
***Tasgius globulifer*** (Fourcroy, 1785)
7 SU16: Marlborough area, 1869-1938 (Meyrick, 1938)
8 ST82: Oysters Coppice, 6-Oct-02 (MDD)
SU03: Wylye Down NNR, 1-Oct-06 to 4-Sep-07 (MDD) (Darby, 2009)
SU22: Langley Wood NNR, 22-Apr-04 to 24-Mar-05 (MDD) (Darby, 2005)
Resident. Local; in litter and at roots of plants in woodland and grassland.

***Tasgius melanarius*** (Heer, 1839)
8 SU03: Wylye Down NNR, 1 to 16-Oct-06 (MDD) (Darby, 2009); Grovely Wood, 8-Apr-72, 30-Oct-75 (DRN)
SU21: Hamptworth Estate, 15-Apr-74, 20-Aug-76 (DRN)
SU22: Langley Wood NNR, 12-Apr-74 (DRN), 8-Dec-06 to 5-Jan-07 (MDD) (Darby, 2008)
Resident. Very local; in litter and at roots of plants in woodland and grassland.

***Tasgius morsitans*** (Marsham, 1802)

7 SU26: Savernake Forest, 15-Jun-73 (SAW); 21-Sep-93 (EN det SAW); 30-May-73 (DBS)
8 SU01: Martin Down NNR, 1999 (EN, 1999)
SU03: Grovely Wood, 19-Aug-95 (MAS); Little Langford, 3-Jul-96 (MAS); Wylye Down NNR, 3-Aug to 4-Sep-07 (MDD) (Darby, 2008a)
SU05: Ell Barrow (SPTAC), 3-Jun - 12-Aug-08 (Boyce & Wright, 2009)
SU22: Langley Wood NNR, 6-Aug-04 to 11-Feb-05, 5-Aug to 26-Oct-06 (MDD) (Darby, 2005, 2008)
SU23: Porton Down, 15-Aug-93 (PTW det MDD)
Resident. Widespread; at roots of plants, in litter, etc. in woodland and grassland.

***Tasgius winkleri*** (Bernhauer, 1906)
8 SU01: Martin Down NNR, 1999 (EN, 1999)
SU03: Wylye Down NNR, 22-Feb to 10-Oct-07 (MDD) (Darby, 2009)
Resident. Very local; at roots of plants in calcareous grassland.

TASGIUS Stephens, 1829

***Tasgius ater*** (Gravenhorst, 1802)
7 SU16: Marlborough area, 1869-1938 (Meyrick, 1938)
Resident. Very local; under stones or running in open. [Synanthropic in houses and outbuildings].

Tasgius pedator

***Tasgius pedator*** (Gravenhorst, 1802)
**Na, LCGR-Fh Cu**
7 SU16: Marlborough area, 1869-1938 (Meyrick, 1938)
Resident. Very local; under stones or in open. [at roots of herbage, etc. in limestone and chalk grassland]

Tribe XANTHOLININI

**GAUROPTERUS Thomson, C.G, 1860**

***Gauropterus fulgidus*** (Fabricius, 1787)
7 SU26: Savernake Forest, 23-Oct-01 (MDD)
8 SU24; Newton Tony, Oct-01 (DG det MDD)
Resident. Very local; in litter and a garden.

**GYROHYPNUS Leach, 1819**

***Gyrohypnus angustatus*** Stephens, 1833
8 SU21: Hamptworth Estate, 15-Apr-74 (DRN)
SU22: Langley Wood NNR, 14-May-04 (MDD) (Darby, 2005)
Resident. Very local; in litter and straw.

***Gyrohypnus fracticornis*** (Müller, O.F, 1776)
7 SU26: Savernake Forest, 3-Jul to 4-Aug-95 (EN det MDD)
8 ST91: Rushmore Park, 23-Sep-01 (AJWA)
SU03: Grovely Wood, 29-Oct-73 (DRN)
SU13: Salisbury, 13-Apr-00 (MDD) (Darby, 2001); Boscombe Down, 9-Oct-94 (MDD)
Resident. Very local; in straw and litter.

***Gyrohypnus punctulatus*** (Paykull, 1789)
7 SU09: Ashton Keynes, 26-Apr-01 (JKG det MDD)
SU16: Marlborough area, 1869-1938 (Meyrick, 1938)
SU26: Savernake Forest, 15-Jul to 3-Aug-95 (EN det MDD)
8 ST93: Great Ridge Wood, 18-Apr-74 (DRN)
SU03: Wylye Down NNR, 22-Feb-07 (MDD) (Darby, 2009)
SU21: Hamptworth Estate, 15-Apr-74 (DRN)
SU22: Langley Wood NNR, 6 to 26-Oct-06 (MDD) (Darby, 2008)
SU23: Porton Down, 1970-2001 (Anon, 2001)
Resident. Widespread; in litter, under dry dung and in straw, in grassland and woodland.

**HYPNOGYRA Casey, 1906**

***Hypnogyra angularis*** (Ganglbauer, 1895)
**Na**
8 SU03: Dinton Park, 15-Aug-01 (KNAA) (Alexander, 2004)
Resident. Very local; in white-rotted heartwood debris in a hollow ancient ash.

**MEGALINUS Mulsant and Rey, 1877**

***Megalinus glabratus*** (Gravenhorst, 1802)
8 SU03: Dinton Park, 15-Aug-01 (KNAA); Wylye Down NNR, 3-Aug to 4-Sep-07 (MDD) (Darby, 2009)
Resident. Very local; at roots of plants in grassland.

**NUDOBIUS Thomson, C.G, 1860**

***Nudobius lentus*** (Gravenhorst, 1806)
8 ST84: Longleat, 12-Apr-02 (MB)
ST93: Great Ridge Wood, 29-Oct-75 (DRN)
SU21: Burnt Ground Wood, 31-Oct-76 (DRN)
Hamptworth Estate, 31-Oct-73 (DRN)

Resident. Very local; under pine and beech bark.

**PHACOPHALLUS Coiffait, 1956**
***Phacophallus parumpunctatus*** (Gyllenhal, 1827)
8 ST91: Rushmore Park, 23-Sep-01 (AJWA)
Resident. Very local; in a dung heap.

**XANTHOLINUS Dejean, 1821**
PURROLINUS Coiffait, 1956
***Xantholinus elegans*** (Olivier, 1795)
8 SU01: Martin Down NNR, 1999 (EN, 1999)
Resident. Very local; at roots of vegetation in grassland.

XANTHOLINUS Dejean, 1821
***Xantholinus linearis*** (Olivier, 1795)
7 SU06: Roundway Camp, Jul-40 (FDB) (Buck, 1956)
SU09: Ashton Keynes, 7-May-01 (JKG det MDD)
SU16: Gopher Wood, 13-Jul-96 (MA det MDD); Marlborough area, 24 Feb 1879 (WFB) (Meyrick, 1880); 1869-1938 (Meyrick, 1938)
SU26: Savernake Forest, 1870-1873 (Meyrick, 1874); 15-Sep to 10-Oct-95 (EN det MDD
8 SU01: Martin Down NNR, 1999 (EN, 1999)
SU02: Knighton Wood, Sep, Oct-06, Aug-07 (AJWA)
SU03: Wylye Down NNR, 2-Dec-06 to 4-Sep-07 (MDD) (Darby, 2009)
SU13: Salisbury, Apr-00 (MAS) (Darby, 2001)
SU21: Landford Bog, 9-Jun-01 (APF) (Foster, 2001)
SU22: Bentley Wood, 27-May-97; Langley Wood NNR, 22-Apr-04 to 24-Mar-05, 26-Oct-06 to 2-Jul-07 (MDD) (Darby, 2005, 2008)
SU26: Bedwyn Brail, 5-Nov-03 (MDD)
Resident. Widespread; in litter and amongst roots in woodland and grassland.

***Xantholinus longiventris*** Heer, 1839
7 SU06: Roundway Camp, Jul-40 (FDB) (Buck, 1956)
SU08: Somerford Common, 23-Apr-01 (JKG det
SU16: Marlborough area, 1869-1938 (Meyrick, 1938)
8 ST85: Bullen Hill Farm, 25-Mar-02 (EGS det MDD)
ST93: Great Ridge Wood, 18-Apr-74 (DRN)
SU02: Knighton Wood, Mar-07 (AJWA)
SU03: Grovely Wood, 12-Jun-00; Wylye Down NNR, 29-Mar to 4-May-07 (MDD) (Darby, 2008a)
SU13: Salisbury, Bemerton Folly, 7-Mar - 15-Jun-02 (MDD) (Darby, 2002d)
SU22: Langley Wood NNR, 17-Nov to 8-Dec-06 (MDD) (Darby, 2008)
Resident. Widespread; in litter and at roots of vegetation in grassland and woodland.

## Superfamily SCARABAEOIDEA
## Family LUCANIDAE
### Subfamily SYNDESINAE

**SINODENDRON Hellwig, 1792**
**Rhinocerus Beetle**
***Sinodendron cylindricum*** (Linnaeus, 1758)
**SQS-2**
7 ST96: Spye Park, 24-Apr-87 (KNAA)
SU16: Marlborough area, 11 May 1879 (WFB) (Meyrick, 1880); 1869-1938 (Meyrick, 1938)
SU26: Savernake Forest, 29-May-76 (DHM) (Murray, 1978); 22-Jun-90 (APFw); 23-Oct-90 (TGW); 16-Jun-93 (EN det SAW); 30-May-04 (KNAA)
8 ST73: Stourhead, before 4-Jul-95 (KNAA) (Alexander, 1996)
ST84: Longleat, 15-Jun-73 (PM); 29-Aug-99 (MB)
ST91: Cranborne Chase, 16 to 25-Jun-80 (NCC); 4-Jul-97 (MAS); 22-Jun-99, 19-Apr-06 (MDD)
ST92: Sutton Mandeville, 29-Jan-98 (MDD); Wardour, 5-Jun-96 (MAS)
ST93: Great Ridge Wood, 8-May-00 (MDD)
ST95: West Lavington, 7-Jun-39 (PLB) n(DFL, 1948)
SU03: Grovely Wood, 10-Apr, 13-Aug-70, 18-Jul-73, 8-Apr-74, 7-Aug-85, 4-Aug-87 (DRN); 14-Jun-01 (WANHS det MDD)
SU15: New Mill, nr Pewsey, 2-Apr-02 (IJG)
SU16: Jones's Mill, 19-Jun-01 (APF) (Foster, 2002)
SU22: Blackmoor Copse, 13-Jun-94 (PM); Bentley Wood, 29-Jun-90 (PM); 11-Aug-01 (MDD); Blackmoor Copse, 20-May-99 (PM); 17-Mar-02 (APF) (Foster, 2004)
SU23: Porton Down, 23-Sep-90 (PTW det MDD), 14-May-00 (BL det MDD); 20-Jul-97 (MAS)
SU26: Bedwyn Brail, 2-Jun-03 (MDD)
Resident. Widespread; in rotten wood of deciduous trees.

### Subfamily LUCANINAE

**LUCANUS Scopoli, 1763**
**Stag Beetle**
***Lucanus cervus*** (Linnaeus, 1758)
**Nb, SQS-8**
7 SU16: Marlborough area, 1869-1938 (Meyrick, 1938)
SU27: Ramsbury, 1870-1873 (Meyrick, 1874)
8 ST95: West Lavington, 1-Jun-39 (PLB) (DFL, 1948)
SU03: Wilton, 17-Jun-00 (DS); 1-Jun-06 (SG)
SU12: Downton, 5-Jul-98 (ID-W)
SU13: Salisbury, 7-Jul-54 (RT); 11-Jul-83 (EG)
*Note: Many other recent records not confirmed*

Resident. Local; at roots of rotten tree stumps and in flight or walking on the ground, near habitations and not in woodland.

### Subfamily DORCINAE

Dorcus parallelopipedus

**DORCUS MacLeay, 1819**
**Lesser Stag Beetle**
***Dorcus parallelipipedus*** (Linnaeus, 1758)
**SQS-2**

7 ST96: Spye Park, 24-Apr-87 (KNAA)
SU19: Hannington, 15-Jul-96 (MEN)
SU26: Savernake Forest, 30-May-73 (DBS); 6-Jun-73 (APF); 6-Dec-73 (RCS); 6-Dec-73; 22-Jun-90 (APFw); 23-Oct-90 (TGW); 21-May-01 (IJG)
SU29: Watchfield, 10-Jul-99 (MEN)
8 ST84: Longleat, 5-Sep-97 (PM); 7, 25-Mar, 8-May, Jul, 29-Aug-99 (MB)
ST85: Green Lane Wood, 7-Jun-95 (WANHS det MDD); 19-Jun-04 (MDD); Bullen Hill Farm, 1-Jan-82 (EGS det MDD)
ST91: Cranborne Chase, 30-Jul-97 (MAS); 21-May-02 (MDD)
ST92: Sutton Mandeville, 28-Jun, 12-Jul-95 (MDD) and many other occasions.
ST93: Great Ridge Wood, 6-Aug-85 (DRN); 10-May-00 (MDD)
ST95: West Lavington, Jun-39 (ITH) (DFL, 1948); 22-May-45 (JMD) (DFL, 1948); 6-Jun-45 (PLJ) (DFL, 1948)
SU02: Knighton Wood, Jun, Jul-07 (AJWA)
SU03: Barford St. Martin, 12-Jun-89 (SP); Grovely Wood, 3, 13-Aug-71, 18-Jul-73, 4-Aug-87, 1- Aug-88 (DRN); 6-Jul-01 (KNAA) (Alexander, 2004); 14-Jun-01 (WANHS det MDD); Dinton Park, 21-Aug-01 (MDD)
SU04: Shrewton, 18-Aug-40 (KJA) (DFL, 1948)
SU06: Devizes, 6-Jul-99 (BG)
SU12: Coombe Bissett Down, 9-Jun-04 (JN); Downton, 4-May-98 (ID-W)
SU13: Idmiston, 5-Aug-82 (DRN); Little Durnford Down, 21-Oct-97, 5-Jun-06 (PM); Salisbury, 19-Jun-73 (PM)
SU22: Langley Wood NNR, 6-Jun-02 (MDD)
SU23: Porton Down, 11-Jul-93 (PTW det MDD); 15-Oct-95 (MAS)
SU26: Bedwyn Brail, 5-Sep, 5-Nov-03 (MDD)

Resident. Widespread; in rotten wood of deciduous trees and walking on the ground.

## Family TROGIDAE

**TROX Fabricius, 1775**
***Trox Scaber*** (Linnaeus, 1767)

7 SU16: Marlborough area, Jun-1873 (EWJ) (Meyrick, 1874); 1869-1938 (Meyrick, 1938)
8 SU23: West Winterslow, 9-Jul-01 (NB)

Resident. Very local; at light [on old bones].

## Family BOLBOCERATIDAE

Odonteus armiger

**ODONTEUS Samouelle, 1819**
***Odonteus armiger*** (Scopoli, 1772)
**Na**

7 SU16: Avebury, 3-Jul-07 (DB)
8 SU23: Porton Down, 4-Jun-93 (SP)

Resident. Very local; at light [under dry dung, on calcareous soils]

## Family GEOTRUPIDAE

**TYPHAEUS Leach, 1815**
**Minotaur Beetle**
***Typhaeus typhoeus*** (Linnaeus, 1758)

7 SU16: Marlborough area, 7-Mar-1879 (EKC) (Meyrick, 1880); 1869-1938 (Meyrick, 1938)
8 ST84: Longleat, 17-Jun-73, 8-Jun-96 (PM)
SU21: Hamptworth Estate, 19-Apr-74 (DRN); Landford Bog, 28-May-96, 22-Apr, 22-Oct-97, 25-Jun-98 (PM)
SU22: Langley Wood NNR, 1 to 22-Apr-04, 26-Oct-06 to 2-Jul-07 (MDD) (Darby, 2005, 2008)

Resident. Local; in burrows under deer and rabbit dung on sandy soils. A winter insect.

Typhaeus typhoeus

**ANOPLOTRUPES Jekel, 1866**
***Anoplotrupes stercorosus*** (Hartmann in Scriba, 1791)
7 SU16: Savernake Forest, 1-Jul to 9-Aug-94 (EN det SAW); 14-Sep-01 (IJG); West Woods, 11-Oct-99 (IJG); Marlborough area, 11 May 1879 (WFB) (Meyrick, 1880), 1869-1938 (Meyrick, 1938)
8 ST84: Longleat, 11-Aug-05 (DJG)
ST92: Sutton Row, 14-Sep-95 (MDD)
ST95: West Lavington, 1938 (PLP) (DFL, 1948)
SU21: Burnt Ground Wood, 21-Jul-73 (DRN)
SU22: Bagfield Copse, 19-Aug-88 (DRN); Langley Wood, 6-Jun-02, 22-Apr-04 to 29-Oct-04, 5-Aug-06 to 2-Jul-07 (MDD) (Darby, 2005, 2008)
Resident. Widespread and frequent; under dung and walking in woodland. (A 12 month pitfall trapping exercise in Langley Wood NNR recorded 19,793 specimens).

**GEOTRUPES Latreille, 1796**
***Geotrupes spiniger*** (Marsham, 1802)
7 SU16: Marlborough area, 1869-1938 (Meyrick, 1938)
8 ST95: West Lavington, 16-Sep-39 (ITH) (DFL, 1948)
ST85: Bullen Hill Farm, 1-Jan-8231-Dec-02 (EGS det MDD)
SU02: Knighton Wood, Sep-07 (AJWA)
SU03: Grovely Wood, 13-Aug-80, 11-Aug-87 (DRN); Wylye Down NNR, 4-Sep to 10-Oct-07 (MDD) (Darby, 2009)
SU13: Stratford sub Castle, 21-Aug-70 (DRN), Cockey Down, 6-Oct, 6-Dec-73 (PM); Woodford, 18-Aug-93 (PM)
SU24: Sidbury Hill (SPTAE), 4-Sep-96 (MG)
Resident. Local; in herbivore dung.

**Dumble Dor, Clock Lousy Watchman**
***Geotrupes stercorarius*** (Linnaeus, 1758)
7 ST99: Flisteridge Wood, 25-Apr-82 (KNAA)
SU16: Marlborough area, Jun-1873 (EWJ) (Meyrick, 1874); 11 May 1879 (WFB) (Meyrick, 1880); 1869-1938 (Meyrick, 1938); Lockeridge, 1994 (JEO)
8 ST84: Longleat, 29-Aug-99 (MB)
ST91: Cranborne Chase, 4-Jun-97 (MAS)
ST92: Sutton Mandeville, Jun-99 (MDD)
ST93: Great Ridge Wood, 2, 10-May-00 (MDD); Stockton Wood, 7-Jul-96 (MAS)
ST95: Potterne, 27-Sep-40 (KJA) (DFL, 1948); West Lavington, 1938 (PLP) (DFL, 1948); 20-Mar-41 (AME) (DFL, 1948)
SU03: Wylye Down NNR, 4-Sep to 10-Oct-07 (MDD) (Darby, 2009)
SU04: Shrewton, 18-Aug-40 (KJA) (DFL, 1948)
SU13: Old Sarum, 31-May-97 (MAS)
SU22: Langley Wood NNR, 22-Apr to 29-Oct-04 (MDD) (Darby, 2005)
SU23: West Winterslow, 1-Sep-01 (NB)
Resident. Widespread; in herbivore dung.

**TRYPOCOPRIS Motschulsky, 1860**
***Trypocopris pyrenaeus*** (Charpentier, 1825)
**Na**
8 ST93: Stockton Wood, 28-Sep-95 (PJH) (Hodge, 1995)
Resident. Very local, on wing and crawling amongst grass. Unusual because almost all records in GB from sandy heaths between May and July.

Trypocopris pyraeneus

## Family SCARABAEIDAE
### Subfamily AEGIALIINAE

**AEGIALIA Latreille, 1807**
***Aegialia arenaria*** (Fabricius, 1787)
7 SU16: Figheldean, 1869-1938 (Meyrick, 1938)
Resident. Very local; in dung. A surprising record for a species usually found on coastal sand dunes. There are five specimens without data under this name in MCC. (conf, not *sabuleti* MDD).

### Subfamily APHODIINAE

Tribe APHODIINI

**APHODIUS Illiger, 1798**

ACROSSUS Mulsant, 1842

***Aphodius depressus*** (Kugelann, 1792)

7 SU16: Marlborough area, Jun-1873 (EWJ) (Meyrick, 1874); 4-May-1879 (WFB) (Meyrick, 1880); 1869-1938 (Meyrick, 1938)

8 ST92: Donhead St. Mary, 18, 19-Jul-95 (HK det MDD)

SU03: Wylye Down NNR, 3-Aug to 4-Sep-07 (MDD) (Darby, 2009)

Resident. Very local; in sheep and cow dung.

***Aphodius luridus*** (Fabricius, 1775)

7 SU16: Marlborough area, Jun-1873 (EWJ) (Meyrick, 1874); 1869-1938 (Meyrick, 1938)

SU26: Savernake Forest, 4-Aug-95 (EN)

8 SU02: Knighton Wood, Oct-06 (AJWA); Middleton Down, 23-May-94 (BL)

Resident. Very local; in sheep and cow dung.

***Aphodius rufipes*** (Linnaeus, 1758)

7 SU16: Marlborough area 1869-1938 (Meyrick, 1938)

SU17: Avebury, 20-May-03 (DB)

SU26: Savernake Forest, 7-Aug-69 (DRN); 28-Aug-93, 9-Jun to 1-Jul-94 (EN det SAW); 4 to 24-Aug-95 (EN det MDD)

8 ST95: West Lavington, 1938 (PLP) (DFL, 1948)

ST92: Donhead St. Mary, 18, 19-Jul-95 (HK det MDD)

SU02: Middleton Down, 23-May-94 (BL)

SU03: Wylye Down NNR, 10-Jul to 3-Aug-07 (MDD) (Darby, 2009); Grovely Wood 3, 5-Aug-71 (DRN)

SU13: Cockey Down, 6-Oct-73 (PM)

SU22: Langley Wood NNR,16-Jul to 6-Aug-04 (MDD) (Darby, 2005)

SU23: Bentley Wood, 6-Dec-73 (PM); 31-May-97 (MDD)

Resident. Widespread; in cow and horse dung, and often at light

AGRILINUS Mulsant and Rey, 1870

***Aphodius ater*** (De Geer, 1774)

7 ST99: Flisteridge Wood, 25-Apr-82 (KNAA)

SU09: Cotswold Water Park, 9-Apr-97 (APF)

SU16: Marlborough area, 1869-1938 (Meyrick, 1938)

SU26: Savernake Forest, 30-Aug to 20-Sep-94 (EN det SAW)

8 ST91: Cranborne Chase, 16 to 25-Jun-80 (NCC)

ST93: Great Ridge Wood, 18-Apr-74 (DRN)

SU01: Martin Down NNR, 1-Jan-7931-Dec-92 (RGB)

SU02: Knighton Wood, May-08 (AJWA); Middleton Down, 4-May-95, 3-Apr-97 (MDD)

SU03: Grovely Wood, 13-Apr-71, 16-Apr-84 (DRN); Wylye Down NNR, 4-May to 5-Jun-07 (MDD) (Darby, 2009)

SU13: Cockey Down, 6-Oct-73 (PM)

SU16: Jones's Mill, 21-Jun-97 (MDD)

SU21: Burnt Ground Wood, 19-Apr-74 (DRN); Hamptworth Estate, 29-May-74 (DRN)

SU22: Langley Wood NNR, 25-Jun to 16-Jul-04 (MDD) (Darby, 2005); Bentley Wood, 31-May-97 (MDD)

Resident. Widespread; in dung and putrefying vegetation.

***Aphodius constans*** Duftschmid, 1805

7 SU16: Marlborough area, 1869-1938 (Meyrick, 1938)

8 SU03: Grovely Wood, 10-Apr-70 (DRN)

SU21: Pound Bottom, 9-Apr-74 (DRN)

Resident. Very local; in cow and horse dung.

***Aphodius rufus*** (Moll, 1782)

7 SU16: Marlborough area, 1869-1938 (Meyrick, 1938)

8 SU02: Knighton Wood, May-07 (AJWA)

SU03: Grovely Wood, 5-Aug-71, 11-Aug-87, 1-Aug-88 (DRN)

SU23: Bentley Wood, 6-Dec-73 (PM)

Resident. Very local; in cow dung and one in a dead hare.

***Aphodius sordidus*** (Fabricius, 1775)

**Na, LCGR-Fm Cu**

7 SU16: Marlborough area, 1869-1938 (Meyrick, 1938)

Resident. Very local; [in various sorts of dung in dry, sandy or chalky areas, often near the coast, prefers exposed habitats] No specimen in MCC.

APHODIUS Illiger, 1798

***Aphodius fimetarius*** (Linnaeus, 1758)

7 SU16: Marlborough area, 30 May 1879 (RAW) (Meyrick, 1880); 1869-1938 (Meyrick, 1938)

SU26: Savernake Forest, 20-Sep to 11-Oct-94 (EN det SAW)

8 ST84: Longleat, 4-May-98 (MB)

ST85: Bullen Hill Farm, 1982-2002 (EGS det MDD)

ST93: Great Ridge Wood, 18-Apr-74 (DRN)

SU02: Knighton Wood, Apr-07 (AJWA); Middleton Down, 21-May-95 (MDD)

SU03: Grovely Wood, 1-Aug-88 (DRN); Wylye Down NNR, 4-May to 5-Jun-07 (MDD) (Darby, 2009); Grovely Wood, 18-Aug-69 (DRN)

SU04: Winterbourne Stoke, 30-Mar-97 (BL); Parsonage Down, 20-Feb-98 (BG)

SU13: Cockey Down, 6-Oct-73 (PM)

SU21: Landford Bog, 14-May-93 (PM)

*A.pedellus* has recently (2006) been separated from *A. fimetarius* so that all previous records must be considered indeterminate. In dung and decomposing vegetable material.

***Aphodius foetens*** (Fabricius, 1787)
8 SU03: Grovely Wood, 11-Aug-87 (DRN)
Resident. Very local; in cow dung.

***Aphodius pedellus*** (De Geer, 1774)
8 SU15: Beach's Plantation (SPTAE), 10-Jun-97 (MGT)

CALAMOSTERNUS Motschulsky, 1860

***Aphodius granarius*** (Linnaeus, 1767)
7 SU16: Marlborough area, 1869-1938 (Meyrick, 1938)
8 ST85: Bullen Hill Farm, 1982-2002 (EGS det MDD)
ST92: Sutton Mandeville, 20-Apr, 3-May-92 (MDD)
ST93: Great Ridge Wood, 18-Apr-74 (DRN)
SU03: Grovely Wood, 17-Apr-71 (DRN)
SU15: Beach's Plantation (SPTAE), 16-Jun-97, 1-May-05 (MGT)
SU22: Langley Wood NNR, 25-Apr to 6-Jun-07 (MDD) (Darby, 2008); Bentley Wood, 31-May-97 (MDD)
SU24: Sidbury Hill (SPTAE), 29-May-97 (MGT)
Resident. Widespread; in cow dung and decomposing vegetable material.

CHILOTHORAX Motschulsky, 1860

***Aphodius distinctus*** (Müller, O.F, 1776)
**Nb, LCGR-Fl Cu**
7 SU16: Marlborough area, 23-Mar-1879 (PCP) (Meyrick, 1880); 1869-1938 (Meyrick, 1938)
Resident. Very local; [in dung of various sorts and decomposing vegetable material]. Five specimens in MCC (det as *inquinatus)* without data.

***Aphodius paykulli*** Bedel, 1908
**Nb, LCGR-Fl Cu**
8 SU03: Grovely Wood, 10-Apr-70 (DRN)
Resident. Very local; in dry horse dung.

COLOBOPTERUS Mulsant, 1842

***Aphodius erraticus*** (Linnaeus, 1758)
7 SU16: Marlborough area, 1869-1938 (Meyrick, 1938)
8 SU02: Knighton Wood, Sep-07 (AJWA); Middleton Down, 4-May-95 (MDD)
SU03: Grovely Wood, 11-Aug-87 (DRN); Wylye Down NNR, 10-Apr to 5-Jun-07 (MDD) (Darby, 2009)
Resident. Very local; in cow and sheep dung.

ESYMUS Mulsant and Rey, 1870

***Aphodius merdarius*** (Fabricius, 1775)
7 SU16: Marlborough area, 1869-1938 (Meyrick, 1938)
8 ST91: Cranborne Chase, 7-May-97 (MAS)
Resident. Very local; [in various sorts of dung, often fresh, prefers exposed conditions].

***Aphodius pusillus*** (Herbst, 1789)
7 SU16: Pewsey Downs, 22-Jun-90 (APFw)
8 SU03: Wylye Down NNR, 4-May to 5-Jun-07 (MDD) (Darby, 2009)
SU21: Hamptworth Estate, Peaked Wood, 29-May-74 (DRN)
Resident. Very local; in horse, cow and sheep dung.

EURODALUS Dellacasa, G, 1983

?***Aphodius coenosus*** (Panzer, 1798)
**Nb**
7 SU16: Marlborough area, Jun-1873 (EWJ) (Meyrick, 1874); 1869-1938 (Meyrick, 1938)
Resident. Very local; in dung [of sheep, cow and rabbit on sandy soil]. A single specimen under this name in the MCC without data is *A. ater.*

EUPLEURUS Mulsant, 1842

***Aphodius subterraneus*** (Linnaeus, 1758)
**RDB1**
7 SU16: Marlborough area, Jun-1873 (EWJ) (Meyrick, 1874); 6-Jun-1879 (WFB) (Meyrick, 1880); 1869-1938 (Meyrick, 1938)
Resident. Very local; in dung [of many kinds, also debris, compost and carrion]. Once widely distributed (in GB) but not recorded for 50 years. A single specimen in the MCC is without data.

LIMARUS Mulsand and Rey, 1870

***Aphodius zenkeri*** Germar, 1813
**Nb**
7 SU26: Savernake Forest, 15-Jun-73 (SAW); 8-Jul to 28-Aug-93, 28-Apr to 18-May-94 (EN det SAW)
8 SU02: Knighton Wood, Sep-06, 6-Jul, Sep-07 (AJWA)
Resident. Very local; in deer dung in woodland.

MELINOPTERUS Mulsant, 1842

***Aphodius prodromus*** (Brahm, 1790)
7 ST99: Flisteridge, 25-Apr-82 (KNAA)
SU09: Ashton Keynes, 26-Apr-01 (JKG det MDD)
SU16: Marlborough area, 30 May 1879 (RAW) (Meyrick, 1880); 1869-1938 (Meyrick, 1938)
SU18: Swindon, 30-Mar, 7-May-01 (JKG det MDD)
SU26: Savernake Forest, 20-Sep to 11-Oct-94 (EN det SAW); 11 to 31-May-95 (EN det MDD)
8 ST84: Longleat, 4-May-98, 7-Mar-99 (MB)
ST91: Cranborne Chase, 24-Apr-97 (MAS)
ST92: Sutton Mandeville, 2-May-92 (MDD)
SU02: Knighton Wood, Oct-06, Mar, Apr-07, May-08 (AJWA); Middleton Down, 21-May-95, 3-Apr-97 (MDD)
SU03: Wylye Down NNR, 4-May to 5-Jun-07 (MDD) (Darby, 2009); Grovely Wood, 10-

Apr-70, 16-Apr-84 (DRN)
SU21: Burnt Ground Wood, 19-Apr-74 (DRN)
SU22: Langley Wood NNR, 22-Apr-04 to 19-Nov-04, 26-Oct to 17-Nov-06 (MDD) (Darby, 2005, 2008); Harestock, 13-Apr-72 (DRN)
Resident. Widespread; in various sorts of dung.

***Aphodius sphacelatus*** (Panzer, 1798)
7 ST99: Flisteridge Wood, 25-Apr-82 (KNAA)
SU08: Webbs Wood, 22-Feb-01 (JKG det MDD)
SU16: Marlborough area, 1869-1938 (Meyrick, 1938)
8 ST84: Longleat, 4-May-98, 7-Mar-99 (MB)
ST92: Sutton Mandeville, 20-Apr-92 (MDD)
ST93: Stockton Wood, 7-Jul-96 (MAS)
SU02: Middleton Down, 3-Apr-97 (MDD)
SU03: Wylye Down NNR, 4-May to 5-Jun-07 (MDD) (Darby, 2009)
SU22: Langley Wood NNR, 21-Jan to 3-Mar-05, 8-Dec-06 to 5-Jan-07 (MDD) (Darby, 2005, 2008)
Resident. Widespread; in various kinds of dung and rotting vegetation.

NIMBUS Mulsand and Rey, 1870
***Aphodius contaminatus*** (Herbst, 1783)
7 SU16: Marlborough area, Jun-1873 (EWJ) (Meyrick, 1874); 1869-1938 (Meyrick, 1938)
SU26: Savernake Forest, 30-Aug to 20-Sep-94 (EN det SAW)
8 SU03: Wylye Down NNR, 20-Sep-05, 4-Sep to 10-Oct-07 (MDD) (Darby, 2009)
SU13: Salisbury, Avon Vslley Reserve, Apr-00 (MAS) (Darby, 2001); Cockey Down, 6-Oct-73 (PM)
Resident. Local; in dung particularly in autumn. Probably widespread.

***Aphodius obliteratus*** Panzer, 1823
7 SU16: Marlborough area, 1869-1938 (Meyrick, 1938)
8 SU02: Knighton Wood, Oct, Nov-06 (AJWA); Langley Wood NNR, 26-Oct-06 to 17-Nov-06 (MDD) (Darby, 2008)
Resident. Local; in dung in wooded areas.

OTOPHORUS Mulsant, 1842
***Aphodius haemorrhoidalis*** (Linnaeus, 1758)
7 SU16: Marlborough area, Jun-1873 (EWJ) (Meyrick, 1874); Marlborough area, 1873-1938 (Meyrick, 1938)
8 SU02: Middleton Down, 21-May-95 (MDD)
SU03: Wylye Down NNR, 4-May to 5-Jun-07 (MDD) (Darby, 2009); Grovely Wood, 12-Jun-00 (MDD)
SU16: Jones's Mill, 21-Jun-97 (MDD)
Resident. Local; in sheep, horse and cow dung. Probably widespread.

PLAGIOGONUS Mulsant, 1842
***Aphodius arenarius*** (Olivier, 1789)
**Nb, LCGR-Fm Cu**
8 SU14: Figheldean, Jun-1870 (RH) (Hislop, 1871)
Resident. Very local; [under vegetable debris and in dung of sheep, horse and cow, in dry, sandy and chalky areas, in open habitats] Formerly *A. putridus.*

TEUCHESTES Mulsant, 1842
***Aphodius fossor*** (Linnaeus, 1758)
7 SU09: Emmett Hill, 1995 (MA det MDD)
SU16: Marlborough area, 16-May-1879 (NNM) (Meyrick, 1880); 1869-1938 (Meyrick, 1938
SU19: Hannington, 27-Jul-96 (MEN)
8 ST73: Stourhead, 4-Jul-95 (KNAA)
ST84: Cley Hill, 6-Jul-82 (KNAA)
SU02: Middleton Down, 4, 21-May-95 (MDD)
SU13: Old Sarum, 31-May-97 (MAS)
SU15: Beach's Plantation (SPTAE), 10-Jun-97 (MGT)
SU16: Huish, Draycott Hill, 17-Aug-01 (IJG); Jones's Mill, 1997 (MAS)
Resident. Widespread; in cow dung.

VOLINUS Mulsant and Rey, 1870
***Aphodius sticticus*** (Panzer, 1798)
7 ST99: Flisteridge Wood, 1-May-97 (APF)
SU16: Marlborough area, Jun-1873 (EWJ) (Meyrick, 1874); 1869-1938 (Meyrick, 1938)
8 ST91: Cranborne Chase, 16 to 25-Jun-80 (NCC)
SU02: Knighton Wood, May, Jul-08 (AJWA)
SU22: Langley Wood NNR, 3 to 25-Jun-04, 25-Apr to 6-Jun-07 (MDD) (Darby, 2005, 2008); Bentley Wood, 31-May-97 (MDD)
Resident. Local; in dung in woodland. Formerly *A. equestris.*

**EUHEPTAULACUS Dellacasa, G, 1938**
***Euheptaulacus villosus*** (Gyllenhal, 1806)
**Na, LCGR-Fm Cm**
8 SU25: Ludgershall, Apr- Jul-1915 (PAH) (Harwood, 1915; Welch, 1978)
Resident. Very local; crawling on bare ground. [In vegetable matter, dung and under stones. In the south (of GB) mostly on inland sites on chalk in exposed and isolated areas].

## Subfamily SCARABAEINAE
### Tribe ONTHOPHAGINI

**ONTHOPHAGUS Latreille, 1802**
PALEONTHOPHAGUS Zunino, 1979
***Onthophagus coenobita*** (Herbst, 1783)
7 SUX9: Colerene Park, 2-Jun-04 (AH der KNAA)
ST98: Malmesbury, Jul-99 (JW det MDD)
SU09: Ashton Keynes (Millenium Green), 26-Apr-01 (JKG det MDD); Upper Waterhay, 9-

Sep-07 (JKG det MDD)
8 SU02: Knighton Wood, May-08 (AJWA)
SU12: Coombe Bissett Down, 1-Apr-97 (MDD)
SU22: Langley Wood NNR, 25-Jun to 16-Jul-04, 5-Aug-06 to 2-Jul-07 (MDD) (Darby, 2005, 2008)
Resident. Local; in dung and carrion.

***Onthophagus joannae*** Goljan, 1953
7 SU16: Marlborough area, Jun-1873 (EWJ) (Meyrick, 1874); 1869-1938 (Meyrick, 1938)
8 SU01: Martin Down NNR, 1999 (Anon, 1999)
SU02: Middleton Down, 4, 21-May-95 (MDD)
SU13: Little Durnford Down, 4-Jun-90 (PM)
SU14: Haxton (SPTAE), 24-Jun-02 (PJH) (Hodge, 2002)
SU22: Langley Wood NNR, 14-May-04 to 6-Aug-04, 25-Apr to 6-Jun-07 (MDD) (Darby, 2005, 2008)
SU23: Porton Down, 3-Jul-99 (PJH)
SU24: Bulford Ranges (SPTAE), 3-Jun - 12-Aug-08 (Boyce & Wright, 2009); North Tidworth, Devils Ditch, 30-May-04 (BLe)
Resident. Widespread; in horse and sheep dung on chalky and sandy soils.

?***Onthophagus nuchicornis*** (Linnaeus, 1758)
**Na**
7 SU16: Marlborough area, Jun-1873 (EWJ) (Meyrick, 1874); 1869-1938 (Meyrick, 1938)
Resident. Very local; [in sheep, horse, rabbit and dog dung in sandy areas inland]. The single specimen without data in the MCC under this name is *O. joannae*.

***Onthophagus similis*** (Scriba, 1790)
8 SU03: Grovely Castle, 19-Aug-95 (MAS)
SU13: Salisbury, 27-Jun-90 (PM)
SU21: Burnt Ground Wood, 19-Apr-74 (DRN)
SU22: Langley Wood NNR, 25-Apr to 6-Jun-07 (MDD) (Darby, 2008)
Resident. Very local; in dung.

## Subfamily MELOLONTHINAE

### Tribe HOPLIINI

**HOPLIA Illiger, 1803**
***Hoplia philanthus*** (Füessly, 1775)
7 SU16: Marlborough area, Jun-1873 (EWJ) (Meyrick, 1874); 1869-1938 (Meyrick, 1938)
8 SU12: Whaddon Common, 22-Jul-73 (DRN)
Resident. Very local; on vegetation.

### Tribe MELOLONTHINI

**MELOLONTHA Fabricius, 1775**
**Cockchafer, May Bug**
***Melolontha melolontha*** (Linnaeus, 1758)
7 ST96: Prickly Sevens Meadow, 7-Jun-00 (GT)
ST99: Crudwell, 3, 15, 16-May-97 (APF)
SU16: Marlborough area, 24 May 1879 (WFB) (Meyrick, 1880); 1869-1938 (Meyrick, 1938); Lockeridge, 1994 (JEO); Oare, 2-Jun-00 (IJG)
SU17: Avebury, 7-May-03 (DB)
SU19: Hannington, 14-Jun-96 (MEN)
8 ST84: Longleat, Jun-97 (MB)
ST85: Green Lane Wood, 7-Jun-95 (WANHS det MDD)
ST92: Sutton Mandeville, 5-Jun-93 (MDD)
ST93: Ham Cross, 11-May-89 (SP)
ST95: West Lavington, 1939-1941 (DFL, 1948)
SU02: Middleton Down, 17-Jun-73 (BL)
SU03: Dinton, 1983-1993 (SP); Berwick St. James, 6-May-02 (BL)
SU04: Orcheston Down (SPTA), 18-May-02 (PJH) (Hodge, 2002); Shrewton Folly, 29-May-94 (SP)
SU06: Roundway Gardens, 26-May-08 (BG) Devizes, 19-May-99 (BG); 15-May-99 (GH det BG); 12-May-08 (PD)
SU14: Haxton (SPTAE), 17-May-02 (PJH) (Hodge, 2002)
SU15: Beach's Plantation (SPTAE), 1-May-05 (MGT); Everleigh (SPTAE), 18-May-02 (PJH) (Hodge, 2002)
SU22: Langley Wood NNR, 6-Sep-73 (GT); 22-Apr to 14-May-04 (MDD) (Darby, 2005)
SU24: Newton Tony, 13-May-01 (DG)
Resident. Widespread; walking on the ground, on vegetation and many records to light. Most records of singletons.

*Melolontha melolontha*

### Tribe RHIZOTROGINI

**AMPHIMALLON Latreille, 1825**
**Summer Chafer**
***Amphimallon solstitiale*** (Linnaeus, 1758)
7 SU06: Calstone Down, 6-Jul-95 (KNAA)

SU16: Marlborough area, 1869-1938 (Meyrick, 1938)
SU17: Avebury, 7-May-05 (DB)
8 ST95: West Lavington, Jul-44 (ITH) (DFL, 1948)
ST85: Trowbridge, 17-Jul-83 (EGS det MDD)
SU03: Wylye Down NNR, 5-Jun to 10-Jul-07 (MDD) (Darby, 2009)
SU05: Market Lavington, Jul-40 (ITH, EleC, AOC) (DFL, 1948)
SU13: Cockey Down, 14-Jul-07 (PM); Salisbury 6-Aug-85 (DRN)
Resident. Widespread; on vegetation and swarming in evening.

Tribe SERICINI

**OMALOPLIA Schönerr, 1817**
***Omaloplia ruricola*** (Fabricius, 1775)
**Nb, LCGR-Fh Ch**
7 SU16: Marlborough area, 1869-1938 (Meyrick, 1938)
8 SU02: Middleton Down, 1-Jul-93 (BL); 25-Jun-93, 14-Aug-01, 31-Aug-02 (PM)
SU12: Coombe Bissett Down, 19-Aug-99, 22-Jul-04, 9-Aug-07 (PM)
SU13: High Post, 20-Jun-95 (MAS); Cockey Down, 10-Jul-96 (PM); Old Sarum, 31-May-97 (MAS); Figsbury Ring, 1-Jun-92 (PM)
SU22: Witherington Down, 24-Jun-93 (PM)
SU23: Porton Down, 2-Jul-94 (PM)
SU25: Ludgershall, Apr- Jul-1915 (PAH) (Harwood, 1915); Sidbury Hill (SPTAE), 17-Jun-97 (MGT)
Resident. Widespread; on vegetation and in flight.

Omaloplia ruricola

**SERICA Macleay, 1819**
**Brown Chafer**
***Serica brunnea*** (Linnaeus, 1758)
7 SU06: Morgan's Hill, 14-Jul-08 (PM)
SU16: Marlborough area, Jun-1873 (EWJ) (Meyrick, 1874)
SU17: Avebury, 20-Jul-03, 30-Jul-06 (DB)
SU26: Savernake Forest, 30-May-73 (DBS); 15-Jun-73 (SAW); 27-May to 28-Jul-93, 1-Jul to 30-Aug-94 (EN det SAW); 15-Jul to 3-Aug-95 (EN det MDD)
8 SU01: Martin Down NNR, 1999 (EN, 1999)
SU02: Knighton Wood, Aug-07 (AJWA)
SU03: Wylye Down NNR, 3-Aug-07 (MDD) (Darby, 2008); Grovely Wood, 11-Aug-71 (DRN)
SU13: Cockey Down, 1-Aug-88 (PM)
SU22: Langley Wood NNR, 5-Aug-06 (MDD) (Darby, 2007)
SU23: West Winterslow, 9-Aug-01 (NB det MDD)
SU25: Collingbourne Forest, 25-Aug-72 (DRN)
Resident. Widespread; In vegetation, pitfall and flight intercept traps and at light.

Subfamily RUTELINAE
Tribe ANOMALINI

**ANOMOLA Samouelle, 1819**
***Anomala dubia*** (Scopoli, 1763)
7 SU26: Savernake Forest, 30-May-73 (DBS)
Resident. Very local; [in sandy areas at the foot of plants and on leaves, or buried in the sand].

**PHYLLOPERTHA Stephens, 1830**
**Fernshaw Beetle, Chovey, June Bug, Field Chafer, Garden Chafer, Bracken Clock, Fernweb**
***Phyllopertha horticola*** (Linnaeus, 1758)
7 ST86: Neston Park, 3-Jun-04 (KNAA)
ST96: Spye Park, 31-May-04 (KNAA)
SU06: Morgan's Hill, 1-Jun-94, 8-Jun-01, 8-Jun-04 (PM)
SU16: Marlborough, 1994 (JEO); Marlborough area, 18-Jun-1879 (JA) (Meyrick, 1880); 1869-1938 (Meyrick, 1938); Lockeridge, 1994 (JEO)
SU17: Avebury, 26-May-07 (DB)
SU26: Savernake Forest, 12-Jun-00 (IJG)
SU27: High Clear Down, 1-Jun-99, 7-Jun-05, 29-May-08 (PM)
SU36: Ham Hill , 31-May-95, 8-Jun-00, 11-Jun-02 (PM)
8 ST93: Park Bottom, Jun-97 (BL det MDD); Great Ridge Wood 25-May-00 (MDD)
ST94: Bowls Barrow (SPTAW), 22-Jun-02 (PJH) (Hodge, 2002); Ladywell (SPTAW), 3-Jun – 12-Aug-08 (Boyce & Wright, 2009); Salisbury Plain, 19-Jun-88, 12-Jun-93, 29-May-94 (SP); Tinker's Track (SPTAW), 3-Jun - 12-Aug-08 (Boyce & Wright, 2009); Wadman's Coppice (SPTAW) 3-Jun - 12-Aug-08 (Boyce & Wright, 2009)
SU02: Fovant Down, 1993 (SP); Middleton Down 22-Jun-88 (BL det PM)
SU03: Little Langford Down, 26-May-99, 6-Jun-08 (PM)
SU04: Fox Covert, 3-Jun - 12-Aug-08 (Boyce & Wright, 2009); Orcheston Down (SPTA), 22-Jun-02 (PJH) (Hodge, 2002); Salisbury

Plain, 25-Jun-02 (PJH) (Hodge, 2002) Shrewton Folly, 15-Jun-02 (BG); Yarnbury Castle, 1-Jun-99 (BL det MDD)
SU05: Black Heath (SPTAC), 3-Jun - 12-Aug-08 (Boyce & Wright, 2009)Ell Barrow (SPTAC), 3-Jun - 12-Aug-08 (Boyce & Wright, 2009); Lavington Down, 7-Jun-03 (EGS det MDD)
SU06: Knoll Down, 9-Jun-99 (BG)
SU13: Boscombe Down, 4-Jun-93 (SP); Cockey Down 1-Jun-93 to 21-May-08 (PM); Middle Woodford Down 23-May-90; 24-May-97 (PM); Figsbury Ring, 31-May-92, 7-Jun-93, 1-Jun-94 (PM); Little Durnford Down, 4-Jun-90, 20-May-92, 19-May-93, 30-May-94, 2-Jun-95 (PM)
SU14: Boscombe Down, 25-May-98 (SP); Haxton (SPTAE), 24-Jun-02 (PJH) (Hodge, 2002); 3-Jun - 12-Aug-08 (Boyce & Wright, 2009); Larkhill west, 2-Jun-97 (MGT) ); Salisbury Plain, 31-May-99 (BG); Wexland Hanging (SPTAC), 22-Jun-02 (PJH) (Hodge, 2002); 3-Jun - 12-Aug-08 (Boyce & Wright, 2009)
SU15: Beach's Plantation (SPTAE), 10, 16-Jun-97 (MGT); Charlton, Nr Rushall, 28-Jun-00 (IJG); Bruce Down, 27-Jun-00 (IJG); Everleigh (SPTAE), 24-Jun-02 (PJH) (Hodge, 2002); West Everleigh Down, Jun-01 (IJG)
SU16: Jones's Mill, 13-Jun-00 (IJG); 19-Jun-01 (APF) (Foster, 2002);
SU23: Porton Down, 14-Jun-9, 21-May-92, 1-Jun-93, 28-May-94 (PTW det MDD)
SU24: Beacon Hill (SPTAE), 3-Jun - 12-Aug-08 (Boyce & Wright, 2009)Bulford Ranges (SPTAE), 3-Jun - 12-Aug-08 (Boyce & Wright, 2009)Driver Training Area (SPTAE), 24-Jun-02 (PJH) (Hodge, 2002); Dunch Hill (SPTAE), 3-Jun - 12-Aug-08 (Boyce & Wright, 2009); Sidbury Hill (SPTAE), 18, 29-May-97 (MGT)
SU25: Haxton Down, 17-Jun-00 (BG); Salisbury Plain (Weather Hill Firs), 21-Jun-01 (MDD); Sidbury Hill (SPTAE), 28-May, 4-Jun-97 (MGT)

Resident. Widespread; on vegetation sometimes in numbers

### Subfamily CETONIINAE

**CETONIA Fabricius, 1775**
**Rose Chafer**
***Cetonia aurata*** (Linnaeus, 1758)
7 SU16: Lockeridge 1994 (JEO)
SU18: Swindon, 5-Jun-96 (MEN)

Resident. Very local; on flowers.

## Family CLAMBIDAE

### Subfamily CALYPTOMERINAE

**CALYPTOMERUS Redtenbacher, 1849**
***Calyptomerus dubius*** (Marsham, 1802)
7 SU16: Marlborough area, Jun-1873 (EWJ) (Meyrick, 1874); 1869-1938 (Meyrick, 1938)
8 SU26: Cobham Frith, 24-Jul-82 (DRN)

Resident. Very local; in a grass pile and [almost invariably in bottom of haystacks and amongst decaying straw].

### Subfamily CLAMBINAE

**CLAMBUS Fischer von Waldheim, 1821**
***Clambus armadillo*** (De Geer, 1774)
7 SU16: Marlborough area, Jun-1873 (EWJ) (Meyrick, 1874); 1869-1938 (Meyrick, 1938)
8 ST92: Sutton Mandeville, 25-Jun-07 (MDD)
SU04: Shrewton, 1-Aug-85 (DRN)
SU16: Jones's Mill, 21-Jun-97 (MDD)

Resident. Very local; in decaying vegetation including a garden compost heap.

***Clambus gibbulus*** (LeConte, 1850)
8 SU26: Wilton Brail, 19-Jun-03 (MDD); Bedwyn Brail, 19-Jun-03 (MDD)

Resident. Very local; in grass piles.

***Clambus pubescens*** Redtenbacher, 1849
8 ST92: Sutton Mandeville, 6-Jul-93 (MDD)
ST96: Rowde, 5-Aug-83 (DRN)
SU04: Shrewton, 28-Jul-82 (DRN)
SU15: Fittleton, 29-Jul-83 (DRN); North Newnton, 5- Aug-83 (DRN); Widdington, 5-Aug-83 (DRN)
SU17: Hallam, 6-Aug-83 (DRN)
SU22: Redlynch, 4-Aug-82 (DRN)

Resident. Very local; in old straw, weed refuse and leaf litter.

***Clambus punctulum*** (Beck, 1817)
8 ST93: Great Ridge Wood, 16-May-00 (MDD)

Resident. Very local; in rotting vegetation

***Clambus simsoni*** Blackburn, 1902
8 ST92: Sutton Mandeville, 25-Jun-07 (MDD), 25-Sep-07 (MDD)
SU02: Knighton Wood, Sep-06, Jul-08 (AJWA)

Resident. Very local; in a flight intercept trap and garden compost. Australian species first detected in GB in 1996 and appears to have spread rapidly. Easily confused with *C. pallidulus*.

## Family SCIRTIDAE

**ELODES Latreille, 1796**
***Elodes elongata*** Tournier, 1868

**RDBi**
8 SU13: Great Durnford, 6-Aug-71 (DRN)
Resident. Very local; on vegetation near the R. Avon.

***Elodes minuta*** (Linnaeus, 1767)
8 ST91: Cranborne Chase, 15-May-97 (MAS)
SU16: Jones's Mill, 11-Jun-97 (MAS); Marlborough area, 1869-1938 (Meyrick, 1938)
Resident. Very local; on vegetation near water. Marlborough record is indeterminate because until 1992 four species were confused under this name.

***Elodes pseudominuta*** Klausnitzer, 1970
8 SU16: Jones's Mill, 21-Jun-97 (MDD), 19-Jun-01 (APF) (Foster, 2002)
Resident. Very local; on vegetation near water.

**ODELES Klaustnitzer, 2004**
***Odeles marginata*** (Fabricius, 1798)
8 SU02: Bishopstone, Jul-91 (MDD)
SU13: Salisbury, Apr-00 (MAS) (Darby, 2001)
Resident. Very local; on vegetation near water.

**MICROCARA Thomson, C.G, 1859**
***Microcara testacea*** (Linnaeus, 1767)
7 SU09: Upper Waterhay, 5 - 11-Jun-07 (JKG det MDD); Blakehill Farm, 11-May-01 (JKG det MDD)
8 SU16: Green L\ne Wood, 31-May-04 (KNAA); Jones's Mill, 19-Jun-01 (APF) (Foster, 2002); 11-Jun-97 (MAS); 2000 (MDD)
SU21: Hamptworth Estate, 5-Aug-75 (DRN)
SU22: East Grimstead, 8-Jun-05 (JSD)
Resident. Local; in moss and on vegetation near water.

**CYPHON Paykull, 1799**
***Cyphon coarctatus*** Paykull, 1799
7 ST87: Castle Combe, 1934 (HBNH)
SU09: Upper Waterhay, 17-May, 4-Jun-05 (JKG det MDD)
SU13: Salisbury, Apr-00 (MDD) (Darby, 2001)
SU16: Marlborough area, Jun-1873 (EWJ) (Meyrick, 1874); 1869-1938 (Meyrick, 1938)
SU18: Swindon, 7-May-01 (JKG det MDD)
SU26: Savernake Forest, 1 to 31-May-95 (EN det MDD)
8 ST84: Longleat, 7-Jul-99 (MDD)
SU12: Petersfinger, 25-Jul-91 (DRN)
SU13: Salisbury, Apr-00 (MAS); 12-May, 1-Jun-00 (MDD); Great Durnford, 6-Aug-71 (DRN)
SU16: Jones's Mill, 19-Jun-01 (APF) (Foster, 2002); 21-Jun-97 (MDD)
SU21: Landford Bog, 25-May-98 (MDD); 9, 24-Jun-01 (APF) (Foster, 2001)
Resident. Widespread; on vegetation near water.

***Cyphon hilaris*** Nyholm, 1944
8 SU21: Landford Bog, 24-Jun-01 (APF) (Foster, 2001)
SU22: Landford, 12-Aug-87 (DRN)
Resident. Very local; on vegetation near water.

***Cyphon ochraceus*** Stephens, 1830
8 SU22: Glazier's Copse, 16-Aug-88 (DRN)
Resident. Very local; on sallow/alder.

***Cyphon padi*** (Linnaeus, 1758)
8 SU13: Salisbury, Apr-00 (MAS) (Darby, 2001)
SU16: Jones's Mill, 19-Jun-01 (APF) (Foster, 2002)
SU21: Landford Bog, 9-Jun-01 (APF) (Foster, 2001)
Resident. Very local; on vegetation near water.

***Cyphon variabilis*** (Thunberg, 1787)
7 SU16: Marlborough area, Jun-1873 (EWJ) (Meyrick, 1874); 1869-1938 (Meyrick, 1938)
Resident. Very local; [on vegetation in damp woods and meadows].

**PRIONOCYPHON Redtenbacher, 1858**
***Prionocyphon serricornis*** (Müller, P.W.J, 1821)
**Nb, IEC-3, SQS-8**
7 SU16: Cadley, 2-May-00 (JSD)
SU26: Savernake Forest, 1 to 19-Jul-94 (EN det SAW); 2-May-00 (JSD)
8 ST92: Ansty Combe, 21-Sep-00 (JSD)
SU02: Fovant Wood, 21-Sep-00 (JSD)
SU03: Bemerhills, 11-Aug-76 (DRN); Grovely Wood, 11-Aug-71 (DRN)
SU22: Bagfield Copse, 7, 22-Aug-86 (DRN); Langley Wood, 15-Jun-00 (JSD)
Resident. Very local; on oaks [and beech, the larvae in water holes formed by the roots].

**SCIRTES Illiger, 1807**
***Scirtes hemisphaericus*** (Linnaeus, 1758)
7 SU16: Marlborough area, 1869-1938 (Meyrick, 1938)
8 SU12: Petersfinger, 25-Jul-91 (DRN)
SU13: Lower Woodford, 1-Aug-83 (DRN)
SU26: Bedwyn Brail, 19-Jun-03 (MDD)
Resident. Local; on vegetation near water.

## Family DASCILLIDAE

**DASCILLUS Latreille, 1796**
**Orchid Beetle**
***Dascillus cervinus*** (Linnaeus, 1758)
**LCGR-Fl Cl**
7 SU06: Cherhill Down, 15-Jun-82 (KNAA); 31-May-04 (BLe); Morgan's Hill, 17-Jun-93, 14-Jun-99, 11-Jun-02 (PM)
SU16: Marlborough area, 20 Jun 1879 (PHM) (Meyrick, 1880); 1869-1938 (Meyrick,

1938)
SU27: High Clear Down, 1-Jun-99; 2-Jun-00 (PM)
8 ST83: Whitesheet Hill, 21-Jun-93 (PM)
ST91: Cranborne Chase, 16 to 25-Jun-80 (NCC), 4-Jun-97 (MAS)
ST94: Bowls Barrow (SPTAW), 22-Jun-02 (PJH) (Hodge, 2002); Ladywell (SPTAW), 23-Jun-02 (PJH) (Hodge, 2002); Norton Down (SPTAW), 23-Jun-02 (PJH) (Hodge, 2002);
SU01: Martin Down NNR, 1979-1992 (RGB); 1999 (EN, 1999)
SU02: Chickengrove Bottom, 6-Jun-98, 9-Jun-07 (BL det MDD); Middleton Down, 15-Jun-93, 5-Jun-95, 9-Jun-05 (PM)
SU03: Berwick St. James, 26-May-99 (MDD); Grovely Wood, 19-Jun-95, 11-Jun-00 (MDD); 14-Jun-01 (WANHS det MDD) 31-May-97 (MAS); Little Langford, 3-Jul-96 (MAS); Wylye Down NNR, 5-Jun to 10-Jul-07 (MDD) (Darby, 2009)

*Dascillus cervinus*

SU04: Fox Covert, 3-Jun - 12-Aug-08 (Boyce & Wright, 2009); Orcheston Down (SPTA), 22-Jun-02 (PJH) (Hodge, 2002); Salisbury Plain, 25-Jun-02 (PJH) (Hodge, 2002);
SU12: Coombe Bissett Down, 31-May-95, 8-Jun-99, 1-Jun-02 (PM)
SU13: Camp Down, 6-Nov-73 (PM); Little Durnford Down, 18-Jun-92, 30-May-94 (PM); Old Sarum, 31-May-97 (MAS); Figsbury Ring, 31-May-92, 10-Jun-94 (PM); Cockey Down, 13-Jun-73, 6-Aug-73, 10-Jun-90, 14-Jun-94, 19-Jun-01 (PM)
SU14: Haxton (SPTAE), 24-Jun-02 (PJH) (Hodge, 2002); 3-Jun - 12-Aug-08 (Boyce & Wright, 2009); Larkhill west, 2-Jun-97 (MGT); Wexland Hanging (SPTAC), 22-Jun-02 (PJH) (Hodge, 2002); 3-Jun - 12-Aug-08 (Boyce & Wright, 2009)
SU15: Beach's Plantation (SPTAE), 13-Jun-97 (MGT); Manningford Common, 20-Jun-02 (IJG); Everleigh (SPTAE), 24-Jun-02 (PJH) (Hodge, 2002); West Everleigh Down, 19-Jun-02 (IJG)
SU16: Jones's Mill, 1-Jan-9731-Dec-97 (MAS)
SU24: Driver Training Area (SPTAE), 24-Jun-02 (PJH) (Hodge, 2002); Dunch Hill (SPTAE), 3-Jun - 12-Aug-08 (Boyce & Wright, 2009); Sidbury Hill (SPTAE), 4, 5,13-Jun-97 (MGT)
Resident. Widespread; on vegetation in calcareous grassland.

## Superfamily BUPRESTOIDEA
## Family BUPRESTIDAE
### Subfamily BUPRESTINAE

**BUPRESTIS Linnaeus, 1758**
***Buprestis aurulenta*** Linnaeus, 1758
7 ST97: Chippenham, 16-Apr-00 (IR)
Foreign Introduction. Found in clothing. Regularly imported into GB in conifer wood from Western N. America. (Shaw, 1961).
.

### Subfamily AGRILINAE
Tribe AGRILINI

**AGRILUS Curtis, 1825**
AGRILUS Curtis, 1825
***Agrilus viridis*** (Linnaeus, 1758)
**Na, SQS-24**
8 SU21: Hamptworth Common, 28-Jul-77 (DRN)
SU22: Holmere Common, 17-Aug-92 (DRN)
Resident. Very local; on willow and sallow, and flying.

ANAMBUS Thomson C.G., 1864
***Agrilus angustulus*** (Illiger, 1803)
**Nb, SQS-8**
8 ST85: Green Lane Wood, 26-Jun-98 (JN); 2-Jun-04 (AH det KNAA); 19-Jun-04 (MDD)
ST91: Cranborne Chase, 30-Jun-99 (MDD)
SU03: Grovely Wood, 16-Aug-71, 21-Aug-72 (DRN)
Resident. Very local; on dying branches of oaks with leaves still attached.

***Agrilus biguttatus*** (Fabricius, 1777)
**Na, SQS-8**
7 ST87: Corsham Court Park, 3-Jun-04 (KNAA)
8 ST85: Green Lane Wood, 11-Apr-97 (GWG det CT); Jun-97 (GWG); 2-Jun-04 (LD); 19-Jun-04 (MDD)
SU22: Bentley Wood, 31-Jul-07 (JSN grandson); Blackmoor Copse, 2-Jun-92 (PM); 13-Jul-02 (PM)
Resident. Very local; on cut oak stumps and logs. Formerly *A. pannonicus*.

Agrilus biguttatus

***Agrilus laticornis*** (Illiger, 1803)
**Nb, SQS-8**
7 ST87: Corsham Court Park, 3-Jun-04 (KNAA)
8 ST85: Green Lane Wood, 26-Jun-98 (JN))
ST91: Cranborne Chase, 7-May-97 (MAS)
SU03: Grovely Wood, 11-Aug-70, 13, 16-Aug-71, 23-Jul-73 (DRN)
SU22: Bagfield Copse, 7-Aug-86, 6-Aug-86 (DRN); Landford, 12-Aug-87 (DRN); Blackmoor Copse, 12-Jul-92 (PM); Langley Wood NNR, 16-Jul to 6-Aug-04 (MDD) (Darby, 2005); Bentley Wood, 1-Aug-90 (PM), 11-Aug-01 (MDD); 24-Jul-90 (PM)
Resident. Local; on oaks and sometimes willows.

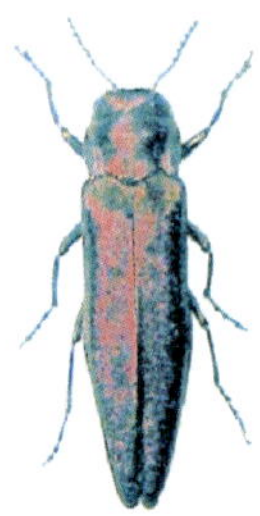

Agrilus sinuatus

**Hawthorn Jewel Beetle**
***Agrilus sinuatus*** (Olivier, 1790)
**Na, SQS-4**
7 ST96: Upper Selves Wood, 31-May-04 (KNAA)
ST86: Neston Park, 3-Jun-04 (KNAA)
ST87: Corsham Court Park, 3-Jun-04 (DG det KNAA)
SU27: High Clear Down, 1-Jun-99 (PM)
8 ST87: Green Lane Wood Meadow, 31-May-04 (KNAA)
SU02: Middleton Down, 2-Jul-99 (PM)
SU04: Parsonage Down NNR, 22-Jul-08 (MDD)
SU12: Coombe Bissett Down, 28-Apr-99 (PM)
SU13: Cockey Down, 14-Jun-73, 29-Apr-99 (PM); Salisbury, 26-Jul-51 (HGS) (Stokes, 1952); Little Durnford Down, 12-Jun-99 (PM)
Resident. Local; on Hawthorns. Several records are of the distinctive larval borings.

Tribe APHANISTICINI

**APHANISTICUS Latreille, 1829**
***Aphanisticus pusillus*** (Olivier, 1790)
**Nb, LCGR-Fm Ch**
8 SU03: Wylye Down NNR, 29-Mar to 10-Apr-07 (MDD) (Darby, 2009)
SU14: Wexland Hanging (SPTAC), 22-Jun-02 (PJH) (Hodge, 2002)
SU24: Perham Down, 30-May-04 (JW det KNAA)
Resident. Very local; in pitfall traps and suction samples in calcareous grassland

Tribe TRACHYINI

**TRACHYS Fabricius, 1801**
**Ground-ivy Jewel Beetle**
***Trachys scrobiculatus*** Kiesenwetter, 1857
**Na, LCGR-Fh Ch**
8 SU03: Grovely Wood, 28-Jul-83 (DRN)
SU14: Stonehenge, summer-03 (PJH) (Hodge, 2003); Haxton (SPTAE), 17-May-02 (PJH) (Hodge, 2002)
SU24: Driver Training Area (SPTAE), 24-Jun-02 (PJH) (Hodge, 2002); Perham Down, 1-Jun-04 (BLe)
Resident. Very local; on vegetation and in suction samples on calcareous grassland.

**Devil's-bit Jewel Beetle**
***Trachys troglodytes*** Gyllenhal in Schönherr, 1817
**LCGR-Fm Ch**
8 SU06: Cherhill Down, 1-Jun-04 (BLe)
SU02: Woodminton Down, 1974 (MGM) (Morris, 1974)
SU03: Wylye Down NNR, 10-Apr to 5-Jun-07 (MDD) (Darby, 2009)
SU04: Orcheston Down, 22-Jun-02 (PJH) (Hodge, 2002)
SU05: Black Heath (SPTAC), 3-Jun - 12-Aug-08 (Boyce & Wright, 2009)
SU12: Coombe Bissett Down, 13-May-99 (PM)
SU13: Little Durnford Down, 14-May-99 (PM)
SU14: Haxton (SPTAE), 3-Jun - 12-Aug-08 (Boyce & Wright, 2009); Wexland Hanging (SPTAC) 22-Jun-02 (PJH) (Hodge, 2002)
SU22: Blackmoor Copse, 17-May-99 (PM)
SU24: Bulford Ranges (SPTAE), 3-Jun - 12-Aug-08 (Boyce & Wright, 2009)
Resident. Local. Many records are of the distinctive mines made by the larvae in Devil's Bit Scabious. Adults in pitfall traps and suction samples.

Trachys troglodytes

## Family BYRRHIDAE
Subfamily BYRRHINAE

**SIMPLOCARIA Stephens, 1829**
***Simplocaria semistriata*** (Fabricius, 1794)

7 SU26: Savernake Forest, 21-Sep to 11-Oct-93, 20-Sep to 11-Oct-94 (EN det SAW)
8 ST91: Cranborne Chase, 12-May-97 (MAS)
ST92: Sutton Mandeville, 22, 23-Nov-92 (MDD)
ST93: Stockton Wood, 28-Sep-95 (PJH) (Hodge, 1995); Stockton Wood, 28-Sep-95 (PJH) (Hodge, 1995)
SU01: Martin Down NNR, 1999 (EN, 1999)
SU03: Wylye Down NNR, 31-Dec-06 to 27-Jan-07 (MDD) (Darby, 2009)
SU13: Boscombe Down, 8, 9-Oct-94 (MDD)
SU16: Jones's Mill, 1997 (MAS)
SU24: Driver Training Area (SPTAE), 1-May-05 (MGT); Dunch Hill (SPTAE), 3-Jun - 12-Aug-08 (Boyce & Wright, 2009); Sidbury Hill (SPTAE), 29-May-97 (MGT)

Resident. Widespread; at roots of plants and walking on the ground, and twice on house wall (attracted by light?)

**CYTILUS Erichson, 1846**
***Cytilus sericeus*** (Forster, 1771)

7 SU16: Marlborough area, Jun-1873 (EWJ) (Meyrick, 1874); 1-Apr-1879 (RAW) (Meyrick, 1880); 1869-1938 (Meyrick, 1938)
8 ST91: Cranborne Chase, 16 to 25-Jun-80 (NCC)
ST92: Sutton Mandeville, 2-May-93 (MDD)
SU03: Grovely Wood, 12-Jun-00 (MDD)
SU21: Hamptworth Estate, 6-Apr-77 (DRN); Pound Bottom, 9-Apr-74 (DRN); Hamptworth Estate, 12-Aug-86 (DRN)
SU22: Langley Wood NNR, 14-May to 3-Jun-04 (MDD) (Darby, 2005); Blackmoor Copse, May-92 (PM)
SU24: Sidbury Hill (SPTAE), 29-May-97 (MGT)

Resident. Widespread; in moss and litter, and under stones.

**BYRRHUS Linnaeus, 1767**
***Byrrhus fasciatus*** (Forster, 1771)

7 SU16: Marlborough area, Jun-1873 (EWJ) (Meyrick, 1874); 1869-1938 (Meyrick, 1938)
8 SU21: Pound Bottom, 9-Apr-74 (DRN)

Resident. Very local; under stones [at plant roots].

Byrrhus pilula

**Pill Beetle**
***Byrrhus pilula*** (Linnaeus, 1758)

7 ST87: Rack Hill, 1995 (MA det MDD)
SU09: Ashton Keynes, 7-May-01 (JKG det MDD)
SU16: Marlborough area, Jun-1873 (EWJ) (Meyrick, 1874); 6-Feb-1879 (GTKM) (Meyrick, 1880); 1869-1938 (Meyrick, 1938); Savernake Forest, 1870-1873 (Meyrick, 1874)
SU17: Avebury, 1-Feb-07 (DB det MDD)
8 ST93: Great Ridge Wood, 7, 25-May-00 (MDD)
SU01: Martin Down NNR, 1999 (EN, 1999)
SU02: Middleton Down, 14-May-89 (BL)
SU03: Wylye Down NNR, 4-May to 5-Jun-07 MDD) (Darby, 2009); Grovely Wood, 14-Apr-71 (DRN)
SU04: Tilshead Down, 1-Jun-97 (MGT)
SU05: Black Heath (SPTAC), 3-Jun - 12-Aug-08 (Boyce & Wright, 2009); Ell Barrow (SPTAC), 3-Jun - 12-Aug-08 (Boyce & Wright, 2009)
SU13: Cockey Down, 13-Jun-73, 6-Oct-73 (PM); Salisbury, 22-May-77 (KNAA)
SU14: Bourne Bottom (SPTAE), 13-Jun-97 (MGT); Larkhill west, 2-Jun-97 (MGT)
SU15: Beach's Plantation (SPTAE), 13-Jun-97 (MGT)
SU21: Pound Bottom, 19-Apr-74 (DRN) 24-Jul-73 (DRN)
SU22: Langley Wood NNR, 14-May-04 to 24-Mar-05, 24-Mar to 2-Jul-07 (MDD) (Darby, 2005, 2008)
SU24: Sidbury Hill (SPTAE), 13 to 17-Jun-97 (MCT)

Resident. Widespread; at plant roots, in litter and under stones.

***Byrrhus pustulatus*** (Forster, 1771)

8 SU22: Langley Wood NNR, 22-Apr to 14-May-04 (MDD) (Darby, 2005)

Resident. Very local; in plant roots and litter in damp area.

**PORCINOLUS Mulsant and Rey, 1869**
***Porcinolus murinus*** (Fabricius, 1794)
**Nb**
7 SU16: Marlborough area, 1869-1938 (Meyrick, 1938)
Resident. Very local; [at roots of heather and possibly other plants on heathland and other open habitats on sandy soils]. No specimen in MCC.

Subfamily SYNCALYPTINAE

**CHAETOPHORA Kirby and Spence, 1823**
***Chaetophora spinosa*** (Rossi, 1794)
8 ST94: Ladywell (SPTAW), 3-Jun - 12-Aug-08 (Boyce & Wright, 2009); Wadman's Coppice (SPTAW) 3-Jun - 12-Aug-08 (Boyce & Wright, 2009)
SU03: Grovely Wood, 8-Apr-74 (DRN)
SU14: Wexland Hanging (SPTAC) 22-Jun-02 (PJH) (Hodge, 2002)
SU24: Dunch Hill (SPTAE), 3-Jun - 12-Aug-08 (Boyce & Wright, 2009)
Resident. Local. In na suction sample and under stones with meadow ants.

**CURIMOPSIS Ganglbauer, 1902**
***Curimopsis maritima*** (Marsham, 1802)
**LCGR-Fm Ch**
8 SU01: Martin Down NNR, 1999 (EN, 1999)
SU24: Newton Tony, 25-Aug to 20-Sep-06, (MA det MGT)
Resident. Very local; at roots of plants in calcareous grassland. Formerly *Syncalypta stratiopunctata.*

## Family ELMIDAE

**ELMIS Latreille, 1802**
***Elmis aenea*** (Müller, P.W.J, 1806)
7 SU06: Marden, Quemerford, 27-Mar-07 (DJL); Sprays Farm, 27-Mar-07 (DJL); Rivers Brook, 27-Mar-07 (DJL)
SU16: R. Kennet, Manton, 1975 (DAC) (Cooling, 1981); R. Kennet, Preshute, 1975 (DAC) (Cooling, 1981); R. Og, Marlborough, 1975 (DAC)
SU26: R. Kennet, Savernake, 1974-1976 (DAC) (Cooling, 1981)
SU37: R. Kennet, Littlecote, 19871-1980 (DAC) (Cooling, 1981)
8 ST94: R. Wylye, Norton Bavant, 1975 (DAC) (Cooling, 1981)
ST95: Erlestoke, 1991-1992 (RC) (Carr, 1997)
SU03: R. Wylye, 5-Aug-91 (RC) (Carr, 1997)
SU04: R.Till, Winterbourne Stoke, 1991-1992 (RC) (Carr, 1997)
SU13: R. Bourne, St Thomas' Bridge, Apr-81 (APF); Middle Woodford, 27-Jul-77 (DRN); Great Durnford, 16, 19-Aug-69, 7-Aug-70, 18-Aug-72, 27-Jul-77 (DRN); Stratford sub Castle, 5-Apr-72 (DRN);
SU14: R. Avon, Figheldean (SPTAE), 20-Jul-91(RC) (Carr, 1997)
Resident. Widespread; in rivers clinging to stones, also on riverside vegetation.

**LIMNIUS Illiger, 1802**
***Limnius volckmari*** (Panzer, 1793)

7 SU06: Marden, 27-Mar-07 (DJL)
SU07: Cherhill, Rivers Brook, 27-Mar-07 (DJL)
SU37: R. Kennet, Littlecote, 1971-1980 (DAC) (Cooling, 1981)
8 ST94: R. Wylye, Norton Bavant, 1975 (DAC) (Cooling, 1981)
SU13: Lower Woodford, 30-Jul, 2-Aug-82 (DRN); Salisbury, 15-Aug-88 (DRN)
SU14: Stonehenge, 22-Apr-03 (PJH) (Hodge, 2003)
Resident. Very local; in rivers and on riverside vegetation.

**OULIMNIUS des Gozis, 1886**
***Oulimnius tuberculatus*** (Müller, P.W.J, 1806)
7 SU37: R. Kennet, Littlecote, 1971-1980 (DAC) (Cooling, 1981)
8 ST94: R. Wylye, 1975 (DAC) (Cooling, 1981)
SU03: R. Wylye, 5-Aug-91 (RC) (Carr, 1997)
SU13: R. Bourne, St Thomas' bridge, Apr-81 (APF)
SU18: Lower Woodford, 27-Jul-82 (DRN)
SU22: Homan's Copse, 31-Jul-85 (DRN)
Resident. Local; in rivers and one on a hot car.

**RIOLUS Mulsant and Rey, 1872**
***Riolus subviolaceus*** (Müller, P.W.J, 1817)
**Nb,**
7 SU06: Cherhill, Rivers Brook,27-Mar-07 (DJL); Marden, 27-Mar-07 (DJL)
SU26: R. Kennet, Savernake, 1974-1976 (DAC) (Cooling, 1981)
SU37: R. Kennet, Littlecote, 1971-1980 (DAC) (Cooling, 1981)
8 ST94: R. Wylye, Norton Bavant, 1975 (DAC)
SU13: River Bourne, ST Thomas' bridge, Aug-81 (APF)
Resident. Very local; in rivers.

## Family DRYOPIDAE

**DRYOPS Olivier, 1791**
DRYOPS Olivier, 1791
***Dryops auriculatus*** (Fourcroy, 1785)
**Nb, LRnt**
8 SU14: Bulford (SPTAE), muddy pond, 19-Jul-91(RC) (Carr, 1997)

Resident. Very local.

***Dryops ernesti*** des Gozis, 1886
8 SU13: Stratford sub Castle, 11-Apr-71 (DRN); Lower Woodford, 2-Aug-82 (DRN)
SU14: Bulford, Nine Mile R. (SPTAE), 23-Jun-93 (CMD)
Resident. Very local; in flood refuse and under a stone by R. Avon.

**?*Dryops griseus*** (Erichson, 1847)
**RDB3, VU (B12ab)**
7 SU16: Marlborough area, 1869-1938 (Meyrick, 1938)
This is almost certainly not a reference to the true *D. griseus* Er. but to *D. griseus* sensu auctt. Brit. which is largely referable to *D.similaris* Bollow, 1936, found in moss at the edge of fens and man-made pools. The record is still very much open to doubt, however, given the difficulty of determining species with the knowledge available in 1938, and the absence of a specimen under this name in the MCC.

***Dryops luridus*** (Erichson, 1847)
7 ST99: R. Thames, side stream, 9-May-99 (APF)
SU09: Cotswold Water Park, 30-May-97 (JSD)
8 ST94: Codford St. Peter, 1991-1992 (RC) (Carr, 1997)
SU14: Bulford Ranges (SPTAE), pond, 3-Jun - 12-Aug-08 (Boyce & Wright, 2009)
Resident. Very local; in ponds.

YRDOPS Steffan, 1961
***Dryops nitidulus*** (Heer, 1841)
**RDB3, VU (B12ab)**
8 SU13: Salisbury, 15-Aug-88 (DRN det CJ)
Resident. Very local; on area washed by R. Avon.

## Family HETEROCERIDAE

**HETEROCERUS Fabricius, 1792**
***Heterocerus fenestratus*** (Thunberg, 1784)
8 ST94: Ladywell (SPTAW), 3-Jun - 12-Aug-08 (Boyce & Wright, 2009); Wadman's Coppice (SPTAW), 3-Jun - 12-Aug-08 (Boyce & Wright, 2009)
SU13: Lower Woodford, 28-Jul-78, 30-Jul-82 (DRN); Salisbury, 25-Jul-78 (DRN)
Resident. Very local; on mud at edge of R. Avon.

***Heterocerus marginatus*** (Fabricius, 1787)
8 ST92: Wardour, 5-Jun-96 (MDD)
SU13: Middle Woodford, 5-Aug-82 (DRN)
Resident. Very local; on mud at edge of a pond and R. Avon.

## Family PSEPHENIDAE
Subfamily EUBRIINAE

**EUBRIA Germar, 1818**
***Eubria palustris*** Germar, 1818
**RDB3**
8 SU16: Jones's Mill, 19-Jun-01 (APF) (Foster, 2002)
Resident. Very local; in a fen. [In flushes and hollows on brookweed *Samolus valerandi* and sallow blossoms].

Superfamily ELATEROIDEA
## Family EUCNEMIDAE
Subfamily MELASINAE
Tribe MELASINI

**MELASIS Olivier, 1790**
***Melasis buprestoides*** (Linnaeus, 1761)
**Nb, IEC-3, SQS-4**
7 ST99: Flisteridge Wood, 19-May-02 (APF)
8 SU03: Grovely Wood, 21-Aug-72 (DRN), 18-Jul-73 (DRN)
Resident. Very local; in rotten wood of deciduous trees, particularly beech.

Tribe EPIPHANINI

**HYLIS des Gozis, 1886**
***Hylis olexai*** (Palm, 1955)
**RDB3, SQS-24**
8 ST93: Great Ridge Wood, 8-May-00 (MDD)
SU22: Langley Wood NNR, 16-Jul to 6-Aug-04 (MDD) (Darby, 2005)
Resident. Very local; in rotten wood, particularly beech.

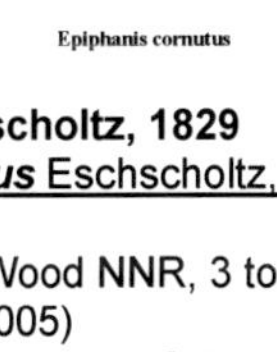

Epiphanis cornutus

**EPIPHANIS Eschscholtz, 1829**
***Epiphanis cornutus*** Eschscholtz, 1829
**SQS-8**
8 SU22: Langley Wood NNR, 3 to 25-Jun-04 (MDD) (Darby, 2005)
Resident. Very local; in a pitfall trap [in rotten wood of sycamore, spruce and other trees].

Tribe DIRHAGINI

**MICRORHAGUS Dejean, 1833**
***Microrhagus pygmaeus*** (Fabricius, 1792)
**RDB3, SQS-8**
7 ST97: North Draycot Park, 22-May-00 (MDD)
8 ST91: Cranborne Chase, 4-Jul-97 (MAS); 22-Jun-99 (MDD)
SU03: Grovely Wood, 21-Aug-72 (DRN)
SU22: Langley Wood NNR, 3 to 25-Jun-04 (MDD) (Darby, 2005)
Resident. Local; in rotten wood of beech and oak, and on vegetation under beech trees.

## Family THROSCIDAE

**TRIXAGUS Kugelann, 1794**
***Trixagus carinifrons*** (Bonvouloir, 1859)
7 SU26: Savernake Forest, 31-May to 23-Jun-95 (EN det MDD)
8 ST92: Sutton Mandeville, 25-Sep-07 (MDD)
SU21: Franchises Common, 24-Jul-73 (DRN)
SU22: Langley Wood NNR, 3-Sep-97 (WANHS det MDD); 14-May to 3-Jun-04, 5 to 25-Aug-06 (MDD) (Darby, 2005, 2008)
Resident. Local; on vegetation under trees including pines in woodland.

***Trixagus dermestoides*** (Linnaeus, 1767)
7 SU16: Marlborough area, Jun-1873 (EWJ) (Meyrick, 1874); 1869-1938 (Meyrick, 1938)
SU26: Savernake Forest, 15-Jun-73 (SAW), 26-May to 7-Jul-93, 9-Jun to 1-Jul-94 (EN det SAW)
8 SU02: Knighton Wood, May-07 (AJWA)
SU03: Grovely Wood, 11-Aug-71, 24-Aug-72, 22-Jul-73 (DRN); 12-Jun-00 (MDD)
SU21: Landford Bog, 2003 (APF) (Foster, 2003)
SU22: Landford, 12-Aug-87 (DRN); Langley Wood NNR, 3-Jun to 16-Jul-04, 25-Apr to 2-Jul-07 (MDD) (Darby, 2005, 2008); Blackmoor Copse, 4 to 18-May-02, 3-Aug-03 (APF) (Foster, 2004)
SU26: Bedwyn Brail, 2-Jun-03 (MDD)
Resident. Widespread; on vegetation under trees in woodland.

## Family ELATERIDAE
Subfamily AGRYPNINAE

**AGRYPNUS Eschscholtz, 1829**
***Agrypnus murinus*** (Linnaeus, 1758)
**LCGR-Fm Ch**
7 ST87: Rack Hill, 1995 (MA det MDD)
SU06: Cherhill Down, 1-Jun-04 (BLe)
SU16: Marlborough area, Jun-1873 (EWJ) (Meyrick, 1874); 12-Jun-1879 (AP) (Meyrick, 1880); 1869-1938 (Meyrick, 1938)

Agrypnus murinus

SU01: Martin Down NNR, 1999 (EN, 1999)
SU02: Middleton Down, 24-May-92 (BL); 9-Jun-96 (PM)
SU03: Wylye Down NNR, 4-May to 10-Jul-07 (MDD) (Darby, 2009); Grovely Wood, 13-Apr-71, 8, 10-Apr-72 (DRN); 14-Jun-01 (WANHS det MDD)
SU04: Fox Covert, 3-Jun - 12-Aug-08 (Boyce & Wright, 2009); Orcheston Down (SPTA), 22-Jun-02 (PJH) (Hodge, 2002); Salisbury Plain, 16-May, 25-Jun-02 (PJH) (Hodge, 2002)
SU05: Black Heath (SPTAC), 3-Jun - 12-Aug-08 (Boyce & Wright, 2009)
SU13: Little Durnford Down, 19-May-91 (PM)
SU14: Bourne Bottom (SPTAE), 13-Jun-97 (MGT); Haxton (SPTAE), 17-May-02 (PJH) (Hodge, 2002); 3-Jun - 12-Aug-08 (Boyce & Wright, 2009); Larkhill west, 2-Jun-97 (MGT); Stonehenge, 22-Apr-03 (PJH) (Hodge, 2003); Wexland Hanging (SPTAC), 22-Jun-02 (PJH) (Hodge, 2002); 3-Jun - 12-Aug-08 (Boyce & Wright, 2009)
SU15: Beach's Plantation (SPTAE), 13 to 17-Jun-97 (MGT); Everleigh (SPTAE), 24-Jun-02 (PJH) (Hodge, 2002); West Everleigh Down, 19-Jun-02 (IJG);
SU23: Porton Down, 3-Jul-99 (PJH) (Hodge, 2002)
SU24: Sidbury Hill (SPTAE), 18-May, 13 to 17-Jun-97 (MGT)
SU25: Sidbury Hill (SPTAE), 4, 13-Jun-97 (MGT)
Resident. Widespread; on vegetation in grassland

Subfamily HYPNOIDINAE

**HYPNOIDUS Dillwyn, 1829**
***Hypnoidus riparius*** (Fabricius, 1792)
7 VC7, after 1950 (Mendel & Clarke, 1996)
SU16: Marlborough, spring-1894 (GCC) (Jebb, 1895); Marlborough area, 1869-1938

(Meyrick, 1938)
SU27: Ramsbury, spring-1894 (GCC) (Jebb, 1895);
8 SU13: Salisbury, 13-Apr-00 (MDD) (Darby, 2001)
Resident. Very local; in litter at the edge of water.

## Subfamily DENTICOLLINAE
### Tribe CTENICERINI

**ACTENICERUS Kiesenwetter, 1858**
**Marsh Click Beetle**
***Actenicerus sjaelandicus*** (Müller, O.F, 1764)
8 SU21: Landford Bog, 22-Jun-03 (APF) (Foster, 2003)
SU16: Jones's Mill, 1997 (MAS); 19-Jun, 2-Jul-01 (APF) (Foster, 2002); 30-May-04 (KNAA)
Resident. Very local; on vegetation in damp areas.

**CTENICERA Latreille, 1829**
***Ctenicera cuprea*** (Fabricius, 1775)
7 SU16: Marlborough area, 1869-1938 (Meyrick, 1938)
SU17: Avebury, 1869-1938 (Meyrick, 1938)
SU27: Upper Upham, 1869-1938 (Meyrick, 1938)
Resident. Very local; on vegetation in dry grassland. Three specimens in the MCC without data. No recent records and the species appears to have declined in the county.

***Ctenicera pectinicornis*** (Linnaeus, 1758)
**Na**
7 SU16: Marlborough area, 1-Jul-1879 (AGC) (Meyrick, 1880); 1869-1938 (Meyrick, 1938)
Resident. Very local; [in lush grassland in old hay meadows]. Breeds in soil at the roots of the plants. Almost all records in GB are to the north of Wiltshire. Two specimens in MCC without data.

**CALAMBUS Thomson, C.G, 1859**
***Calambus bipustulatus*** (Linnaeus, 1767)
**Nb, IEC-3, SQS-8**
7 SU16: Marlborough area, 1869-1938 (Meyrick, 1938)
SU26: Savernake Forest, 6-Dec-73 (RCS); 11-Apr-90 (TGW); May-95 (MDD); 1-Jun-04 (MH det KNAA)
8 SU22: Langley Wood NNR, 12-Apr-74 (DRN) (Nash, 1979)
Resident. Very local; in dead wood and under bark of various deciduous trees.

**APLOTARSUS Stephens, 1830**
***Aplotarsus incanus*** (Gyllenhal, 1827)
7 ST96: Conigre Meadow, 20-May-01 (JKG det MDD)
SU18: Rushy Platt, 20-May-01 (JKG det MDD)
8 ST91: Cranborne Chase, 24-May-97 (MAS)
ST95: West Lavington, 13-May-45 (DCL) (DFL, 1948)
SU21: Burnt Ground Wood, 31-May-74 (DRN)
Resident. Local; on vegetation and young pines, often in damp areas.

**PARAPHOTISTUS Kishii, 1966**
***Paraphotistus impressus*** (Fabricius, 1792)
**Nb**
7 SU16: Marlborough area, 1869-1938 (Meyrick, 1938)
Resident. Very local; [in woodland]. The biology is unknown but adults have been found on pine, birch and oak. Two specimens without data in the MCC.

**PROSTERNON Latreille, 1834**
**Chequered Click Beetle**
***Prosternon tessellatum*** (Linnaeus, 1758)
**LCGR-FI CI**
7 SU16: Marlborough area, 1869-1938 (Meyrick, 1938)
SU26: Savernake Forest, 30-May-73 (DBS)
8 ST93: Great Ridge Wood, 29-Jul-82 (DRN)
ST94: Ladywell (SPTAW), 3-Jun - 12-Aug-08 (Boyce & Wright, 2009); Norton Down (SPTAW), 15-May-02 (PJH) (Hodge, 2002)
ST95: West Lavington, 1938 (PLP) (DFL, 1948)
SU01: Martin Down NNR, 1979-1992 (RGB)
SU03: Grovely Wood, 27-May-74 (DRN)
SU14: Haxton (SPTAE), 17-May, 24-Jun-02 (PJH) (Hodge, 2002); Wexland Hanging (SPTAC), 22-Jun-02 (PJH) (Hodge, 2002)
SU23: Porton Down, 3-Jul-99 (PJH) (Hodge, 2002)
Resident. Widespread; on vegetation in grassland.

### Tribe DENTICOLLINI

**CIDNOPUS Thgomson, C.G, 1859**
***Cidnopus aeruginosus*** (Olivier, 1790)
8 VC8, after 1950 (Mendel & Clarke, 1996)
Resident. Very local.

**KIBUNEA Kishii, 1966**
***Kibunea minuta*** (Linnaeus, 1758)
7 ST87: Corsham Court Park, 3-Jun-04 (KNAA)
SU09: Ashton Keynes, 7-May-01 (JKG det MDD)
SU16: Marlborough area, Jun-1873 (EWJ) (Meyrick, 1874); 1869-1938 (Meyrick 1938)
SU18: Swindon, 7, 15-May-01 (JKG det MDD)
SU26: Savernake Forest, 9-Jun to 1-Jul-94 (EN det SAW); 25-Apr to 11-May-95 (EN det MDD); 30-May-04 (KNAA)
8 ST93: Great Ridge Wood, 25-May-00 (MDD)
ST94: Norton Down (SPTAW), 15-May-02 (PJH) (Hodge, 2002); Ladywell (SPTAW), 15-May-02 (PJH) (Hodge, 2002); Bowls Barrow (SPTA), 22-Jun-02 (PJH) (Hodge, 2002)

SU01: Martin Down NNR, 1999 (EN, 1999)
SU03: Grovely Wood, 27-May-74 (DRN)
SU04: Orcheston Down (SPTA), 18-May, 22-Jun-02 (PJH) (Hodge, 2002); Salisbury Plain, 16-May, 25-Jun-02 (PJH) (Hodge, 2002)
SU14: Haxton (SPTAE), 17-May-02 (PJH) (Hodge, 2002); Stonehenge, 22-Apr-03 (PJH) (Hodge, 2003); Wexland Hanging (SPTAC), 16-May, 22-Jun-02 (PJH) (Hodge, 2002)
SU15: Everleigh (SPTAE), 18-May-02 (PJH) (Hodge, 2002)
SU23: Porton Down, 3-Jul-99 (PJH) (Hodge, 2002)
Resident. Widespread; on vegetation in grassland.

Denticollis linearis

**DENTICOLLIS Piller & Mitterpacher, 1783**
***Denticollis linearis*** (Linnaeus, 1758)
**SQS-1**
7 ST96: Spye Park, 24-Apr-87, 31-May-04 (KNAA); Silver Street Wood, 31-May-04 (KNAA)
ST99: Flisteridge Wood, 13-Jun-82 (KNAA)
SU16: Gopher Wood, 13-Jul-96 (MA det MDD); Marlborough area, Jun-1873 (EWJ) (Meyrick, 1874); 18-Jun-1879 (JA) (Meyrick, 1880); 1869-1938 (Meyrick, 1938)
SU17: Avebury, 31-May, 11-Jun-03, 2-Jun-05, 2-Jun-07 (DB det MDD)
SU26: Savernake Forest, 25-Apr to 28-Jul-93, 9-Jun to 1-Jul-94 (EN det SAW); 11-May to 23-Jun-95 (EN det MDD); 31-May-00 (WANHS det MDD)
8 ST84: Longleat, 28-May-93 (PM)
ST91: Cranborne Chase, 24-May-97 (MAS)
ST92: Wardour, 5-Jun-96 (MAS), Sutton Row, 21-Sep-95 (MDD); Sutton Mandeville, May-92 (MDD)
ST93: Great Ridge Wood, 10, 25-May-00 (MDD)
SU02: Knighton Wood, May, Jun-07, May-08 (AJWA)
SU03: Grovely Wood, 8-Apr-72, 27-May-74, 4-Aug-87 (DRN), 19-Jun-95 (MDD), 14-Jun-01 (WANHS det MDD)
SU13: Old Sarum, 31-May-97 (MAS)
SU16: Jones's Mill, 19-Jun-01 (APF) (Foster, 2002), 21-Jun-97 (MDD), 11-Jun-97 (MAS); 30-May-04 (KNAA)
SU21: Burnt Ground Wood, 31-May-74 (DRN)
SU22: Blackmoor Copse, 22-May-91, 21-May-93 (PM), 31-May-96 (MDD), 17-Mar-02, 3-Aug-03 (APF) (Foster, 2004); Bentley Wood, 2-Jun-91 (PM), 27-May-97, 27-May-98 (MDD), 23-May-04 (AGD); Langley Wood NNR, 28-May-74 (DRN); 9-Oct-82 (KNAA) 29-May-03, 14-May to 3-Jun-04, 25-Apr to 6-Jun-07 (MDD) (Darby, 2005, 2008)
SU23: Porton Down, 15-Oct-95, 12-Nov-96 (MAS)
Resident. Widespread; on vegetation in woodland.

**ATHOUS Eschscholtz, 1829**
ATHOUS Eschscholtz, 1829
***Athous haemorrhoidalis*** (Fabricius, 1801)
7 ST84: Smallbrook Meadows, 2-Jun-01 (JKG det MDD)
ST96: Spye Park, 31-May-04 (KNAA)
ST97: North Draycot Park, 22-May-00 (MDD)
ST99: Flisteridge Wood, 16-May-82 (KNAA)
SU06: Cherhill Down, 1-Jun-04 (BLe)
SU08: Somerford Common, 11-May-02 (JKG det MDD)
SU09: Upper Waterhay, 17-May-054-Jun-05 (JKG det MDD)
SU16: Marlborough area, Jun-1873 (EWJ) (Meyrick, 1874); 28 May 1879 (EEdeS) (Meyrick, 1880); 1869-1938 (Meyrick, 1938); Piggledene, 14-Jun-82 (KNAA); West Woods, May-02 (MDD)
SU17: Avebury, 14-Jun-82 (KNAA)
SU19: Hannington, 8-Jun-96 (MEN)
SU26: Savernake Forest, 17, 31-May-90 (TGW); 25-Apr to 28-Jul-93, 28-Apr to 20-Sep-94 (EN det SAW), 31-May to 23-Jun-95 (EN det MDD), 31-May-00 (WANHS det MDD)
8 ST84: Longleat, 15-Jun-73 (PM), Jun-97 (MB)
ST85: Green Lane Wood, 7-Jun-95 (WANHS det MDD), 26-Jun-98 (JN)
ST91: Cranborne Chase, 16 to 25-Jun-80 (NCC), 7-May-97 (MAS)
ST92: Wardour, 5-Jun-96 (MAS)
ST93: Great Ridge Wood, 10, 16, 25-May-00 (MDD)
ST94: Bowls Barrow (SPTAW), 16-May, 22-Jun-02 (PJH) (Hodge, 2002); Ladywell (SPTAW), 15-May, 23-Jun-02 (PJH) (Hodge, 2002); 3-Jun - 12-Aug-08 (Boyce & Wright, 2009); Norton Down (SPTAW), 15-May, 23-Jun-02 (PJH) (Hodge, 2002); Tinker's Track (SPTAW), 3-Jun - 12-Aug-08 (Boyce & Wright, 2009)
ST95: West Lavington, 1-Jun-39 (PLB) (DFL, 1948); 24-May-44, 13-May-45 (DCL) (DFL, 1948)
SU01: Martin Down NNR, 1979-1992 (RGB); 1999 (EN, 1999)

SU02: Knighton Wood, Jun-07, May, Jul-08 (AJWA)
SU03: Grovely Wood, 11-Aug-71, 27, 30-May-74 (DRN); 19-Jun-95 (MDD); 31-May-97 (MAS); 14-Jun-01 (WANHS det MDD); Wylye Down NNR, 5-Jun to 10-Jul-07 (MDD) (Darby, 2009)
SU04: Salisbury Plain, 16-May, 25-Jun-02 (PJH) (Hodge, 2002); Orcheston Down (SPTA), 18-May, 22-Jun-02 (PJH) (Hodge, 2002)
SU05: Black Heath (SPTAC), 3-Jun - 12-Aug-08 (Boyce & Wright, 2009)Ell Barrow (SPTAC), 3-Jun - 12-Aug-08 (Boyce & Wright, 2009)
SU13: Cockey Down, 21-May-91 (PM); Salisbury, 1-Jun-00 (MDD) (Darby, 2001); Little Durnford Down, 19-May-91 (PM)
SU14: Bourne Bottom (SPTAE), 13-Jun-97 (MGT); Haxton (SPTAE), 17-May, 24-Jun-02 (PJH) (Hodge, 2002); 3-Jun - 12-Aug-08 (Boyce & Wright, 2009); Larkhill west, 2-Jun-97 (MGT); Stonehenge, 22-Apr-03 (PJH) (Hodge, 2003); Wexland Hanging (SPTAC), 16-May, 22-Jun-02 (PJH) (Hodge, 2002); 3-Jun - 12-Aug-08 (Boyce & Wright, 2009)
SU15: Beach's Plantation (SPTAE), 10, 12-Jun-97 (MGT); Everleigh (SPTAE), 18-May, 24-Jun-02 (PJH) (Hodge, 2002)
SU16: Jones's Mill, 21-Jun-97 (MDD), 1997 (MAS), 3-May-00 (JKG det MDD);
SU22: Bentley Wood, 27-May-97 (MDD); Blackmoor Copse, 22-May-91, 13-May-93 (PM), 3-Aug-03 (APF) (Foster, 2004); Langley Wood, 28-May-74 (DRN); May-99 (BL); Jun-02 (DRN); 29-May-03, 14-May to 27-Aug-04, 6-Jun to 2-Jul-07 (MDD) (Darby, 2005, 2008)
SU23: Porton Down, 6 Jun 93 (PTW det MDD); 3-Jul-99 (MAS)
SU24: Beacon Hill (SPTAE), 3-Jun - 12-Aug-08 (Boyce & Wright, 2009); Bulford Ranges (SPTAE), 3-Jun - 12-Aug-08 (Boyce & Wright, 2009) Driver Training Area (SPTAE), 17-May-02 (PJH) (Hodge, 2002); Dunch Hill (SPTAE), 3-Jun - 12-Aug-08 (Boyce & Wright, 2009)
SU25: Salisbury Plain (Weather Hill Firs), 21-Jun-01 (MDD); Sidbury Hill (SPTAE), 4-Jun-97 (MGT)
SU26: Bedwyn Brail, 2-Jun-03 (MDD)

Resident. Widespread and frequent on vegetation in woodland and grassland.

***Athous vittatus*** (Fabricius, 1792)

7 SU16: Marlborough area, Jun-1873 (EWJ) (Meyrick, 1874); 18-Jun-1879 (TNH-S) (Meyrick, 1880), 1869-1938 (Meyrick, 1938); West Woods, May-02 (MDD)
SU26: Savernake Forest, 25-Apr to 28-Jul-93, 9-Jun to 20-Sep-94 (EN det SAW), 31-May to 23-Jun-95 (EN det MDD), 31-May-04 (BLe)
8 ST91: Cranborne Chase, 16 to 25-Jun-80 (NCC), 23-May-97 (MAS), 22-Jun-99 (MDD)
ST93: Great Ridge Wood, 16-May-00 (MDD)
SU02: Knighton Wood, May, Jul-08 (AJWA)
SU03: Grovely Wood, 11-Aug-71 (DRN), 19-Jun-95, 12-Jun-00 (MDD)
SU22: Bagfield Copse, 6, 7-Aug-86 (DRN); Bentley Wood, 31-May-97 (MDD); Langley Wood NNR, 8-Jun-02, 14-May-04, 6-Jun-07 (MDD) (Darby, 2005, 2008)
SU23: Porton Down, 3-Jul-99 (MAS)
SU26: Bedwyn Brail, 2-Jun-03 (MDD)

Resident. Widespread; on vegetation in woodland and grassland.

ORTHATHOUS Reitter, 1905

***Athous bicolor*** (Goeze, 1777)

7 SU16: Marlborough area, Jun-1873 (EWJ) (Meyrick, 1874), 1869-1938 (Meyrick, 1938)
8 ST83: Whitesheet Hill, 7-Jul-82 (KNAA)
ST85: Bullen Hill Farm, 1982-2002 (EGS det MDD)
ST91: Cranborne Chase, 16 to 25-Jun-80 (NCC)
ST93: Pits Wood, Chilmark, 12-Jul-95 (MDD); Stockton Wood, 7-Jul-96 (MAS)
ST94: Bowls Barrow (SPTAW), 22-Jun-02 (PJH) (Hodge, 2002)
SU01: Martin Down NNR, 1999 (EN, 1999)
SU02: Bishopstone, 17-Jul-89 (MDD)
SU03: Grovely Wood, 29-Jul-78 (DRN); Wylye Down NNR, 10-Jul to 3-Aug-07 (MDD) (Darby, 2009); Grovely Wood, 16-Aug-71, 4-Aug-78 (DRN)
SU04: Shrewton, 1-Aug-85 (DRN); Parsonage Down, 2, 22-Jul-08 (MDD); Tilshead Park (SPTAW), 14-Jul-99 (PJH) (Hodge, 1999);
SU06: Devizes, 5-Aug-83 (DRN); Horton, 5-Aug-83 (DRN)
SU13: Middle Woodford, 27-Jul-77 (DRN); Lower Woodford, 28-Jul-78 (DRN); Great Durnford, 17-Jul-73 (DRN)
SU14: Haxton (SPTAE), 24-Jun-02 (PJH) (Hodge, 2002); Stonehenge, 22-Apr-03 (PJH) (Hodge, 2003); Wexland Hanging (SPTAC), 22-Jun-02 (PJH) (Hodge, 2002)
SU15: Haxton Down, 14-Jul-99 (PJH) (Hodge, 1999)
SU16: Jones's Mill, 1997 (MAS)
SU23: Porton Down, 3-Jul-99 (PJH) (Hodge, 2002)
SU24: Driver Training Area (SPTAE), 24-Jun-02 (PJH) (Hodge, 2002)

Resident. Widespread; on vegetation in grassland and woods.

**STENAGOSTUS Thomson, C.G, 1859**

***Stenagostus rhombeus*** (Olivier, 1790)

**SQS-4**

7 VC7, after 1950 (Mendel & Clarke)
ST87: Corsham Court Park, 3-Jun-04 (KNAA)
ST96: Spye Park, 24-Apr-87 (KNAA)
ST99: Flisteridge Wood, 6-May-89 (KNAA)
8 SU03: Grovely Wood, 11-Aug-87, 1-Aug-88 (DRN)
SU21: Hamptworth Estate, 2-Nov-73 (DRN)
Resident. Local; under beech bark and on hazel.

**HEMICREPIDIUS Germar, 1839**

***Hemicrepidius hirtus*** (Herbst, 1784)

7 SU06: Cherhill Down, 15-Jun-82 (KNAA)
SU16: Marlborough area, 1869-1938 (Meyrick, 1938)
8 ST83: Whitesheet Down, 7-Jul-82 (KNAA)
ST91: Cranborne Chase, 16 to 25-Jun-80 (NCC), 22, 30-Jun-99 (MDD)
SU04: Parsonage Down NNR, 2-Jul-08 (MDD); Tilshead Park (SPTAW), 14-Jul-99 (PJH) (Hodge, 1999)
SU15: Everleigh (SPTAE), 24-Jun-02 (PJH) (Hodge, 2002)
SU21: Pound Bottom, 24-Jul-73 (DRN)
SU22: Bentley Wood, 18-Jul-93 (PM), 2-Jul-96 (MDD)
SU24: Bulford Ranges (SPTAE), 3-Jun - 12-Aug-08 (Boyce & Wright, 2009); Driver Training Area (SPTAE), 24-Jun-02 (PJH) (Hodge, 2002)
Resident. Widespread; on vegetation in grassland and woodland.

## Subfamily ELATERINAE

### Tribe ADRASTINI

**ADRASTUS Eschscholtz, 1829**

***Adrastus pallens*** (Fabricius, 1792)

7 SU16: Marlborough area, 1869-1938 (Meyrick, 1938)
8 SU12: Petersfinger, 25-Jul-91 (DRN)
SU14: Stonehenge, 22-Apr-03 (PJH) (Hodge, 2003)
SU16: Woodborough, 4-Aug-83 (DRN)
SU21: Hamptworth Estate, 3-Aug-73 (DRN)
SU22: Bagfield Copse, 7-Aug-86 (DRN)
Resident. Local; on vegetation in various habitats.

### Tribe AGRIOTINI

**AGRIOTES Eschscholtz, 1829**

***Agriotes acuminatus*** (Stephens, 1830)

7 SU16: Marlborough area, Jun-97 (EWJ) (Meyrick, 1874); 4-May-1879 (WFB) (Meyrick, 1880); 1869-1938 (Meyrick, 1938)
SU26: Savernake Forest, 17, 31-May-90 (TGW); 16-Jun to 7-Jul-93, 28-Apr to 18-May-94 (EN det SAW), 31-May to 23-Jun-95 (EN det MDD), 31-May-04 (BLe)
8 ST94: Ladywell (SPTAW), 15-May-02 (PJH) (Hodge, 2002)
SY99: Flisteridge Wood, 13-Jun-82 (KNAA)
SU03: Grovely Wood, 27-May-74 (DRN), 19-Jun-95, 12-Jun-00 (MDD)
SU04: Salisbury Plain, 16-May-02 (PJH) (Hodge, 2002)
SU13: Salisbury, Bemerton Folly, 7-Mar - 15-Jun-02 (MDD) (Darby, 2002d)
SU15: Everleigh (SPTAE), 18-May-02 (PJH) (Hodge, 2002)
SU22: Langley Wood NNR, 28-May-74 (DRN); Bentley Wood, 27-May-97 (MDD)
Resident. Widespread and frequent on vegetation and trees in woodland.

***Agriotes lineatus*** (Linnaeus, 1767)

7 SU16: Marlborough area, Jun-97 (EWJ) (Meyrick, 1874); 1869-1938 (Meyrick, 1938)
8 ST95: West Lavington, 14-May-44 (NR) (DFL, 1948)
SU01: Martin Down NNR, 1999 (EN, 1999)
SU03: Wylye Down NNR, 5-Jun-07 (MDD) (Darby, 2008a)
SU04: Orcheston Down (SPTA), 22-Jun-02 (PJH) (Hodge, 2002); Tilshead Down, 4-Jun-97 (MGT)
SU13: Salisbury, 22-May-77 (DRN)
SU15: Beach's Plantation (SPTAE), 13 to 17-Jun-97 (MGT)
SU22: Langley Wood NNR, 8-Jun-02 (MDD)
SU24: Driver Training Area (SPTAE), 1-May-05 (MGT)
Resident. Widespread; on vegetation in grassland.

***Agriotes obscurus*** (Linnaeus, 1758)

7 ST85: Green Lane Wood, 31-May04 (KNAA)
SU09: Ashton Keynes, 7-May-01 (JKG det MDD)
SU16: Marlborough area, Jun-97 (EWJ) (Meyrick, 1874); 1869-1938 (Meyrick, 1938)
SU18: Swindon, 7-May-01 (JKG det MDD)
SU26: Savernake Forest, 30-May-04 (KNAA)
8 ST93: Great Ridge Wood, 16-May-00 (MDD)
ST94: Norton Down (SPTAW), 23-Jun-02 (PJH) (Hodge, 2002); Ladywell (SPTAW), 23-Jun-02 (PJH) (Hodge, 2002)
SU02: Middleton Down, 6-Nov-73 (BL)
SU03: Wylye Down NNR, 27-Jan to 10-Jul-07 (MDD) (Darby, 2008); Grovely Wood, 13-Apr-71, 8-Apr-72, 8-Apr-74 (DRN)
SU04: Salisbury Plain, 16-May-02 (PJH) (Hodge, 2002)
SU13: Salisbury, 11-Apr-71, 26-May-74 (DRN), 13-Apr-00 (MDD) (Darby, 2001); Cockey Down, 24-Apr-93 (PM)
SU14: Larkhill west, 2-Jun-97 (MGT); Stonehenge, 22-Apr-03 (PJH) (Hodge, 2003); Wexland Hanging (SPTAC), 16-May, 22-Jun-02 (PJH) (Hodge, 2002)
SU24: Beach's Plantation (SPTAE), 1-May-05

(MGT)
SU22: Langley Wood NNR, 25-Jun to 6-Aug-04, 6-Jun to 2-Jul-07 (MDD) (Darby, 2005, 2008)
SU23: Porton Down, 19-Apr-92 (PTW det MDD)
Resident. Widespread; on vegetation in grassland and woodland.

***Agriotes pallidulus*** (Illiger, 1807)
7 ST96: Spye Park, 30-May-04 (KNAA)
ST97: North Draycot Park, 22-May-00 (MDD)
SU16: Marlborough area, Jun-1873 (EWJ) (Meyrick, 1874); 31 May 1879 (TNH-S) (Meyrick, 1880); 1869-1938 (Meyrick, 1938)
SU26: Savernake Forest, 25-Apr to 7-Jul-93, 28-Apr to 20-Sep-94 (EN det SAW), 25-Apr to 3-May-95 (EN det MDD); 30-May-04 (KNAA)
8 ST91: Cranborne Chase), 16 to 25-Jun-80 (NCC), 7-May-97 (MAS), 22-Jun-99 (MDD)
SU02: Knighton Wood, Apr-07, Jul-08 (AJWA)
SU03: Grovely Wood, 11-Aug-71, 27-May-74 (DRN), 19-Jun-95 (MDD)
SU13: Stratford sub Castle, 7-Aug-71 (DRN); Old
SU16: Jones's Mill, 1997 (MAS); Sarum, 31-May-97 (MAS)
SU21: Hamptworth Estate, 29-May-74 (DRN)
SU22: Bagfield Copse, 7-Aug-86 (DRN); Blackmoor Copse, 4 to 18-May-02, 3-Aug-03 (APF) (Foster, 2004); Langley Wood NNR, 28-May-74 (DRN); 8-Jun-02, 29-May-03, 22-Apr to 6-Aug-04, 5-Aug-06 to 2-Jul-07 (MDD) (Darby, 2005, 2008); 25-May-94 (WANHS det MDD); Whiteparish Common, 2-Jul-95 (MAS)
SU24: Driver Training Area (SPTAE), 17-May-02 (PJH) (Hodge, 2002)
SU26: Bedwyn Brail, 2-Jun-03 (MDD)
Resident. Widespread and frequent; on vegetation and in litter in woodland and on grassland

***Agriotes sputator*** (Linnaeus, 1758)
7 SU06: Cherhill Down, 1-Jun-04 (BLe)
SU09: Somerford Keynes, 4-Jun-04 (BLe)
SU16: Marlborough area, 1869-1938 (Meyrick, 1938)
8 ST93: Great Ridge Wood, 18-Apr-74 (DRN)
ST94: Ladywell (SPTAW), 23-Jun-02 (PJH) (Hodge, 2002); 3-Jun - 12-Aug-08 (Boyce & Wright, 2009); Tinker's Track (SPTAW), 3-Jun - 12-Aug-08 (Boyce & Wright, 2009); Wadman's Coppice (SPTAW), 3-Jun - 12-Aug-08 (Boyce & Wright, 2009)
ST95: West Lavington, 24-May-44 (DCL) (DFL, 1948)
SU01: Martin Down NNR, 1999 (EN, 1999)
SU03: Dinton Park, 6-Jul-00 (KNAA); Grovely Wood, 10-Apr-72 (DRN); Wylye Down NNR, 22-Feb to 10-Jul-07 (MDD) (Darby, 2008a)
SU04: Orcheston Down (SPTA), 18-May, 22-Jun-02 (PJH) (Hodge, 2002); Fox Covert, 25-Jun-02 (PJH) (Hodge, 2002)
SU05: Ell Barrow (SPTAC), 3-Jun - 12-Aug-08 (Boyce & Wright, 2009)
SU13: Salisbury, 22-May-77 (KNAA)
SU14: Haxton, 24-Jun-02 (PJH) (Hodge, 2002); Wexland Hanging (SPTAC), 22-Jun-02 (PJH) (Hodge, 2002)
SU15: Beach's Plantation (SPTAE), 10, 13 to 17-Jun-97, 1-May-05 (MGT)
SU23: Porton Down, 3-Jul-99 (PJH) (Hodge, 2002)
SU24: Beacon Hill (SPTAE), 3-Jun - 12-Aug-08 (Boyce & Wright, 2009); Driver Training Area (SPTAE), 1-May-05 (MGT); Dunch Hill (SPTAE), 16-Jun-97 (MGT); 3-Jun - 12-Aug-08 (Boyce & Wright, 2009); Sidbury Hill (SPTAE), 18-May, 13 to 17-Jun-97 (MGT)
SU25: Sidbury Hill (SPTAE), 28-May-97 (MGT)
Resident. Widespread; on vegetation and under stones and a cowpat, in grassland.

**DALOPIUS Eschscholtz, 1829**
***Dalopius marginatus*** (Linnaeus, 1758)
7 ST96: Chittoe, 9-May-00 (EGS det MDD); Spye Park, 31-May-04 (KNAA)
SU16: Marlborough area, 1869-1938 (Meyrick, 1938)
SU26: Savernake Forest, 25-Apr to 26-May-93, 28-Apr to 1-Jul-94 (EN det SAW); 25-Apr to 31-May-95 (EN det MDD); 31-May-04 (BLe)
8 ST84: Longleat, 28-May-93 (PM)
ST91: Cranborne Chase, 16 to 25-Jun-80 (NCC), 7- May-97 (MAS), 22-Jun-99 (MDD)
ST92: Wardour, 5-Jun-96 (MAS)
SU03: Grovely Wood, 16-Aug-71 (DRN), 19-Jun-95 (MDD)
SU13: Old Sarum, 31-May-97 (MAS)
SU22: Langley Wood NNR, 6-Jun-07 (MDD) (Darby, 2008)
SU22: Bagfield Copse, 7-Aug-86 (DRN); Blackmoor Copse, 12-May-93 (PM); Langley Wood NNR, 29-May-03 (MDD)
Resident. Local; on vegetation and trees in woods.

Tribe AMPEDINI

**AMPEDUS Dejean, 1833**
***Ampedus balteatus*** (Linnaeus, 1758)
**SQS-2, IEC-possible addition**
7 SU16: Marlborough area, Jun-1873 (EWJ) (Meyrick, 1874), 1869-1938 (Meyrick, 1938)
8 ST84: Longleat, 17-Jun-73 (PM)
SU21: Burnt Ground Wood, 15-May-03 (MDD); Franchises Common, 25, 27-Jul-73 (DRN); Hamptworth Estate, 29-May-74 (DRN); Pound Bottom, 31-May-74 (DRN)

Resident. Very local; in rotten wood of pines in woodland.

Ampedus balteatus

***Ampedus cinnabarinus*** (Eschscholtz, 1829)
**RDB3**, **IEC-1**, **SQS-16**
8 SU22: Bentley Wood, 31-May-96 (TWED conf. HM)
Resident. Very local; under bark of a cut oak log in a pile of logs.

Ampedus cinnabarinus

***Ampedus elongantulus*** (Fabricius, 1787)
**Na**, **IEC-3**, **SQS-8**
8 SU13: Cockey Down, 29-Apr-99 (PM)
SU21: Pound Bottom, 25-Jul-73, 31-May-74 (DRN)
SU22: Blackmoor Copse, 20-May-99, 4-May-03 (PM)
SU25: Ludgershall, Apr - Jul-1915 (PAH) (Harwood, 1915)
Resident. Local; on vegetation and pine trees.

***Ampedus quercicola*** (du Buysson, 1887)
**Nb**, **IEC-1**, **SQS-8**
8 SU21: Burnt Ground Wood, 31-Oct-75 (DRN); Franchises Common, 25-Jul-73 (DRN); Hamptworth Estate, 2-Nov-73 (DRN) Landford Bog, 9-Jun-01 (APF) (Foster, 2001)
SU22: Bagfield Copse, 7-Aug-86 (DRN); Bentley Wood, 2-Jun-04 (PJC et KNAA); Blackmoor Copse, 17-May-02 (APF) (Foster, 2004); Langley Wood NNR, 29-Aug-72 (DRN), 8-Jun-02, 29-May-03 (MDD)
Resident. Very local; in rotten wood, mainly birch.

The Marlborough specimen of Ampedus sanguineus

***Ampedus sanguineus*** (Linnaeus, 1758)
**Extinct**
8 SU12/13: Salisbury July c.1830 (WEL) (Stephens, 1830, Mendel & Clarke, 1996)
This is probably the specimen in the NHM. The only other extant specimens recorded are in the Dale Collection at Oxford (Hope Department) and in the Manchester Museum both from the New Forest in the 19thC. A correctly identified specimen under this name in the MCC without data (conf. HM), is, therefore, a fourth addition to the list. Stephens described the species as occurring 'in the utmost profusion on oaks' in the New Forest c. 1828, so that the lack of specimens is surprising but probably explained by confusion at the time between this species and other *Ampedus* spp. with red elytra.

***Ampedus sanguinolentus*** (Schrank, 1776)
**Na**, **SQS-16**
8 SU12/13: Salisbury c.1830 (WEL *as Elater ephippium*) (Stephens, 1830)
SU21: Pound Bottom, 31-May-74 (DRN)
Resident. Very local; flying by and on young pines.

Tribe MEGAPENTHINI

**PROCRAERUS Reitter, 1905**
***Procraerus tibialis*** (Boisduval & Lacordaire, 1835)
**RDB3**, **IEC-1**, **SQS-16**
7 SU26: Savernake Forest, mid-Jun-70 (AAA) (Allen, 1971)
8 VC8, before 1950 (Mendel & Clarke, 1996)
Allen's record is presumably the one cited in Savernake ISR, 1980-82. He found 'a thorax and part of an elytron in a large and very rotten standing trunk of beech... but a thorough search failed to reveal living examples in any stage'. [Widely distributed but local in southern England in dead wood of oak, beech and hornbeam].

Tribe ELATERINI

**SERICUS Eschscholtz, 1829**
***Sericus brunneus*** (Linnaeus, 1758)
7 SU16: Marlborough area, 1869-1938 (Meyrick, 1938)
8 SU21: Pound Bottom, 31-May-73 (DRN)
Resident. Very local; Nash's specimen was on a young pine. [Vegetation on heaths].

Subfamily MELANOTINAE

**MELANOTUS Eschscholz, 1829**
***Melanotus villosus*** (Fourcroy, 1785)
**SQS-1**
7 ST96: Spye Park, 24-Apr-87 (KNAA)
ST97: North Draycot Park, 22-May-00 (MDD)
ST99: Flisteridge Wood, 2-May-82 (KNAA)
SU16: Marlborough area, 29-May-1879 (EEdeS) (Meyrick, 1880), 1869-1938 (Meyrick, 1938)
SU26: Savernake Forest, spring-1894 (GCC) (Jebb, 1895); 17, 31-May-90 (TGW); 31-May-00 (WANHS det MDD); 30-May-04 (KNAA)
8 ST84: Longleat, Jun-97 (MB); 25-May-05 (DJG)
ST92: Wardour, 18-Jun-02 (MDD)
ST93: Great Ridge Wood, 19-Apr-83 (DRN), 14-Jun-96, 18-Mar, 2 to 16-May-00 (MDD)
ST95: West Lavington, 1938 (PLP) (DFL, 1948)
SU02: Knighton Wood, 6-Jul-07 (AJWA)
SU03: Grovely Wood, 10-Apr-70 (DRN), 19-Jun-95 (MDD), 14-Jun-01 (WANHS det MDD)
SU13: Little Durnford Down, 20-May-03 (MDD); Salisbury, Bemerton Folly, 7-Mar - 15-Jun-02 (MDD) (Darby, 2002d)
SU22: Blackmoor Copse, 17-Mar-02 (APF) (Foster, 2004); Langley Wood NNR, 28-May-74 (DRN), 9-Oct-82 (KNAA); 6-Jun-02, 29-May-03 (MDD)
SU26: Bedwyn Brail, 2-Jun, 5-Nov-03 (MDD)
Resident. Widespread; in rotten wood, some records as larvae.

Subfamily NEGASTRIINAE

**OEDOSTETHUS LeConte, 1853**
***Oedostethus quadripustulatus*** (Fabricius, 1792)
**Na**
7 VC7, since 1950 (Mendel & Clarke, 1996)
SU16: Wooton Rivers, 1905-1938 (Meyrick 1938)
Resident. Very local; [at roots of grass on edges of fields, etc. near water].

## Family DRILIDAE

**DRILUS Olivier, 1790**
***Drilus flavescens*** (Fourcroy, 1785)
**Na, LCGR-Fh Cm**
8 SU13: Cockey Down, 2-Jun-92 (PM), 10-Jun-90 (PM)
SU23: Porton Down, 3-Jul-99 (BL)
Resident. Very local; on vegetation in calcareous grassland. The female of this species is larviform and apterous.

Drilus flavescens

## Family LYCIDAE

Subfamily EROTINAE

**PLATYCIS Thomson, C.G, 1859**
***Platycis minutus*** (Fabricius, 1787)
**Nb, IEC-3, SQS-8**
7 SU08: Ravensroost Wood, 21-Aug-08 (PD)
SU26: Savernake Forest, 28-Jul to 28-Sep-93 (EN det SAW); 25-Aug to 14-Sep-95 (EN det MDD)
8 ST92: Sutton Mandeville, 22-Aug-01 (MDD)
SU13: Middle Woodford, 5-Aug-82 (DRN)
SU22: Blackmoor Copse, 24-Aug-01 (APF) (Foster, 2004)
Resident. Local; on rotten logs of birch and beech, flying in garden in evening, and on herbage.

Platycis minuta

## Family LAMPYRIDAE

**LAMPYRIS Müller, O.F, 1764**
**Glow Worm**

***Lampyris noctiluca*** (Linnaeus, 1758)
**LCGR-Fm Ch**
7 ST86: Bradford on Avon, 4-Aug-07 (WRQ); Corsham, Jul-01 (BM); Turleigh, 10-Jul-06 (HT); Wadswick Common, 13-Jul-98, 26-Jun-01 (ML)
ST87: Ford, 1988-1989 (MrsPT); Giddeahall Verge, 18-Jun-73 (CD); Long Dean, 24-Apr-87 (JHB); Stanton Park, 18-Jun-99 (HK det MDD); West Kington, 17-Mar-07 (MLu)
ST96: Chittoe, 1-Jan-01 (WC); Kennet and Avon Canal, 12-Aug-97 (PM); Sells Green, 15-Jul-05 (JED)
ST98: Corston, Jul-04 (AS)
SU06: Oliver's Castle, 6-Dec-73, 1990-2000 (PM)
SU16: Lockeridge, 25-May-73 (JEO); Marlborough area, 10-Jul-1879 (BJR) (Meyrick, 1880); 14-Jul-1879 (TNH-S) (Meyrick, 1880); 1869-1938 (Meyrick, 1938)
SU26: Cadley, May-01 (RD); Great Bedwyn, 2002 (KF); Savernake Forest, 6-Sep-73 (MrT)
SU27: Aldbourne, 19-Jun-96 (GW); High Clear Down, 13-Apr-02 (PM)
8 ST84: Black Dog Woods, Aug-99 (VH); Longleat, 14-Jun-07 (SB); Warminster, 1-May-01 (MrsPS)
ST85: Biss Wood, 15-Jun-02 (GWG); Green Lane Wood, 8-Jul-92 (PM); 24, 29-Jul-07 (GHa); Heywood, 2000 (ERBS)
ST92: Crockerton Down, 30-May-86; Donhead St. Mary, Jun, Jul-02 (HK det MDD); Fovant, 25-May-96 (MDD); Sutton Mandeville, 11-May-96 (MDD); Tollard Down, 1991 (RS);
ST93: Hawking Down, Chicklade, 14-Aug-04 (RM); Chilmark Down, 10-May-87 (PM); Great Ridge Wood, 18-Apr-74 (DRN)
ST95: West Lavington: May, Jun-36, 38, 46, 48 (ITH, BWHC, TFG) (DFL, 1948)
ST96: Seend, 1991(RS); Seend Cleeve, 1991 (RD)
SU01: Martin Down NNR, 1999 (EN, 1999)
SU02: Chickengrove Bottom, Apr-92 (PM); Middleton Down, May-88, Jul-89, Aug-89 (BL), Jun-90 (SP), 20-May-01 (BL)
SU03: Dinton, 17-Jul-87 (SP); Grovely Wood, 11-Apr, 28-May-71, 1974 (DRN); Jun-06 (MDD); Little Langford, 18-Jun-06 (DS); Wylye, Jul-01 (JF)
SU04: White Barrow, Tilshead, 25-Aug-01 (MA det MDD); Horse Down, Tilshead, 5-Aug-01 (MA det MDD)
SU13: Cockey Down, 15-Jul-94 (PM); Figsbury Ring, 4-Jul-91, 13-Jul-94 (PM); Quidhampton, 6-Dec-73 (PM)
SU15: New Mill, nr Pewsey, Jul-02 (HK det MDD)
SU16: Jones's Mill, 28-Jun-87, 22-Mar-92 (PM), 12-May-99 (BH); 6-Mar-99, 6-Mar-00, Jul-02 (HK det MDD)
SU22: Bentley Wood, 28-May, 23-Aug-90, 22-May, 24-Apr, 2-Sep-91 (PM); Blackmoor Copse, 27-May-73 (HFC), 13-Jun-73 (SP), 12-Jul-92, 3, 14-Jun-94 (SWa), 21-Jul-96 (PM), 13-Jul-97, 9-Jul-00 (SWa); Langley Wood NNR, 22-May-04 (MDD) (Darby, 2005)
SU23: West Winterslow, 5-Jul-01 (NB)
SU24: Beacon Hill (SPTAE), 3-Jun - 12-Aug-08 (Boyce & Wright, 2009); Perham Ranges, 18-Jun-73 (JM); Shipton Bellinger, 1996 (BL)
SU26: Bedwyn Brail, 12-Jun-03 (RMC)
Resident. Widespread; on vegetation and crawling on roads. Majority of records of ten species or less.

## Family CANTHARIDAE
**Soldier and Sailor Beetles**
### Subfamily CANTHARINAE
#### Tribe PODABRINI

***Podabrus alpinus*** (Paykull, 1798)
7 SU16: Marlborough area, 1869-1938 (Meyrick, 1938)
SU26: Savernake Forest, 11 to 30-May-95 (EN det MDD), 8-May-02 (IJG)
8 ST91: Cranborne Chase, 16 to 25-Jun-80 (NCC)
SU22: Blackmoor Copse, 12-May-93 (PM); Holmere Common, 29-May-90 (DRN); Langley Wood NNR, 29-May-03 (MDD), 28-May-74 (DRN)
SU26: Bedwyn Brail, 2-Jun-03 (MDD)
Resident. Local; on vegetation usually in woodland.

Podabrus alpinus

#### Tribe CANTHARINI

**CANTHARIS Linnaeus, 1758**
***Cantharis cryptica*** Ashe, 1947
7 ST96: Spye Park, 31-May-04 (KNAA)
SU19: Hannington, 8-Jun-96 (MEN)
SU26: Savernake Forest, 25-Apr to 26-May-93, 9-Jun to 19-Jul-94 (EN det SAW); 31-May to

23-Jun-95, 15-Jul to 3-Aug-95 (EN det MDD); 30-May-04 (KNAA); 31-May-04 (BLe)
8 ST84: Longleat, 15-Jun-73 (PM)
SU02: Fovant Down, 16-Jun-95 (MDD)
SU03: Grovely Wood, 27-May-74 (DRN)
SU16: Jones's Mill, 19-Jun-01 (APF) (Foster, 2002); 30-May-04 (KNAA)
SU21: Landford Bog, 9-Jun-01 (APF) (Foster, 2001)
SU26: Bedwyn Brail, 2-Jun-03 (MDD)
Resident. Widespread; on vegetation in grassland and woodland. Often confused with *C. pallida* from which it was separated by Ashe in 1946.

***Cantharis decipiens*** Baudi, 1871
7 ST86: Neston Park, 3-Jun-04 (KNAA)
ST87: Corsham Court Park, 3-Jun-04 (KNAA, PJC det KNAA)
SU19: Hannington, 26-May-97(MEN)
8 ST84: Longleat, 28-May-93 (PM)
ST93: Great Ridge Wood, 25-May-00 (MDD)
SU02: Knighton Wood, Jul-08 (AJWA)
SU13: Little Durnford Down, 18-May-93 (PM)
SU14: Stonehenge, 22-Apr-03 (PJH) (Hodge, 2003)
SU21: Landford Bog, 9-Jun-01 (APF) (Foster, 2001)
SU22: Blackmoor Copse, 17-May-02 (APF) (Foster, 2004)
Resident. Local; on vegetation in woodland and grassland.

***Cantharis figurata*** Mannerheim, 1843
7 SU09: Upper Waterhay, 13-Jun-04 (JKG det MDD)
Resident. Very local; on vegetation near water.

***Cantharis fusca*** Linnaeus, 1758
**RDB3**
7 SU16: Marlborough area, 28-May-1879 (WFB) (Meyrick, 1880); 1869-1938 (Meyrick, 1938)
SU19: Hannington, 26-May-97 (MEN)
8 ST95: West Lavington, 25-May-46 (DCL) (DFL, 1948)
SU12: Britford, Jun – 15-Sep-02 (CMD, 2008)
SU15: Beach's Plantation (SPTAE), 12-Jun-97 (MGT)
Resident. Very local; [on vegetation in grassland]. Perhaps not as rare as its RDB status suggests.

***Cantharis lateralis*** Linnaeus, 1758
7 SU09: Somerford Keynes, 4-Jun-04 (BLe)
8 ST94: Ladywell (SPTAW), 3-Jun - 12-Aug-08 (Boyce & Wright, 2009); Wadman's Coppice (SPTAW) 3-Jun - 12-Aug-08 (Boyce & Wright, 2009); Warminster Ranges (SPTAW), 3-Jun - 12-Aug-08 (Boyce & Wright, 2009)
SU04: Fox Covert, 3-Jun - 12-Aug-08 (Boyce & Wright, 2009); Salisbury Plain, 25-Jun-02 (PJH) (Hodge, 2002)
SU14: Haxton (SPTAE), 3-Jun - 12-Aug-08 (Boyce & Wright, 2009)
Resident. Very local; on vegetation in grassland.

***Cantharis livida*** Linnaeus, 1758
7 SU16: Marlborough area, Jun-1873 (EWJ) (Meyrick, 1874); 24 Jun 1879 (FSA) (Meyrick, 1880); 1869-1938 (Meyrick, 1938)
8 ST94: Bowls Barrow (SPTA), 16-May, 22-Jun-02 (PJH) (Hodge, 2002); Ladywell (SPTAW), 15-May, 23-Jun-02 (PJH) (Hodge, 2002); Norton Down (SPTAW), 15-May-02 (PJH) (Hodge, 2002); Warminster Ranges (SPTAW), 3-Jun -12-Aug-08 (Boyce & Wright, 2009)
ST95: West Lavington, 24-May-44 (DCL) (DFL, 1948)
SU04: Salisbury Plain, 16-May, 25-Jun-02 (PJH) (Hodge, 2002); Tilshead Down, 4-Jun-97 (MGT)
SU14: Haxton (SPTAE), 24-Jun-02 (PJH) (Hodge, 2002) (Hodge, 2002); 3-Jun - 12-Aug-08 (Boyce & Wright, 2009); Wexland Hanging (SPTA), 16-May-22-Jun-02 (PJH) (Hodge, 2002)
SU15: Beach's Plantation (SPTAE), 10-Jun-97 (MGT); Everleigh (SPTAE), 18-May, 24-Jun-02 (PJH) (Hodge, 2002)
SU24: Bulford Ranges (SPTAE), 3-Jun - 12-Aug-08 (Boyce & Wright, 2009); Driver Training Area (SPTAE), 17-May, 24-Jun-02 (PJH) (Hodge, 2002)
SU25: Sidbury Hill (SPTAE), 4-Jun-97 (MGT)
Resident. Widespread; on vegetation in grassland.

***Cantharis nigra*** (De Geer, 1774)
7 SU08: Distillery Meadows, 1-Jul-00, 14-Jun-73 (PM)
SU09: Upper Waterhay, 13-Jun-04, 6-Jul-07 (JKG det MDD); Emmett Hill, 1-Jun-92 (PM); Stoke, 19-Jun-95, 24-Jun-99 (PM)
SU16: Caen Hill Locks, 11-Jun-96 (EGS det MDD); Marlborough area, 1869-1938 (Meyrick, 1938)
SU19: Hannington, 26-May-97, 22-Jun-96 (MEN)
8 ST85: Bullen Hill Farm, 1982-2002 (EGS det MDD)
ST92: Sutton Mandeville, 7-Jul-96 (MDD)
ST94: Bowls Barrow (SPTAW), 22-Jun-02 (PJH) (Hodge, 2002); Ladywell (SPTAW), 23-Jun-02 (PJH) (Hodge, 2002); Norton Down (SPTAW), 23-Jun-02 (PJH) (Hodge, 2002)
SU04: Salisbury Plain, 25-Jun-02 (PJH) (Hodge, 2002) (Hodge, 2002)
SU14: Wexland Hanging (SPTAC), 22-Jun-02 (PJH) (Hodge, 2002)
SU16: Jones's Mill, 21-Jun-97 (MDD), 2-Jul-01

(APF) (Foster, 2002
SU22: Bentley Wood, 2-Jul-96 (MDD); Blackmoor Copse, 2-Jul-96 (MDD); Holmere Common, 29-May-90 (DRN)
SU26: Bedwyn Brail, 19-Jun-03 (MDD)
Resident. Widespread; on vegetation in grassland and woodland.

***Cantharis nigricans*** (Müller, O.F. 1776)
7 ST84: Smallbrook Meadows, 2-Jun-01 (JKG det MDD)
SU06: Roundway, Devizes, 17-May-00 (BG)
SU08: Grone Farm, Grittenham, 14-Jun-98 (BLe); Red Lodge Pond, 31-May-01 (JKG det MDD)
SU09: Upper Waterhay, 17-May to 4-Jun-05, 5 to 11-Jun-07 (JKG det MDD)
SU16: Marlborough area, Jun-1873 (EWJ) (Meyrick, 1874); 7-Jun-1879 (WFB) (Meyrick, 1880); 1869-1938 (Meyrick, 1938)
SU19: Hannington, 27-May, 22-Jun-96 (MEN)
SU24: Barham Down, 30-May-04 (AES det KNAA)
SU26: Savernake Forest, 31-May-04 (BLe), 26-May to 16-Jun-93 (EN det SAW); 31-May to 23-Jun-95 (EN det MDD)
8 ST84: Longleat, 15-Jun-73 (PM)
ST85: Green Lane Wood, 7-Jun-95 (WANHS det MDD); 31-May-04 (KNAA)
ST91: Cranborne Chase, 16 to 25-Jun-80 (NCC), 12-May-97 (MAS)
ST93: Great Ridge Wood, 10, 16, 25-May-00 (MDD)
ST94: Bowls Barrow (SPTAW), 16-May-02 (PJH) (Hodge, 2002); Ladywell (SPTAW), 3-Jun – 12-Aug-08 (Boyce & Wright, 2009) Norton Down (SPTAW), 15-May-02 (PJH) (Hodge, 2002)
SU02: Bishopstone, 27-May-91 (MDD)
SU03: Grovely Wood, 27, 30-May-74 (DRN)
SU04: Orcheston Down (SPTA), 18-May-02 (PJH) (Hodge, 2002)
SU13: Salisbury, 1-Jun-00 (Darby, 2001); Salisbury, Bemerton Folly, 7-Mar - 15-Jun-02 (MDD) (Darby, 2002d)
SU14: Wexland Hanging (SPTAC), 16-May-02 (PJH) (Hodge, 2002)
SU15: Everleigh (SPTAE), 18-May-02 (PJH) (Hodge, 2002)
SU16: Jones's Mill, 2-Jul-01 (APF) (Foster, 2002)
SU22: Blackmoor Copse, 13, 23-May-93 (PM)
SU25: Sidbury Hill (SPTAE), 4-Jun-97 (MGT)
Resident. Widespread; on vegetation in grassland and woodland.

***Cantharis pallida*** Goeze, 1777
7 SU16: Lockeridge, 1994 (JEO); Marlborough area, Jun-1873 (Meyrick, 1874); 7-Jun-1879 (WFB) (Meyrick, 1880);1869-1938 (Meyrick, 1938)
SU26: Savernake Forest, 31-May-04 (BLe)
8 ST85: Green Lane Wood, 7-Jun-95 (WANHS det MDD)
ST95: West Lavington, 25-Apr-46 (DCL) (DFL, 1948)
SU12: Charlton-All-Saints, 9-May-96 (MAS)
SU13: Great Durnford, 17-Jul-73 (DRN)
SU16: Jones's Mill, 19-Jun-01 (APF) (Foster, 2002); 1-Jun-04 (BLe)
SU22: Langley Wood NNR, 28-May-74 (DRN)
Resident. Widespread; on vegetation sometimes near water. See note under *C. cryptica*

***Cantharis pellucida*** Fabricius, 1792
7 SU16: Marlborough area, 31 May 1879 (TNH-S) (Meyrick, 1880); 1869-1938 (Meyrick, 1938)
SU19: Hannington, 26-May-97(MEN)
SU26: Savernake Forest, 25-Apr to 26-May-93, 9-Jun to 1-Jul, 30-Aug to 20-Sep-94 (EN det SAW); 30-May-04 (KNAA); 31-May-04 (BLe)
8 ST91: Cranborne Chase, 16 to 25-Jun-80 (NCC)
ST95: West Lavington, 25-Apr-44 (DCL) (DFL, 1948)
SU03: Grovely Wood, 27, 30-May-74 (DRN)
SU04: Orcheston Down (SPTA), 18-May-02 (PJH) (Hodge, 2002)
SU14: Stonehenge, 22-Apr-03 (PJH) (Hodge, 2003)
SU15: Everleigh (SPTAE), 18-May-02 (PJH) (Hodge, 2002)
SU16: Jones's Mill, 30-May-04 (KNAA); 1-Jun-04 (BLe)
SU21: Landford Bog, 24-Jun-01 (APF) (Foster, 2001); Peaked Wood, 29-May-74 (DRN)
SU22: Bentley Wood, 23-May-04 (AGD); Langley Wood NNR, 28-May-74 (DRN)
Resident. Widespread; on vegetation in woodland and grassland.

***Cantharis rufa*** Linnaeus, 1758
7 ST85: Green Lane Wood, 31-May-04 (KNAA)
ST87: Corsham Park, 3-Jun-04 (BLe, KNAA)
SU09: Somerford Keynes, 4-Jun-04 (BLe); Upper Waterhay, 17-May to 4-Jun-05 (JKG det MDD)
SU16: Marlborough area, 1869-1938 (Meyrick, 1938)
SU97: Lacock, 29-May-04 (AES det KNAA)
8 ST94: Bowls Barrow (SPTAW), 16-May, 22-Jun 02 (PJH) (Hodge, 2002); Norton Down (SPTAW), 23-Jun-02 (PJH) (Hodge, 2002); Warminster Ranges (SPTAW), 3-Jun -12-Aug-08 (Boyce & Wright, 2009)
SU14: Wexland Hanging (SPTAC), 22-Jun-02 (PJH) (Hodge, 2002)
SU15: Beach's Plantation (SPTAE), 10, 12-Jun-97 (MGT); Everleigh (SPTAE), 18-May, 24-Jun-02 (PJH) (Hodge, 2002)

SU24: Sidbury Hill (SPTAE), 28-May, 4-Jun-97 (MGT)
Resident. Widespread; on vegetation in grassland.

***Cantharis rustica*** Fallén, 1807
7 ST85: Green Lane Wood, 31-May-04 (KNAA)
ST86: Neston Park, 3-Jun-04 (BLe)
ST87: Corsham Park, 3-Jun-04 (KNAA)
ST94: Wadman's Coppice (SPTAW) 3-Jun - 12-Aug-08 (Boyce & Wright, 2009); ST94: Warminster Ranges (SPTAW), 3-Jun -12-Aug-08 (Boyce & Wright, 2009)
ST96: Spye Park, 31-May-04 (KNAA); Upper Selves Wood, 31-May-04 (KNAA)
SU04: Fox Covert, 3-Jun - 12-Aug-08 (Boyce & Wright, 2009)
SU05: Ell Barrow (SPTAC), 3-Jun - 12-Aug-08 (Boyce & Wright, 2009)
SU14: Haxton (SPTAE), 3-Jun - 12-Aug-08 (Boyce & Wright, 2009); Wexland Hanging (SPTAE), 3-Jun - 12-Aug-08 (Boyce & Wright, 2009)
SU16: Gopher Wood, May-96 (MA det MDD); Marlborough area, 1869-1938 (Meyrick, 1938)
SU19: Hannington, 26-May-97 (MEN)
SU24: Barham Down, 30-May-04 (KNAA); Bulford Ranges (SPTAE), 3-Jun - 12-Aug-08 (Boyce & Wright, 2009); Dunch Hill (SPTAE), 3-Jun - 12-Aug-08 (Boyce & Wright, 2009)
SU26: Savernake Forest, 30-May-04 (KNAA)
SU26/27: Axford, Jun-1873 (EWJ) (Meyrick, 1874)
8 ST84: Berkley Woods, 21-May-00 (AGD)
ST92: South Down Farm, 30-Jun-97 (EGS det MDD)
ST93: Sutton Mandeville, 2-Jun-95 (MDD)
ST94: Bowls Barrow (SPTAW), 16-May-02 (PJH) (Hodge, 2002); Ladywell (SPTAW), 15-May-02 (PJH) (Hodge, 2002); Norton Down (SPTA), 15-May-02 (PJH) (Hodge, 2002)
ST95: West Lavington, 1938, 1939 (PLP) (DFL, 1948); 24-May-46 (DCL) (DFL, 1948)
SU02: Middleton Down, 24-May-92 (BL)
SU03: Wylye Down NNR, 4-May to 5-Jun-07 (MDD) (Darby, 2009)
SU04: Orcheston Down (SPTA), 18-May-02 (PJH) (Hodge, 2002); Salisbury Plain, 16-May-02 (PJH) (Hodge, 2002)
SU13: Old Sarum, 31-May-97 (MAS)
SU14: Bourne Bottom (SPTAE), 13-Jun-97 (MGT); Haxton (SPTAE), 17-May-02 (PJH) (Hodge, 2002); Larkhill west, 2-Jun-97 (MGT); Stonehenge, 22-Apr-03 (PJH) (Hodge, 2003); Wexland Hanging (SPTAC), 16-May, 22-Jun-02 (PJH) (Hodge, 2002)
SU15: Beach's Plantation (SPTAE), 10, 13-Jun-97 (MGT); Everleigh (SPTAE), 18-May, 24-Jun-02 (PJH) (Hodge, 2002)
SU22: Bentley Wood, 27-May-97 (MDD); Langley Wood, 29-May-03 (MDD)
SU23: Porton Down, 31-May-92 (PTW det MDD, 6-Jun-93 (PTW det MDD)
SU24: Driver Training Area (SPTAE), 17-May-02 (PJH) (Hodge, 2002); North Tidworth, Devils Ditch, 0-May-04 (BLe); Sidbury Hill (SPTAE), 29-May-97 (MGT)
SU25: Sidbury Hill (SPTAE), 28-May, 4-Jun-97 (MGT)
Resident. Widespread; on vegetation in grassland and woodland

***Cantharis thoracica*** (Olivier, 1790)
7 SU16: Marlborough area, 1869-1938 (Meyrick, 1938)
8 SU12: Petersfinger, 25-Jul-91 (DRN); Salisbury, 30-Jul-73 (DRN)
SU13: Great Durnford, 7-Aug-70 (DRN)
SU14: Figheldean, 4-Aug-88 (DRN)
SU15: Fittleton, 29-Jul-83 (DRN)
Resident. Local; on vegetation near R. Avon and R. Bourne.

**RHAGONYCHA Eschscholtz, 1830**
***Rhagonycha fulva*** (Scopoli, 1763)
7 SU08: Ravensroost Wood, 3-Aug-96 (MEN); Somerford Common, 11-Aug-98 (MEN)
SU16: Lockeridge, 1994 (JEO); Marlborough area, 1869-1938 (Meyrick, 1938); Wootton Rivers, 10-Aug-81 (DRN)
SU19: Hannington, 26, 31-Jul, 3-Aug-96 (MEN)
8 ST02: Vernditch Chase, 17-Jul-95 (MDD)
ST84: Longleat, 15-Jun-73 (PM); 11-Aug-05 (DJG)
ST85: Bullen Hill Farm, 1982-2002 (EGS det MDD); Green Lane Wood, 0-Jul-92 (PM)
ST92: Sutton Mandeville, 7-Jul-96 (MDD)
ST93: Fonthill, 19-Jul-95 (MDD); Pits Wood, Chilmark, 12-Jul-95 (MDD)
ST94: Tinker's Track (SPTAW), 3-Jun - 12-Aug-08 (Boyce & Wright, 2009); Wadman's Coppice (SPTAW) 3-Jun - 12-Aug-08 (Boyce & Wright, 2009); Warminster Ranges (SPTAW), 3-Jun - 12-Aug-08 (Boyce & Wright, 2009)
ST95: West Lavington, 18-Jul-40 (ITH) (DFL, 1948)
SU01: Martin Down NNR, 1999 (EN, 1999)
SU02: Chickengrove Bottom, 17-Jul-95 (MDD), 15-Jul-02 (BL); Middleton Down, 24-Jul-89, 21-Jul-95 (BL)
SU03: Dinton, 9-Jul-95 (MDD); Grovely Wood, 7-Aug-85 (DRN); Wylye Down NNR, 10-Jul, 3-Aug 07, 4-Sep, 10-Oct-07 (MDD) (Darby, 2008a)
SU04: Parsonage Down, 2-Jul-08 (MDD); Tilshead Park, 14-Jul-99 (PJH) (Hodge, 2002) (Hodge, 1999); Shrewton, 28-Jul, 6-Aug-82

(DRN); Yarnbury Castle, 8-Jul-08 (MDD)
SU05: Black Heath (SPTAC), 3-Jun - 12-Aug-08 (Boyce & Wright, 2009); Ell Barrow (SPTAC), 3-Jun - 12-Aug-08 (Boyce & Wright, 2009)
SU12: Petersfinger, 7-Aug-88, 25-Jul-91 (DRN)
SU13: Cockey Down, 13-Jun-73 (PM); Great Durnford, 18-Aug-72, 17-Jul-73 (DRN); Idmiston, 5-Aug-82 (DRN); Little Durnford Down, 6-Dec-73 (PM); Lower Woodford, 5-Aug-82 (DRN); Middle Woodford, 27-Jul-77 (DRN); Salisbury, 1-Apr-00 (MAS) (Darby, 2001)
SU14: Figheldean, 28-Jul-82 (DRN); Haxton (SPTAE), 3-Jun - 12-Aug-08 (Boyce & Wright, 2009); Netheravon, 29-Jul-83 (DRN); Stonehenge, 8-Jul-01(AGD); 22-Apr-03 (PJH) (Hodge, 2003); Wexland Hanging (SPTAE), 3-Jun - 12-Aug-08 (Boyce & Wright, 2009)
SU15: Fittleton, 29-Jul-83 (DRN); Haxton Down (SPTAE), 14-Jul-99 (PJH) (Hodge, 1999)
SU16: Jones's Mill, 2-Jul-01 (APF) (Foster, 2002); Woodborough, 4-Aug-83 (DRN)
SU22: Bentley Wood, 1996 (JN); 11-Aug-01(MDD); 8-Jul-04 (AGD); Blackmoor Copse, 3-Aug-03 (APF) (Foster, 2004); East Grimstead, 1-Aug-77 (DRN); Langley Wood NNR, 25-Aug-06 (MDD) (Darby, 2008)
SU23: Porton Down, 7-Jul-91 (PTW det MDD); 3-Jul-99 (PJH); West Winterslow, 7-Jul-01 (NB)
SU24: Bulford Ranges (SPTAE), 3-Jun - 12-Aug-08 (Boyce & Wright, 2009); Dunch Hill (SPTAE), 3-Jun - 12-Aug-08 (Boyce& Wright, 2009)

Resident. Widespread and frequent; on vegetation in grassland and woodland.

***Rhagonycha lignosa*** (Müller, O.F, 1764)

7 ST86: Neston Park, 3-Jun-04 (KNAA)
ST96: Spye Park, 31-May-04 (KNAA)
SU16: Marlborough area, 1869-1938 (Meyrick, 1938)
SU26: Savernake Forest, 26-May to 16-Jun-93, 28-Apr to 19-Jul, 30-Aug to 20-Sep-94 (EN det SAW), 11 to 31-May, 15-Jul to 3-Aug-95 (EN det MDD); 30-May-04 (KNAA)
8 ST84: Longleat, 28-May-93 (PM)
ST85: Green Lane Wood, 7-Jun-95 (WANHS det MDD)
ST91: Cranborne Chase, 16 to 25-Jun-80 (NCC); 7-May-97 (MAS); 22-Jun-99 (MDD)
ST93: Great Ridge Wood, 16-May-00 (MDD)
SU02: Middleton Down, 24-May-95 (MDD)
SU03: Grovely Wood, 27-May-74, 1-Aug-88 (DRN)
SU04: Orcheston Down (SPTA), 18-May-02 (PJH) (Hodge, 2002)
SU11: Burnt Ground Wood, 31-May-74 (DRN)
SU13: Little Durnford Down, 18-May-93 (PM); Old Sarum, 31-May-97 (MAS)
SU14: Stonehenge, 22-Apr-03 (PJH) (Hodge, 2003)
SU15: Everleigh (SPTAE), 18-May-02 (PJH) (Hodge, 2002)
SU22: Bentley Wood, 31-May-97 (MDD); Langley Wood, 28-May-74 (DRN)
SU24: North Tidworth, Devils Ditch, 30-May-04 (BLe)
SU26: Bedwyn Brail, 2-Jun-03 (MDD)

Resident. Widespread; on vegetation in grassland and woodland

***Rhagonycha limbata*** Thomson, C.G., 1864

7 ST84: Smallbrook Meadows, 2-Jun-01 (JKG det MDD)
ST87: Castle Combe, 3-Jun-04 (BLe, KNAA)
ST95: Cherhill Down, 30-May-04 (AES det KNAA)
SU09: Upper Waterhay, 17-May to 11-Jun-05 (JKG det MDD)
SU16: Marlborough area, 28 May 1879 (TNH-S) (Meyrick, 1880); 1869-1938 (Meyrick, 1938)
SU26: Savernake Forest, 25-Apr to 26-May-93 (EN det SAW); 3-Jun-04 (BLe)
8 ST91: Cranborne Chase, 16 to 25-Jun-80 (NCC), 24-May-97 (MAS)
ST94: Bowls Barrow (SPTAW), 16-May-02 (PJH) (Hodge, 2002); Ladywell (SPTAW), 23-Jun-02 (PJH) (Hodge, 2002) (Hodge, 2002); Norton Down (SPTA), 15-May, 23-Jun-02 (PJH) (Hodge, 2002)
SU13: Old Sarum, 31-May-97 (MAS)
SU14: Wexland Hanging (SPTAC), 23-Jun-02 (PJH) (Hodge, 2002)
SU15: Everleigh (SPTAE), 24-Jun-02 (PJH) (Hodge, 2002)
SU25: Sidbury Hill (SPTAE), 6-Jun-97 (MGT)
SU26: Bedwyn Brail, 2-Jun-03 (MDD)

Resident. Widespread; on vegetation in grassland and woodland. Formerly *R. femoralis* sensu auctt Brit..

***Rhagonycha lutea*** (Müller, O.F, 1764)
**Nb**

8 SU03: Wylye Down NNR, 4-May to 10-Jul-07 (MDD) (Darby, 2009)
SU13: Cockey Down, 13-Jun-73 (PM)
SU23: Porton Down, 3-Jul-99 (MAS)

Resident. Very local; on vegetation in calcareous grassland.

***Rhagonycha testacea*** (Linnaeus, 1758)

7 SU16: Marlborough area, 1869-1938 (Meyrick, 1938)
8 ST85: Green Lane Wood, 7-Jun-95 (WANHS det MDD); 31-May-04 (KNAA); 2-Jun-04 (PJC

det KNAA)
SU13: Salisbury, Apr-00 (MDD) (Darby, 2001)
SU14: Figheldean, 8-Jun-97 (MGT); Stonehenge, 22-Apr-03 (PJH) (Hodge, 2003)
SU16: Jones's Mill, 21-Jun-97 (MDD), 11-Jun-97 (MAS); 30-May-04 (KNAA); 1-Jun-04 (BLe)
SU22: Langley Wood NNR, 28-May-74 (DRN)
Resident. Local; on vegetation in wooded areas.

***Rhagonycha translucida*** (Krynicki, 1832)
**Nb**
7 SU26: Savernake Forest, 9-Jun to 1-Jul-94 (EN det SAW); 31-May to 23-Jun-95 (EN det MDD)
8 ST91: Cranborne Chase, 16 to 25-Jun-80 (NCC)
SU16: Jones's Mill, 21-Jun-97 (MDD)
Resident. Very local; on vegetation in wooded areas.

## Subfamily SILINAE

Silis ruficollis

**SILIS Charpentier, 1825**
***Silis ruficollis*** (Fabricius, 1775)
**Nb**
8 SU12: Petersfinger, 25-Jul-91 (DRN)
SU13: Great Durnford, 27-Jul-77 (DRN); Lower Woodford, 1-Aug-83 (DRN); Jun – 15-Sep-02 (CMD); Middle Woodford, 27-Jul-77 (DRN)
Resident. Very local; on vegetation at edge of R. Avon and a lake.

## Subfamily MALTHININAE
### Tribe MALTHININI

**MALTHINUS Latreille, 1806**
***Malthinus balteatus*** Suffrian, 1851
**Nb, SQS-8**
7 SU16: Marlborough area, 1869-1938 (Meyrick, 1938)
8 ST91: Cranborne Chase, 30-Jun-99 (MDD), 4-Jun-97 (MAS)
SU03: Grovely Wood, 13-Aug-71 (DRN)
SU04: Parsonage Down NNR, 22-Jul-08 (MDD)
SU23: Porton Down, 3-Jul-99 (MAS)
SU24: Driver Training Area (SPTAE), 24-Jun-02 (PJH) (Hodge, 2002)
Resident. Local; on foliage in calcareous grassland.

Malthinus balteatus

***Malthinus flaveolus*** (Herbst, 1786)
**SQS-1**
7 ST86: Neston Park, 3-Jun-04 (KNAA)
SU16: Marlborough area, 1869-1938
SU26: Savernake Forest, 16-Jun to 28-Jul-93 (EN det SAW)
8 SU21: Landford Bog, 24-Jun-01, 22-Jun-03 (APF) (Foster, 2001, 2003)
SU22: Blackmoor Copse, 13-Jul-03 (APF) (Foster, 2004)
Resident. Local; on shrubs and trees in wooded areas.

***Malthinus frontalis*** (Marsham, 1802)
**Nb, SQS-8**
8 SU03: Dinton Park, 6-Jul-01 (KNAA) (Alexander, 2004)
Resident. Very local; on vegetation in parkland.

***Malthinus seriepunctatus*** Kiesenwetter, 1852
**SQS-2**
7 ST86: Neston Park, 3-Jun-04 (KNAA)
SU16: Marlborough area, 1-Jan-0031-Dec-05 (Meyrick, 1906), 1869-1938 (Meyrick, 1938)
8 ST85: Bullen Hill Farm, 1982 - 2002 (EGS det MDD)
ST91: Cranborne Chase, 30-Jun-99 (MDD)
ST93: Stockton Wood, 7-Jul-96 (MAS)
SU03: Grovely Wood, 26-Aug-72 (DRN), 12-Jun-00 (MDD)
SU04: Parsonage Down NNR, 2-Jul-08 (MDD)
SU22: Landford, 12-Aug-87 (DRN); Glazier's Copse, 16-Aug-88 (DRN)
SU26: Bedwyn Brail, 2, 19-Jun-03 (MDD)
Resident. Widespread; on vegetation in wooded areas.

### Tribe MALTHODINI

**MALTHODES Kiesenwetter, 1852**
***Malthodes dispar*** (Germar, 1824)
**SQS-2**
8 SU03: Grovely Wood, 14-Jun-01 (WANHS det MDD)
Resident. Very local; on vegetation in woodland.

***Malthodes fibulatus*** Kiesenwetter, 1852
**Nb, SQS-8**
7 SU16: Marlborough area, Jun-1873 (EWJ) (Meyrick, 1874); 1869-1938 (Meyrick, 1938)
SU26: Savernake Forest, 22-Jun-90 (APFw)
8 SU22: Langley Wood NNR, 2-Jul-07 (MDD) (Darby, 2008)
Resident. Very local; on trees in woodland.

***Malthodes guttifer*** Kiesenwetter, 1852
**Nb, SQS-8**
8 ST84: Longleat, 7-Jul-99 (MDD)
Resident. Very local; on trees in woodland.

***Malthodes marginatus*** (Latreille, 1806)
**SQS-1**
7 ST86: Neston Park, 3-Jun-04 (KNAA)
ST87: Castle Combe, 3-Jun-04 (BLe)
ST96: Spye Park, 31-May-04 (KNAA)
SU16: Marlborough area, 1869-1938 (Meyrick, 1938)
SU26: Savernake Forest, 8 to 28-Jul-93, 28 Apr to 18-May, 9-Jun to 1-Jul-94 (EN det SAW); 1-May to 23-Jun-95 (EN det MDD), 30-May-04 (KNAA); 31-May-04 (BLe)
8 ST91: Cranborne Chase, 7-May-97 (MAS)
SU02: Middleton Down, 24-May-95 (MDD)
SU13: Salisbury, 1-Apr-00 (MAS), 1-Jun-00 (MDD) (Darby, 2001)
SU14: Stonehenge, 22-Apr-03 (PJH) (Hodge, 2003)
SU16: Jones's Mill, 30-May-04 (AH det KNAA)
SU21: Landford Bog, 24-Jun-01 (APF) (Foster, 2001)
SU22: Bentley Wood, 2-Jul-96, 31-May-97 (MDD); Blackmoor Copse, 17-May-02 (APF) (Foster, 2004)
SU26: Bedwyn Brail, 2-Jun-03 (MDD)
Resident. Widespread; on trees in woodland.

***Malthodes maurus*** (Laporte de Castelnau, 1840)
**Nb, SQS-16**
7 SU26: Savernake Forest, 6-Dec-73 (APFw)
Resident. Very local; [on trees in damp woodland].

***Malthodes minimus*** (Linnaeus, 1758)
**SQS-1**
7 ST86: Neston Park, 3-Jun-04 (KNAA, DJG)
ST87: Corsham Park, 3-Jun-04 (BLe, KNAA); Castle Combe, 3-Jun-04 (BLe, KNAA)
SU16: Marlborough area, 1869-1938 (Meyrick, 1938)
SU17: Avebury, 13-Jun-03 (DB det MDD)
SU26: Savernake Forest, 8 to 28Jul-93 (EN det SAW)
8 ST84: Longleat, 7-Jul-99 (MDD)
ST91: Cranborne Chase, 4-Jun-97 (MAS), 22,, 30-Jun-99 (MDD)
SU02: Fovant Down, 16-Jun-95 (MDD)
SU22: Bentley Wood, 27-May-98 (MDD)
Resident. Widespread; on vegetation in woodland and wooded areas.

***Malthodes mysticus*** Kiesenwetter, 1852
**SQS-2**
7 SU16: Marlborough area, 1869-1938 (Meyrick, 1938)
Resident. Very local; [on vegetation in woodland].

***Malthodes pumilus*** (Brébisson, 1835)
**SQS-2**
7 ST96: Spye Park, 31-May-04 (KNAA)
SU16: Marlborough area, Jun-1873 (EWJ) (Meyrick, 1874); 1869-1938 (Meyrick, 1938)
Resident. Very local; [on vegetation in fields and hedgerows].

## Superfamily BOSTRICHOIDEA
## **Family DERMESTIDAE**
### Subfamily DERMESTINAE

**DERMESTES Linnaeus, 1758**
DERMESTINUS Zhantiev, 1967
**Hide Beetle**
***Dermestes murinus*** Linnaeus, 1758
7 SU16: Marlborough area, Jun-1873 (EWJ) (Meyrick, 1874); 17-Feb-1879 (WFB) (Meyrick, 1880); 1869-1938 (Meyrick, 1938)
8 SU03: Grovely Wood, 13-Aug-69, 27-Oct-71, 11-Apr-72, 8-Apr-74 (DRN)
Resident. Very local; [on dead animals and birds].

DERMESTES Linnaeus, 1758
**Larder Beetle, Bacon Beetle**
***Dermestes lardarius*** Linnaeus, 1758
7 SU16: Marlborough area, 1869-1938 (Meyrick, 1938)
SU27: Ramsbury, 1870-1873 (Meyrick, 1874)

### Subfamily ATTAGENINAE

**ATTAGENUS Latreille, 1802**
**Two-spotted Carpet Beetle, Fur Beetle**
***Attagenus pellio*** (Linnaeus, 1758)
7 ST86: Neston Park, 3-Jun-04 (KNAA)
SU16: Marlborough area, Jun-1873 (EWJ) (Meyrick, 1874); 1869-1938 (Meyrick, 1938)

Resident. Very local; [in houses, nests, and on flowers].

### Subfamily MEGATOMINAE

**CTESIAS Stephens, 1830**
**Cobweb Beetle**
***Ctesias serra*** (Fabricius, 1792)
**Nb, SQS-4**
7 ST96: Spye Park, 24-Apr-87 (KNAA)
ST96/97: Bowood Park, 24-Aug-86 (KNAA)
ST99: Flisteridge Wood, 6-May-89 (KNAA)
SU26: Savernake Forest, 6-Dec-73, 22-Jun-90 (APFw), 6-Dec-73 (PS); 5-May-90 (KNAA)
8 ST95: West Lavington, Jun-41 (det GJK) (DFL, 1948); 30-Jun-41 (ITH) (DFL, 1948)
SU03: Dinton Park, 6-Jul-01 (KNAA) (Alexander, 2004)
SU13: Stratford sub Castle, 11-Apr-71 (DRN)
Resident. Local; under bark and in rotting trees and stumps. [In spiders' webs in crevices in old oaks]. Often recorded as larvae.

### Subfamily ANTHRENINAE

**ANTHRENUS Muller, O.F, 1764**
HELOCERUS
***Anthrenus fuscus*** Olivier, 1789
7 SU16: Marlborough area, Jun-1873 (EWJ) (Meyrick, 1874), 1869-1938 (Meyrick, 1938)
8 SU03: Grovely Wood, 5-Aug-71 (DRN)
SU22: Bentley Wood, 31-Jul-91 (DRN)
Resident. Very local; on Umbelliferae.

NATHRENUS
**Varied Carpet Beetle, Varied Cabinet Beetle, Small Cabinet Beetle**
***Anthrenus verbasci,*** (Linnaeus, 1767)
7 SU09: Ashton Keynes, 20-Mar-01 (JKG det MDD)
SU16: Marlborough area, Jun-1870 (RH) (Hislop, 1871); 1869-1938 (Meyrick, 1938); Oare, 2002 (IJG)
8 SU03: Berwick St. James, 29-Apr-97 (BL det MDD); Wilton, 23-May-04 (AGD)
SU12: Alderbury, 24-Jun-03(DG det MAS)
SU13: Salisbury, Aug-79 (DRN); Apr-00 (MAS) (Darby, 2001)
SU23: West Winterslow, 7-Jul-01 (NB)
SU24: Newton Tony, Jun-03 (DG det MAS)
Resident. Widespread; on flowers and in houses.

## Family BOSTRICHIDAE
### Subfamily LYCTINAE

**LYCTUS Fabricius, 1793**
***Lyctus brunneus*** (Stephens, 1830)
**IEC-3, SQS-4**
8 SU03: Dinton Park, 6-Jul-01 (KNAA) (Alexander, 2004)
ST92: Sutton Mandeville, 22-Jul-08 (MDD)
Resident. Very local; flying in kitchen [in old oak stumps, occasionally oak palings].

***Lyctus linearis*** (Goeze, 1777)
**Nb, SQS-8**
7 SU16: Marlborough area, Jun-1873 (EWJ) (Meyrick, 1874); 1869-1938 (Meyrick, 1938)
Resident. Very local; [in dead wood, particularly oak, beech and ash. Often found in fresh oak palings. In ancient woodland, timber yeards and houses].

## Family ANOBIIDAE
### Subfamily EUCRADINAE.

Hedobia imperialis

**HEDOBIA Dejean, 1821**
PTINOMORPHUS Mulsand & Rey, 1868
***Hedobia imperialis*** (Linnaeus, 1767)
**Nb, SQS-8**
7 SU16: Manton, nr. Marlborough, 1900-1905 (Meyrick, 1906); Marlborough area, 1873-1938 (Meyrick, 1938)
SU26: Savernake Forest, 9-Jun to 1-Jul-94 (EN det SAW)
8 ST92: Fovant, 27-May-74 (DRN)
SU02: Fovant Down, 16-Jun-95 (MDD)
Resident. Local; in dead wood in old hedgerows and woodland.

### Subfamily PTININAE
Tribe PTININI

**NIPTUS Boieidieu, 1856**
***Niptus hololeucus*** (Faldermann, 1835)
7 SU16: Marlborough area, 1900 (Meyrick, 1906); 1869-1938 (Meyrick, 1938)

Resident. Very local; in old houses [especially in cupboards] among seeds, etc.. One specimen without data in MCC.

**TIPNUS Thomson, C.G, 1859**
***Tipnus unicolor*** (Piller & Mitterpacher, 1783)
7 SU16: Marlborough area, 1869-1938 (Meyrick, 1938)
Resident. Very local; among old wood. [Occasionally in bird nests and about houses]. One specimen without data in MCC.

**PTINUS Linnaeus, 1767**
**White-marked Spider Beetle**
***Ptinus fur*** (Linnaeus, 1758)
7 SU16: Marlborough area, Jun-1873 (Meyrick, 1874); 1869-1938 (Meyrick, 1938)
SU26: Savernake Forest, 28-Apr to 18-May-94 (EN det SAW)
8 ST92: Sutton Mandeville, 11-May-94, 6-Jun-01 (MDD)
Resident. Very local; generally indoors [often found in bird nests].

***Ptinus subpilosus*** Sturm, 1837
**Nb, IEC-2, SQS-8**
7 SU26: Savernake Forest, 8 to 28-Jul-93 (EN det SAW); 27-Oct to 16-Nov-95 (EN det MDD)
8 SU02: Knighton Wood, 6-Jul-07 (AJWA)
SU22: Langley Wood NNR, 25-Jun to 16-Jul-04 (MDD) (Darby, 2005)
Resident. Very local; in old hollow oak tree [under bark and in ant nests] in woodland.

**Australian Spider Beetle**
***Ptinus tectus*** Boieldieu, 1856
7 SU16: Clyffe Pypard, 1905-1938 (Meyrick, 1938)
Resident or casual introduction. Very local; in granaries. No specimens under this name in MCC.

## Subfamily DRYOPHILINAE

**GRYNOBIUS Thomson, C.G, 1859**
***Grynobius planus*** (Fabricius, 1787)
7 ST87: Corsham Park, 3-Jun-04 (BLe)
SU09: Upper Waterhay, 17-May, Jun-05 (JKG det MDD)
SU16: Marlborough area, Jun-1873 (EWJ) (Meyrick, 1874); 1869-1938 (Meyrick, 1938)
8 ST91: Cranborne Chase, 23-May-97 (MAS)
ST93: Great Ridge Wood, 16, 25-May-00 (MDD)
SU02: Knighton Wood, Apr, 6-Jul-07 (AJWA)
SU03: Grovely Wood, 23-Jul-73, 27-May-74 (DRN)
SU22: Blackmoor Copse, 3-Aug-03, 17-Mar-02 (APF) (Foster, 2004); Langley Wood, 28-May-74 (DRN)
Resident. Widespread; sits on old dead stumps in woodland.

**DRYOPHILUS Chevrolat, 1832**
***Dryophilus pusillus*** (Gyllenhal, 1808)
**SQS-2**
7 SU16: Marlborough area, Jun-1873 (EWJ) (Meyrick, 1874); 1869-1938 (Meyrick, 1938)
8 ST91: Cranborne Chase, 16 to 25-Jun-80 (NCC)
SU03: Grovely Wood, 11-Aug-71 (DRN)
Resident. Very local; on coniferous trees in woodland.

## Subfamily ERNOBIINAE

**OCHINA Dejean, 1821**
***Ochina ptinoides*** (Marsham, 1802)
**SQS-2**
7 ST86: Neston Park, 3-Jun-04 (KNAA)
ST87: Corsham Park, 3-Jun-04 (BLe)
SU16: Marlborough area, Jun-1873 (EWJ) (Meyrick, 1874); 1869-1938 (Meyrick, 1938)
8 ST92: Sutton Mandeville, 25-Jul-07 (MDD)
Resident. Very local; in dead ivy.

**XESTOBIUM Motschulsky, 1845**
**Death Watch Beetle**
***Xestobium rufovillosum*** (De Geer, 1774)
**IEC-3, SQS-4**
7 ST96: Spye Park, 31-May-04 (KNAA)
ST73: Stourhead, before 4-Jul-95 (KNAA) (Alexander, 1996)
ST84: Longleat, 12-Apr-02 (MB)
ST92: Sutton Mandeville, 23-May-92, 25-May-01 (MDD)
SU03: Dinton Park, 6-Jul-01 (KNAA) (Alexander, 2004)
SU13: Salisbury, 20-Jun-73 (PM)
Resident. Very local; in old dead wood, particularly oak in buildings.

**ERNOBIUS Thomson, C.G, 1859**
***Ernobius mollis*** (Linnaeus, 1758)
**SQS-2**
7 SU16: Marlborough area, 1869-1938 (Meyrick, 1938)
Resident. Very local; [in old dead wood] sometimes in houses.

## Subfamily ANOBIINAE

**ANOBIUM Fabricius, 1775**
***Anobium fulvicorne*** (Sturm, 1837)
7 ST86: Neston Park, 3-Jun-04 (BLe, DJG)
ST87: Corsham Park, 3-Jun-04 (BLe)
8 SU22: Blackmoor Copse, 3-Aug-03 (APF) (Foster, 2004); Langley Wood NNR, 29-May-03 (MDD)
SU26: Bedwyn Brail, 2-Jun-03 (MDD)
Resident. Local; on or near dead trees in woodland and parkland.

***Anobium inexspectatum*** Lohse, 1954
**Nb**, **SQS-8**
8 SU22: Blackmoor Copse, 13-Jul-03 (APF) (Foster, 2004)
Resident. Very local; [in dead branches of ivy].

**Common Furniture Beetle**
***Anobium punctatum*** (De Geer, 1774)
**SQS-1**
7 SU16: Marlborough area, Jun-1873 (EWJ) (Meyrick, 1874); 1869-1938 (Meyrick, 1938)
8 ST84: Longleat, 28-Aug-00 (MB)
SU02: Bishopstone, 1-Jul-91 (MDD)
SU03: Berwick St. James, May-01 (MrsAL det MDD); Grovely Wood, 16-Aug-71, 23-Aug-72 (DRN), 27-May-02 (MDD); Little Langford Down, 14-Jul-04 (MDD)
SU13: Great Durnford, 3-Aug-83, 8-Aug-85 (DRN)
SU22: Blackmoor Copse, 3-Aug-03 (APF) (Foster, 2004)
SU23: West Winterslow, 1-Aug-01 (NB)
Resident. Widespread; in or near dead wood in trees, building timber and furniture..

## Subfamily PTILININAE

**PTILINUS Geoffroy, 1762**
**Fan-bearing Wood Borer**
***Ptilinus pectinicornis*** (Linnaeus, 1758)
**SQS-1**
7 SU16: Marlborough area, 1869-1938 (Meyrick, 1938)
SU17: Avebury, 17-Jun-06 (DB det MDD)
SU26: Savernake Forest, 16-Jun to 7-Jul-93, 1 to 19-Jul-94 (EN det SAW); Tottenham Deer Park, 19-Dec-87 (KNAA)
8 ST91: Cranborne Chase, 8-Jul-97 (MAS)
SU03: Grovely Wood, 3-Aug-71, 21, 24-Aug-72, 4-Aug-78 (DRN)
SU16: Jones's Mill, 19-Jun, 17-Oct-01 (APF) (Foster, 2002)
SU22: Blackmoor Copse, 13-Jul-03 (APF) (Foster, 2004)
Resident. Local; under bark, on trunks and in dead wood of trees, and on vegetation nearby.

## Subfamily DORCATOMINAE

**DORCATOMA Herbst, 1792**
***Dorcatoma chrysomelina*** Sturm, 1837
**IEC-3**, **SQS-4**
8 SU03: Grovely Wood, 6-Apr-72, 23-Jul-73 (DRN)
Resident. Very local; in dead, red-rotten oak trees.

# Superfamily LYMEXYLOIDEA

# Family LYMEXYLIDAE
## Subfamily HYLECOETINAE

**HYLECOETUS Latreille, 1806**
***Hylecoetus dermestoides*** (Linnaeus, 1761)
**IEC-3**, **SQS-4**
8 ST95: West Lavington, 24-May-44 (DCL) (DFL, 1948)
Resident. Very local; [in oak, conifers and birch].

# Family TROGOSSITIDAE
## Subfamily PELTINAE

Thymalus limbatus

**THYMALUS Latreille, 1802**
***Thymalus limbatus*** (Fabricius, 1787)
**Nb**, **IEC-2**, **SQS-8**
8 ST73: Stourhead, 4-Jul-95 (KNAA) (Alexander, 1996)
ST84: Longleat, 28-Aug-00 (MB)
SU22: Bagfield Copse, 31-Jul-85, 6-Aug-86 (DRN)
Resident. Very local; under bark of oak.

# Family CLERIDAE
## Subfamily TILLINAE

Tillus elongatus

**TILLUS Olivier, 1790**
***Tillus elongatus*** (Linnaeus, 1758)
**Nb**, **IEC-3**, **SQS-8**
7 ST86: Neston Park, 3-Jun-04 (BLe det KNAA)
SU16: Marlborough area, 1869-1938 (Meyrick, 1938)

SU26: Savernake Forest, 30-May-04 (KNAA)
8 ST91: Cranborne Chase, 8-Jul-97 (MAS)
Resident. Very local; on beech in company with *Ptilinus* and on hawthorn blossom. Breeds in hardwoods where larvae parasitise Anobiidae.

Subfamily CLERINAE

Thanasimus formicarius

**THANASIMUS Latreille, 1806**
**Ant Beetle**
***Thanasimus formicarius*** (Linnaeus, 1758)
**IEC-3, SQS-4**
7 ST87: Corsham Court Park, 3-Jun-04 (BLe det KNAA)
SU26: Savernake Forest, 6-Dec-73 (PS, APF); Ju-84 (APF)
SU36: Ham Hill , 23-May-01 (PM)
8 ST84: Longleat, Aucombe Bottom, 25-May-05 (DJG)
ST92: Sutton Mandeville, 20-Apr-03 (MDD)
SU13: Little Durnford Down, 12-Jun-99 (PM)
SU14: Wexland Hanging (SPTAC), 22-Jun-02 (PJH) (Hodge, 2002)
SU21: Cloven Hill, 25-Apr-84 (DRN); Hamptworth Common, 30-Mar-03 (MDD); Hamptworth Estate, 30-Jul-85 (DRN)
SU23: Winterslow, 21-Jan-90 (PM)
Resident. Widespread; under bark and in dead and dying wood, usually coniferous, often with Scolytidae.

Subfamily KORYNETINAE

**KORYNETES Herbst, 1792**
***Korynetes caeruleus*** (De Geer, 1775)
**Nb, IEC-8, SQS-3**
7 SU16: Marlborough, 22-Jun-1902 (EEM) (Meyrick, 1938)
SU27: Ramsbury, 1869-1938 (Meyrick, 1938)
Resident. Very local; about old timber. [On old bones and carrion, and in dead wood] The larvae prey on Anobiidae and possibly Scolytidae. Seven specimens in the MCC one of which from Marlborough bears data as above.

Korynetes caeruleus

**NECROBIA Olivier, 1795**
***Necrobia violacea*** (Linnaeus, 1758)
7 SU16: Marlborough area, 1869-1938 (Meyrick, 1938)
8 SU03: Grovely Wood, 13-Aug-69 (DRN)
Resident. Very local; on carcases and old bones. One specimen without data in MCC.

## Family DASYTIDAE
Subfamily DASYTINAE

**DASYTES Paykull, 1799**
***Dasytes aeratus*** Stephens, 1830
**SQS-2**
7 ST96: Spye Park, 31-May-04 (KNAA)
ST86: Neston Park, 3-Jun-04 (BLe, KNAA)
ST87: Corsham Court Park, 3-Jun-04 (BLe, KNAA)
ST97: North Draycot Park, 22-May-00 (MDD)
SU16: Marlborough area, 1869-1938 (Meyrick, 1938)
SU26: Savernake Forest, 6-Dec-73 (RCS); 31-May-90 (TGW); 28-Apr to 18-May, 30-Aug to 20-Sep-94 (EN det SAW); 30-May-04 (KNAA); 31-May-04 (BLe)
8 ST85: Bullen Hill Farm, 1982-2002 (EGS det MDD)
ST93: Great Ridge Wood, 25-May-00 (MDD)
SU02: Knighton Wood, May, Jul-08 (AJWA)
SU03: Grovely Wood, 27, 30-May-74 (DRN)
SU13: Salisbury, 26-May-74 (DRN); Salisbury, Bemerton Folly, 7-Mar - 15-Jun-02 (MDD) (Darby, 2002d)
SU14: Haxton (SPTAE), 17-May-02 (PJH) (Hodge, 2002)
SU21: Burnt Ground Wood, 31-May-74 (DRN)
SU22: Blackmoor Copse, 17-May-02 (APF) (Foster, 2004); Langley Wood NNR, 28-May-74 (DRN), 25-May-94 (WANHS det MDD); 14-May to 16-Jul-04 (MDD) (Darby, 2005)
Resident. Widespread; on vegetation.

***Dasytes niger*** (Linnaeus, 1761)
**Na, SQS-16**
7 SU16: Marlborough area, Jun-1873 (EWJ) (Meyrick, 1874); 1869-1938 (Meyrick, 1938)
8 ST93: Great Ridge Wood, 29-Jul-85 (DRN)
SU23: Porton Down, 3-Jul-99 (PJH) (Hodge, 2002)

Resident. Very local; on flowers including thistles. Larvae probably develop in dead wood.

***Dasytes plumbeus*** (Müller, O.F, 1776)
**Nb, SQS-8**
8 SU12: Petersfinger, 25-Jul-91 (DRN)
Resident. Very local; on herbage at edge of lake.

## Family MALACHIIDAE

**AXINOTARSUS Motschulsky, 1853**
***Axinotarsus marginalis*** (Laporte de Castelnau, 1840)
8 SU22: Blackmoor Copse, 2-Jul-96 (MDD det CJ)
SU23: Porton Down, 3-Jul-99 (PJH) (Hodge, 2002)
Resident. Very local; in litter in woodland and [on vegetation in damp meadows].

**CORDYLEPHERUS Evers, 1985**
***Cordylepherus viridis*** Fabricius, 1787
7 SU16: Marlborough area, Jun-1873 (EWJ) (Meyrick, 1874); 1869-1938 (Meyrick, 1938)
SU26: Savernake Forest, 30-May-04 (KNAA)
8 ST94: Norton Down (SPTAW), 23-Jun-02 (PJH) (Hodge, 2002)
SU02: Middleton Down, 24-May-95 (MDD)
SU04: Orcheston Down (SPTA), 22-Jun-02 (PJH) (Hodge, 2002); Tilshead Park (SPTAW), 14-Jul-99 (PJH) (Hodge, 1999)
SU14: Haxton (SPTAE), 17-May-02 (PJH) (Hodge, 2002); Stonehenge, 22-Apr-03 (PJH) (Hodge, 2003); Wexland Hanging (SPTAC), 22-Jun-02 (PJH) (Hodge, 2002)
SU15: Beach's Plantation (SPTAE), 10-Jun-97 (MGT)
SU16: Jones's Mill, 21-Jun-97 (MDD)
SU22: Bentley Wood, 27, 31-May-97 (MDD)
SU23: Porton Down, 3-Jul-99 (PJH) (Hodge, 2002)
SU24: Driver Training Area (SPTAE), 24-Jun-02 (PJH) (Hodge, 2002); North Tidworth, Devils Ditch, 30-May-04 (BLe)
SU25: Sidbury Hill (SPTAE), 4-Jun-97 (MGT)
Resident. Widespread; on vegetation in grassland and woodland.

**MALACHIUS Fabricius, 1775**
***Malachius aeneus*** (Linnaeus, 1758)
**RDB3**
7 SU16: Chilton, 1869-1938 (Meyrick, 1938)
8 SU02: Knighton Wood, 6-Jul-07 (AJWA)
Resident. Very local; by sweeping and in a flight intercept trap in woodland. [On flowers, especially *Ranunculus* spp. in grassland and grassy areas in woodland].

**Malachite Beetle**
***Malachius bipustulatus*** (Linnaeus, 1758)
**SQS-1**
7 ST84: Smallbrook Meadows, 2-Jun-01 (JKG det MDD)
ST86: Neston Park, 3-Jun-04 (KNAA)
ST95: Erlestoke Park Woods, 1-Jun-04 (KNAA)
ST96: Conigre Meadow, 20-May-01 (JKG det MDD); Spye Park, 31-May-04 (KNAA)
ST97: Chippenham, Lackham College, 29-May-04 (BLe); North Draycot Park, 22-May-00 (MDD)
ST98: Braydon Pond Plantation, 27-May-91 (KNAA)
ST99: Flisteridge Wood, 13-Jun-82 (KNAA)
SU08: Great Wood, 27-May-91 (KNAA); Red Lodge Pond, 31-May-01 (JKG det MDD)
SU09: Somerford Keynes, Swillbrook Lakes, 4-Jun-04 (BLe); Upper Waterhay, 17-May to 11-Jun-05, 5 to 11-Jun-07 (JKG det MDD)
SU16: Marlborough area, Jun-1873 (EWJ) (Meyrick, 1874); 18-Jun-1879 (AMF) (Meyrick, 1880); 1869-1938 (Meyrick, 1938); Piggledene, 14-Jun-82 (DBS det KNAA)
SU17: Avebury, 24-May-05 (DB det MDD); Clout's Wood, Wroughton, 31-May-04 (BLe)
SU19: Hannington, 8-Jun-96, 26-May-97, 8-May-99 (MEN)
SU27: Aldbourne Chase, 19-Jun-96 (BL)
8 ST85: Bullen Hill Farm, 1982-2002 (EGS det MDD); Green Lane Wood, 19-May-92 (PM)
ST91: Cranborne Chase, 16 to 25-Jun-80 (NCC), 12-May-97 (MAS)
ST92: Sutton Mandeville, 24-Apr-92 (MDD); Wardour, 5-Jun-96 (MAS)
ST94: Wadman's Coppice (SPTAW) 3-Jun - 12-Aug-08 (Boyce & Wright, 2009), Warminster Ranges (SPTAW), 3-Jun -12-Aug-08 (Boyce & Wright, 2009)
ST95: West Lavington, 1-Jun-39 (PLB) (DFL, 1948); 25-May-44, 25-May-46 (DCL one det MissH)
SU02: Knighton Wood, 6-Jul-07 (AJWA)
SU13: Little Durnford Down, 20-May-92 (PM); Old Sarum, 31-May-97 (MAS); Salisbury, Apr-00 (MAS) (Darby, 2001)
SU14: Stonehenge, 22-Apr-03 (PJH) (Hodge, 2003)
SU15: Charlton, 11-Jun-02 (IJG)
SU16: Green Lane Wood, 31-May-04 (KNAA)
SU21: Hamptworth Estate, 31-May-74 (DRN); Landford Bog, 14-May-93 (PM)
SU22: Bentley Wood, 25-May-90 (PM), 31-May-96, 2-Jul-96 (MDD); Blackmoor Copse, 21-May-92, 23-May-93 (PM); 2-Jul-96 (MDD); 9-Jun-01 (APF) (Foster, 2004); Langley Wood, 28-May-74 (DRN), May-99 (BL det MDD); 29-May-03 (MDD)

SU23: Hound Wood, 1-Jun-98 (BL det MDD)
Porton Down, 3-Jul-99 (PJH) (Hodge, 2002)
SU24: Beacon Hill (SPTAE), 3-Jun - 12-Aug-08 (Boyce & Wright, 2009); Bulford Ranges (SPTAE), 3-Jun - 12-Aug-08 (Boyce & Wright, 2009)
Resident. Ubiquitous; on vegetation in grassland and woodland.

**ANTHOCOMUS Erichson, 1840**
***Anthocomus fasciatus*** (Linnaeus, 1758)
**SQS-4**
7 ST98: Lacock, 29-May-04 (BLe)
SU16: Marlborough area, 1869-1938 (Meyrick, 1938)
SU27: Ramsbury, 1869-1938 (Meyrick, 1938)
8 SU03: Berwick St. James, May-01 (MrsAL det MDD); Grovely Wood, 5-Aug-71 (DRN)
SU24: Newton Tony, Jun-03 (DG det MAS)
SU36: Bagshot, 1869-1938 (Meyrick, 1938)
Resident. Local; on vegetation, in a spider's web. [In damp meadows].

***Anthocomus rufus*** (Herbst, 1784)
7 SU16: Chilton, 1869-1938 (Meyrick, 1938)
Resident. Very local; by sweeping [vegetation in fens].

## Superfamily CUCUJOIDEA
## Family SPHINDIDAE

**SPHINDUS Dejean, 1821**
***Sphindus dubius*** (Gyllenhal, 1808)
**Nb, 8**
8 SU03: Grovely Wood, 5, 13-Aug-71 (DRN)
Resident. Very local; in myxomycete fungus on beech and pine, and in horse chestnut stump.

**ASPIDIPHORUS Dejean, 1821**
***Aspidiphorus orbiculatus*** (Gyllenhal, 1808)
**SQS-2**
7 SU26: Savernake Forest, 15-Jun to 28-Aug-73, 9-Jun to 19-Jul-94 (EN det SAW), 31-May to 23-Jun-95 (EN det MDD)
8 SU02: Knighton Wood, May-07, May, Jul-08 (AJWA)
SU03: Grovely Wood, 5-Aug-71 (DRN)
SU23: Porton Down, 3-Jul-99 (BL)
SU25: Ludgershall, Apr- Jul-1915 (PAH) (Harwood, 1915)
Resident. Local; in chestnut stump, a flight intercept trap and [in fungi].

## Family KATERETIDAE

**KATERETES Herbst, 1793**
***Kateretes pedicularius*** (Linnaeus, 1758)
8 SU13: Great Durnford, 17-Jul-73 (DRN)
Salisbury, 26-May-74 (DRN)
Resident. Very local; on vegetation by the R. Avon.

***Kateretes pusillus*** (Thunberg, 1794)
8 SU12: Charlton-All-Saints, 9-May-96 (MAS)
SU13: Great Durnford, 17-Jul-73 (DRN); Salisbury, 26-May-74 (DRN); 12-May-00 (MDD) (Darby, 2001)
Resident. Very local; on vegetation in a reedbed and by the R. Avon. Formerly *K. bipustulatus*.

***Kateretes rufilabris*** (Latreille, 1807)
7 SU16: Marlborough area, 1869-1938 (Meyrick, 1938)
8 SU12: Charlton-All-Saints, 9-May-96 (MAS); Whaddon, 22-Jul-73 (DRN)
SU13: Middle Woodford, 27-Jul-77 (DRN)
SU21: Hamptworth Common, 28-Jul-77 (DRN)
Resident. Local; on vegetation, on a damp part of the common and by the R. Avon.

**BRACHYPTERUS Kugelann, 1794**
***Brachypterus glaber*** (Stephens, 1835)
7 SU06: Roundway Camp, Devizes, Jul-40 (FDB) (Buck, 1956)
SU16: Marlborough area, Jun-1873 (EWJ) (Meyrick, 1874); 1869-1938 (Meyrick, 1938)
8 ST91: Cranborne Chase, 24-May-97 (MAS)
SU04: Shrewton, 1-Aug-85 (DRN)
SU12: Salisbury, 30-Jul-73 (DRN); Apr-00 (MAS) (Darby, 2001)
SU13: Old Sarum, 31-May-97 (MAS)
SU16: Jones's Mill, 1-Jan-9731-Dec-97 (MAS)
SU22: Bentley Wood, 27, 31-May-97 (MDD)
Resident. Widespread; on stinging nettles, *Urtica dioica.*

**Nettle Pollen Beetle**
***Brachypterus urticae*** (Fabricius, 1792)
7 SU16: Marlborough area, Jun-1873 (EWJ) (Meyrick, 1874); 1869-1938 (Meyrick, 1938)
8 ST91: Cranborne Chase, 30-Jun-99 (MDD)
SU03: Grovely Wood, 1-Aug-88 (DRN)
SU12: Great Durnford, 17-Jul-73 (DRN);
SU13: Salisbury, 30-Jul-73 (DRN), 1-Jun-00 (MDD) (Darby, 2001)
SU16: Jones's Mill, 21-Jun-97 (MDD); Woodborough, 4-Aug-83 (DRN)
SU26: Bedwyn Brail, 2-Jun-03 (MDD)
Resident. Local; on stinging nettles, *Urtica dioica.*

**BRACHYPTEROLUS Grouvelle, 1913**
***Brachypterolus linariae*** (Stephens, 1830)
8 SU03: Grovely Wood, 4-Aug-78 (DRN)
SU04: Orcheston Down (SPTA), 22-Jun-02 (PJH) (Hodge, 2002)
SU14: Stonehenge, 22-Apr-03 (PJH) (Hodge, 2003)

Resident. Very local; on vegetation and in a harebell *Campanula rotundifolia.*

***Brachypterolus pulicarius*** (Linnaeus, 1758)
8 SU04: Orcheston Down (SPTA), 22-Jun-02 (PJH) (Hodge, 2002)
SU14: Haxton (SPTAE), 24-Jun-02 (PJH) (Hodge, 2002); Stonehenge, 22-Apr-03 (PJH) (Hodge, 2003)
Resident. Very local; on vegetation [in flowers of yellow toadflax, *Linaria vulgaris.*

***Brachypterolus vestitus*** (Kiesenwetter, 1850)
8 ST91: Cranborne Chase, 7-May-97 (MAS)
Resident. Very local; [on antirrhinum flowers]

## Family NITIDULIDAE

### Subfamily CARPOPHILINAE

#### Tribe CARPOPHILINI

**CARPOPHILUS Stephens, 1829**

***Carpophilus marginellus*** Motschulsky, 1858
8 SU02: Knighton Wood, Sep-06 (AJWA)
SU13: Old Sarum, 4-Aug-71 (DRN) (Nash, 1975)
SU14: Netheravon, 29-Jul-83 (DRN)
SU15: Widdington, 5-Aug-83 (DRN)
SU21: Burnt Ground Wood, 31-Oct-73 (DRN) (Nash, 1975)
Resident. Local; in fungus, under sappy bark, in cut grass pile, in a flight intercept trap.

***Carpophilus sexpustulatus*** (Fabricius, 1792)
**IEC-3, SQS-8**
8 SU03: Grovely Wood, 7-Aug-85 (DRN) (Nash, 1977)
SU21: Hamptworth Common, 24-Mar-03 (MDD)
Resident. Very local; under bark.

#### Tribe EPURAEINI

**EPURAEA Erichson, 1843**

EPURAEA Erichson, 1843

***Epuraea aestiva*** (Linnaeus, 1758)
7 SU16: Marlborough area, Jun-1873 (EWJ) (Meyrick, 1874); 1-Jun-1879 (TNH-S) (Meyrick, 1880); 1869-1938 (Meyrick, 1938)
SU26: Savernake Forest, 25-Apr to 26-May-93 (EN det SAW); 31-May to 23-Jun-95 (EN det MDD)
8 ST91: Cranborne Chase, 15-May-97 (MAS)
ST94: Ladywell (SPTAW), 15-May-02 (PJH) (Hodge, 2002)
SU02: Knighton Wood, Mar-07 (AJWA)
SU04: Salisbury Plain, 16-May-02 (PJH) (Hodge, 2002)
SU22: Blackmoor Copse, 17-May-02 (APF) (Foster, 2004)
SU24: Driver Training Area (SPTAE), 17-May-02 (PJH) (Hodge, 2002)
Resident. Local; under bark, in bumble bee's nest, on vegetation.

***Epuraea biguttata*** (Thunberg, 1784)
**SQS-2**
8 SU02: Knighton Wood, Sep, Oct-06, Mar to May-07 (AJWA)
SU03: Grovely Wood, 16-Apr-84 (DRN)
Resident. Very local; one on a cut oak log, in woodland.

***Epuraea distincta*** (Grimmer, 1841)
**Na, SQS-8**
8 SU13: Salisbury, 24-May-00 (MDD) (Darby, 2001)
Resident. Very local; in fungus on a dead willow in a boggy area.

***Epuraea fuscicollis*** (Stephens, 1832)
**Nb, SQS-8**
8 SU02: Knighton Wood, Oct-06, Oct-07 (AJWA) (Allen, 2008)
Resident. Very local; in a flight intercept trap. [Almost always associated with trees damaged by*Cossus* larvae in woodland].

***Epuraea longula*** Erichson, 1845
**Nb, SQS-8**
7 SU26: Savernake Forest, 28-Jul to 28-Aug-93 (EN det SAW)
Resident. Very local; in a malaise trap [on flowers in woodland].

***Epuraea marseuli*** Reitter, 1872
**SQS-1**
8 SU02: Knighton Wood, Sep-06, Apr-07 (AJWA)
SU03: Grovely Wood, 25-Oct-71 (DRN)
SU21: Cloven Hill, 25-Apr-84 (DRN); Hamptworth Common, 27-May-74 (DRN), 30-Mar-03 (MDD)
SU22: East Grimstead, 1-Aug-78 (DRN)
Resident. Local; on a pine, on vegetation under trees in woodland.

***Epuraea melina*** Erichson, 1843
8 SU22: East Grimstead, 11-Aug-86 (DRN)
Resident. Very local; on Cow Parsley, *Anthriscus syvestris* near a pond

***Epuraea pallescens*** (Stephens, 1835)
**SQS-2**
7 SU16: Marlborough area, 28-May-1879 (TNH-S) (Meyrick, 1880); 1869-1938 (Meyrick, 1938)
SU26: Savernake Forest, 20 to 25-Apr-93 (EN det SAW)
Resident. Very local; under bark, at sap or in fungi [in woodland].

***Epuraea silacea*** (Herbst, 1783)
**SQS-1**
7 SU16: Marlborough area, 1-Jul-1879 (TNH-S) (Meyrick, 1880); 1869-1938 (Meyrick, 1938)
Resident. Very local; under bark, at sap or in fungi.

EPURAEANELLA Crotch, 1874
***Epuraea limbata*** (Fabricius, 1787)
**SQS-2**
8 SU15: Widdington, 5-Aug-83 (DRN)
Resident. Very local; teeming in fungus on elm.

MICRURIA Reitter, 1875
***Epuraea melanocephala*** (Marsham, 1802)
7 SU16: Marlborough area, 1869-1938 (Meyrick, 1938)
SU26: Savernake Forest, 25-Apr to 26-May-93 (EN det SAW)
8 ST91: Cranborne Chase, 7-May-97 (MAS)
SU03: Dinton Park, 6-Jul-00 (KNAA); Grovely Wood, 1-Aug-88 (DRN)
SU14: Stonehenge, 22-Apr-03 (PJH) (Hodge, 2003)
Resident. Local; in flowers, on trees, on vegetation under trees.

## Subfamily MELIGETHINAE

**PRIA Stephens, 1829**
***Pria dulcamarae*** (Scopoli, 1763)
8 SU13: Great Durnford, 27-Jul-77, 3-Aug-83 (DRN); Lower Woodford, 28-Jul-78 (DRN)
SU16: Woodborough, 4-Aug-83 (DRN)
Resident. Very local; on vegetation by R. Avon.

**MELIGETHES Stephens, 1829**
**Common Pollen Beetle**
***Meligethes aeneus*** (Fabricius, 1775)
7 ST86: Box, 26-May-56 (AME)
ST87: Colerne Park Lane, 26-May-56 (AME) Rudloe, 26-May-56 (AME); Slaughterford, 26-May-56 (AME);
ST96: Conigre Meadow, 20-May-01 (JKG det MDD); Sandy Lane, 20-Oct-56 (AME);
ST97: Chippenham, 10-Jul-48 (AME); North Draycot Park, 22-May-00 (MDD)
SU06: Roundway Camp, Jul-40 (FDB) (Buck, 1956)
SU08: Somerford Common, 23-Apr-01 (JKG det MDD)
SU09: Ashton Keynes, 26-Apr-01 (JKG det MDD); Upper Waterhay, 13-Jun-04, 6-Jul-07 (JKG det MDD)
SU16: Marlborough area, Jun-1873 (EWJ) (Meyrick, 1874); 12 Mar 1879 (WFB) (Meyrick, 1880); 1869-1938 (Meyrick, 1938)
SU18: Rushy Platt, 20-May-01 (JKG det MDD)
SU26: Savernake Forest, 25-May-78 (MIR det AME), 20 to 25-Apr, 16-Jun to 20-Sep-93, 28-Apr to 18-May, 1 to 19-Jul-94 (EN det SAW); 31-May to 23-Jun-95 (EN det MDD)
SU26/27: Axford, 25-May-78 (MIR det AME)
SU27: Knighton, 18-Aug-63 (AME)
8 ST91: Cranborne Chase, 24-Apr-97 (MAS); 22, 30-Jun-99 (MDD)
ST93: Great Ridge Wood, 4-May-00 (MDD); Stockton Down, 7-Jul-56 (AME)
ST94: Bowls Barrow (SPTAW), 16-May, 22-Jun-02 (PJH) (Hodge, 2002); Ladywell (SPTAW), 15-May, 23-Jun-02 (PJH) (Hodge, 2002); Norton Down (SPTAW), 15-May, 23-Jun-02 (PJH) (Hodge, 2002)
SU00: Dean Hill, 30-Jul-50 (AME)
SU01: Martin Down NNR, 1979-1992, 21-Aug-89 (RGB)
SU02: Knighton Wood, Mar, Apr-07 (AJWA); Middleton Down, 14-May-89, 9-Jun-93 (BL)
SU04: Fox Covert, 3-Jun - 12-Aug-08 (Boyce & Wright, 2009); Orcheston Down (SPTA), 18-May-02 (PJH) (Hodge, 2002); Parsonage Down NNR, 22-Jul-08 (MDD); Salisbury Plain, 25-Jun-02 (PJH) (Hodge, 2002); Tilshead Park (SPTAW), 14-Jul-99 (PJH) (Hodge, 1999)
SU12: Alderbury, 24-Jun-03 (DG det MAS)
SU13: Salisbury, Apr-00 (MAS) (Darby, 2001); Salisbury, Bemerton Folly, 7-Mar - 15-Jun-02 (MDD) (Darby, 2002d)
SU14: Stonehenge, 22-Apr-03 (PJH) (Hodge, 2003); Wexland Hanging (SPTAC), 22-Jun-02 (PJH) (Hodge, 2002)
SU15: Everleigh (SPTAE), 18-May, 24-Jun-02 (PJH) (Hodge, 2002); Haxton Down, 14-Jul-99 (PJH) (Hodge, 1999)
SU16: Jones's Mill, 1997 (MAS)
SU21: East Grimstead, 17-Jun-64 (AME)
SU22: Mean Wood, 30-Jul-50 (AME); Whiteparish, 28-Apr-56 (AME); Whiteparish Common, 2-Jul-95 (MAS); Witherington Down, 4-Oct-53 (AME)
SU23: Pitton, 4-Oct-53 (AME); Porton Down, 3-Jul-99 (PJH) (Hodge, 2002); Winterslow, 17-Jun-64 (AME)
SU24: Driver Training Area (SPTAE), 17-May, 24-Jun-02 (PJH) (Hodge, 2002); Dunch Hill (SPTAE), 3-Jun - 12-Aug-08 (Boyce& Wright, 2009); Newton Tony, Jun-03 (DG det MAS)
SU26: Bedwyn Common, 26-May-56 (AME); Cobham Frith, 26-May-56 (AME)
Resident. Widespread and abundant especially on yellow flowers.

***Meligethes atramentarius*** Foerster, 1849
**Notable**
8 SU22: Whiteparish, 28-Apr-56 (AME)

SU26: Cobham Frith, 26-May-56 (AME)
Resident. Very local; [in shaded deciduous woodland coppice on heavy calcareous soils]

***Meligethes atratus*** (Olivier, 1790)
7 ST87: Slaughterford, 26-May-56 (AME)
SU16: Marlborough area, Jun-1873 (EWJ) (Meyrick, 1874); 1869-1938 (Meyrick, 1938)
SU26: Savernake Forest. NE, 22-Aug-90 (AGD)
8 ST91: Cranborne Chase, 7-May-97 (MAS)
SU03: Grovely Wood, 31-May-97 (MAS)
SU13: Old Sarum, 31-May-97 (MAS); Salisbury, Apr-00 (MAS) (Darby, 2001)
SU16: Jones's Mill, 19-Jun-01 (APF) (Foster, 2002)
SU22: East Grimstead, 17-Jun-64 (AME); 31-Jul-91 (DRN); Bentley Wood, 27-May-98 (MDD); Glazier's Copse, 16-Aug-88 (DRN); Mean Wood, 30-Jul-50 (AME)
SU23: Winterslow, 17-Jun-64 (AME)
SU26: Bedwyn, Sep-42 (KGB); Bedwyn Common, 26-May-56 (AME); Cobham Frith, 26-May-56 (AME)
Resident. Widespread; [woodland margins, hedgerows, roadsides, rough grassy places].

***Meligethes bidens*** Brisout, 1863
**LCGR-Fh Cu**
7 ST87: Colerne Park, 19-Aug-56 (AME)
8 ST93: Stockton Down, 7-Jul-56 (AME)
SU00: Dean Hill, 30-Jul-50 (AME)
SU14: Stonehenge, 22-Apr-03 (PJH) (Hodge, 2003)
SU23: Porton Down, 4-Oct-53 (AME); 3-Jul-99 (BLe)
Resident. Local; in open woodlands, scrub, hedgerows and dry grassy places, especially on calcareous soils.

***Meligethes bidentatus*** Brisout, 1863
**RDBK, Fl Cm**
7 ST98: Broydon, 19-Aug-56 (AME)
SU04: Fox Covert, 3-Jun - 12-Aug-08 (Boyce & Wright, 2009)
Resident. Very local; on Dyer's Greenwood, *Genista tinctoria,* in [meadows, road-verges and banks, heaths and open woods].

***Meligethes brunnicornis*** Sturm, 1845
7 ST86: Box, 26-May-56 (AME)
ST87: Slaughterford, 26-May-56 (AME)
SU16: Marlborough area, Jun-1873 (EWJ) (Meyrick, 1874); 1869-1938 (Meyrick, 1938)
8 SU22: Mean Wood, 30-Jul-50 (AME)
SU26: Bedwyn Common, 26-May-56 (AME)
Resident. Local; in [shadowy, deciduous woodlands, hedgerows, banks ans margins of cultivated land.

***Meligethes carinulatus*** Förster, 1849
7 ST86: Box, 26-May-56 (AME)
ST87: Slaughterford, 26-May-56 (AME)
ST98: Broydon, 19-Aug-56 (AME)
SU08: Callow Hill, 19-Aug-56 (AME)
8 ST93: Bowls Barrow (SPTAW), 16-May-02 (PJH) (Hodge, 2002); Stockton Down, 7-Jul-56 (AME)
ST94: Ladywell (SPTAW), 3-Jun - 12-Aug-08 (Boyce & Wright, 2009)Norton Down (SPTAW), 15-May, 23-Jun-02 (PJH) (Hodge, 2002); Warminster Ranges (SPTAW), 3-Jun -12-Aug-08 (Boyce & Wright, 2009)
SU00: Dean Hill, 30-Jul-50 (AME)
SU04: Orcheston Down (SPTAC), 18-May-02 (PJH) (Hodge, 2002); Salisbury Plain, 16-May-02 (PJH) (Hodge, 2002)
SU22: Whiteparish, 28-Apr-56 (AME)
SU23: Lopcombe Corner, 17-Jun-64 (AME)
SU26: Cobham Frith, 26-May-56 (AME)
Resident. Widespread; in [short grassy habitats, road verges, and heathland].

***Meligethes difficilis*** (Heer, 1841)
7 ST86: Box, 28-May-56 (AME)
ST97: Chippenham, 27-May-56 (AME)
8 SU22: West Dean, 4-Oct-53 (AME)
SU23: Pitton, 4-Oct-53 (AME)
SU26: Bedwyn Common, 26-May-56 (AME)
Resident. Local; [grassy places, woodland margins, pathways, parks hedgerows, banks, farmyards and waste areas].

***Meligethes erichsonii*** Brisout, 1863
**Notable, Fh Cu**
7 ST86: Box, 27-May-56 (AME)
ST87: Slaughterford, 26-May-56 (AME)
8 SU23: Lopcombe Corner, 17-Jun-64 (AME)
Resident. Very local; on short turf on chalk and limestone.

***Meligethes exilis*** Sturm, 1845
**Notable**
8 SU05: Black Heath (SPTAC), 3-Jun - 12-Aug-08 (Boyce & Wright, 2009)
Resident. Very local; [in sandy areas on thyme].

***Meligethes flavimanus*** Stephens, 1830
7 ST86: Box, 26-May-56 (AME)
SU26: Savernake Forest, 28-Jul to 28-Aug-93 (EN det SAW)
8 SU14: Haxton (SPTAE), 17-May-02 (PJH) (Hodge, 2002); Stonehenge, 22-Apr-03 (PJH) (Hodge, 2003)
SU21: East Grimstead, 17-Jun-64 (AME)
Resident. Local; [in woodland margins, hedgerows, roadsides, rough grassy places].

***Meligethes gagathinus*** Erichson, 1845

**Notable**
7 ST86: Box, 26-May-56 (AME)
ST87: Colerne Park Lane, 26-May-56 (AME); Rudloe, 26-May-56 (AME)
8 SU00: Dean Hill, 30-Jul-50 (AME)
SU22: Whiteparish, 1950, 30-Jul-50 (AME)
SU23: Pitton, 1953, 4-Oct-53 (AME)
Resident. Local; [damp habitats, open woodland, grassy roadsides, arable fields and ditches].

***Meligethes incanus*** Sturm, 1845
**Notable**
7 SU29: Frias Hill, nr Sevenhampton, 19-Aug-56 (AME)
Resident. Very local; on catmint, *Nepeta cataria*. [Hedgerows, roadsides, gardens, banks and rocky places, usually on calcareous soils].

***Meligethes kunzei*** Erichson, 1845
8 SU22: Whiteparish, 28-Apr-56 (AME)
SU26: Cobham Frith, 26-May-56 (AME)
Resident. Very local; on dandelions, *Taraxacum offinalis,* and yellow archangel, *Lamiastrum galeobdolon* [in shady deciduous coppice woodlands, on heavy calcareous soils].

***Meligethes lugubris*** Sturm, 1845
**Notable, LCGR-FI Cu**
7 ST86: Box Hill, 7-Jul-56 (AME)
8 SU00: Dean Hill, 1950, 30-Jul-50 (AME)
SU22: Farley, 26-Apr-36 (PAH)
Resident. Very local; on Breckland thyme, *Thymus serpyllum* and dwarf thistle, *Cirsium acaule* [in dry grassy habitats, heaths, grassland, banks, close-grazed pasture].

***Meligethes morosus*** Erichson, 1845
7 ST86: Box, 26-May-56 (AME)
8 SU22: Whiteparish, 30-Jul-50 (AME); West Dean, 28-Apr-56 (AME)
SU23: Pitton, 4-Oct-53 (AME)
SU26: Bedwyn Common, 26-May-56 (AME)
Resident. Local; on white dead-nettle, *Lamium album* [in grassy places, woodland margins, pathways, hedgerows, banks, farmyards, parks and waste areas].

***Meligethes nigrescens*** Stephens, 1830
7 ST86: Box, 26-May-56 (AME)
ST87: Slaughterford, 26-May-56 (AME)
ST98: Broydon, 19-Aug-56 (AME)
SU08: Callow Hill, 19-Aug-56 (AME)
SU16: Marlborough area, Jun-1873 (EWJ) (Meyrick, 1874); 1869-1938 (Meyrick, 1938)
SU26: Savernake Forest, 20 to 25-Apr, 28-Jul to 28-Aug-93, 28-Apr to 18-May-94 (EN det SAW)
SU27: Knighton, 18-Aug-63 (AME)
8 ST93: Stockton Down, 7-Jul-56 (AME)
ST94: Tinker's Track (SPTAW), 3-Jun - 12-Aug-08 (Boyce & Wright, 2009)
SU02: Knighton Wood, Jul-08 (AJWA)
SU04: Parsonage Down NNR, 22-Jul-08 (MDD)
SU06: Devizes, May-25-19? (*Jones* det AME)
SU22: Blackmoor Copse, 17-May-02 (APF) (Foster, 2004); West Dean, 28-Apr-56 (AME); Whiteparish, 28-Apr-56 (AME); Witherington Down, 4-Oct-53 (AME)
SU23: Pitton, 4-Oct-53 (AME)
SU26: Bedwyn Common, 26-May-56 (AME); Cobham Frith, 26-May-56 (AME)
Resident. Widespread; on a wide range of flowering plants and shrubs in [grassy places, meadows and pastures, lawns, banks, etc. on a wide variety of soils].

***Meligethes obscurus*** Erichson, 1845
7 ST96: Spye Park, Sep-56 (AME)
8 SU01: Martin Down NNR, 1999 (EN, 1999)
ST73: Gasper Street, 27-May-60 (Anon)
SU22: Knightswood Copse, 4-Oct-53 (AME)
Resident. Local; on devil's bit scabious, *Succisa pratensis,* and yellow Compositae [in open woods and on sand dunes, on dry non-calcareous soils]

***Meligethes ochropus*** Sturm, 1845
**Notable**
8 SU23: Pitton, 4-Oct-53 (AME)
Resident. Very local; on charlock, *Sinapis arvensis* [in damp habitats such as the margins of rivers, ditches and ponds].

***Meligethes ovatus*** Sturm, 1845
7 ST96: Sandy Lane, 19-Aug-56 (AME)
SU16: Marlborough area, 1869-1938 (Meyrick, 1938)
8 SU00: Dean Hill, 30-Jul-50 (AME)
SU22: Blackmoor Copse, 24-Aug-01 (APF) (Foster, 2004); Mean Wood, 30-Jul-50 (AME); Whiteparish, 28-Apr-56 (AME)
SU26: Bedwyn Common, 26-May-56 (AME)
Resident. Local; mainly on ground vy, *Glechoma hederacea* [in woodlands, coppices, hedgerows, grassland and waste places, usually on damp soils].

***Meligethes pedicularius*** (Gyllenhal, 1808)
7 SU16: Marlborough area, Jun-1873 (EWJ) (Meyrick, 1874); 1869-1938 (Meyrick, 1938)
8 ST94: Norton Down (SPTAW), 15-May-02 (PJH) (Hodge, 2002)
Resident. Very local; [in open woods, heaths and arable land].

***Meligethes persicus*** Faldermann, 1837
7 ST98: Broydon, 19-Aug-56 (AME)
8 ST94: Bowls Barrow (SPTAW), 22-Jun-02 (PJH) (Hodge, 2002)
SU22: Mean Wood, 30-Jul-50 (AME)

SU26: Cobham Frith, 26-May-56 (AME)
Resident. Local; on various plants [in permanent grassland, heaths, banks and open woodland on light soils].

***Meligethes planiusculus*** (Heer, 1841)
**LCGR-Fm Cu**
7 ST87: Colerne Park, 19-Aug-56 (AME)
8 ST94: Bowls Barrow (SPTAW), 16-May, 22-Jun – 02 (PJH) (Hodge, 2002); Ladywell (SPTAW), 23-Jun-02 (PJH) (Hodge, 2002); 3-Jun - 12-Aug-08 (Boyce & Wright, 2009); Tinker's Track (SPTAW), 3-Jun - 12-Aug-08 (Boyce & Wright, 2009); Wadman's Coppice (SPTAW) 3-Jun - 12-Aug-08 (Boyce & Wright, 2009)
SU00: Dean Hill, 30-Jul-50 (AME)
SU04: Fox Covert, 3-Jun - 12-Aug-08 (Boyce & Wright, 2009); Salisbury Plain, 25-Jun-02 (PJH) (Hodge, 2002)
SU14: Haxton (SPTAE), 17-May, 24-Jun-02 (PJH) (Hodge, 2002); 3-Jun - 12-Aug-08 (Boyce & Wright, 2009); Wexland Hanging (SPTAC), 22-Jun-02 (PJH) (Hodge, 2002); 3-Jun - 12-Aug-08 (Boyce & Wright, 2009)
SU24: Driver Training Area (SPTAE), 17-May, 24-Jun-02 (PJH) (Hodge, 2002); Dunch Hill (SPTAE), 3-Jun - 12-Aug-08 (Boyce& Wright, 2009)
Resident. Widespread; on viper's bugloss, *Echium vulgare* [in dry open places].

***Meligethes ruficornis*** (Marsham, 1802)
7 ST87: North Wraxall, 7-Jul-56 (AME)
SU16: Marlborough area, Jun-1873 (EWJ) (Meyrick, 1874), 1869-1938 (Meyrick, 1938)
8 SU22: Whiteparish, 30-Jul-50 (AME)
Resident. Very local; on black horehound, *Ballota nigra*. [In woodland margins and paths, hedgerows, road verges and waste ground].

***Meligethes serripes*** (Gyllenhal, 1827)
[**RDBK** but omitted in error by Hyman and Parsons, 1992,1994]
8 SU23: Winterslow, 4-Oct-53 (AME)
Resident. Very local; on red hemp nettle, *Galeopsis angustifolia* [on arable land and in bare places].

***Meligethes solidus*** (Kugelann, 1794)
**Notable, LCGR-Fh Cu**
7 ST87: Slaughterford, 26-May-56 (AME)
8 ST94: Ladywell (SPTAW), 15-May-02 (PJH) (Hodge, 2002); Norton Down (SPTAW), 15-May-02 (PJH) (Hodge, 2002); Warminster Ranges (SPTAW), 3-Jun -12-Aug-08 (Boyce & Wright, 2009)
SU01: Martin Down NNR, 1979-1992, 1994 (RGB)
SU02: Wilton, 18-Aug-47 (AME)
SU04: Orcheston Down (SPTA), 18-May-02 (PJH) (Hodge, 2002)
SU05: Black Heath (SPTAC), 3-Jun - 12-Aug-08 (Boyce & Wright, 2009); Ell Barrow (SPTAC), 3-Jun - 12-Aug-08 (Boyce & Wright, 2009)
SU14: Wexland Hanging (SPTAE), 3-Jun - 12-Aug-08 (Boyce & Wright, 2009)
SU15: Everleigh (SPTAE), 18-May-02 (PJH) (Hodge, 2002)
SU23: Porton Down, 18-Aug-05 (KNAA)
SU24: Driver Training Area (SPTAE), 24-Jun-02 (PJH) (Hodge, 2002)
Resident. Widespread; on vegetation but particularly rock-rose, *Helianthemum nummularium*, in calcareous grassland.

***Meligethes viridescens*** (Fabricius, 1787)
7 ST86: Box, 26-May-56 (AME)
ST87: Rudloe, 26-May-56 (AME)
ST96: Sandy Lane, 19-Aug-56, 20-Oct-57 (AME)
SU16: Marlborough area, 1869-1938 (Meyrick, 1938)
SU27: Knighton, 18-Aug-63 (AME)
8 SU00: Dean Hill, 30-Jul-50 (AME)
SU22: Mean Wood, 30-Jul-50 (AME); Whiteparish, 28-Apr-56 (AME)
SU23: Winterslow, 17-Jun-64 (AME)
SU26: Bedwyn, Sep-42 (Anon); Bedwyn Common, 26-May-56 (AME)
Resident. Widespread; on various plants [in a wide range of habitats].

## Subfamily NITIDULINAE

**NITIDULA Fabricius, 1775**
***Nitidula bipunctata*** (Linnaeus, 1758)
7 SU16: Marlborough area, Jun-1873 (EWJ) (Meyrick, 1880); 1869-1938 (Meyrick, 1938)
8: SU03: Grovely Wood, 9-Aug-71 (DRN)
Resident. Very local; on old carrion and flowers, and a dead hare.

**OMOSITA Erichson, 1843**
***Omosita colon*** (Linnaeus, 1758)
7 SU16: Marlborough area, 1869-1938 (Meyrick, 1938)
8 ST91: Rushmore Park, 23-Sep-01 (AJWA)
Resident. Very local; [on carrion, chiefly old dry bones] and in a dung heap. One specimen without data in MCC.

***Omosita depressa*** (Linnaeus, 1758)
8 ST92: Sutton Mandeville, 20-Apr-92 (MDD)
SU03: Grovely Wood, 9-Mandeville, 20-Apr-92 (MDD)
SU22: Langley Wood NNR, 5-Aug to 15-Sep-06 (MDD) (Darby, 2008)
Resident. Very local; on dry carrion.

***Omosita discoidea*** (Fabricius, 1775)
7 SU16: Marlborough area, Jun-1873 (EWJ) (Meyrick, 1874); 1869-1938 (Meyrick, 1938)
8 SU02: Knighton Wood, Oct, Nov-06, Apr-07 (AJWA)
SU03: Grovely Wood, Oct-71, 31-Jul-79 (DRN)
SU22: Langley Wood NNR, 26-Oct to 17-Nov-06 (MDD) (Darby, 2008)
Resident. Local; on dry carrion.

**SORONIA Erichson, 1843**
***Soronia grisea*** (Linnaeus, 1758)
**SQS-2**
7 SU09: Upper Waterhay, 17-May to 4-Jun to 11-Jun-05, Jun-07 (JKG det MDD)
SU16: Marlborough area, Jun-1873 (EWJ) (Meyrick, 1874); 14 May 1879 (TAP) (Meyrick, 1880); 1869-1938 (Meyrick, 1938)
8 SU14: Figheldean, 28-Jul-82 (DRN)
Resident. Very local; on flowers, at sap of trees damaged by *Cossus* larvae, swept off vegetation by R. Avon.

***Soronia punctatissima*** (Illiger, 1794)
**SQS-2**
7 SU26: Savernake Forest, 15-Jun-73 (SAW); 8 to 28-Jul-93 (EN det SAW); 4 to 24-Aug-95 (EN det MDD)
8 ST91: Cranborne Chase, 16 to 25-Jun-80 (NCC)
SU02: Knighton Wood, Apr, May, Sep-07 (AJWA)
SU13: Boscombe Down, 9-Oct-94 (MDD)
Resident. Local; in flight intercept traps and an old bird's nest [under bark of old oak and at sap of trees damaged by *Cossus* larvae].

**CYCHRAMUS Kugelann, 1794**
***Cychramus luteus*** (Fabricius, 1787)
7 SU16: Marlborough area, Jun-1873 (EWJ) (Meyrick, 1874); 1869-1938 (Meyrick, 1938)
SU26: Savernake Forest, 20-Aug-70 (DRN); 31-May-00 (WANHS det MDD)
8 ST91: Cranborne Chase, 29-Apr-97 (MAS)
ST93: Great Ridge Wood, 16, 25-May-00 (MDD); Stockton Wood, 7-Jul-96 (MAS)
SU02: Knighton Wood, Sep, Nov-06, Apr-07 (AJWA)
SU03: Grovely Wood, 27-May-74 (DRN); 11-Jun-00 (MDD)
SU16: Jones's Mill, 1997 (MAS), 19-Jun-01 (APF) (Foster, 2002),
SU22: Bentley Wood, 12-Feb-01 (NB det MDD)
Resident. Widespread; on vegetation and fungi in wooded areas.

**POCADIUS Erichson, 1843**
***Pocadius adustus*** Reitter, 1888
7 SU26: Savernake Forest, 31-May to 23-Jun-95 (EN det MDD)
Resident. Very local; in a malaise trap. [In fungi in woodland]. Formerly *P. Lanuginosus*, see note below.

***Pocadius ferrugineus*** (Fabricius, 1775)
7 SU16: Marlborough area, 1869-1938 (Meyrick, 1938)
SU17: Fyfield Down, 30-Dec-1899 (APF) (Meyrick, 1938)
SU26: Savernake Forest, 25-Apr to 26-May, 21-Sep to 11-Oct-93, 1 to 19-Jul-94 (EN det SAW); 6 to 26-Oct-95 (EN det MDD)
8 ST92: Sutton Mandeville, Aug-92 (MDD)
SU02: Knighton Wood, Sep-07 (AJWA)
SU03: Grovely Wood, 3-Aug-71 (DRN)
SU21: Franchises Wood, 3-May-95 (MAS)
SU22: Bentley Wood, 31-May-96 (MDD)
Resident. Widespread; in woodland on fungi. *P. lanuginosus* was not separated from *adustus* until 1969 so that determinations before that date are not reliable.

**THALYCRA Erichson, 1843**
***Thalycra fervida*** (Olivier, 1790)
**Nb**
8 SU02: Knighton Wood, Sep-07 (AJWA)
Resident. Very local; in a flight intercept trap.

## Subfamily CRYPTARCHINAE

**CRYPTARCHA Shuckard, 1839**
***Cryptarcha strigata*** (Fabricius, 1787)
**Nb, SQS-8**
7 SU16: Marlborough area, Jun-1873 (EWJ) (Meyrick, 1874); 1869-1938 (Meyrick, 1938)
8 SU02: Knighton Wood, Apr, May, 6-Jul, Sep-07 (AJWA) (Allen, 2008)
Resident. Very local; at sap exuding from damage by *Cossus* larvae and in a flight intercept trap. One specimen without data in MCC.

**GLISCHROCHILUS Reitter, 1873**
GLISCHROCHILUS Reitter, 1873
***Glischrochilus quadripunctatus*** (Linnaeus, 1758)
7 SU26: Savernake, spring-1894 (GCC) (Jebb, 1895)
Resident. Very local; at sap and under bark.

LIBRODOR Reitter, 1884
***Glischrochilus hortensis*** (Geoffroy in Fourcroy, 1785)
7 SU16: Marlborough area, 1869-1938 (Meyrick, 1938)
8 ST92: Sutton Mandeville, 24-May-92 (MDD)
SU02: Knighton Wood, Mar, Apr-07 (AJWA)
SU21: Hamptworth Estate, 19-Apr-74 (DRN)
SU22: Langley Wood NNR, 16-Jul to 6-Aug-04 (MDD) (Darby, 2005)

Resident. Local; at sap and under bark.

***Glischrochilus quadriguttatus*** (Fabricius, 1777)
**SQS-2**
7 SU16: Marlborough area, 1 869-1938 (Meyrick, 1938)
SU26: Savernake Forest, spring-1894 (GCC) (Jebb, 1895); 3, 7-Apr-70 (DRN)
8 ST84: Longleat, 26-Mar-89 (AGD)
ST92: Sutton Mandeville, 1-May-97 (MDD)
SU02: Knighton Wood, Mar-07 (AJWA)
SU03: Grovely Wood, 6-Apr-72, 27-May-74, 4-Apr, 4-Aug-78 (DRN)
SU21: Burnt Ground Wood, 19-Apr-74 (DRN); Hamptworth Estate, 3-Apr-78 (DRN)
SU22: Langley Wood NNR, 15-Mar to 22-Apr, 3 to 25-Jun-04 (MDD) (Darby, 2005)
SU23: Porton Down, 1970-2001 (Anon, 2001), 12-Nov-96 (MAS)
Resident. Widespread; at sap and under bark of beech an oak.

**PITYOPHAGUS Shuckard, 1839**
***Pityophagus ferrugineus*** (Linnaeus, 1761)
8 SU21: Hamptworth Estate, 25-Apr-84 (DRN); Hamptworth Common, 30-Mar-03 (MDD)
Resident. Very local; under pine bark.

## Family MONOTOMIDAE
### Subfamily RHIZOPHAGINAE

**RHIZOPHAGUS Herbst, 1793**
RHIZOPHAGUS Herbst, 1793
***Rhizophagus bipustulatus*** (Fabricius, 1792)
**SQS-1**
7 ST96: Spye Park, 24-Apr-87 (KNAA)
ST99: Flisteridge Wood, 2-May-82 (KNAA)
SU16: Marlborough area, Jun-1873 (EWJ) (Meyrick, 1874); 6-Mar-1879 (PHM) (Meyrick, 1880); 1869-1938 (Meyrick, 1938); Martinsell, 1870-1873 (Meyrick, 1874)
SU26: Savernake Forest, 8-Jan-90 (AGD; 16-Jun to 7-Jul-93 (EN det SAW); Tottenham Deer Park, 19-Dec-87 (KNAA)
8 ST84: Longleat, 7-Mar-99 (MB), 7-Jul-99 (MDD)
ST91: Cranborne Chase, 21-May-02 (MDD)
SU02: Knighton Wood, Mar, Apr-07 (AJWA)
SU03: Grovely Wood, 25-Oct-71, 23-Jul-73, 16-Apr-84 (DRN)
SU21: Burnt Ground Wood, 31-May-74 (DRN)
SU22: Blackmoor Copse, 3-Aug-03 (APF) (Foster, 2004); Redlynch, 13-Apr-74 (DRN)
SU23: Porton Down, 12-Nov-96 (MAS)
Resident. Widespread; under bark in woodland.

***Rhizophagus dispar*** (Paykull, 1800)
**SQS-1**
7 ST96: Spye Park, 24-Apr-87 (KNAA)
ST99: Flisteridge Wood, 6-May-89 (KNAA)
SU16: West Woods, May-02 (MDD)
SU17: Bincknoll Dip Wood, 22-Feb-87 (KNAA)
SU26: Savernake Forest, 2-Apr, 20-Aug-70 (DRN); 29-May-83 (KNAA); 22-Jun-90 (APFw); 23-Oct-90 (TGW; 20-Sep to 11-Oct-94; (EN det SAW) 11-May to 30-May-95 (EN det MDD)
8 ST73: Tucking Mill Hanging, 11-Feb-84 (KNAA)
ST82: Oysters Coppice, 6-Oct-02 (MDD)
ST84: Longleat, 7-Mar-99 (MB), 7-Jul-99 (MDD)
ST85: Green Lane Wood, 26-Jun-98 (JN)
ST92: Sutton Mandeville, 1-May-97 (MDD); Wardour, 13-Nov-95 (MDD), 3-Apr-96 (EN)
ST93: Fonthill, 26-Jun-02 (MDD)
SU02: Gardiner Forest, 1-Sep-72 (DRN); Knighton Wood, 6-Jul-07 (AJWA)
SU03: Grovely Wood, 6-Aug-70, 6, 8-Apr, 23-Aug-72, 16-Apr-74, 3, 8-Apr, 30-Oct-75, 4, 8-Aug-80, 7-Jul-85 (DRN); 21-Aug-01 (MDD)
SU13: Little Durnford Down, 13-Sep-99 (MDD)
SU14: Netheravon, 29-Jul-83 (DRN)
SU21: Hamptworth Common, 30-Mar-03 (MDD)
SU22: Bentley Wood, 5-May-03 (MDD); Blackmoor Copse, 3-Aug-03 (APF) (Foster, 2004); Langley Wood NNR, 6-Jun-02, 24-Mar-05 (MDD) (Darby, 2005)
SU26: Bedwyn Brail, 2-Jun-03 (MDD); Bedwyn Common, 24-Jul-82 (DRN)
Resident. Widespread; under bark of both coniferous and deciduous trees and in fungi in woodland.

***Rhizophagus ferrugineus*** (Paykull, 1800)
**SQS-2**
7 SU26: Savernake Forest, 31-May to 23-Jun, 15-Jul to 3-Aug-95 (EN det MDD)
8 SU03: Grovely Wood, 18-Jul-73 (DRN)
SU21: Hamptworth Estate, 3-Aug, 31-Oct-73, 31-May-74, 31-Oct-75 (DRN)
SU26: Bedwyn Brail, 19-Jun-03 (MDD)
Resident. Local; under bark of both coniferous and deciduous trees in woodland.

***Rhizophagus nitidulus*** (Fabricius, 1798)
**Nb, SQS-4**
7 SU16: West Woods, May-02 (MDD)
8 ST93: Great Ridge Wood, 13-Aug-80 (DRN)
SU03: Grovely Wood, 6-Apr, 25-Oct-71, 23-Jul-73, 8-Apr-75 (DRN)
Resident. Very local; under beech bark in woodland.

**Graveyard Beetle**
***Rhizophagus parallelocollis*** Gyllenhal, 1827
**SQS-2**
7 ST99: Crudwell, 19-Jun-73 (APF)
Resident. Very local; [frequently swarming in graves, graveyards, on tombstones, and on corpses in coffins

(from 10 months to 2 years old) buried at some depth below ground].

***Rhizophagus perforatus*** Erichson, 1845
**SQS-2**
7 SU16: Marlborough area, Jun-1873 (EWJ) (Meyrick, 1874); 1869-1938 (Meyrick, 1938)
8 SU02: Bishopstone, 19-May-91 (MDD)
SU21: Burnt Ground Wood, 31-Oct-75 (DRN)
Resident. Very local; in garden compost and on sappy birch [often in company with *R. parallelocollis*, rarely under bark].

EURHIZOPHAGUS Méquignon, 1909
***Rhizophagus depressus*** (Fabricius, 1792)
**SQS-2**
8 SU21: Cloven Hill, 25-Apr-84 (DRN); Hamptworth Common, 30-Mar-03 (MDD)
Resident. Very local; under bark.

## Subfamily MONOTOMINAE

**MONOTOMA Herbst, 1793**
MONOTOMA Herbst, 1793
***Monotoma bicolor*** Villa, 1835
7 SU06: Roundway Camp, Jul-40 (FDB) (Buck, 1956)
8 SU03: Grovely Wood, 29-Oct-73 (DRN)
SU21: Hamptworth Estate, 1-Apr-75, 1-Aug-79 (DRN)
Resident. Very local; in straw and other decaying vegetable matter. *M. quadricollis* has recently been separated from *bicolor* (Booth, 2001) so that these records should be treated as indeterminate.

***Monotoma brevicollis*** Aubé, 1837
8 ST92: Sutton Mandeville, 25-Jul-07 (MDD)
SU03: Grovely Wood, 2-Aug-79 (DRN)
Resident. Very local; in mouldy grass and garden compost.

***Monotoma longicollis*** (Gyllenhal, 1827)
8 ST92: Sutton Mandeville, 6-Jul-93 (MDD)
SU03: Burcombe, 23-Jun-95 (MDD); Grovely Wood, 2-Aug-79 (DRN)
SU13: Lower Woodford, 30-Jul-82 (DRN)
SU22: East Grimstead, 12-Aug-87 (DRN)
Resident. Very local; in mouldy grass and other rotting vegetable matter.

***Monotoma picipes*** Herbst, 1793
7 SU16: Marlborough area, Jun-1873 (EWJ) (Meyrick, 1874); 1869-1938 (Meyrick, 1938)
SU26: Froxfield, 1870-1873 (Meyrick, 1874)
8 ST92: Sutton Mandeville, 25-Jun-07 (MDD)
SU03: Grovely Wood, 29-Oct-73, 2-Aug-79 (DRN)
SU21: Hamptworth Estate, 29-Mar, 3-Apr-78 (DRN)
SU22: East Grimstead, 12-Aug-87 (DRN); Redlynch, 4-Aug-82 (DRN)
Resident. Local; in grass piles and other rotting vegetation.

***Monotoma spinicollis*** Aubé, 1837
8 SU22: Redlynch, 4-Aug-82 (DRN)
Resident. Very local; in sileage grass.

***Monotoma testacea*** Motschulsky, 1845
8 SU03: Burcombe, 23-Jun-95 (MDD); Grovely Wood, 2-Aug-79 (DRN)
Resident. Very local; in mouldy grass piles.

# Family SILVANIDAE
## Subfamily BRONTINAE

**PSAMMOECUS Latreille, 1829**
***Psammoecus bipunctatus*** (Fabricius, 1792)
7 SU16: Marlborough area, 1869-1938 (Meyrick, 1938)
SU27: Ramsbury, spring-1894 (GCC) (Jebb, 1895)
8 SU12: Petersfinger, 25-Jul-91 (DRN)
SU13: Great Durnford, 16, 19-Aug-69, 7-Aug-70, 27-Jul-77, 3-Aug-83 (DRN); Salisbury, Apr-00 (MAS) (Darby,2001); Stratford sub Castle, 26-May-74 (DRN)
SU15: Fittleton, 29-Jul-83 (DRN)
SU16: Jones's Mill, 1997 (MAS)
Resident. Local; in litter and on vegetation beside R. Avon.

## Subfamily SILVANINAE

**AHASVERUS des Gozis, 1881**
**Foreign Grain Beetle**
***Ahasverus advena*** (Waltl, 1834)
8 SU03: Grovely Wood, 2-Aug-79 (DRN)
SU14: Netheravon, 29-Jul-83 (DRN)
SU21: Hamptworth Estate, 29-Mar-78 (DRN)
Resident, introduced synanthropic species. Very local; in mouldy grass and straw, and fungus at the base of a poplar. [Originally associated with stored grain but now spread to other habitats].

**SILVANUS Latreille, 1804**
***Silvanus bidentatus*** (Fabricius, 1792)
**Nb, IEC-2**
8 ST84: Longleat, 12-Apr-02 (MB)
ST92: Wardour, 29-Oct-95 (MDD)
ST93: Fonthill, 14-Apr-85 (MDD)
SU02: Gardiner Forest, 1-Sep-72 (DRN) (Nash, 1975)
SU03: Grovely Wood, 7-Jul-85 (DRN)
SU21: Hamptworth Common, 12-Aug-86 (DRN)
SU22: Blackmoor Copse, 3-Aug-03 (APF) (Foster, 2004)

Resident. Widespread; under bark of oaks and conifers

***Silvanus unidentatus*** (Olivier, 1790)
**IEC-3**
7 SU16: Marlborough area, Jun-1873 (EWJ) (Meyrick, 1874); 1869-1938 (Meyrick, 1938); 'Marlborough Forest', Jun-1870 (RH) (Hislop, 1871)
SU26: Savernake Forest, spring-1894 (GCC) (Jebb, 1895); 3-Apr-70 (DRN); 29-May-73 (CMJ); 6-Jun-73 (APF); 30-May-04 (KNAA)
8 ST84: Longleat, Jun-97 (MB)
ST91: Cranborne Chase, 21-May-02 (MDD)
ST93: Great Ridge Wood, 13-Aug-81 (DRN)
SU03: Grovely Wood, 25-Oct-71, 8,16, 19-Apr-74, 4-Aug-78, 16-Apr-84 (DRN) (DRN); Little Langford, 3-Jul-96 (MAS)
SU21: Burnt Ground Wood, 21-Jul-73, 31-Oct-76 (DRN); Hamptworth Estate, 19-Apr-74 (DRN)
SU22: Redlynch, 13-Apr-74 (DRN); 12-Feb-01 (NB); 11-Aug, 10-Oct-01 (MDD)
Resident. Widespread; under bark of beech and conifers.

## Family CUCUJIDAE

**PEDIACUS Shuckard, 1839**
***Pediacus depressus*** (Herbst, 1797)
**Na, IEC-2, SQS-16**
7 SU26: Savernake Forest, 15-Jun-73 (SAW); 25-Apr to 26-May-93 (EN det SAW)
8 SU02: Knighton Wood, Jul-08 (AJWA)
Resident. Very local; under bark of conifers

***Pediacus dermestoides*** (Fabricius, 1792)
**IEC-3, SQS-4**
7 ST86: Neston Park, 3-Jun-04 (KNAA)
ST99: Flisteridge Wood, 25-Apr-82, 6-May-89 (KNAA)
SU26: Savernake Forest, 2, 7-Apr-70 (DRN); 29-May-73 (CM-J); 15-Jun-73, 30-ay-04 (KNAA); 15-Jun-73 (SAW); 20-Apr to 16-Jun-93, 28-Apr to 18-May-94 (EN det SAW)
8 ST73: Stourhead, before 4-Jul-95 (KNAA) (Alexander, 1996)
ST84: Little Bradley Wood, 5-May-91 (KNAA); Longleat, 7-Mar-99 (MB)
ST93: Fonthill, 26-Jun-02 (MDD); Great Ridge Wood, 19-Apr-84 (DRN)
SU02: Knighton Wood, Mar-07, May-08 (AJWA)
SU03: Grovely Wood, 10-Apr-70, 25-Oct-71, 6-Apr-72, 23,Jul-73, 4-Apr-78, 29-Jul-79, 16-Apr-84 (DRN); 19-Aug-95 (MAS)
SU13: Little Durnford Down, 13-Sep-99, 20-May-03 (MDD)
SU21: Burnt Ground Wood, 19-Apr-74 (DRN); Hamptworth Common, 30-Mar-03 (MDD)
SU22: Bagfield Copse, 4-Aug-86 (DRN); Blackmoor Copse, 31-May-96 (MDD); Langley Wood, 28-May-74 (DRN)
SU26: Bedwyn Brail, 2-Jun, 5-Sep, 5-Nov-03 (MDD)
Resident. Widespread; on trees and under bark of beech, oak and birch.

Pediacus dermestoides

## Family LAEMOPHLOEIDAE

**CRYPTOLESTES Ganglbauer, 1899**
***Cryptolestes ferrugineus*** (Stephens, 1831)
**IEC-addition, SQS-2**
7 SU16: Marlborough area, Jun-1873 (EWJ) (Meyrick, 1874); 1869-1938 (Meyrick, 1938)
8 ST92: Sutton Mandeville, 1-Sep-96 (JAO); 14-Jul-03 (MDD)
SU03: Grovely Wood, 30-Oct-75, 4-Aug-78 (DRN)
SU15: Widdington, 5-Aug-83 (DRN)
SU21: Burnt Ground Wood, 21-Jul-73, 31-Oct-75 (DRN)
SU22: Bentley Wood, 12-Feb-01 (NB); 10-Oct-01, 5-May-03 (MDD)
SU26: Bedwyn Brail, 2-Jun, 5-Sep, 5-Nov-03 (MDD)
Resident. Widespread; under bark of deciduous and coniferous trees.

***Cryptolestes spartii*** (Curtis, 1834)
**Na**
8 SU21: Burnt Ground Wood, 19-Apr, 21-Jul, 31-Oct-73 (DRN)
Resident. Very local; under beech bark.

## Family PHALACRIDAE

**PHALACRUS Paykull, 1800**
***Phalacrus caricis*** Sturm, 1807
7 SU16: Marlborough area, 1869-1938 (Meyrick, 1938)
SU27: Ramsbury, spring-1894 (GCC) (Jebb, 1895)
Resident. Very local; by sweeping [smutted sedges,

*Carex* spp.]. One specimen without data in MCC.

***Phalacrus championi*** Guillebeau, 1892
**Na**
8 ST94: Bowls Barrow (SPTAW), 16-May-02 (PJH) (Hodge, 2002)
SU22: Bentley Wood, 27-May-97 (MDD)
Resident. Very local; on vegetation [smutted sedges *Carex* spp.]. Previously *P. brunnipes*.

***Phalacrus corruscus*** (Panzer, 1796)
7 SU16: Marlborough area, Jun-1873 (EWJ) (Meyrick, 1874); 1869-1938 (Meyrick, 1938)
Resident. Very local; by sweeping flowers.

***Phalacrus fimetarius*** (Fabricius, 1775)
**LCGR-Fh Cu**
8 ST94: Bowls Barrow (SPTAW), 22-Jun-02 (PJH) (Hodge, 2002); Ladywell (SPTAW), 23-Jun-02 (PJH) (Hodge, 2002)
SU04: Orcheston Down (SPTA), 22-Jun-02 (PJH) (Hodge, 2002); Salisbury Plain, 16-May-02 (PJH) (Hodge 2002); Shrewton, 6-Aug-82, 2-Aug-83 (DRN)
SU14: Haxton (SPTAE), 17-May-02 (PJH) (Hodge, 2002); Wexland Hanging (SPTAC), 16-May, 22-Jun-02 (PJH) (Hodge, 2002)
SU22: Langley Wood NNR, 15-Jul to 5-Aug-06 (MDD) (Darby, 2008)
SU25: Salisbury Plain (Weather Hill Firs), 21-Jun-01 (MDD)
Resident. Local; on vegetation in calcareous grassland.

**OLIBRUS Erichson, 1845**
***Olibrus aeneus*** (Fabricius, 1792)
7 SU16: Marlborough area, Jun-1873 (EWJ) (Meyrick, 1874); 1869-1938 (Meyrick, 1938)
8 ST93: Stockton, 28-Sep-95 (PJH) (Hodge, 1995)
SU04: Shrewton, 1-Aug-85 (DRN)
Resident. Very local; on vegetation.

***Olibrus affinis*** (Sturm, 1807)
7 SU06: Cherhill Down, 1-Jun-04 (BLe)
8 ST93: Stockton, 28-Sep-95 (PJH) (Hodge, 1995)
ST94: Bowls Barrow (SPTAW), 22-Jun-02 (PJH) (Hodge, 2002); Ladywell (SPTAW), 23-Jun-02 (PJH) (Hodge, 2002); Norton Down (SPTAW), 23-Jun-02 (PJH) (Hodge, 2002)
SU01: Martin Down NNR, 21-Aug-89 (RGB)
SU04: Fox Covert, 25-Jun-02 (PJH) (Hodge, 2002) Salisbury Plain, 25-Jun-02 (PJH) (Hodge, 2002)
SU14: Haxton (SPTAE), 24-Jun-02 (PJH) (Hodge, 2002); Wexland Hanging (SPTAC), 22-Jun-02 (PJH) (Hodge, 2002)
SU23: Porton Down, 3-Jul-99 (BLe)
Resident. Local; on vegetation in calcareous grassland.

***Olibrus corticalis*** (Panzer, 1796)
8 SU21: Burnt Ground Wood, 31-May-74 (DRN)
SU23: Porton Down, 20-Jul-97 (MAS)
Resident. Very local; on vegetation.

***Olibrus flavicornis*** (Sturm, 1807)
**RDBK**
8 ST94: Ladywell (SPTAW), 3-Jun - 12-Aug-08 (Boyce & Wright, 2009); Tinkers Track (SPTAW), 3-Jun - 12-Aug-08 (Boyce & Wright, 2009); Wadman's Coppice (SPTAW), 3-Jun - 12-Aug-08 (Boyce & Wright, 2009); Warminster Ranges (SPTAW), 3-Jun - 12-Aug-08 (Boyce & Wright, 2009)
SU14: Wexland Hanging (SPTAC), 3-Jun - 12-Aug-08 (Boyce & Wright, 2009)
Resident. Local; in grassland [on the continent associated with autumn hawkbit, *Leontodon autumnalis*].

***Olibrus liquidus*** Erichson, 1845
7 SU16: Marlborough area, Jun-1873 (EWJ) (Meyrick, 1874); 1869-1938 (Meyrick, 1938)
8 ST94: Warminster Ranges (SPTAW), 3-Jun - 12-Aug-08 (Boyce & Wright, 2009)
SU03: Wylye Down NNR, 20-Sep-05 (MDD) (Darby, 2009)
SU05: Black Heath (SPTAC), 3-Jun - 12-Aug-08 (Boyce & Wright, 2009)
SU13: Salisbury, 25-Jul-78 (DRN)
SU14: Wexland Hanging (SPTAC), 3-Jun - 12-Aug-08 (Boyce & Wright, 2009)
SU23: West Winterslow, 1-Aug-01 (NB)
SU24: Beacon Hill (SPTAE), 3-Jun - 12-Aug-08 (Boyce & Wright, 2009); Dunch Hill (SPTAE), 3-Jun - 12-Aug-08 (Boyce & Wright, 2009)
Resident. Widespread; on vegetation in calcareous grassland and weeds in a former tip.

**STILBUS Seidlitz, 1872**
***Stilbus oblongus*** (Erichson, 1845)
7 SU09: Somerford Keynes, 4-Jun-04 (BLe)
8 SU03: Grovely Wood, 11-Aug-71 (DRN)
SU14: Netheravon, 29-Jul-83 (DRN)
Resident. Very local; on vegetation [associated with bulrushes, *Typha latifolia*].

***Stilbus testaceus*** (Panzer, 1796)
7 SU16: Marlborough area, 1869-1938 (Meyrick, 1938)
SU26: Savernake Forest, 15-Jul to 3-Aug-95 (EN det MDD)
8 ST92: Sutton Mandeville, 2-Aug-95 (MDD), 1-Sep-96 (JAO), 29-Jan-98 (MDD)
ST93: Stockton Wood, 28-Sep-95 (PJH) (Hodge, 1995)
ST94: Ladywell (SPTAW), 3-Jun - 12-Aug-08

(Boyce & Wright, 2009)
SU02: Knighton Wood, Mar-07 (AJWA)
SU03: Grovely Wood, 12-Jun-00 (MDD)
SU26: Bedwyn Brail, 20-Sep-04 (MDD)
Resident. Local; on vegetation in woodland. This appears to represent an extension of its habitat range; Thompson (1958) cites dry grass and hay.

## Family CRYPTOPHAGIDAE

### Subfamily CRYPTOPHAGINAE

#### Tribe CRYPTOPHAGINI

**TELMATOPHILUS Heer, 1841**

***Telmatophilus caricis*** (Olivier, 1790)
7 SU16: Marlborough area, Jun-1873 (EWJ) (Meyrick, 1874), 1869-1938 (Meyrick, 1938) (One specimen without data in MCC conf. CJ)
8 SU13: Salisbury, 26-May-74 (DRN)
SU16: Woodborough, 4-Aug-83 (DRN)
Resident. Very local; on water plants, *Typha* etc..

***Telmatophilus typhae*** (Fallén, 1802)
8 SU13: Great Durnford, 6-Aug-71 (DRN)
Resident. Very local; on vegetation at edge of R. Avon at dusk.

**CRYPTOPHAGUS Herbst, 1792**

***Cryptophagus acutangulus*** Gyllenhal, 1827
8 ST92: Sutton Mandeville, summer 1993, (MDD conf AD)
SU04: Shrewton, 28-Jul-82 (DRN)
Resident. Very local; in straw.

***Cryptophagus dentatus*** (Herbst, 1793)
**SQS-1**
7 SU06: Roundway Camp, Jul-40 (FDB) (Buck, 1956)
SU16: Marlborough area, 1869-1938 (Meyrick, 1938)
8 SU02: Knighton Wood, Apr-07 (AJWA)
SU03: Grovely Wood, 6-Ap, 27-Oct-71, 6-Apr-72, 30-Oct-75 (DRN)
SU13: Salisbury, Bemerton Folly, 7-Mar - 15-Jun-02 (MDD) (Darby, 2002d)
SU16: Jones's Mill, 11-May-97 (MDD det CJ)
SU21: Hamptworth Estate, 3-Apr-78 (DRN)
SU22: Langley Wood NNR, 12-Apr-74 (DRN)
Resident. Local; in rotting vegetation and fungi. A single specimen with out data in the MCC conf. CJ.

***Cryptophagus denticulatus*** Heer, 1841
8 ST93: Great Ridge Wood, 29-Oct-75 (DRN)
SU02: Knighton Wood, Nov-07, Sep-08 (AJWA)
SU03: Grovely Wood, 27-Oct, 4-Aug-78, 2-Aug-79, 8-Aug-80 (DRN); Stratford sub Castle, 5-Apr-72 (DRN)
SU13: Lower Woodford, 28-Jul-78 (DRN)
SU15: Widdington, 5-Aug-83 (DRN)
SU21: Hamptworth Estate, 29-Mar, 3-Apr-78 (DRN)
Resident. Local; in rotting vegetation, straw, fungi and under bark. Formerly *C. pilosus*

***Cryptophagus distinguendus*** Sturm, 1845
8 SU02: Wilton, 6-Sep-95 (MDD conf. CJ)
SU13: Salisbury, Bemerton Folly, 7-Mar - 15-Jun-02 (MDD det CJ) (Darby, 2002d)
Resident. Very local; in old bird's nest and woodland litter.

***Cryptophagus laticollis*** Lucas, 1846
7 SU16: Marlborough area, Jun-1873 (EWJ) (Meyrick, 1874); 1869-1938 (Meyrick, 1938) (One specimen without data in MCC conf. CJ)
8 SU03: Grovely Wood, 2-Aug-79 (DRN)
SU21: Hamptworth Estate, 15-Apr-74 (DRN)
SU22: Landford, 12-Aug-87 (DRN)
Resident. Local; in straw, a mouldy grass heap and beaten from hazel.

***Cryptophagus lycoperdi*** (Scopoli, 1763)
8 ST92: Wardour, 29-Oct-95 (MDD)
SU21: Hamptworth Estate, 28-Oct-75 (DRN)
Resident. Very local; in *Scleroderma* fungus.

***Cryptophagus pubescens*** Sturm, 1845
8 SU03: Grovely Wood, 16-Aug-71 (DRN)
Resident. Very local; sweeping grass at dusk.

***Cryptophagus punctipennis*** Brisout, 1863
7 SU26: Savernake Forest, 22-Aug-90 (AGD)
8 SU03: Wylye Down NNR, 29-Mar to 4-May-07 (MDD conf JC) (Darby, 2009)
SU21: Hamptworth Estate, 29-Mar-78 (DRN)
Resident. Very local; in a pitfall trap and in straw near a pond. Previously *C. pilosus.*

***Cryptophagus reflexus*** Rey, 1889
8 SU13: Devenish Reserve, 24-May-05 (MDD det CJ)
Resident. Very local; sweeping.

***Cryptophagus ruficornis*** Stephens, 1830
**Notable, SQS-8**
8 SU02: Knighton Wood, 6-Jul-07 (AJWA)
SU03: Grovely Wood, 16-Apr-74 (DRN)
Resident. Very local; in a flight intercept trap and on an old beech branch.

***Cryptophagus scanicus*** (Linnaeus, 1758)
7 SU16: Marlborough area, Jun-1873 (EWJ) (Meyrick, 1874); 1869-1938 (Meyrick, 1938)
8 ST92: Sutton Mandeville, 24-Oct-95 (MDD)
SU03: Grovely Wood, 1-Aug-80 (DRN)

Resident. Very local; in garden compost, and fungi in woodland.

***Cryptophagus scutellatus*** Newman, 1834
8 SU03: Grovely Wood, 29-Oct-73 (DRN)
Resident. Very local; in straw in woodland.

***Cryptophagus setulosus*** Sturm, 1845
7 SU16: Marlborough area, Jun-1873 (EWJ) (Meyrick, 1874); 1869-1938 (Meyrick, 1938)
8 SU03: Grovely Wood, 11-Aug-71, 24-Aug-72 (DRN)
SU13: Salisbury, 9-Aug-87 (DRN)
Resident. Very local; on a window at dusk and by sweeping at dusk.

**MICRAMBE Thomson, C.G, 1863**
MICRAMBE Thomson, C.G., 1863
***Micrambe ulicis*** (Stephens, 1830)
7 SU16: Marlborough area, Jun-1873 (EWJ) (Meyrick, 1874); 1869-1938 (Meyrick, 1938)
SU27: Ramsbury, 1870-1873 (Meyrick, 1874)
8 SU02: Knighton Wood, Sep-06 (AJWA)
SU04: Parsonage Down NNR, 2, 22-Jul-08 (MDD conf CJ)
SU21: Hamptworth Estate, 24-Apr-84 (DRN)
SU23: Thorny Down woods, 16-Oct-00 (MDD)
Resident. Local; on gorse. Previously *M. vini.*

**ANTHEROPHAGUS Dejean, 1821**
***Antherophagus pallens*** (Linnaeus, 1758)
7 SU26: Savernake Forest, 31-May to 23-Jun-95 (EN det MDD)
8 ST73: Stourhead, 4-Jul-95 (KNAA)
SU03: Wylye Down NNR, 4-May to 5-Jun-07 (MDD) (Darby, 2009)
SU23: Porton Down, 3-Jul-99 (BLe)
Resident. Very local; in malaise and pitfall traps in woodland and grassland [probably near Bumble bee nests]. Previously *A. nigricornis.*

***Antherophagus similis*** (Curtis, 1835)
8 SU13: Great Durnford, 27-Jul-77 (DRN)
Resident. Very local; on vegetation near R. Avon. Previously *A. pallens.*

Tribe CAEOSCELINI

**CAENOSCELIS Thomson, C.G, 1863**
***Caenoscelis ferruginea*** (Sahlberg, C.R, 1820)
7 SU16: Marlborough area, 1869-1938 (Meyrick, 1938)
Resident. Very local; by evening sweeping [in litter and fungi in woodland]. No specimen under this name in MCC.

**ATOMARIA Stephens, 1829**
ATOMARIA Stephens, 1829

**Pygmy Mangel Beetle**
***Atomaria linearis*** Stephens, 1830
7 SU16: Marlborough area, Jun-1873 (EWJ) (Meyrick, 1874); 1869-1938 (Meyrick, 1938)
Resident. Very local; on mangolds. A farmland pest [in heaps of cut vegetation and haystacks]. One specimen in MCC without data, conf. CJ.

***Atomaria nigrirostris*** Stephens, 1830
7 VC7, n.d. (Johnson, 1993)
8 SU21: Hamptworth Estate, 25-Apr-84 (DRN)
Resident. Very local; under pine bark. Woodland species [mostly associated with broadleaved trees, on ground in litter, etc.].

***Atomaria nigriventris*** Stephens, 1830
**Notable**
8 SU04: Shrewton, 1-Aug-85 (DRN) (Johnson, 1993)
Resident. Very local; in old straw. Grassland species especially around farms.

***Atomaria pulchra*** Erichson, 1846
8 SU02: Knighton Wood, May, Jul-08 (AJWA)
SU21: Hamptworth Estate, 29-May-74 (DRN) (Johnson, 1993)
Resident. Very local; on vegetation. [Mostly associated with coniferous woodland].

***Atomaria punctithorax*** Reitter, 1887
**Notable**
8 SU04: Shrewton, 1-Aug-85 (DRN) (Johnson, 1993)
SU21: Hamptworth Estate, 1-Aug-79 (DRN) (Nash, 1981; Johnson, 1993)
Resident. Very local; in old straw. Grassland species associated with farms and gardens.

ANCHICERA
***Atomaria apicalis*** Erichson, 1846
7 VC7, n.d. (Johnson, 1993)
SU16: Marlborough area, 1869-1938 (Meyrick, 1938)
SU17: Wootton Rivers, 10-Aug-81 (DRN)
8 ST91: Cranborne Chase, 30-Jun-99 (MDD det CJ)
ST93: Great Ridge Wood, 29-Jul-82 (DRN)
SU03: Grovely Wood, 4, 8-Aug-80 (DRN)
SU05: Market Lavington, 2-Aug-83 (DRN)
SU06: Horton, 5-Aug-83 (DRN)
SU13: Lower Woodford, 27-Jul, 5-Aug-82 (DRN); Salisbury, 13-Aug-81 (DRN)
SU14: Figheldean, 7-Aug-81 (DRN)
SU15: Fittleton, 29-Jul-83 (DRN); Widdoington, 5-Aug-83 (DRN)
SU16: Woodborough, 4-Aug-83 (DRN)
SU21: Hamptworth Estate, 15-Apr-74, 3-Apr-78 (DRN)
SU22: East Grimstead, 12-Aug-87
Resident. Widespread; in cut vegetation, straw and on sappy spruce. Grassland species.

***Atomaria atricapilla*** Stephens, 1830
7 VC7 n.d. (Johnson, 1993)
SU16: Marlborough area, Jun-1873 (EWJ) (Meyrick, 1874); 1869-1938 (Meyrick, 1938)
SU17: Wootton Rivers, 10-Aug-81 (DRN)
SU26: Savernake Forest, 28-Jul to 28-Aug-93 (EN det SAW)
8 ST93: Great Ridge Wood, 19-Apr-84 (DRN)
ST96: Rowde, 5-Aug-83 (DRN)
SU02: Knighton Wood, Apr-07, Jul-08 (AJWA)
SU03: Grovely Wood, 8-Aug-80 (DRN)
SU04: Shrewton, 1-Aug-85 (DRN)
SU13: Idmiston, 5-Aug-82 (DRN)
SU14: Netheravon, 30-Jul-85 (DRN)
SU15: Manningford, 13-Aug-87 (DRN); Widdington, 5-Aug-83 (DRN)
SU22: Redlynch, 4-Aug-82 (DRN)
Resident. Local; in moss, cut grass and on vegetation.

***Atomaria basalis*** Erichson, 1846
7 SU16: Marlborough area, Jun-1873 (EWJ) (Meyrick, 1874) No specimens in MCC.
8 SU13: Lower Woodford, 1-Aug-83 (DRN)
SU15: Fittleton, 29-Jul-83 (DRN) (Johnson, 1993)
Resident. Very local; in weed refuse and cut grass. Wetland species.

***Atomaria fuscata*** (Schönherr, 1808)
7 VC7, n.d. (Johnson, 1993)
8 SU02: Knighton Wood, Sep-06, Apr, Sep-07 (AJWA)
SU03: Grovely Wood, 11-Aug-71 (DRN)
SU04: Shrewton, 1-Aug-85 (DRN)
SU06: Horton, 5-Aug-83 (DRN)
SU13: Great Durnford, 27-Jul-77 (DRN); Lower Woodford, 27, 30-Jul, 5-Aug-82 (DRN)
SU14: Netheravon, 30-Jul-85 (DRN)
SU15: Widdington, 5-Aug-83 (DRN)
SU16: Woodborough, 4-Aug-83 (DRN)
SU21: Hamptworth Estate, 3-Apr-78 (DRN)
SU22: Landford, 12-Aug-87 (DRN)
Resident. Widespread; in fungus and straw and on vegetation in a range of habitats.

***Atomaria gutta*** Newman, 1834
8 SU14: Amesbury n.d. (PS) (Johnson, 1993)
SU13: Lower Woodford, 5-Aug-82 (DRN) (Johnson, 1993)
Resident. Very local; [in damp litter near water].

***Atomaria lewisi*** Reitter, 1877
7 VC7 n.d. (Johnson, 1993)
SU17: Wootton Rivers, 10-Aug-81 (DRN)
8 ST91: Rushmore Park, 23-Sep-01 (AJWA)
ST92: Sutton Mandeville, 25-Jun-07 (MDD)
ST93: Great Ridge Wood, 29-Jul-82 (DRN)
ST96: Rowde, 5-Aug-83 (DRN)
SU02: Knighton Wood, Jul, Sep-08 (AJWA)
SU03: Grovely Wood, 1, 8-Aug-80 (DRN)
SU04: Shrewton, 1-Aug-83 (DRN)
SU06: Horton, 5-Aug-83 (DRN)
SU13: Idmiston, 5-Aug-82 (DRN); Lower Woodford 27-Jul, 5-Aug-82 (DRN)
SU14: Netheravon, 30-Jul-85 (DRN)
SU15: Widdington, 5-Aug-83 (DRN)
SU16: Woodborough, 4-Aug-83 (DRN)
SU17: Hallam, 6-Aug-83 (DRN)
SU21: Hamptworth Estate, 29-Mar, 3-Apr-78 (DRN)
SU22: East Grimstead, 12-Aug-87 (DRN); Landford, 12-Aug-87 (DRN); Redlynch, 4-Aug-82 (DRN)
Resident. Local; in straw, garden compost and rotting vegetation generally. Grassland species. Probably widespread.

***Atomaria mesomela*** (Herbst, 1792)
7 VC7, n.d. (Johnson, 1993)
8 SU13: Middle Woodford, 5-Aug-82 (DRN); Lower Woodford, 30-Jul-82 (DRN) Great Durnford, 27-Jul-77 (DRN)
Resident. Very local; by R. Avon sweeping and in weed refuse. Wetland species.

***Atomaria munda*** Erichson, 1846
7 ST96: Devizes n.d. (Johnson, 1993); Rowde, 5-Aug-83 (DRN)
SU16: Wootton Rivers, 10-Aug-81 (DRN)
8 SU03: Grovely Wood, 8-Aug-80 (DRN)
SU04: Shrewton, 28-Jul-82 (DRN)

Resident. Very local; in old straw and a grass heap. Grassland species [especially indoors in barns].

***Atomaria nitidula*** (Marsham, 1802)
7 SU16: Marlborough area, 1869-1938 (Meyrick, 1938)
8 SU02: Knighton Wood, May-07 (AJWA)
SU03: Wylye Down NNR, 5-Jun to 10-Jul-07 (MDD det CJ)
SU04: Parsonage Down NNR, 22-Aug-08 (MDD det CJ)
SU05: Market Lavington, 2-Aug-85 (DRN)
Resident. Very local; in vegetable refuse and by general sweeping [in a wide range of habitats]. One specimen under this name in MCC without data.

***Atomaria pusilla*** (Paykull, 1798)
7 SU16: Marlborough area, Jun-1873 (EWJ) (Meyrick, 1874); 1869-1938 (Meyrick, 1938) (One specimen under this name in MCC without data conf. CJ)
8 SU03: Grovely Wood, 2-Aug-79 (DRN)
Resident. Very local; in mouldy cut grass. Grassland species [especially around farms and gardens]..

***Atomaria rubella*** Heer, 1841

7 VC7 n.d. (Johnson, 1993)
8 SU06: Horton, 5-Aug-83 (DRN)
SU13: Great Durnford, 3-Aug-83 (DRN); Idmiston, 5-Aug-82 (DRN); Lower Woodford, 5-Aug-82 (DRN)
SU15: Fittleton, 29-Jul-83 (DRN); Widdington, 5-Aug-83 (DRN)
SU16: Woodborough, 4-Aug-83
Resident. Very local; in fungus on elm. Probably widespread especially in damp situations.

***Atomaria rubricollis*** Brisout de Barneville, 1863
**RDBi**
8 SU01: Martin Down NNR, n.d. (LJW) (Johnson, 1993)
SU03: Wylye Down NNR, 10-Jul-07 (MDD det CJ)
Resident. Very local; Grassland species [ in wet areas on base-rich soils]. Nationally very rare species with only four other records since 1969.

***Atomaria testacea*** Stephens, 1830
7 VC7, n.d. (Johnson, 1993)
SU16: Wootton Rivers, 10-Aug-81 (DRN)
SU27: Rabley Wood, 1939 (det KGB) (Knight, 1940)
8 ST83: Chicklade, 30-Jul-82 (DRN)
ST93: Great Ridge Wood, 29-Jul-82 (DRN)
ST96: Rowde, 5-Aug-83 (DRN)
SU03: Grovely Wood, 4-Aug-78, 8-Aug-80, 5-Aug-81 (DRN); Wylye Down NNR, 10-Jul to 3-Aug-07 (MDD conf CJ)
SU04: Shrewton, 28-Jul-82, 1-Aug-85 (DRN)
SU05: Market Lavington, 2-Aug-83 (DRN)
SU06: Horton, 5-Aug-83 (DRN)
SU13: Lower Woodford, 27, 30-Jul, 5-Aug-82 (DRN); Middle Woodford, 8-Aug-79 (DRN); Salisbury, Bemerton Folly, 7-Mar - 15-Jun-02 (MDD) (Darby, 2002d)
SU14: Netheravon, 29-Jul-83, 30-Jul-85 (DRN)
SU15: Fittleton, 29-Jul-83 (DRN); North Newnton, 5-Aug-83; Widdington, 5-Aug-83 (DRN)
SU16: Woodborough, 4-Aug-83 (DRN)
SU21: Hamptworth Estate, 29-Mar-78, 5-Aug-85 (DRN)
SU22: East Grimstead, 12-Aug-87; Landford, 12-Aug-87 (DRN); Redlynch, 4-Aug-82 (DRN)
Resident. Local; in grass heap, straw, swan's nest, and rotten beech.

**OOTYPUS Ganglbauer, 1899**
***Ootypus globosus*** (Waltl, 1838)
8 SU21: Landford Bog, 2003 (APF) (Foster, 2003)
SU22: Bentley Wood, 3-Sep-97 (WANHS det MDD)
Resident. Very local; in rotting vegetation and dung.

**EPHISTEMUS Stephens, 1829**
***Ephistemus globulus*** (Paykull, 1786)
7 VC7 n.d. (Johnson, 1993)
SU16: Marlborough area, 1869-1938 (Meyrick, 1938); Wootton Rivers, 10-Aug-81 (DRN)
8 ST92: Sutton Mandeville, 1-Jun-93 (MDD)
SU03: Grovely Wood, 2-Aug-79 (DRN)
SU04: Shrewton, 28-Jul-82 (DRN)
SU06: Horton, 5-Aug-83 (DRN)
SU13: Idmiston, 5-Aug-82 (DRN); Lower Woodford, 27-Jul, 5-Aug-82 (DRN); Salisbury, 13- Aug-81 (DRN)
SU14: Netheravon, 30-Jul-85 (DRN)
SU15: Widdington, 5-Aug-83 (DRN)
SU21: Hamptworth Estate, 15-May, 29-Mar-78, 3-Apr-78, 6-Aug-79 (DRN)
SU22: Redlynch, 4-Aug-82 (DRN)
Resident. Local; in garden compost, old straw and mouldy grass.

## Family EROTYLIDAE
### Subfamily EROTYLINAE
#### Tribe DACNINI

**DACNE Latreille, 1796**
***Dacne bipustulata*** (Thunberg, 1781)
**SQS-2**
7 SU16: Marlborough area, 1869-1938 (Meyrick, 1938)
8 SU03: Grovely Wood, 5, 11-Aug, 25-Oct-71, 11-Apr-72 (DRN)
SU13: Stratford sub Castle, 5-Apr-72 (DRN)
SU16: Jones's Mill, 21-Jun-97 (MDD), 1-Sep-97 (HK det MDD); 19-Jun-01 (APF) (Foster, 2002)
Resident. Very local; on fungi. No specimens in MCC.

***Dacne rufifrons*** (Fabricius, 1775)
**SQS-2**
8 SU03: Grovely Wood, 27-Oct-71 (DRN)
SU13: Stratford sub Castle, 5-Apr-72 (DRN)
SU16: Jones's Mill, 19-Jun-01 (APF) (Foster, 2002)
SU26: Wilton Brail, 19-Jun-03 (MDD)
Resident. Local; in fungi on trees.

#### Tribe TRITOMINI

**TRIPLAX Herbst, 1793**
***Triplax aenea*** (Schaller, 1783)
**IEC-possible addition, SQS-2**
8 ST84: Longleat, 29-Aug-99, 28-Aug-00 (MB)
SU03: Grovely Wood, 14 records 4-Aug-70 – 26-Jul-82 (DRN)
SU22: Homan's Copse, 31-Jul-85 (DRN)
SU23: Porton Down, 1970-2001 (Anon, 2001), 15-Oct-95, 12-Nov-96 (MAS)
Resident. Local; in fungi on trees.

**TRITOMA Fabricius, 1775**
***Tritoma bipustulata*** Fabricius, 1775
**Na, IEC-3, SQS-16**
8 SU03: Grovely Wood, 4-Aug-70, 3-Aug-71 (DRN)
SU25: Ludgershall, Apr- Jul-1915 (PAH) (Harwood, 1915)
Resident. Very local; in fungi on trees.

Tritoma bipustulata

## Family BYTURIDAE

**BYTURUS Latreille, 1796**
***Byturus ochraceus*** (Scriba, 1790)
7 SU16: Marlborough area, 1869-1938 (Meyrick, 1938)
SU26: Savernake Forest, 31-May to 23-Jun-95 (EN det MDD)
8 ST85: Green Lane Wood, 26-Jun-98 (JN)
ST91: Cranborne Chase, 30-Jul-97 (MAS)
ST93: Stockton Wood and Do, 7-Jul-96 (MAS)
SU02: Knighton Wood, Apr, 6-Jul-07, May, Jul-08 (AJWA)
SU03: Grovely Wood, 27, 30-May-74 (DRN)
SU13: Little Durnford Down, 1-May-93 (PM); Old Sarum, 31-May-97 (MAS); Salisbury, Apr-00 (MAS) (Darby, 2001)
SU15: Everleigh (SPTAE), 18-May-02 (PJH) (Hodge, 2002)
SU16: Jones's Mill, 1997 (MAS)
SU21: Hamptworth Estate, 29-May-74 (DRN)
SU26: Bedwyn Brail, 2-Jun-03 (MDD)
Resident. Widespread; on vegetation in woodland. Easily confused with *B. Tomentosus* but scarcer.

**Raspberry Beetle**
***Byturus tomentosus*** (De Geer, 1774)
7 ST96: Conigre Meadow, 20-May-01 (JKG det MDD)
ST97: North Draycot Park, 22-May-00 (MDD)
SU16: Marlborough area, Jun-1873 (EWJ) (Meyrick, 1874); 31-May-1879 (TNH-S) (Meyrick, 1880); 1869-1938 (Meyrick, 1938); 19-Jun-01 (APF) (Foster, 2002)
SU17: Avebury, 24-May-03, 12-Jun-06, 19-May-07 (DB det MDD)
SU26: Savernake Forest, 26-May to 16-Jun-93, 28-Apr to 18-May, 19-Jul to 9-Aug-94 (EN det SAW); 31-May to 23-Jun, 4 to 24-Aug-95 (EN det MDD)
ST91: Cranborne Chase, 16 to 25-Jun-80 (NCC); 24-Jun-97 (MAS)
ST92: Wardour, 5-Jun-96 (MDD)
ST93: Fonthill, 27-May-95 (MDD); Great Ridge Wood, 14-Jun-96, 16, 25-May-00 (MDD)
ST94: Ladywell (SPTAW), 15-May-02 (PJH) (Hodge, 2002)
SU02: Middleton Down, 24-May-95 (MDD); Wilton, 12-May-97 (MDD)
SU03: Grovely Wood, 27-May-74 (DRN); 11-Jun-00 (MDD)
SU04: Parsonage Down NNR, 22-Jul-08 (MDD); Salisbury Plain, 16-May-02 (PJH) (Hodge, 2002)
SU13: Old Sarum, 31-May-97 (MAS); Salisbury, Apr-00 (MAS) (Darby, 2001); Salisbury, Bemerton Folly, 7-Mar - 15-Jun-02 (MDD) (Darby, 2002d)
SU14: Haxton (SPTAE), 17-May-02 (PJH) (Hodge, 2002)
SU16: Jones's Mill, 21-Jun-97 (MDD)
SU22: Bentley Wood, 31-May-97 (MDD); Blackmoor Copse, 17-May-02 (APF) (Foster, 2004); Langley Wood NNR, 8-Jun-02 (MDD); Whiteparish Common, 2-Jul-95 (MAS)
Resident. Widespread; on vegetation in woodland and grassland. Easily confused with *B. ochraceus.*

## Family BIPHYLLIDAE

**BIPHYLLUS Dejean, 1821**
***Biphyllus lunatus*** (Fabricius, 1787)
**IEC-3, SQS-4**
8 ST84: Longleat, 12-Apr-02 (MB)
SU03: Grovely Wood, 11-Apr-72, 16-Apr-74 (DRN)
Resident. Very local; in *Daldinia* fungus on ash.

**DIPLOCOELUS Guérin-Méneville, 1844**
***Diplocoelus fagi*** Guérin-Méneville, 1844
**Nb, IEC-3, SQS-8**
7 SU26: Savernake Forest, 9-Jun to 1-Jul-94 (EN det SAW)
8 ST92: Wardour, 29-Oct-95, 5-Jun-96 (MDD)
SU02: Knighton Wood, Jun, Sep-07 (AJWA)
SU03: Grovely Wood, 13-Aug-70, 16-Apr-74, 3-Apr, 30-Oct-75, 4, 11-Aug-87 (DRN)
SU22: Bentley Wood, 11-Aug-01 (MDD); Homan's Copse, 31-Jul-85 (DRN)
SU23: Porton Down, 15-Oct-95, 12-Nov-96 (MAS)
Resident. Local; in rotten wood and under bark of beech, and in fungus on beech.

## Family BOTHRIDERIDAE
Subfamily ANOMMATINAE

**ANOMMATUS Wesmael, 1835**
***Anommatus duodecimstriatus,*** (Müller, P.W.J. 1821)
**Na**
7 ST99: Crudwell, 19-Jun-73 (APF)
Resident. Very local; [in decaying wood and rotting seed potatoes underground; also under half buried stones in damp places and occasionally in vegetable and haystrack refuse].

## Family CERYLONIDAE
Subfamily CERYLONINAE

Cerylon fagi

**CERYLON Latreille, 1802**
***Cerylon fagi*** Brisout, 1867
**Nb, IEC-2, SQS-8**
7 SU16: West Woods, May-02 (MDD)
SU26: Savernake Forest, 3-Apr-70 (DRN)
8 ST91: Cranborne Chase, 19-Apr-06 (MDD)
ST92: Wardour, 13-Nov-95 (MDD)
ST93: Great Ridge Wood, 18-Apr-74 (DRN)
SU03: Grovely Wood, 6-Aug-70, 11, 13-Aug-71, 6-Apr-72, 16-Apr-74 (DRN)
SU22: Langley Wood NNR, 8-Jun-02 (MDD)
Resident. Widespread; under bark, mainly beech but also oak and birch.

***Cerylon ferrugineum*** Stephens, 1830
**IEC-addition, SQS-2**
7 SU16: Marlborough area, 1869-1938 (Meyrick, 1938)
SU26: Savernake Forest, 2-Apr-70 (DRN); 6-Jun-73 (APF); 15-Jun-73, 29-May-83, 5-May-90 (KNAA); 15-Jun-73 (SAW); 6-Dec-73, 22-Jun-90 (APFw); 8 to 28-Jul-93 (EN det SAW)
8 ST73: Stourhead, 11-Feb-84 (KNAA) (Alexander, 1996); Tucking Mill Hanging, 11-Feb-84 (KNAA)
ST84: Longleat, 7-Jul-99 (MDD); 12-Apr-02 (MB)
ST91: Cranborne Chase, 21-May-02 (MDD)
ST92: Sutton Row, 21-Sep-95 (MDD); Wardour, 3-Apr-96 (EN); 18-Jun-02 (MDD)
ST93: Fonthill, 26-Jun-02 (MDD); Great Ridge Wood, 18-Apr-74 (DRN); 18-Mar, 2-May-00 (MDD)
ST96: Spye Park, 24-Apr-87 (KNAA)
ST99: Flisteridge Wood, 6-May-89 (KNAA)
SU02: Knighton Wood, 6-Jul-07 (AJWA)
SU03: Grovely Wood, 11 records 6-Aug-70 - 4-Aug-87 (DRN); 21-Aug-01 (MDD)
SU13: Little Durnford Down, 13-Sep-99 (MDD); Salisbury, Bemerton Folly, 7-Mar - 15-Jun-02 (MDD) (Darby, 2002d)
SU21: Hamptworth Common, 24-Mar-03 (MDD)
SU22: Bentley Wood, 5-May-03 (MDD)
SU23: Thorny Down Woods, 16-Oct-00 (MDD); Porton Down, 1970-2001 (Anon, 2001)
SU26: Bedwyn Brail, 5-Nov-03 (MDD); Wilton Brail, 19-Jun-03 (MDD)
Resident. Widespread; under bark of mainly beech and sometimes in fungi.

***Cerylon histeroides*** (Fabricius, 1792)
**IEC-addition, SQS-4**
7 ST99: Flisteridge Wood, 21-Mar, 5-May-82 (KNAA)
SU16: Marlborough area, Jun-18738 (EWJ) (Meyrick, 1874); 1869-1938 (Meyrick, 1938)
SU26: Savernake Forest, 6-Dec-73, 22-Jun-90 (APFw); 29-May-83 (KNAA)
8 ST84: Longleat, 5-May-86 (AGD)
ST91: Cranborne Chase, 19-Apr-06 (MDD)
ST93: Great Ridge Wood, 19-Apr-84 (DRN); 18-Mar-00 (MDD)
SU02: Knighton Wood, Apr-07, May-08 (AJWA)
SU03: Grovely Wood, 6-Aug-70, 5-Apr-71, 16-Apr-74, 30-Oct-75, 1-Aug-80, 26-Jul-824-Aug-87 (DRN); 17-Feb-98, 27-May-02 (MDD)
SU13: Little Durnford Down, 13-Sep-99 (MDD)
SU22: Blackmoor Copse, 3-Aug-03 (APF) (Foster, 2004); Langley Wood NNR, 12-Apr-74 (DRN); 6-Jun-02 (MDD)
SU26: Bedwyn Brail, 2-Jun-03, 20-Sep-04 (MDD)
Resident. Local; under bark of beech and oak.

## Family ALEXIIDAE

**SPHAEROSOMA Stephens, 1832**
***Sphaerosoma pilosum*** (Panzer, 1793)
8 SU03: Grovely Wood, 8-Aug-80 (DRN)
Resident. Very local; beaten from sappy conifer strippings.

## Family ENDOMYCHIDAE
Subfamily ENDOMYCHINAE

**ENDOMYCHUS Panzer, 1795**
***Endomychus coccineus*** (Linnaeus, 1758)
7 SU16: Marlborough area, 1869-1938 (Meyrick, 1938)
SU26: Savernake Forest, 2-Apr-70 (DRN)
8 ST82: Oysters Coppice, 6-Oct-02 (MDD)

ST85: Bullen Hill Farm, 1982-2002 (EGS det MDD)
ST94: Norton Down (SPTAW), 23-Jun-02 (PJH) (Hodge, 2002)
SU02: Bishopstone, 3-Aug-91 (MDD)
SU03: Dinton, 8-Aug-91 (SP); Grovely Wood, 10 records 10-Apr-70 – 4-Aug-78 (DRN)
SU13: Great Durnford, 6-Apr-70 (DRN); Little Durnford Down, 6-Dec-73 (PM); Salisbury, 23-May-93 (PM)
SU21: Landford Bog, 21-Oct-97 (PM)
SU22: Bentley Wood, 23-Jul-90 (PM)
SU23: Allington, 1-Dec-97 (MrsP); Porton Down, 15-Oct-95 (MAS)

On *Chondrostereum* fungus on rotten logs and stumps.

Endomychus coccineus

## Subfamily ANAMORPHINAE

**SYMBIOTES Redtenbacher, 1849**
***Symbiotes latus*** Redtenbacher, 1849
**Nb, IEC-3, SQS-8**
8 SU13: Middle Woodford, 8-Aug-79 (DRN)

Resident. Very local; under fungoidal bark of upright elm.

## Subfamily MEROPHYSIINAE

**HOLOPARAMECUS Curtis, 1833**
**Holoparamecus caularum** (Aubé, 1843)
8 SU02: Wilton, 6-Sep-95 (MDD)

Resident. Very local; in pile of rotting vegetation.

## Subfamily LYCOPERDININAE

**LYCOPERDINA Latreille, 1807**
***Lycoperdina bovistae*** (Fabricius, 1792)
**RDB3, LCGR-FI C?**
8 SU01: Martin Down NNR, 1999 (EN, 1999)
SU03: Grovely Wood, 5, 6-Apr-71 (DRN) (Nash, 2004)
SU22: Langley Wood NNR, 27-Aug to 17-Sep-04 (MDD) (Darby, 2005)

Resident. Very local; in pitfall traps and in *Lycoperdon*, puff-ball, fungi.

Lycoperdina bovistae

## Subfamily MYCETAEINAE

**MYCETAEA Stephens, 1829**
**Hairy Cellar Beetle**
***Mycetaea subterranea*** (Fabricius, 1801)
7 SU16: Marlborough area, Jun-1873 (EWJ) (Meyrick, 1874); 1869-1938 (Meyrick, 1938)
8 SU03: Wylye Down NNR, 20-Sep-05 (MDD) (Darby, 2009)
SU13: Little Durnford Down, 13-Sep-99, 20-May-03 (MDD)
SU21: Hamptworth Estate, 1-Apr-75, 1-Aug-79 (DRN)

Resident. Local; in straw and litter, and sweeping in calcareous grassland. Formerly *M. hirta.*

# Family COCCINELLIDAE

## Subfamily COCCIDULINAE

**COCCIDULA Kugelann in Illiger, 1798**
***Coccidula rufa*** (Herbst, 1783)
7 SU09: Somerford Keynes, 4-Jun-04 (BLe)
SU16: Marlborough area, 12-Feb-0531 Dec 1873 (Meyrick, 1874); 1869-1938 (Meyrick, 1938)
8 ST91: Cranborne Chase, 4-Jul-97 (MAS)
ST94: Ladywell (SPTAW), 3-Jun - 12-Aug-08 (Boyce & Wright, 2009); Wadman's Coppice (SPTAW), 3-Jun - 12-Aug-08 (Boyce & Wright, 2009)
SU01: Martin Down NNR, 1999 (EN, 1999)
SU12: Charlton-All-Saints, 9-May-96 (MAS); Salisbury, 30-Jul-73 (DRN); Stratford-Sub-Castle, 10-Aug-70 (DRN)
SU13: Great Durnford, 16-Aug-69, 7-Aug-70, 27-Jul-77 (DRN); Idmiston, 5-Aug-82 (DRN); Lower Woodford, 28-Jul-78, 2, 5-Aug-82 (DRN); Middle Woodford, 27-Jul-77 (DRN); Old Sarum, 31-May-97 (MAS); Stratford sub

Castle, 26-May-74 (DRN)
SU14: Netheravon, 29-Jul-83 (DRN); Nine Mile R. pond (SPTAE), 5-Jun-97 (MGT)
SU15: Fittleton, 29-Jul-83 (DRN)
SU16: Jones's Mill, 21-Jun-97 (MDD)
Resident. Widespread; on vegetation in damp areas.

**RHYZOBIUS Stephens, 1829**
***Rhyzobius litura*** (Fabricius, 1787)
7 SU06: Cherhill Down, 31-May-04 (BLe); Roundway Camp, Jul-40 (FDB) (Buck, 1956)
SU09: Somerford Keynes, 4-Jun-04 (BLe)
SU16: Marlborough area, Jun-1873 (EWJ) (Meyrick, 1874); 1869-1938 (Meyrick, 1938); Wootton Rivers, 10-Aug-81 (DRN)
SU26: Savernake Forest, 20-Sep to 11-Oct-94 (EN det SAW)
8 ST91: Cranborne Chase, 16 to 25-Jun-80 (NCC)
ST93: Fonthill, 27-May-95 (MDD); Great Ridge Wood, 18-Apr-74 (DRN)
ST94: Bowls Barrow (SPTAW), 16-May, 22-Jun-02 (PJH) (Hodge, 2002); Ladywell (SPTAW), 15- May, 23-Jun-02 (PJH) (Hodge, 2002); Norton Down (SPTAW), 15-May, 23-Jun-02 (PJH) (Hodge, 2002); Warminster Ranges (SPTAW), 3-Jun - 12-Aug-08 (Boyce & Wright, 2009)
ST95: West Lavington, 20-May-44 (DCL) (DFL, 1948)
SU01: Martin Down NNR, 21-Aug-89 (RGB); 1999 (EN, 1999)
SU03: Grovely Wood, 11-Aug-71, 1-Aug-73, 1-Aug-88 (DRN)
SU04: Shrewton, 28-Jul-82 (DRN); Yarnbury Castle 21-Jun-90 (APFw)
SU05: Black Heath (SPTAC), 3-Jun - 12-Aug-08 (Boyce & Wright, 2009)
SU12: Coombe Bissett Down, 1-Apr-97 (MDD); Salisbury, 30-Jul-73 (DRN)
SU13: Boscombe Down, 8, 9-Oct-94 (MDD); Great Durnford, 19-Aug-69, 7-Aug-70, 3-Aug-83 (DRN); Idmiston, 5-Aug-82 (DRN); Middle Woodford, 27-Jul-72 (DRN); Salisbury, 15-Aug-88 (DRN); Stratford-Sub-Castle, 10-Aug- 70 (DRN)
SU14: Haxton, 17-May, 24-Jun-02 (PJH) (Hodge, 2002); 3-Jun - 12-Aug-08 (Boyce & Wright, 2009); Netheravon, 29-Jul-83 (DRN); Stonehenge, 22-Apr-03 (PJH) (Hodge, 2003); Wexland Hanging (SPTAC), 22-Jun-02 (PJH) (Hodge, 2002)
SU15: Beach's Plantation (SPTAE), 13 to 17-Jun-97 (MGT); Everleigh (SPTAE), 18-May, 24-Jun-02 (PJH) (Hodge, 2002)
SU16: Woodborough, 4-Aug-83 (DRN)
SU22: Bentley Wood, 31-May-97 (MDD); Holmere Common, 19-Aug-88 (DRN); Redlynch, 4-Aug-82 (DRN)
SU23: Porton Down, 3-Jul-99 (PJH) (Hodge, 2002)
SU24: Sidbury Hill (SPTAE), 13 to 17-Jun-97 (MGT)
SU25: Sidbury Hill (SPTAE), 28-May, 4, 6-Jun-97 (MGT)
SU26: Bedwyn Brail, 19-Jun-03 (MDD)
Resident. Widespread; on vegetation in grassland. .
Two other similar species of *Rhzobius, chrysomeloides* and *lophanthae*, have recently been added to the British list (Booth, 2000 and Hawkins, 2001) so that these records should be treated as indeterminate.

**SCYMNUS Kugelann, 1794**
SCYMNUS Kugelann, 1794
***Scymnus femoralis*** (Gyllenhal, 1827)
**LCGR-Fm Cu**
8 SU24: Bulford Ranges (SPTAE), 3-Jun - 12-Aug-08 (Boyce & Wright, 2009)
Resident. Very local: [at plant roots on dry sandy or stony soils].

***Scymnus nigrinus*** Kugelann, 1794
8 SU21: Pound Bottom, 24-Jul-73 (DRN); Franchises Common, 25-Jul-73 (DRN)
Resident. Very local; beating pines and sweeping gorse, etc..

PULLUS Mulsant, 1846
***Scymnus auritus*** Thunberg, 1795
7 SU16: Marlborough area, Jun-1873 (EWJ) (Meyrick, 1874); 1869-1938 (Meyrick, 1938)
8 SU03: Grovely Wood, 13-Aug-70 (DRN)
SU26: Bedwyn Brail, 2-Jun-03 (MDD)
Resident. Very local; on oak trees.

***Scymnus suturalis*** Thunberg, 1795
7 SU16: Marlborough area, 1869-1938 (Meyrick, 1938)
8 SU04: Orcheston Down (SPTA), 18-May-02 (PJH) (Hodge, 2002)
Resident. Very local; beaten off pines.

NEOPULLUS Sasaji, 1971
***Scymnus haemorrhoidalis*** Herbst, 1797
7 SU09: Somerford Keynes, 4-Jun-04 (BLe)
8 SU13: Salisbury, 11-Apr-71 (DRN)
SU21: Hamptworth Estate, 24-Apr-84 (DRN)
SU24: Sidbury Hill (SPTAE), 29-May-97 (MGT)
Resident. Local; in moss and under stone. [In many habitats but particularly damp areas].

**NEPHUS Mulsant, 1846**
***Nephus redtenbacheri*** (Mulsant, 1846)
8 SU21: Landford Bog, 24-Jun-01 (APF) (Foster, 2001)
SU24: North Tidworth, Devils Ditch, 30-May-04 (BLe)
Resident. Very local; [in low growing vegetation in a wide variety of habitats].

**HYPERASPIS Dejean, 1836**
***Hyperaspis pseudopustulata*** Mulsant, 1853
**Nb**
8 SU21: Hamptworth Estate, 31-May-74 (DRN) (Nash, 1981)
Resident. Very local; on gravel path by pines. [In a diverse range of habitats including deciduous woodland and acid heathland].

## Subfamily CHILOCORINAE

**Heather Ladybird**
***Chilocorus bipustulatus*** (Linnaeus, 1758)
7 SU16: Marlborough area, Jun-1873 (EWJ) (Meyrick, 1874); 1869-1938 (Meyrick, 1938)
8 SU21: Landford Bog, 9, 24-Jun-01 (APF) (Foster, 2001)
Resident. Very local; on heather, *Erica*, and ling, *Calluna*.

**Kidney-spot Ladybird**
***Chilocorus renipustulatus*** (Scriba, 1791)
7 SU06: Morgan's Hill, 28-Mar-96 (PM)
SU16: Jones's Mill, 2-Jul-01 (APF) (Foster, 2002); Marlborough area, 1869-1938 (Meyrick, 1938)
SU17: Clout's Wood, 29-Dec-00 (PM)
SU26: Savernake Forest, 8-Sep-69 (DRN), 15-Sep to 5-Oct-95 (EN det MDD)
8 ST85: Green Lane Wood, 7-Jun-95 (WANHS det MDD)
ST93: Great Ridge Wood, 14-Jun-96 (MDD)
SU03: Grovely Wood, 4-Aug-70, 13-Aug-71, 8, 16-Apr-74 (DRN)
SU12: Petersfinger, 25-Jul-91 (DRN)
SU13: Idmiston, 13-Aug-87 (DRN); Great Durnford, 11-Apr-71 (DRN); Little Durnford Down, 19-Jul-90, 25-Mar-91 (PM)
SU14: Figheldean, 4, 19-Aug-88 (DRN)
SU15: Everleigh (SPTAE), 18-May-02 (PJH) (Hodge, 2002)
SU20: Gurston Down, Broad Chalke, 13-Aug-95 (MDD)
SU21: Hamptworth Estate, Aug-86 (DRN)
SU22: Bagfield Copse, 31-Dec-42 (DRN); Bentley Wood, 18-Mar, 29-Jul-90 (PM); Holmere Common, 19-Aug-88 (DRN); Landford, 12-Aug-87 (DRN); Langley Wood, 6-Aug-86 (DRN)
Resident. Widespread; on trees particularly ash, willows, sallows and birches.

**EXOCHOMUS Redtenbacher, 1843**
**Pine Ladybird**
***Exochomus quadripustulatus*** (Linnaeus, 1758)
8 ST95: West Lavington, Jun-41 (PVS) (DFL, 1948 'unconfirmed record')
SU03: Grovely Wood, 13-Aug-71, 18-Jul-73, 16-Apr-74 (DRN)
SU04: Orcheston Down (SPTA), 18-May-02 (PJH) (Hodge, 2002); Parsonage Down NNR, 8-Jul-08 (MDD)
SU13: Little Durnford Down, 25-Mar-91 (PM); Salisbury, 7-Apr-74 (DRN)
SU21: Burnt Ground Wood, 31-May-74, 1-Apr-75 (DRN); Pound Bottom, 24-Jul-73 (DRN)
SU22: Bentley Wood, 28-Jun-90, 13-Mar-91 (PM); PepperBox Hill, 9-May-93 (PM)
SU23: Porton Down, 3-Jul-99 (BLe)
Resident. Local; mainly on needled conifers [but in some localities on deciduous trees also].

## Subfamily COCCINELLINAE

**ANISOSTICTA Dejean, 1836**
**Water Ladybird**
***Anisosticta novemdecimpunctata*** (Linnaeus, 1758)
7 SU16: Marlborough area, 1869-1938 (Meyrick, 1938)
8 SU12: Charlton-All-Saints, 9-Aug-95 (MAS); Petersfinger, 25-Jul-91 (DRN)
SU13: Great Durnford, 16-Aug-69, 7-Aug-70, 6-Aug-71, 27-Jul-77 (DRN); Lower Woodford, 28-Jul-78 (DRN)
Resident. Very local; on vegetation by water.

**TYTTHASPIS Crotch, 1874**
**16-Spot Ladybird**
***Tytthaspis sedecimpunctata*** (Linnaeus, 1761)
7 SU09: Ashton Keynes, 27-Apr-00 (JKG det MDD); Upper Waterhay, 17-May to 11-Jun-05 (JKG det MDD)
SU16: Marlborough area, 6-May-1879 (PHM) (Meyrick, 1880)
SU18: Haydon Meadow, 10-Apr-02 (JKG det MDD)
8 ST95: West Lavington, 13-May-45 (DCL) (DFL, 1948)
ST94: Bowls Barrow (SPTAW), 22-Jun-02 (PJH) (Hodge, 2002); Ladywell (SPTAW), 3-Jun – 12-Aug-08 (Boyce & Wright, 2009); Norton Down (SPTAW), 23-Jun-02 (PJH) (Hodge, 2002)
SU01: Martin Down NNR, 1999 (EN, 1999)
SU04: Orcheston Down (SPTA), 22-Jun-02 (PJH) (Hodge, 2002)
SU05: Ell Barrow (SPTAC), 3-Jun - 12-Aug-08 (Boyce & Wright, 2009)
SU13: Great Durnford, 7-Aug-70 (DRN); Stratford sub Castle, 26-May-74 (DRN)_
SU14: Bourne Bottom (SPTAE), 13-Jun-97 (MGT); Figheldean, 28-Jul-82 (DRN); Larkhill west, 2-Jun-97 (MGT); Stonehenge, 22-Apr-03 (PJH) (Hodge, 2003)
SU15: Beach's Plantation (SPTAE), 10-Jun-97 (MGT); Fittleton, 29-Jul-83 (DRN)

SU16: Woodborough, 4-Aug-83 (DRN)
SU21: Hamptworth Estate, 16-Apr-74, 3-Apr-78 (DRN); Landford Bog, 9-Jun-01 (APF) (Foster, 2001)
SU24: Dunch Hill (SPTAE), 16-Jun-97 (MGT); North Tidworth, Devils Ditch, 30-May-04 (BLe)
SU25: Sidbury Hill (SPTAE), 28-May, 4-Jun-97 (MGT)

Resident. Widespread; on vegetation in grassland, wet places and heathland.

**MYZIA Mulsant, 1846**

***Myzia oblongoguttata*** (Linnaeus, 1758)

7 SU16: Marlborough area, 1869-1938 (Meyrick, 1938)
8 ST85: Bullen Hill Farm, 1982-2002 (EGS det MDD)
SU04: Orcheston Down (SPTA), 18-May-02 (PJH) (Hodge, 2002)
SU21: Woodfalls, 19-Jun-86 (EGS det MDD)

Resident. Local; [on needled conifers, particularly mature or semi-senile Scots pine].

Myzia oblongoguttata

**MYRRHA Mulsant, 1846**

**18-Spot Ladybird**

***Myrrha octodecimguttata*** (Linnaeus, 1758)

7 SU16: Marlborough area, 17-Feb-1879 (HMP) (Meyrick, 1880)
8 SU03: Grovely Wood, 27-Oct-71 (DRN)

Resident. Very local; on Scots pine.

**PROPYLEA Mulsant, 1846**

**14-Spot Ladybird**

***Propylea quattuordecimpunctata*** (Linnaeus, 1758)

7 ST96: Chittoe, 9-May-00 (EGS det MDD)
SU06: Roundway. Home Cover, 13-May-00 (BG)
SU08: Red Lodge Planatatio, 9-May-00 (BG)
SU09: Upper Waterhay, 5 to 11-Jun-05, 5 to 11-Jun-07 (JKG det MDD)
SU16: Caen Hill Locks, 11-Jun-96 (EGS det MDD); Lockeridge, 1994 (JEO); Marlborough area, Jun-1873 (EWJ) (Meyrick, 1874); 1-Apr-1879 (RAW) (Meyrick, 1880); 16-Jul-1879 (GH) (Meyrick, 1880); 1869-1938 (Meyrick, 1938)
SU17: Avebury, 25-Apr-04, 29-Jul-05 (DB)
SU19: Hannington, 26, 31-Jul-95, 26-May-97(MEN)
SU26: Savernake Forest, 22-Jun-90 (APFw); 25-Apr to 16-Jun, 29-Aug to 11-Oct-93, 28-Apr to 18-May, 19-Jul to 30-Aug-94 (EN det SAW); 11 to 30-May-95, 4 to 24-Aug-95 (EN det MDD)
8 ST02: Vernditch Chase, 17-Jul-95 (MDD)
ST84: Longleat, 15-Jun-73 (PM); 29-Aug-99 (MB); 25-May-05 (DJG)
ST85: Green Lane Wood, 8-Jul-92 (PM); 7-Jun-95 (MDD)
ST91: Cranborne Chase, 24-Apr-97 (MAS)
ST92: Sutton Mandeville, 7-Jul-96 (MDD); Wardour, 5-Jun-96 (MDD)
ST93: Fonthill, 19-Jul-95 (MDD); Great Ridge Wood, 14-Jun-96 (MDD); Stockton Wood, 28-Sep-95 (PJH) (Hodge, 1995)
ST94: Bowls Barrow (SPTAW), 16-May-02 (PJH) (Hodge, 2002);
Ladywell, 10-Aug-86 (EGS det MDD)
ST95: West Lavington, 25, 26-May-46 (DCL) (DFL, 1948)
SU02: Broad Chalke, 23-May-94 (BL); Chicken-grove Bottom, 21-May-93 (PM), 17-Jul-95 (MDD); Knighton Wood, 2006-2007 (AJWA);
Middleton Down, 1-Aug-92, 23-May-94, 21-May-95 (MDD); Wilton, 12-May-97 (MDD)
SU03: Dinton, 21-Jul-95 (MDD); Grovely Wood, 25-Oct-71, 24-Aug-72, 1-Aug-73, 4-Aug-81, 1- Aug-88 (DRN); 19-Aug-95 (MAS), 11, 12-Jun-00 (MDD)
SU04: Orcheston Down (SPTA), 18-May-02 (PJH) (Hodge, 2002); Shrewton, 6-Aug-82, 2-Aug-83 (DRN)
SU06: Devizes, 6-May-00 (BG)
SU13: Cockey Down, 1-May-90 (PM);Great Durnford , 18-Aug-72, 17-Jul-73, 3-Aug, 83 (DRN); Little Durnford Down, 6-Dec-73 (PM); Lower Woodford, 5-Aug-82 (DRN); Salisbury, 15-Aug-88 (DRN); 1-Jun-00 (Darby, 2001);Salisbury, Bemerton Folly, 7-Mar - 15-Jun-02 (MDD) (Darby, 2002d)
Stratford sub Castle, 26-May-74 (DRN)
SU14: Figheldean, 28-Jul-82; Haxton (SPTAE), 17-May, 24-Jun-02 (PJH) (Hodge, 2002); Stonehenge, 22-Apr-03 (PJH) (Hodge, 2003)
SU15: Beach's Plantation (SPTAE), 10, 16-Jun-97 (MGT); Everleigh (SPTAE), 18-May-02 (PJH) (Hodge, 2002); Fittleton, 29-Jul-83 (DRN)
SU16: Jones's Mill, 2-Jul-01 (APF) (Foster, 2002); Woodborough, 4-Aug-83 (DRN)
SU21: Landford Bog, 13-Jun-73 (SP); 24-Jun-01 (APF) (Foster, 2001)

SU22: Bentley Wood, 25-Mar, 28-Jun, 1-Aug-90 (PM); 31-May-96, 11-Aug-01 (MDD); Blackmoor Copse, 13-Jun-73 (SP), 4-May to 18-May-02, 3-Aug-03 (APF) (Foster, 2004); East Grimstead, 5-Apr-77 (DRN); Glaziers Copse, 16-Aug-88 (DRN); PepperBox Hill, 9-May-93 (PM)
SU23: Porton Down, 3-Jul-99 (PJH) (Hodge, 2002)
SU24: Sidbury Hill (SPTAE), 4-Jun-97 (MGT)

Resident. Ubiquitous; in most habitats.
Some records may be confusions with *Calvia quattuordecimguttata.*

**CALVIA Mulsant, 1846**
**Cream-spot Ladybird**
***Calvia quattuordecimguttata*** (Linnaeus, 1758)

7 SU09: Upper Waterhay, 17-May, Jun-05 (JKG det MDD)
SU17: Avebury, 30-May-03 (DB)
8 ST91: Cranborne Chase, 19-Apr-06 (MDD)
ST92: Sutton Mandeville, 2-May-02 (MDD)
ST93: Stockton Wood, 28-Sep-95 (PJH) (Hodge, 1995)
SU02: Knighton Wood, Sep-06, Mar-07 (AJWA)
SU03: Grovely Wood, 11-Aug-71, 23-Jul, 1-Aug-73 (DRN)
SU12: Petersfinger, 7-Aug-88 (DRN)
SU13: Great Durnford, 19-Aug-69, 17-Jul-73 (DRN); Middle Woodford, 27-Jul-77 (DRN)
SU15: Beach's Plantation (SPTAE), 1-May-05 (MGT)
SU22: Bagfield Copse, 7-Aug-86 (DRN); Langley Wood, 28-May-74 (DRN)
SU26: Bedwyn Brail, 20-Sep-04 (MDD)

Resident. Widespread; in deciduous woodland and hedgerows.

Halyzia sedecimguttata

**HALYZIA Mulsant, 1846**
**Orange Ladybird**
***Halyzia sedecimguttata*** (Linnaeus, 1758)

7 ST87: Castle Combe, 3-Jun-04 (BLe)
SU06: Home Covert, Devizes, 16-Dec-94 (BG); Roundway Hill Covert, 7-Aug-96 (BG); Marlborough area, Jun-1873 (Meyrick, 1874); 1869-1938 (Meyrick, 1938)
SU17: Avebury, 17-May-07 (DB)
SU18: Swindon, 8-Aug-05 (JWo)
SU26: Savernake Forest, 31-May-04 (BLe)
8 ST84: Longleat, 15, 17-Jun-73 (PM); 25-May-05 (DJG)
ST85: Green Lane Wood, 7-Jun-95 (WANHS det MDD)
ST86: Bradford on Avon, 20-Dec-07 (SG)
ST91: Cranborne Chase, 19-Apr-06 (MDD)
ST92: Wardour, 5-Jun-96 (MDD, MAS)
ST93: Great Ridge Wood, 16-May-00 (MDD)
ST94: Imber village, 18-Oct-97 (EGS det MDD) Salisbury Plain, 9-Oct-93 (SP)
SU02: Middleton Down, 1-Nov-04 (PM)
SU03: Little Langford, 3-Jul-91 (PM)
SU04: Orcheston Down (SPTA), 18-May-02 (PJH) (Hodge, 2002); Parsonage Down NNR, 2-Jul-08 (MDD)
SU06: Devizes, 10-Jun-99 (BG)
SU13: Boscombe Down, 8, 9-Oct-94 (MDD); Little Durnford Down, 19-Jul-90, 25-Mar-91 (PM)
SU16: Jones's Mill, 21-Jun-97 (MDD)
SU21: Franchises Wood, 3-May-95 (MAS); Bentley Wood, 12-Aug-87 (DRN); 25-Mar-90 (PM)
SU22: Bagfield Copse, 7-Aug-86, 19-Aug-88 (DRN); Blackmoor Copse, 21-Jul-90 (SP), 27-Mar-91, 31-May-93 (PM); Langley Wood NNR, 29-May-03 (MDD)
SU23: Porton Down, 20-Aug-94 (PM); 3-Jul-99 (PMP); Winterslow, 13-Jul-90 (PM)
SU26: Bedwyn Brail, 2-Jun-03 (MDD)

Resident. Ubiquitous; on deciduous trees. Appears to be increasing.

**PSYLLOBORA Dejean, 1836**
**22-Spot Ladybird**
***Psyllobora vigintiduopunctata*** (Linnaeus, 1758)

7 SU09: Blakehill Farm, 11-May-01 (JKG det MDD)
SU16: Lockeridge, 1994 (JEO); Marlborough area, Jun-1873 (EWJ) (Meyrick, 1874); 1869-1938 (Meyrick, 1938); Oare, 18-Aug, Sep-02 (IJG)
SU17: Avebury: 26-Jan, 20-Jun-07 (DB)
SU19: Hannington, 26-Jul-95(MEN)
8 ST85: Bullen Hill Farm, 1982-2002 (EGS det MDD)
ST94: Bowls Barrow (SPTAW), 16-May, 22-Jun-02 (PJH) (Hodge, 2002); Salisbury Plain, 11-Aug-90, 27, 31-May-92, 27-Aug, 4-Sep-93 (SP)
ST95: Keevil, 6-Oct-00 (BG); West Lavington, 26-May-46 (DCL) (DFL, 1948)
SU03: Grovely Wood, 14-Jun-01 (WANHS det MDD)
SU04: Orcheston Down (SPTA), 18-May-02 (PJH) (Hodge, 2002); Salisbury Plain, 16-May, 25-Jun-02 (PJH) (Hodge, 2002); Shrewton, 6-Aug-82 (DRN)
SU13: Cockey Down, 6-Oct-73 (PM); Great Durnford, 8-Aug-85 (DRN); Salisbury, 1-Jun-00 (MDD) (Darby, 2001)
SU15: Beach's Plantation (SPTAE), 16-Jun-97

(MGT); Everleigh (SPTAE), 24-Jun-02 (PJH) (Hodge, 2002)
SU23: West Winterslow, 7-Jul-01 (NB) Winterslow, 2-Sep-90 (PM)
Resident. Widespread; on low vegetation in both woodland and grassland.

Anatis ocellata

**ANATIS Mulsant, 1846**
**Eyed Ladybird**
***Anatis ocellata*** (Linnaeus, 1758)
7 ST96: Prickmore Wood, 14-May-99 (BL)
SU06: Roundway, Home Covert, 27-Apr-00 (BG)
SU16: Marlborough area, Jun-1873 (EWJ) (Meyrick, 1874); 1869-1938 (Meyrick, 1938)
8 ST82: Oysters Coppice, 9-Jan-93 (LB)
ST84: Longleat, 17-Jun-73, 8-Jun-96 (PM)
ST91: Cranborne Chase, 16 to 25-Jun-80 (NCC)
ST92: Sutton Mandeville, 21-May-98 (MDD)
SU03: Grovely Wood, 11-Aug-71 (DRN)
SU04: Orcheston Down (SPTA), 18-May-02 (PJH) (Hodge, 2002)
SU21: Franchises Wood, 4-Aug-77 (DRN)
SU22: Bentley Wood, 3-Sep-90 (PM); 31-May-96, 27-May-97 (MDD)
SU23: Porton Down, 15-Jun-73n (J.Walker)
SU26: Bedwyn Brail, 2-Jun-03 (MDD)
Resident. Widespread; on conifers.

**APHIDECTA Weise, 1899**
**Larch Ladybird**
***Aphidecta obliterata*** (Linnaeus, 1758)
7 SU16: Marlborough area, Jun-1873 (EWJ) (Meyrick, 1874); 873-1938 (Meyrick, 1938)
SU27: Ramsbury, 1870-1873 (Meyrick, 1874)
8 SU03: Grovely Wood, 11-Aug, 25-Oct-71, 18-Jul-73, 3-Aug-79 (DRN)
SU22: Bagfield Copse, 7-Aug-86 (DRN)
Resident. Very local; on needled conifers.

**HIPPODAMIA Dejean, 1836**
HIPPODAMIA Dejean, 1836
**13-Spot Ladybird**
***Hippodamia tredecimpunctata*** (Linnaeus, 1758)
8 ST95: West Lavington, Jun-39 (PLB) (DFL, 1948)
Possible continental immigrant. Very local; [on reeds in damp and marshy places]. This is a previously unnoticed record of this ladybird of which there have been no British records since 1950. (Majerus, 1994)

**COCCINELLA Linnaeus, 1758**
**Scarce 7-spot Ladybird**
***Coccinella magnifica*** Redtenbacher, 1843
**Na**
7 SU16: Marlborough area, 1869-1938
8 SU12: Whaddon Common, 22-Jul-73 (DRN) (Nash, 1982)
Resident. Very local; in nests of the wood ant, *Formica rufa,* or on vegetation nearby.

Coccinella magnifica

**5-Spot Ladybird**
***Coccinella quinquepunctata*** Linnaeus, 1758
**RDB3**
7 SU16: Marlborough area, 1869-1938 (Meyrick, 1938) (Darby, 2002, 2005)
Resident. Very local; only on or by unstable river shingle. There is a single specimen without data correctly identified under this name in the MCC.

The Marlborough College specimen of Coccinella quinequepunctata

**7-Spot Ladybird**
***Coccinella septempunctata*** Linnaeus, 1758
7 ST87: Slaughterford, 14, 15-Mar-07 (KC)
SU08: Distillery Meadows, 9-Jun-91 (SP)
SU16: Lockeridge, 1994 (JEO); Marlborough area, Jun-1873 (EWJ) (Meyrick, 1874); 1869-1938 (Meyrick, 1938); Pewsey Downs, 22-Jun-90 APFw); Wootton Rivers, 10-Aug-81 (DRN)
SU17: Avebury, 11-Feb-04 (DB)
SU19: Hannington, 26, 31-Jul-95, 26-May-97 (MEN)
SU26: Savernake Forest, 22-Jun-90 (APFw); 26-May to 16-Jun-93, 18-May to 11-Oct-93 (EN

det SAW); 31 May to 16-Nov-95 (EN det MDD)
8 ST82: Oysters Coppice, 15-Jun-73 (SP)
ST84: Longleat, 15-Jun-73 (PM); 29-Aug-99 (MB)
ST85: Biss Wood, 15-Aug-00 (BG); Green Lane Wood, 8-Jul-92 (PM), 7-Jun-95 (MDD)
ST91: Cranborne Chase, 16 to 25-Jun-80 (NCC); 25-May-05 (DJG); 19-Apr-06 (MDD)
ST92: Sutton Mandeville, 7-Jul-96 (MDD); Wardour, 5-Jun-96 (MDD)
ST93: Fonthill, 19-Jul-95 (MDD); Great Ridge Wood, 4-May-00 (MDD); Stockton Wood, 28-Sep-95 (PJH) (Hodge, 1995)
ST94: Ladywell (SPTAW), 3-Jun - 12-Aug-08 (Boyce & Wright, 2009); Norton Down (SPTAW), 15-May, 23-Jun-02 (PJH) (Hodge, 2002);Tinkers Track (SPTAW), 3-Jun - 12-Aug-08 (Boyce & Wright, 2009); Wadman's Coppice (SPTAW), 3-Jun - 12-Aug-08 (Boyce & Wright, 2009); Warminster Ranges (SPTAW), 3-Jun - 12-Aug-08 (Boyce & Wright, 2009)
ST95: West Lavington 25-Jan, Apr to Sep-41 (ITH) (DFL, 1948)
SU01: Martin Down NNR, 1999 (EN, 1999)
SU02: Bishopstone, 7-May-89 (MDD); Middleton Down, 24-Apr-89, 24-Apr-90, 23-May-94, 21-May-95 (MDD); Wilton, 12-May-97 (MDD)
SU03: Barford Down, 10-Apr-72 (DRN); Dinton, R. Nadder, 21-Jul-95 (MDD); Grovely Wood, 25-Oct-71, 8, 10-Apr, 24-Aug-72, 1-Aug-73, 4-Apr-77, 1-Aug-88 (DRN; Wylye Down NNR, 20-Sep-05, 4-Sep to 10-Oct-07 (MDD) (Darby, 2008)
SU04: Fox Covert (SPTAW), 3-Jun - 12-Aug-08 (Boyce & Wright, 2009); Orcheston Down (SPTA), 18-May-02 (PJH) (Hodge, 2002); Salisbury Plain, 16-May, 25-Jun-02 (PJH) (Hodge, 2002); Shrewton 2- Aug-82, 6-Aug-83 (DRN)
SU12: Coombe Bissett Down, 1-Apr-97 (MDD)
SU13: Boscombe Down, 9-Oct-94 (MDD); Cockey Down, 1-May-90 (PM); Great Durnford, 3-Aug-83 (DRN); Little Durnford Down, 6-Dec-73 (PM); Lower Woodford, 5-Aug-82; Salisbury, 15-Aug-88 (DRN); 13-Apr-00 (MDD) (Darby, 2001)
SU14: Haxton (SPTAE), 17-May-02 (PJH) (Hodge, 2002); 3-Jun - 12-Aug-08 (Boyce & Wright, 2009); Larkhill west, 2-Jun-97 (MGT); Netheravon, 29-Jul-83 (DRN); Stonehenge, 22-Apr-03 (PJH) (Hodge, 2003); Wexland Hanging (SPTAC), 22-Jun-02 (PJH) (Hodge, 2002); 3-Jun - 12-Aug-08 (Boyce & Wright, 2009)
SU15: Beach's Plantation (SPTAE), 13 to 17-Jun-97, 1-May-05 (MGT); Everleigh (SPTAE), 24-Jun-02 (PJH) (Hodge, 2002)
SU16: Woodborough, 4-Aug-83 (DRN)
SU21: Landford Bog, 13-Jun-73 (SP); 25-May-98 (MDD)
SU22: Bentley Wood, 19-Jun-73 (JN); 25-Mar-90 (PM); 31-May-96 (MDD); Blackmoor Copse, 2-Jun-86 (BG); 31-May-96 (MDD); 24, 31-Aug-01, 17-Mar-02 (APF) (Foster, 2004); PepperBox Hill, 9-May-93 (PM); Redlynch, 4-Aug-82 (DRN)
SU23: Porton Down, 15-Jun-73 (J.Walker); West Winterslow, 24-Jul-01 (NB)
SU24: Beacon Hill (SPTAE), 3-Jun - 12-Aug-08 (Boyce & Wright, 2009); Bulford Ranges (SPTAE), 3-Jun - 12-Aug-08 (Boyce & Wright, 2009); Driver Training Area (SPTAE), 17-May-02 (PJH) (Hodge, 2002); Dunch Hill (SPTAE), 3-Jun - 12-Aug-08 (Boyce & Wright, 2009); Sidbury Hill (SPTAE), 29-May, 13 to 17-Jun-97 (MGT)
SU25: Sidbury Hill (SPTAE), 28-May, 5-97 (MGT)
Resident. Widespread and frequent in a wide range of habitats.

**11-Spot Ladybird**
***Coccinella undecimpunctata*** Linnaeus, 1758
7 SU06: Roundway Camp, Jul-40 (FDB) (Buck, 1956)
SU09: Somerford Keynes, 4-Jun-04 (BLe)
SU16: Marlborough area, 1869-1938 (Meyrick, 1938)
8 ST93: Stockton area, 28-Sep-95 (PJH) (Hodge, 1995)
SU01: Martin Down NNR, 21-Aug-89 (RGB)
SU03: Grovely Wood, 25-Oct-71 (DRN)
SU04: Shrewton, 2-Aug-83 (DRN)
SU13: Cockey Down, 2-Sep-02 (PM); Lower Woodford, 28-Jul-78 (DRN); Salisbury, 24-Jul-92 (PM)
SU23: Winterslow, 19-Aug-90 (PM)
Resident. Widespread; on vegetation.

**ADALIA Mulsant, 1846**
**2-Spot Ladybird**
***Adalia bipunctata*** (Linnaeus, 1758)
7 ST97: North Draycot Park, 22-May-00 (MDD)
SU06: Calstone and Cherhil, 2-Jul-00 (BG); Calstone, R. Mard, 9-Jun-99 (BG) Tam Hill, 1-Oct-00 (BG)
SU16: Lockeridge, 1994 (JEO); Marlborough area, Jun-1873 (EWJ) (Meyrick, 1874); 10 Feb 1879 (GTKM) (Meyrick, 1880); 1873-1938 (Meyrick, 1938)
SU18: Swindon, 7-May-01 (JKG det MDD)
SU19: Hannington, 26, 31-Jul-95, 26-May-97(MEN)
SU26: Savernake Forest, 28-Aprl to 18-May-94 (EN det SAW)
SU29: Watchfield, 15-May-99 (MEN)

8 ST84: Longleat, 25-May-05 (DJG)
ST91: Cranborne Chase, 24-Apr-97 (MAS)
ST92: Fovant, 25-Apr-95 (MDD)
ST93: Stockton area, 28-Sep-95 (PJH) (Hodge, 1995)
ST94: Ladywell (SPTAW), 15-May-02 (PJH) (Hodge, 2002); Salisbury Plain, 30-Dec-90, 27-Aug-93 (SP)
ST95: West Lavington, Apr-Sep-41 (ITH) (DFL, 1948)
SU03: Grovely Wood, 25-Oct-71, 4-Apr-77 (DRN)
SU04: Fox Covert (SPTAW), 3-Jun - 12-Aug-08 (Boyce & Wright, 2009)
SU12: Petersfinger, 7-Aug-88 (DRN)
SU13: Cockey Down, 7-Aug-92 (PM); Great Durnford, 18-Aug-72, 17-Jul-73, 27-Jul-74, 27-Jul-77, 3-Aug-83 (DRN); Lower Woodford, 28-Jul-78 (DRN); Middle Woodford, 27-Jul-77; Old Sarum, 31-May-97 (MAS); Salisbury, 18-Jun-73 (PM); 7-Apr-74 (DRN); 1-May-00 (MDD) (Darby, 2001)
SU14: Haxton (SPTAE), 3-Jun - 12-Aug-08 (Boyce & Wright, 2009); Netheravon, 29-Jul-83 (DRN); Stonehenge, 22-Apr-03 (PJH) (Hodge, 2003)
SU15: Beach's Plantation (SPTAE), 1-May-05 (MGT)
SU16: Jones's Mill, 1-Jan-9731-Dec-97 (MAS)
SU21: Cloven Hill, 19-Aug-88 (DRN); Hamptworth Estate, 16-Apr-74 (DRN)
SU22: Bentley Wood, 23-Jul-90 (PM)
SU23: Porton Down, 26-Apr-91, 24, 25-Apr-93 (PTW det MDD); 3-Jul-99 (PJH) (Hodge, 2002)
SU24: Beacon Hill (SPTAE), 3-Jun - 12-Aug-08 (Boyce & Wright, 2009)
SU25: Sidbury Hill (SPTAE), 28-May-97 (MGT)

Resident. Widespread in a wide range of situations, particularly urban and suburban.

**10-Spot Ladybird**
***Adalia decempunctata*** (Linnaeus, 1758)

7 SU09: Upper Waterhay, 13-Jun-04 (JKG det MDD)
SU16: Marlborough area, 1869-1938
SU17: Avebury, 28-Dec-05, 4, 13-Sep-06 (DB)
SU26: Savernake Forest, 31-May to 23-Jun, 5 to 26-Oct-95 (EN det MDD); 22-Jun-90 (APFw)
8 ST84: Longleat, 31-Mar-85 (AGD)
ST85: Green Lane Wood, 8-Jul-92 (PM)
ST91: Cranborne Chase, 16 to 25-Jun-80 (NCC), 22-Jun-99 (MDD)
ST92: Wardour, 5-Jun-96 (MDD)
ST93: Great Ridge Wood, 14-Jun-96 (MDD) Stockton, 28-Sep-95 (PJH) (Hodge, 1995)
ST94: Salisbury Plain, 30-Dec-90 (SP)
ST95: West Lavington, 11-May-43 (ITH) (DFL, 1948)
SU02: Middleton Down, 20-May-01 (BL)
SU03: Grovely Wood, 13-Aug-70, 11, 13-Aug-71, 24-Aug-72, 1-Aug-73 (DRN); Wylye Down NNR, 20-Sep-05 (MDD) (Darby, 2009)
SU04: Shrewton Folly, 29-May-94 (SP)
SU13: Cockey Down, 1-May-90 (PM); Salisbury, Apr-00 (MAS) (Darby, 2001); Stratford-Sub-Castle, 7-Aug-71 (DRN)
SU14: Stonehenge, 22-Apr-03 (PJH) (Hodge, 2003)
SU16: Jones's Mill, 19-Jun-01 (APF) (Foster, 2002); Woodborough, 4-Aug-83 (DRN)
SU22: Bentley Wood, 11-Aug-01 (MDD); Glazier's Copse, 16-Aug-88 (DRN); Langley Wood, 3 to 24-Mar-05 (MDD) (Darby, 2005)
SU23: Porton Down, 24-Apr, 3-May-93 (PTW det MDD) West Winterslow, 9-Jul-01 (NB)

Resident. Widespread in woodlands and hedgerows on many trees and shrubs.

**HARMONIA Mulsant, 1846**
**Harlequin Ladybird**
***Harmonia axyridis*** (Pallas, 1773)

7 SU17: Avebury, 16-Jun-08 (DB)
8 ST92: Sutton Mandeville, 8-Dec-06 (MDD)
ST93: Hindon, 8-Sep-06 (ColH)
SU04: Tilshead, 24-Jan-07 (TJ)
SU06: Devizes, 15-Oct-07 (RL)
SU12: Harnham, 12-Oct-07 (JLE)
SU23: Allington, 24-Oct-06 (Mrspp)

Resident. Widespread; recent introduction reported from in houses in particular. See more references to Wiltshire on www.harlequin-survey.org

**Cream-streaked Ladybird**
***Harmonia quadripunctata*** (Pontoppidan, 1763)

8 SU13: Little Durnford Down, 18-Mar-91 (PM); Salisbury, 5-Jun-06 (PM)

Resident. Very local, on Scots pine. First recorded in GB in the 1937 and now reported to be spreading rapidly.

## Subfamily EPILACHNINAE

**SUBCOCINELLA Guérin-Méneville, 1844**
***Subcoccinella vigintiquattuorpunctata*** (Linnaeus, 1758)

7 ST98: Conygre Mead, Malmesbury, 5-Apr-02 (JKG det MDD)
SU09: Blakehill Farm, 11-May-01 (JKG det MDD)
SU16: Marlborough area, 1869-1938
SU18: Swindon, 7-May-01 (JKG det MDD)
8 ST82: Oysters Coppice, 15-Jun-73 (SP)
ST85: Green Lane Wood, 31-May-04 (KNAA)
ST92: Wardour, 5-Jun-96 (MDD)
ST94: Wadman's Coppice (SPTAW), 3-Jun - 12-Aug-08 (Boyce & Wright, 2009)
SU01: Martin Down NNR, 21-Aug-89 (RGB); 1999

(EN, 1999)
SU02: Chickengrove Bottom, Apr-92 (PM); 17-Jul-95 (MDD); Middleton Down, 26-May-92 (PM), 8-May-95 (PM)
SU04: Orcheston Down (SPTA), 22-Jun-02 (PJH) (Hodge, 2002)
SU14: Stonehenge, 22-Apr-03 (PJH) (Hodge, 2003); Wexland Hanging (SPTAC), 16-May-02 (PJH) (Hodge, 2002)
SU15: Beach's Plantation (SPTAE), 1-May-05 (MGT); Everleigh (SPTAE), 18-May, 24-Jun-02 (PJH) (Hodge, 2002)
SU16: Jones's Mill, 19-Jun, 2-Jul-01 (APF) (Foster, 2002)
SU21: Hamptworth Estate, 29-Mar-78 (DRN)
SU23: Porton Down, 3-Jul-99 (PJH) (Hodge, 2002)
SU24: Driver Training Area (SPTAE), 24-Jun-02 (PJH) (Hodge, 2002)

Resident. Widespread; in grasslands, meadowlands, road and woodland verges.

## Family CORYLOPHIDAE
### Subfamily CORYLOPHINAE

**SERICODERUS Stephens, 1829**
***Sericoderus lateralis*** (Gyllenhal, 1827)
8 ST92: Sutton Mandeville, 14-Aug-07 (MDD), 25-Sep-07 (MDD); Wardour, 5-Jun-96 (MAS)
SU03: Grovely Wood, 2-Aug-79, 8-Aug-80 (DRN)
SU14: Netheravon, 30-Jul-85 (DRN); Wexland Hanging (SPTAC), 22-Jun-02 (PJH) (Hodge, 2002)
SU21: Hamptworth Estate, 29-Mar-78 (DRN)

Resident. Local; in piles of mouldt grass and straw A second species of *Sericoderus , brevicornis* has recently been added to the British list by Galsworthy and Booth , 2007 so that these records require checking.

### Subfamily ORTHOPERINAE

**ORTHOPERUS Stephens, 1829**
***?Orthoperus atomus*** (Gyllenhal, 1808)
7 SU16: Marlborough area, 1906 (Meyrick, 1938)

Resident. Very local; in vegetable refuse. Probably a misidentification. A single specimen under this name in the MCC has been det. by SB as *O.nigrescens.*

***Orthoperus corticalis*** (Redtenbacher, 1849)
**SQS-4**
7 SU26: Savernake Forest, 5-Aug-86 (DRN)
8 ST93: Fonthill, 26 Jun-02 (MDD det SB)
SU03: Grovely Wood, 8-Apr-72 (DRN)
SU16: Jones's Mill, 21-Jun-97 (MDD det SB); Langley Wood NNR, 6-Jun-02 (MDD det SB)

Resident. Very local. Beating oaks, under beech bark and in fungus.

***Orthoperus nigrescens*** Stephens, 1829
**Nb**
8 ST92: Sutton Mandeville, 13-Aug-95, 6-Jul-07 (MDD det SB)
SU02: Wilton, 6-Sep-95 (MDD det SB)
SU13: Great Durnford, 18-Aug-72 (DRN); Salisbury, May-00; Salisbury, Bemerton Folly, 7-Mar - 15-Jun-02 (MDD det SB) (Darby, 2002d)
SU16: Jones's Mill, 21-Jun-97 (MDD det SB)
SU21: Burnt Ground Wood, 15-May-03 (MDD)
SU22: Bentley Wood, 11-Aug-01 (MDD det SB); Langley Wood NNR, 6-Jun-02 (MDD det SB)

Resident. Local; beating oaks and sweeping *Clematis*, etc. by the R. Avon.

## Family LATRIDIIDAE
### Subfamily LATRIDIINAE

**STEPHOSTETHUS LeConte, 1878**
***Stephostethus alternans*** (Mannerheim, 1844)
**SQS-4**
8 SU02: Knighton Wood, Apr-07, May-08 (AJWA) (Allen, 2008)

Resident. Very local; in a flight intercept trap.

***Stephostethus angusticollis*** (Gyllenhal, 1827)
7 SU16: Marlborough area, 1869-1938 (Meyrick, 1938)

Resident. Very local, in vegetable refuse. There is a single specimen under this name in the MCC, inscribed under the card *3205.*

***Stephostethus lardarius*** (De Geer, 1775)
7 SU16: Marlborough area, Combe Farm, 1873-1938 (Meyrick, 1938)
8 ST92: Fovant, 25-Apr-95 (MDD)
SU03: Grovely Wood, 27-May-74 (DRN)
SU04: Shrewton, 6-Aug-82 (DRN)
SU13: Great Durnford, 16-Aug-69 (DRN)
SU16: Woodborough, 4-Aug-83 (DRN)
SU22: Blackmoor Copse, 17-Mar-02 (APF) (Foster, 2004); West Grimstead, 12-Aug-87 (DRN)

Resident. Local; in vegetable refuse.

**CARTODERE Thomson, C.G, 1859**
ARIDIUS Motschulsky, 1866
***Cartodere bifasciata*** (Reitter, 1877)
7 SU26: Savernake Forest, 16-Jun to 28-Jul-93 (EN det SAW); 4 to 24-Aug-95 (EN det MDD)
8 ST84: Longleat, 24-Sep-89 (AGD)

ST93: Stockton Wood and Do, 28-Sep-95 (PJH) (Hodge, 2002)
SU01: Martin Down NNR, 21-Aug-89 (RGB)
SU02: Knighton Wood, Apr-07, May-08 (AJWA)
SU03: Grovely Wood, 1-Aug-73, 8, 16-Apr-74 (DRN)
SU05: Market Lavington, 2-Aug-83 (DRN)
SU06: Horton, 5-Aug-83 (DRN)
SU13: Great Durnford, 18-Aug-72, 27-Jul-77 (DRN); Lower Woodford, 5-Aug-82 (DRN); Middle Woodford, 27-Jul-77 (DRN); Salisbury, 15-Aug-88 (DRN)
SU14: Netheravon, 29-Jul-83 (DRN); Stonehenge, 22-Apr-03 (PJH) (Hodge, 2003)
SU15: Fittleton, 29-Jul-83 (DRN); Widdington, 5-Aug-83 (DRN)
SU21: Hamptworth Estate, 29-Mar-78 (DRN)
SU22: Blackmoor Copse, 3-Aug-03 (APF) (Foster, 2004); Glazier's Copse, 16-Aug-88 (DRN); Holmere Common, 19-Aug-88 (DRN); Redlynch, 4-Aug-82 (DRN)
SU26: Bedwyn Brail, 20-Sep-04 (MDD)
Resident. Widespread; on vegetation, in litter and rotting vegetation, straw, fungi, etc..

***Cartodere nodifer*** (Westwood, 1839)
7 ST99: Crudwell, 19-Jun-73 (APF)
SU26: Savernake Forest, 27-May to 21-Sep-93, 18-My to 11-Oct-94 (EN det SAW)
8 ST91: Cranborne Chase, 30-Jun-99 (MDD)
SU02: Knighton Wood, Sep-06, Apr, Aug, Sep-07 (AJWA)
SU03: Grovely Wood, 6-Aug-70, 5-Aug-71, 24-Aug-72, 8-Apr, 30-Oct-75, 4-Aug-78, 2-Aug-79 (DRN), 3-Dec-00, 3-Aug-04 (MDD)
SU13: Lower Woodford, 28-Jul-78 (DRN); Salisbury, Bemerton Folly, 7-Mar - 15-Jun-02 (MDD) (Darby, 2002d)
SU16: Jones's Mill, Sep-97 (HK det MDD); 17-Oct-01 (APF) (Foster, 2002)
SU21: Hamptworth Estate, 3-Apr-78, 6-Aug-79 (DRN); Hamptworth Common, 24, 30-Mar-03 (MDD)
SU22: Bentley Wood, 31-May-96, 11-Aug-01 (MDD); Blackmoor Copse, 17-Mar-02, 3-Aug-03 (APF) (Foster, 2004); Holmere Common, 19-Aug-88 (DRN); Langley Wood, 29-May-03, 25-Jun to 6-Aug-04 (MDD) (Darby, 2005)
SU23: Porton Down, 1970-2001 (Anon, 2001)
Resident. Widespread; in rotting vegetation, straw, under bark, on trees, etc..

**LATRIDIUS Herbst, 1793**
***Latridius minutus*** (Linnaeus, 1767)
7 SU16: Marlborough area, 24-Mar-1879 (WFB) (Meyrick, 1880); 1900-1938 (Meyrick, 1938)
Resident. Very local; 'in almost any situation'. *minutus* was recognised in 1972 as representing three species (Tozer, 1972) so that this record must be considered indeterminate. Two specimens in MCC, one labelled *Marlborough 31/7/02.*

***Latridius porcatus*** Herbst, 1793
8 SU03: Grovely Wood, 3-Aug-04 (MDD)
Resident. Very local; beating oaks. Formerly *L. anthracinus.*

**ENICMUS Thomson, C.G, 1859**
***Enicmus brevicornis*** (Mannerheim, 1844)
**Notable, IEC-3, SQS-8**
8 ST92: Sutton Mandeville, 1-Sep-96 (JAO)
ST93: Great Ridge Wood, 29-Oct-75 (DRN)
SU03: Grovely Wood, 23-Aug-72, 23-Jul-73, 30-Oct-75, 4-Aug-78, 1-Aug-88 (DRN); Little Langford Down, 14-Jul-04 (MDD)
SU13: Little Durnford, 14-Jul-04 (MDD)
Resident. Local; under beech bark and on an indeterminate mouldy log.

***Enicmus histrio*** Joy & Tomlin, 1910
8 ST92: Sutton Mandeville, 24-Oct-95 (MDD)
SU02: Knighton Wood, Sep-06 (AJWA)
SU03: Grovely Wood, 4-Aug-87 (DRN)
SU13: Boscombe Down, 8, 9-Oct-94 (MDD); Salisbury, Apr-00 (MAS) (Darby, 2001)
SU21: Hamptworth Estate, 28-Oct-75, 29-Mar-78 (DRN)
SU22: Langley Wood NNR, 14-May to 3-Jun-04 (MDD) (Darby, 2005)
Resident. Very local; in leaf litter, in straw, on a barn and in a compost heap.

***Enicmus testaceus*** (Stephens, 1830)
**SQS-2**
7 SU26: Savernake Forest, 1-Jul (EN det SAW); 4 to 24-Aug-95 (EN det MDD)
8 ST93: Great Ridge Wood, 29-Oct-75 (DRN)
SU03: Grovely Wood, 4-Aug-80 (DRN)
SU21: Hamptworth Estate, 29-Mar-78 (DRN)
Resident. Local; on fungus and in litter.

***Enicmus transversus*** (Olivier, 1790)
7 SU06: Roundway Camp, Jul-40 (FDB) (Buck, 1956)
SU16: Marlborough area, 1869-1938 (Meyrick, 1938)
SU26: Savernake Forest. NE, 22-Aug-90 (AGD)
8 SU01: Martin Down NNR, 21-Aug-89 (RGB)
SU02: Knighton Wood, Apr-07 (AJWA)
SU03: Grovely Wood, 6-Apr-71, 16-Aug-71, 21, 22, 24-Aug-72, 2, 3-Aug-79, 4-Aug-87 (DRN); Wylye Down NNR, 10-Jul to 3-Aug-07 (MDD) (Darby, 2009)
SU13: Great Durnford, 18-Aug-72 (DRN); Idmiston, 5-Aug-82 (DRN); Salisbury, 2-May-00 (MDD) (Darby, 2001); Stratford sub Castle, 7-Aug-71 (DRN)

SU14: Wexland Hanging (SPTAC) 22-Jun-02 (PJH) (Hodge, 2002)
SU21: Hamptworth Estate, 29-Mar-79 (DRN)
SU22: Blackmoor Copse, 17-Mar-02 (APF) (Foster, 2004); Langley Wood NNR, 16-Jul to 6-Aug-04 (MDD) (Darby, 2005); West Grimstead, 12-Aug-87 (DRN)
SU23: Porton Down, 1970-2001 (Anon, 2001)
SU26: Bedwyn Brail, 2-Jun-03 (MDD)
Resident. Widespread; in grass piles, litter, fungus and on vegetation, sometimes near water.

**DIENERELLA Reitter**
CARTODEREMA Reitter, 1911
***Dienerella clathrata*** (Mannerheim, 1844)
8 SU21: Burnt Ground Wood, 21-Jul-73 (DRN) (Krell *et al.*, 2005)
SU22: Langley Wood NNR, 3 to 25-Jun-04 (MDD) (Darby, 2005)
SU26: Cobham Frith, 24-Jul-82 (DRN) (Krell *et al*, 2005)
Resident. Very local; in grass pile, woodland litter and under bark.

***Dienerella ruficollis*** (Marsham, 1802)
7 SU16: Marlborough area, Jun-1873 (EWJ) (Meyrick, 1874); 1869-1938 (Meyrick, 1938)
8 SU03: Dinton Park, 15-Aug-08 (KNAA)
SU04: Shrewton, 28-Jul-82 (DRN)
SU22: Redlynch, 4-Aug-82 (DRN)
Resident. Very local; in vegetable refuse , fungi and sileage grass.

## Subfamily CORTICARIINAE

**CORTICARIA Marsham, 1802**
***Corticaria crenulata*** (Gyllenhal, 1827)
7 SU16: Marlborough area, Jun-1873 (EWJ) (Meyrick, 1874); 1869-1938 (Meyrick, 1938)
Resident. Very local; in vegetable refuse, etc..

***Corticaria elongata*** (Gyllenhal, 1827)
7 SU06: Roundway Camp, Jul-40 (FDB) (Buck, 1956)
SU16: Marlborough area, 17-Feb-1879 (WFB) (Meyrick, 1880); 1869-1938 (Meyrick, 1938)
8 SU03: Grovely Wood, 2, 3-Aug-79, 8-Aug-80 (DRN)
SU21: Hamptworth Estate, 29-Mar, 3-Apr-78, 6-Aug-79 (DRN)
SU22: Redlynch, 4-Aug-82 (DRN)
Resident. Very local; in grass piles and old straw.

***Corticaria impressa*** (Olivier, 1790)
8 SU03: Grovely Wood, 2-Aug-79 (DRN)
SU04: Shrewton, 28-Jul-82 (DRN)
SU05: Market Lavington, 2-Aug-83 (DRN)
SU15: Fittleton, 29-Jul-83 (DRN); Widdington, 5-Aug-83 (DRN)
SU16: Woodborough, 4-Aug-83 (DRN)
Resident. Very local; in bracket fungi, mouldy grass and on vegetation.

***Corticaria punctulata*** Marsham, 1802
7 SU16: Marlborough area, 1869-1938 (Meyrick, 1938)
SU26: Savernake Forest, 4 to 24-Aug-95 (EN det MDD)
Resident. Very local; in vegetable refuse, etc..

***Corticaria umbilicata*** (Beck, 1817)
8 SU21: Landford Bog, 24-Aug-01 (APF) (Foster, 2001)
Resident. Very local; [in moss in woods].

**CORTICARINA Reitter, 1880**
***Corticarina minuta*** (Fabricius, 1792)
7 ST96: Rowde, 5-Aug-83 (DRN)
SU06: Roundway Camp, Jul-40 (FDB) (Buck, 1956)
SU16: Marlborough area, 1869-1938 (Meyrick, 1938)
SU17: Hallam, 6-Aug-83 (DRN)
8 ST93: Great Ridge Wood, 29-Oct-75 (DRN)
ST94: Bowls Barrow (SPTAW), 16-May-02 (PJH) (Hodge, 2002)
SU01: Martin Down NNR, 21-Aug-89 (RGB)
SU02: Knighton Wood, Sep-06, Apr-07 (AJWA)
SU03: Grovely Wood, 11-Aug-70, 9-Aug, 25-Oct-7, 3-Apr-74, 3-Aug-79, 1-Aug-80 (DRN); Wylye Down NNR, 4-May to 10-Jul-07 (MDD) (Darby, 2008)
SU04: Salisbury Plain, 16-May-02 (PJH) (Hodge, 2002)
SU13: Boscombe Down, 8, 9-Oct-94 (MDD); Lower Woodford, 27, 30-Jul, 5-Aug-82 (DRN); Salisbury, 10-May-00 (MAS det RT)
SU14: Netheravon, 30-Jul-85; Stonehenge, 22-Apr-03 (PJH) (Hodge, 2003)
SU15: Fittleton, 29-Jul-83 (DRN); Widdington, 5-Aug-83 (DRN)
SU16: Woodborough, 4-Aug-83 (DRN)
SU21: Hamptworth Estate, 6-Apr-77, 29-Mar-78, 6-Aug-79 (DRN); Landford Bog, 24-Jun-01 (APF) (Foster, 2001)
SU22: Glaziers Copse, 16-Aug-88 (DRN); Redlynch, 4-Aug-82 (DRN)
Resident. Widespread; in cut grass, fungi, moss, straw, etc.. Previously *C. fuscula.*

***Corticarina similata*** (Gyllenhal, 1827)
8 SU22: Bentley Wood, 3-Sep-97 (WANHS det MDD)
Resident. Very local; on pines.

**CORTINICARA Johnson, 1975**
***Cortinicara gibbosa*** (Herbst, 1793)

7 ST97: North Draycot Park, 22-May-00 (MDD)
SU16: Marlborough area, 1869-1938
SU26: Savernake Forest, 20-Apr to 25-May-93, 30-Aug to 20-Sep-94 (EN det SAW)
8 ST92: Sutton Mandeville, 29-Jan-98 (MDD)
ST93: Great Ridge Wood; 29-Jul-82 (DRN); Stockton Wood, 28-Sep-95 (PJH) (Hodge, 1995)
SU01: Martin Down NNR, 21-Aug-89 (RGB)
SU02: Knighton Wood, Oct-06, Mar, Nov-07, Jul-08 (AJWA)
SU03: Grovely Wood, 29-Oct-70 (DRN); 17-Feb-98, 3-Dec-00 (MDD)
SU04: Parsonage Down NNR, 2-Jul-08 (MDD); Fox Covert, 25-Jun-02 (PJH) (Hodge, 2002)
SU13: Boscombe Down, 8, 9-Oct-94 (MDD); Salisbury, 10-May-00 (MAS det RT); Stratford sub Castle, 10-Aug-70 (DRN)
SU14: Stonehenge, 22-Apr-03 (PJH) (Hodge, 2003)
SU16: Woodborough, 4-Aug-83 (DRN)
SU22: Bentley Wood, 27-May-97 (MDD); 3-Sep-97 (WANHS det MDD)
SU23: Porton Down, 1970-2001 (Anon, 2001); Thorny Down woods, 16-Oct-00 (MDD)
Resident. Widespread and frequent at roots of vegetation, under stone, in tussocks, etc..

## Family MYCETOPHAGIDAE

**TRIPHYLLUS Dejean, 1821**
***Triphyllus bicolor*** (Fabricius, 1777)
**IEC-2, SQS-4**
7 SU26: Savernake Forest, 20-Aug-70 (DRN)
8 SU03: Grovely Wood, 11-Aug-71, 23-Aug-72 (DRN)
Resident. Very local; on bracket fungi in woodland

**LITARGUS Erichson, 1846**
***Litargus connexus*** (Fourcroy, 1785)
**SQS-2**
7 ST86: Neston Park, 3-Jun-04 (KNAA)
ST96: Spye Park, 31-May-04 (KNAA)
SU16: Marlborough area, Jun-1873 (EWJ) (Meyrick, 1874); 1869-1938 (Meyrick, 1938)
SU26: Savernake Forest, 6-Jun-73, 13-Jun-85 (APF)
8 ST91: Cranborne Chase, 21-May-02 (MDD)
ST93: Great Ridge Wood, 29-Oct-75 (DRN)
SU03: Grovely Wood, 6-Apr, 5, 9-Aug-71, 11-Apr, 23-Aug-72, 30-Oct-75, 4-Aug-78, 1-Aug-80 (DRN)
SU21: Burnt Ground Wood, 31-Oct-73 (DRN); Hamptworth Estate, 19-Apr-74 (DRN)
SU22: Bentley Wood, 12-Feb-01 (NB det MDD); 11-Aug-01 (MDD)
SU23: Porton Down, 15-Oct-95 (MAS)
SU26: Bedwyn Brail, 2-Jun-03 (MDD)
Resident. Widespread; under bark, especially beech, and in fungus in woodland.

**MYCETOPHAGUS Hellwig, 1792**
***Mycetophagus atomarius*** (Fabricius, 1787)
**IEC-3, SQS-2**
7 SU16: Marlborough area, Jun-1873 (EWJ) (Meyrick, 1874); 1869-1938 (Meyrick, 1938)
8 ST84: Longleat, 12-Apr-02 (MB)
ST92: Sutton Mandeville, 3-May-92 (MDD); Wardour, 5-Jun-96 (MDD)
ST93: Great Ridge Wood, 29-Jul-85 (DRN)
SU03: Grovely Wood, 10-Apr, 4, 6-Aug-70, 25-Oct-71, 16-Apr-74, 1-Aug-80, 16-Apr-84 (DRN)
SU13: Little Durnford Down, 20-May-03 (MDD)
SU15: Everleigh (SPTAE), 18-May-02 (PJH) (Hodge, 2002)
SU26: Bedwyn Brail, 2-Jun-03 (MDD)
Resident. Widespread; in fungi and under bark, especially of beech.

***Mycetophagus multipunctatus*** Fabricius, 1792
**SQS-2**
7 SU16: Marlborough area, 1869-1938
SU26: Savernake Forest, 1970-2001 (Anon)
8 SU03: Grovely Wood, 6, 11, 12-Apr-72 (DRN)
SU13: Salisbury, Apr-00 (MAS); 24-May-00 (MDD) (Darby, 2001); Stratford sub Castle, 5-Apr-72 (DRN)
SU14: Netheravon, 29-Jul-83 (DRN)
SU16: Jones's Mill, 21-Jun-97 (MDD)
Resident. Local; in fungi on trees.

***Mycetophagus quadriguttatus*** Müller, P.W.J, 1821
**Na, IEC-2, SQS-16**
8 SU13: Little Durnford Down, 20-May-03 (MDD)
Resident. Very local. A synanthropic species normally found in granaries and foodstores, but occasionally, like this, in fungi on trees.

***Mycetophagus quadripustulatus*** (Linnaeus, 1761)
**SQS-2**
7 ST99: Flisteridge Wood, 2-May-82 (KNAA)
SU16: Marlborough area, Jun-1873 (EWJ) (Meyrick, 1874); 1869-1938 (Meyrick, 1938)
SU26: Savernake Forest, 2-Apr-70 (DRN)
8 ST73: Tucking Mill Hanging, 11-Feb-84 (KNAA)
ST84: Longleat, 29-Aug-99 (MB); Southleigh Wood, 24-Apr-87 (RSK)
ST85: Bullen Hill Farm, 1982-2002 (EGS det MDD); Green Lane Wood, 31-May-04 (KNAA)
ST93: Great Ridge Wood, 19-Apr-84 (DRN)
SU03: 4-Aug-70, 3, 5, 11-Aug, 25-Oct-71, n12-Apr-72 (DRN)
SU13: Stratford sub Castle, 11-Apr-71, 5-Apr-72 (DRN)
SU14: Netheravon, 29-Jul-83 (DRN)

SU15: Widdington, 5-Aug-83 (DRN)
SU16: Jones's Mill, 21-Jun-97 (MDD); Sep-97 (HK det MDD); 19-Jun-01 (APF) (Foster, 2002)
Resident. Local; on fungus.

**TYPHAEA Stephens, 1829**
***Typhaea stercorea*** (Linnaeus, 1758)
7 SU06: Roundway Camp, Jul-40 (FDB) (Buck, 1956)
8 SU13: Great Durnford, 16-Aug-69 (DRN)
SU21: Hamptworth Estate, 6-Aug-79 (DRN)
SU22: Redlynch, 4-Aug-82 (DRN)
Resident. Very local; in cut grass and on vegetation.

## Family CIIDAE
### Subfamily CIINAE
#### Tribe OROPHIINI

**OCTOTEMNUS Mellié, 1847**
***Octotemnus glabriculus*** (Gyllenhal, 1827)
7 SU16: Marlborough area, Jun-1873 (EWJ) (Meyrick, 1874); 5-Jun-1879 (WFB) (Meyrick, 1880); Marlborough area, 1873-1938 (Meyrick, 1938)
8 ST84: Longleat, 2-Jan-99, 5-Sep-00(GO)
ST92: Fovant, 25-Apr-95 (MDD)
ST93: Fonthill, 26-Jun-02 (MDD); Great Ridge Wood, 29-Jul-85 (DRN); 28-Apr-00 (MDD)
SU02: Knighton Wood, Apr-07, Jul-08 (AJWA)
SU03: Dinton Park, 15-Aug-01 (KNAA); Grovely Wood, 10-Apr, 29-Oct-70 (DRN); 17-Feb-98 (MDD)
SU13: Salisbury, Bemerton Folly, 7-Mar - 15-Jun-02 (MDD) (Darby, 2002d)
SU16: Jones's Mill, 1997 (MAS); 21 Jun 97 (MDD); Sep-97 (HK det MDD)
SU21: Landford Bog, 9-Jun-01 (APF) (Foster, 2001); Peaked Wood, 15-Apr-74 (DRN)
SU22: Bentley Wood, 10-Oct-01 (MDD), 12-Feb-01 (NB); Langley Wood NNR, 6-Jun-02, 29-May-03 (MDD)
SU26: Bedwyn Brail, 20-Sep-04 (MDD)
Resident. Widespread; in fungus, particularly *Trametes* .

#### Tribe CIINI

**SULCACIS Dury, 1917**
***Sulcacis affinis*** (Gyllenhal, 1827)
**SQS-2**
8 ST84: Longleat, 2-Jan-99 (GO)
ST93: Great Ridge Wood, 29-Jul-85 (DRN)
SU03: Grovely Wood, 5-Apr-71 (DRN)
SU12: Whaddon Common, 22-Jul-73 (DRN)
SU21: Peaked Wood, 15-Apr-74 (DRN)
SU26: Bedwyn Brail, 20-Sep-04 (MDD)
Resident. Local; in fungus on trees.

**STRIGOCIS Dury, 1917**
***Strigocis bicornis*** (Mellié, 1848)
**Nb, SQS-8**
8 ST84: Longleat, 5-Sep-00 (GO)
Resident. Very local; [in fungus].

**ORTHOCIS Casey, 1898**
***Orthocis alni*** (Gyllenhal, 1813)
**SQS-2**
8 ST92: Sutton Mandeville, 20-Jan-99 (MDD)
ST93: Great Ridge Wood, 29-Jul-85 (DRN)
SU22: Blackmoor Copse, 17-Mar-02 (APF) (Foster, 2004)
Resident. Very local; in fungi on rotten wood.

**CIS Latreille, 1796**
***Cis bidentatus*** (Olivier, 1790)
**SQS-2**
8 ST93: Great Ridge Wood, 4-Oct-04 (MDD)
SU03: Grovely Wood, 3-Apr-75 (DRN)
SU13: Great Durnford, 6-Apr-70 (DRN)
Resident. Very local; on fungus in woodland

***Cis bilamellatus*** Wood, 1884
8 ST84: Longleat, 2-Jan-99 (GO)
ST85: Biss Wood, 19-Jun-73 (GO)
ST91: Cranborne Chase, 16 to 25-Jun-80 (NCC)
ST93: Great Ridge Wood, 4-Oct-04 (MDD)
SU03: Grovely Wood, 5,9-Aug-71, 6, 10-Apr, 21-Aug-72, 16-Apr-74 (DRN)
SU08: Ravensroost Wood, Oct-96 (KNAA)
SU13: Great Durnford, 6-Apr-70 (DRN); Salisbury, Apr-00 (MAS) (Darby, 2001)
SU15: Widdington, 5-Aug-83 (DRN)
SU22: Bentley Wood, 10-Oct-01 (MDD); Blackmoor Copse, 9-Jun-01, 17-Mar-02 (APF) (Foster, 2004); Langley Wood NNR, 12-Apr-74 (DRN; 6-Jun-02 (MDD)
SU26: Bedwyn Brail, 2-Jun-03 (MDD)
Resident. Widespread; in fungi, particularly *Piptoporus* on birch. Recently introduced species from Australia.

***Cis boleti*** (Scopoli, 1763)
**SQS-1**
7 SU16: Marlborough area, Jun-1873 (EWJ) (Meyrick, 1874); 1869-1938 (Meyrick, 1938)
SU26: Savernake Forest, 16-Jun to 20-Sep-93 (EN det SAW); Tottenham Deer Park, 19-Dec-87 (KNAA)
SU27: Rabley Wood, 1870-1873 (Meyrick, 1874)
8 ST73: Stourhead Woods, 11-Feb-84 (KNAA)
ST84: Longleat, 5-Sep-00 (GO)
ST91: Cranborne Chase, 16 to 25-Jun-80 (NCC); 30-Jun-99, 21-May-02 (MDD)
ST93: Great Ridge Wood, 29-Jul-85 (DRN); 4-Oct-04 (MDD)
SU03: Grovely Wood, 5-Apr-71, 8-Apr-74, 1-Aug-80, 5-Aug-81, 11-Aug-87 (DRN); Wylye

Down NNR, 4-May to 5-Jun-07 (MDD) (Darby, 2009)
SU13: Salisbury, 24-May-00 (MDD) (Darby, 2001)
SU16: Jones's Mill, 21-Jun-97 (MDD), 17-Oct-01 (APF) (Foster, 2002)
SU21: Burnt Ground Wood, 31-Oct-73 (DRN); Landford Bog, 25-May-98 (MDD); 9-Jun-01 (APF) (Foster, 2001)
SU22: Bentley Wood, 12-Feb-01 (NB det MDD); Blackmoor Copse, 24, 31-Aug-01 (APF) (Foster, 2004)
Resident. Widespread; on fungi in woodland, particularly on beech.

***Cis fagi*** Waltl, 1839
**SQS-2**
8 SU26: Bedwyn Brail, 20-Sep-04 (MDD)
Resident. Very local; in a bracket fungus.

***Cis festivus*** (Panzer, 1793)
**Nb, SQS-2**
8 SU02: Gardiner Forest, 1-Sep-72 (DRN)
SU03: Grovely Wood, 4-Aug-80 (DRN)
Resident. Very local; in fungoid cypress bark and in *Stereum* fungi on hazel.

***Cis hispidus*** (Paykull, 1798)
**SQS-4**
8 ST84: Longleat, 5-Sep-00 (GO)
Resident. Very local; [in bracket fungi].

***Cis micans*** (Fabricius, 1792)
**SQS-4**
7 SU16: Marlborough area, Jun-1873 (EWJ) (Meyrick, 1874); 1869-1938 (Meyrick, 1938)
8 SU21: Franchises Wood, 3-May-95 (MAS)
Resident. Very local; in fungi on wood.

***Cis nitidus*** (Fabricius, 1792)
**SQS-2**
7 SU16: Marlborough area, 1869-1938 (Meyrick, 1938)
8 ST84: Longleat, 2-Jan-99 (GO)
SU03: Wylye Down NNR, 20-Sep-05 (MDD) (Darby, 2009); Grovely Wood, 5, 6-Apr-71 4-Aug-80 (DRN)
SU13: Little Durnford Down, 20-May-03 (MDD)
Resident. Local; in fungi on wood and by sweeping.

***Cis pygmaeus*** (Marsham, 1802)
**SQS-2**
8 SU03: Dinton Park, 6-Jul-00 (KNAA); Grovely Wood, 1-Aug-88 (DRN)
Resident. Very local; in fungus on beech.

**ENNEARTHRON Mellié, 1847**
***Ennearthron cornutum*** (Gyllenhal, 1827)
**SQS-2**
8 SU03: Grovely Wood, 16-Apr-74 (DRN); 3-Aug-04 (MDD)
SU21: Landford Bog, 9-Jun-01 (APF) (Foster, 2001)
Resident. Very local; in fungi and under bark.

## Family TETRATOMIDAE
### Subfamily HALLOMENINAE

**HALLOMENUS Panzer, 1793**
***Hallomenus binotatus*** (Quensel, 1790)
**Nb**
8 ST84: Longleat, 29-Aug-99 (MB)
SU21: Hamptworth Estate, 6-Apr-77 (DRN)
Resident. Very local; [in fungi].

### Subfamily TETRATOMINAE

**TETRATOMA Fabricius, 1790**
***Tetratoma ancora*** Fabricius, 1791
**Nb, IEC-3, SQS-8**
8 SU03: Grovely Wood, 3-Aug-04 (MDD)
Resident. Very local; under fungoid bark.

***Tetratoma fungorum*** Fabricius, 1790
**IEC-possible addition, SQS-2**
7 SU16: Marlborough area, 1869-1938 (Meyrick, 1938)
SU26: Savernake Forest, 3-Apr-70 (DRN); 1970-2001 (Anon); 6-Oct-96 (KNAA)
8 SU02: Hare Warren, Wilton, 8-Apr-70 (DRN)
SU03: Grovely Wood, 27-Oct-71, 6, 11-Apr-72 (DRN)
SU13: Great Durnford, 6-Apr-70 (DRN)
SU22: Langley Wood, 12-Apr-74 (DRN)
SU23: Porton Down. Wood at, 15-Oct-95 (MAS)
Resident. Local; in fungi on wood.

## Family MELANDRYIDAE
### Subfamily MELANDRYINAE

**ORCHESIA Latreille, 1807**
***Orchesia micans*** (Panzer, 1793)
**Nb, IEC-addition, SQS-4**
8 ST85: Iford Manor, 28-Aug-02 (DJG)
SU16: Jones's Mill, 17-Oct-01 (APF) (Foster, 2002)
SU03: Grovely Wood, 11-Aug-70 (DRN); Little Langford Down, 14-Jul-04 (MDD)
Resident. Very local; in fungus on wood, and once on a live beech.

***Orchesia minor*** Walker, 1836
**Nb, IEC-addition, SQS-8**
7 SU16: Marlborough area, Jun-1873 (EWJ)

(Meyrick, 1874); 1869-1938 (Meyrick, 1938)
8 ST93: Great Ridge Wood, 4-Oct-04 (MDD)
SU03: Grovely Wood, 3-Aug-04 (MDD)
SU16: Jones's Mill, 21-Jun-97 (MDD)
SU22: Blackmoor Copse, 17-Mar-02, 3-Aug-03 (APF) (Foster, 2004)
SU26: Bedwyn Brail, 20-Sep-04 (MDD)
Resident. Local; in fungi on trees.

***Orchesia undulata*** Kraatz, 1853
**IEC-4, SQS-3**
7 ST96: Spye Park, 24-Apr-87 (KNAA)
ST99: Flisteridge Wood, 25-Apr-82 (KNAA); 20-May-02 (APF)
SU26: Savernake Forest, 6-Jun-73 (APF); 9-Jun to 1-Jul, 20-Sep to 11-Oct-94 (EN det SAW); 25-Aug-to 14-Sep-95 (EN det MDD)
8 ST73: Stourhead, 11-Feb-84, 4-Jul-95 (KNAA); Tucking Mill Hanging, 11-Feb-84 (KNAA)
ST84: Longleat, 28-Aug-00 (MB)
ST93: Great Ridge, 19-Apr-84 (DRN)
SU02: Gardiner Forest, 1-Sep-72 (DRN); Knighton Wood, May to Jul-08 (AJWA); Wilton, 8-Apr-70 (DRN)
SU03: Grovely Wood, ten records 10-Apr-70 - 8-Apr-75 (DRN)
SU16: Jones's Mill, 17-Oct-01 (APF) (Foster, 2002)
SU21: Hamptworth Estate, 16-Apr-74, 31-Jul-85 (DRN)
SU22: Blackmoor Copse, 3-Aug-03 (APF) (Foster, 2004); Langley Wood NNR, 6-Aug-86 (DRN); 6-Jun to 2-Jul-07 (MDD) (Darby, 2008)
SU26: Bedwyn Brail, 20-Sep-04 (MDD)
Resident. Widespread; under bark, particularly beech

**ABDERA Stephens, 1832**
***Abdera biflexuosa*** (Curtis, 1829)
**Nb, IEC-3, SQS-8**
7 SU16: Marlborough area, 1869-1938 (Meyrick, 1938)
SU36: Stype Wood, Jun-1870 (RH) (Hislop, 1871)
8 SU03: Dinton Park, 6-Jul-00 (KNAA) (Alexander, 2004); Grovely Wood, 16-Aug-71 (DRN)
Resident. Very local; on fungi on trees.

**PHLOIOTRYA Stephens, 1832**
***Phloiotrya vaudoueri*** Mulsant, 1856
**Nb, IEC-2, SQS-8**
7 SU26: Savernake Forest, 1970-2001 (Anon); 29, 30-May-76 (MacKechnie-Jarvis,1976)
**Resident. Very local; [in rotten wood].**

**MELANDRYA Fabricius, 1801**
***Melandrya caraboides*** (Linnaeus, 1761)
**Nb, IEC-3, SQS-4**
7 SU16: Marlborough area, 12-Feb-05 (Meyrick, 1874); 5-Jun-1879 (WFB) (Meyrick, 1880); 1869-1938 (Meyrick, 1938)
SU26: Savernake Forest, 24-May-73 (PJC); 26-Jul-85 (DRN); 30-May-04 (DS det KNAA)
Resident. Very local; on old oak or willow stumps, and on live beech.

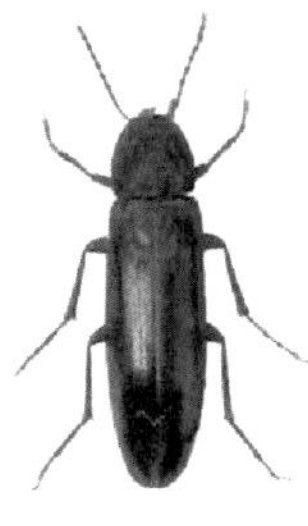

Phloiotrya vaudoueri

## Subfamily OSPHYINAE

**CONOPALPUS Gyllenhal, 1810**
***Conopalpus testaceus*** (Olivier, 1790)
**Nb, IEC-3, SQS-8**
7 SU26: Savernake Forest, 15-Jun-73 (SAW); 16-Jun to 7-Jul-93, 1 to 19-Jul-94 (EN det SAW)
8 ST73: Stourhead, 4-Jul-95 (KNAA) (Alexander, 1996)
SU03: Grovely Wood, 11-Aug-71 (DRN)
Resident. Very local; on oaks.

**OSPHYA Illiger, 1807**
***Osphya bipunctata*** (Fabricius, 1775)
**RDB3, SQS-16**
7 ST87: Castle Combe, 3-Jun-04 (BLe det KNAA)
Resident. Very local; [on flowering hawthorn in wooded areas].

# Family MORDELLIDAE
## Subfamily MORDELLINAE
### Tribe MORDELLINI

Tomoxia bucephala

**TOMOXIA Costa, A, 1853**

***Tomoxia bucephala*** Costa, 1853
**Na, IEC-3**
7 SU26: Savernake Forest, 6-Jun-73 (APF); 6-Dec-73 (PS); 1, 5-Aug-86 (DRN)
8 ST91: Cranborne Chase, 8-Jul-97 (MAS)
SU21: Landford Bog, 22-Jun-03 (APF) (Foster, 2003)
SU25: Ludgershall, Apr - Jul-1915 (PAH) (Harwood, 1915)
Resident. Local, on felled timber in woodland.

***Mordella holomelaena*** Apfelbeck, 1914
**RDBK**
8 SU22: Bentley Wood, 27-May-97 (MDD)
Resident. Very local; on Umbelliferae.

***Mordella leucaspis*** Küster, 1849
**RDBK**
8 SU22: East Grimstead, 11-Aug-86 (DRN)
Resident. Very local; on Umbelliferae.

**VARIIMORDA Méquignon**
***Variimorda villosa*** (Schrank, 1781)
**Nb**
7 SU16: Oare, 2-Aug-02 (IJG)
8 SU03: Grovely Wood, 18-Aug-69, 3-Aug-71, 26-Jul-82 (DRN); 15-Aug-01 (MDD)
SU04: Parsonage Down NNR, 2-Jul-08 (MDD); Salisbury Plain Jul-1915 (PH)
SU16: Jones's Mill, 2-Jul-01 (APF) (Foster, 2002)
SU21: Landford Bog, 24-Jun-01, 22-Jun-03 (APF) (Foster 2001, 2004)
SU22: Bentley Wood, 1-Aug-78, 11-Aug-86 (DRN); 1996, 20-Aug-06 (JN); 2-Jul-96, 11-Aug-01 (MDD); PepperBox Hill, 29-Jul-01 (BL)
SU23: Porton Down, 3-Jul-99 (MAS); West Winterslow, 7-Jul-01 (NB)
SU24: Bulford Ranges (SPTAE), 3-Jun - 12-Aug-08 (Boyce & Wright, 2009)
SU25: Collingbourne Wood, Jun-1915 (PH); Ludgershall, Apr - Jul-1915 (PAH) (Harwood, 1915)
Resident. Widespread; on Umbelliferae and other flowering vegetation.

Variimorda villosa

Tribe MORDELLISTENINI

**MORDELLISTENA Costa, A, 1854**
MORDELLISTENA Costa, A., 1854
***Mordellistena brevicauda*** (Boheman, 1849)
**RDBK, LCGR-Fh C?**
7 SU16: Marlborough area, Jun-1873 (EWJ) (Meyrick, 1874); 1869-1938 (Meyrick, 1938)
8 SU24: Driver Training Area (SPTAE), 24-Jun-02 (PJH) (Hodge, 2002)
Resident. Very local; on vegetation in calcareous grassland. No specimen under this name in MCC.

***Mordellistena humeralis*** (Linnaeus, 1758)
**RDBK, IEC-possible addition**
8 SU22: East Grimstead, 11-Aug-86 (DRN)
SU26: Easton Royal, 13-Jul-02 (IJG)
Resident. Very local; on Umbelliferae in broadleaved woodland.

***Mordellistena neuwaldeggiana*** (Panzer, 1796)
**RDBK, IEC-3**
8 ST91: Cranborne Chase, 16 to 25-Jun-80 (NCC)
SU22: Bagfield Copse, 7-Aug-86 (DRN)
Resident. Very local; [on hogweed at edges of woodland and pasture woodland].

***Mordellistena parvula*** (Gyllenhal, 1827)
**RDBK, LCGR-Fm C?**
7 SU16: Marlborough area, 1869-1938 (Meyrick, 1938)
SU36: Stype Wood, Jun-1871 (RH) (Hislop, 1871)
8 SU14: Stonehenge, Summer-03 (PJH) (Hodge, 2003)
Resident. Very local; the Stonehenge example in a pitfall trap. [Mainly associated with mugwort, but also recorded from yarrow]. Recent records from only two other counties. No specimen under this name in MCC.

***Mordellistena pumila*** (Gyllenhal, 1810)
7 ST86: Neston Park, 3-Jun-04 (BLe)
SU16: West Woods, 9-Jul-46 (Holmes & Tennant, 1947)
8 ST91: Cranborne Chase, 4-Jun-97 (MAS)
ST94: Salisbury Plain, Imber Firs, 24-Jun-99 (EGS det MDD); Ladywell (SPTAW), 23-Jun-02 (PJH) (Hodge, 2002)
SU16: Jones's Mill, 19-Jun-01 (APF) (Foster, 2002)
SU24: Driver Training Area (SPTAE), 24-Jun-02 (PJH) (Hodge, 2002); North Tidworth, Devils Ditch, 30-May-04 (BLe)
Resident. Local; on flowers in woodland and grassland.

***Mordellistena variegata*** (Fabricius, 1798)
8 SU03: Wylye Down NNR, 5-Jun to 10-Jul-07 (MDD) (Darby, 2009)

SU14: Stonehenge, 22-Apr-03 (PJH) (Hodge, 2003)
Resident. Very local; in a flight intercept trap on calcareous grassland.

**MORDELLOCHROA Emery, 1876**
***Mordellochroa abdominalis*** (Fabricius, 1775)
7 ST87: Castle Combe, 3-Jun-04 (BLe); Corsham Park, 3-Jun-04 (BLe)
ST89: Colerene Park, 2-Jun-04 (AH det KNAA)
ST97: Chippenham, Lackham College, 29-May-04 (BLe)
SU08: Grone Farm, Grittenham, 14-Jun-98 (BLe)
8 ST83: Stourhead, 4-Jul-95 (KNAA)
ST91: Cranborne Chase, 4-Jun-97 (MAS)
SU02: Knighton Wood, 6-Jul-07, May-08 (AJWA)
SU13: Old Sarum, 31-May-97 (MAS)
SU22: Bentley Wood, 31-May-97 (MDD)
Resident. Widespread; on flowering vegetation.

## Family RIPIPHORIDAE

Metoecus paradoxus

**METOECUS Dejean, 1834**
***Metoecus paradoxus*** (Linnaeus, 1761)
8 ST85: Bullen Hill Farm, 1982-2002 (EGS det MDD)
SU21: Harnham, Salisbury, 2-Oct-99 (MDD)
Resident. Very local; in or near wasps' nests.

## Family COLYDIIDAE
Subfamily COLYDIINAE
Tribe SYNCHITINI

**SYNCHITA Hellwig, 1792**
***Synchita humeralis*** (Fabricius, 1792)
**Nb, IEC-3, SQS-8**
8 SU22: Blackmoor Copse, 13-Jul-03 (APF) (Foster, 2004)
Resident. Very local; [under fungus infected bark in broadleaved woodland].

**BITOMA Herbst, 1793**
***Bitoma crenata*** (Fabricius, 1775)
7 ST86: Neston Park, 3-Jun-04 (KNAA)
ST96: Spye Park, 31-May-04 (KNAA)
SU16: Marlborough area, Jun-1873 (EWJ) (Meyrick, 1874); 1869-1938 (Meyrick, 1938)
SU26: Savernake Forest, spring-1894 (GCC) (Jebb, 1895); 1970-2001 (Anon), 2- Apr-70 (DRN); 15-Jun-73, 5-May-90, 30-May-04 (KNAA); 6-Dec-73 (APFw); 6-Dec-73 (RCS); 9-May, 4-Sep-90 (TGW); 22-Jun-90 (APFw); 22-Aug-90 (AGD); 14-Sep-01 (IJG)
8 ST73: Stourhead, before 4-Jul-95 (KNAA) (Alexander, 1996)
ST82: Oysters Coppice, 6-Oct-02 (MDD)
ST84: Longleat, 29-Aug-99 (MB)
ST85: Green Lane Wood, 7-Jun-95 (MDD); 26-Jun-98 (JN)
ST91: Cranborne Chase, 21-May-02 (Anon)
ST92: Wardour, 5-Jun-96 (MDD)
ST93: Fonthill, 26-Jun-02 (MDD); Great Ridge Wood, 12-Apr-77, 13-Aug-81, 29-Jul-82, 29-Jul-85 (DRN); 2, 10-May-00 (MDD)
SU02: Wilton, 16-Aug-88 (DRN)
SU03: Grovely Wood, 23 records Aug-70 – Aug-88 (DRN); 19-Aug-95 (MAS); 11-May-00 (MDD)
SU12: Burnt Ground Wood, 31-Oct-76 (DRN)
SU13: Little Durnford Down, 13-Sep-99 (MDD)
SU21: Hamptworth Estate, 16, 19-Apr-74, 30-Jul-85 (DRN)
SU22: Bentley Wood, 31-May-96 (MDD); 12-Feb-01 (NB); Blackmoor Copse, 31-May-96 (MDD)
SU23: Porton Down, 12-Nov-96 (MAS)
SU25: Ludgershall, Apr - Jul-1915 (PAH) (Harwood, 1915)
SU26: Bedwyn Brail, 2-Jun, 5-Sep, 5-Nov-03 (MDD)
Resident. Ubiquitous; under bark in both deciduoius and coniferous woodland.

Tribe COLYDIINI

Colydium elongatum

**COLYDIUM Fabricius, 1793**
***Colydium elongatum*** (Fabricius, 1787)

**RDB3, IEC-possible addition, SQS-16**
7 ST96: Spye Park, 31-May-04 (KNAA)
SU26: Savernake Forest, 1970-2001 (Anon); 29, 30-May-76 (MacKechnie-Jarvis,1976)
8 SU03: Grovely Wood, 10-Apr, 10-May-70, 8-Apr-23, 30-Aug-72, 23-Jul-73, 16-Apr, 27-May-74, 8-Apr-75 (DRN)
SU21: Burnt Ground Wood, 31-May-74 (DRN)
SU22: Langley Wood NNR, 12-Apr-74 (DRN)
SU26: Bedwyn Brail, 2-Jun-03 (MDD)
Resident. Local; under bark of beech and oak. Seems to have been plentiful in 1970s but I have only encountered it once since.

## Family TENEBRIONIDAE
### Subfamily LAGRIINAE

**LAGRIA Fabricius, 1775**
***Lagria hirta*** (Linnaeus, 1758)
7 SU16: Marlborough area, Jun-1873 (EWJ) (Meyrick, 1874); 1869-1938 (Meyrick, 1938)
SU17: Avebury, 12-Jul-03 (DB det MDD)
SU27: High Clear Down, 21-Aug-95 (MA det MDD)
8 ST85: Bullen Hill Farm, 1982-2002 (EGS det MDD)
SU16: Jones's Mill, 7-Jul-05 (MDD); Woodborough, 4-Aug-83 (DRN)
SU23: Bentley Wood, 6-Dec-73 (PM)
Resident. Local; on vegetation. Probably widespread.

### Subfamily TENEBRIONINAE
Tribe BOLITOPHAGINI

**ELEDONA Latreille, 1796**
***Eledona agricola*** (Herbst, 1783)
**Nb, IEC-3, SQS-4**
7 ST86: Neston Park, 3-Jun-04 (DJG det KNAA)
SU26: Savernake Forest, 29, 30-May-76 (MacKechnie-Jarvis, 1976)
8 ST92: Wardour, 13-Nov-95 (MDD)
SU03: Grovely Wood, 3-Aug-71 (DRN)
SU12: Burnt Ground Wood, 31-Oct-75 (DRN)
Resident. Local; on fungus in woodland.

Tribe TENEBRIONINI

**TENEBRIO Linnaeus, 1758**
**Meal-Worm Beetle**
***Tenebrio molitor*** Linnaeus, 1758
7 SU06: Roundway Camp, Jul-40 (FDB) (Buck, 1956)
SU16: Marlborough area, 1869-1938 (Meyrick, 1938)
8 ST95: West Lavington, 1938 (PLP) (DFL, 1948)
ST92: Sutton Mandeville, 14-Jul-03 (MDD)
SU13: Salisbury, 28-Aug-72 (DRN)
Resident. Local; in and around buildings where it breeds in bird nests.

**Dark Meal-Worm Beetle**
***Tenebrio obscurus*** Fabricius, 1792
7 SU16: Marlborough area, 1869-1938 (Meyrick, 1938)
Resident. Very local; in College and in old flour mills.

Tribe ALPHITOBIINI

**ALPHITOBIUS Stephens, 1829**
**Lesser Mealworm Beetle**
***Alphitobius diaperinus*** (Panzer, 1796)
8 SU02: Knighton Wood, 7-Oct-06 (AJWA) (Allen, 2008)
Resident. Very local; in aerial litter, bird nests, etc.. [Associated with a wide range of stored commodities eg. grains, nuts, leather, bones, and sometimes in large numbers in deep litter poultry houses].

Tribe HELOPINI

**NALASSUS Mulsant, 1854**
***Nalassus laevioctostriatus*** (Goeze, 1777)
7 SU16: Marlborough area, Jun-1873 (EWJ) (Meyrick, 1874); 1869-1938 (Meyrick, 1938)
SU26: Savernake Forest, 13-May-91 (EGS det MDD)
8 ST84: Longleat, 4-May-98, 8-May, 1-Jul-99 (MB)
SU21: Hamptworth Estate, 16,19-Apr-74 (DRN)
SU22: Blackmoor Copse, 31-Aug-01 (APF) (Foster, 2004); Langley Wood NNR, 12-Apr-74, 6-Aug-86 (DRN); 6-Jun-02, 29-May-03, 1 to 22-Apr-04, 20-Feb to 24-Mar-07 (MDD) (Darby, 2008)
Resident. Local; under bark and in rotten wood.

Tribe BLAPTINI

**BLAPS Fabricius, 1775**
?***Blaps mortisaga*** (Linnaeus, 1758)
7 SU16: Marlborough area, 24-May-1879 (GTKM) (Meyrick, 1880); 1869-1938 (Meyrick, 1938)
Possibly a misidentification. There have been no verified records of this insect in GB since 1864 and no specimen exists in the MCC.

**Cellar Beetle, Churchyard Beetle**
***Blaps mucronata*** Latreille, 1804
7 SU16: Marlborough area, Jun-1873 (EWJ) (Meyrick, 1874); 1869-1938 (Meyrick, 1938)
8 ST85: Bullen Hill Farm, 1982-2002 (EGS det MDD)
ST92: Sutton Mandeville, numerous sightings

1992-2008 (MDD)
ST95: West Lavington, 12-May-45 (DCL) (DFL, 1948)
Resident. Local; in or near buildings.

## Subfamily DIAPERINAE
### Tribe HYPOPHLAEINI

**CORTICEUS Piller& Mitterpacher, 1783**
***Corticeus bicolor*** (Olivier, 1790)
8 SU03: Grovely Wood, 4-Aug-78 (DRN)
Resident. Very local, under bark. [Thrived as a result of Dutch Elm disease but there have been fewer records since].

### Tribe SCAPHIDEMINI

**SCAPHIDEMA Redtenbacher, 1849**
***Scaphidema metallicum*** (Fabricius, 1792)
**Nb**
7 SU27: Rabley, 1939 (det KGB) (Knight, 1940)
8 SU13: Salisbury, Bemerton Folly, 7-Mar - 15-Jun-02 (MDD) (Darby, 2002d)
SU14: Stonehenge, Summer-03 (PJH) (Hodge, 2003)
SU16: Jones's Mill, Sep-97 (HK det MDD); 17-Oct-01 (APF) (Foster, 2002)
Resident. Local; in woodland litter and under fungoid bark.

## Subfamily ALLECULINAE
### Tribe ALLECULINI

**PRIONYCHUS Solier, 1835**
***Prionychus ater*** (Fabricius, 1775)
**Nb, IEC-3, SQS-8**
7 ST87: Corsham Court Park, 3-Jun-04 (KNAA)
SU17: Avebury, 11-Aug-03 (DB)
8 ST84: Longleat, 18-Mar - 24-Apr-90 (AGD)
Resident. Very local; at light and as a larva in a tree hollow.

Prionychus ater

**GONODERA Mulsant, 1856**
***Gonodera luperus*** (Herbst, 1783)
**SQS-2**
7 SU16: Marlborough area, Jun-1873 (EWJ) (Meyrick, 1874); 1869-1938 (Meyrick, 1938)
8 ST93: Great Ridge Wood, 25-May-00 (MDD)
SU23: Porton Down, 3-May-93 (PTW det MDD)
Resident. Very local; on vegetation.

**PSEUDOCISTELA Crotch, 1873**
***Pseudocistela ceramboides*** (Linnaeus, 1758)
**Nb, IEC-2, SQS-8**
7 SU16: Marlborough area, 21-Apr-73 (Meyrick, 1938)
SU26: Savernake Forest, 1939 (det KGB) (Knight, 1940); 1970-2001 (Anon)
Resident. Very local; [in dead and dying trees in ancient broadleaved woodland]. No specimens in the MCC

**ISOMIRA Mulsant, 1856**
***Isomira murina*** (Linnaeus, 1758)
7 SU06: Cherhill Down, 31-May-04 (BLe); 1-Jun-94 (PM)
SU16: Marlborough area, 1869-1938 (Meyrick, 1938)
8 ST91: Cranborne Chase, 15-May-97 (MAS)
ST92: Sutton Mandeville, Jun-96 (MDD); Win Green, 3-Jul-77 (BLe)
ST93: Great Ridge Wood, 25-May-00 (MDD)
ST94: Bowls Barrow (SPTAW), 22-Jun-02 (PJH) (Hodge, 2002); Ladywell (SPTAW), 23-Jun-02 (PJH) (Hodge, 2002); Norton Down (SPTAW), 23-Jun-02 (PJH) (Hodge, 2002); Warminster Ranges (SPTAW), 3-Jun - 12-Aug-08 (Boyce & Wright, 2009)
SU01: Martin Down NNR, 1999 (EN, 1999)
SU02: Middleton Down, 22-Jun-98 (BL)
SU03: Grovely Wood, 27-May-74 (DRN); Wylye Down NNR, 4-May to 10-Jul-07 (MDD). (Darby, 2009)
SU04: Fox Covert (SPTAW), 3-Jun - 12-Aug-08 (Boyce & Wright, 2009)
SU12: Salisbury, 29-May-40 (FTG)
SU13: Cockey Down, 15-Jun-73 (PM); Old Sarum, 31-May-97 (MAS); Salisbury, Apr-00 (MAS) (Darby, 2001)
SU14: Stonehenge, 22-Apr-03 (PJH) (Hodge, 2003); Wexland Hanging (SPTAC), 22-Jun-02 (PJH) (Hodge, 2002)
SU23: Porton Down, 3-Jul-99 (PJH) (Hodge, 2002)
SU24: Bulford Ranges (SPTAE), 3-Jun - 12-Aug-08 (Boyce & Wright, 2009); Driver Training Area (SPTAE), 24-Jun-02 (PJH) (Hodge, 2002)
SU24: Sidbury Hill (SPTAE), 28-May, 4-Jun-97 (MGT)
Resident. Widespread; on vegetation

**MYCETOCHARA Berthold, 1827**
***Mycetochara humeralis*** (Fabricius, 1787)

**Na**, **IEC-2**, **SQS-16**
7 SU26: Savernake Forest, 1-Aug-86 (DRN); 29, 30-May-76 (MacKechnie-Jarvis,1976)
Resident. Very local; on felled bech.

## Family OEDEMERIDAE
### Subfamily OEDEMERINAE
#### Tribe NACERDINI

**NACERDES Dejean, 1834**
**Wharf Borer**
***Nacerdes melanura*** (Linnaeus, 1758)
7 SU16: Marlborough area, Jun-1873 (EWJ) (Meyrick, 1874); 1869-1938 (Meyrick, 1938)
Resident. Very local; on old timber. Two specimens in MCC without data.

#### Tribe ASCLERINI

**ISCHNOMERA Stephens, 1832**
***Ischnomera caerulea*** (Linnaeus, 1758)
**RDB3**, **IEC-1**, **SQS-24**
7 SU16: Marlborough area, 1869-1938 (Meyrick, 1938); West Woods, 1900-1905 (Meyrick, 1906)
SU26: Savernake Forest, 29, 30-May-76 (MacKechnie-Jarvis,1976); 15-Jun-73 (SAW); 26-May to 16-Jun-93 (EN det SAW)
8 ST85: Bullen Hill Farm, 1982-2002 (EGS det MDD)
ST92: Sutton Mandeville, 24-May-99 (MDD)
Resident. Very local; on vegetation in woodland. British records before 1988 comprised both *caerula* and *cyanea*, so must be considered indeterminate.

***Ischnomera cyanea*** (Fabricius, 1787)
**Nb**
7 ST87: Castle Combe, 3-Jun-04 (BLe, KNAA); Corsham Court Park, 3-Jun-04 (BLe, DJG det KNAA)
ST96: Clink Lane, 31-May-04 (KNAA); Upper Selves Wood, 31-May-04 (KNAA)
ST99: Flisteridge Wood, 19-May-02 (APF)
SU26: Savernake Forest, 29, 30-May-76 (MacKechnie-Jarvis, 1976)
8 ST85: Green Lane Wood, 31-May-04 (KNAA)
SU22: Blackmoor Copse, 17-May-02 (APF) (Foster, 2004)
Resident. Local; [on flowering vegetation in woodland].

***Ischnomera sanguinicollis*** (Fabricius, 1787)
**Nb**, **IEC-1**, **SQS-8**
7 ST96: Upper Selves Wood, 31-May-04 (KNAA)
SU26: Savernake Forest, 1970-2001 (Anon)
Resident. Very local; [on dead or dying trees].

Ischnomera sanguinicollis

#### Tribe OEDEMERINI

**OEDEMERA Olivier, 1789**
ONCOMERA Stephens, 1832
***Oedemera femoralis*** (Olivier, 1803)
**Nb**
8 SU02: Knighton Wood, Nov-06, Mar-07 (AJWA) (Allen, 2008)
Resident. Very local; in a flight intercept trap. [Associated with dead ivy and beaten from sallow bloom in spring].

OEDEMERA Olivier, 1789
***Oedemera lurida*** (Marsham, 1802)
7 SU09: Clattinger Farm, 15-Jun-97 (PM); Somerford Keynes, Swillbrook Lakes, 4-Jun-04 (BLe)
SU16: Marlborough area, Jun-1873 (EWJ) (Meyrick, 1874), 5-Jun-1879 (WFB) (Meyrick, 1880); 1869-1938 (Meyrick, 1938)
SU36: Ham Hill , 8-Aug-93 (PM)
8 ST84: Longleat, 25-May-05 (DJG)
ST91: Cranborne Chase, 14-Jun-97 (MAS); 22-Jun-99 (MDD)
ST94: Bowls Barrow (SPTAW), 22-Jun-02 (PJH) (Hodge, 2002); Ladywell (SPTAW), 3-Jun – 12-Aug-08 (Boyce & Wright, 2009); Norton Down (SPTAW), 15-May, 23-Jun-02 (PJH) (Hodge, 2002); Tinkers Track (SPTAW), 3-Jun - 12-Aug-08 (Boyce & Wright, 2009)
ST95: Bratton, 22-Mar-08 (AJM)
SU01: Martin Down NNR, 1979-1992 (RGB)
SU03: Grovely Wood, 29-Jul-78; Wylye Down NNR, 5-Jun to 3-Aug-07 (MDD) (Darby, 2008a)
SU04: Orcheston Down (SPTA), 22-Jun-02 (PJH) (Hodge, 2002); Salisbury Plain, 16-May-02 (PJH) (Hodge, 2002)
SU05: Ell Barrow (SPTAC), 3-Jun - 12-Aug-08 (Boyce & Wright, 2009)
SU12: Coombe Bissett Down, 31-May-95 (PM)

SU13: Cockey Down, 14-Jun-96 (PM); Figsbury Ring, 13-Jul-94 (PM); Little Durnford Down, 25-Jul-90 (PM); Old Sarum, 31-May-97 (MAS); Salisbury, 20-Jun-73 (PM); Apr-00 (MAS) (Darby, 2001)
SU14: Bourne Bottom (SPTAE), 13-Jun-97 (MGT); Haxton (SPTAE), 24-Jun-02 (PJH) (Hodge, 2002); Stonehenge, 22-Apr-03 (PJH) (Hodge, 2003); Wexland Hanging (SPTAC), 22-Jun-02 (PJH) (Hodge, 2002); 3-Jun - 12-Aug-08 (Boyce & Wright, 2009)
SU15: Beach's Plantation (SPTAE), 10-Jun-97 (MGT); Everleigh (SPTAE), 18-May, 24-Jun-02 (PJH) (Hodge, 2002)
SU16: Jones's Mill, 1997 (MAS); 19-Jun, 2-Jul-01 (APF) (Foster, 2002); 30-May-04 (KNAA)
SU21: Landford Bog, 9-Jun-01 (APF) (Foster, 2001)
SU22: Bentley Wood, 2-Jul-96, 27, 31-May-97 (MDD); Blackmoor Copse, 31-May-93 (PM) Langley Wood NNR, 25-Apr to 6-Jun-07 (MDD) (Darby, 2008)
SU23: Newton Tony (Allington Farm), 26-Jun-03 (DG); Porton Down, 3-Jul-99 (PJH) (Hodge, 2002); West Winterslow, 16-Jul-01 (NB)
SU24: Beacon Hill (SPTAE), 3-Jun - 12-Aug-08 (Boyce & Wright, 2009); Bulford Ranges (SPTAE), 3-Jun - 12-Aug-08 (Boyce & Wright, 2009); Driver Training Area, 24-Jun-02 (PJH) (Hodge, 2002); Dunch Hill (SPTAE), 16-Jun-97 (MGT); 3-Jun - 12-Aug-08 (Boyce & Wright, 2009); North Tidworth, Devils Ditch, 30-May-04 (BLe); Sidbury Hill (SPTAE), 18-May-97 (MGT)
SU25: Sidbury Hill (SPTAE), 4-Jun-97 (MGT)

Resident. Ubiquitous; on vegetation in grassland and open areas in woodland.

**Swollen-thighed Beetle**
***Oedemera nobilis*** (Scopoli, 1763)

7 ST86: Neston Park, 3-Jun-04 (KNAA)
SU08: Distillery Meadows, 14-Jun-04 (PM); Echo Lodge Meadows, 13-Jun-01 (PM); 5-Jun-07 (MS)
SU09: Clattinger Farm, 15-Jun-97 (PM); Upper Waterhay, 6-Jul-07 (JKG det MDD)
8 ST84: Longleat, 15-Jun-73, 8-Jun-96 (PM); 7-Jul-99 (MDD); 25-May-05 (DJG)
ST85: Biss Wood, 16-May-00 (BG); Green Lane Wood, 7-Jun-95, 19-Jun-04 (MDD)
ST91: Cranborne Chase, 24-May-97 (MAS)
ST92: Sutton Mandeville, 7-Jul-96 (MDD)
SU02: Chickengrove Bottom, 15-Jun-02 (BL); Middleton Down, 9, 10-Aug-96 (PM)
SU03: Wylye Down NNR, 4-May to 3-Aug-07 (MDD) (Darby, 2009)
SU04: Parsonage Down NNR, 2-Jul-08 (MDD); Yarnbury Castle, 11-Jun-02 (BL); 8-Jul-08 (MDD)
SU12: Coombe Bissett Down, 31-May-95 (PM); Petersfinger, 7-Aug-88 (DRN)
SU13: Cockey Down, 6-Nov-73, 27-Jun-90 (PM); Little Durnford Down, Jul-90 (PM); Old Sarum, 31-May-97 (MAS); Salisbury, 20-Jun-73 (PM)
SU14: Haxton (SPTAE), 3-Jun - 12-Aug-08 (Boyce & Wright, 2009); Stonehenge, 22-Apr-03 (PJH) (Hodge, 2003)
SU22: Bentley Wood, 31-May, 2-Jul-96; 27-May-97 (MDD); Blackmoor Copse, 2-Jul-96 (MDD); 9-Jun-01, 17-May-02 (APF) (Foster, 2004); Langley Wood NNR, May-99 (BL)
SU23: West Winterslow, 9-Jul-01 (NB)
SU24: Beacon Hill (SPTAE), 3-Jun - 12-Aug-08 (Boyce & Wright, 2009); Bulford Ranges (SPTAE), 3-Jun - 12-Aug-08 (Boyce & Wright, 2009); Driver Training Area (SPTAE), 24-Jun-02 (PJH) (Hodge, 2002)
SU26: Wilton Water slope, 20-Jun-01 (IJG)

Resident. Widespread on vegetation in grassland and open areas in woodland.

## Family MELOIDAE
### Subfamily MELOINAE

**MELOE Linnaeus, 1758**
**Common Oil Beetle**
***Meloe proscarabaeus*** Linnaeus, 1758

7 SU16: Lockeridge, 1994 (JEO); Marlborough area, 27 Mar 1879 (AWM) (Meyrick, 1880); 1869-1938 (Meyrick, 1938)
8 ST84: Cley Hill, 1-May-05 (RD)
ST92: Upper Chicksgrove, 29-Mar-08 (MDD)
ST93: Whitesheet Down, 10-May-08 (ET, BT)
ST94: Imber Firs, 26-May-86 (EGS det MDD)
ST95: West Lavington: 1, 24-Mar-41 (KJA) (DFL, 1948)
SU00: Barbury Castle, 2006-2007 (HS)
SU02: Broad Chalke, May-99 (MDD); 6-May-99 (BM); Ebbesbourne Wake, 18-Apr-03 (DJG); Middleton Down, 28-Mar-89, 28-Mar-92, 23-May-04 (BL); 2-May-94, 10-May-96, 1-Apr-99, 31- Aug-02, 20-Mar-03, 22-Apr-08 (PM); 4-Apr-02 (BL); 17-Mar-04 (CH)
SU03: Barford Down, 10-Apr-72 (DRN); Grovely Wood, 14-Apr-71 (DRN); Little Langford Down, 8-Apr-07 (PM)
SU13: Figsbury Ring, 25-Apr-91 (PM)
SU14: Haxton (SPTAE), 17-May-02 (PJH) (Hodge, 2002); Wexland Hanging (SPTAC), 22-Jun-02 (PJH) (Hodge, 2002)
SU15: Everleigh (SPTAE), 18-May-02 (PJH) (Hodge, 2002)
SU23: Porton, 1970-2001 (Anon); 13-Apr-02 (AM)

Resident. Widespread; wandering in the open on heathland, grassland and a road. Numbers appear to have increased in recent years.

**Rugose Oil Beetle**
***Meloe rugosus*** Marsham, 1802
**RDB3, LCGR-Fm Ch**
8 ST95: Bratton Hill, 24-Sep-84 (EGS det MDD)
Resident. Very local; on vegetation on a grassy hiilside.

Meloe violaceus

**Violet Oil Beetle**
***Meloe violaceus*** Marsham, 1802
**Nb**
7 SU16: Marlborough area, 1869-1938 (Meyrick, 1938)
Resident. Very local; on grassy banks and heaths. Two specimens in MCC, one labelled *05*.

## Family PYROCHROIDAE

**PYROCHROA Geoffroy, 1762**
**Black-Headed Cardinal Beetle**
***Pyrochroa coccinea*** (Linnaeus, 1761)
**Nb, IEC-3, SQS-4**
7 SU16: West Woods, May-02 (MDD); Marlborough area, 1869-1938 (Meyrick, 1938)
SU26: Savernake Forest, 1970-2001 (Anon), 6-Jan-73 (PCT); 15-Jun-73 (SAW); 6-Nov-73 (SB); 6-Dec-73 (RCS); 17-May-90 (TGW); 27-May to 16-Jun-93, 9-Jun to 1-Jul-94 (EN det SAW); 31-May to 23-Jun-95 (EN det MDD); 31-May-00 (WANHS det MDD); 14-Jun-02 (IJG)
8 SU03: Grovely Wood, 12-Apr-72, 16-Apr, 27-May-74, 4, 11-Aug-81, 26-Jul-82, 1-Aug-88 (DRN); Little Langford Down, 26-May-99 (PM)
SU16: Jones's Mill, 19-Jun, 2-Jul, 17-Oct-01 (APF) (Foster, 2002)
SU22: Bentley Wood, 20, 22-May-90 (PM); 27-May-98 (MDD); Blackmoor Copse, 14-May-92, 13, 23-May-93, 16-Jun-97, 20-May-99, 17-May-04 (PM); 29-May-00, 17-May-02, 18-May-03, 3-Aug-03 (APF) (Foster, 2004) (PM); Langley Wood NNR, 28-May-84 (DRN); 6-Jun-02 (MDD)
SU24: Bulford Ranges (SPTAE), 3-Jun - 12-Aug-08 (Boyce & Wright, 2009); Butley Wood, 8-Jul-90 (SJG)
Resident. Widespread; on logs, stumps and vegetation in woodland.

Pyrochroa coccinea

**Common Cardinal Beetle**
***Pyrochroa serraticornis*** (Scopoli, 1763)
7 ST84: Smallbrook Meadows, 9-Jun-00 (PM)
ST87: Castle Combe, 16-Jun-54 (JET); 3-Jun-04 (KNAA)
ST97: Chippenham, Lackham College, 29-May-04 (BLe)
SU06: Roundway Down, 26-May-99 (BG)
SU08: Brockhurst Meadows, 9-Jun-03 (PM); Echo Lodge Meadows, 24-May-00 (MS); Great Wood, 27-May-91 (KNAA)
SU09: Blakehill Farm, 8-Jun-05 (PM); Clattinger Farm, 15-Jun-97 (PM); Upper Waterhay, 17-May-05 (JKG det MDD)
SU16: Avebury, 29-May-92 (PM); Lockeridge, 16-Jun-73 (JEO); Marlborough area, 21 May 1879 (EEdeS) (Meyrick, 1880); 1869-1938 (Meyrick, 1938)
SU17: Avebury, 24-May-03 (DB); Clout's Wood, 13-May-00 (PM)
SU18: Swindon, 4-Jun-01 (JKG det MDD)
SU19: Hannington, 8, 22-Jun-96, 26-May-97, 30-May-98 (MEN)
SU26: Savernake Forest, 31-May-00 (WANHS det MDD); 31-May to 23-Jun-95 (EN det MDD); 30-May-04 (KNAA)
SU36 Ham Hill , 31-May-95 (PM)
8 ST84: Longleat, Jun-1997 (MB)

ST85: Bullen Hill Farm, 1982-2002 (EGS det MDD); Green Lane Wood, 19-May-92 (PM)
ST91: Cranborne Chase, 16 to 25-Jun-80 (NCC); 22-Jun-99 (MDD)
ST92: Sutton Mandeville, 11-Jul-96, 4-May-97 (MDD)
ST93: Great Ridge Wood, 10-May-00 (MDD)
ST94: Warminster Ranges (SPTAW), 3-Jun - 12-Aug-08 (Boyce & Wright, 2009)
ST95: West Lavington, 11-Jul-34 -17-May-45 (ITH, BWHC, KJA, JA) (DFL, 1948 various records probably this species )
SU02: Chickengrove Bottom, 26-May-92 (PM); Knighton Wood, May-08 (AJWA) Middleton Down, 5-Jun-95, 31-May-00 (PM)
SU03: Grovely Wood, 27-May-74 (DRN)
SU04: Salisbury Plain, 25-Jun-02 (PJH) (Hodge, 2002)
SU12: Clearbury Ring, May-928 (PM)
SU13: Cockey Down, 6-Aug-73, 23-May-90, 1-Jun-93, 14-Jun-96 (PM); Figsbury Ring, 17-May-88, 1-Jun-94 (PM); Little Durnford Down, 27-May-97 (MDD); 3-May-93 (PM); Old Sarum, 31-May-97 (MAS); 21-Apr-97 (PM); Salisbury, 1-Jun-00 (MDD) (Darby, 2001); 20-Jun-73 (PM); Stratford sub Castle, 5-Apr-72 (DRN)
SU16: Jones's Mill, 19-Jun-01 (APF) (Foster, 2002)
SU21: Landford Bog, 9-Jun-01 (APF) (Foster, 2001)
SU22: Blackmoor Copse, 8-Jun-97, 20-May-99 (PM); 9-Jun-01 (APF) (Foster, 2004)
SU23: Bentley Wood, 22-May-90 (PM)
SU25: Sidbury Hill (SPTAE), 18-May-97 (MGT)

Resident. Ubiquitous; on logs, stumps and vegetation, particularly in woodland.

## Family SALPINGIDAE

### Subfamily SALPINGINAE

**LISSODEMA Curtis, 18333**

***Lissodema denticolle*** (Gyllenhal, 1813)

**Nb, SQS-8**

7 SU16: Marlborough area, Jun-1873 (EWJ) (Meyrick, 1874); 1869-1938 (Meyrick, 1938)
8 SU03: Little Langford Down, 14-Jul-04 (MDD)
SU22: Blackmoor Copse, 17-Mar-02, 13-Jul-03 (APF) (Foster, 2004)

Resident. Very local; under bark in woodland. Formerly *L. quadripustulata.*

**VINCENZELLUS Reitter, 1911**

***Vincenzellus ruficollis*** (Panzer, 1794)

**SQS-2**

7 SU26: Savernake Forest, 2, 3-Apr-70 (DRN)
8 SU02: Knighton Wood, Apr, 6-Jul-07 (AJWA)
SU03: Grovely Wood, 8, 16-Apr-74, 3-Apr-75 (DRN)
SU13: Cockey Down, 6-Nov-73 (PM); Great Durnford, 6-Apr-70 (DRN); Little Durnford Down, 13-Sep-99 (MDD)
SU21: Hamptworth Estate, 24-Apr-84 (DRN)

Resident. Local; in moss and under bark.

**SALPINGUS Illiger, 1801**

***Salpingus planirostris*** (Fabricius, 1787)

**SQS-1**

7 SU16: Marlborough area, Jun-1873 (EWJ) (Meyrick, 1874); 17-Feb-1879 (WFB) (Meyrick, 1880); Marlborough area, 1873-1938 (Meyrick, 1938)
SU26: Savernake Forest, 22-Aug-90 (AGD); 16-Jun to 21-Sep-93, 28-Apr to 19-Jul-94 (EN det SAW), 15-Sep to 10-Oct-95 (EN det MDD)
8 ST84: Longleat, 4-May-98 (MB)
ST85: Green Lane Wood, 17-Aug-04 (MDD)
SU03: Grovely Wood, 8-Apr-74, 30-Oct-75, 4-Aug-78, 4-Aug-81, 1-Aug-88 (DRN); 3-Aug-04 (MDD); Little Langford Down, 14-Jul-04 (MDD)
SU04: Shrewton, 1-Aug-85 (DRN)
SU13: Boscombe Down, 8, 9-Oct-94 (MDD)
SU16: Jones's Mill, 19-Jun-01 (APF) (Foster, 2002)
SU22: Blackmoor Copse, 31-Aug-01, 17-May-02, 13-Jul, 3-Aug-03 (APF) (Foster, 2004); Langley Wood, 9-Oct-82 (KNAA)
SU26: Bedwyn Brail, 20-Sep-04 (MDD)

Resident. Widespread; on and under bark of deciduous trees.

***Salpingus ruficollis*** (Linnaeus, 1767)

7 SU16: Marlborough area, Jun-1873 (EWJ) (Meyrick, 1874); 6-Feb-1879 (WFB) (Meyrick, 1880); 1869-1938 (Meyrick, 1938)
8 ST91: Cranborne Chase, 12-May-97 (MAS)
SU03: Grovely Wood, 10-Apr, 13-Aug-70, 9-Aug-71, 6-Apr, 21-Aug-72, 3, 16-Apr-75, 1-Aug-88 (DRN); 19-Aug-95 (MAS)
SU16: Jones's Mill, 19-Jun-01 (APF) (Foster, 2002)
SU26: Bedwyn Brail, 20-Sep-04 (MDD)

Resident. Local; on and under bark of deciduous trees.

## Family ANTHICIDAE

### Subfamily ANTHICINAE

#### Tribe NOTOXINI

**NOTOXUS Geoffroy, 1762**

***Notoxus monoceros*** (Linnaeus, 1761)

7 ST96: Spye Park, 31-May-04 (KNAA) (Alexander,

2005)
Resident. Very local; [at plant roots on sandy soils].

Tribe ANTHICINI

**ANTHICUS Paykull, 1798**
***Anthicus antherinus* (Linnaeus, 1761)**
7 SU09: Somerford Keynes, Jun-04 (BLe)
SU16: Marlborough area, 1869-1938 (Meyrick, 1938)
8 ST92: Sutton Mandeville, 1-May-97 (MDD)
SU15: Beach's Plantation (SPTAE), 16-Jun-97 (MGT)
Resident. Local; on vegetation.

**OMONADUS Mulsant and Rey, 1866**
***Omonadus bifasciatus* (Rossi, 1792)**
**Nb**
8 ST96: Rowde, 5-Aug-83 (DRN)
Resident. Very local; abundant in an old farmyard dungheap.

***Omonadus floralis* (Linnaeus, 1758)**
7 ST98: Lacock, 29-May-04 (BLe)
SU16: Marlborough area, Jun-1873 (EWJ) (Meyrick, 1874)
SU26: Savernake Forest, 20-Sep to 11-Oct-94 (EN det SAW)
8 ST91: Rushmore Park, 23-Sep-01 (AJWA)
ST92: Sutton Mandeville, 25-Sep-07 to 3-Jan-08 (MDD)
SU02: Knighton Wood, Oct-06 (AJWA)
SU03: Grovely Wood, 11, 16-Aug-71, 23-Jul-73, 3-Aug-79 (DRN)
SU12: Whaddon Common, 22-Jul-73 (DRN)
SU21: Hamptworth Estate, 15-Apr-74, 28-Oct-75, 29-Mar-78 (DRN)
Resident. Widespread; in decaying vegetation, straw, garden compost, an old dungheap, etc..

***Omonadus formicarius* (Goeze, 1777)**
8 ST91: Rushmore Park, 23-Sep-01 (AJWA)
Resident. Very local; in a dung heap.

**CORDICOMUS Pic, 1894**
**[*Cordicomus instabilis* Schmidt, 1842**
7 SU16: Marlborough area, Jun-1873 (EWJ) (Meyrick, 1874)]
Meyrick (1900) noted that 'this record is probably ... erroneus'.

## Family ADERIDAE

**ADERUS Stephens, 1829**
***Aderus oculatus* (Paykull, 1798**
**Nb, IEC-3, SQS-8**
8 SU03: Dinton Park, 6-Jul-01 (KNAA) (Alexander, 2004)
Resident. Very local; in dead wood of broadleaved trees.

## Family SCRAPTIIDAE
Subfamily ANASPIDINAE

**ANASPIS Geoffroy, 1762**
ANASPIS Geoffroy, 1762
***Anaspis frontalis* (Linnaeus, 1758)**
**SQS-1**
7 SU09: Somerford Keynes, Swillbrook Lakes, 4-Jun-04 (BLe)
SU16: Marlborough area, Jun-1873 (EWJ) (Meyrick, 1874); 4-Jun-1879 (TNH-S) (Meyrick, 1880); 1869-1938 (Meyrick, 1938)
SU26: Savernake Forest, 25-Apr to 8-Jul-93 (EN det SAW); 31-May to 23-Jun-95 (EN det MDD); 31-May-04 (BLe)
8 ST91: Cranborne Chase, 16 to 25-Jun-80 (NCC)
ST93: Great Ridge Wood, 16, 25-May-00 (MDD)
SU03: Grovely Wood, 11-Aug-71 (DRN); 12-Jun-00 (MDD)
SU13: Old Sarum, 31-May-97 (MAS); Salisbury, Apr-00 (MAS) (Darby, 2001)
SU16: Jones's Mill, 19978 (MAS)
SU22: Bagfield Copse, 6-Aug-86 (DRN); Bentley Wood, 31-May-97 (MDD); Blackmoor Copse, 9-Jun-01, 17-May-02, 3-Aug-03 (APF) (Foster, 2004)
SU26: Bedwyn Brail, 19-Jun-03 (MDD)
Resident. Widespread; on flowers, especially Hawthorn.

***Anaspis fasciata* (Forster, 1771)**
**SQS-2**
7 ST86: Neston Park, 3-Jun-04 (BLe, KNAA)
SU16: Marlborough area, 1869-1938 (Meyrick, 1938)
SU26: Savernake Forest, 31-May-90 (TGW)
8 ST85: Green Lane Wood, 31-May-04 (KNAA)
ST91: Cranborne Chase, 16 to 25-Jun-80 (NCC); 7-May-97 (MAS)
SU13: Old Sarum, 31-May-97 (MAS); Salisbury, Apr-00 (MAS) (Darby, 2001)
SU16: Jones's Mill, 1997 (MAS)
SU22: Blackmoor Copse, 9-Jun-01 (APF) (Foster, 2004); Whiteparish Common, 2-Jul-95 (MAS)
SU24: North Tidworth, Devils Ditch, 3-Jun-04 (BLe)
Resident. Widespread; on flowers, especially hawthorn.

***Anaspis garnesyi* Fowler, 1889**
7 ST87: Corsham Park, 3-Jun-04 (BLe)

ST97: Chippenham, Lackham College, 29-May-04 (BLe), 22-May-00 (MDD)
ST98: Lacock, 29-May-04 (BLe)
SU09: Somerford Keynes, Swillbrook Lakes, 4-Jun-04 (BLe); Upper Waterhay, 5 to 11-Jun-05 (JKG det MDD)
SU26: Savernake Forest, 9-Jun to 1-Jul-94 (EN det SAW); 31-May-04 (BLe)
8 ST91: Cranborne Chase, 16 to 25-Jun-80 (NCC)
SU13: Salisbury, Apr-00 (MAS) (Darby, 2001)
SU23: Porton Down, 3-Jul-99 (MAS)
Resident. Widespread; on flowers, especially Hawthorn.

***Anaspis lurida*** Stephens, 1832
**SQS-2**
7 ST87: Corsham Park (ST8770), 3-Jun-04 (BLe)
ST97: Chippenham, Lackham College, 29-May-04 (BLe)
SU16: Jones's Mill, 19-Jun-01 (APF) (Foster, 2002)
8 SU22: Blackmoor Copse, 9-Jun-01, 17-May-02 (APF) (Foster, 2004)
Resident. Local; on flowers in wooded areas.

***Anaspis maculata*** (Geoffroy in Fourcroy, 1785)
7 ST85: Green Lane Wood, 31-May-04 (KNAA)
ST86: Neston Park, 3-Jun-04 (KNAA)
ST87: Corsham Park, 3-Jun-04 (BLe)
ST96: Spye Park, 31-May-04 (KNAA)
ST97: Chippenham, Lackham College, 29-May-04 (BLe); North Draycot Park, 22-May-00 (MDD)
ST98: Lacock, 29-May-04 (BLe)
SU09: Somerford Keynes, Swillbrook Lakes, 4-Jun-04 (BLe); Upper Waterhay, 17-May- to 4-Jun-05 (JKG det MDD)
SU16: Marlborough area, Jun-1873 (EWJ) (Meyrick, 1874); 31 May 1879 (TNH-S) (Meyrick, 1880); 1869-1938 (Meyrick, 1938)
SU17: Avebury, 11-Jun-03 (DB det MDD)
SU26: Savernake Forest, 26-May to 16-Jun-93 (EN det SAW), 31-May to 23-Jun-95 (EN det MDD)
8 ST91: Cranborne Chase, 16 to 25-Jun-80 (NCC); 7-May-97 (MAS); 22-Jun-99 (MDD)
ST93: Great Ridge Wood, 16, 25-May-00 (MDD)
ST94: Bowls Barrow (SPTAW), 16-May-02 (PJH) (Hodge, 2002); Ladywell (SPTAW), 15-May-02 (PJH) (Hodge, 2002); Norton Down (SPTA), 15-May-02 (PJH) (Hodge, 2002)
SU04: Parsonage Down NNR, 2-Jul-08 (MDD); Salisbury Plain, 16-May-02 (PJH) (Hodge, 2002)
SU14: Haxton (SPTAE), 17-May-02 (PJH) (Hodge, 2002); Wexland Hanging (SPTAC), 16-May-02 (PJH) (Hodge, 2002)
SU15: Everleigh (SPTAE), 18-May-02 (PJH) (Hodge, 2002); Stonehenge, 22-Apr-03 (PJH) (Hodge, 2003)
SU16: Jones's Mill, 19-Jun, 2-Jul-01 (APF) (Foster, 2002)
SU22: Blackmoor Copse, 9-Jun-01 (APF) (Foster, 2004); Langley Wood NNR, 29-May-03 (MDD)
SU23: West Winterslow, 1-Aug-01 (NB)
SU24: North Tidworth, Devils Ditch, 30-May-04 (BLe)
SU26: Bedwyn Brail, 2-Jun-03 (MDD)
Resident. Widespread and often abundantly on flowers, especially hawthorn, in woods and grassland.

***Anaspis pulicaria*** Costa, 1854
**SQS-1**
7 ST87: Castle Combe, 3-Jun-04 (BLe)
SU09: Somerford Keynes, Swillbrook Lakes, 4-Jun-04 (BLe)
SU16: Marlborough area, 1869-1938 (Meyrick, 1938)
8 ST91: Cranborne Chase, 16 to 25-Jun-80 (NCC)
SU04: Parsonage Down NNR, 2-Jul-08 (MDD)
SU12: Petersfinger, 25-Jul-91 (DRN)
SU23: West Winterslow, 1-Aug-01 (NB)
Resident. Local; sometimes in numbers on flowers, particularly hawthorn.

***Anaspis regimbarti*** Schilsky, 1895
7 ST86: Neston Park, 3-Jun-04 (BLe, KNAA)
ST87: Corsham Park, 3-Jun-04 (BLe)
ST97: Chippenham, Lackham College, 29-May-04 (BLe)
SU16: Marlborough area, 1869-1938 (Meyrick, 1938)
SU26: Savernake Forest, 16-Jun to 7-Jul-93, 28-Apr to 18-May-94 (EN det SAW); 31-May to 23-Jun-95 (EN det MDD)
8 ST91: Cranborne Chase, 16 to 25-Jun-80 (NCC)
SU14: Haxton (SPTAE), 17-May-02 (PJH) (Hodge, 2002); Stonehenge, 22-Apr-03 (PJH) (Hodge, 2003)
SU15: Everleigh (SPTAE), 18-May-02 (PJH) (Hodge, 2002)
SU16: Jones's Mill, 19-Jun-01 (APF) (Foster, 2002)
SU22: Holmere Common, 19-Aug-88 (DRN)
SU23: Porton Down, 3-Jul-99 (BLe); 3-Jul-99 (PJH) (Hodge, 2002)
Resident. Widespread; on flowers, especially hawthorn.

NASSIPA Emery, 1876

***Anaspis rufilabris*** (Gyllenhal, 1827)
**SQS-1**
7 SU09: Upper Waterhay, 17-May to 4-Jun-05 (JKG det MDD)
SU16: Marlborough area, 31-May-1879 (TNH-S) (Meyrick, 1880); 1869-1938 (Meyrick, 1938)
SU26: Savernake Forest, 17, 31-May-90 (TGW)

27-May to 16-Jun-93, 9-Jun to 1-Jul-94 (EN det SAW); 31-May to 23-Jun-95 (EN det MDD); 31-May-04 (BLe)
8 ST85: Bullen Hill Farm, 1982-2002 (EGS det MDD)
ST91: Cranborne Chase, 7-May-97 (MAS)
SU13: Old Sarum, 31-May-97 (MAS)
SU22: Bentley Wood, 31-May-97 (MDD); Langley Wood, 25-Jun-04 (MDD) (Darby, 2005); Whiteparish Common, 2-Jul-95 (MAS)
SU23: Porton Down, 3-Jul-99 (MAS)
Resident. Widespread; on flowers, especially hawthorn, and occasionally overwintering in dead wood.

## Superfamily CHRYSOMELOIDEA
## **Family CERAMBYCIDAE***
### Subfamily PRIONINAE

Prionus corarius

**PRIONUS Geoffroy, 1762**
**Tanner Beetle**
***Prionus coriarius*** (Linnaeus, 1758)
**Na**, **IEC-3**, **SQS-16**
7 VC7, n.d. (RRUK)
SU26: Savernake Forest, 1869-1938 (Meyrick, 1938); before 1994 (ISR (Sav), 1994)
8 VC8, n.d. (RRUK)
Resident. Very local; [in dead wood of various broadleaved and sometimes coniferous trees]. Two specimens in the MCC, one labelled *Sav Forest.* Not recorded in recent years. One of our largest and most impressive insects. Adults are crepuscular and occasionally seen at dusk flying around the upper branches of trees. Attracted to light.

*Distributional publications by R.R.Uhthoff-Kaufmann which include Wiltshire (but no more precise data) have not been included except where there is no other evidence of a species' existence in the county. Uhthoff-Kaufmann informed me *in litt.* that all his detailed records up to 1948, including over a hundred personal lists sent in by his correspondents, were accidentally destroyed in a fire in his house, and were irreplaceable.

### Subfamily LEPTURINAE
#### Tribe RHAGIINI

**RHAGIUM Fabricius, 1775**
RHAGIUM Fabricius, 1775
**[*Rhagium inquisitor*** (Linnaeus, 1758)
**Nb**
7 SU16: Marlborough area, Jun-1873 (EWJ) (Meyrick, 1874); 1869-1938 (Meyrick, 1938)**]**
3 specimens under this name in the MCC are all *R. mordax* which was cited as. *R.inquisitor* in 19th century. *R. inquisitor* is a Scottish species and records from other parts of Britain are considered as introductions or adventives.

HAGRIUM Villiers, 1978
**Two-banded Longhorn Beetle**
***Rhagium bifasciatum*** Fabricius, 1775
**SQS-1**
7 VC7, n.d. (RRUK)
ST99: Flisteridge Wood, 1982 (KNAA)
SU16: Marlborough area, Jun-1873 (EWJ) (Meyrick, 1874); 1869-1938 (Meyrick, 1938)
SU26: Savernake Forest, spring-1894 (GCC) (Jebb, 1895); 22-Jun-90 (APFw); 25-Apr to 27-May-93 (EN det SAW); 25-Apr-to 30-May-95 (EN det MDD); 31-May-00 (WANHS det MDD); 30-May-04 (KNAA)
8 ST84: Longleat, 28-May-93 (PM); Jun-97 (MB); 25-May-05 (DJG)
ST85: Biss Wood, 20-Mar-06 (GWG)
ST92: Sutton Row, 21-Sep-95 (MDD)
ST93: Great Ridge Wood, 8, 16-May-00 (MDD)
SU03: Grovely Wood, 27-May-02 (MDD)
SU21: Hamptworth Estate, 19-Apr-74 (DRN)
SU22: Bentley Wood, 27-May-97 (MDD); Langley Wood, 29-Aug-72, 12-Apr, 28-May-74 (DRN); 6-Jun-02 (MDD)
Resident. Widespread; around dead wood stumps and logs, larvae and pupae under bark.

MEGARHAGIUM Reitter, 1913
**Black-spotted Longhorn Beetle**
***Rhagium mordax*** (De Geer, 1775)
**SQS-1**
7 ST73: Tucking Mill Hanging, 11-Feb-84 (KNAA)

ST86: Box, 17-Jun-24 (JVB)
ST96: Spye Park, 24-Apr-87 (KNAA)
ST99: Flisteridge Wood, 2-May-82 (KNAA)
SU08: Echo Lodge Meadows, 12-Jun-03 (MS); Stonehill Wood, Minety, 13-May-02 (JKG det MDD)
SU16: Marlborough area, 1869-1938 (Meyrick, 1938); West Woods, May-02 (MDD); 20-May-05 (IJG)
SU17: Bincknoll Dip Wood, 22-Feb-87 (KNAA)
SU26: Bedwyn Common, 15-Jun-06 (MEN); Savernake Forest, 2-Apr-70 (DRN); 22-Jun-90 (APFw); 25-Apr to 27-May-93, 28-Apr to 18-May, 9-Jun to 1-Jul-94 (EN det SAW); 31-May to 23-Jun-95 (EN det MDD); 31-May-00 (WANHS det MDD); 30-May-04 (KNAA)

8 ST82: Oysters Coppice, 6-Oct-02 (MDD)
ST85: Biss Wood, 10-Jun-82 (EGS det MDD); Brokerswood County Park, 13-May-06 (MCH); Green Lane Wood, 19-May-92 (PM); 7-Jun-95 (WANHS det MDD); 4-Apr-99 (EGS det MDD)
ST93: Great Ridge Wood, 2-May-00 (MDD)
SU02: Chickengrove Bottom, 21-May-93 (PM); Wilton, 16-Aug-88 (DRN)
SU03: Dinton, 1-May-89 (SP); Grovely Wood, 12 records 10-Apr-70 – 4-Aug-81 (DRN)
SU15: Everleigh, 2006 (IJG)
SU21: Hamptworth Estate, 16-Apr-74 (DRN); Landford Bog, 14-May-93 (PM)
SU22: Bentley Wood, 20-Jun-73, 1-Jan-96 (JN); 22-Jun-73 (EG); 22-May-91, 13-May-93 (PM); 31-May-96 (MDD); 9-Jun, 31-Aug-01, 17-May-02 (APF) (Foster, 2004); 23-May-04 (AGD); Langley Wood NNR, 28-May-74 (DRN); 6-Sep-73 (Anon)
SU26: Bedwyn Brail, 19-Jun-03 (MDD)

Resident. Widespread; around dead wood stumps and logs, larvae and pupae under bark.

**STENOCORUS Geoffroy, 1762**
**Variable Longhorn Beetle**
***Stenocorus meridianus*** (Linnaeus, 1758)
**SQS-2**

7 VC7, n.d. (RRUK)
ST86: Neston Park, 3-Jun-04 (KNAA)
ST89: Colerne Park, 2-Jun-04 (AH det KNAA)
ST96: Spye Park, 31-May-04 (KNAA)
ST97: Chippenham, 11-Jun-19 (JVB)
SU08: Brockhurst Meadows, 9-Jun-03 (PM)
SU16: Marlborough area, Jun-03 (EWJ) (Meyrick, 1874); 14 Jun 1879 (FSA) (Meyrick, 1880); 1869-1938 (Meyrick, 1938); West Woods, 1870-1873 (Meyrick, 1874)
SU18: Coate Water, 4-Jul-98 (MEN)
SU19: Hannington, 30-May, 6-Jun-98 (MEN); Highworth, Pentylands Lane, 16-Jul-03, 5-Jun-04, 17-Jun, 1-Jul-06 (MEN)
SU26: Savernake Forest, 6-Dec-73, 22-Jun-90 (APFw)
SU26/27: Axford, 2006 (IJG)
SU29: Round Robin Farm, 3-Jul-98 (MEN)

8 ST85: Green Lane Wood, 31-May-04 (KNAA)
ST91: Cranborne Chase, 16 to 25-Jun-80 (NCC)
ST93: Pits Wood, Chilmark, 12-Jul-95 (MDD)
SU03: Little Langford Down, 22-Jun-99 (PM); Steeple Langford, 8-Jun-02 (SC det MDD)
SU12/13: Salisbury, Jun-1947 (HGS)
SU13: Little Durnford Down, 17-Jun-05 (PM)
SU22: Blackmoor Copse, 31-May-92, 16-Jun-97 (PM); 17-May-02 (APF) (Foster, 2004)

Resident. Widespread; on flowering vegetation at woodland edges, larvae in underground roots. Appears to be rarer than in the past.

**GRAMMOPTERA Audinet-Serville, 1835**
**Common Grammoptera**
***Grammoptera ruficornis*** (Fabricius, 1781)
**SQS-1**

7 VC7, n.d. (RRUK)
ST86: Neston Park, 3-Jun-04 (BLe, KNAA); Box, 7-Jun-24 (JVB)
ST87: Corsham Park, 3-Jun-04 (BLe)
ST96: Spye Park, 31-May-04 (KNAA)
ST97: Chippenham, Lackham College, 29-May-04 (BLe); North Draycot Park, 22-May-00 (MDD)
ST99: Flisteridge Wood, 13-Jun-82 (KNAA); 1, 15-May-97 (APF)
SU05: Drews Pond, 8-Jun-06 (MEN)
SU06: Devizes, 20-Jun-06 (MEN); Oliver's Castle, 4-May-05 (IJG)
SU08: Brockhurst Meadows, 25-Jun-06, 9-Jun-07, 25-Jun-08 (MEN); Minety (Stonehill wood), 13-May-02 (JKG det MDD); Ravensroost Wood, 29-Jun-05 (IJG); Somerford Common, 20-Jun-05 (IJG)
SU09: Cotswold Water Park, 27-Apr-97 (APF) Somerford Keynes, Swillbrook Lakes, 4-Jun-04 (BLe); Upper Waterhay, 17-May to 11-Jun-05, 5 to 11-Jun-07 (JKG det MDD)
SU16: Furze Hill West, 22-Jun-05 (IJG); Marlborough area, Jun-03 (EWJ) (Meyrick, 1874); 1869-1938 (Meyrick, 1938); West Woods, May-02 (MDD)
SU17: Clout's Wood, 2-Jul-99 (MEN)
SU19: Hannington, 8-May, 3-Jul-99 (MEN); Highworth, Pentylands Lane, 5-Jun-04 (MEN)
SU25: Sidbury Hill (SPTAE), 28-May-97 (MGT)
SU26: Bedwyn Common, 28-Jun-06 (MEN); Savernake Forest, 26-May to 8-Jul-93 (EN det SAW); 31-May to 23-Jun-95 (EN); 31-May-00 (WANHS det MDD); 31-May-05 (IJG)
SU27: Ramsbury, 3-Jun-05 (IJG)

SU29: Highworth, 18-May-07, 3, 29-Jun-06 (MEN); Raglans Wood, 23-Jun-06 (MEN); Starveall Barn, 23-Jun-07 (MEN); Watchfield, 15-May-99 (MEN)
8 ST85: Green Lane Wood, 7-Jun-95 (MDD); Bullen Hill Farm, 1-Jan-82 (EGS det MDD); Green Lane Wood, 15-Jun-92 (PM)
ST91: Cranborne Chase, 16 to 25-Jun-80 (NCC); 7-May-97 (MAS); 22-Jun-99 (MDD)
ST92: Wardour, 5-Jun-96 (MAS)
ST93: Fonthill, 26-Jun-02 (MDD); Great Ridge Wood, 14-Jun-96, 16, 25-May-00 (MDD)
ST93: Sutton Mandeville, 2-Jun-95 (MDD)
ST94: Norton Down (SPTAW), 15-May-02 (PJH) (Hodge, 2002)
SU02: Chickengrove Bottom, 21-May-93 (PM); Middleton Down, 21-May-93 (PM)
SU03: Grovely Wood, 8-Apr-72, 27-May-74 (DRN); 11-Jun-00 (MDD)
SU13: Cockey Down, 4-Jun-91 (PM); Little Durnford Down, 13-Jun-73, 28-Apr-93 (PM); Old Sarum, 31-May-97 (MAS); Salisbury, Apr-00 (Darby, 2001); Salisbury, Bemerton Folly, 7-Mar - 15-Jun-02 (MDD) (Darby, 2002d)
SU14: Haxton (SPTAE), 17-May-02 (PJH) (Hodge, 2002); Stonehenge, 22-Apr-03 (PJH) (Hodge, 2003)
SU16: Jones's Mill, 21-Jun-97 (MDD); 12-Jun-99 (HK det MDD); 19-Jun-01, 2-Jul-01 (APF) (Foster, 2002)
SU21: Burnt Ground Wood, 31-May-74 (DRN)
SU22: Bentley Wood, 31-May-97 (MDD); Blackmoor Copse, 2-Jul-96 (MDD); 9-Jun-01, 17-May-02 (APF) (Foster, 2004); Langley Wood NNR, 28-May-74 (DRN); 8-Jun-02, 2-Jul-07 (MDD) (Darby, 2008) PepperBox Hill, 9-May-93 (PM); White-parish Common, 2-Jul-95 (MAS)
SU23: Porton Down, 3-Jul-99 (PJH) (Hodge, 2002)
SU25: Linden Copse, Everleigh Ashes, 19-May-05 (IJG)
SU26: Bedwyn Brail, 2-Jun-03 (MDD)

Resident. Ubiquitous; on woodland flowers. Larvae in fungus-infected small branches of various broadleaved trees.

Tribe LEPTURINI

**LEPTURA Linnaeus, 1758**
**Four-banded Longhorn Beetle**
***Leptura quadrifasciata*** Linnaeus, 1758
**IEC-3, SQS-2**

7 SU26: Savernake Forest, 3-Aug-89, 15-Jul to 3-Aug-95 (EN det MDD)
8 SU01: Martin Down, 16-Jun-95 (MCH)
SU03: Grovely Wood, 4-Aug-70, 23-Aug-72, 1-Aug-88 (DRN); Steeple Langford, Jul-05 (MDD)
SU13: Salisbury, 26-May-03, 9-Jul-05, 8, 9-Jun-06 (PM)
SU21: Franchises Wood, 4-Aug-77 (DRN); Hamptworth Estate, 26-Jul-77 (DRN); Landford Bog, 22-Jun-03 (APF) (Foster, 2003)

Leptura quadrifasciata

SU22: Blackmoor Copse, 3-Aug-03 (APF) (Foster, 2004); 10-Jul-05 (PM); Bentley Wood, 1, 11-Aug-86 (DRN); 1996 (JN)
SU25: Linden Copse, Everleigh Ashes, 10-Jul-02 (IJG)

Resident. Widespread; on flowers in wooded areas and on a log, larvae in dead wood of various broadleaved trees, especially birch. More frequent than in the past. Should not be confused with *L.aurulenta* which is also becoming more frequent but not yet found in the county.

**PARACORYMBIA Miroshnikov, 1998**
**Tawny Longhorn Beetle**
***Paracorymbia fulva*** (De Geer, 1775)
**RDB3, SQS-24**

8 ST92: Sutton Mandeville, 2-Jul-07 (MDD)
SU02: Bishopstone, Jul-91 (MDD); Middleton Down, 6-Jul-06 (PM)
SU03: Steeple Langford, 14-Jul-05 (MDD)
SU04: Yarnbury Castle, 22-Jul-08 (MDD)
SU12: Coombe Bissett Down, 31-Jul-03, 13-Jul-06 (PM); Milford, 29, 31-Jul, 1-Aug-91 (DRN)
SU13: Salisbury, Jul-47 (HGS); 2003, 2004, 2005, Jul-o6, 1-Jul-08 (PM)
SU23: Newton Tony, 26-Jun-03 (DG det MDD)

Resident. Local. A southern Wiltshire speciality. Nationally regarded as rare but found here frequently on umbelliferae and thistles usually in grassland. The biology is unknown but is probably subterranean.

Paracorymbia fulva

**ANOPLODERA Mulsant, 1839**
**Six-spotted Longhorn Beetle**
***Anoplodera sexguttata*** (Fabricius, 1775)
**RDB3, IEC-2, SQS-24**

7 SU26: Savernake Forest, 1972 (PJC); 1976 (CMJ); 27-May to 16-Jun-93 (EN det SAW); 1-Jun-04 (MH det KNAA)
8 SU12/13: Clarendon Wood, Salisbury, c.1802 (DL) (Marsham, 1802)

Resident. Very local; on Umbelliferae and Hawthorn, larvae in dead wood of oak and beech. Recent records from Savernake only.

**PACHYTODES Pic, 1891**
**Speckled Longhorn Beetle**
***Pachytodes cerambyciformis*** (Schrank, 1781)
**SQS-2**

7 VC7, n.d. (RRUK)
SU16: Marlborough area, 1869-1938 (Meyrick, 1938)

Resident. Very local; on flowers in summer, larvae in rotten wood, especially exposed roots of fallen trees. No recent records which is surprising given that the insect is not uncommon in neighbouring counties.

**ALOSTERNA Mulsant, 1863**
**Tobacco-coloured Longhorn Beetle**
***Alosterna tabacicolor*** (De Geer, 1775)

7 VC7, n.d. (RRUK)
ST97: Chippenham, Lackham College, 29-May-04 (BLe)
ST98: Braydon Pond Plantation, 27-May-91 (KNAA)
ST99: Flisteridge Wood, 13, 27-Jun-82 (KNAA); 15-May-97 (APF)
SU16: Marlborough area, 1869-1938 (Meyrick, 1938)
SU26: Savernake Forest, 30-May-73 (DBS); 6-Jun-73 (APF); 15-Jun-73 (SAW); 22-Jun-90 (APFw); 26-May to 16-Jun-93, 9-Jun to 1-Jul-94 (EN det SAW); 31-May to 23-Jun-95 (EN det MDD); 30-May-04 (KNAA) 8-Jun-00, 31-May-05 (IJG)
8 ST85: Green Lane Wood, 7-Jun-95 (MDD)
ST91: Cranborne Chase, 16 to 25-Jun-80 (NCC); 23-Jun-97 (MAS); 22-Jun-99 (MDD)
ST93: Great Ridge Wood, 16-May-00 (MDD)
SU03: Grovely Wood, 16-Apr-84 (DRN); 11-Jun-00 (MDD); Little Langford Down, 26-May-99 (PM)
SU13: Old Sarum, 31-May-97 (MAS)
SU22: Bentley Wood, 4-Jul-90, 15-Jun-93 (PM); 27, 31-May-97 (MDD); Blackmoor Copse, 31-May-93, 13-Jun-94 (PM); 9-Jun-01 (APF) (Foster, 2004); Langley Wood NNR, 6-Jun-02 (MDD)
SU26: Bedwyn Brail, 2-Jun-03 (MDD)

Resident. Widespread; adults on flowers, larvae in damp rotten stumps and branches.

**PSEUDOVADONIA Lobanov, Danilevsky & Murzin, 1981**
**Fairy-ring Longhorn Beetle**
***Pseudovadonia livida*** (Fabricius, 1777)
**SQS-8**

8 SU13: Little Durnford, Jun-03 (MDD)
SU14: Wexland Hanging (SPTAC), 3-Jun - 12-Aug-08 (Boyce & Wright, 2009)
SU24: Beacon Hill (SPTAE), 3-Jun - 12-Aug-08 (Boyce & Wright, 2009)

Resident. Very local; on Umbelliferae in grassland. Associated with the fairy ring fungus *Marasmius oreades*.

**STRANGALIA Audinet-Serville**
[***Strangalia attenuata*** (Linnaeus, 1758)
**Extinct**

8 SU12/13: Salisbury, c.1830 (Stephens, 1831, 1839)]

Similar in appearance to *R.maculata* and sometimes said to be a variant of that species. Stephens cites it as 'Very rare: several specimens have been captured at different periods near Salisbury...'. Included in the latest British catalogue (Duff, 2008) but the last confirmed record seems to be from Windsor in 1845, another, from Arundel in 1991, being doubtful. Its status in Britain is reviewed by Kauffmann (1995).

**RUTPELA Nakane & Ohbayashi, 1959**
**Black-and-yellow Longhorn Beetle**
***Rutpela maculata*** Poda, 1761
**SQS-1**

7 ST99: Flisteridge Wood, 13, 27-Jun, 17-Jul-82 (KNAA)
SU08: Brockhurst Meadows, 25-Jun-06, 9-Jun-07 (MEN); Ravensroost Wood, 29-Jun-05 (IJG); Echo Lodge Meadows, 12-Jun-02 (MS det MDD)
SU16: Marlborough area, Jun-03 (EWJ) (Meyrick, 1874), Aug-1879 (MC) (Meyrick, 1880);

Oare, 26-Jun-05 (IJG)
SU26: Bedwyn Common, 28-Jun-06 (MEN)
SU29: Highworth, 29-Jun-06 (MEN)
8 ST84: Longleat, 11-Aug-05 (DJG)
ST92: Sutton Mandeville, 7-Jul-96 (MDD)
ST93: Fonthill, 19-Jul-95 (MDD)
ST95: West Lavington, 1937 (HGM) (DFL, 1948)
SU01: Martin Down, 16-Jun-95 (MCH)
SU02: Chickengrove Bottom, May-97 (BL)
SU03: Grovely Wood, 21, 23-Aug-72, 30, 31-Jul-79 (DRN)
SU12: Petersfinger, 25-Jul-91 (DRN)
SU21: Landford Bog, 24-Jun-01 (APF) (Foster, 2001); 22-Jun-03 (APF) (Foster, 2003);
SU22: Bentley Wood, 18-Jul-04 (AGD); Blackmoor Copse, 2-Jul-96 (MDD); East Grimstead, 1-Aug-78, 11-Aug-86 (DRN)Hamptworth Estate, 31-Jul-85 (DRN)
SU23: Bentley Wood, 1996 (JN); 12-Aug-05 (IJG);
Resident. Widespread; on flowers in summer, larvae in damp rotting wood of a wide range of trees.

**STENURELLA Villiers, 1974**
**Black-striped Longhorn Beetle**
***Stenurella melanura*** (Linnaeus, 1758)
**SQS-2**
7 VC7, n.d. (RRUK)
ST99: Flisteridge Wood 13, 27-Jun-82 (KNAA)
SU08: Brockhurst Meadows, 9-Jun-07, 25-Jun-08 (MEN); Somerford Common, 3-Aug-96 (MEN); 11-Jul-01 (JKG det MDD); 20-Jun-05 (IJG)
SU16: Marlborough area, 1-Jun-1873 (EWJ) (Meyrick, 1874); 30 Jun 1879 (EEDeS) (Meyrick, 1880); 1869-1938 (Meyrick, 1938)
SU17: Clout's Wood, 2-Jul-99 (MEN)
SU19: Highworth, 16-Jun-02, 16-Jul-03, 17-Jun-06 (MEN)
SU26: Bedwyn Common, 28-Jun-06 (MEN); Savernake Forest, 30-May-73 (DBS); 15-Jun-73 (SAW); 16-Jun to 7-Jul-93 (EN det SAW)
SU27: Aldbourne, 23-Jun-05 (IJG)
8 ST84: Longleat, 15-Jun-73 (PM); 1-Jun-97 (MB); 7-Jul-99 (MDD)
ST91: Cranborne Chase, 16 to 25-Jun-80 (NCC); 14-Jun-97 (MAS)
ST93: Chilmark, 21-Jun-99 (BL)
ST95: Great Cheverell Hill, 17-Jun-94 (PM)
SU01: Martin Down, 16-Jun-95 (MCH)
SU03: Grovely Wood, 30-Jul-79, 28-Jul-83 (DRN); Wylye Down, 14-Jun-97 (MGT)
SU04: Orcheston Down (SPTA), 22-Jun-02 (PJH) (Hodge, 2002)
SU06: Devizes, 19-Jun-06 (MEN); Morgan's Hill, 5-Jun-06 (MEN)
SU13: Cockey Down, 29-Jun-90 (PM)
SU14: Wexland Hanging (SPTAC), 3-Jun - 12-Aug-08 (Boyce & Wright, 2009)
SU15: Everleigh (SPTAE), 24-Jun-02 (PJH) (Hodge, 2002)
SU16: Jones's Mill, 1997 (MAS)
SU21: Landford Bog, 24-Jun-01 (APF) (Foster, 2001), 22-Jun-03 (APF) (Foster, 2004)
SU22: Bentley Wood, 28, 29-May-90 (PM); 2-Jul-96, 31-May-97 (MDD); 1-Jun-97 (MGT); Blackmoor Copse, 2-Jun-92 (PM); 2-Jul-96 (MDD)
SU23: Porton Down, 3-Jul-99 (PJH) (Hodge, 2002); West Winterslow, 7-Jul-01 (NB)
SU24: Furze Hill, Tidworth, 22-Jun-05 (IJG)
SU25: Ludgershall, Apr - Jul-1915 (PAH) (Harwood, 1915)
Resident. Ubiquitous; on flowers in the summer. Larvae develop in smaller pieces of dead wood usually in close contact with the ground.

**Small Black Longhorn Beetle**
***Stenurella nigra*** (Linnaeus, 1758)
**Na, IEC-possible addition, SQS-24**
8 VC8 (Hyman and Parsons, 1992)
SU22: Bentley Wood, 8-Jun-97; 4-Jul-90 (PM); East Grimstead, 1-Aug-78 (DRN)
Very local; on buttercups in a ride and on a thistle at the edge of a conifer wood.Easily confused with dark specimens of *S. Melanura.*

## Subfamily SPONDYLIDINAE
### Tribe ASEMINI

**ASEMUM Eschscholtz, 1830**
***Asemum striatum*** (Linnaeus, 1758)
**SQS-2**
7 SU16: Marlborough, 8-Jun-45 (Holmes & Tennant, 1947)
Resident. Very local; around dead wood and roots in coniferous woods. The single specimen under this name in the MCC is inscribed *WPGT Pyrford June '19* which I assume is Pyrford in Surrey. A native of northern Scotland which has spread south. Mainly crepuscular.

**TETROPIUM Kirby, 1837**
**Black Spruce Longhorn Beetle**
***Tetropium castaneum*** (Linnaeus, 1758)
8 ST92: Tisbury, 21-Aug-24 (NHJ) (Joy, 1925)
Naturalised. A surprising record given that the species is only known from the north of England and Scotland but, given the recorder, one assumes it was not mis-identified!

**Larch Longhorn Beetle**
**Tetropium gabrieli** Weise, 1905
8 VC8, n.d. (RRUK)
Naturalised. No recent records. Rather local in Larch plantations.

**ARHOPALUS Audinet-Serville, 1834**
**Burnt Pine Longhorn Beetle**
***Arhopalus ferus*** (Mulsant, 1839)
8 SU06: Devizes, Jul-00 (BG)
SU23: Porton Down, 2000 (PTW det MDD)
SU24: Newton Tony, 23-May-01 (DG)
Naturalised. Very local; in pine and rarely spruce. Strongly attracted to scorched trees soon after a fire. Mainly nocturnal and known to come to light.

**Dusky Longhorn Beetle**
***Arhopalus rusticus*** (Linnaeus, 1758)
**SQS-2**
8 SU06: Devizes, Jul-00 (BG)
SU23: Porton Down, 20-Jul-97 (MAS)
Resident. Very local; in coniferous trees. Mainly nocturnal and sometimes attracted to lights

## Subfamily CERAMBYCINAE

### Tribe GRACILIINI

**Basket Longhorn**
***Gracilia minuta*** (Fabricius, 1781)
**RDB2**
7 VC7, n.d. (RRUK)
SU16: Marlborough area, 1869-1938 (Meyrick, 1938)
Resident. Very local; in basketwork [and dry dead twigs particularly of rosaceous species such as bramble]. One of the three MCC specimens is labelled *Poulton 25/6/01*.

Marlborough College specimen of Gracilia minuta

### Tribe OBRIINI

**OBRIUM Dejean, 1821**
**Brown Longhorn Beetle**
***Obrium brunneum*** (Fabricius, 1793)
7 SU16: Marlborough, 13-Jul-02 (EEM) (insect in MCC but not listed by Meyrick, 1938)
8 SU22: Bentley Wood, 27, 31-May-97 (MDD)
Resident, naturalised. Very local. The Bentley Wood examples on flowers near conifers, their usual habitat. This is the most westerly record of this beetle in Britain.

### Tribe MOLORCHINI

**MOLORCHUS Fabricius, 1793**
**Spruce Shortwing Beetle**
***Molorchus minor*** (Linnaeus, 1758)
7 VC7, n.d. (RRUK)
SU29: Raglans Wood, 23-Jun-07 (MEN)
8 VC8, n.d. (RRUK)
Resident, naturalised introduction. Very local; on flowers associated with coniferous woodland.

### Tribe CALLICHROMATINI

**AROMIA Audinet-Serville, 1833**
**Musk Beetle**
***Aromia moschata*** (Linnaeus, 1758)
**Nb, SQS-8**
7 VC7, n.d. (RRUK)
SU16: Marlborough area, Jun-1873 (Meyrick, 1874); 1869-1938 (Meyrick, 1938)
SU27: Ramsbury, 1870-1873 (Meyrick, 1874)
8 ST95: West Lavington, 11-Jul-46 (JA) (DFL, 1948)
SU03: Berwick St. James, 15-Sep-07 (BL); Deptford Ponds, 2002 (JF); Steeple Langford, 2005 (Larval borings) (MDD)
Resident. Local; in young and healthy wood of willows and sallows, often near the ground. Adults fly in sunshine but again near to the ground. Probably more frequent than these records suggest because of its reclusive habits.

Aromia moschata

### Tribe CALLIDIINI

**CALLIDIUM Fabricius, 1864**
**Violet Longhorn Beetle**
***Callidium violaceum*** (Linnaeus, 1758)
7 VC7, n.d. (RRUK)
SU16: Marlborough area, Jun-1873 (EWJ) (Meyrick, 1874); 22 Jun 1879 (FSA)

(Meyrick, 1880); 1869-1938 (Meyrick, 1938)
Resident, naturalised introduction. No recent records in GB. Meyrick noted in 1896 'it is in abundance in a summer house of un-dressed fir timber (with the bark on) erected some four years at the bottom of the master's garden; the larvae are burrowing through the wood in all directions, and apparently intend to destroy it completely' (*Rep.Marlboro.Coll.nat.Hist. Soc.* 45, 1897)

**PHYMATODES Mulsant, 1839**
**Tanbark Borer**
***Phymatodes testaceus* (Linnaeus, 1758)**
**IEC-3, SQS-4**
7 VC7, n.d. (RRUK)
ST86: Neston Park, 3-Jun-04 (BLe)
ST87: Corsham Court Park, 3-Jun-04 (KNAA)
SU16: Marlborough area, 1869-1938
SU26: Savernake Forest, 31-May-04 (BLe)
8 ST85: Green Lane Wood, 19-Jun-04 (MDD)
ST95: West Lavington, 20-May-44 (DCL) (DFL, 1948)
SU02: Knighton Wood, Jul-08 (AJWA)
SU03: Dinton Park, 6-Jul-01 (KNAA) (Alexander, 2004)
SU22: Blackmoor Copse, 17-Mar-02 (APF) (Foster, 2004)
Resident. Widespread; in recently dead wood (such as broken branches) of various trees, particularly oak, in ancient woodland.

Marlborough College specimen of Callidium violaceum

**POECILIUM Fairmaire, 1864**
**White-banded Longhorn Beetle**
***Poecilium alni* (Linnaeus, 1767)**
**Nb, IEC-addition, SQS-16**
7 VC7, n.d. (RRUK)
SU16: Marlborough area, Jun-1873 (EWJ) (Meyrick, 1874); 1869-1938 (Meyrick, 1938); 1-Jun-04 (JWe det KNAA)
8 VC8, n.d. (RRUK)
Resident. Very local; in dead wood [under bark of recently dead or dying twigs of various broadleaved trees, particularly oak]. Much declined in GB in recent years. Single specimen in MCC without data.

Tribe CLYTINI

**CLYTUS Laicharting, 1784**
**Wasp Beetle**
***Clytus arietis* (Linnaeus, 1758)**
**SQS-1**
7 VC7, n.d. (RRUK)
ST87: Castle Combe, 3-Jun-04 (KNAA)
ST97: Chippenham, 29-May-04 (BLe)
ST98: Lacock, 29-May-04 (BLe)
SU08: Minety (Stonehill wood), 17-May-02 (JKG det MDD); Somerford Common, 16-May-00 (JKG det MDD)
SU09: Upper Waterhay, 5 to 11-Jun-07 (JKG det MDD)
SU15: Manningford Bruce, 2006 (IJG)
SU16: Lockeridge, 1994 (JEO); Marlborough area, Jun-1873 (EWJ) (Meyrick, 1874); 2-Jun-1879 (WFB) (Meyrick, 1880); 1873-1938 (Meyrick, 1938)
SU19: Hannington, 22-Jun-96, 26-May-97, 30-May-98, 3-Jul-99 (MEN); Highworth, 8-May-94, 7-May-95, 3-Jul-98, 16-Jun-02, 7, 15, 16, 21-Jun-03, 18-May-07 (MEN)
SU26: Savernake Forest, 17-May-90 (TGW); 25-Apr to 7-Jul-93 (EN det SAW)
8 ST85: Bullen Hill Farm, 1982-2002 (EGS det MDD); Green Lane Wood, 5-Jun-96 (RB), 19-Jun-04 (MDD)
ST91: Cranborne Chase, 16 to 25-Jun-80 (NCC); 7-May-97 (MAS)
ST92: Sutton Mandeville, 24-May-92 (MDD); Wardour, 5-Jun-96 (MAS)
ST94: Salisbury Plain, 12-Jun-93 (SP)
ST95: West Lavington, 2-Jun-40, 15-Jun-41, 13-Jun-46 (ITH) (DFL, 1948)
SU03: Berwick St. James, 14-Jun-96 (BL); 26-May-99 (MDD); Dinton, 5-Jun-93 (SP); Grovely Wood, 27-May-74 (DRN)
SU04: Fox Covert (SPTAW), 3-Jun - 12-Aug-08 (Boyce & Wright, 2009)
SU13: Cockey Down, 21-May-91, 2-Jun-92 (PM); Little Durnford Down, 18-May-93 (PM); Salisbury, 15-Jun-73 (PM)
SU14: Stonehenge, 22-Apr-03 (PJH) (Hodge, 2003)
SU16: Jones's Mill, 4-Jun-04 (BLe)
SU22: Blackmoor Copse, 21-May-92 (PM); 31-May, 2-Jul-96 (MDD); 17-May-02, 3-Aug-03 (APF) (Foster, 2004); Langley Wood NNR, 28-May-74 (DRN); May-99 (BL)
SU23: Bentley Wood, 22-May-90 (PM)

Resident. Widespread; in dead wood of various trees, especially fence posts. Sometimes seen on the wing and often the first longhorn to appear in spring.

Tribe ANAGLYPTINI

**ANAGLYPTUS Mulsant, 1839**
**Rufous-shouldered Longhorn Beetle**
***Anaglyptus mysticus*** (Linnaeus, 1758)
**Nb, SQS-4**
7 SU08: Great Wood, 27-May-91 (KNAA)
SU09: Upper Waterhay, 17-May to 4-Jun-05 (JKG det MDD)
SU19: Highworth, Pentylands Lane, 17-Jun-06
SU26: Savernake Forest, 6-Jun-73 (APF); 6-Dec-73 (PS); 31-May to 23-Jun-95 (EN det MDD)
8 ST85: Bullen Hill Farm, 24-May-03 (EGS det MDD); Green Lane Wood, 7-Jun-95 (WANHS det MDD)
SU09: Clattinger Farm, 2-Jun-06 (PM)
SU22: Blackmoor Copse, 17-May-02 (APF) (Foster, 2004)
Resident. Local; adults often seen on hawthorn, larvae in dead wood of various broadleaved trees particularly oak.

Subfamily LAMIINAE
Tribe MESOSINI

**MESOSA Latreille, 1829**
**White-clouded Longhorn Beetle**
***Mesosa nebulosa*** (Fabricius, 1781)
**RDB3, IEC-2, SQS-24**
8 SU22: Blackmoor Copse, 14-Jun-05 (PM)
Resident. Very local; crawling on a stump. Larvae develop in rotten branches of various broadleaved trees.

Mesosa nebulosa

Tribe AGAPANTHIINI

**AGAPANTHIA Audinet-Serville, 1835**
**Golden-bloomed Grey Longhorn Beetle**
***Agapanthia villosoviridescens*** (De Geer, 1775)
7 SU06: Beckhampton, Jul-00 (BG); Roundway Down, 22-May-04 (AC det GO'D)
SU08: Somerford Common, 27-May-01 (MDD)
SU09: Upper Waterhay, 6-Jul-07 (JKG det MDD)
SU17: Clouts Wood, 18-Jun-07 (MEN)
SU18: Wroughton, 6-May-08 (JSm)
SU19: Hannington, 26-May-97 (MEN)
SU26: Hat Gate Reserve, 24-May-00 (IJG)
SU27: High Clear Down, 27-Jul-99 (PM)
8 SU04: Fox Covert (SPTAW), 3-Jun - 12-Aug-08 (Boyce & Wright, 2009); Yarnbury Castle, 2-Jul-08 (MDD)
SU15: Everleigh (SPTAE), 24-Jun-02 (PJH) (Hodge, 2002)
SU16: Pewsey, Jones's Mill, 1-Jun-04 (BLe)
SU24: Bulford Ranges (SPTAE), 3-Jun - 12-Aug-08 (Boyce & Wright, 2009)Resident.
Widespread; usually on hogweed. Extended its range in recent years in GB.

Agapanthia villosoviridescens

Tribe POGONOCHERINI

**POGONOCHERUS Dejean, 1821**
**Greater Thorn-tipped Longhorn Beetle**
***Pogonocherus hispidulus*** (Piller & Mitterpacher, 1783)
**SQS-2**
7 SU16: Oare, 2006 (IJG); Marlborough area, 1873-1938 (Meyrick, 1938)
8 ST92: Sutton Mandeville, 22-Nov-92 (MDD)
SU22: Bentley Wood, 30-Oct-04 (DT det MDD); Blackmoor Copse, 3-Aug-03 (APF) (Foster, 2004)
Resident. Local; on dead twigs of broadleaved trees in woods, and one at light.

**Lesser Thorn-tipped Longhorn Beetle**
***Pogonocherus hispidus*** (Linnaeus, 1758)
7 VC7, n.d. (RRUK)

SU16: Marlborough area, Jun-1873 (EWJ) (Meyrick, 1874)
SU26: Savernake Forest, 30-May-04 (KNAA)
SU29: Highworth, 3-Jun-06 (MEN)
8 ST91: Cranborne Chase, 16 to 25-Jun-80 (NCC)
SU13: Salisbury, 26-Apr-98 (PM)
SU22: Bagfield Copse, 6, 7-Aug-86 (DRN)
Resident. Local; in shrubs including holly and ivy in woods and old hedgerows.

Tribe ACANTHOCININI

**Black-clouded Longhorn Beetle**
***Leiopus nebulosus*** (Linnaeus, 1758)
**SQS-2**
7 VC7, n.d. (RRUK)
ST86: Neston Park, 3-Jun-04 (BLe)
ST87: Corsham Court Park, 3-Jun-04 (KNAA)
SU16: Marlborough area, Jun-1873 (EWJ) (Meyrick, 1874); 27 May 1879 (CTS) (Meyrick, 1880); 1869-1938 (Meyrick, 1938)
SU26: Savernake Forest, 28-Jul to 20-Sep-93 (EN SAW); 31-May to 23-Jun--95 (EN det MDD)
8 ST85: Green Lane Wood, 7-Jun-95 (WANHS det MDD); 2-Jun-04 (AH det KNAA)
ST91: Cranborne Chase, 16 to 25-Jun-80 (NCC)
ST93: Pits Wood, Chilmark, 2-Jun-95 (MDD); Stockton Wood, 7-Jul-96 (MAS)
ST95: Earlstoke Wood, 9-May-00 (EGS det MDD)
SU03: Grovely Wood, 11, 16-Aug-71, 29-Jul, 4-Aug-78, 1- Aug-80 (DRN)
SU16: Jones's Mill, 19-Jun-01 (APF) (Foster, 2002)
SU21: Burnt Ground Wood, 31-May-74 (DRN)
SU22: Bagfield Copse, 7-Aug-86 (DRN); Langley Wood, 12-Apr-74 (DRN); 6-Aug-04 (MDD) (Darby, 2005)
Resident. Widespread; in woods.

Tribe SAPERDINI

**SAPERDA Fabricius, 1775**
**Small Poplar Borer**
***Saperda populnea*** (Linnaeus, 1758)
7 SU08: Red Lodge Pond, 31-May-01 (JKG det MDD)
8 ST85: Picket Wood, 17-Jun-83 (EGS det MDD)
SU21: Hamptworth Estate, 30-Mar-78 (DRN) (Nash, 1979), 24-Apr-84 (DRN)
SU22: Bentley Wood, 30-May-90 (PM)
Resident. Local; on aspen in particular. Presence is often given away by the galls which the larvae make in the living branches.

**STENOSTOLA Dejean, 1835**
**Lime Longhorn Beetle**
***Stenostola dubia*** (Laicharting, 1784)
**Nb, IEC-addition, SQS-8**
7 VC7, n.d. (RRUK)
8 SU22: Blackmoor Copse, 9-Jun-01 (APF) (Foster, 2004)
Resident. Very local; [known to eat the leaves of limes, larvae in dead branches of various broadleaved trees but usually limes].

Tribe PHYTOECIINI

Phytoecia cylindrica

**PHYTOECIA Dejean, 1835**
**Umbellifer Longhorn Beetle**
***Phytoecia cylindrica*** (Linnaeus, 1758)
**Nb**
7 ST98: Braydon Pond Plantation, 27-May-91 (KNAA); Conygre Mead, Malmesbury, 20-Apr-02 (JKG det MDD)
SU16: Oare, 13-Jun-00 (IJG); West Woods, 19-Jul-00 (IJG)
SU19: Hannington, 3-May-99 (MEN); Highworth, 27-May-96 (MEN)
SU26: Savernake Forest, 30-May-04 (KNAA)
8 VC7, n.d. (RRUK)
ST85: Green Lane Wood, 1-Jun-06 (SC det MDD)
ST93: Fonthill, 27-May-95 (MDD)
SU13: Cockey Down, 29-Jun-90 (PM)
SU14: Haxton (SPTAE), 17-May-02 (PJH) (Hodge, 2002); Stonehenge, summer-03 (PJH) (Hodge, 2003)
SU15: Everleigh, 24-Jun-02 (PJH) (Hodge, 2002)
Resident. Widespread; on umbelliferae particularly cow parsley (often inside the supporting stems of the flower head). Larvae develop in the upright stems.

**OBEREA Dejean, 1835**
**Twin-spotted Longhorn Beetle**
***Oberea oculata*** (Linnaeus, 1758)
**RDB1**
8 ST95: West Lavington (Strawberry Hill) 20-May-44 (DCL) (DFL, 1948)
This is a remarkable record of a spectacular insect . The beetle was not uncommon in the 19[th] century from the fens of East Anglia and Cambridgeshire but

is only known since 1900 from one site in the Cam Washes of Cambridgeshire.

Tribe TETROPIINI

**TETROPS Stephens, 1829**
**Plum Longhorn Beetle**
***Tetrops praeustus*** (Linnaeus, 1758)
**SQS-2**
7 SU16: Marlborough area, 1869-1938 (Meyrick, 1938)
SU19: Highworth, Pentylands Lane 30-May-98 (MEN)
8 SU04: Parsonage Down NNR, 2-Jul-08 (MDD)
SU25: Sidbury Hill (SPTAE), 28-May-97 (MGT)
Resident. Very local; in dead hedges and clumps of trees [in dying branches of Rosaceae in particular].

## Family ORSODACNIDAE

**ORSODACNE Latreille, 1802**
***Orsodacne cerasi*** (Linnaeus, 1758)
7 ST97: Chippenham, Lackham College, 29-May-04 (BLe)
SU26: Savernake Forest, 31-May-95 (EN)
8 SU16: Milton Lilbourne, 5-Jul-02 (IJG)
Resident. Very local; on flowers of herbaceous plants, trees and shrubs in mixed and deciduous woodland.

***Orsodacne humeralis*** Latreille, 1802 **Nb**
8 SU22: Bentley Wood, 27-May-98 (MDD)
Resident. Very local; on Hawthorn in woodland ride.

## Family CHRYSOMELIDAE

Subfamily BRUCHINAE

**BRUCHUS Linnaeus, 1767**
***Bruchus atomarius*** (Linnaeus, 1761)
**Nb**
7 ST97: Chippenham, Lackham College, 29-May-04 (BLe)
SU06: Cherhill Down, 31-May-04 (BLe)
SU16: Marlborough area, 1869-1938 (Meyrick, 1938)
8 ST91: Cranborne Chase, 4-Jun-97 (MAS)
ST94: Norton Down (SPTAW), 23-Jun-02 (PJH) (Hodge, 2002)
SU16: Jones's Milll, 1-Jun-04 (BLe); Woodborough, 4-Aug-83 (DRN)
Resident. Widespread; in a variety of habitats..Adults are probably pollen feeders.

***Bruchus loti*** Paykull, 1800
7 SU16: Marlborough area, 1-Jan-0031-Dec-05 (Meyrick, 1906);1869-1938 (Meyrick, 1938)
8 ST94: Bowls Barrow (SPTAW), 22-Jun-02 (PJH) (Hodge, 2002); Ladywell (SPTAW), 23-Jun-02 (PJH) (Hodge, 2002); Norton Down (SPTAW), 15-May-02 (PJH) (Hodge, 2002)
SU04: Fox Covert (SPTAW), 3-Jun - 12-Aug-08 (Boyce & Wright, 2009)
SU05: Black Heath (SPTAC), 3-Jun - 12-Aug-08 (Boyce & Wright, 2009)
SU13: Salisbury, 26-May-74 (DRN)
SU14: Haxton (SPTAE), 24-Jun-02 (PJH) (Hodge, 2002); Stonehenge, 22-Apr-03 (PJH) (Hodge, 2003); Tinkers Track (SPTAW), 3-Jun - 12-Aug-08 (Boyce & Wright, 2009); Wexland Hanging (SPTAC), 3-Jun - 12-Aug-08 (Boyce & Wright, 2009)
SU15: Everleigh (SPTAE), 18-May, 24-Jun-02 (PJH) (Hodge, 2002)
SU16: Jones's Mill Reserve, 30-May-04 (KNAA)
SU22: Bentley Wood, 31-May, 2-Jul-96 (MDD)
SU24: Dunch Hill (SPTAE), 3-Jun - 12-Aug-08 (Boyce & Wright, 2009)
Resident. Local; on vegetation in woodland and grassland. Adults are probably pollen feeders.

**Bean Beetle**
***Bruchus rufimanus*** Boheman, 1833
7 SU16: Marlborough area, Jun-1873 (EWJ) (Meyrick, 1874); 1869-1938 (Meyrick, 1938)
SU26: Savernake Forest, 20-Sep to 11-Oct-94 (EN det SAW)
8 ST85: Bullen Hill Farm, 1982-2002 (EGS det MDD)
ST92: Sutton Mandeville, 12-Feb-01 (MDD)
ST95: West Lavington, 13-May-45 (DCL) (DFL, 1948)
SU04: Orcheston Down (SPTA), 18-May-02 (PJH) (Hodge, 2002)
SU14: Stonehenge, 22-Apr-03 (PJH) (Hodge, 2003)
SU22: Bentley Wood, 31-May-97 (MDD)
Resident. Widespread; on different plants in a wide variety of habitats. Adults are probably pollen feeders.

***Bruchus rufipes*** Herbst, 1783
7 SU16: Marlborough area, Jun-1873 (EWJ) (Meyrick, 1874); 1869-1938 (Meyrick, 1938)
8 SU04: Shrewton, 6-Aug-82 (DRN)
SU05: Ell Barrow (SPTAC), 3-Jun - 12-Aug-08 (Boyce & Wright, 2009)
SU25: Sidbury Hill (SPTAE), 4-Jun-97 (MGT)
Resident. Very local; by sweeping and on Weld, *Reseda luteola.* Adults are probably pollen feeders.

**BRUCHIDIUS Schilsky, 1905**
***Bruchidius cisti*** (Fabricius, 1775)
**LCGR-Fh Cl**
7 SU16: Pewsey Downs an, 22-Jun-90 (APFw)
8 ST83: Whitesheet Down, 7-Jul-82 (KNAA)

ST94: Ladywell (SPTAW), 23-Jun-02 (PJH) (Hodge, 2002)
SU03: Wylye Down NNR, 10-Jul-07 (MDD) (Darby, 2009)
SU05: Black Heath (SPTAC), 3-Jun - 12-Aug-08 (Boyce & Wright, 2009)
SU14: Haxton (SPTAE), 24-Jun-02 (PJH) (Hodge, 2002); Wexland Hanging (SPTAC), 22-Jun-02 (PJH) (Hodge, 2002); 3-Jun - 12-Aug-08 (Boyce & Wright, 2009)
SU23: Porton Down, 3-Jul-99 (PJH) (Hodge, 2002)
SU24: Dunch Hill (SPTAE), 3-Jun - 12-Aug-08 (Boyce & Wright, 2009); North Tidworth, Devils Ditch, 30-May-04 (BLe)
SU25: Linden Copse, Everleigh Ashes, 6-Jun-01 (IJG); Sidbury Hill (SPTAE), 18-May-97 (MGT)
Resident. Widespread; on various plants in calcareous grassland.

**CALLOSOBRUCHUS Pic, 1902**
**Cowpea Weevil, Cowpea Seed Beetle**
***Callosobruchus maculatus*** (Fabricius, 1775)
7 SU27: Ramsbury, 20-Oct-06 (CM det MDD)
Resident. Very local; in stored pulses. A major pest of stored foodstuffs but only scattered records in GB since 1925.

## Subfamily DONACIINAE

***Donacia aquatica*** (Linnaeus, 1758)
**RDB3**
8 SU12: Salisbury, 1830 (AI)
Resident. Very local; [on sedges, grasses and reeds by water].

Donacia aquatica

***?Donacia bicolora*** Zschach, 1788
**RDB2**
7 SU16: Marlborough area, Jun-1873 (EWJ) (Meyrick, 1874); 1869-1938 (Meyrick, 1938)
Resident. Very local; on *Sparganium* and sedges besides water. No specimen in the MCC and considered doubtful because of the difficulties of determination.

***Donacia cinerea*** Herbst, 1784
**Nb**
7 SU09: Cotswold Water Park, 28-Apr-82 (GNF)
Resident. Very local; [on bulrushes on which they feed].

Donacia dentata

***Donacia dentata*** Hoppe, 1795
**Na**
8 SU12/13: Salisbury, c.1830 (AI) (Stephens, 1830)
Resident. Very local; [usually associated with arrowhead and possibly also water-plantains, *Alisma*].

***Donacia semicuprea*** Panzer, 1796
7 ST96: Kennet & Avon Canal, 2-Jul-97 (MNS)
SU16: Marlborough area, 1869-1938 (Meyrick, 1938)
8 ST93: Sutton Mandeville, 2-Jun-95 (MDD)
SU12: Charlton-All-Saints, 2-Jun-95, 9-May-96 (MAS)
SU13: Lower Woodford, 2-Aug-82 (DRN)
Resident. Local; at edge of water [where there is sweetgrass, *Glyceria*, on which the adults feed].

***Donacia simplex*** Fabricius, 1775
7 ST96: Kennet & Avon Canal, 2-Jul-97 (MNS)
SU16: Marlborough area, Jun-1873 (EWJ) (Meyrick, 1874); 1869-1938 (Meyrick, 1938)
8 SU13: Lower Woodford, 28-Jul-78 (DRN)
SU14: Figheldean, R. Avon, 8-Jun-97 (MGT); Netheavon, 8-Jun-97 (MGT)
SU16: Pewsey, Kennet & Avon Canal, 1-Jun-04 (BLe); Wootton Rivers, 8-Jun-97 (MGT det MLC)
Resident. Local; on vegetation [usually bur-reeds, *Sparganium*,] at edge of water. The adults eat the surface of the leaves.

***Donacia thalassina*** Germar, 1811
**Nb**
7 SU16: Marlborough area, Jun-1873 (EWJ) (Meyrick, 1874); 1869-1938 (Meyrick, 1938)

8 SU15: New Mill, nr Pewsey, 12-Jun-99 (HK det MDD)
Resident. Very local; on water plants, chiefly *Carex,* and vegetation at edge of a pond.

***Donacia versicolorea*** (Brahm, 1791)
7 SU26: Bedwyn, 1869-1938 (Meyrick, 1938)
8 SU22: Bentley Wood, 1-Aug-78 (DRN)
Resident. Very local; chiefly on broadleaved pondweed, *Potamogeton*, and on/around arrowhead, *Sagittaria*.

***Donacia vulgaris*** Zschach, 1788
7 SU09: Cotswold Water Park, 21-Jun-90 (APFw); 30-May-97 (JSD)
Resident. Very local; [usually amongst stands of common reed, *Phragmites australis,* and bulrushes, *Typha*].

**PLATEUMARIS Thomson, C.G, 1859**
***Plateumaris affinis*** (Kunze, 1818)
**Nb**
7 SU16: Marlborough area, Jun-1871 (RH) (Hislop, 1871); 1869-1938 (Meyrick, 1938)
SU36: Stype Wood, Jun-1870 (RH) (Hislop, 1871)
8 ST93: Fonthill, 27-May-95 (MDD)
SU12: Britford, Jun - 15-Sep-02 (CMD) (Drake, 2008) (this is presumably the species referred to by Drake (2008) as *rustica*)
SU13: Lower Woodford, Jun - 15-Sep-02 (CMD) (Drake, 2008) see above; Middle Woodford, 27-Jul-77, 26-May-74 (DRN)
SU16: Jones's Mill, 19-Jun-01 (APF) (Foster, 2002)
Resident. Local; on vegetation [mainly sedges] at the edge of rivers.

Plateumaris braccata

***Plateumaris braccata*** (Scopoli, 1772)
**Na**
8 SU16: Jones's Mill, 30-May-04 (KNAA)
Resident. Very local; [in beds of common reed, *Phragmites australis*, growing in water at the margins of lakes, canals and dykes in fens].

***Plateumaris discolor*** (Panzer, 1795)
7 ST96: Lacock, 16-Jun-82 (KNAA)
SU16: Jones's Mill, 19-Jun-01 (APF) (Foster, 2002); 30-May-04 (KNAA)
Resident. Very local; [on vegetation in boggy areas and around ponds, rivers and canals].

***Plateumaris sericea*** (Linnaeus, 1758)
7 SU16: Marlborough area, 12 Jun 1879 (RAW) (Meyrick, 1880); 1869-1938 (Meyrick, 1938)
8 ST93: Fonthill, 27-May-95 (MDD)
SU13: Lower Woodford, 28-Jul-78 (DRN); Salisbury, 26-May-74 (DRN)
SU16: Jones's Mill, 30-May-04 (KNAA)
SU26: Wolfhall Bridge, Kennett and Avon canal, 13-Jun-01 (IJG)
Resident. Local; on vegetation at the edge of ponds, rivers and a canal.

## Subfamily CRIOCERINAE

**LEMA Fabricius, 1798**
***Lema cyanella*** (Linnaeus, 1758)
7 SU09: Upper Waterhay, 5-Jun-05 (JKG det MDD)
SU16: Marlborough area, Jun-1873 (EWJ) (Meyrick, 1874); 31 Jan 1879 (EBB) (Meyrick, 1880); 1869-1938 (Meyrick, 1938)
SU26: Savernake Forest, 22-Aug-90 (AGD)
8 ST84: Longleat, 4-May-98 (MB)
SU01: Martin Down, 25-Jun-80 (DAS)
SU02: Bishopstone, 19-May-91 (MDD)
SU03: Grovely Wood, 29-Jul-78 (DRN)
SU13: Cockey Down, 14-May-91 (PM)
SU14: Bourne Bottom (SPTAE), 13-Jun-97 (MGT det MLC)
SU15: Beach's Plantation (SPTAE), 13-Jun-97 (MGT det MLC)
SU16: Woodborough, 4-Aug-82 (DRN)
Resident. Widespread; on vegetation [adults feed on leaves of creeping thistle, *Cirsium arvense*].

**OULEMA des Gozis, 1886**
**Cereal Leaf Beetle**
***Oulema melanopus*** (Linnaeus, 1758)
7 SU09: Ashton Keynes (Millenium Green), 26-Apr-01 (JKG det MDD); Upper Waterhay, 17-May-05 (JKG det MDD)
SU16: Marlborough area, Jun-1873 (EWJ) (Meyrick, 1874); 1869-1938 (Meyrick, 1874)
SU26: Savernake Forest, 9 to 30-Aug, 20-Sep to 11-Oct-94 (EN det SAW)
8 SU03: Berwick St. James, Feb-01 (BL)
SU13: Salisbury, Bemerton Folly, 7-Mar - 15-Jun-02 (MDD) (Darby, 2002d)

SU16: Jones's Mill, 2-Jul-01 (APF) (Foster, 2002)
Resident. Local; on vegetation [feeds on cereal crops] in a wide range of habitats. '*O.melanopus*' was recognised in 1990s to consist of two distinct species: *O. melanopus* and *O. rufocyanea* so that unchecked records before that date are indeterminate.

Oulema melanopus

***Oulema obscura*** (Stephens, 1831)
7 SU09: Somerford Keynes, Swillbrook Lakes, 4-Jun-04 (BLe)
SU16: Marlborough area, 1869-1938 (Meyrick, 1938)
8 ST85: Bullen Hill Farm, 25-Feb-02 (EGS det MDD)
ST92: Wardour, 5-Jun-96 (MAS)
ST93: Great Ridge Wood, 19-Apr-84 (DRN); 14-Jun-96 (MDD)
ST94: Bowls Barrow (SPTAW), 16-May, 22-Jun-02 (PJH) (Hodge, 2002)
SU02: Middleton Down, 6-Nov-73 (BL)
SU03: Grovely Wood, 4-Aug-70, 11-Aug-71, 21-Aug-72, 8-Apr-75, 28-Jul-83 (DRN)
SU12: Charlton-All-Saints, 9-May-96 (MAS); Petersfinger, 25-Jul-91 (DRN)
SU13: Great Durnford, 16-Aug-69, 6-Aug-71, 18-Aug-72, 27-Jul-77, 3-Aug-88 (DRN); Lower Woodford, 8-Apr-70, 5-Aug-82 (DRN)
SU14: Netheravon, 29-Jul-83 (DRN)
SU15: Everleigh (SPTAE), 18-May-02 (PJH) (Hodge, 2002)
SU16: Woodborough, 4-Aug-83 (DRN)
SU21: Hamptworth Estate, 1-Apr-75 (DRN)
SU22: Bentley Wood, 11-Aug-01 (MDD); Glaziers Copse, 16-Aug-88 (DRN)
SU23: Porton Down, 3-Jul-99 (PJH) (Hodge, 2002)
SU24: North Tidworth, Devils Ditch, 30-May-04 (BLe)
SU26: Bedwyn Brail, 5-Nov-03 (MDD)
Resident. Widespread; on vegetation [adults and larvae feed on Poaceae] in a wide variety of habitats.

***Oulema rufocyanea*** (Suffrian, 1847)
8 ST93: Great Ridge Wood, 19-Apr-84 (DRN); Stockton Wood, 28-Sep-95 (PJH) (Hodge, 1995)
ST94: Norton Down (SPTAW), 23-Jun-02 (PJH) (Hodge, 2002)
SU03: Grovely Wood, 24-Aug-72 (DRN)
SU13: Great Durnford, 16-Aug-69 (DRN); Lower Woodford, 27-Jul, 5-Aug-82 (DRN)
SU14: Netheravon, 29-Jul-83 (DRN); Stonehenge, 22-Apr-03 (PJH) (Hodge, 2003)
SU15: Everleigh (SPTAE), 18-May-02 (PJH) (Hodge, 2002)
SU16: Woodborough, 4-Aug-83 (DRN)
SU22: Glaziers Copse, 16-Aug-88 (DRN)
Resident. Local; on vegetation [adults and larvae feed on cereal crops and grasses] in a wide range of habitats.

Crioceris asparagi

**CRIOCERIS Geoffroy, 1762**
**Asparagus Beetle**
***Crioceris asparagi*** (Linnaeus, 1758)
7 SU16: Marlborough area, 1869-1938 (Meyrick, 1938)
8 ST85: Bullen Hill Farm, 1982-2002 (EGS det MDD)
ST92: Sutton Mandeville, 6-Jun-96 (MDD)
ST95: West Lavington, 1937 (HGM), 5-Jun-39 (PLB), 6-Apr-40 (ITH) all (DFL, 1948)
SU02: Bishopstone, 20-Jul-91 (MDD)
Resident. Widespread; on asparagus. One specimen in MCC with data: *Marlborough 22.Jun.*[19]*02.*

Liliocerus lilii

**LILIOCERUS Reitter, 1912**
**Lily Beetle**
***Lilioceris lilii*** (Scopoli, 1763)
7 ST86: Neston Park, 3-Jun-04 (BLe); Bradford-on-Avon, 8-Aug-99 (GY)
ST98: Malmesbury Abbey, 27-Jul-04 (DK)
SU09: Clattinger Farm, 9-May-06 (PM)
SU16: Pewsey, 17-Jun-73 (HK det MDD)
SU17: Avebury, 24-Jun-06 (DB)

8 SU23: Allington, nr Salisbury, 23-May-01(DG); 26-Mar-07(PP)
SU24: Newton Tony, 26-Jun-0, 8-May-Jun-06 (DG)
Resident. Widespread; on various plants but particularly lilies and fritallaries in private and public gardens and commercial situations. [Rarely in natural situations].

## Subfamily CRYPTOCEPHALINAE

### Tribe CLYTRINI

**CLYTRA Laicharting, 1781**
***Clytra quadripunctata*** (Linnaeus, 1758)
7 SU16: Marlborough area, Jun-1873 (EWJ) (Meyrick, 1874); 29-May-1879 (PHM) (Meyrick, 1880); 1869-1938 (Meyrick, 1938)
Resident. Very local; by beating bushes near nests of the ant *Formica rufa*. Larvae are associated with ants, requiring the nests to complete their development. They probably feed on plant debris in the nest. 3 specimens without data in MCC.

Clytra quadripunctata

### Tribe CRYPTOCEPHALINI

**CRYPTOCEPHALUS Geoffroy, 1762**
***Cryptocephalus aureolus*** Suffrian, 1847
**Nb, LCGR-Fm Cm**
7 SU06: Morgan's Hill, 17-Jul-99 (BL det MDD)
SU16: Marlborough area, 4-Jun-1879 (RAW); 1869-1938 (Meyrick, 1938)
8 ST94: Norton Down (SPTAW), 23-Jun-02 (PJH) (Hodge, 2002); Wadman's Coppice (SPTAW), 3-Jun - 12-Aug-08 (Boyce & Wright, 2009)
SU04: Fox Covert (SPTAW), 3-Jun - 12-Aug-08 (Boyce & Wright, 2009)
SU14: Haxton (SPTAE), 17-May-02 (PJH) (Hodge, 2002)
SU23: Porton Down, 3-Jul-99 (PJH) (Hodge, 2002)
SU24: Newton Tony, Jun-03 (DG det MAS); Sidbury Hill (SPTAE), 18-May-97 (MGT det MLC)
SU25: Sidbury Hill (SPTAE), 18-May-97 (MGT det MLC)
Resident. Widespread; on vegetation in grassland. Adults feed on pollen of yellow flowers in particular.

***Cryptocephalus bipunctatus*** (Linnaeus, 1758)
**Nb, LCGR-Fm Ch**
8 SU21: Landford Bog, 24-Jun-01 (APF) (Foster, 2001); 22-Jun-93 (PM)
SU23: Porton Down, 3-Jul-99 (PMP)
Resident. Very local; [on trees in a wide variety of habitats including woodland and downland].

***Cryptocephalus fulvus*** Goeze, 1777
**LCGR-Fm Cl**
8 SU22: Holmere Common, 19-Aug-88 (DRN); Blackmoor Copse, 31-Aug-01 (APF) (Foster, 2004)
SU22: Glazier's Copse, 16-Aug-88 (DRN)
SU23: West Winterslow, 1-Aug-01 (NB); Porton Down, 3-Jul-99 (BLe)
Resident. Very local; on vegetation in mainly open habitats.

***Cryptocephalus hypochaeridis*** Dufour, 1843
**LCGR-Fm Cl**
7 SU16: Marlborough area, 1869-1938 (Meyrick, 1938)
SU27: High Clear Down, 1-Jun-99, 27-May-03, 7-Jun-05 (PM) (PM)
8 SU02: Middleton Down, 8-Jun-93 (BL)
SU16: Clench Common, 4-Jun-02 (IJG)
Resident. Local; on vegetation in open downland.

***Cryptocephalus labiatus*** (Linnaeus, 1761)
7 SU16: Marlborough area, Jun-03 (EWJ) (Meyrick, 1874); 1869-1938 (Meyrick, 1938)
8 ST93: Great Ridge Wood, 6-Aug-85 (DRN)
SU03: Grovely Wood, 16-Aug-71 (DRN)
SU21: Hamptworth Common, 28-Jul-77 (DRN); Landford Bog, 24-Jun-01, 22-Jun, 13-Jul-03 (APF) (Foster, 2004)
SU22: Glazier's Copse, 16-Aug-88 (DRN)
SU23: Porton Down, 20-Jul-97 (MAS); 3-Jul-99 (PJH) (Hodge, 2002); West Winterslow, 1-Aug-01 (NB)
SU25: Sidbury Hill (SPTAE), 28-May-97 (MGT det MLC)
Resident. Widespread; usually on Birches in a wide range of habitats.

***Cryptocephalus moraei*** (Linnaeus, 1758)
**LCGR-Fl Cl**
8 ST91: Cranborne Chase, 16 to 25-Jun-80 (NCC); 4-Jun-97 (MAS)
ST94: Norton Down, 23-Jun-02 (PJH) (Hodge, 2002); Warminster Ranges (SPTAW), 3-Jun - 12-Aug-08 (Boyce & Wright, 2009)
SU15: Everleigh, 24-Jun-02 (PJH) (Hodge, 2002)
SU23: Porton Down, 3-Jul-99 (MAS)

Resident. Local; in calcareous grassland [on low vegetation].

Cryptocephalus parvulus

***Cryptocephalus parvulus*** Müller, O.F, 1776
**Nb**
7 SU16: Marlborough area, 1869-1938 (Meyrick, 1938)
Resident. Very local; on young trees [usually birches in both downland and woodland]. One specimen in MCC without data.

***Cryptocephalus punctiger*** Paykull, 1799
**Na**
7 SU16: Marlborough area, Jun-03 (EWJ) (Meyrick, 1874); 1869-1938 (Meyrick, 1938)
Resident. Very local; on young trees [usually on birches in broadleaved woodland and commons]. Only a few scattered records in GB mainly in the SE. No specimens in the MCC.

Cryptocephalus punctiger

***Cryptocephalus pusillus*** Fabricius, 1777
7 SU16: Marlborough area, 1869-1938 (Meyrick, 1938)
ST93: Great Ridge Wood, 6-Aug-85 (DRN)
8 SU03: Grovely Wood, 11, 16-Aug-71 (DRN)
SU16: Jones's Mill, 1997 (MAS)
SU21: Hamptworth Estate, 3-Aug-73 (DRN); Landford Bog, 24-Jun-01 (APF) (Foster, 2001)
SU22: Glazier's Copse, 16-Aug-88 (DRN); Langley Wood NNR, 16-Jul to 6-Aug-04 (MDD) (Darby, 2005)
Resident. Widespread; on various broadleaved trees in a range of habitats.

***Cryptocephalus sexpunctatus*** (Linnaeus, 1758)
**RDB2**
8 SU25: Ludgershall, Apr - Jul-1915 (PAH) (Harwood, 1915)
Resident. Very local; on vegetation [in calcareous grassland with dense scrub, especially on west-facing hillsides, and broadleaved woodland].

## Subfamily CHRYSOMELINAE

**TIMARCHA Samouelle, 1819**
***Timarcha goettingensis*** (Linnaeus, 1758)
**LCGR-Fm Ch**
7 SU16: Marlborough area, Jun-1873 (EWJ) (Meyrick, 1874); 10-Mar-1879 (HCB) (Meyrick, 1880); Woodborough, 1900-1905 (Meyrick, 1905)
SU17: Marlborough Downs, 6-Oct-45 (DCL) (DFL, 1948)
8 ST93: Great Ridge Wood, 18-Apr-74 (DRN)
SU01: Martin Down NNR, 1999 (EN, 1999)
SU05: Black Heath (SPTAC), 3-Jun - 12-Aug-08 (Boyce & Wright, 2009)
SU12: Coombe Bissett Down, 1, 9-Mar-00 (MDD)
SU16: Huish, Draycott Hill, 4-Oct-02 (IJG);
SU23: Porton Down, Jul-91, 4-May-92, 3-May-93 (PTW det MDD)
SU24: Beacon Hill (SPTAE), 3-Jun - 12-Aug-08 (Boyce & Wright, 2009); Driver Training Area (SPTAE), 1-May-05 (MGT)
SU25: Sidbury Hill (SPTAE), 18-May-97 (MGT det MLC)
Resident. Widespread; in open situations, particularly calcareous grassland.

**Bloody-nosed Beetle**
***Timarcha tenebricosa*** (Fabricius, 1775)
**LCGR-Fl Cm**
7 SU06: Beckhampton, 25-May-94 (BG); Morgan's Hill, 22-May-04 (AC); 30-Aug-91, 30-Aug-92 (PM); Tam Hill, 10-Mar-00 (BG)
SU16: Alton, 30-Jun-00 (IJG); Furze Hill West, 6-May-98 (GT); Lockeridge, 1994 (JEO); Marlborough area, Jun-1873 (EWJ) (Meyrick, 1874); 12 Mar 1879 (GTKM) (Meyrick, 1880); 1869-1938 (Meyrick, 1938); Martinsell Hill East, 2-Mar-99 (GT)
SU19: Hannington, 2-Sep-95 (MEN)
8 ST91: Quarry Bottom, 14-Mar-00 (GT)
ST92: Sutton Mandeville, 19-Aug, 12-Sep-98 (MDD)
ST93: Great Ridge Wood, 2-Aug-82 (DRN); 7-May-00 (MDD); Chilmark, 18-Aug-98 (MDD)
ST94: Knook Down Coombe, 24-Feb-98 (PD); Ladywell (SPTAW), 15-May-02 (PJH) (Hodge, 2002); Norton Down (SPTAW), 15-May-02 (PJH) (Hodge, 2002); Salisbury Plain, 1987-1994 (SP); Summer Down

(SPTA), 7-Aug-92 (SGB det JB); Wadman's Coppice (SPTAW), 3-Jun - 12-Aug-08 (Boyce & Wright, 2009)
ST95: Great Cheverell Hill, 17-Jun-94 (PM); West Lavington, May-Jun, 1934-44 (ITH, HGM)
SU01: Martin Down NNR, 21-Aug-89 (RGB); 1999 (EN, 1999)
SU02: Bishopstone, 12, 19, 27-May-91 (MDD) Broad Chalke, 26-Aug-94 (BL); Chickengrove Bottom, 1-Apr-92 (PM); Gardiner Forest, 1-Sep-72 (DRN); Middleton Down, 28-Mar-89, 4-Apr, 17-May-90, 23-Mar-92, 2-May, 26-Aug-94 (BL)
SU03: Barford Down, 3, 10-Apr-72 (DRN); Bu-co mbe, 1-Apr-92 (PM); Deptford Bank, 19-Mar-98 (PD); Grovely Wood, 11-Apr-72, 17-Apr-74 (DRN); 25-May-92 (SP); 17-Feb-98, 11-Jun-00 (MDD); Middle Hills, 31-Mar-98 (PD)
SU04: Can Down (SPTA), 4-Mar-00 (BG); Yarnbury Castle path, 8-Jul-08 (MDD)
SU05: Black Heath (SPTAC), 3-Jun - 12-Aug-08 (Boyce & Wright, 2009); Redhorn Hill, 5-Sep-98 (BG); New Copse (SPTA), 25-May-98 (BG)
SU12: Coombe Bissett Down, 24-Jul-04 (PM); Great Yews, 19-May-68 (CA)
SU14: Well Bottom, 7-May-89 (SP)
SU15: Charlton, 28-Jun-00 (IJG); West Everleigh Down, 27-Jun-00 (IJG)
SU23: Porton Down, 19-Apr-92 (PTW det MDD), 4-May-92, 12-Jun-9, 15-Oct-95 (MAS), 3-Jul-99 (PJH) (Hodge, 2002)

Resident. Widespread; walking in the open in a wide range of habitats. Adults and larvae feed on the leaves and stems of bedstraws, *Galium spp.*.

Timarcha tenebricosa

**CHRYSOLINA Motschulsky, 1860**

***Chrysolina banksi*** (Fabricius, 1775)

7 SU16: Marlborough area, Jun-1873 (EWJ) (Meyrick, 1874); 1869-1938 (Meyrick, 1938)

8 ST94: Imber Long Beeches, 16-Jun-85 (EGS det MDD); South Down Sleight (SPTA), 7-Aug-92 (JB det ACW)
SU05: Black Heath (SPTAC), 3-Jun - 12-Aug-08 (Boyce & Wright, 2009)
SU15: Everleigh (SPTAE), 22-Aug-00 (IJG); Upavon, 23-May-94 (BL); West Everleigh Down, 14-Aug-02 (IJG)

Resident. Local; in open habitats on various plants.

***Chrysolina fastuosa*** (Scopoli, 1763)

7 SU16: Rabley, Marlborough area, 26-May-1879 (EdeS) (Meyrick, 1880); 1869-1938 (Meyrick, 1938)

Resident. Very local; on *Galeopsis tetrahit*. [Hemp nettle]. One specimen in MCC without data.

**[*Chrysolina haemoptera*** (Linnaeus, 1758)

**Nb, LCGR-Fm Ch**

7 SU16: Henswood, 1873-1895 (Jebb, 1895)**]**

Most records in GB from coastal habitats, perhaps the reason why Meyrick later noted that the record was 'almost certainly erroneous, and should be struck out' (Meyrick, 1900).

**Mint Leaf Beetle**

***Chrysolina herbacea*** (Duftschmid, 1825)

7 ST87: Castle Combe, 3-Jun-04 (BLe)
SU09: Cotswold Water Park, 9-Sep-84 (KNAA), 21-Jun-90 (APFw)
SU16: Clatford, New Arboretum, May-97 (JEO); Lockeridge, 1994 (JEO)
SU26: Savernake Forest, 19-Jun-98 (EGS det MDD)
SU27: Ramsbury Meadow, 5-Aug-93 (PM)

8 ST73: Stourhead. Ponds, 2000 (DBG)
ST82: Oysters Copplce, Jun-89 (PM); 6-Oct-02 (MDD)
ST85: Green Lane Wood, 8-Jul-92 (PM)
ST93: Fonthill, 27-May-95 (MDD)
SU06: Devizes, 16-Jun-97 (JEO)
SU13: Great Durnford, 16, 19-Aug-69, 18-Aug-72, 27-Jul-77, 3-Aug-83, 3-Aug-88 (DRN); Lower Woodford, 28-Jul-78, 2, 5-Aug-82 (DRN); 27-Jun-89 (PM); Salisbury, 9-Jul-06 (PM)
SU14: Figheldean, 28-Jul-82; Netheravon, 29-Jul-83 (DRN); Stonehenge, 22-Apr-03 (PJH) (Hodge, 2003)
SU15: East Sharcott, 31-Aug-00 (IJG)
SU16: Jones's Mill, 1997 (MAS); 18-Jun-97 (PM); 21-Jun-97 (MDD); 19-Jun-01 (APF) (Foster, 2002); 30-May-04 (KNAA)
SU22: Bentley Wood, 5, 17-Jun-90, 2-Jun-91 (PM); 2-Jul-96 (MDD); Blackmoor Copse, 18-Jun-93 (PM); East Grimstead, 12-Aug-87 (DRN); Homan's Copse, 31-Jul-85 (DRN); Langley Wood NNR, 6-Sep-73 (Anon); Whiteparish Common, 2-Jul-95

(MAS)
Resident. Widespread; on vegetation in wet habitats. Formery *C. menthastri.*

***Chrysolina hyperici*** (Forster, 1771)
**LCGR-FI CI**
8 ST91: Cranborne Chase, 16 to 25-Jun-80 (NCC)
ST94: Tinkers Track (SPTAW), 3-Jun - 12-Aug-08 (Boyce & Wright, 2009)
SU03: Grovely Wood, 29-Jul-78 (DRN)
SU04: Tilshead Down, 4-Jun-97 (MGT det MLC)
SU13: Old Sarum, 31-May-97 (MAS)
SU14: Haxton (SPTAE), 22, 24-Jun-02 (PJH) (Hodge, 2002); Wexland Hanging (SPTAC), 3-Jun - 12-Aug-08 (Boyce & Wright, 2009)
SU15: Everleigh (SPTAE), 24-Jun-02 (PJH) (Hodge, 2002)
SU22: Blackmoor Copse, Jan-93 (PM)
SU23: Porton Down, 3-Jul-99 (PJH) (Hodge, 2002)
SU24: Driver Training Area (SPTAE), 24-Jun-02 (PJH) (Hodge, 2002); 1-May-05 (MGT); Dunch Hill (SPTAE), 3-Jun - 12-Aug- 08 (Boyce & Wright, 2009)
SU25: Sidbury Hill (SPTAE), 28, 29-May-97 (MGT det MLC)

***Chrysolina oricalcia*** (Müller, O.F. 1776)
**Nb**
7 SU16: Marlborough area, 1869-1938 (Meyrick, 1938)
8 ST84: Longleat, Jun-97, 7-Mar, 29-Aug-99 (MB)
ST94: Ladywell (SPTAW), 3-Jun - 12-Aug-08 (Boyce & Wright, 2009); South Down Sleight (SPTA), 7- Aug-92 (DP det ACW)
SU04: Tilshead Down, 1 - 4-Jun-97 (MGT det MLC)
Resident. Local; on Umbelliferae and by general sweeping. [Occurs in various habitats including woodland rides and edges, hedgerows, calcareous grassland, chalk pits, and edges of water].

Chrysolina oricalcia

***Chrysolina polita*** (Linnaeus, 1758)
7 ST98: Lacock, 29-May-04 (BLe)
SU08: Red Lodge Pond, 31-May-01 (JKG det MDD)
SU09: Blakehill Farm, 11-May-01 (JKG det MDD)
SU16: Marlborough area, Jun-1873 (EWJ) (Meyrick, 1874); 15-Feb-1879 (WFB) (Meyrick, 1880); 1869-1938 (Meyrick, 1938)
SU19: Hannington, 3-May-99 (MEN); Highworth, 15-Aug-99 (MEN)
SU26: Savernake Forest, 16-Jun to 17-Jul-93 (EN det SAW)
8 ST85: Bullen Hill Farm, 1982-2002 (EGS det MDD); Green Lane Wood, 7-Jun-95 (WANHS det MDD)
ST91: Cranborne Chase, 16 to 25-Jun-80 (NCC); 22-Jun-99 (MDD)
ST92: Sutton Mandeville, 7-Jul-96 (MDD)
ST93: Fonthill, 19-Jul-95 (MDD); Great Ridge Wood, 18-Apr-74, 19-Apr-84 (DRN); 14-Jun-96 (MDD)
ST94: Ladywell (SPTAW), 15-May-02 (PJH) (Hodge, 2002)
SU03: Grovely Wood, 29-Oct-70, 11-Aug-71, 8-Apr-75, 4-Apr-77, 19-Apr-84 (DRN); 12-Jun-00, 15-Jun-01 (MDD)
SU13: Great Durnford, 6-Apr-70, 17-Jul-73 (DRN); High Post, 3-May-95 (MAS); Lower Woodford, 2, 5-Aug-82 (DRN); Salisbury, Apr-00 (MAS) (Darby, 2001)
SU14: Figheldean, 28-Jul-82 (DRN); Stonehenge, 22-Apr-03 (PJH) (Hodge, 2003)
SU16: Jones's Mill, 21-Jun-97 (MDD); 19-Jun-01 (APF) (Foster, 2002)
SU22: Bagfield Copse, 7-Aug-86 (DRN); Bentley Wood, 5-Jul-90 (PM); 2-Jul-96 (MDD); Blackmoor Copse, 4-May-92 (PM); 15-Jun-93 (PM); 2-Jul-96 (MDD); 17-May-02 (APF) (Foster, 2004); Langley Wood NNR, 6-Aug-86 (DRN); 25-Jun to 16-Jul-04 (MDD) (Darby, 2005)
SU26: Bedwyn Brail, 19-Jun-03 (MDD)
Resident. Widespread; on vegetation in a great variety of habitats.

Chrysolina sanguinolenta

***Chrysolina sanguinolenta*** (Linnaeus, 1758)
**Na**
8 ST94: Ladywell (SPTAW), 15-May-02 (PJH) (Hodge, 2002); Imber Firs, 17-May-82 (EGS det MDD)
SU15: Beach's Plantation (SPTAE), 1-May-05

(MGT)
Resident. Very local; [adults usually on common toadflax, *Linaria vulgaris*, and sometimes on snapdragon, *Antirrhinum majus*]

***Chrysolina staphylaea*** (Linnaeus, 1758)
7 SU16: Marlborough area, Jun-1873 (EWJ) (Meyrick, 1874); 26 Feb 1879 (HW) (Meyrick, 1880); 1869-1938 (Meyrick, 1938)
8 ST84: Longleat, 29-Aug-99 (MB) (Meyrick, 1938)
SU01: Martin Down NNR, 1999 (EN, 1999)
SU13: Great Durnford, 18-Aug-72 (DRN)
SU15: Manningford, 4-Aug-88 (DRN)
SU16: Woodborough, 4-Aug-83 (DRN)
Resident. Widespread; in a great variety of habitats.

***Chrysolina varians*** (Schaller, 1783)
7 SU08: Red Lodge Pond, 31-May-01 (JKG det MDD)
SU16: Marlborough area, 1869-1938 (Meyrick, 1938)
8 ST91: Cranborne Chase, 16 to 25-Jun-80 (NCC)
ST93: Great Ridge Wood, 18-Apr-74 (DRN)
SU02: Chickengrove Bottom, 26-May-92 (PM)
SU03: Grovely Wood, 11-Aug-71 (DRN); 11, 12, 15-Jun-01 (MDD)
SU22: Bentley Wood, 3, 4-Jul-90, 9-May-91 (PM)
SU23: Porton Down, 3-Jul-99 (PMP)
Resident. Widespread. Adults are nocturnal so usually found hiding under leaves, stones, etc..They feed on foliage of St. John's Worts, *Hyepricum* spp.

***Chrysolina violacea*** (Müller, O.F, 1776)
**Nb, LCGR-Fm Cm**
7 SU16: Marlborough area, 1895-1938 (Meyrick, 1938); West Woods, 12-May-46 (Holmes and Tennant, 1947)
8 ST91: Cranborne Chase, 1-Jan-99 (MDD)
ST92: Sutton Mandeville, Mar, 5-Aug-95, Aug-97 (MDD)
SU03: Grovely Wood, 6, 13-Apr-71, 6-Apr-72, 8-Apr-72, 30-May-74, 3-Apr-75 (DRN)
SU13: Cockey Down, May-93 (PM)
SU24: Sidbury Hill (SPTAE), 29-May-97 (MGT)
Resident. Very local; usually walking around in the open as singletons. Adults feed on ground ivy, *Glechoma hederacea*, on chalk downland with scrub or young trees, and unimproved calcareous grassland, heaths, scrub and disturbed ground on base-rich soils.

**GASTROPHYSA Dejean, 1836**
***Gastrophysa polygoni*** (Linnaeus, 1758)
7 ST97: Chippenham, 1999 (MW)
SU06: Roundway Camp, Jul-40 (FDB) (Buck, 1956)
SU16: Marlborough area, Jun-1873 (EWJ) (Meyrick, 1874); 1869-1938 (Meyrick, 1938)
SU18: Rushy Platt, 20-May-01 (JKG det MDD)
8 ST92: Wardour, 5-Jun-96 (MAS)
ST94: Bowls Barrow (SPTAW), 16-May-02 (PJH) (Hodge, 2002)
ST95: West Lavington 22-May-41 (KJA), 20-May-44 (DCL) both (DFL, 1948)
SU02: Wilton, 12-May-97 (MDD)
SU03: Grovely Wood, 13-Apr-71, 27-May-74 (DRN)
SU13: Salisbury, 26-May-74 (DRN)
SU15: Beach's Plantation (SPTAE), 16, 19-Jun-97, 1-May-05 (MGT det MLC)
SU16: Jones's Mill, 3-May-00 (JKG det MDD); 30-May-00, 16-May-02 (IJG)
SU22: Langley Wood NNR, 14-May to 3-Jun-04 (MDD) (Darby, 2005); Redlynch, 4-Aug-82 (DRN)
Resident. Widespread; on vegetation in a wide variety of natural and disturbed habitats. Leaves of Polygonaceae are often riddled with large holes.

**Green Dock Beetle**
***Gastrophysa viridula*** (De Geer, 1775)
7 ST96: Conigre Meadow, 20-May-01 (JKG det MDD)
SU09: Dance Common, 15-Apr-02 (JKG det MDD); Upper Waterhay, 13-Jun-04, 17-May to 11-Jun-05, 5-Jun to 11-Jun-07, 6-Jul-07 (JKG det MDD); North Meadow, Apr-92 (PM) (Meyrick, 1938)
SU16: Marlborough area, 1869-1938 (Meyrick, 1938)
8 SU14: Stonehenge, 22-Apr-03 (PJH) (Hodge, 2003)
SU15: Beach's Plantation (SPTAE), 1-May-05 (MGT); Charlton, 11-Jun-02 (IJG); Everleigh (SPTAE), 18-May-02 (PJH) (Hodge, 2002)
SU16: Jones's Mill, 2-Jul-01 (APF) (Foster, 2002)
Resident. Widespread; on vegetation in a great variety of habitats. Not many records but probably one of our most common leaf beetles. Adults perforate leaves of Polygonaceae with circular holes c.1cm in diameter.

**PHAEDON Latreille, 1829**
***Phaedon armoraciae*** (Linnaeus, 1758)
7 SU09: Somerford Keynes, Swillbrook Lakes, 4-Jun-04 (BLe); Upper Waterhay, 17-May to 4-Jun-05 (JKG det MDD)
SU16: Marlborough area, 1869-1938 (Meyrick, 1938)
8 SU12: Charlton-All-Saints, 2-Jun-95 (MAS)
SU13: Great Durnford, 27-Jul-77 (DRN)
SU14: Nine Mile R., pond (SPTAE), 5-Jun-97 (MGT); Sheep Bridge, Bulford, 5-Jun-97 (MGT det MLC); Stonehenge, 22-Apr-03 (PJH) (Hodge, 2003)
SU16: Jones's Mill, 1997 (MAS); 19-Jun, 17-Oct-01 (APF) (Foster, 2002)

Resident. Local; on various water plants in wetland habitats.

***Phaedon cochleariae*** (Fabricius, 1792)
7 SU16: Marlborough area, Jun-1873 (EWJ) (Meyrick, 1874); 1869-1938 (Meyrick, 1938)
8 ST92: Wardour, 5-Jun-96 (MAS)
SU12: Charlton-All-Saints, 2-Jun-95 (MAS)
SU13: Great Durnford, 18-Aug-72 (DRN); Salisbury, Apr-00 (MAS) (Darby, 2001)
SU14: Nine Mile R., pond (SPTAE), 5-Jun-97 (MGT); Sheep Bridge, Bulford, 5-Jun-97 (MGT det MLC); Stonehenge, 22-Apr-03 (PJH) (Hodge, 2003)
SU15: Everleigh (SPTAE), 18-May-02 (PJH) (Hodge, 2002)
SU16: Jones's Mill, 19-Jun, 17-Oct-01 (APF) (Foster, 2002)
SU25: Sidbury Hill (SPTAE), 4-Jun-97 (MGT)
Resident. Widespread; on Brassicaceae in a wide range of habitats.

**Celery Leaf Beetle**
***Phaedon tumidulus*** (Germar, 1824)
7 ST98: Conygre Mead, Malmesbury, 5-Apr-02 (JKG det MDD); Rodbourne, 29-Mar-02 (JKG det MDD)
SU09: Ashton Keynes, 27-Apr-00 (JKG det MDD)
SU16: Marlborough area, Jun-18738 (EWJ) (Meyrick, 1874); 1869-1938 (Meyrick, 1938)
SU18: Swindon, 7-May-01 (JKG det MDD)
SU26: Savernake Forest, 30-Aug to 20-Sep-94 (EN det SAW)
8 ST94: Bowls Barrow (SPTAW), 22-Jun-02 (PJH) (Hodge, 2002)
SU01: Martin Down NNR, 1999 (EN, 1999)
SU03: Grovely Wood, 24-Aug-72, 27-May-74, 4-Aug-78 (DRN); 12-Jun-00 (MDD)
SU04: Fox Covert, 25-Jun-02 (PJH) (Hodge, 2002); Salisbury Plain, 25-Jun-02 (PJH) (Hodge, 2002); Shrewton, 1-Aug-85 (DRN)
SU13: Salisbury, 15-Aug-88 (DRN); 2-May-00 (MDD ) (Darby, 2001); Stratford sub Castle, 5-Apr-72 (DRN)
SU14: Haxton (SPTAE), 17-May-02 (PJH) (Hodge, 2002); Stonehenge, 22-Apr-03 (PJH) (Hodge, 2003); Wexland Hanging (SPTAC), 3-Jun - 12-Aug-08 (Boyce & Wright, 2009)
SU15: Everleigh (SPTAE), 18-May-02 (PJH)
SU16: Woodborough, 4-Aug-83 (DRN)
SU21: Hamptworth Estate, 15-Apr-74 (DRN)
SU22: Harestock, 13-Apr-72 (DRN)
SU23: Porton Down, 3-Jul-99 (BLe); West Winterslow, 1-Aug-01 (NB)
SU24: Driver Training Area (SPTAE), 17-May, 24-Jun-02 (PJH) (Hodge, 2002)
SU25: Sidbury Hill (SPTAE), 19-May, 6-Jun-97 (MGT det MLC)
Resident. Widespread; in a wide range of habitats on the foliage of Apiaceae including ground elder and cow parsley, on which the adults feed.

**HYDROTHASSA Thomson, C.G, 1859**
***Hydrothassa glabra*** (Herbst, 1783)
7 ST98: Rodbourne, 29-Mar-02 (JKG det MDD)
Resident. Very local; [on Ranunculaceae in various habitats]. Leaves of the plants are riddled with holes.

***Hydrothassa marginella*** (Linnaeus, 1758)
7 SU16: Marlborough area, Jun-1873 (EWJ) (Meyrick, 1874); 31-May-1879 (TNH-S) (Meyrick, 1880); 1869-1938 (Meyrick, 1938); Poulton, 1870-1873 (Meyrick, 1874
8 ST92: Wardour, 5-Jun-96 (MAS)
SU12: Charlton-All-Saints, 9-May-96 (MAS)
SU13: Great Durnford, 27-Jul-77 (DRN); Lower Woodford, 27-Jul, 2, 5-Aug-82 (DRN); Stratford sub Castle, 7-Aug-71 (DRN)
SU15: New Mill, nr Pewsey, 8-Jun-01 (IJG)
SU16: Jones's Mill, 1-Jan-97 (MAS); 21-Jun-97 (MDD); 19-Jun-01 (APF) (Foster, 2002); 1-Jun-04 (BLe); Woodborough, 4-Aug-83 (DRN)
Resident. Widespread; on Ranunculaceae in a wide range of habitats.

**PRASOCURIS Latreille, 1802**
***Prasocuris junci*** (Brahm, 1790)
7 SU09: Ashton Keynes, 7-May, 26-Apr-01 (JKG det MDD); Upper Waterhay, 17-May to 4-Jun-05 (JKG det MDD)
SU16: Marlborough area, 1869-1938 (Meyrick, 1938)
8 ST84: Longleat, 7-Mar-99 (MB)
SU02: Bishopstone, Jun-918 (MDD)
SU13: Idmiston, 13-Aug-87 (DRN); Lower Woodford, 28-Jul-78, 2-Aug-82 (DRN); Middle Woodford, 27-Jul-77 (DRN); Salisbury, 15- Aug-88 (DRN)
SU14: Stonehenge, 22-Apr-03 (PJH) (Hodge, 2003)
SU15: Fittleton, 29-Jul-83 (DRN)
SU16: Jones's Mill, 3-May-00 (JKG det MDD); 21-Jun-97 (MDD)
Resident. Very local; on various plants in damp situations. Adults and larvae feed on the leaves of brooklime, *Veronica beccabunga*.

***Prasocuris phellandrii*** (Linnaeus, 1758)
7 SU09: Somerford Keynes, Swillbrook Lakes, 4-Jun-04 (BLe)
SU16: Marlborough area, Jun-1873 (EWJ) (Meyrick, 1874); 30-May-1879 (RAW) (Meyrick, 1880); 1869-1938 (Meyrick, 1938)
Resident. Very local; on dropwort, *Oenanthe,* [also on other wetland plants]. 5 specimens without data in MCC .

**PLAGIODERA Dejean, 1836**
***Plagiodera versicolora*** (Laicharting, 1781)
7 SU09: Upper Waterhay, 13-Jun-04, 17-May to 4-Jun-05 (JKG det MDD)
SU16: Marlborough area, 1869-1938 (Meyrick, 1938)
SU26: Savernake Forest, 30-May-73 (DBS)
8 ST91: Cranborne Chase, 16 to 25-Jun-80 (NCC)
SU12: East Harnham, 1-Aug-82 (DRN)
SU13: Idmiston, 5-Aug-82, 13-Aug-87 (DRN); Great Durnford, 6-Aug-87 (DRN); Lower Woodford, 28-Jul-78 (DRN); Middle Woodford, 27-Jul-77 (DRN); Stratford sub Castle, 11-Apr, 7-Aug-71 (DRN)
SU14: Netheravon, 29-Jul-83 (DRN)
SU15: Fittleton, 29-Jul-83 (DRN)
SU16: Avebury, 4-May-03 (MGT); Pewsey, Jones's Mill, 1-Jun-04 (BLe)
SU26: Bedwyn Brail, 19-Jun-03 (MDD)
Resident. Widespread; on willows near water.

**CHRYSOMELA Linnaeus, 1758**
***Chrysomela populi*** Linnaeus, 1758
7 SU09: Upper Waterhay, 30-Apr-00 (JKG det MDD)
SU16: Marlborough area, Jun-1873 (EWJ) (Meyrick, 1874); 1869-1938 (Meyrick, 1938)
8 SU21: Hamptworth Estate, 6-Aug-79 (DRN); Landford Bog, Jun-90 (SP det DC)
SU16: Jones's Mill, 3-May-00 (JKG det MDD)
SU22: Blackmoor Copse, 4-May-92, 13-May, 7-Jun-93 (PM); Whiteparish Common, 22-Jul-73 (DRN)
Resident. Local; on aspens and willows.

**GONIOCTENA Dejean, 1836**
***Gonioctena olivacea*** (Forster, 1771)
8 ST94: Bowls Barrow (SPTAW), 22-Jun-02 (PJH) (Hodge, 2002); Ladywell (SPTAW), 23-Jun-02 (PJH) (Hodge, 2002)
SU14: Haxton (SPTAE), 17-May, 24-Jun-02 (PJH) (Hodge, 2002)
Resident. Very local; [mainly on broom, *Cytisus* spp.where adults and larvae feed on the leaves and stems].

***Gonioctena pallida*** (Linnaeus, 1758)
7 SU16: Marlborough area, Pumphrey's Wood, 1869-1938 (Meyrick, 1938)
8 SU14: Haxton (SPTAE), 3-Jun - 12-Aug-08 (Boyce & Wright, 2009)
Resident. Very local; on sallow and hazel; adults and larvae feed on the underside of the leaves.

***Gonioctena viminalis*** (Linnaeus, 1758)
7 SU26: Durley Heath, Savernake Forest, 4-Jun-01 (IJG); Savernake Forest, 7-Aug-69 (DRN)
8 SU22: Glaziers Copse, 31-Jul-85 (DRN)
Resident. Very local; on *Salix Spp.* where adults and larvae feed on the underside of the leaves.

**PHRATORA Dejean, 1836**
***Phratora laticollis*** (Suffrian, 1851)
7 ST97: Chippenham, 29-May-04 (BLe)
ST98: Lacock, 29-May-04 (BLe)
8 SU13: Stratford sub Castle, 11-Apr-71 (DRN)
SU14: Figheldean, 28-Jul-82 (DRN)
Resident. Local; on poplars and under willow bark. Adults and larvae feed on the underside of leaves.

***Phratora vitellinae*** (Linnaeus, 1758)
7 SU09: Upper Waterhay, 6-Jul-07 (JKG det MDD)
SU16: Oare, 20-May-01 (IJG); Marlborough area, 1869-1938 (Meyrick, 1938)
8 SU13: Great Durnford, 19-Aug-69 (DRN)
SU14: Figheldean, 4-Aug-88 (DRN)
SU22: Blackmoor Copse, 3-Aug-03 (APF) (Foster, 2004); Glazier's Copse, 16-Aug-88 (DRN)
Resident. Local; on sallows and poplars. Adults and larvae feed on the underside of the leaves.

**Blue Willow Beetle**
***Phratora vulgatissima*** (Linnaeus, 1758)
7 SU09: Somerford Keynes, Swillbrook Lakes, 4-Jun-04 (BLe)
SU16: Lockeridge, 1994 (JEO); Marlborough, 1994 (JEO)
8 SU16: Jones's Mill, 2-Jul-01 (APF) (Foster, 2002)
SU21: Landford Bog, 24-Jun-01 (APF) (Foster, 2001)
SU22: Blackmoor Copse, 24-Aug-01 (APF) (Foster, 2004)
Resident. Local; mainly on willows where the adults and larvae feed on the underside of the leaves.

## Subfamily GALERUCINAE
### Tribe GALERUCINI

**GELERUCELLA Crotch, 1873**
***Galerucella lineola*** (Fabricius, 1781)
7 ST96: Conigre Meadow, 20-May-01 (JKG det MDD)
SU09: Somerford Keynes, 4-Jun-04 (BLe)
SU15: Fittleton, 29-Jul-83 (DRN)
SU16: Marlborough area, Jun-1873 (Meyrick, 1874); 1869-1938 (Meyrick, 1938)
8 SU15: Fittleton, 29-Jul-83 (DRN)
Resident. Local on waterside vegetation. [Adults and larvae feed on alder, hazel and willows.

***Galerucella nymphaeae*** (Linnaeus, 1758)/***Galerucella sagittariae*** (Gyllenhal, 1813)
These species are linked because there are few morphological characters to separate them.
7 SU16: Caen Hill Locks, 21-Aug-00 (EGS det MDD); Marlborough area, 1869-1938 (Meyrick, 1938)
8 ST93: Fonthill Bishop, 3-Aug-78 (DRN)
SU13: Salisbury, 26-May-74, 15-Aug-88 (DRN);

Apr-00 (MAS) (Darby, 2001)
SU14: Figheldean, 2, 28-Jul-82 (DRN)
SU16: Jones's Mill, 18-Dec-01 (IJG); 1-Jun-04 (BLe); Woodborough, 4-Aug-83 (DRN)
SU23: Porton Down, 3-Jul-03 (EGS det MDD)
Resident. Local; on vegetation near water. Adults of the usually darker, larger form (*nymphaeae*) occur on water lilies, whilst those of the usually smaller paler form (*sagittariae*) occur on a wide range of semi-aquatic and terrestrial plants.

***Galerucella pusilla*** (Duftschmid, 1825)
8 SU13: Great Durnford, 16-Aug-69 (DRN)
Resident. Very local; on vegetation at edge of R. Avon.

***Galerucella tenella*** (Linnaeus, 1761)
7 SU16: Marlborough area, 1869-1938 (Meyrick, 1938)
SU12: Charlton-All-Saints, 9-May-96 (MAS)
8 SU13: Great Durnford, 19-Aug-69 (DRN); Middle Woodford, 27-Jul-77 (DRN); Salisbury, 26-May-74, 15-Aug-88 (DRN); 12-May-00 (MDD) (Darby, 2001)
SU14: Stonehenge, 22-Apr-03 (PJH) (Hodge, 2003)
SU15: Fittleton, 4-Aug-88 (DRN)
SU16: Jones's Mill, 1997 (MAS); 11-Jun-97 (MAS); 2-Jul-01 (APF) (Foster, 2002); 1-Jun-04 (BLe)
Resident. Widespread; on vegetation in wetland habitats.

**PYRRHALTA Joannis, 1865**
***Pyrrhalta viburni*** (Paykull, 1799)
8 SU03: Grovely Wood, 15, 18-Aug-69, 1-Aug-73 (DRN)
SU16: Jones's Mill, 19-Jun-01 (APF) (Foster, 2002)
SU22: Blackmoor Copse, 31-Aug-01 (APF) (Foster, 2004)
Resident. Local; in woodland and wetland on viburnums and hazel.

**GALERUCA Geoffroy, 1762**
***Galeruca tanaceti*** (Linnaeus, 1758)
**LCGR-Fm Cl**
7 SU06: Morgan's Hill, 24-Sep-97 (PM)
SU16: Marlborough area, 20-Sep-1879 (WFB) (Meyrick, 1880); 1869-1938 (Meyrick, 1938)
SU26: Savernake Forest, 15-Jun-73 (SAW); 28-Jul to 28-Aug-93 (EN det SAW)
8 SU02: Middleton Down, 8-Jan-89, 15-Nov-05 (PM)
SU13: Cockey Down, 2-Oct-89 (PM); Little Durnford Down, 31-Oct-92 (PM)
SU23: Porton Down, 3-Jul-99 (BLe)
Resident. Very local; usually in grassland.

**LOCHMAEA Weise, 1883**
**Willow Leaf Beetle**
***Lochmaea caprea*** (Linnaeus, 1758)
7 SU16: Chilton, 1900-1905 (Meyrick, 1906) Marlborough area, 1869-1938 (Meyrick, 1938); Oare, 7-May-00 (IJG)
8 SU16: Jones's Mill, 1-Jun-04 (BLe); Pewsey, Kennet & Avon Canal, 1-Jun-04 (BLe)
SU21: Landford Bog, 9-Jun-01 (APF) (Foster, 2001); Peaked Wood, 29-May-74 (DRN)
SU22: Bentley Wood, 28-Jun-90 (PM); 2-Jul-96 (MDD); Blackmoor Copse, 14-Jun-73 (PM); 3-Aug-03 (APF) (Foster, 2004); Holmere Common, 19-Aug-88 (DRN); Homan's Copse, 29-May-90 (DRN); Langley Wood NNR, 22-Apr to 14-May-04 (MDD) (Darby, 2005)
Resident. Widespread; on trees mainly in wet areas. Adults feed on the underside of the leaves of willows, birches and alders.

**Hawthorn Leaf Beetle**
***Lochmaea crataegi*** (Forster, 1771)
7 ST96: Chittoe, 9-May-00 (EGS det MDD)
ST97: North Draycot Park, 22-May-00 (MDD)
SU16: Marlborough area, Jun-1873 (EWJ) (Meyrick, 1874); 1869-1938 (Meyrick, 1938); 'Marlborough Forest', Jun-1870 (RH) (Hislop, 1871)
SU18: Swindon, 30-Mar-01 (JKG det MDD)
SU26: Savernake Forest, 17, 31-May-90 (TGW); 31-May-04 (BLe)
8 ST91: Cranborne Chase, 4-Jun-97 (MAS)
ST93: Great Ridge, 19-Apr-84 (DRN)
ST94: Bowls Barrow (SPTAW), 16-May-02 (PJH) (Hodge, 2002);
Norton Down (SPTAW), 15-May-02 (PJH) (Hodge, 2002)
SU02: Middleton Down, 27-Apr-93 (PM)
SU03: Grovely Wood, 27, 30-May-74, 28-Jul-83 (DRN)
SU04: Orcheston Down (SPTA), 18-May-02 (PJH) (Hodge, 2002); Salisbury Plain, 16-May-02 (PJH) (Hodge, 2002); Shrewton, 2-Aug-83, 1- Aug-85 (DRN)
SU13: Cockey Down, May-93 (PM); Little Durnford Down, 14-May, 28-Apr-93 (PM); Old Sarum, 31-May-97 (MAS); Salisbury, Apr-00 (MAS) (Darby, 2001)
SU14: Haxton (SPTAE), 17-May-02 (PJH) (Hodge, 2002); Stonehenge, 22-Apr-03 (PJH) (Hodge, 2003); Wexland Hanging (SPTAC), 16-May-02 (PJH) (Hodge, 2002)
SU15: Everleigh (SPTAE), 18-May-02 (PJH) (Hodge, 2002)
SU16: Jones's Mill, 1-Jan-97 (MAS); 21-Jun-97 (MDD)
SU24: Driver Training Area (SPTAE), 17-May-02 (PJH) (Hodge, 2002)

SU26: Bedwyn Brail, 2-Jun-03 (MDD)
Resident. Widespread; mainly on hawthorn on which the adults and larvae feed, the latter mining the fruits.

Lochmaea crataegi

**Heather Beetle**
***Lochmaea suturalis*** (Thomson, C. G, 1866)
7 SU16: Marlborough area, 1869-1938 (Meyrick, 1938)
8 SU21: Hamptworth Estate, 24-Apr-84 (DRN); Landford Bog, 24-Jun-01 (APF) (Foster, 2001); Pound Bottom, 9-Apr-74 (DRN)
SU22: Langley Wood NNR, 14-May to 3-Jun-04 (MDD) (Darby, 2005)
SU26: Bedwyn Common, 1900-1905 (Meyrick, 1905)
Resident. Local; on heather and more rarely heaths, *Erica* spp. Requires a damp understorey of moss or litter for oviposition and pupation.

**PHYLLOBROTICA Dejean, 1836**
**Skullcap Leaf Beetle**
***Phyllobrotica quadrimaculata*** (Linnaeus, 1758)
8 ST73: Stourhead. Ponds, 1-Jan-00 (DBS)
SU13: Great Durnford, 19-Aug-69 (DRN)
SU22: Bentley Wood, 2-Jul-96 (MDD)
Resident. Very local; [on skullcaps, *Scutellaria* spp.] the leaves of which are eaten by the adult beetles.

Luperus flavipes

**LUPERUS Geoffroy, 1762**
***Luperus flavipes*** (Linnaeus, 1767)
**Nb**
7 SU16: Marlborough area, Jun-1873 (EWJ) (Meyrick, 1874); 1869-1938 (Meyrick, 1938)
Resident. Very local; on willows, birch, etc.. 2 specimens in MCC without data.

***Luperus longicornis*** (Fabricius, 1781)
7 SU16: Marlborough area, 1869-1938 (Meyrick, 1938)
8 ST91: Cranborne Chase, 15-May-97 (MAS)
SU21: Hamptworth Estate, 31-May-74 (DRN); Landford Bog, 14-May-93 (PM); 25-May-98 (MDD); 9, 24-Jun-01, 22-Jun-03 (APF) (Foster, 2001, 2003); Pound Bottom, 24-Jul-73 (DRN)
Resident. Local; on young bushes and trees.

**SERMYLASSA Reitter, 1912**
***Sermylassa halensis*** (Linnaeus, 1767)
**LCGR-Fm Cl**
7 SU06: Morgan's Hill, 30-Aug-92 (PM); Ranscombe Bottom, 3-Sep-94 (PM)
SU16: Marlborough area, Jun-1873 (EWJ) (Meyrick, 1874); 1869-1938 (Meyrick, 1938)
SU26: Savernake Forest, 15-Jun-73 (SAW); 28-Aug to 11-Oct-93 (EN det SAW)
8 SU01: Martin Down NNR, 21-Aug-89 (RGB); 1999 (EN, 1999)
SU03: Grovely Wood, 1-Aug-73 (DRN)
SU04: Shrewton, 1-Aug-85 (DRN)
SU05: Black Heath (SPTAC), 3-Jun - 12-Aug-08 (Boyce & Wright, 2009)
SU13: Cockey Down, 14-Jun-73, 10-Jul-95 (PM); Great Durnford, 16-Aug-69 (DRN); Old Sarum, 31-May-97 (MAS)
SU14: Ablington Down (SPTA), 7-Aug-92 (JHB); Stonehenge, 22-Apr-03 (PJH) (Hodge, 2003)
SU16: Woodborough, 4-Aug-83 (DRN)
SU23: Porton Down, 20-Aug-94 (PM); 15-Oct-95, 20-Jul-97 (MAS)
SU25: Collingbourne Forest, 25-Aug-72 (DRN)
Resident. Widespread; on bedstraws, *Galium* spp., the foliage of which is eaten by the adult beetles.

Tribe ALTICINI

**PHYLLOTRETA Dejean, 1836**
***Phyllotreta atra*** (Fabricius, 1775)
7 SU06: Calstone Down, 22-Aug-95 (KNAA det MLC)
SU16: Marlborough area, 1869-1938 (Meyrick, 1938)
8 ST95: West Lavington, 30-May-44 (PJW) (DFL, 1948)
ST94: Norton Down (SPTAW), 15-May-02 (PJH)

(Hodge, 2002)
SU03: Wylye Down NNR, 29-Mar to 10-Apr-07 (MDD) (Darby, 2009)
SU04: Shrewton, 6-Aug-82 (DRN)
SU13: Great Durnford, 7-Aug-70 (DRN); Idmiston, 5-Aug-82 (DRN); Old Sarum, 31-May-97 (MAS)
SU14: Stonehenge, 22-Apr-03 (PJH) (Hodge, 2003); Wexland Hanging (SPTAC), 22-Jun-02 (PJH) (Hodge, 2002)
SU22: Blackmoor Copse, 17-Mar-02 (APF) (Foster, 2004)
SU23: Thorny Down Woods, 16-Oct-00 (MDD)
Resident. Local; on Brassicaceae. Adults feed on the leaves causing 'shot-holing'.

Phyllotreta cruciferae

***Phyllotreta cruciferae*** (Goeze, 1777)
**Nb**
8 SU02: Knighton Wood, Jun-07 (AJWA)
SU13: Lower Woodford, 28-Jul-78 (DRN)
Resident. Very local; on vegetation [Brassicaceae] at edge of the R. Avon and in a flight intercept trap. Adults eat the leaves causing pits and 'shot-holing'.

***Phyllotreta diademata*** Foudras, 1860
7 SU06: Cherhill Down, 31-May-04 (BLe)
8 SU02: Knighton Wood, Jul-08 (AJWA)
SU13: Salisbury, Avon Valley Reserve, 2-May-00 (MDD) (Darby, 2001)
Resident. Very local; [on Brassicaceae]. Adults eat the leaves causing 'shot-holing'.

***Phyllotreta nemorum*** (Linnaeus, 1758)
7 SU06: Roundway Camp, Jul-40 (FDB) (Buck, 1956)
SU16: Marlborough area, Jun-1873 (EWJ) (Meyrick, 1874); 1869-1938 (Meyrick, 1938)
8 ST95: West Lavington, 30-May-44 (PJW) (DFL, 1948)
ST94: Norton Down (SPTAW), 23-Jun-02 (PJH) (Hodge, 2002)
SU14: Netheravon, 29-Jul-83 (DRN)
Resident. Local; on vegetation [Brassicaceae]. Adults eat the leaves causing 'shot-holing'.

***Phyllotreta nigripes*** (Fabricius, 1775)
7 ST87: Castle Combe, 3-Jun-04 (BLe)
SU06: Calstone Down, 22-Aug-95 (KNAA det MLC); Roundway Camp, Jul-40 (FDB) (Buck, 1956)
SU16: Marlborough area, 1869-1938 (Meyrick, 1938)
SU26: Savernake Forest, 15-Sep to 27-Oct-95 (EN det MDD); 31-May-04 (BLe)
8 ST93: Setting Copse, 28-Sep-95 (PJH) (Hodge, 1995); Stockton Wood, 28-Sep-95 (PJH) (Hodge, 1995)
ST94: Bowls Barrow (SPTAW), 16-May, 22-Jun-02 (PJH) (Hodge, 2002); Ladywell (SPTAW), 15-May-02 (PJH) (Hodge, 2002); Norton Down (SPTA), 15-May, 23-Jun-02 (PJH) (Hodge, 2002)
SU02: Knighton Wood, Sep to Nov-06, Mar, Oct-07 (AJWA)
SU03: Dinton Park, 15-Aug-01 (KNAA); Grovely Wood, 28-Jul-83 (DRN); Wilton, 14-Aug-76 (DRN); Wylye Down NNR, 22-Feb to 29-Mar-07 (MDD) (Darby, 2009)
SU04: Salisbury Plain, 25-Jun-02 (PJH) (Hodge, 2002)
SU14: Haxton (SPTAE), 17-May, 24-Jun-02 (PJH) (Hodge, 2002); Larkhill west, 2-Jun-97 (MGT det MLC); Stonehenge, 22-Apr-03 (PJH) (Hodge, 2003); Wexland Hanging (SPTA), 22-Jun-02 (PJH) (Hodge, 2002)
SU15: Everleigh (SPTAE), 18-May, 24-Jun-02 (PJH) (Hodge, 2002)
SU22: Bentley Wood, 31-May-97 (MDD); Langley Wood, 11-Feb to 3-Mar-05 (MDD) (Darby, 2005)
SU23: Driver Training Area (SPTAE), 17-May-02 (PJH) (Hodge, 2002); 24-Jun-02 (PJH) (Hodge, 2002); North Tidworth, Devils Ditch, 30-May-04 (BLe); Porton Down, 3-Jul-99 (PJH) (Hodge, 2002)
SU25: Sidbury Hill (SPTAE), 4-Jun-97 (MGT)
Resident. Widespread; on Brassicaceae, the leaves of which are eaten by the adults causing 'shot-holes'. One of the three commonest species on oil seed rape.

***Phyllotreta nodicornis*** (Marsham, 1802)
**LCGR-FI CI**
8 ST94: Bowls Barrow (SPTAW), 16-May-02 (PJH) (Hodge, 2002); Ladywell (SPTAW), 15-May-88 (EGS det MDD); 15-May-02 (PJH) (Hodge, 2002); 3-Jun - 12-Aug-08 (Boyce & Wright, 2009); Norton Down (SPTAW), 15-May, 23-Jun-02 (PJH) (Hodge, 2002); Tinkers Track (SPTAW), 3-Jun - 12-Aug-08 (Boyce & Wright, 2009); Warminster Ranges (SPTAW), 3-Jun - 12-Aug-08 (Boyce & Wright, 2009)
SU03: Grovely Wood, 28-Jul-83 (DRN)
SU04: Salisbury Plain, 16-May-02 (PJH) (Hodge, 2002)
SU05: Ell Barrow (SPTAC), 3-Jun - 12-Aug-08

(Boyce & Wright, 2009)
SU14: Haxton (SPTAE), 3-Jun - 12-Aug-08 (Boyce & Wright, 2009); Larkhill west, 2-Jun-97 (MGT); Stonehenge, 22-Apr-03 (PJH) (Hodge, 2003); Wexland Hanging (SPTAC), 3-Jun - 12-Aug-08 (Boyce & Wright, 2009)
SU15: Everleigh (SPTAE), 18-May-02 (PJH) (Hodge, 2002)
SU22: Blackmoor Copse, 31-Aug-01 (APF) (Foster, 2004)
SU24: Bulford Ranges (SPTAE), 3-Jun - 12-Aug-08 (Boyce & Wright, 2009); Dunch Hill (SPTAE), 3-Jun - 12-Aug-08 (Boyce & Wright, 2009); North Tidworth, Devils Ditch, 30-May-04 (BLe)
Resident. Widespread; on vegetation in a range of habitats including woodland and grassland. [Adults usually occur on the mignonette family, Resedeaceae].

***Phyllotreta ochripes*** (Curtis, 1837)
7 SU16: Marlborough area, 1869-1938
8 SU14: Figheldean meadows, Jun-1870 (RH) (Hislop, 1871); Stonehenge, 22-Apr-03 (PJH) (Hodge, 2003)
Resident. Very local; [adults usually on Brassicaceae] causing 'shot-holing' of the leaves.

***Phyllotreta tetrastigma*** (Comolli, 1837)
8 ST91: Cranborne Chase, 16 to 25-Jun-80 (NCC)
Resident. Very local; [in wet areas on water-cress and other cresses in particular. Adults cause 'shot-holing' of the leaves.

***Phyllotreta undulata*** Kutschera, 1860
7 SU06: Roundway Camp, Jul-40 (FDB) (Buck, 1956)
SU16: Marlborough area, 17-Feb-1879 (WFB) (Meyrick, 1880); 1869-1938 (Meyrick, 1938)
SU26: Savernake Forest, 4 to 25-Aug-95 (EN det MDD)
8 ST93: Great Ridge Wood, 16-May-00 (MDD)
ST94: Bowls Barrow (SPTAW), 22-Jun-02 (PJH) (Hodge, 2002); Norton Down (SPTAW), 15-May-02 (PJH) (Hodge, 2002)
SU02: Knighton Wood, Apr, Sep-07 (AJWA)
SU03: Grovely Wood, 4-Aug-78 (DRN); Wilton, 14-Aug-76 (DRN)
SU04: Salisbury Plain, 25-Jun-02 (PJH) (Hodge, 2002)
SU13: Great Durnford, 6-Apr-70 (DRN); Idmiston, 5-Aug-82 (DRN); Salisbury, Apr-00 (MAS) (Darby, 2001)
SU14: Stonehenge, 22-Apr-03 (PJH) (Hodge, 2003)
SU15: Everleigh (SPTAE), 18-May, 24-Jun-02 (PJH) (Hodge, 2002)
Resident. Widespread; on vegetation, especially Brassicaceae, and sometimes on trees, in a wide range of habitats. Adults feed on the leaves causing 'shot-holing'.

***Phyllotreta vittula*** (Redtenbacher, 1849)
7 SU16: Marlborough area, Jun-1873 (EWJ) (Meyrick, 1874); 1869-1938 (Meyrick, 1938)
8 SU02: Knighton Wood, Sep, Oct-06, Nov-07 (AJWA)
SU03: Grovely Wood, 4-Aug-70 (DRN)
SU04: Shrewton, 30-Jul-82 (DRN)
SU13: Idmiston, 5-Aug-82 (DRN)
Resident. Local; on water-cress, grass and other vegetation [particularly Brassicaceae, in various habitats].

**APHTHONA Dejean, 1836**
***Aphthona ?atratula*** Allard, 1859
7 SU16: Marlborough area, 1869-1938 (Meyrick, 1938)
SU36: Stype Wood, Jun-1870 (RH) (Hislop, 1871)
8 SU03: Grovely Wood, 1-Aug-78, 28-Jul-83 (DRN)
This is the *A atrovirens* of Pope (1977) but this species does not occur in GB and these specimens are probably referable to *atratula*. However *atratula* has recently been synonymised by several continental authors with *A euphorbiae*. Cox, however, believes that the species are distinct.
Resident. Very local; on herbage in woodland.

***Aphthona atrocaerulea*** (Stephens, 1831)
7 ST97: North Draycot Park, 22-May-00 (MDD)
SU26: Savernake Forest, 31-May-04 (BLe)
8 ST93: Great Ridge Wood, 25-May-00 (MDD)
SU03: Little Langford, 3-Jul-96 (MAS)
SU22: Blackmoor Copse, 31-Aug-01, 4-May-02, 13-Jul-03, 3-Aug-03 (APF) (Foster, 2004)
Resident. Local; on Spurge in woodland. Adult feeding causes 'shot-holing' of the leaves.

***Aphthona euphorbiae*** (Schrank, 1781
7 SU18: Swindon, 7-May-01 (JKG det MDD)
SU26: Savernake Forest, 15-Jun-73 (SAW); 22-Jun-90 (APFw); 20-Apr to 26-May, 8-Jul to 11-Oct-93, 28-Apr to 11-Oct-94 (EN det SAW); 11 to 30-May-95, 4-Aug to 27-Oct-95 (EN det MDD)
ST85: Bullen Hill Farm, 1982-2002 (EGS det MDD)
8 ST91: Cranborne Chase, 22-Jun-99 (MDD)
ST93: Great Ridge Wood, 4-May-00 (MDD), 16-May-00 (MDD); Stockton area, 28-Sep-95 (PJH) (Hodge, 1995)
ST94: Bowls Barrow (SPTAW), 16-May-02 (PJH) (Hodge, 2002); Ladywell (SPTAW), 15-May-02 (PJH) (Hodge, 2002); Tinkers Track (SPTAW), 3-Jun - 12-Aug-08 (Boyce & Wright, 2009)
SU01: Martin Down NNR, 1979-1992 (RGB); 1999 (EN, 1999)

SU02: Knighton Wood, Sep, 7-Oct-06, Mar, Sep, Nov-07 (AJWA)
SU03: Wylye Down NNR, 20-Sep-05, 4-May to 10-Jul, 4-Sep-07 (MDD) (Darby, 2009)
SU04: Fox Covert (SPTAW), 3-Jun - 12-Aug-08 (Boyce & Wright, 2009); Orcheston Down (SPTA), 18-May-02 (PJH) (Hodge, 2002)
SU05: Ell Barrow (SPTAC), 3-Jun - 12-Aug-08 (Boyce & Wright, 2009)
SU15: Everleigh (SPTAE), 18-May-02 (PJH) (Hodge, 2002)
SU16: Jones's Mill, 17-Oct-01 (APF) (Foster, 2002)
SU21: Franchises Wood, 3-May-95 (MAS)
SU22: Langley Wood NNR, 14-May-04 to 24-Mar-05, 25-Aug-06 to 6-Jun-07 (MDD) (Darby, 2005, 2008)
SU23: Thorny Down Woods, 16-Oct-00 (MDD)
SU24: Bulford Ranges (SPTAE), 3-Jun - 12-Aug-08 (Boyce & Wright, 2009)
SU25: Sidbury Hill (SPTAE), 4-Jun-97 (MGT det MLC)

Resident. Widespread and frequent on a wide range of plants in a very wide range of habitats. Much more common than formerly. Adult feeding causes typical 'shot-holing' of leaves.

***Aphthona herbigrada*** (Curtis, 1837)
**LCGR-Fh Cl**

8 SU01: Martin Down NNR, 1979-1992 (RGB)
SU23: Porton Down, 3-Jul-99 (MAS); 18-Aug-05 (KNAA)
SU24: Beacon Hill (SPTAE), 3-Jun - 12-Aug-08 (Boyce & Wright, 2009)

Resident. Very local; [usually on rock-roses, but sometimes gorse and thymes]. Adult feeding causes 'shot-holing' of the leaves.

***Aphthona lutescens*** (Gyllenhal, 1813)

7 SU09: Somerford Keynes, Swillbrook Lakes, 4-Jun-04 (BLe)
SU16: Chilton, 1900-1905 (Meyrick, 1906); Marlborough area, 1869-1938 (Meyrick, 1938)
8 SU13: Great Durnford, 19-Aug-69 (DRN); Middle Woodford, 27-Jul-77 (DRN); Salisbury, Avon Valley Reserve, 12- May, 1-Jun-00 (MDD) (Darby, 2001)

Resident. Very local; on vegetation in wetlands. Adults feed on leaves of purple-loosestrife, *Lythrum salicaria.*

***Aphthona melancholica*** Weise, 1888

7 SU16: Marlborough area, 4-May-1879 (WFB) (Meyrick, 1880); 1869-1938 (Meyrick, 1938)
8 ST91: Cranborne Chase, 7-May-97 (MAS)
SU03: Grovely Wood, 1-Aug-88 (DRN)

Resident. Very local; on spurge, *Euphorbia* spp. in woodland.

***Aphthona nigriceps*** (Redtenbacher, 1842)
**Na**

8 SU15: Manningford, 13-Aug-87 (DRN)
SU16: Woodborough, 4-Aug-83 (DRN)

Resident. Very local; on meadow crane's-bill, *Geranium pratense* [and other plants]. Adult feeding causes 'shot-holing' of the leaves.

***Aphthona nonstriata*** (Goeze, 1777)

7 SU16: Oare, 16-Mar-02 (IJG)
8 SU13: Middle Woodford, 27-Jul-77 (DRN); Great Durnford, 17-May-73, 27-Jul-77 (DRN) Salisbury, 15-Aug-88 (DRN); Apr-00 (MAS); 1-May-00 (MDD) (Darby, 2001)
SU16: Jones's Mill, 21-Jun-97 (MDD); 19-Jun-01 (APF) (Foster, 2002); 1-Jun-04 (BLe)

Resident. Local; exclusively on yellow iris, *Iris pseudacorus.* Adults eat longitudinal strips in the leaves and larvae mine the leaves causing long, white feeding marks usually near the extremities.

**LONGITARSUS Berthold, 1827**

***Longitarsus agilis*** (Rye, 1868)
**Na**

8 SU13: Great Durnford, 6-Aug-87 (DRN)

Resident. Very local; on vegetation near the R. Avon. [Adults usually on figworts causing 'shot-holing' of the leaves].

Longitarsus anchusae

***Longitarsus anchusae*** (Paykull, 1799)
**Nb, LCGR-Fm Cl**

8 ST94: Ladywell (SPTAW), 3-Jun - 12-Aug-08 (Boyce & Wright, 2009)
SU04: Fox Covert (SPTAW), 3-Jun - 12-Aug-08 (Boyce & Wright, 2009)
SU23: Porton Down, 3-Jul-99 (PMP)
SU24: Bulford Ranges (SPTAE), 3-Jun - 12-Aug-08 (Boyce & Wright, 2009); North Tidworth, Devils Ditch, 30-May-04 (BLe)

Resident. Very local. [On various members of the borage family (Boraginaceae)]. Adult feeding causes 'shot-holing' of the leaves.

***Longitarsus atricillus*** (Linnaeus, 1761)
7 SU16: Marlborough area, Jun-1873 (EWJ) (Meyrick, 1874); 1869-1938 (Meyrick, 1938)
8 SU03: Wylye Down NNR, 20-Sep-05, 27-Jan to 5-Jun-07 (MDD) (Darby, 2009)
Resident. Very local; on plants of several families in a wide range of habitats. Adult feeding causes 'shot-holing' of the leaves.

***Longitarsus brunneus*** (Duftschmid, 1825)
**Nb**
8 SU14: Stonehenge, 22-Apr, 19-Jul-03 (PJH) (Hodge, 2003)
Resident. Very local; on common meadow-rue, *Thalictrum flavum* growing besides the R. Avon. Adult feeding causes 'shot-holing' of leaves.

***Longitarsus curtus*** (Allard, 1860)
**Na, LCGR-Fm Cl**
8 SU24: Bulford Ranges (SPTAE), 3-Jun - 12-Aug-08 (Boyce & Wright, 2009)
Residen. Very local: [on various Boraginaceae. Adult feeding causes 'shot-holing' of the leaves].

Longitarsus dorsalis

***Longitarsus dorsalis*** (Fabricius, 1781)
**Nb, LCGR-Fm Cl**
7 SU06: Cherhill Down, 31-May-04 (BLe)
8 SU02: Knighton Wood, Jun-07 (AJWA)
SU03: Wylye Down NNR, 22-Feb to 29-Mar-07 (MDD) (Darby, 2009)
SU04: Fox Covert (SPTAW), 3-Jun - 12-Aug-08 (Boyce & Wright, 2009)
SU12: Coombe Bissett Down, 1, 9-Mar-00 (MDD)
SU14: Stonehenge, 22-Apr, 20-May-03 (PJH) (Hodge, 2003); Wexland Hanging (SPTAC), 16-May-02 (PJH) (Hodge, 2002)
SU15: Beach's Plantation (SPTAE), 1-May-05 (MGT)
SU22: Bentley Wood, 27-May-98 (MDD); Langley Wood, 14-May-04 to 21-Jan-05 (MDD) (Darby, 2005)
SU24: Dunch Hill (SPTAE), 3-Jun - 12-Aug-08 (Boyce & Wright, 2009); North Tidworth, Devils Ditch, 30-May-04 (BLe)
Resident. Very local; in calcareous grassland, mainly on ragworts, *Senecio* spp.. Adult feeding causes 'shot-holes' in the leaves.

***Longitarsus exoletus*** (Linnaeus, 1758)
**LCGR-Fm Cl**
8 ST94: Ladywell (SPTAW), 23-Jun-02 (PJH) (Hodge, 2002); Tinkers Track (SPTAW), 3-Jun - 12-Aug-08 (Boyce & Wright, 2009)
SU01: Martin Down NNR, 1979-1992 (RGB)
SU04: Fox Covert (SPTAW), 3-Jun - 12-Aug-08 (Boyce & Wright, 2009); Salisbury Plain, 25-Jun-02 (PJH) (Hodge, 2002); Tilshead Park (SPTAW), 14-Jul-99 (PJH) (Hodge, 1999)
SU05: Black Heath (SPTAC), 3-Jun - 12-Aug-08 (Boyce & Wright, 2009)
SU14: Haxton, 24-Jun-02 (PJH) (Hodge, 2002); Wexland Hanging (SPTAC), 22-Jun-02 (PJH) (Hodge, 2002); 3-Jun - 12-Aug-08 (Boyce & Wright, 2009)
SU15: Everleigh (SPTAE), 18-May-02 (PJH) (Hodge, 2002); Haxton Down, 14-Jul-99 (PJH) (Hodge, 1999)
SU23: Porton Down, 3-Jul-99 (MAS)
SU24: Driver Training Area (SPTAE), 17-May, 24-Jun-02 (PJH) (Hodge, 2002)
SU25: Sidbury Hill (SPTAE), 28-May, 5-Jun-97 (MGT det MLC)
Resident. Local; [usually on borages, Boraginaceae]. Adult feeding causes 'shot-holes' in the leaves.

***Longitarsus flavicornis*** (Stephens, 1831)
7 SU06: Calstone Down, 22-Aug-95 (KNAA det MLC)
SU26: Savernake Forest, 20-Sep to 11-Oct-94 (EN det SAW)
8 SU01: Martin Down NNR, 21-Aug-89 (RGB); 1999, (EN, 1999)
SU03: Grovely Wood, 9-Aug-71, 4-Aug-78 (DRN)
SU13: Lower Woodford, 28-Jul-78 (DRN); Salisbury, 15-Aug-88 (DRN)
SU14: Wexland Hanging (SPTAC) 22-Jun-02 (PJH) (Hodge, 2002)
Resident. Widespread; on various plants including scabious and mustard, some by the R. Avon. [Usually on ragworts. Adult feeding causes 'shot-holes' in the leaves

***Longitarsus fowleri*** Allen, 1967
**Na, LCGR-Fh Ch**
8 SU24: Beacon Hill (SPTAE), 3-Jun - 12-Aug-08 (Boyce & Wright, 2009)
Resident. Very local; on [the undersides of young ] teasel rosettes [and on the upper surface of older plants later in the year]. Adult feeding causes 'shot-holes' in the leaves.

Longitarsus ganglbaueri

***Longitarsus ganglbaueri*** Heikertinger, 1912
**Na**
7 SU16: Marlborough area, 1869-1938 (Meyrick, 1938)
Resident. Very local; on *Senecio jacobaea* [Ragwort]

***Longitarsus gracilis*** Kutschera, 1864
8 ST93: Stockton Wood, 28-Sep-95 (PJH) (Hodge, 1995)
SU22: Bentley Wood, 31-May-96 (MDD)
Resident. Very local; on ragworts, *Senecio* spp., adult feeding causes 'shot-holes' in the leaves.

***Longitarsus holsaticus*** (Linnaeus, 1758)
7 SU16: Granham, Marlborough area, 1939 (det KGB) (Knight, 1940)
Resident. Very local; [adults feed on marsh lousewort, *Pedicularis palustris*, and possibly speedwells, *Veronica* spp.].

***Longitarsus kutscherae*** (Rye, 1872)
8 SU03: Grovely Wood, 4-Aug-70 (DRN)
Resident. Very local; on vegetation in a ride. [Adults feed on plantains] causing 'shot-holes' in the leaves.

***Longitarsus luridus*** (Scopoli, 1763)
7 SU16: Marlborough area, 1869-1938 (Meyrick, 1938)
SU26: Savernake Forest, 28 –Apr to 18-May, 1-Jul to 11-Oct-94 (EN det SAW); 4-Aug to 27-Oct-95 (EN det MDD)
8 ST91: Cranborne Chase, 4-Jun-97 (MAS)
ST93: Stockton area, 28-Sep-95 (PJH) (Hodge, 1995)
ST94: Bowls Barrow (SPTAW), 16-May-02 (PJH) (Hodge, 2002)
SU01: Martin Down NNR, 21-Aug-89 (RGB); 1999 (EN, 1999)
SU02: Knighton Wood, Sep-08 (AJWA)
SU03: Dinton Park, 15-Aug-01 (KNAA det ML); Little Langford, 19-Aug-95 (MAS)
SU13: Great Durnford, 16-Aug-69 (DRN); Old Sarum, 31-May-97 (MAS); Salisbury, 15-Aug-88 (DRN); Apr-00 (MAS); 7-Apr, 12-May-00 (MDD) (Darby, 2001)
SU22: Bentley Wood, 31-May-97 (MDD); 3-Sep-97 (WANHS det MDD); Blackmoor Copse, 31-Aug-01 (APF) (Foster, 2004)
Resident. Widespread; on plants of many families in almost any habitat. Adult feeding causes 'shot-holes' in the leaves.

***Longitarsus lycopi*** (Foudras, 1860)
**Nb, LCGR-Fm C?**
8 SU24: North Tidworth, Devils Ditch, 30-May-04 (BLe)
Resident. Very local; [on various species of the mint family]. Adult feeding causes 'shot-holes' in the leaves.

***Longitarsus melanocephalus*** (De Geer, 1775)
7 SU16: Marlborough area, 1869-1938 (Meyrick, 1938)
8 SU01: Martin Down NNR, 21-Aug-89 (RGB)
SU02: Knighton Wood, May-07 (AJWA)
SU13: Salisbury, Avon Valley Reserve, Apr-00 (MAS) (Darby, 2001)
Resident. Very local; in various habitats on plantains. Adult feeding causes 'shot-holes' in the leaves.

***Longitarsus membranaceus*** (Foudras, 1860)
7 SU26: Mildenhall, 1939 (det KGB) (Knight, 1940)
Resident. Very local; [usually in woodland in clearings and rides. Adults feed on wood sage causing 'shot-holes' in the leaves.

Longitarsus nasturtii

***Longitarsus nasturtii*** (Fabricius, 1792)
**Nb, LCGR-Fl Cl**
7 SU16: Marlborough area, Jun-1873 (Meyrick, 1874); 1869-1938 (Meyrick, 1938)
Resident. Very local; on Boraginaceae. Adult feeding causes 'shot-holes' in the leaves.

***Longitarsus nigrofasciatus*** (Goeze, 1777)
**Na, LCGR-Fh Cl**
8 ST91: Cranborne Chase, 7-May-97 (MAS)
Resident. Very local; [on mulleins and figworts]. Adult feeding causes 'shot-holes' in the leaves.

***Longitarsus obliteratus*** (Rosenhauser, 1847)
**LCGR-Fh Cm**
7 SU06: Calstone Down, 22-Aug-95 (KNAA)

Resident. Very local; [on Lamiaceae in various habitats]. Adult feeding causes 'shot-holes' in the leaves.

***Longitarsus parvulus*** (Paykull, 1799)
**Na**
7 ST99: Crudwell, 19-Jun-73 (APF)
SU26: Savernake Forest, 15-Jun-73 (SAW); 20 to 25-Apr, 29-Aug to 11-Oct-93 (EN det SAW); 11 to 30-May, 4-Aug to 10-Oct-95 (EN det MDD)
8 ST91: Cranborne Chase, 24-Apr-97 (MAS)
ST92: Sutton Mandeville, 1-Sep-96 (JAO)
ST93: Stockton Wood, 28-Sep-95 (PJH) (Hodge, 1995)
SU01: Martin Down NNR, 1979-1992 (RGB)
SU02: Knighton Wood, Oct, Nov-06, Mar, Sep-07 (AJWA)
SU04: Fox Covert (SPTAW), 3-Jun - 12-Aug-08 (Boyce & Wright, 2009)
SU13: Salisbury, Apr-00 (MAS) (Darby, 2001)
SU15: Everleigh (SPTAE), 18-May-02 (PJH) (Hodge, 2002)
SU22: Whiteparish Common, 2-Jul-95 (MAS)
SU23: Porton Down, 3-Jul-99 (MAS)
Resident. Widespread; in spring and early summer mainly on Linseed, *Linum usitatissimum*, but in late summer, after the linseed has been harvested, on many plants and trees. Adult feeding causes 'shot-holes' in the leaves. Appears to be no longer deserving of its Na status.

***Longitarsus pratensis*** (Panzer, 1794)
8 SU06: Calstone Down, 22-Aug-95 (KNAA det MLC)
ST93: Stockton Wood, 28-Sep-95 (PJH) (Hodge, 2002)
ST94: Norton Down (SPTAW), 15-May, 23-Jun-02 (PJH) (Hodge, 2002)
SU01: Martin Down NNR, 21-Aug-89 (RGB)
SU03: Dinton Park, 15-Aug-01 (KNAA det MCL); Grovely Wood, 9-Aug-71 (DRN); Wylye Down NNR, 20-Sep-05, 29-Mar-07 (MDD) (Darby, 2009)
SU04: Orcheston Down, 22-Jun-02 (PJH) (Hodge, 2002)
SU14: Haxton (SPTAE), 17-May-02 (PJH) (Hodge, 2002); Stonehenge, 22-Apr-03 (PJH) (Hodge, 2003); Wexland Hanging (SPTAC), 16-May-02 (PJH) (Hodge, 2002)
SU15: Everleigh (SPTAE), 18-May-02 (PJH) (Hodge, 2002)
SU23: Porton Down, 3-Jul-99 (PMP
SU24: Driver Training Area (SPTAE), 17-May-02 (PJH) (Hodge, 2002)
Resident. Very local; [in a wide variety of habitats, usually on plantains, *Plantago* spp.]. Adult feeding causes 'shot-holes' in the leaves.

***Longitarsus reichei*** (Allard, 1860)
8 SU23: Porton Down, 3-Jul-99 (BLe)
Resident. Very local; [in a variety of habitats on plantains, *Plantago* spp.]. Adult feeding cuses 'shot-holes' in the leaves.

***Longitarsus succineus*** (Foudras, 1860)
8 SU01: Martin Down NNR, 21-Aug-89 (RGB)
SU03: Grovely Wood, 4-Aug-78 (DRN)
SU04: Shrewton, 2-Aug-83 (DRN)
SU23: Porton Down, 20-Jul-97 (MAS); 3-Jul-99 (BLe)
Resident. Very local; [usually on many species of the daisy family (Asteraceae) including *Achillea*, *Artemisia*, chamomiles, mugworts, oxeye daisies, etc.]. Adult feeding causes 'shot-holes' in the leaves.

***Longitarsus suturellus*** (Duftschmid, 1825)
7 SU16: Marlborough area, 1869-1938 (Meyrick, 1938)
8 ST94: Ladywell (SPTAW), 3-Jun - 12-Aug-08 (Boyce & Wright, 2009)
SU12: Coombe Bissett Down, 1-Mar-00 (MDD)
Resident. Very local; [usually on many species of the daisy family (Asteraceae) including ragworts, tansy, colt's foot, etc.]. Adult feeding causes 'shot-holes' in the leaves.

***Longitarsus tabidus*** (Fabricius, 1775)
**Nb**
8 SU03: Grovely Wood, 16-Aug-84 (DRN)
SU24: North Tidworth, Devils Ditch, 30-May-04 (BLe)
Resident. Very local; on mulleins, *Verbascum* spp.. Adult feeding causes 'shot-holes' in the leaves.

Altica brevicollis

**ALTICA Geoffroy, 1762**
***Altica brevicollis*** Foudras, 1860
**Na**
8 ST91: Cranborne Chase, 4-Jun-97 (MAS)
Resident. Very local; [probably feeds only on hazel]. Adult feeding causes small holes in the leaves and larvae feed on the undersurface causing similar damage.

***Altica carinthiaca*** Weise, 1888
8 ST94: Bowls Barrow (SPTAW), 16-May-02 (PJH) (Hodge, 2002); Ladywell (SPTAW), 15-May-02 (PJH) (Hodge, 2002)
SU04: Fox Covert (SPTA), 16-May-02 (PJH) (Hodge, 2002)
SU14: Haxton (SPTAE), 17-May-02 (PJH) (Hodge, 2002); Stonehenge, 20-May-03 (PJH) (Hodge, 2003); Stonehenge, 22-Apr-03 (PJH) (Hodge, 2003); Wexland Hanging (SPTA), 16-May-02 (PJH) (Hodge, 2002)
Resident. Local; adults and larvae feed only on meadow vetchling, *Lathyrus pratensis,* making longitudinal slits in the leaves. Newly recognised as a British species.

***Altica helianthemi*** (Duftschmid, 1825)
7 SU16: Marlborough area, 1869-1938 (Meyrick, 1938); Pewsey Downs, 22-Jun-90 (APFw)
8 SU23: Porton Down, 3-Jul-99 (PJH) (Hodge, 2002)
SU24: Beacon Hill (SPTAE), 3-Jun - 12-Aug-08 (Boyce & Wright, 2009)
Resident. Very local; [usually on species of the rock-rose family, Cistaceae].

***Altica longicollis*** (Allard, 1859)
**Nb**
8 SU22: Langley Wood NNR, 25-Jun to 16-Jul-04 (MDD) (Darby, 2005)
SU23: Porton Down, 3-Jul-99 (MAS)
Resident. Very local; on heather. Can cause extensive damage. This is listed as *A ericeti* in Duff (2008) but has since been changed.

***Altica lythri*** Aubé, 1843
7 ST96: Conigre Meadow, 20-May-01 (JKG det MDD)
SU09: Somerford Keynes, 4-Jun-04 (BLe); Upper Waterhay, 18-Apr-01, 13-Jun-04 (JKG det MDD)
SU16: Marlborough area, 1869-1938
SU18: Rushy Platt, 20-May-01 (JKG det MDD)
SU36: Stype Wood, Jun-1870 (RH) (Hislop, 1871)
8 ST94: Ladywell (SPTAW), 3-Jun - 12-Aug-08 (Boyce & Wright, 2009)
SU02: Fovant. Water Cress, 31-May-95 (MDD)
SU05: Lavington Down, 6-May-03 (EGS det MDD)
SU12: Petersfinger, 25-Jul-91 (DRN)
SU13: Great Durnford, 16-Aug-69, 3-Aug-83 (DRN); Idmiston, 5-Aug-82 (DRN); Salisbury, 27-Jul-77 (DRN); Salisbury, Avon Valley Reserve, Apr-00 (MAS); 12-Apr-00 (MDD) (Darby, 2001)

SU14: Stonehenge, 22-Apr-03 (PJH) (Hodge, 2003)
SU16: Jones's Mill, 11-Jun-97 (MAS); 19-Jun-01 (APF) (Foster, 2002); 1-Jun-04 (BLe)
SU22: Blackmoor Copse, 17-May-02 (APF) (Foster, 2004)
Resident. Widespread; on members of the willowherb family (Onagraceae) in a wide variety of habitats.

***Altica oleracea*** (Linnaeus, 1758)
7 SU06: Calstone Down, 27-Aug-95 (KNAA der MLC)
SU08: Somerford Common, 23-Apr-01 (JKG det MDD)
SU16: Marlborough area, Jun-1873 (EWJ) (Meyrick, 1874); 1869-1938 (Meyrick, 1938)
SU98: Great Somerford, 12-Mar-95 (BG det MDD)
8 ST91: Cranborne Chase, 24-Apr-97 (MAS)
SU01: Martin Down NNR, 1999 (EN, 1999)
SU03: Wylye Down NNR, 20-Sep-05, 22-Feb to 29-Mar, 5-Jun to 10-Jul-07 (MDD) (Darby, 2008a)
SU23: Porton Down, 20-Jul-97, 3-Jul-99 (MAS)
SU24: Driver Training Area (SPTAE), 17-May-02 (PJH) (Hodge, 2002)
Resident. Widespread; in various habitats on a wide range of plants and trees. Adults feed on the edges of leaves causing notches and holes.

***Altica palustris*** Weise, 1888
8 ST93: Great Ridge Wood, 16-May-00 (MDD)
ST94: Ladywell (SPTAW), 3-Jun - 12-Aug-08 (Boyce & Wright, 2009)
SU03: Grovely Wood, 9-Aug-71 (DRN)
Resident. Very local; in various habitats on various plants. Adults eat small holes and notches in the leaves of willowherbs.

**HERMAEOPHAGA Foudras, 1860**
***Hermaeophaga mercurialis*** (Fabricius, 1792)
7 SU16: Marlborough area, Jun-1873 (EWJ) (Meyrick, 1874); 1869-1938 (Meyrick, 1938)
8 ST91: Cranborne Chase, 24-Apr-97 (MAS); 22, 30-Jun-99 (MDD)
ST93: Great Ridge Wood, 14-Jun-96, 4, 10-May-00 (MDD); 18-Apr-74 (DRN); Stockton Wood, 28-Sep-95 (PJH) (Hodge, 2002) (Hodge, 1995)
SU02: Knighton Wood, Sep, Oct-06, Aug to Sep-07, May-08 (AJWA)
SU03: Grovely Wood, 9 records 11-Aug-71 - 4-Aug-87 (DRN); 11, 12-Jun, 3-Dec-00, 21-Aug-01 (MDD)
SU22: Bentley Wood, 31-May-97 (MDD); Blackmoor Copse, 4-May-0318-May-03 (APF) (Foster, 2004)
Resident. Widespread; in woodland on dogs mercury, *Mercurialis perennis.* Adult feeding causes small holes in the leaves.

**BATOPHILA Foudras, 1860**
***Batophila rubi*** (Paykull, 1799)
7 SU16: Marlborough area, 1869-1938 (Meyrick, 1938)

Resident. Very local; chiefly on *Rubus* [brambles].

**OCHROSIS Foudras, 1860**
***Ochrosis ventralis*** (Illiger, 1807)
**RDB3**
8 SU22: Langley Wood NNR, 14-May to 3-Jun-04 (MDD) (Darby, 2005)
Resident. Very local; in a pitfall trap. .Adults feed on the leaves of nightshades, *Solanum* spp., causing typical 'shot-holes'. Very local and scattered in GB. Alexander (2006) notes that there have been very few records of this species in recent years.

Ochrosis ventralis

**NEOCREPIDODERA Heikertinger, 1911**
***Neocrepidodera ferruginea*** (Scopoli, 1763)
7 SU06: Roundway Camp, Jul-40 (FDB) (Buck, 1956)
SU16: Marlborough area, 18-Jun-1879 (FED-H) (Meyrick, 1880); 1869-1938 (Meyrick, 1938)
8 ST94: Ladywell, 23-Jun-02 (PJH) (Hodge, 2002)
SU03: Grovely Wood, 24-Aug-72, 1-Aug-73 (DRN); Wylye Down NNR, 5-Jun to 10-Jul-07 (MDD) (Darby, 2009)
SU04: Parsonage Down NNR, 2, 22-Jul-08 (MDD); Salisbury Plain, 23-Jun-02 (PJH) (Hodge, 2002); Shrewton, 6-Aug-82, 1-Aug-85 (DRN)
SU05: Ell Barrow (SPTAC), 3-Jun - 12-Aug-08 (Boyce & Wright, 2009)
SU12: Petersfinger, 25-Jul-91 (DRN)
SU13: Great Durnford, 19-Aug-69, 7-Aug-70 (DRN); Lower Woodford, 28-Jul-78 (DRN); Middle Woodford, 27-Jul-77 (DRN); Old Sarum, 31-May-97 (MAS)
SU14: Wexland Hanging (SPTAC) 23-Jun-02 (PJH) (Hodge, 2002)
SU15: Manningford, 4-Aug-88 (DRN)
SU16: Woodborough, 4-Aug-83 (DRN)
SU21: Hamptworth Common, 28-Jul-77 (DRN); Hamptworth Estate, 3-Aug-73 (DRN)
SU23: Porton Down, 3-Jul-99 (PJH) (Hodge, 2002)
SU24: Sidbury Hill (SPTAE), 13 to 17-Jun-97 (MGT)
Resident. Widespread; on many different plants in a wide range of habitats. Adult feeding causes 'shot-holes' in the leaves.
***Neocrepidodera transversa*** (Marsham, 1802)
8 ST94: Ladywell (SPTAW), 3-Jun - 12-Aug-08 (Boyce & Wright, 2009);Tinkers Track (SPTAW), 3-Jun - 12-Aug-08 (Boyce & Wright, 2009); Wadman's Coppice (SPTAW), 3-Jun - 12-Aug-08 (Boyce & Wright, 2009); Warminster Ranges (SPTAW), 3-Jun - 12-Aug-08 (Boyce & Wright, 2009)
SU05: Black Heath (SPTAC), 3-Jun - 12-Aug-08 (Boyce & Wright, 2009)
SU13: Salisbury, 15-Aug-88 (DRN)
SU15: Beach's Plantation (SPTAE), 10-Jun-97 (MGT)
SU24: Beacon Hill (SPTAE), 3-Jun - 12-Aug-08 (Boyce & Wright, 2009)
Resident. Local; On several plant families in a wide range of habitats. Adult feeding causes 'shot-holing' in leaves.

**DEROCREPIS Weise, 1886**
***Derocrepis rufipes*** (Linnaeus, 1758)
7 ST97: Chippenham, Lackham College, 29-May-04 (BLe)
SU06: Cherhill Down, 31-May-04 (BLe)
SU16: Marlborough area, Jun-18738 (EWJ) (Meyrick, 1874); 1869-1938 (Meyrick, 1938); West Kennet, 4-May-03 (MGT)
8 ST94: Bowls Barrow (SPTAW), 16-May, 22-Jun-02 (PJH) (Hodge, 2002); Ladywell (SPTAW), 15-May, 23-Jun-02 (PJH) (Hodge, 2002)
ST95: Cheverell, 30-May-97 (MGT det MLC)
SU03: Grovely Wood, 11-Jun-00 (MDD)
SU04: Parsonage Down NNR, 2-Jul-08 (MDD)
SU23: Porton Down, 3-Jul-99 (MAS)
Resident. Local; on species of the pea family (Fabaceae) including vetches, broom, gorse, etc. as well as cultivated varieties.

**HIPPURIPHILA Foudras, 1860**
***Hippuriphila modeeri*** (Linnaeus, 1761)
8 SU14: Netheravon, 29-Jul-83 (DRN)
Resident. Very local; on vegetation near the R. Avon. Adults feed on the leaves and stems of horsetails, *Equisetum* spp..

**CREPIDODERA Dejean, 1836**
***Crepidodera aurata*** (Marsham, 1802)
7 SU16: Oare, 21-Apr-02 (IJG)
8 SU12: Petersfinger, 25-Jul-91 (DRN)
SU13: Salisbury, May-00 (MAS, MDD) (Darby, 2001)
SU15: Charlton, 11-Jun-06 (IJG)
SU22: Blackmoor Copse, 4 to 18-May-02 (APF) (Foster, 2004); Holmere Common, 19-Aug-88 (DRN); Homan's Copse, 31-Jul-85 (DRN); Langley Wood NNR, 28-May-74 (DRN)
Resident. Local; on willows and more rarely poplars. Adults feed on the leaves causing typical 'shot-holes'.

***Crepidodera aurea*** (Fourcroy, 1785)
7 ST98: Lacock, 29-May-04 (BLe)
8 SU03: Dinton Park, 15-Aug-01 (KNAA)
SU13: Salisbury, May-00 (MDD) (Darby, 2001)
SU22: Homan's Copse, 29-May-90 (DRN)
Resident. Very local; [on poplars and less frequently willows] Adults feed on poplar leaves causing typical 'shot-holes'.

***Crepidodera fulvicornis*** (Fabricius, 1792)
7 ST87: Castle Combe, 3-Jun-04 (BLe)
8 SU21: Hamptworth Common, 28-Jul-77 (DRN)
SU22: Glazier's Copse, 16-Aug-88 (DRN); Holmere Common, 19-Aug-88 (DRN); Homan's Copse, 29-May-90 (DRN); Langley Wood NNR, 28-May-74, 19-Aug-88 (DRN)
Resident. Local; on willows and more rarely poplars. Adults feed on the leaves of willows causing typical 'shot-holes'.

Crepidodera nitidula

***Crepidodera nitidula*** (Linnaeus, 1758)
**Nb**
7 SU16: Marlborough area, Jun-1873 (EWJ) (Meyrick, 1874)
8 ST94: Wadman's Coppice (SPTAW), 3-Jun - 12-Aug-08 (Boyce & Wright, 2009)
Resident. Very local. Stype Wood etc. on aspen. Adults feed on the leaves of poplars causing typical 'shot-holes'.

***Crepidodera plutus*** (Latreille, 1804)
8 SU12: Petersfinger, 25-Jul-91 (DRN)
Resident. Very local; [on willows and more rarely poplars]. Adults feed by removing the lower epidermis from the leaf leaving the upper surface intact.

**EPITRIX Foudras, 1860**
***Epitrix atropae*** Foudras, 1860
**Nb**
8 SU23: Porton Down, 15-Oct-95, 12-Nov-96, 3-Jul-99 (MAS)
Resident. Very local; [on undersides of large leaves of nightshades especially deadly nightshade, *Atropa belladonna*].

Epitrix atropae

***Epitrix pubescens*** (Koch, J.D.W, 1803)
7 ST96: Spye Park, 31-May-04 (KNAA)
ST98: Lacock, 29-May-04 (BLe)
SU09: Somerford Keynes, Swillbrook Lakes, 4-Jun-04 (BLe); Upper Waterhay, 17-May to 4-Jun-05 (JKG det MDD)
8 ST94: Norton Down (SPTAW), 15-May, 23-Jun-02 (PJH) (Hodge, 2002)
SU13: Great Durnford, 27-Jul-77, 6-Aug-87 (DRN); Middle Woodford, 27-Jul-77 (DRN); Salisbury, Avon Valley Reserve, Apr-00 (MAS) (Darby, 2001); Salisbury, Bemerton Folly, 7-Mar - 15-Jun-02 (MDD) (Darby, 2002d)
SU14: Stonehenge, 22-Apr-03 (PJH) (Hodge, 2003); Wexland Hanging (SPTAC) 22-Jun-02 (PJH) (Hodge, 2002)
SU16: Jones's Mill, 30-May-04 (KNAA); 4-Jun-04 (BLe)
SU22: Langley Wood NNR, 29-May-03 (MDD)
Resident. Widespread; [on various nightshades in a wide variety of habitats, especially bittersweet, *Solanum dulcamara*].

**PODAGRICA Dejean, 1836**
***Podagrica fuscicornis*** (Linnaeus, 1767)
**Nb**
7 SU16: Marlborough area, 26 May 1879 (CTS) (Meyrick, 1880); 1869-1938 (Meyrick, 1938)
Resident. Very local; on Malva[ceae, mallows]. Adults feed on the leaves causing typical 'shot-holes'.

**MANTURA Stephens, 1831**
***Mantura matthewsii*** (Curtis, 1833)
**LCGR-Fh Cm**
7 SU06: Cherhill Down, 1-Jun-04 (BLe)
8 SU24: Driver Training Area (SPTAE), 24-Jun-02 (PJH) (Hodge, 2002)
Resident. Very local; [adults usually on common rock-rose, *Helianthemum mummularium*].

Mantura rustica

***Mantura rustica*** (Linnaeus, 1767)
**Nb**
7 SU16: Marlborough area, 1869-1938 (Meyrick, 1938)
8 SU13: Lower Woodford, 1-Aug-83 (DRN)
SU21: Hamptworth Estate, 29-Mar-78 (DRN)
Resident. Very local; on vegetation near the R. Avon and in straw by a fishpond. [Usually on broadleaved dock, *Rumex obtusifolius*, and other docks].

**CHAETOCNEMA Stephens, 1831**
***Chaetocnema concinna*** (Marsham, 1802)
7 SU06: Roundway Camp, Jul-40 (FDB) (Buck, 1956)
SU09: Ashton Keynes, 16-Apr-01 (JKG det MDD); Somerford Keynes, Swillbrook Lakes, 4-Jun-04 (BLe); Upper Waterhay, 17-May to 4-Jun-05 (JKG det MDD)
SU16: Marlborough area, Jun-1873 (EWJ) (Meyrick, 1874); 6-Feb-1879 (WFB) (Meyrick, 1880); 1869-1938 (Meyrick, 1938)
SU26: Savernake Forest, 15-Sep to 5-Oct-95 (EN det MDD)
SU27: Ramsbury, 1870-1873 (Meyrick, 1874)
8 SU01: Martin Down NNR, 1979-1992 (RGB)
SU02: Knighton Wood, Oct-06, Oct-07 (AJWA)
SU03: Dinton Park, 15-Aug-01 (KNAA det MLC); Grovely Wood, 24-Aug-72 (DRN); 21-Aug-01 (MDD)
SU13: Salisbury, 15-Aug-88 (DRN), Apr-00 (MAS) (Darby, 2001)
SU21: Landford Bog, 25-May-98 (MDD)
SU22: Langley Wood NNR, 14-May-04 (MDD) (Darby, 2005); Redlynch, 4-Aug-82 (DRN)
SU23: Porton Down, 1970-2001 (Anon, 2001)
Resident. Widespread; on a wide range of plants in almost any habitat.*C. picipes* was not separated from *C. concinna* until 1997 so that earlier records of this species are indeterminate.

***Chaetocnema confusa*** (Boheman, 1851)
8 SU03: Wylye Down NNR, 20-Sep-05 (MDD) (Darby, 2009)
Resident. Very local; in a pitfall trap. A surprising record for a species usually associated with sedges and possibly rushes in wetland habitats.
***Chaetocnema hortensis*** (Fourcroy, 1785)
8 ST93: Stockton Wood, 28-Sep-95 (PJH) (Hodge, 1995)
SU01: Martin Down NNR, 21-Aug-89 (RGB)
SU03: Grovely Wood, 25-Oct-71 (DRN); Wylye Down NNR, 5-Jun to 3-Aug-07 (MDD) (Darby, 2009)
SU04: Salisbury Plain, 16-May-02 (PJH) (Hodge, 2002)
SU12: Charlton-All-Saints, 9-May-96 (MAS)
SU21: Hamptworth Estate, 6-Apr-77 (DRN); Landford Bog, 24-Jun-01 (APF) (Foster, 2001)
SU22: Bentley Wood, 3-Sep-97 (WANHS det MDD); Blackmoor Copse, 4-May-03 (APF) (Foster, 2004); Glaziers Copse, 16-Aug-88 (DRN)
SU23: Porton Down, 1970-2001 (Anon, 2001)
Resident. Widespread; in a range of habitats on grasses and cultivated crops. Adults feed on the leaves causing elongate holes.

**SPHAERODERMA Stephens, 1831**
***Sphaeroderma rubidum*** (Graëlls, 1858)
7 SU16: Marlborough area, 1869-1938 (Meyrick, 1938); Pewsey Downs, 22-Jun-90 (APFw)
8 SU04: Salisbury Plain, 25-Jun-02 (PJH) (Hodge, 2002); Shrewton, 1-Aug-85 (DRN)
SU13: Boscombe Down, 9-Oct-94 (MDD)
SU14: Haxton (SPTAE), 24-Jun-02 (PJH) (Hodge, 2002)
SU15: Beach's Plantation (SPTAE), 13 to 17-Jun-97 (MGT det MLC); Everleigh, 24-Jun-02 (PJH) (Hodge, 2002)
SU25: Sidbury Hill (SPTAE), 5-Jun-97 (MGT det MLC)
Resident. Local; [on various species of the daisy family (Asteraceae) especially knapweeds and thistles]. Adults feed on the leaves causing 'shot-holes'.

***Sphaeroderma testaceum*** (Fabricius, 1775)
7 SU09: Ashton Keynes, 7-May-01 (JKG det MDD)
SU16: Marlborough area, Jun-1873 (EWJ) (Meyrick, 1874); 1869-1938 (Meyrick, 1938)
8 SU01: Martin Down NNR, 21-Aug-89 (RGB)
SU03: Wilton, 14-Aug-76 (DRN); Grovely Wood, 11-Aug-71, 8-Apr-72, 29-Jul-78 (DRN)
SU04: Parsonage Down, 2-Jul-08 (MDD)
SU05: Ell Barrow (SPTAC), 3-Jun - 12-Aug-08 (Boyce & Wright, 2009)
SU12: Petersfinger, 25-Jul-91 (DRN)
SU14: Stonehenge, 22-Apr-03 (PJH) (Hodge, 2003)
SU15: Beach's Plantation (SPTAE), 13 - 17-Jun-97 (MGT det MLC)
SU16: Woodborough, 4-Aug-83 (DRN)
SU21: Franchises Common, 25-Jul-73 (DRN)
SU22: Homan's Copse, 31-Jul-85 (DRN)
SU23: Porton Down, 3-Jul-99 (PJH) (Hodge, 2002)

SU24: Driver Training Area (SPTAE), 24-Jun-02 (PJH) (Hodge, 2002)
Resident. Widespread; in a wide range of habitats on thistles and knapweeds.

**APTEROPEDA Dejean, 1836**
***Apteropeda globosa*** (Illiger, 1794)
**Nb, LCGR-Fl Cl**
7 SU16: Marlborough area, Jun-1873 (EWJ) (Meyrick, 1874); 1869-1938 (Meyrick, 1938)
Resident. Very local; in moss and by evening sweeping [on species of the dead nettle family, Lamiaceae, also on speedwells, *Veronica*. Usually on calcareous soils]. Adult feeding causes 'shot-holes' in the leaves. One specimen in MCC without data.

***Apteropeda orbiculata*** (Marsham, 1802)
7 SU16: Marlborough area, 1869-1938 (Meyrick, 1938); Pewsey Downs, 22-Jun-90 (APFw)
8 SU01: Martin Down NNR, 1999 (EN, 1999)
SU03: Grovely Wood, 11-Aug-71, 10-Apr-72 (DRN)
SU04: Orcheston Down (SPTA), 18-May-02 (PJH) (Hodge, 2002)
SU14: Wexland Hanging (SPTAC) 22-Jun-02 (PJH) (Hodge, 2002)
SU15: Everleigh (SPTAE), 18-May-02 (PJH) (Hodge, 2002)
SU22: Bentley Wood, 27-May-97 (MDD); Blackmoor Copse, 17-Mar-02 (APF) (Foster, 2004); Langley Wood NNR, 17-Sep-04 (MDD) (Darby, 2005)
SU23: Porton Down, 3-Jul-99 (PJH) (Hodge, 2002)
SU24: Dunch Hill (SPTAE), 16-Jun-97 (MGT det MLC)
Resident. Widespread; [on plantains and members of the daed-nettle family, Lamiaceae]. Adult feeding causes 'shot-holes' in the leaves.

**PSYLLIODES Berthold, 1827**
***Psylliodes affinis*** (Paykull, 1799)
7 SU26: Mildenhall, 1939 (det KGB) (Knight, 1940)
8 SU13: Great Durnford, 27-Jul-77, 7-Aug-70 (DRN); Middle Woodford, 27-Jul-77 (DRN)
SU14: Bulford Ranges (SPTAE), 3-Jun - 12-Aug-08 (Boyce & Wright, 2009); Haxton (SPTAE), 17-May-02 (PJH) (Hodge, 2002)
Resident. Local; [on members of the nightshade family, Solanaceae, in various habitats]. Adults feed on bittersweet causing 'shot-holes' in the leaves.

***Psylliodes chalcomera*** (Illiger, 1807)
**Nb**
8 ST94: Norton Down (SPTAW), 15-May-02 (PJH) (Hodge, 2002)
SU01: Martin Down NNR, 1979-1992 (RGB)
SU03: Grovely Wood, 4-May-78 (DRN)
SU04: Fox Covert (SPTAW), 3-Jun - 12-Aug-08 (Boyce & Wright, 2009)
SU14: Haxton (SPTAE), 3-Jun - 12-Aug-08 (Boyce & Wright, 2009)
Resident. Very local; [usually on thistles but sometimes on other Asteraceae]. Adult feeding on the leaves causes 'shot-holes'.

***Psylliodes chrysocephala*** (Linnaeus, 1758)
7 SU09: Ashton Keynes, 16-Apr-01 (JKG det MDD)
SU26: Savernake Forest, 20-Sep to 11-Oct-93, 20-Sep to 11-Oct-94 (EN det SAW); 15-Sep to 10-Oct-95 (EN det MDD)
8 ST93: Setting Copse, 28-Sep-95 (PJH) (Hodge, 1995); Stockton Wood, 28-Sep-95 (PJH) (Hodge, 1995)
SU02: Knighton Wood, Sep-06, May, Sep-07 (AJWA); Wilton, 6-Sep-95 (MDD)
SU03: Wylye Down NNR, 20-Sep-05 (MDD) (Darby, 2009)
SU14: Stonehenge, 22-Apr-03 (PJH) (Hodge, 2003)
Resident. Widespread; [in almost any habitat where Brassicaceae grow wild or are cultivated]. Adult feeding causes 'shot-holes' in the leaves.

***Psylliodes cuprea*** (Koch, J.D.W, 1803)
7 SU26: Savernake Forest, 15-Jun to 7-Jul, 29-Aug to 20-Sep, 11-Oct to 9-Nov-93 (EN det SAW)
8 SU02: Knighton Wood, Aug-07 (AJWA)
SU03: Wylye Down NNR, 10-Apr-07 (MDD) (Darby, 2009)
Resident. Local; in pitfall and flight intercept traps. [On various species of the cabbage family, and on oaks and ash in the autumn].

***Psylliodes dulcamarae*** (Koch, J.D.W, 1803)
8 SU13: Great Durnford, 27-Jul-77 (DRN); Middle Woodford, 27-Jul-77 (DRN)
SU14: Stonehenge, 22-Apr-03 (PJH) (Hodge, 2003)
Resident. Very local; [adults usually on bittersweet, *Solanum dulcamara*] their feeding causing 'shot-holes' in the leaves.

***Psylliodes laticollis*** Kutschera, 1860
8 SU13: Idmiston, 5-Aug-82 (DRN) (Nash, 1979); Salisbury, 26-May-74 (DRN) (Nash, 1979)
SU14: Netheravon, 29-Jul-83 (DRN) (Nash, 1979)
Resident. Very local; on vegetation including water-cress at the edge of the R. Avon and the R.Bourne.

***Psylliodes luteola*** (Müller, O.F, 1776)
**RDBK**
8 ST93: Great Ridge Wood, 6-Aug-85 (DRN); Stockton Wood, 28-Sep-95 (PJH) (Hodge, 1995)
SU02: Knighton Wood, Sep-06, Aug to Nov-07 (AJWA)
SU03: Dinton Park, 15-Aug-01 (KNAA); Little

Langford, 3-Jul-96 (MAS); Wylye Down NNR, 20-Sep-05 (MDD) (Darby, 2009)
SU14: Stonehenge, 22-Apr-03, 19-Jul-03 (PJH) (Hodge, 2003)
SU22: Blackmoor Copse, 31-Aug-01 (APF) (Foster, 2004)
SU24: Newton Tony, 25-Aug to 20-Sep-06, (MA det MGT)
Resident. Widespread; in a wide range of habitats on cultivated wheat, barleys and rye-grass, and on wild grasses. First recorded in GB in 1912 and known from only a handful of sites until 1984 when it appears to have rapidly increased its range.

Psylliodes luteola

***Psylliodes napi*** (Fabricius, 1792)
7 SU09: Somerford Keynes, Swillbrook Lakes, 4-Jun-04 (BLe)
SU16: Lockeridge, 1939 (det KGB) (Knight, 1940)
8 ST93: Stockton Wood, 28-Sep-95 (PJH) (Hodge, 1995)
SU01: Martin Down NNR, 1999 (EN, 1999)
SU04: Shrewton, 6-Aug-82 (DRN)
Resident. Local; [mainly on various species of the Brassicaceae]. Adult feeding causes 'shot-holes ' in the leaves.

***Psylliodes picina*** (Marsham, 1802)
8 SU13: Great Durnford, 27-Jul-77, 6-Aug-87 (DRN)
SU16: Woodborough, 4-Aug-83 (DRN)

## Subfamily CASSIDINAE

Pilemostoma fastuosa

**PILEMOSTOMA Desbrochers, 1891**
***Pilemostoma fastuosa*** (Schaller, 1783)
**Na, LCGR-Fh Ch**
8 SU22: Bentley Wood, 20-Aug-06 (EG)
Resident. Very local; [on various species of the daisy family, Asteraceae, particularly ploughman's spikenard, *Inula conyzae,* and common fleabane, *Pulicaria dysenterica*]. Adults eat variably-sized holes in the leaves.

**CASSIDA Linnaeus, 1758**
***Cassida flaveola*** Thunberg, 1794
7 SU16: Marlborough area, 1869-1938 (Meyrick, 1938)
8 ST83: Chicklade, 14-Aug-87 (DRN)
SU03: Grovely Wood, 14-Apr-71, 27-May-74 (DRN)
SU13: Great Durnford, 16-Aug-69 (DRN)
SU22: Bentley Wood, 31-May-97 (MDD)
Resident. Local; [adults usually on species of the pink family, Caryophyllaceae].

**Fleabane Tortoise Beetle**
***Cassida murraea*** Linnaeus, 1767
8 ST93: Fonthill, 3-Aug-78 (DRN)
SU03: Berwick St. James, 26-May-99 (MDD); Dinton, 9-Jul-95 (MDD)
SU13: Lower Woodford, 27-Jul, 2-Aug-82 (DRN); Salisbury, 15-Aug-88 (DRN); 23-May-93 (PM)
SU22: Bentley Wood, 18-Jul-04(AGD); Langley Wood NNR, 28-May-74 (DRN) (Nash, 1979)
SU23: Bentley Wood, 20-Aug-06 (EG)
Resident. Local; on fleabane, *Pulicaria dysenterica* [and marsh thistle *Cirsium palustre*] in damp habitats.

***Cassida nebulosa*** Linnaeus, 1758
**RDBi**
8 SU04: Salisbury Plain, 1950 (MGT) (Telfer, 2005)
SU15: Beach's Plantation (SPTAE), 16-Jun-97 (MGT det MLC)
Resident. Very local; in a flax field [on various species of the goosefoot family, Chenopodiaceae, and knotgrasses]. Adults eat irregular holes in the leaves.

***Cassida nobilis*** Linnaeus, 1758
**Nb**
8 SU22: Hawks Grove, 5-Apr-77 (DRN)
Resident. Very local; in moss in a damp area. [On various plants usually on sandy and chalky soils].

***Cassida prasina*** Illiger, 1798
**Nb, LCGR-Fh Ch**
8 ST94: Warminster Ranges (SPTAW), 3-Jun - 12-Aug-08 (Boyce & Wright, 2009)
SU03: Wylye Down NNR, 22-Feb to 29-Mar-07 (MDD) (Darby, 2009)
Resident. Very local; in a pitfall trap on calcareous grassland. [Adults usually occur on yarrow, *Achillea millefolium*, and sneezewort, *A. ptarmica*].

Cassida prasina

**Thistle Tortoise Beetle**
***Cassida rubiginosa*** Müller, O.F. 1776

7 ST96: Conigre Meadow, 20-May-01 (JKG det MDD)
SU09: Cotswold Water Park, 14-Apr-97 (APF); Upper Waterhay, 17-May-05 (JKG det MDD)
SU16: Marlborough area, 1869-1938 (Meyrick, 1938)
SU18: Rushy Platt, 20-May-01 (JKG det MDD)
8 ST84: Longleat, 29-Aug-99 (MB) SU02: Middleton Down, 23-Mar-92 (PM)
ST91: Cranborne Chase, 4-Jun-97 (MAS); 30-Jun-99 (MDD)
ST92: South Down Farm, 4-May-96 (EGS det MDD); Sutton Mandeville, 5-Jul-95 (MDD); Wardour, 5-Jun-96 (MAS)
ST94: Bowls Barrow (SPTAW), 16-May-02 (PJH) (Hodge, 2002); Ladywell (SPTAW), 15-May- 02 (PJH) (Hodge, 2002); Norton Down (SPTA), 15-May-02 (PJH) (Hodge, 2002)
SU02: Middleton Down, 19-Apr-89 (BL)
SU03: Wylye Down NNR, 20-Sep-05 (MDD) (Darby, 2009)
SU04: Fox Covert (SPTAW), 3-Jun - 12-Aug-08 (Boyce & Wright, 2009)Salisbury Plain, 16-May, 25-Jun-02 (PJH) (Hodge, 2002); Stonehenge, 22-Apr-03 (PJH) (Hodge, 2003)
SU12: Charlton-All-Saints, 9-May-96 (MAS)
SU13: Boscombe Down, 9-Oct-94 (MDD); Little Durnford Down, 25-May-92 (PM); Salisbury, 26-May-74, 15-Aug-88 (DRN)
SU14: Haxton, 17-May-02; 24-Jun-02 (PJH) (Hodge, 2002); 3-Jun - 12-Aug-08 (Boyce & Wright, 2009); Weland Hanging (SPTAC), 16-May-02 (PJH) (Hodge, 2002)
SU15: Everleigh (SPTAE), 24-Jun-02 (PJH) (Hodge, 2002)
SU16: Woodborough, 4-Aug-83 (DRN)
SU21: Hamptworth Estate, 1-Apr-75 (DRN)
SU22: Bentley Wood, 31-May, 2-Jul-96 (MDD); Blackmoor Copse, 31-May-92 (PM)
SU23: Porton Down, 3-Jul-99 (PJH) (Hodge, 2002)
SU24: Beacon Hill (SPTAE), 3-Jun - 12-Aug-08 (Boyce & Wright, 2009); Bulford Ranges (SPTAE), 3-Jun - 12-Aug-08 (Boyce & Wright, 2009); Driver Training Area (SPTAE), 17-May, 24-Jun-02 (PJH) (Hodge, 2002); Dunch Hill (SPTAE), 3-Jun – 12-Aug-08 (Boyce & Wright, 2009)

Resident. Ubiquitous; mainly on thistles, *Cirsium* and *Carduus,* in a great variety of habitats.

***Cassida vibex*** Linnaeus, 1758

8 ST85: Green Lane Wood, 31-May-04 (KNAA)
ST95: West Lavington, May-32, May-46 (ITH) (DFL, 1948)
SU13: Salisbury, 15-Aug-88 (DRN)

Resident. Very local [in a wide range of habitats on members of the daisy family, Asteraceae, in particular]. Adults feed on common knapweed, *Centaurea nigra*.

**Green Tortoise Beetle**
***Cassida viridis*** Linnaeus, 1758

7 ST96: Conigre Meadow, 20-May-01 (JKG det MDD)
SU15: Everleigh Ashes, 2-Jun-04 (KM det KNAA)
SU16: Marlborough area, Jun-1873 (EWJ) (Meyrick, 1874); 29 May 1879 (EEdeS) (Meyrick, 1880); 1869-1938; (Meyrick, 1938); West Woods, 11-Aug-02 (BLe)
SU26: Savernake Forest, 30-May-73 (DBS); 30-May-04 (KNAA)
8 ST93: Fonthill,, 19-Jul-95 (MDD)
ST95: West Lavington, 18-May-40 (PLB) (DFL, 1948); Jul-40 (ITH) (DFL, 1948)
SU03: Berwick St. James, 20-Jun-95 (MDD)
SU13: Great Durnford, 19-Aug-69, 27-Jul-77 (DRN); Salisbury, 15-Aug-88 (DRN); Apr-00 (MAS); 1-Jun-00 (MDD (Darby, 2001)

Resident. Widespread; on various plants usually in damp situations. Adult feeding causes irregular holes in the leaves.

Superfamily CURCULIONOIDEA
Division ORTHOCERI
## Family ANTHRIBIDAE
Subfamily ANTHRIBINAE
Tribe ANTHRIBINI

**ANTHRIBUS Geoffroy, 1762**
***Anthribus nebulosus*** Forster, 1771
**Nb**

7 ST87: Corsham Park, 3-Jun-04 (BLe)

Resident. Very local; [in woods, hedges and plantations on a variety of broadleaved and coniferous trees].

Tribe PLATYRHININI

**PLATYRHINUS Clairville, 1798**
***Platyrhinus resinosus*** (Scopoli, 1763)

**Nb**, **IEC-3**, **SQS-4**
8 ST84: Longleat, 12-Apr-02 (MB)
Resident. Very local; [in woods, particularly on dead ash but also recorded from dead wood of other trees. Larvae in the fungus *Daldinia concentrica*].

Platyrhinus resinosus

Tribe PLATYSTOMINI

**PLATYSTOMOS Schneider, 1791**
***Platystomos albinus*** (Linnaeus, 1758)
**Nb, IEC-3, SQS-8**
8 ST91: Cranborne Chase, 16 to 25-Jun-80 (NCC)
ST93: Great Ridge Wood, 4-Oct-04 (MDD)
SU03: Grovely Wood, 4-Aug-87 (DRN)
SU21: Pound Bottom, Jul-08 (DB det MDD)
SU22: Blackmoor Copse, May-92 (PM)
Resident. Local; in woods on dead and dying trees.

Platystomos albinus

## Subfamily CHORAGINAE
### Tribe CHORAGINI

**CHORAGUS Kirby, 1819**
***Choragus sheppardi*** Kirby, 1819
**Na, SQS-16**
8 SU22: Hamptworth Estate, 12-Apr-74 (DRN)
Resident. Very local; [in hedges and woods on old, dead ivy; the larvae in the rotten, fungus-infested wood].

# Family RHYNCHITIDAE
## Subfamily RHYNCHITINAE
### Tribe RHYNCHITINI

**INVOLVULUS Schrank, 1798**
**Twig Cutting Weevil**
***Involvulus caeruleus*** (De Geer, 1775)
7 SU16: West Woods, 9-Jul-46 (Holmes & Tennant, 1947)
SU18: Haydon Meadow, 10-Apr-02 (JKG det MDD)
8 SU14: Wexland Hanging (SPTAC), 22-Jun-02 (PJH) (Hodge, 2002)
Resident. Very local; [in woods and hedgerows on rosaceous trees and shrubs] A minor pest of fruit trees.

**LASIORHYNCHITES Jekel, 1860**
***Lasiorhynchites cavifrons*** Gyllenhal, 1833
**Nb**
7 SU16: Marlborough area, Jun-1873 (EWJ) (Meyrick, 1874); 1869-1938 (Meyrick, 1938) (One specimen in MCC insc *5568* underneath card)
SU26: Savernake Forest, 30-May-73 (DBS) (Meyrick, 1938)
8 ST95: West Lavington, 22-May-41 (KJA) (DFL, 1948)
SU22: Blackmoor Copse, May-92 (PM) (Meyrick, 1938)
SU23: Thorny Down, Porton, 1974-1978 (MGM) (Morris, 1979)
Resident. Local; in woods, etc. on oaks.

Lasiorhynchites cavifrons

**NEOCOENORRHINUS Voss, 1951**
***Neocoenorrhinus aeneovirens*** (Marsham, 1802)
7 SU26: Savernake Forest, 3-Jun-04 (BLe)
8 ST85: Green Lane Wood, 7-Jun-95 (MDD)
SU21: Burnt Ground Wood, 31-May-74 (DRN)
SU22: Bentley Wood, 3-Sep-97 (WANHS det MDD); Langley Wood NNR 25-May-94 (WANHS det MDD); Whiteparish Common, 1965-1974 (MGM) (Morris, 1974, 1975)
Resident. Local; in woods, etc. on oaks.

**Apple Fruit Rhynchites**
***Neocoenorrhinus aequatus*** (Linnaeus, 1767)

7 ST97: North Draycot Park, 22-May-00 (MDD)
SU16: Marlborough area, Jun-1873 (EWJ) (Meyrick, 1874)
SU26: Savernake Forest, 1869-1938 (Meyrick, 1938); 17-May-90 (TGW); 8 to 28-Jul-93 (EN det SAW); 31-May-00 (WANHS det MDD)
8 ST92: Sutton Mandeville, May-95 (MDD)
ST93: Great Ridge Wood, 6-Aug-85 (DRN); 16, 25-May-00 (MDD)
SU02: Bishopstone, 23, 28-May-91 (MDD)
SU03: Grovely Wood, 30-May-74 (DRN); Wylye Down NNR, 3-Aug-07 (MDD) (Darby, 2008a)
SU04: Orcheston Down (SPTA), 18-May-02 (PJH) (Hodge, 2002)
SU12: Coombe Bissett Down, 1-Apr-97 (MDD)
SU13: Cockey Down, 1-May-93 (PM); Little Durnford Down, 21-May-91 (PM); Old Sarum, 31-May-97 (MAS); Salisbury, 26-May-74 (DRN); Apr-00 (MAS); 2-May-00 (MDD) (Darby, 2001)
SU14: Haxton (SPTAE), 17-May-02 (PJH) (Hodge, 2002); Wexland Hanging (SPTAC), 16-May-02 (PJH) (Hodge, 2002)
SU22: PepperBox Hill, 9-May-93 (PM)
SU23: Porton Down, 6-Jun-93 (PTW det MDD); 3-Jul-99 (PJH) (Hodge, 2002); Thorny Down, 1965-1974 (MGM) (Morris, 1974, 1975)
SU25: Sidbury Hill (SPTAE), 28-May-97 (MGT)
Resident. Widespread; on rosaceous trees and shrubs in hedgerows and woodland edges.

**Strawberry Rhynchites**
***Neocoenorrhinus germanicus*** Herbst, 1797
7 SU09: Cotswold Water Park, 13-Apr-97 (APF)
SU16: Marlborough area, Jun-1873 (EWJ) (Meyrick, 1874); Wernham, 1869-1938 (Meyrick, 1938)
SU26: Savernake Forest, 15-Jun-73 (SAW); 25-Apr to 27-May, 11-Oct to 9 Nov-93 (EN det SAW)
8 ST94: Ladywell (SPTAW), 15-May, 23-Jun-02 (PJH) (Hodge, 2002)
SU02: Knighton Wood, 6-Jul-07 (AJWA)
SU03: Grovely Wood, 16-Aug-71, 1-Aug-88 (DRN)
SU15: Everleigh (SPTAE), 18-May-02 (PJH) (Hodge, 2002)
Resident. Widespread; [on a wide variety of herbaceous and shrubby Rosaceae, also on oaks and other trees including, apparently, willows]. A pest of strawberry, blackberry, raspberry, loganberry, etc..

***Noecoenorrhinus interpunctatus*** Stephens, 1831
**Nb**
7 SU26: Savernake Forest, 3-Jun-04 (BLe)
8 SU22: Whiteparish Common, 1974-1978 (MGM) (Morris, 1979)
SU23: Thorny Down, 1974-1978 (Morris, 1979)
Resident. Very local; on oaks. The biology is uncertain.

**TEMNOCERUS Thunberg, 1815**
***Temnocerus longiceps*** Thomson, C. G, 1888
**Nb**
8 SU04: Parsonage Down NNR, 2-Jul-08 (MDD)
SU12: Whaddon Common, 22-Jul-73 (DRN)
SU21: Hamptworth Common, 28-Jul-77 (DRN)
SU22: Bentley Wood, 27-May-98 (MDD)
Resident. Local; on sallows, birches, [and pear trees].

***Temnocerus nanus*** (Paykull, 1792)
8 ST91: Cranborne Chase, 15-May-97 (MAS)
ST93: Great Ridge Wood, 6-Aug-85 (DRN)
SU21: Franchises Common, 25-Jul-73 (DRN); Franchises Wood, 4-Jul-77 (DRN)
SU22: Homan's Copse, 31-Jul-85 (DRN); Whiteparish Common, 1974-1978 (MGM) (Morris, 1979)
Resident. Local; on birch and hazel [less frequently willows and alder].

***Temnocerus tomentosus*** Gyllenhal, 1839
**Nb**
8 SU21: Landford Bog, 22-Jun-03 (APF) (Foster, 2003)
SU22: Whiteparish Common, 1974-1978 (MGM) (Morris, 1979)
SU26: Bedwyn Common, 30-May-46 (Holmes & Tennant, 1947)
Resident. Very local; [on willows and poplars, and possibly other trees].

Tribe BYCTISCINI

**BYCTISCUS Thomson, C.G., 1859**
***Byctiscus betulae*** (Linnaeus, 1758)
**Nb**
8 SU22: Blackmoor Copse, 17-May-02 (APF) (Foster, 2004); Whiteparish Common, 1965-1979 (MGM) (Morris, 1974, 1975, 1979)
Resident. Very local; [on a variety of trees, most frequently Hazel and birch]. The larvae make characteristic ‘cigars’out of the leaves.

Tribe DEPORAINI

**DEPORAUS Samouelle, 1819**
***Deporaus betulae*** (Linnaeus, 1758)
7 SU16: Marlborough area, 18-Jun-1879 (AMF) (Meyrick, 1880); Pumphrey's Wood, 1873-1938 (Meyrick, 1938)
SU17: Fyfield Down, 1-Jan-73 (MGM) (Morris, 1974, 1975, 1979)
SU26: Savernake Forest, 3-May-02 (IJG)
8 ST93: Great Ridge Wood, 6-Aug-85 (DRN);

Stockton Wood, 1965-1979 (MGM)
SU03: Grovely Wood, 21-Aug-72, 30-May-74, 1-Aug-88 (DRN)
SU21: Burnt Ground Wood, 31-May-74 (DRN)
SU22: Bagfield Copse, 7-Aug-86 (DRN); Homan's Copse, 29-May-90 (DRN); Landford, 13-Aug-87 (DRN); Langley Wood NNR, 28-May-74, 6-Aug-86 (DRN); Whiteparish Common, 1974-1978 (Morris, 1979)
SU23: Thorny Down, 1965-1979 (MGM) (Morris, 1974, 1975, 1979)
Resident. Widespread; on birch, alder and hazel. The females cut the leaves to make 'conical' rolls in which the larvae develop.

## Family ATTELABIDAE
### Subfamily ATTELABINAE

**ATTELABUS Linnaeus, 1758**
***Attelabus nitens*** (Scopoli, 1763)
7 SU26: Savernake Forest, 6-Jun-73 (APF); 6-Dec-73 (PS)
8 SU22: Bentley Wood, 31-May-97 (MDD); Whiteparish Common, 22-Jul-73 (DRN); 1974-1978 (MGM) (Morris, 1979)
Resident. Very local; [in woods and coppice on oak and less commonly chestnut]. The female makes leaf rolls in which the larvae develop especially on regrowth and 'scrub oak'.

### Subfamily APODERINAE

Apoderus coryli

**APODERUS Olivier, 1807**
**Hazel Leaf-roller Weevil**
***Apoderus coryli*** (Linnaeus, 1758)
7 ST96: Chittoe, 22-Jul-00 (EGS det MDD)
SU08: Ravensroost Wood, 11-Aug-98 (MEN); Red Lodge Pond, 31-May-01 (JKG det MDD); Somerford Common, 11-May-02 (JKG det MDD)
SU16: Marlborough area, 18-Jun-1879 (AMF) (Meyrick, 1880); West Woods, 1869-1938 (Meyrick, 1938)
SU27: Ramsbury, 1869-1938 (Meyrick, 1938)
8 ST91: Cranborne Chase, 24-Apr-97 (MAS)
ST92: Sutton Mandeville, Jun-96 (MDD)
ST93: Great Ridge Wood, 6-Aug-85 (DRN)
SU03: Grovely Wood, 1-Aug-88 (DRN)
SU13: Little Durnford Down, 28-Apr-93 (PM)
SU21: Burnt Ground Wood, 31-May-74 (DRN)
SU22: Bagfield Copse, 7-Aug-86 (DRN); Bentley Wood, 28-Jun, 4-Jul-90 (PM); Blackmoor Copse, 31-May-92, 19-May-94 (PM); 9-Jun, 24-Aug-01, 13-Jul, 3-Aug-03 (APF) (Foster, 2004); Glaziers Copse, 31-Jul-85 (DRN); Langley Wood, 6-Aug-86; Whiteparish Common, 22-Jul-73 (DRN); 1974-1978 (MGM) (Morris, 1979)
Resident. Widespread; on hazels in woods. The female makes characteristic, rather untidy, leaf rolls in which the larvae develop.

## Family APIONIDAE
### Subfamily APIONINAE
#### Tribe APIONINI

**APION Herbst, 1797**
ERYTHRAPION Schilsky, 1906
***Apion cruentatum*** Walton, J, 1844
8 SU23: Hamptworth Estate, 31-Oct-73 (DRN)
Resident. Very local; in moss on gravel. Larvae develop in the dock *Rumex acetosa.*

***Apion frumentarium*** Sharp, 1918
7 ST96: Conlgre Meadow, 20-May-01 (JKG det MDD)
ST98: Conygre Mead, Malmesbury, 5-Apr-02 (JKG det MDD)
SU09: Ashton Keynes, 7-May-01 (JKG det MDD)
SU16: Marlborough area, 1869-1938 (Meyrick, 1938)
8 ST84: Longleat, 29-Aug-99 (MB)
ST92: Sutton Mandeville, 22-May-95 (MDD)
SU03: Barford Down, 10-Apr-72 (DRN); Grovely Wood, 11-Aug-71 (DRN); 11-Jun-00 (MDD); Wylye, 1965-1974 (MGM) (Morris, 1974, 1975)
SU04: Salisbury Plain, 16-May-02 (PJH) (Hodge, 2002); Shrewton, 6-Aug-82 (DRN)
SU12: Coombe Bissett Down, 1-Apr-97 (MDD)
SU13: Stratford sub Castle, 10-Aug-70 (DRN); Great Durnford, 7, 22-Aug-70 (DRN); Idmiston Down, 1974-1978 (MGM) (Morris, 1979)
SU14: Figheldean, 28-Jul-82 (DRN); Stonehenge, 22-Apr-03 (PJH) (Hodge, 2003)
SU15: Everleigh (SPTAE), 18-May-02 (PJH) (Hodge, 2002)
SU16: Jones's Mill, 16-May-02 (IJG)

SU22: Bentley Wood, 27-May-97 (MDD); Redlynch, 4-Aug-82 (DRN)
SU23: West Winterslow, 11-Aug-01 (NB)
Resident. Widespread; on broadleaved docks, *Rumex* spp..

***Apion haematodes*** Kirby, 1808
7 SU16: Marlborough area, 1869-1938 (Meyrick, 1938)
8 ST92: Wardour, 5-Jun-96 (MAS)
Resident. Very local; [on *Rumex acetosella* agg. in areaceous and dry, peaty areas]. Single specimen without data in MCC (conf. MGM).

Tribe APLEMONINI

**PERAPION Wagner, 1907**
PERAPION Wagner, 1907
***Perapion curtirostre*** (Germar, 1817)
7 SU06: Cherhill Down, 15-Jun-82 (KNAA); Roundway Camp, Jul-40 (FDB) (Buck, 1956); 1965-1979 (MGM) (Morris, 1974, 1975, 1979)
SU16: Combe Farm, nr Marlborough, 1869-1938 (Meyrick, 1938); Silbury Hill, 1965-1974 (MGM) (Morris, 1974, 1975)
SU17: Fyfield Down, 1965-1974 (MGM) (Morris, 1974, 1975)
8 ST92: Prescombe Down, 1965-1974 (MGM) (Morris, 1974, 1975)
ST94: Ladywell (SPTAW), 15-May-02 (PJH) (Hodge, 2002); Norton Down (SPTAW), 23-Jun-02 (PJH) (Hodge, 2002)
SU01: Martin Down NNR, 1965-1974 (MGM) (Morris, 1974, 1975); 21-Aug-89 (RGB)
SU02: Middleton Down, 24-May-95 (MDD)
SU04: Orcheston Down (SPTA), 18-May-02 (PJH) (Hodge, 2002)
SU13: Great Durnford, 19-Aug-69 (DRN); Middle Woodford, 5-Aug-82 (DRN); Old Sarum, 1965-1974 (Morris, 1974, 1975); Rox Hill, 1965-1974 (Morris, 1974, 1975); Stratford Sub Castle, 10-Aug-70 (DRN)
SU14: Haxton, 24-Jun-02 (PJH) (Hodge, 2002)
SU15: Everleigh (SPTAE), 18-May-02 (PJH) (Hodge, 2002)
SU22: Langley Wood NNR, 28-May-74 (DRN)
Resident. Widespread; on a wide variety of docks, *Rumex* spp..

***Perapion hydrolapathi*** (Marsham, 1802)
7 SU16: Manton, nr. Marlborough, 1869-1938 (Meyrick, 1938); Gopher Wood, Huish, 1965-1974 (MGM) (Morris, 1974, 1975)
8 ST83: Chicklade, 14-Aug-87 (DRN)
ST84: Longleat, 4-May, 29-Aug-99 (MB)
ST93: Great Ridge Wood, 16-May-00 (MDD det RT)
SU04: Salisbury Plain, 16-May-02 (PJH) (Hodge, 2002); Shrewton, 6-Aug-82 (DRN)
SU13: Great Durnford, 17-Aug-70 (DRN); Idmiston, 1974-1978 (MGM) (Morris, 1979); 15-Aug-88 (DRN); Lower Woodford, 2-Aug-82 (DRN); Old Sarum, 31-May-97 (MAS); Salisbury,15-Aug-88 (DRN); Salisbury, Avon Valley Reserve, Apr-00 (MAS) (Darby, 2001); Stratford sub Castle, 10-Aug-70 (DRN)
SU14: Figheldean, 28-Jul-82 (DRN); Stonehenge, 22-Apr-03 (PJH) (Hodge, 2003)
SU15: Everleigh (SPTAE), 18-May, 24-Jun-02 (PJH) (Hodge, 2002)
SU16: Jones's Mill, 19-Jun-05 (MAS); Woodborough, 4-Aug-83 (DRN)
SU22: Blackmoor Copse, 17-Mar-02 (APF) (Foster, 2004); Redlynch, 4-Aug-82 (DRN)
Resident. Widespread; on various species of docks, *Rumex spp.*.

***Perapion marchicum*** (Herbst, 1797)
7 SU06: Calstone Down, 4-Aug-95 (KNAA)
Resident. Very local; [on dock *Rumex acetosella*]. Larvae make complex galls on the roots.

***Perapion violaceum*** (Kirby, 1808)
8 ST94: Ladywell (SPTAW), 3-Jun - 12-Aug-08 (Boyce & Wright, 2009)
SU13: Middle Woodford, 27-Jul-77 (DRN); Stratford sub Castle, 10-Aug-70 (DRN)
SU22: Langley Wood, 28-May-74 (DRN)
Resident. Very local; [on docks, *Rumex spp.*].

Tribe ASPIDAPIINI

**ASPIDAPION Schilsky, 1901**
ASPIDAPION Schilsky, 1901
***Aspidapion radiolus*** (Marsham, 1802)
8 ST85: Bullen Hill Farm, 1982-2002 (EGS det MDD)
SU03: Barford St. Martin, 1965-1979 (MGM) (Morris, 1974, 1975); Wylye, 1965-1979 (MGM) (Morris 974, 1975)
SU12: East Harnham, 1-Aug-82 (DRN)
SU14: Bulford House, 3-Jun-97 (MGT)
Resident. Very local; [on mallow *Malva sylvestris* and other Malvaceae].

***Aspidapion soror*** (Rey, 1895)
**Na**
8 SU03: Barford St. Martin, 1965-1979 (MGM) (Morris, 1974, 1975); Wylye, 1965-1979 (MGM) (Morris, 1974, 1975)
Resident. Very local; [on marsh mallow, *Althea officinalis*].

KOESTLINIA Alonso-Zarazaga, 1990

***Aspidapion aeneum*** (Fabricius, 1775)

7 SU16: Marlborough area, 1869-1938 (Meyrick, 1938)

Resident. Very local; [chiefly on common mallow, *Malva sylvestris*]. Four specimens without data in MCC (conf. MGM).

Tribe CERATAPIINI

**CERATAPION Schilsky, 1901**

CERATAPION Schilsky, 1901

***Ceratapion carduorum*** (Kirby, 1808)

7 SU08: Somerford Common, 23-Apr-01 (JKG det MDD)
SU09: Ashton Keynes, 7-May-01 (JKG det MDD)
SU16: Gopher Wood, Huish, 1965-1974 (MGM) (Morris, 1974, 1975); Marlborough area, 17-Feb-1879 (WFB) (Meyrick, 1880); 1873-1938 (Meyrick, 1938); Pewsey Downs, 1965-1974 (MGM) (Morris, 1974, 1975)
SU17: Fyfield Down, 1965-1974 (MGM) (Morris, 1974, 1975)
SU18: Swindon, 7-May-01 (JKG det MDD)
8 ST83: Whitesheet Hill, 1965-1974 (MGM) (Morris, 1974, 1975)
ST94: Bowls Barrow, 22-Jun-02 (PJH) (Hodge, 2002); Cotley Hill, 1965-1974 (MGM) (Morris, 1974, 1975)
ST95: Fore Hill, nr West Lavington, 1965-1974 (MGM) (Morris, 1974, 1975)
SU03: Grovely Wood, 13-Aug-70, 23-Jul-73, 29-Jul-78 (DRN)
SU04: Orcheston Down (SPTA), 22-Jun-02 (PJH) (Hodge, 2002); Parsonage Down, 22-Jul-08 (MDD); Yarnbury Castle, 1965-1974 (MGM) (Morris, 1974, 1975)
SU05: Chirton Gorse, 1965-1974 (MGM) (Morris, 1974, 1975)
SU13: Boscombe Down, 8, 9-Oct-94 (MDD); Idmiston, 5-Aug-82 (DRN)
SU14: Haxton (SPTAE), 17-May-02 (PJH) (Hodge, 2002); Stonehenge, 22-Apr-03 (PJH) (Hodge, 2003)
SU22: Bentley Wood, 31-May, 2-Jul-96 (MDD)
SU23: Battery Hill, 1974-1978 (MGM) (Morris, 1979); Porton Down, 3-Jul-99 (PMP)

Resident. Widespread; on thistles , especially *Cirsium vulgare* and *C. arvense*. This species has been confused until recently with *lacertense* and *gibbirostre* so that many records are indeterminate.

***Ceratapion gibbirostre*** (Gyllenhal, 1813)

7 SU06: Calstone Down, 22-Aug-95 (KNAA)
8 ST84: Longleat, 29-Aug-99 (MB)
ST93: Stockton Wood, 28-Sep-95 (PJH) (Hodge, 1995)
SU01: Martin Down NNR, 21-Aug-89 (RGB)
SU03: Grovely Wood, 4-Aug-78, 28-Jul-83, 4-Aug-87 (DRN)
SU04: Shrewton, 6-Aug-82 (DRN)
SU13: Lower Woodford, 1-Aug-83 (DRN); Salisbury, 15-Aug-88 (DRN)
SU15: Everleigh (SPTAE), 18-May, 24-Jun-02 (PJH) (Hodge, 2002)
SU16: Woodborough, 4-Aug-83 (DRN)

Formerly *carduorum*, see above.

ACANEPHODUS Alonso-Zarazaga, 1990

***Ceratapion onopordi*** (Kirby, 1808)

7 ST07: Windmill Hill, nr Avebury, 1965-1974 (MGM) (Morris, 1974, 1975)
SU06: Cherhill Down, 15-Jun-82 (KNAA)
SU09: Somerford Keynes, 4-Jun-04 (BLe)
SU16: Marlborough area, 1869-1938 (Meyrick, 1938); Pewsey Downs, 1965-1974 (MGM) (Morris, 1974, 1975)
8 ST84: Longleat, 29-Aug-99 (MB)
ST93: Stockton Down, 1965-1974 (MGM) (Morris, 1974, 1975)
ST94: Bowls Barrow, 22-Jun-02 (PJH) (Hodge, 2002); Cotley Hill, 1965-1974 (MGM) (Morris, 1974, 1975); Starveall Down, 1965-1974 (MGM) (Morris, 1974, 1975)
ST95: Fore Hill, nr West Lavington, 1965-1974 (MGM) (Morris, 1974, 1975)
SU02: Middleton Down, 24-May-95 (MDD)
SU13: Old Sarum, 1965-1974 (MGM) (Morris, 1974, 1975); Salisbury, Avon Valley Reserve, Apr-00 (MAS) (Darby, 2001)
SU14: Haxton (SPTAE), 17-May, 24-Jun-02 (PJH) (Hodge, 2002); Stonehenge, 22-Apr-03 (PJH) (2003)
SU16: Woodborough, 4-Aug-83 (DRN)
SU23: Porton Down, 12-Nov-96 (MAS); 3-Jul-99 (PJH) (Hodge, 2002); 1965-1974 (MGM) (Morris, 1974, 1975)
SU24: Driver Training Area (SPTAE), 17-May-02 (PJH) (Hodge, 2002)

Resident. Widespread; [in grassland on a wide variety of Compositae, including burdocks, knapweeds and thistles].

**OMPHALAPION Schilsky, 1901**

***Omphalapion hookerorum*** (Kirby, 1808)

7 SU16: Marlborough area, 1869-1938 (Meyrick, 1938)
8 ST92: Prescombe Down, 1965-1974 (MGM) (Morris, 1974, 1975); Sutton Mandeville, 1-Aug-08 (MDD)
ST94: Norton Down (SPTAW), 23-Jun-02 (PJH) (Hodge, 2002)
SU02: Compton Chamberlayne, 1965-1974 (MGM) (Morris, 1974, 1975)
SU12: Little Woodbury, nr Salisbury, 1965-1974 (MGM) (Morris, 1974, 1975)
SU22: Redlynch, 4-Aug-82 (DRN)

Resident. Local; in waste places, etc. on *Tripleuro-spermum* and *Matricaria* spp..

Tribe EXAPIINI

**EXAPION Bedel, 1885**

EXAPION Bedel, 1885

***Exapion difficile*** (Herbst, 1797)

**Na, LCGR- Fl Cl**

8 SU04: Fox Covert (SPTAW), 3-Jun - 12-Aug-08 (Boyce & Wright, 2009)
SU14: Haxton (SPTAE), 3-Jun - 12-Aug-08 (Boyce & Wright, 2009)

Resident. Very local: [chiefly or exclusively on dyers greenweed, *Genista. tinctoria*].

ULAPION Ehret, 1997

**Gorse Weevil**

***Exapion ulicis*** (Forster, 1771)

7 SU16: Marlborough area, 1869-1938 (Meyrick, 1938)
SU17: Fyfield Down, 1965-1974 (MGM) (Morris, 1974, 1975)
SU26: Savernake Forest, 1965-1974 (MGM) (Morris, 1974, 1975)
8 ST83: Brimsdown Hill, 1965-1974 (MGM)
ST92: Prescombe Down, 1965-1974 (MGM)
ST93: Stockton Wood, 1965-1974 (MGM)
ST94: Bowls Barrow (SPTAW), 16-May-02 (PJH) (Hodge, 2002); Ladywell (SPTAW), 15-May-02 (PJH) (Hodge, 2002); Salisbury Plain, 7- Mar-93 (SP)
SU02: Middleton Down, 24-May-95 (MDD)
SU04: Parsonage Down NNR, 2-Jul-08 (MDD)
SU05: Chirton Gorse, 1965-1974 (MGM) (Morris, 1974, 1975)
SU21: Hamptworth Estate, 24-Apr-84 (DRN); Landford Bog, 24-Jun-01 (APF) (Foster, 2001); 22-Jun-03 (APF) (Foster, 2003); Pound Bottom, 24-Jul-73 (DRN)
SU22: Whiteparish Common, 1965-1974 (MGM) (Morris, 1974, 1975)
SU23: Thorny Down, Porton, 1869-1938 (MGM) (Morris, 1979)

Resident. Widespread; on gorse.

Tribe KALCAPIINI

**KALCAPION Schilsky, 1906**

**Dogs-mercury Seed Weevil**

***Kalcapion pallipes*** (Kirby, 1808)

7 SU17: Fyfield Down, 1965-1974 (MGM) (Morris, 1974, 1975)
8 SU22: Bentley Wood, 31-May-97 (MDD)
SU23: Porton Down, 1965-1974 (MGM) (Morris, 1974, 1975)

Resident. Very local; on dog's mercury, *Mercurialis perrenis.*

***Kalcapion semivittatum*** (Gyllenhal, 1833)

**Na**

8 SU13: Salisbury, Aug-87 (DRN)

Resident. Very local; [on annual mercury, *Mercurialis annua,* in waste places, etc.].

**MELANAPION Wagner, 1930**

***Melanapion minimum*** (Herbst, 1797)

**RDB3**

7 SU27: Ramsbury, 1869-1938 (Meyrick, 1938)

Resident. Very local; on sallow [also on a wide variety of other *Salix* spp.]. One specimen in MCC without data (conf. MGM).

**SQUAMAPION Bokor, 1923**

***Squamapion atomarium*** (Kirby, 1808)

7 SU06: Calstone Down, 22-Aug-95 (KNAA)
SU16: Pewsey Downs, 1965-1974 (MGM) (Morris, 1974, 1975)
8 ST83: Brimsdown Hill, 1965-1974 (MGM) (Morris, 1974, 1975); Whitesheet Hill, 1965-1974 (MGM) (Morris, 1974, 1975)
ST92: Prescombe Down, 1965-1974 (MGM) (Morris, 1974, 1975)
ST94: Cotley Hill, 1965-1974 (MGM) (Morris, 1974, 1975)
SU02: Knapp Down, 1965-1974 (MGM) (Morris, 1974, 1975); Knighton Hill, Stoke Farthing, 1965-1974 (MGM) (Morris, 1974, 1975)
SU03: Steeple Langford Cow Down, 1965-1974 (MGM) (Morris, 1974, 1975)
SU23: Porton Down, 1965-1974 (MGM) (Morris, 1974, 1975)

Resident. Widespread; on thyme in dry places.

**TAENIAPION Schilsky, 1906**

***Taeniapion urticarium*** (Herbst, 1784)

7 SU16: Marlborough area, 1869-1938 (Meyrick, 1938)
8 ST94: Ladywell (SPTAW), 15-May-02 (PJH) (Hodge, 2002)
SU12: Coombe Bissett Down, 1-Apr-97 (MDD)
SU13: Great Durnford, 18-Aug-72, 27-Jul-77 (DRN); Salisbury, Bemerton Folly, 7-Mar – 15-Jun-02 (MDD) (Darby, 2002d)
SU14: Haxton (SPTAE), 17-May-02 (PJH) (Hodge, 2002); Stonehenge, 22-Apr-03 (PJH) (Hodge, 2003)
SU16: Jones's Mill, 1997 (MAS)
SU25: Sidbury Hill (SPTAE), 28-May-97 (MGT)

Resident. Widespread; on stinging nettles, *Urtica dioica.*

Tribe MALVAPIINI

**PSEUDAPION Schilsky, 1906**

***Pseudapion rufirostre*** (Fabricius, 1775)

7 SU16: Marlborough area, 1869-1938 (Meyrick, 1938)

8 SU03: Barford St. Martin, 1965-1974 (MGM) (Morris, 1974, 1975); Wylye, 1965-1974 (MGM) (Morris, 1974, 1975)

SU12: Coombe Bissett Down, 1-Apr-97 (MDD); East Harnham, 1-Aug-82 (DRN)

Resident. Very local; [on mallows *Malva sylvestris* and *M. neglecta* on roadsides, waste places, etc.].

Tribe OXYSTOMATINI
Subtribe OXYSTOMTINA

**CYANAPION Bokor, 1923**

CYANAPION Bokor, 1923

***Cyanapion spencii*** (Kirby, 1808)

7 ST07: Windmill Hill, nr Avebury, 1965-1974 (MGM) (Morris, 1974, 1975)

8 SU04: Shrewton, 30-Jul-82, 1-Aug-85 (DRN)

SU14: Haxton (SPTAE), 3-Jun - 12-Aug-08 (Boyce & Wright, 2009)

Resident. Very local; on tufted vetch, *Viccia cracca,* [generally in damp places].

BOTHRYORRHYNCHAPION Bokor, 1923

***Cyanapion afer*** (Gyllenhal, 1833)

7 SU06: Morgan's Hill, 1965-1974 (MGM) (Morris, 1974, 1975)

8 SU02: Woodminton Down, 1965-1974 (MGM) (Morris, 1974, 1975)

SU22: Bentley Wood, 2-Jul-96 (MDD)

Resident. Very local; in woodland [hedges and waysides]. The foodplant requires investigation.

**EUTRICHAPION Reitter, 1916**

EUTRICHAPION Reitter, 1916

***Eutrichapion ervi*** (Kirby, 1808)

7 SU16: Wernham, nr Marlborough, 1869-1938 (Meyrick, 1938)

SU17: Hackpen Hill, 1965-1974 (MGM) (Morris, 1974, 1975); Clout's Wood, 1965-1974 (MGM) (Morris, 1974, 1975)

8 ST93: Stockton Wood, 1965-1974 (MGM) (Morris, 1974, 1975)

ST94: Bowls Barrow (SPTAW), 16-May, 22-Jun-02 (PJH) (Hodge, 2002); Ladywell (SPTAW), 23-Jun-02 (PJH) (Hodge, 2002); Norton Down (SPTA), 23-Jun-02 (PJH) (Hodge, 2002)

ST95: Bratton Castle, 1965-1974 (MGM) (Morris, 1974, 1975)

SU02: Broad Chalke, 1965-1974 (MGM) (Morris, 1974, 1975)

SU03: Grovely Wood, 28-Jul-83 (DRN)

SU04: Orcheston Down (SPTA), 18-May, 22-Jun-02 (PJH) (Hodge, 2002); Shrewton, 30-Jul-82, 2-Aug-83 (DRN)

SU05: Redhorn Hill, 1965-1974 (MGM) (Morris, 1974, 1975)

SU13: Camp Down, Salisbury, 1965-1974 (MGM) (Morris, 1974, 1975); Great Durnford, 16-Aug-69 (DRN); Salisbury, 15-Aug-88 (DRN) Stratford sub Castle, 10-Aug-70 (DRN)

SU14: Haxton (SPTAE), 17-May, 24-Jun-02 (PJH) (Hodge, 2002); Stonehenge, 22-Apr-03 (PJH) (Hodge, 2003); Wexland Hanging (SPTA), 22-Jun-02 (PJH) (Hodge, 2002)

SU15: Manningford, 13-Aug-87 (DRN)

SU16: Jones's Mill, 21-Jun-97 (MDD); Netheravon, 29-Jul-83 (DRN); Woodborough, 4-Aug-83 (DRN)

SU22: West Grimstead, 12-Aug-87 (DRN)

Resident. Widespread; in [waste places, etc., and particularly in hedgerows. Most frequently on vetches and the vetchling, *Lathyrus pratensis.*

***Eutrichapion viciae*** (Paykull, 1800)

7 ST07: Windmill Hill, nr Avebury, 1965-1974 (MGM) (Morris, 1974, 1975)

SU09: Somerford Keynes, 4-Jun-04 (BLe)

SU16: Gopher Wood, Huish, 1965-1974 (MGM) (Morris, 1974, 1975)

8 ST83: Whitesheet Down, 7-Jul-82 (KNAA)

ST94: Norton Down (SPTAW), 23-Jun-02 (PJH) (Hodge, 2002)

SU04: Salisbury Plain, 28-Jul-82 (DRN); 16-May-02 (PJH) (Hodge, 2002); Shrewton, 30-Jul-82 (DRN); 2-Aug-83, 1-Aug-85 (DRN)

SU05: Black Heath (SPTAC), 3-Jun - 12-Aug-08 (Boyce & Wright, 2009); Ell Barrow (SPTAC), 3-Jun - 12-Aug-08 (Boyce & Wright, 2009); Redhorn Hill, 1965-1974 (MGM) (Morris,1974, 1975)

SU12: Charlton-All-Saints, 9-May-96 (MAS)

SU13: Great Durnford, 17-Jul-73 (DRN)

SU14: Haxton (SPTAE), 3-Jun - 12-Aug-08 (Boyce & Wright, 2009); Wexland Hanging (SPTAC), 22-Jun-02 (PJH) (Hodge, 2002);

SU15: Manningford, 13-Aug-87 (DRN)

SU22: Blackmoor Copse, 17-May-02 (APF) (Foster, 2004); Glaziers Common, 19-Aug-88 (DRN); Langley Wood NNR, 28-May-74 (DRN); West Grimstead, 12-Aug-87 (DRN)

SU24: Bulford Down (SPTAE), 1965-1974 (MGM) (Morris, 1974, 1975)

Resident. Widespread; in hedgerows, roadsides, etc. on vetches, particularly *Vicia cracca* and *Lathyrus pratensis.*

CNEMAPION Bokor, 1923

***Eutrichapion vorax*** (Herbst, 1797)

8 SU03: Grovely Wood, 1-Aug-88 (DRN)

SU23: Porton Down, 1965-1974 (MGM) (Morris, 1974, 1975)

Resident. Very local; [in hedgerows and waste places and often in woods on trees, shrubs and ground vegetation. The hosts are vetches, *Vicia* spp.].

PSILOCALYMMA Alonso Zarazaga, 1990

***Eutrichapion punctigerum*** (Paykull, 1792)

7 SU16: West Woods, 1869-1938 (Meyrick, 1938)
SU17: Clout's Wood, 1965-1974 (MGM) (Morris, 1974, 1975)
SU27: Love's Copse, 1869-1938 (Meyrick, 1938)

Resident. Very local; in [waste places, wood edges, roadsides, etc. on species of vetch particularly *Vicia sepium*].

**HEMITRICHAPION Voss, 1959**

TINOCYBA Alonso Zarazaga, 1990

***Hemitrichapion reflexum*** (Gyllenhal, 1833)

**Na, LCGR-Fh Cl**

8 ST95: Bratton Castle, 1974-1978 (MGM) (Morris, 1979)
SU01: Martin Down NNR, 1974-1978 (MGM) (Morris, 1979)
SU04: Fox Covert (SPTAW), 3-Jun - 12-Aug-08 (Boyce & Wright, 2009)
SU14: Wexland Hanging (SPTAC), 3-Jun - 12-Aug-08 (Boyce & Wright, 2009)

Resident. Very local; in calcareous grassland [on sainfoin, *Onobrychis vicifolia*].

***Hemitrichapion waltoni*** (Stephens, 1839)

**LCGR-Fh Cl**

7 SU06: Morgan's Hill, 1965-1974 (MGM) (Morris, 1974, 1975)
SU16: Pewsey Downs, 22-Jun-90 (APFw); 1965-1974 (MGM) (Morris, 1974, 1975)
8 ST83: Whitesheet Hill, 1965-1974 (MGM) (Morris, 1974, 1975)
ST92: Prescombe Down, 1965-1974 (MGM) (Morris, 1974, 1975)
ST94: Starveall Down, 1965-1974 (MGM) (Morris, 1974, 1975); Cotley Hill, 1965-1974 (MGM) (Morris, 1974, 1975); Norton Down, 23-Jun-02 (PJH) (Hodge, 2002)
ST95: Fore Hill, nr West Lavington, 1965-1974 (MGM) (Morris, 1974, 1975); Bratton Castle, 1965-1974 (MGM) (Morris, 1974, 1975)
SU02: Knighton Hill, 1965-1974 (MGM) (Morris, 1974, 1975); Stratford Tony Down, 1965-1974 (MGM) (Morris, 1974, 1975); Woodminton Down, 1965-1974 (MGM) (Morris, 1974, 1975)
SU03: Steeple Langford Cow Down, 1965-1974 (MGM) (Morris, 1974, 1975); Wylye Down, 1965-1974 (MGM) (Morris, 1974, 1975)
SU04: Yarnbury Castle, 1965-1974 (MGM) (Morris, 1974, 1975)
SU14: Bulford Down, 1965-1974 (MGM) (Morris, 1974, 1975); Stonehenge, 22-Apr-03 (PJH) (Hodge, 2003)
SU24: Beacon Hill (SPTAE), 3-Jun - 12-Aug-08 (Boyce & Wright, 2009)

Resident. Widespread; in calcareous grassland [on horse shoe vetch, *Hippocrepis comosa*].

**HOLOTRICHAPION Györffy, 1956**

HOLOTRICHAPION Györffy, 1956

***Holotrichapion ononis*** (Kirby, 1808)

**LCGR-Fm Cl**

8 SU06: Cherhill Down, 15-Jun-82 (KNAA)
ST94: Bowls Barrow, 22-Jun-02 (PJH) (Hodge, 2002); Starveall Down, 1965-1974 (MGM) (Morris, 1974, 1975); Warminster Ranges (SPTAW), 3-Jun - 12-Aug-08 (Boyce & Wright, 2009)
SU02: Stratford Tony Down, 1965-1974 (MGM) (Morris, 1974, 1975)
SU03: Steeple Langford Cow Down, 1965-1974 (MGM) (Morris, 1974, 1975)
SU04: Parsonage Down NNR, 1965-1974 (MGM) (Morris, 1974, 1975)
SU13: Camp Down, Salisbury, 1965-1974 (MGM) (Morris, 1974, 1975)
SU24: North Tidworth, Devils Ditch, 30-May-04 (BLe)

Resident. Widespread; in dry places [on restharrow, *Ononis repens* and *O. spinosa*].

APIOPS Alonso-Zarazaga, 1990

***Holotrichapion pisi*** (Fabricius, 1801)

7 SU17: Marlborough Downs, 1939 (det KGB) (Knight, 1940)
Fyfield Down, 1-Jan-73 (MGM) (Morris, 1974, 1975)
8 ST94: Warminster Ranges (SPTAW), 3-Jun - 12-Aug-08 (Boyce & Wright, 2009)
SU03: Grovely Wood, 16-Aug-71, 24-Aug-72 (DRN)
SU14: Stonehenge, 22-Apr-03 (PJH) (Hodge, 2003)
SU15: Manningford, 13-Aug-87 (DRN)
SU24: Driver Training Area (SPTAE), 24-Jun-02 (PJH) (Hodge, 2002)

Resident. Local; biology in Britain not well known, probably exclusively on *Medicago* spp..

LEGARICAPION Ehret, 1990

***Holotrichapion aethiops*** (Herbst, 1797)

7 SU27: Ramsbury, 1869-1938 (Meyrick, 1938) (3 specimens in MCC one insc. under card *4846*)
8 SU04: Shrewton, 30-Jul-82, 2-Aug-83 (DRN)

Resident. Very local; on tufted vetch, *Vicia cracca* [also on *V. sepium*].

**OXYSTOMA Duméril, 1806**

***Oxystoma cerdo*** (Gerstaecker, 1854)

**Nb**
8 ST94: Bowls Barrow (SPTAW), 22-Jun-02 (PJH) (Hodge, 2002); Imber village, 13-Sep-03 (EGS det MDD); Norton Down (SPTAW), 23-Jun-02 (PJH) (Hodge, 2002); Tinkers Track (SPTAW), 3-Jun - 12-Aug-08 (Boyce & Wright, 2009); Warminster Ranges (SPTAW), 3-Jun - 12-Aug-08 (Boyce & Wright, 2009)
SU04: Orcheston Down (SPTA), 18-May, 22-Jun-02 (PJH) (Hodge, 2002); Salisbury Plain, 16-May, 25-Jun-02 (PJH) (Hodge, 2002); Tilshead Park (SPTAW), 14-Jul-99 (PJH) (Hodge, 1999)
SU05: Ell Barrow (SPTAC), 3-Jun - 12-Aug-08 (Boyce & Wright, 2009)
SU14: Haxton (SPTAE), 17-May-02 (PJH) (Hodge, 2002); Stonehenge, 22-Apr-03, 19-Jul-03 (PJH) (Hodge, 2003); Wexland Hanging (SPTA), 22-Jun-02 (PJH) (Hodge, 2002)
Resident. Local; in [hedgerows and waste places on vetches especially tufted vetch, *Vicia cracca* ]. Used to be known as a northern species but has spread south.

***Oxystoma craccae*** (Linnaeus, 1767)
7 SU09: Upper Waterhay, 6-Jul-07 (JKG det MDD)
SU16: Marlborough area, 1869-1938 (Meyrick, 1938)
8 SU03: Grovely Wood, 1-Aug-73 (DRN)
SU04: Salisbury Plain, 16-May-02 (PJH) (Hodge, 2002)
SU14: Wexland Hanging (SPTAC), 16-May-02 (PJH) (Hodge, 2002)
SU24: Driver Training Area (SPTAE), 17-May-02 (PJH) (Hodge, 2002)
Resident. Widespread; in [hedgerows, wood edges, etc., frequently on trees and shrubs although food plants are a wide variety of vetches, *Vicia* spp.].

***Oxystoma pomonae*** (Fabricius, 1798)
7 SU26: Savernake Forest, 1965-1974 (MGM) (Morris, 1974, 1975)
8 ST94: Bowls Barrow (SPTAW), 16-May, 22-Jun-02 (PJH) (Hodge, 2002); Ladywell (SPTAW), 15-May, 23-Jun-02 (PJH) (Hodge, 2002); Norton Down (SPTAW), 15-May, 23-Jun-02 (PJH) (Hodge, 2002)
SU04: Orcheston Down (SPTA), 18-May, 22-Jun-02 (PJH) (Hodge, 2002); Parsonage Down NNR, 22-Jul-08 (MDD); Salisbury Plain, 16-May, 25-Jun-02 (PJH) (Hodge, 2002)
SU05: Black Heath (SPTAC), 3-Jun - 12-Aug-08 (Boyce & Wright, 2009); Lavington Down, 6-May-03 (EGS det MDD)
SU14: Haxton (SPTAE), 17-May-02 (PJH) (Hodge, 2002); Wexland Hanging (SPTAC), 16-May, 22-Jun-02 (PJH) (Hodge, 2002)
SU15: Everleigh (SPTAE), 18-May-02 (PJH) (Hodge, 2002)
SU23: Porton Down, 1974-1978 (MGM) (Morris, 1979); Thorny Down, Porton, 19741974-1978 (MGM) (Morris, 1979); West Winterslow, 11-Aug-01 (NB)
SU24: Dunch Hill (SPTAE), 3-Jun - 12-Aug-08 (Boyce & Wright, 2009)
SU25: Sidbury Hill (SPTAE), 19-May-97 (MGT)
Resident. Widespread; on vetches, *Vicia* spp. [but adults are often found on trees especially in spring and autumn].

***Oxystoma subulatum*** (Kirby, 1808)
7 SU16: Marlborough area, 1869-1938 (Meyrick, 1938)
SU17: Clout's Wood, 1965-1974 (MGM) (Morris, 1974, 1975)
SU26: Savernake Forest, 21-Sep to 11-Oct-93 (EN det SAW)
8 ST94: Bowls Barrow (SPTAW), 16-May, 22-Jun-02 (PJH) (Hodge, 2002); Norton Down (SPTAW), 23-Jun-02 (PJH) (Hodge, 2002)
SU02: Knighton Hill, Stoke, 1965-1974 (MGM) (Morris, 1974, 1975)
SU04: Orcheston Down (SPTA), 18-May, 22-Jun-02 (PJH) (Hodge, 2002); 1965-1974 (MGM) (Morris, 1974, 1975); Salisbury Plain, 25-Jun-02 (PJH) (Hodge, 2002); Tilshead Park (SPTAW), 14-Jul-99 (PJH) (Hodge, 1999)
SU14: Figheldean, 28-Jul-82 (DRN); Haxton (SPTA), 24-Jun-02 (PJH) (Hodge, 2002)
SU23: Porton Down, 3-Jul-99 (PJH) (Hodge, 2002)
Resident. Widespread; [in hedgerows and wood edges]. The principal and perhaps exclusive foodplant is meadow vetchling, *Lathyrus pratensis.*

**PIRAPION Reitter, 1916**
***Pirapion immune*** (Kirby, 1808)
7 SU16: Martinsell, 1895-1938 (Meyrick, 1938)
Resident. Very local; on broom,*Cytisus scoparius.* One specimen in MCC insc. *15 2/60* (conf. MGM).

Subtribe CATAPIINA

**CATAPION Schilsky, 1906**
***Catapion pubescens*** (Kirby, 1811)
**Nb**
8 SU03: Wylye Down, 1965-1974 (MGM) (Morris, 1974, 1975)
SU22: Holmere Common, 19-Aug-88 (DRN)
Resident. Very local; [foodplant not known].

***Catapion seniculus*** (Kirby, 1808)
7 ST06: Cherhill Down, 1-Jun-04 (BLe); Oldbury Castle, 1965-1974 (MGM) (Morris, 1974, 1975)
SU16: Pewsey Downs, 1965-1974 (MGM) (Morris, 1974, 1975)

SU17: Fyfield Down, 1-Jan-73 (MGM) (Morris, 1974, 1975); Hackpen Hill, 1965-1974 (MGM) (Morris, 1974, 1975)
8 ST83: Brimsdown Hill, 1965-1974 (MGM) (Morris, 1974, 1975); Whitesheet Hill, 1965-1974 (MGM) (Morris, 1974, 1975)
ST94: Bowls Barrow, 22-Jun-02 (PJH) (Hodge, 2002); Cotley Hill, 1965-1974 (MGM) (Morris, 1974, 1975); Starveall Down, 1965-1974 (MGM) (Morris, 1974, 1975)
SU01: Martin Down NNR, 21-Aug-89 (RGB); 1965-1974 (MGM) (Morris, 1974, 1975)
SU02: Knapp Down, 1965-1974 (MGM) (Morris, 1974, 1975); Knighton Hill, 1965-1974 (MGM) (Morris, 1974, 1975); Stratford Tony Down, 1965-1974 (MGM) (Morris, 1974, 1975); Woodminton Down, 1965-1974 (MGM) (Morris, 1974, 1975)
SU03: Grovely Wood, 16-Aug-71 (DRN); Steeple Langford Cow Down, 1965-1974 (MGM) (Morris, 1974, 1975); Wylye Down, 1965-1974 (MGM) (Morris, 1974, 1975)
SU04: Parsonage Down, 1965-1974 (MGM) (Morris, 1974, 1975)
SU05: Ell Barrow (SPTAC), 3-Jun - 12-Aug-08 (Boyce & Wright, 2009)
SU13: Camp Down, Salisbury, 1965-1974 (MGM) (Morris, 1974, 1975); Salisbury, 5-Aug-88 (DRN)
SU14: Bulford Down, 1965-1974 (MGM) (Morris, 1974, 1975); Stonehenge, 22-Apr-03 (PJH) (Hodge, 2003)

Resident. Widespread; [on clovers, *Trifolium* spp. and possibly *Medicago* spp]. Biology not well known.

Subtribe SYNAPIINA

**ISCHNOPTERAPION Bokor, 1923**

ISCHNOPTERAPION Bokor, 1923

***Ischnopterapion loti* (Kirby, 1808)**

7 ST06: Oldbury Castle, 1965-1974 (MGM) (Morris, 1974, 1975); Morgan's Hill, 1965-1974 (MGM) (Morris, 1974, 1975)
SU16: Gopher Wood, 1965-1974 (MGM) (Morris, 1974, 1975); Marlborough area, 1869-1938 (Meyrick, 1938); Pewsey Downs, 1965-1974 (MGM) (Morris, 1974, 1975); 22-Jun-90 (APFw)
SU17: Fyfield Down, 1965-1974 (MGM) (Morris, 1974, 1975); Clout's Wood, 1965-1974 (MGM) (Morris, 1974, 1975)
8 ST83: Brimsdown Hill, 1965-1974 (MGM) (Morris, 1974, 1975)
ST92: Prescombe Down, 1965-1974 (MGM) (Morris, 1974, 1975)
ST93: Stockton Down, 1965-1974 (MGM) (Morris, 1974, 1975)
ST94: Bowls Barrow (SPTAW), 22-Jun-02 (PJH) (Hodge, 2002); Cotley Hill, 1965-1974 (MGM) (Morris, 1974, 1975); Ladywell (SPTA), 15-May, 23-Jun-02 (PJH) (Hodge, 2002); 3-Jun - 12-Aug-08 (Boyce & Wright, 2009); Starveall Down, 1965-1974 (MGM) (Morris, 1974, 1975); Tinkers Track (SPTAW), 3-Jun - 12-Aug-08 (Boyce & Wright, 2009); Warminster Ranges (SPTAW), 3-Jun - 12- Aug-08 (Boyce & Wright, 2009)
SU01: Martin Down NNR, 21-Aug-89 (RGB); 1965-1974 (MGM) (Morris, 1974, 1975)
SU02: Knapp Down, 1965-1974 (MGM) (Morris, 1974, 1975); Knighton Hill, 1965-1974 (MGM) (Morris, 1974, 1975); Stratford Tony Down, 1965-1974 (MGM) (Morris, 1974, 1975)
SU03: Steeple Langford Cow Down, 1965-1974 (MGM) (Morris, 1974, 1975)
SU04: Orcheston Down (SPTA), 18-May, 22-Jun-02 (PJH) (Hodge, 2002); Parsonage Down, 1965-1974 (MGM) (Morris 1974, 1975); Salisbury Plain, 25-Jun-02 (PJH) (Hodge, 2002); Shrewton, 28, 30-Jul, 6-Aug-82, 2-Aug-83, 1-Aug-85 (DRN); Yarnbury Castle, 1965-1974 (MGM) (Morris, 1974, 1975)
SU05: Redhorn Hill, 1965-1974 (MGM) (Morris, 1974, 1975)
SU13: Camp Down, Salisbury, 1965-1974 (MGM) (Morris, 1974, 1975)
SU14: Bulford Down, 1965-1974 (MGM) (Morris, 1974, 1975); Haxton (SPTAE), 17-May-02 (PJH) (Hodge, 2002); 3-Jun - 12-Aug-08 (Boyce & Wright, 2009); Stonehenge, 22-Apr-03 (PJH) (Hodge, 2003); Wexland Hanging (SPTA), 22-Jun-02 (PJH) (Hodge, 2002)
SU23: Porton Down, 1965-1974 (MGM) (Morris, 1974, 1975); 3-Jul-99 (PJH) (Hodge, 2002)
SU24: Driver Training Area (SPTAE), 17-May, 24-Jun-02 (PJH) (Hodge, 2002); Dunch Hill (SPTAE), 3-Jun - 12-Aug-08 (Boyce & Wright, 2009)

Resident. Widespread; in roadsides, grasslands and other dry places [chiefly on birdsfoot-trefoil, *Lotus corniculatus*].

***Ischnopterapion modestum* (Germar, 1817)**

8 SU16: Jones's Mill, 19-Jun, 2-Jul-01 (APF) (Foster, 2002)

Resident. Very local; in wetland [on the trefoil *Lotus uliginosus*]. Added to the GB list in 1976 as *A. sicardi*.

CHLORAPION Györffy, 1956

***Ischnopterapion virens* (Herbst, 1797)**

7 SU16: Gopher Wood, Huish, 1965-1974 (MGM) (Morris, 1974, 1975); Marlborough area, 1869-1938 (Meyrick, 1938)
8 ST92: Prescombe Down, 1965-1974 (MGM)

(Morris, 1974, 1975)
ST94: Cotley Hill, 1965-1974 (MGM) (Morris, 1974, 1975); Starveall Down, 1965-1974 (MGM) (Morris, 1974, 1975)
SU02: Knapp Down, 1965-1974 (MGM) (Morris, 1974, 1975); Knighton Hill, 1965-1974 (MGM) (Morris, 1974, 1975); Stratford Tony Down, 1965-1974 (MGM) (Morris, 1974, 1975); Woodminton Down, 1965-1974 (MGM) (Morris, 1974, 1975)
SU03: Grovely Wood, 28-Jul-83 (DRN); Wylye Down, 1965-1974 (MGM) (Morris, 1974, 1975)
SU04: Fox Covert, 25-Jun-02 (PJH) (Hodge, 2002); Orcheston Down, 22-Jun-02 (PJH) (Hodge, 2002); Parsonage Down NNR, 1965-1974 (MGM) (Morris, 1974, 1975); Salisbury Plain, 25-Jun-02 (PJH) (Hodge, 2002)
SU06: Horton, 5-Aug-83 (DRN)
SU13: Camp Down, Salisbury, 1965-1974 (MGM) (Morris, 1974, 1975)
SU14: Figheldean, 28-Jul-82 (DRN); Haxton (SPTA), 17-May 24-Jun-02 (PJH) (Hodge, 2002); Stonehenge, 22-Apr-03 (PJH) (Hodge, 2003)
SU16: Woodborough, 4-Aug-83 (DRN)
SU24: North Tidworth, Devils Ditch, 30-May-04 (BLe); Tidworth Pennings, 1965-1974 (MGM) (Morris, 1974, 1975)

Resident. Widespread; in a wide variety of open and grassy habitats on clovers, *Trifolium* spp..

**PROTOPIRAPION Alonso-Zarazaga, 1990**

***Protopirapion atratulum*** (Germar, 1817)

7 SU16: Marlborough area, 1869-1938 (Meyrick, 1938)
8 ST83: Brimsdown Hill, 1965-1974 (MGM) (Morris, 1974, 1975)
ST94: Bowls Barrow (SPTAW), 16-May-02 (PJH) (Hodge, 2002)
SU04: Parsonage Down NNR, 2-Jul-08 (MDD); Salisbury Plain, 17-Jul-00 (KNAA)
SU25: Collingbourne Wood, 1869-1938 (Meyrick, 1938)

Resident. Local; [chiefly on broom *Cytisus scoparius*, occasionally on gorse, *Ulex europaeus*, and *Genista* spp.].

**STENOPTERAPION Bokor, 1923**

STENOPTERAPION Bokor, 1923

***Stenopterapion meliloti*** (Kirby, 1808)

7 ST18: Wroughton, 1965-1974 (MGM) (Morris, 1974, 1975)
SU17: Clout's Wood, 1965-1974 (MGM) (Morris, 1974, 1975)
8 ST94: Ladywell (SPTAW), 23-Jun-02 (PJH) (Hodge, 2002); Norton Down (SPTAW), 15-May, 23-Jun-02 (PJH) (Hodge, 2002)
SU04: Salisbury Plain, 16-May-02 (PJH) (Hodge, 2002); Shrewton, 30-Aug-82 (DRN); Tilshead Park (SPTAW), 14-Jul-99 PJH) (Hodge, 1999)
SU15: Haxton Down, 14-Jul-99 (PJH) (Hodge, 1999); Manningford, 13-Aug-87 (DRN)
SU22: Bentley Wood, 27-May-97 (MDD) (Meyrick, 1938)
SU23: Allington Farm, nr Salisbury, 1974-1978 (MGM) (Morris, 1979)

Resident. Widespread; in waste places, etc. on species of *Melilotus*.

***Stenopterapion tenue*** (Kirby, 1808)

7 SU16: Pewsey Downs, 1965-1974 (MGM) (Morris, 1974, 1975)
SU17: Clout's Wood, 1965-1974 (MGM) (Morris, 1974, 1975)
8 ST83: Brimsdown Hill, 1965-1974 (MGM) (Morris, 1974, 1975); Whitesheet Hill, 1965-1974 (MGM) (Morris, 1974, 1975)
ST92: Prescombe Down, 1965-1974 (MGM) (Morris, 1974, 1975)
ST94: Cotley Hill, 1965-1974 (MGM) (Morris, 1974, 1975); Ladywell (SPTAW), 23-Jun-02 (PJH) (Hodge, 2002); Norton Down (SPTAW), 23-Jun-02 (PJH) (Hodge, 2002); Wadman's Coppice (SPTAW), 3-Jun - 12-Aug-08 (Boyce & Wright, 2009)
SU01: Martin Down NNR, 21-Aug-89 (RGB)
SU02: Knapp Down, 1965-1974 (MGM) (Morris, 1974, 1975); Knighton Hill, 1965-1974 (MGM) (Morris, 1974, 1975); Stratford Tony Down, 1965-1974 (MGM) (Morris, 1974, 1975); Woodminton Down, 1965-1974 (MGM) (Morris, 1974, 1975)
SU03: Grovely Wood, 28-Jul-83 (DRN)
SU04: Shrewton, 6-Aug-82, 2-Aug-83 (DRN)
SU05: Redhorn Hill, 1965-1974 (MGM) (Morris, 1974, 1975)
SU13: Salisbury, 25-Jul-78 (DRN)
SU14: Stonehenge, 22-Apr-03 (PJH) (Hodge, 2003)
SU15: Manningford, 13-Aug-87 (DRN)
SU23: Allington Farm, nr Salisbury, 1974-1978 (MGM) (Morris, 1979)

Resident. Widespread; in grasslands, etc. [on *Medicago* spp., particularly *M. lupulina*, and possibly other Papilionaceae, but hosts not well known].

**SYNAPION Schilsky, 1902**

***Synapion ebeninum*** (Kirby, 1808)

7 SU16: Marlborough area, 1869-1938 (Meyrick, 1938)
8 SU22: Barnridge Copse, nr Farley, 1-Jan-79 (MGM) (Morris, 1979); Langley Wood, 14-May to 3-Jun-04 (MDD) (Darby, 2005)

Resident. Very local; in damp places on *Lotus*

*uliginosus*, birdsfoot-trefoil. Single specimen in MCC without data (conf. MGM).

Subtribe TRICHAPIINA

**BETULAPION Ehret, 1994**

***Betulapion simile*** (Kirby, 1811)

8 SU23: Porton Down, 20-Jul-97 (MAS)

Resident. Very local; in [woods, heaths, etc. on birches].

Tribe PIEZOTRACHELINI

**PROTAPION Schilsky, 1908**

***Protapion apricans*** (Herbst, 1797)

7 ST83: Whitesheet Down, 7-Jul-82 (KNAA)
SU06: Cherhill Down, 1-Jun-04 (BLe); Morgan's Hill, 1965-1974 (MGM) (Morris, 1974, 1975); Oldbury Castle, 1965-1974 (MGM) (Morris, 1974, 1975)
SU09: Somerford Keynes, 4-Jun-04 (BLe)
SU16: Gopher Wood, Huish, 1965-1974 (MGM) (Morris, 1974, 1975); Marlborough area, 1869-1938 (Meyrick, 1938); Pewsey Downs, 1965-1974 (MGM) (Morris, 1974, 1975)
SU17: Clout's Wood, 1965-1974 (MGM) (Morris, 1974, 1975); Hackpen Hill, 1965-1974 (MGM) (Morris, 1974, 1975)
ST83: Brimsdown Hill, 1965-1974 (MGM) (Morris, 1974, 1975)
ST94: Cotley Hill, 1965-1974 (MGM) (Morris, 1974, 1975); Ladywell (SPTAW), 3-Jun – 12-Aug-08 (Boyce & Wright, 2009); Norton Down (SPTAW), 15-May-02 (PJH) (Hodge, 2002); Starveall Down, 1965-1974 (MGM) (Morris, 1974, 1975)
ST95: Bratton Castle, 1965-1974 (MGM) (Morris, 1974, 1975)
SU01: Martin Down NNR, 21-Aug-89 (RGB); 1965-1974 (MGM) (Morris, 1974, 1975)
SU02: Knapp Down, 1965-1974 (MGM) (Morris, 1974, 1975); Knighton Hill, 1965-1974 (MGM) (Morris, 1974, 1975); Middleton Down, 24-May-95 (MDD); Stratford Tony Down, 1965-1974 (MGM) (Morris, 1974, 1975); Woodminton Down, 1965-1974 (MGM) (Morris, 1974, 1975)
SU03: Grovely Wood, 11-Aug-70 (DRN); Steeple Langford Cow Down, 1965-1974 (MGM) (Morris, 1974, 1975); Wylye Down NNR, 4-May to 10-Oct-05 (MDD) (Darby, 2009)
SU04: Fox Covert, 25-Jun-02 (PJH) (Hodge, 2002); 3-Jun - 12-Aug-08 (Boyce & Wright, 2009); Orcheston Down (SPTA), 22-Jun-02 (PJH) (Hodge, 2002); Parsonage Down, 1965-1974 (MGM) (Morris, 1974, 1975), 8 to 22-Jul-08 (MDD); Salisbury Plain, 16-May, 25-Jun-02 (PJH) (Hodge, 2002); 17-Jul-00 (KNAA); Shrewton, 1-Aug-85, 30-Jul-82 (DRN)
SU05: Ell Barrow (SPTAC), 3-Jun - 12-Aug-08 (Boyce & Wright, 2009); Redhorn Hill, 1965-1974 (MGM) (Morris, 1974, 1975)
SU13: Camp Down, Salisbury, 1965-1974 (MGM) (Morris, 1974, 1975); Middle Woodford, 5-Aug-82 (DRN); Salisbury, 26-May-74, 15-Aug-88 (DRN); Salisbury, Avon Valley Reserve,1-Jun-00 (MDD) (Darby, 2001)
SU14: Figheldean, 28-Jul-82 (DRN); Stonehenge, 22-Apr-03 (PJH) (Hodge, 2003); Wexland Hanging (SPTAC), 22-Jun-02 (PJH) (Hodge, 2002)
SU15: Beach's Plantation (SPTAE), 13-Jun-97 (MGT); Everleigh (SPTAE), 18-May-02 (PJH) (Hodge, 2002); Manningford, 13-Aug-87 (DRN)
SU16: Woodborough, 4-Aug-83 (DRN)
SU22: Holmere Common, 19-Aug-88 (DRN)
SU23: Porton Down, 3-Jul-99 (MAS)
SU24: Tidworth Pennings, 1965-1974 (MGM) (Morris, 1974, 1975)

Resident. Widespread; in grasslands, waste places, etc. [on clover, *Trifolium pratense*].

***Protapion assimile*** (Kirby, 1808)

7 ST06: Oldbury Castle, 1965-1974 (MGM) (Morris, 1974, 1975); Morgan's Hill, 1965-1974 (MGM) (Morris, 1974, 1975); Roundway Camp, Jul-40 (FDB) (Buck, 1956); 1965-1974 (MGM) (Morris, 1974, 1975)
SU16: Marlborough area, 1869-1938 (Meyrick, 1938); Pewsey Downs, 1965-1974 (MGM) (Morris, 1974, 1975)
SU17: Clout's Wood, 1965-1974 (MGM) (Morris, 1974, 1975); Fyfield Down, 1965-1974 (MGM) (Morris, 1974, 1975)
8 ST83: Brimsdown Hill, 1965-1974 (MGM) (Morris, 1974, 1975)
ST84: Longleat, 29-Aug-99 (MB) (Meyrick, 1938)
ST93: Fonthill, 19-Jul-95 (MDD)
ST94: Cotley Hill, 1965-1974 (MGM) (Morris, 1974, 1975); Norton Down (SPTAW), 15-May-02 (PJH) (Hodge, 2002); Starveall Down, 1965-1974 (MGM) (Morris, 1974, 1975)
ST95: Bratton Castle, 1965-1974 (MGM) (Morris, 1974, 1975)
SU01: Martin Down NNR, 21-Aug-89 (RGB)
SU02: Knapp Down, 1965-1974 (MGM) (Morris, 1974, 1975); Knighton Hill, 1965-1974 (MGM) (Morris, 1974, 1975); Woodminton Down, 1965-1974 (MGM) (Morris, 1974, 1975)
SU03: Steeple Langford Cow Down 1965-1974 (MGM) (Morris, 1974, 1975); Wylye Down, 1965-1974 (MGM) (Morris, 1974, 1975)

SU04: Parsonage Down, 1965-1974 (MGM) (Morris, 1974, 1975)
SU05: Ell Barrow (SPTAC), 3-Jun - 12-Aug-08 (Boyce & Wright, 2009); Redhorn Hill, 1965-1974 (MGM) (Morris, 1974, 1975)
SU13: Camp Down, Salisbury, 1965-1974 (MGM) (Morris, 1974, 1975); Salisbury, 15-Aug-88 (DRN); Stratford sub Castle, 10-Aug-70 (DRN)
SU14: Stonehenge, 22-Apr-03 (PJH) (Hodge, 2003); Wexland Hanging (SPTAC), 22-Jun-02 (PJH) (Hodge, 2002)
SU15: Fittleton, 29-Jul-83 (DRN)
SU16: Woodborough, 4-Aug-83 (DRN)
SU23: Porton Down, 1965-1974 (MGM) (Morris, 1974, 1975)
SU24: Dunch Hill (SPTAE), 3-Jun - 12-Aug-08 (Boyce & Wright, 2009)

Resident. Widespread; in grasslands [also woodland, on clovers, *Trifolium* spp.].

***Protapion difforme*** (Ahrens, 1817)
**Nb**
7 SU16: Gore Copse, nr Marlborough, 1869-1938 (Meyrick, 1938)
8 ST83: Brimsdown Hill, 1965-1974 (MGM) (Morris, 1974, 1975) Whitesheet Down, 7-Jul-82 (KNAA)
ST94: Cotley Hill, 1965-1974 (MGM) (Morris, 1974, 1975)
SU02: Knighton Hill, 1965-1974 (MGM) (Meyrick, 1938); Stratford Tony Down, 1965-1974 (MGM) (Meyrick, 1938)
SU14: Haxton (SPTAE), 17-May-02 (PJH) (Hodge, 2002)
SU15: Everleigh, 24-Jun-02 (PJH) (Hodge, 2002)
SU23: Porton Down, 1965-1974 (MGM) (Morris, 1974, 1975)

Resident. Widespread; in [damp grasslands and marshes]. The biology is unknown.

***Protapion filirostre*** (Kirby, 1808)
**Nb, LCGR-Fm Cl**
8 ST94/SU00/SU01 Salisbury Plain MOD, 17-Jul-00 (KNAA)
ST94: Wadman's Coppice (SPTAW), 3-Jun - 12-Aug-08 (Boyce & Wright, 2009)

Resident. Very local; in [dry grassland, often on calcareous soils]. Biology not well known.

**White Clover Seed Weevil**
***Protapion fulvipes*** (Geoffroy in Fourcroy, 1785)
7 ST97: Chippenham, Lackham College, 29-May-04 (BLe)
SU06: Morgan's Hill, 1965-1974 (MGM) (Morris, 1974, 1975)
SU09: Somerford Keynes, 4-Jun-04 (BLe)
SU16: Pewsey Downs, 1965-1974 (MGM) (Morris, 1974, 1975); West Woods, 1965-1974 (MGM) (Morris, 1974, 1975)
SU17: Fyfield Down, 1-Jan-73 (MGM) (Morris, 1974, 1975)
SU26: Savernake Forest, 28-Jul to 28-Aug-93 (EN det SAW)
8 ST84: Cley Hill, 6-Jul-07 (KNAA); Longleat, 29-Aug-99 (MB)
ST91: Cranborne Chase, 7-May-97 (MAS); 16-Jun-80 (NCC)
ST92: Fovant, 25-Apr-95 (MDD); Prescombe Down, 1965-1974 (MGM) (Morris, 1974, 1975)
ST93: Great Ridge Wood, 16-May-00 (MDD)
ST94: Bowls Barrow (SPTAW), 16-May-02 (PJH) (Hodge, 2002); Starveall Down, 1965-1974 (MGM) (Morris, 1974, 1975)
SU01: Martin Down NNR, 21-Aug-89 (RGB)
SU02: Knapp Down, 1965-1974 (MGM) (Morris, 1974, 1975); Knighton Hill, 1965-1974 (MGM) (Morris, 1974, 1975); Middleton Down, 4, 24-May-95 (MDD); Stratford Tony Down, 1965-1974 (MGM) (Morris, 1974, 1975); Woodminton Down, 1965-1974 (MGM) (Morris, 1974, 1975)
SU03: Little Langford, 3-Jul-96 (MAS) (Morris, 1974, 1975); Steeple Langford Cow Down, 1965-1974 (MGM) (Morris, 1974, 1975); Wylye Down, 1965-1974 (MGM) (Morris, 1974, 1975)
SU04: Orcheston Down, 22-Jun-02 (PJH) (Hodge, 2002); Parsonage Down, 1965-1974 (MGM) (Morris, 1974, 1975); Salisbury Plain, 16-May-02 (PJH) (Hodge, 2002)
SU13: Boscombe Down, 8, 9-Oct-94 (MDD); Old Sarum, 31-May-97 (MAS); Salisbury, Bemerton Folly, 7-Mar - 15-Jun-02 (MDD) (Darby, 2002d)
SU14: Bulford Down, 1965-1974 (MGM) (Morris, 1974, 1975); Figheldean, 28-Jul-82 (DRN); Haxton (SPTAE), 17-May-02 (PJH) (Hodge, 2002); Netheravon, 29-Jul-83 (DRN); Stonehenge, 22-Apr-03 (PJH) (Hodge, 2003); Wexland Hanging (SPTAC), 16-May-02 (PJH) (Hodge, 2002)
SU16: Woodborough, 4-Aug-83 (DRN)
SU22: Blackmoor Copse, 17-May-02 (APF) (Foster, 2004); Homan's Copse, 19-Aug-88 (DRN); Landford, 12-Aug-87 (DRN); Redlynch, 4-Aug-82 (DRN)
SU23: Porton Down, 1965-1974 (MAS); 12-Nov-96 (MGM) (Morris, 1974, 1975)
SU24: North Tidworth, Devils Ditch, 30-May-04 (BLe)

Resident. Widespread; in all kinds of open places and sometimes on trees and shrubs. Host plants are clovers *Trifolium repens* and *T. hybridum*. Formerly *Protapion dichroum*.

***Protapion nigritarse*** (Kirby, 1808)

7 SU09: Somerford Keynes, 4-Jun-04 (BLe)
SU16: Marlborough area, 1869-1938 (Meyrick, 1938)
8 ST93: Great Ridge Wood, 16-May-00 (MDD)
ST94: Cotley Hill, 1965-1974 (MGM) (Morris, 1974, 1975); Norton Down (SPTAW), 15-May-02 (PJH) (Hodge, 2002)
SU02: Knighton Wood, Mar-07 (AJWA); Middleton Down, 24-May-95 (MDD)
SU05: Black Heath (SPTAC), 3-Jun - 12-Aug-08 (Boyce & Wright, 2009); Ell Barrow (SPTAC), 3-Jun - 12-Aug-08 (Boyce & Wright, 2009); Redhorn Hill, 1965-1974 (MGM) (Morris, 1974, 1975)
SU13: Salisbury, Avon Valley Reserve, 1-Jun-00 (MDD) (Darby, 2001)
SU14: Wexland Hanging (SPTAC), 22-Jun-02 (PJH) (Hodge, 2002)
SU22: Hamptworth Estate, 7-Aug-86 (DRN); Langley Wood NNR, 25-May-94 (WANHS det MDD)
Resident. Widespread; in grasslands and dry open places. Biology not well studied.

***Protapion ononidis*** (Bach, 1854)
7 SU16: Wernham, nr Marlborough, 1869-1938 (Meyrick, 1938)
SU26: Bedwyn, 1869-1938 (Meyrick, 1938)
8 ST94: Cotley Hill, 1965-1974 (MGM) (Morris, 1974, 1975); Tinkers Track (SPTAW), 3-Jun - 12-Aug-08 (Boyce & Wright, 2009); Warminster Ranges (SPTAW), 3-Jun - 12-Aug-08 (Boyce & Wright, 2009)
ST95: Bratton Castle, 1965-1974 (MGM) (Morris, 1974, 1975)
SU04: Fox Covert (SPTAW), 3-Jun - 12-Aug-08 (Boyce & Wright, 2009)
SU05: Ell Barrow (SPTAC), 3-Jun - 12-Aug-08 (Boyce & Wright, 2009)
SU13: Little Durnford, 1965-1974 (MGM) (Morris, 1974, 1975)
SU14: Bulford Down, 1965-1974 (MGM) (Morris, 1974, 1975)
Resident. Local; in grassy open places [on restharrow, *Ononis repens* and *O. spinosa*] Previously *Protapion ononicola.*

***Protapion trifolii*** (Linnaeus, 1768)
7 ST06: Oldbury Castle, 1965-1974 (MGM) (Morris, 1974, 1975)
8 SU02: Knighton Hill, 1965-1974 (MGM) (Morris, 1974, 1975); Stratford Tony Down, 1965-1974 (MGM) (Morris, 1974, 1975)
SU03: Wylye Down NNR, 1965-1974 (MGM) (Morris, 1974, 1975)
SU13: Camp Down, Salisbury, 1965-1974 (MGM) (Morris, 1974, 1975)
SU14: Bulford Down, 1965-1974 (MGM) (Morris, 1974, 1975); Haxton (SPTAE), 17-May-02 (PJH) (Hodge, 2002)
SU23: Porton Down, 1965-1974 (MGM) (Morris, 1974, 1975)
SU24: Tidworth Pennings, 1965-1974 (MGM) (Morris, 1974, 1975)
Resident. Widespread; in grasslands, hedgerows and waste places [on clovers *Trifolium pratense* and *T. medium*].

***Protapion varipes*** (Germar, 1817)
**Nb**
7 SU26: Folly Farm nr Marlborough, 1939 (Anon det KGB) (Knight, 1940)
8 SU02: Knighton Hill, 1965-1974 (MGM) (Morris, 1974, 1975)
SU04: Parsonage Down, 1965-1974 (MGM) (Morris, 1974, 1975)
SU13: Salisbury, Avon Valley Reserve, Apr-00 (MAS) (Darby, 2001)
Resident. Local; in grasslands, waste places [and woodland edges, on clover, *Trifolium pratense*, and possibly other *Trifolium* spp.]. Biology not well known.

## Family NANOPHYIDAE

**NANOPHYES Schönherr, 1838**
**Loosestrife Weevil**
***Nanophyes marmoratus*** (Goeze, 1777)
7 ST96/97: Bowood Lake, 24-Aug-86 (KNAA)
SU09: Somerford Keynes, 4-Jun-04 (BLe)
8 SU13: Great Durnford, 18-Aug-72 (DRN); Salisbury, 15-Aug-88 (DRN)
SU14: Netheravon, 29-Jul-83 (DRN)
SU15: Fittleton, 4-Aug-88 (DRN)
SU16: Jones's Mill, 1-Jan-97 (MAS)

## Family DRYOPHTHORIDAE
Subfamily RHYNCHOPHORINAE
Tribe LITOSOMINI

**SITOPHILUS Schönherr, 1838**
**Granary Weevil**
***Sitophilus granarius*** (Linnaeus, 1758)
8 ST95: West Lavington, 16-Aug-40 (KJA) (DFL, 1948)
SU12/13: Salisbury, Dec-77 (SE det KNAA)
Resident. Very local; [in grain stores]. An important pest of grain and also in farinaceous foodstuffs such as biscuits.

## Family ERIRHINIDAE
Subfamily ERIRHININAE
Tribe ERIRHININI

**GRYPUS Germar, 1817**
**Horsetail Weevil**
***Grypus equiseti*** (Fabricius, 1775)
**Nb**
7 SU16: Honey Street, Banks of Kennett and Avon Canal, 1965-1974 (MGM) (Morris, 1974, 1975)
8 SU12/13: Salisbury, 26-May-74 (DRN)
SU16: Jones's Mill, 21-Jun-97 (MDD)
Resident. Very local; on horsetails, *Equisetum* spp. in wet areas.

Grypus equiseti

**NOTARIS Germar, 1817**
***Notaris acridulus*** (Linnaeus, 1758)
7 SU16: Marlborough area, 1869-1938 (Meyrick, 1938)
SU17: Clout's Wood, 1965-1974 (MGM) (Morris, 1974, 1975)
8 ST92: Sutton Mandeville, 1-May-97 (MDD)
ST94: Ladywell (SPTAW), 3-Jun - 12-Aug-08 (Boyce & Wright, 2009)
SU02: Fovant, 31-May-95 (MDD)
SU12: Petersfinger, 25-Jul-91 (DRN)
SU13: Lower Woodford, 2, 5-Aug-82 (DRN); Stratford sub Castle, 10-Aug-70 (DRN)
SU21: Hamptworth Estate, 29-Mar-78 (DRN)
Resident. Widespread; in all kinds of damp and wet places. Also at the sides of rivers, streams, canals, etc..

***Notaris scirpi*** (Fabricius, 1793)
**Nb**
8 ST94: Ladywell (SPTAW), 3-Jun - 12-Aug-08 (Boyce & Wright, 2009); Wadman's Coppice (SPTAW), 3-Jun - 12-Aug-08 (Boyce & Wright, 2009)
Resident. Very local; [in marshes, fens and other wetlands, and at the sides of rivers, canals, etc.]. Biology little known.

**THRYOGENES Bedel, 1884**
***Thryogenes scirrhosus*** (Gyllenhal, 1836)
**Nb**
7 SU36: Stype Wood, Jun-1870 (RH) (Hislop, 1871); 1869-1938 (Meyrick, 1938)
Resident. Very local; [in fens, marshes and wetlands generally]. Host plants not well known but associated with *Sparganium* spp.. There is no specimen under this name in MCC.

Tribe STENOPELMINI

**STENOPELMIS Schönherr, 1836**
***Stenopelmus rufinasus*** Gyllenhal, 1836
7 SU16: Honey Street, Banks of Kennett and Avon Canal, 1965-1974 (MGM) (Morris, 1974, 1975)
8 SU15: New Mill, nr Pewsey, 30-Oct-45 (Holmes & Tennant, 1947)
Resident. Very local; [on water fern, *Azolla filiculoides*, in ponds, lakes ditches, canals and other still-water bodies].

## Family CURCULIONIDAE
### Subfamily CURCULIONINAE
#### Tribe CURCULIONINI

**ARCHARIUS Gistel, 1848**
***Archarius pyrrhoceras*** Marsham, 1802
7 SU16: Gore Copse, nr Marlborough, 1869-1938 (Meyrick, 1938)
SU26: Savernake Forest, 20-Apr to 7-Jul, 21-Sep to 11-Oct-93 (EN det SAW); 31-May-04 (BLe)
8 ST91: Cranborne Chase, 30-Jun-99 (MDD)
ST93: Stockton Down, 1965-1974 (MGM) (Morris, 1974, 1975)
ST94: Norton Down (SPTAW), 15-May-02 (PJH) (Hodge, 2002)
SU03: Grovely Wood, 27-May-74 (DRN); 12-Jun-00 (MDD)
SU13: Salisbury, Bemerton Folly, 7-Mar – 15-Jun-02 (MDD) (Darby, 2002d)
SU15: Everleigh (SPTAE), 18-May-02 (PJH) (Hodge, 2002)
SU22: Bagfield Copse, 6-Aug-86 (DRN); Bentley Wood, 11-Aug-01 (MDD); Blackmoor Copse, 1965-1974 (MGM) (Morris, 1974, 1975); Langley Wood NNR, 3-Jun to 27-Aug-04 (MDD) (Darby, 2005); Whiteparish Common, 1965-1974 (MGM) (Morris, 1974, 1975)
SU23: Bentley Wood, 31-May-97 (MDD); Thorny Down, Porton, 1974-1978 (MGM) (Morris, 1979)
SU26: Bedwyn Brail, 2, 19-Jun-03 (MDD)
Resident. Widespread; on oak, larvae in cynipid galls.

***Archarius salicivorus*** Paykull, 1792
7 SU16: Marlborough area, 1869-1938 (Meyrick, 1938)

8 ST83: Gutch Common, nr Semley, 1965-1974 (MGM) (Morris, 1974, 1975)
ST84: Longleat, 7-Jul-99 (MDD)
SU12: Whaddon, 22-Jul-73 (DRN)
SU13: Salisbury, Avon Valley Reserve, 12-May (Darby, 2001); Salisbury, Bemerton Folly, 7-Mar - 15-Jun-02 (MDD) (Darby, 2002d)
SU14: Stonehenge, 22-Apr-03 (PJH) (Hodge, 2003)
SU21: Hamptworth Common, 28-Jul-77 (DRN); Hamptworth Estate, 6-Apr-77 (DRN)
SU22: Barnridge Copse, nr Farley, 1974-1978 (MGM) (Morris, 1979); Bentley Wood, 11-Aug-01 (MDD); Blackmoor Copse, 17-May-02 (APF) (Foster, 2004); Glazier's Copse, 16-Aug-88 (DRN); Holmere Common, 19-Aug-88 (DRN); Landford, 12-Aug-87 (DRN); Langley Wood NNR, 25-May-94 (WANHS det MDD); Whiteparish, 22-Jul-73 (DRN); Whiteparish Common, 1965-1974 (MGM) (Morris, 1974, 1975)

Resident. Widespread; on willows, sallows and alder. Larvae in sawfly galls (of *Pontania* spp.].

**CURCULIO Linnaeus, 1758**
***Curculio betulae*** (Stephens, 1831)
**Nb**
8 ST93: Great Ridge Wood, 6-Aug-85 (DRN)
Resident. Very local; on birch/hazel.

Curculio glandium

**Acorn Weevil**
***Curculio glandium*** Marsham, 1802
7 ST86: Neston Park, 3-Jun-04 (BLe)
ST87: Corsham Park, 3-Jun-04 (BLe)
SU26: Savernake Forest, 11 to 30-May-95, 15-Sep to 10-Oct-95 (EN det MDD)
SU27: Ramsbury, 1869-1938 (Meyrick, 1938)
8 ST85: Bullen Hill Farm, 1982-2002 (EGS det MDD); Picket Wood, 6-Dec-81 (EGS det MDD)
ST91: Cranborne Chase, 12-May-97 (MAS)
ST93: Great Ridge Wood, 16-May-00 (MDD)
SU02: Knighton Wood, Sep-06, May-08 (AJWA)
SU03: Grovely Wood, 13-Aug-71 (DRN)
SU21: Franchises Wood, 25-Jul-73 (DRN)
SU22: Langley Wood NNR, 8-Jun-02, 16-Jul-04 (MDD) (Darby, 2005)
SU23: Thorny Down Woods, 1965-1974 (MGM) (Morris, 1974, 1975)
SU26: Bedwyn Brail, 2-Jun-03 (MDD)

Resident. Widespread; on oak.

**Nut Weevil**
***Curculio nucum*** Linnaeus, 1758
7 SU16: Marlborough area, 1869-1938 (Meyrick, 1938)
SU19: Hannington, 26-May-97 (MEN)
SU26: Savernake Forest, 28-Apr to 18-May, 1 to 19-Jul, 30-Aug to 20-Sep-94 (EN det SAW); 31-May to 23-Jun-95 (EN det MDD)
8 ST85: Green Lane Wood, 26-Jun-98 (JN)
ST91: Cranborne Chase, 12-May-97 (MAS)
SU03: Grovely Wood, 13-Aug-70, 1-Aug-88 (DRN)
SU22: Langley Wood NNR, 25-May-94 (WANHS det MDD); Whiteparish Common, 1965-1974 (MGM) (Morris, 1974, 1975)

Resident. Widespread; on hazel.

***Curculio venosus*** (Gravenhorst, 1807)
7 ST86: Neston Park, 3-Jun-04 (BLe, KNAA)
ST87: Corsham Park, 3-Jun-04 (BLe)
ST99: Flisteridge Wood, 16-May-82 (KNAA)
SU16: Marlborough area, 1869-1938 (Meyrick, 1938)
SU26: Savernake Forest, 30-May-73 (DBS); 25-Apr to 26-May, 15-Jun to 17-Jul-73 (EN det SAW); 29-May-83 (KNAA)
8 ST91: Cranborne Chase, 30-Jun-99 (MDD)
SU03: Grovely Wood, 18-Jul-73 (DRN); 11-Jun-00 (MDD)
SU21: Burnt Ground Wood, 31-May-74 (DRN)
SU22: Bagfield Copse, 6-Aug-86 (DRN); Bentley Wood, 27-May-97 (MDD); Langley Wood NNR, 28-May-74 (DRN); 25-May-94 (WANHS det MDD); 27-Aug-04 (MDD) (Darby, 2005); Whiteparish Common, 1974-1978 (MGM) (Morris, 1979)
SU23: Thorny Down, Porton, 1974-1978 (MGM) (Morris, 1979)

Resident. Widespread; on oak.

***Curculio villosus*** Fabricius, 1781
**Nb**
7 SU16: Marlborough area, 1869-1938 (Meyrick, 1938)
SU26: Savernake Forest, 11 to 30-May-95 (EN det MDD)
8 SU23: Thorny Down Woods, 1965-1974 (MGM) (Morris, 1974, 1975)

Resident. Very local; [larvae in 'oak apple' galls of *Biorhiza pallida*].

Tribe ANTHONOMINI

**ANTHONOMUS Germar, 1817**

ANTHONOMUS Germar, 1817

***Anthonomus bituberculatus*** Thomson, C. G, 1868

7 SU16: Lockeridge, 1965-1974 (MGM) (Morris, 1974, 1975) Marlborough area, 1869-1938 (Meyrick, 1938)

8 SU15: North Newton, 1965-1974 (MGM) (Morris, 1974, 1975)

SU16: Jones's Mill, 21-Jun-9721-Jun-97 (MDD)

Resident. Very local; on hawthorns. Two specimens in MCC one labelled *Marlborough 18/6/02* and det as *A. ulmi* are this species (conf. MGM)

***Anthonomus pedicularius*** (Linnaeus, 1758)

7 ST99: Crudwell, 18-May-97 (APF)

SU06: Morgan's Hill, 1965-1974 (MGM) (Morris, 1974, 1975)

SU09: Ashton Keynes, 7-May-01 (JKG det MDD)

SU16: Marlborough area, 24-Mar-1879 (WFB) (Meyrick, 1880); 1869-1938 (Meyrick, 1938)

SU17: Fyfield Down, 1-Jan-73 (MGM) (Morris, 1974, 1975)

SU26: Savernake Forest, 27-May to 28-Sep-93 (EN det SAW)

8 ST95: Fore Hill, nr West Lavington, 1965-1974 (MGM) (Morris, 1974, 1975)

SU03: Dinton Park, 6-Jul-00 (KNAA); Grovely Wood, 11-Aug-71, 1-Aug-88 (DRN)

SU04: Orcheston Down, 1965-1974 (MGM) (Morris, 1974, 1975); Parsonage Down NNR, 22-Jul-08 (MDD); Shrewton, 1-Aug-88 (DRN)

SU12: Coombe Bissett Down, 1-Apr-97 (MDD)

SU13: Camp Down, 1965-1974 (MGM) (Morris, 1974, 1975); Salisbury, Avon Valley Reserve, 14-Apr-00 (MAS) (Darby, 2001)

SU14: Stonehenge, 22-Apr-03 (PJH) (Hodge, 2003)

SU21: Burnt Ground Wood, 31-May-74 (DRN)

SU22: Blackmoor Copse, 1965-1974 (MGM) (Morris, 1974, 1975); Langley Wood NNR, 20-May-74 (DRN)

SU24: Driver Training Area (SPTAE), 24-Jun-02 (PJH) (Hodge, 2002)

Resident. Widespread; [on hawthorns] and whitethorn.

**Strawberry Blossom Weevil**

***Anthonomus rubi*** (Herbst, 1795)

7 SU16: GranHam Hill , Marlborough district, 1873-1938 (Meyrick, 1938)

8 ST94: Cotley Hill, 1965-1974 (MGM) (Morris, 1974, 1975)

ST95: Bratton Castle, 1965-1974 (MGM) (Morris, 1974, 1975)

SU13: Great Durnford, 3-Aug-83 (DRN); Salisbury, Avon Valley Reserve, Apr-00 (MAS) (Darby, 2001)

SU14: Haxton (SPTAE), 17-May-02 (PJH) (Hodge, 2002)

SU16: Woodborough, 4-Aug-83 (DRN)

SU22: Blackmoor Copse, 9-Jun-01 (APF) (Foster, 2004)

SU23: Porton Down, 3-Jul-99 (PJH) (Hodge, 2002); 1965-1974 (MGM) (Morris, 1974, 1975)

SU24: Dunch Hill (SPTAE), 3-Jun - 12-Aug-08 (Boyce & Wright, 2009)

Resident. Widespread; on [various rosaceous herbs and shrubs] including strawberries and raspberries.

[***Anthonomus ulmi*** (De Geer, 1775)

**Nb**

7 SU16: Marlborough area, 1869-1938 (Meyrick, 1938) see *A.bituberculatus*]

Tribe CIONINI

**CIONUS Clairville, 1798**

***Cionus alauda*** (Herbst, 1784)

7 SU26: Folly Farm, 1869-1938 (Meyrick, 1938); Mildenhall, 1869-1938 (Meyrick, 1938)

8 ST85: Green Lane Wood, 7-Jun-95 (WANHS det MDD)

ST91: Cranborne Chase, 16 to 25-Jun-80 (NCC)

SU02: Bishopstone, 28-May-91 (MDD)

SU03: Grovely Wood, 4-Aug-78 (DRN)

SU05: Redhorn Hill, 1965-1974 (MGM) (Morris, 1974, 1975)

SU13: Great Durnford, 19-Aug-69, 7-Aug-70, 27-Jul-77, 4-Aug-78, 3-Aug-83, 6-Aug-87 (DRN); Idmiston, 5-Aug-82 (DRN); Lower Woodford, 28-Jul-78, 27-Jul-82 (DRN); Middle Woodford, 27-Jul-77 (DRN); Stratford sub Castle, 8-Aug-70 (DRN); Winterbourne Gunner, 1974-1978 (MGM) (Morris, 1979)

SU14: Figheldean, 28-Jul-82; Stonehenge, 22-Apr-03 (PJH) (Hodge, 2003)

SU15: Fittleton, 29-Jul-83 (DRN)

SU16: Pewsey, Kennet & Avon Canal, 1-Jun-04 (BLe); Woodborough, 4-Aug-83 (DRN)

SU23: Porton Down, 1974-1978 (MGM) (Morris, 1979)

Resident. Widespread; on figworts *Scrophularia nodosa* and *S. auriculata* in woodlands, roadsides and at the edges of water. [Occasionally on *Verbascum* spp.].

***Cionus hortulanus*** (Fourcroy, 1785)

7 SU16: Jones's Mill, 2-Jul-01 (APF) (Foster, 2002);

SU16: Marlborough area, 1965-1974 (MGM) (Morris, 1974, 1975)
8 ST93: Great Ridge Wood, 18-Aug-69 (DRN)
SU03: Grovely Wood, 13-Aug-69, 11-Aug-71, 30-May-74, 29-Jul-78, 4-Aug-87 (DRN)
SU05: Redhorn Hill, 1965-1974 (MGM) (Morris, 1974, 1975)
SU13: Winterbourne Gunner, 1974-1978 (MGM) (Morris, 1979)
SU14: Netheravon, 29-Jul-83 (DRN)
SU15: Fittleton, 29-Jul-83 (DRN)
SU16: Woodborough, 4-Aug-83 (DRN)
SU21: Burnt Ground Wood, 31-May-74 (DRN)
SU22: Bentley Wood, 31-May-96 (MDD); Langley Wood, 5-Aug-85 (DRN)
SU23: Porton Down, 1974-1978 (MGM) (Morris, 1979)
SU24: Driver Training Area (SPTAE), 17-May-02 (PJH) (Hodge, 2002)
SU25: Collingbourne Forest, 25-Aug-72 (DRN)
Resident. Widespread; on figworts *Scrophularia nodosa* and *S. auriculata* in woodlands, roadsides and in waste places. [Occasionally also on *Buddleja* and *Verbascum* spp.].

Cionus scrophulariae

**Figwort Weevil**
***Cionus scrophulariae*** (Linnaeus, 1758)
7 ST96/97: Bowood Lake, 24-Aug-86 (KNAA)
SU16: Jones's Mill, 19-Jun-01 (APF) (Foster, 2002); Wootton Rivers, 8-Jun-97 (MGT)
SU26: Mildenhall, 1869-1938 (MGM) (Morris, (M.G, 1979); Savernake Forest, 25-Apr to 26-May-93 (EN det SAW)
8 ST74: Longleat, 22-Jun-96 (EGS det MDD)
ST85: Green Lane Wood, 7-Jun-95 (WANHS det MDD)
ST91: Cranborne Chase, 16 to 25-Jun-80 (NCC); 8-Jul- 97 (MAS)
ST92: Sutton Mandeville, Jun-96 (MDD)
ST93: Great Ridge Wood, 18-Aug-69 (DRN)
SU02: Bishopstone, 28-May-91 (MDD)
SU03: Dinton, 15-Jun-96 (MDD); Grovely Wood, 29-Jul-78 (DRN)
SU13: Great Durnford, 3-Aug-83 (DRN); Lower Woodford, 28-Jul-78 (DRN); Salisbury, 1869-1938 (MGM) (Morris, 1979); Stratford sub Castle, 8-Aug-70 (DRN)
SU14: Figheldean, 28-Jul-82 (DRN); 8-Jun-97 (MGT)
SU16: Pewsey, Kennet & Avon Canal, 1-Jun-04 (BLe); Woodborough, 4-Aug-83 (DRN)
SU21: Landford Bog, 25-May-98 (MDD)
SU22: Bentley Wood, 5-Jun-04 (AGD); Blackmoor Copse, 31-Aug-01 (APF) (Foster, 2004)
Resident. Widespread; on figworts *Scrophularia nodosa* and *S. aquatica* in woodlands, roadsides and at the edges of water. [Occasionally also on other plants including *Buddleja* spp.].

***Cionus tuberculosus*** (Scopoli, 1763)
7 SU09: Cotswold Water Park, 10-Apr-97 (APF)
SU16: Knighton, nr. Marlborough, 1869-1938 (Anon) (Meyrick, 1938)
8 ST92: Sutton Mandeville, 1-Jun-96 (MDD)
SU02: Bishopstone, 28-May-91 (MDD)
SU03: Wylye, 1965-1974 (MGM) (Morris, 1974, 1975)
SU13: Great Durnford, 19-Aug-69, 22-Aug-70 (DRN); Lower Woodford, 28-Jul-78, 27-Jul-82 (DRN); Salisbury, 26-May-74 (DRN); 1965-1974 (MGM) (Morris, 1974, 1975); Stratford sub Castle, 8, 10-Aug-70 (DRN)
SU14: Figheldean, Jun-1870 (RH) (Hislop, 1871); 28-Jul-82 (DRN); Stonehenge, 22-Apr-03 (PJH) (Hodge, 2003)
SU15: Fittleton, 4-Aug-88 (DRN)
SU16: Jones's Mill, 19-Jun-01 (APF) (Foster, 2002); Woodborough, 4-Aug-83 (DRN)
SU21: Hamptworth Estate, 29-May-74 (DRN)
SU22: Bentley Wood, 12-Aug-87 (DRN)
SU23: Porton Down1974-1978 (MGM) (Morris, 1979)
Resident. Widespread; usually in wet and marshy places on figwort *Scrophularia auriculata*, less frequently in woodlands, roadsides etc..

**CLEOPUS Dejean, 1821**
***Cleopus pulchellus*** (Herbst, 1795)
7 SU16: Marlborough area, 1869-1938 (Meyrick, 1938)
8 ST91: Cranborne Chase, 22-Jul-97 (MAS); 30-Jun-99 (MDD)
SU02: Broad Chalke, 1965-1974 (MGM) (Morris, 1974, 1975)
SU16: Woodborough, 4-Aug-83 (DRN)
Resident. Widespread; on *Scrophularia nodosa* and *S. auriculata* in woodlands, roadsides and at the edges of water. [Occasionally on *Verbascum* spp.].

Tribe ELLESCINI
Subtribe ELLESCINA

**ELLESCUS Dejean, 1821**
***Ellescus bipunctatus*** (Linnaeus, 1758)
**Nb**
8 SU22: Blackmoor Copse, 1965-1974 (MGM) (Morris, 1974, 1975)
Resident. Very local; on various Salix Species, larvae in the catkins.

Subtribe DORYTOMINA

**DORYTOMUS Germar, 1817**
***Dorytomus dejeani*** Faust, 1882
7 SU36: Stype Wood, 1869-1938 (Meyrick, 1938)
8 ST91: Cranborne Chase, 14-Jun-97 (MAS)
SU13: Salisbury, Avon Valley Reserve, Apr-00 (MAS) (Darby, 2001)
SU22: Bentley Wood, 27-May-98 (MDD); Whiteparish Common, 1965-1974 (MGM) (Morris, 1974, 1975)
Resident. Local; on poplar, *Populus tremula.*

***Dorytomus longimanus*** (Forster, 1771)
7 SU16: Lockeridge, 1994 (JEO); Clatford, 1994 (JEO); Marlborough area, 1869-1938 (Meyrick, 1938)
8 SU03: Steeple Langford, Jul-04 (MDD)
SU14: Netheravon, 29-Jul-83 (DRN)
Resident. Very local; under willow bark and on poplar, often close to wetlands.

***Dorytomus rufatus*** (Bedel, 1888)
7 SU17: Clout's Wood, 1965-1974 (MGM) (Morris, 1974, 1975)
8 SU21: Hamptworth Common, 28-Jul-77 (DRN)
SU22: Barnridge Copse, nr Farley, 1869-1938 (MGM) (Morris, 1979); Holmere Common, 29-May-90 (DRN)
Resident. Very local; on sallows and other *Salix Spp.*

***Dorytomus salicinus*** (Gyllenhal, 1827)
**Nb**
7 SU16: Marlborough area, 1869-1938 (Meyrick, 1938)
Resident. Very local; on willow. One nspecimen in MCC without data (con. MGM)

***Dorytomus taeniatus*** (Fabricius, 1781)
7 ST99: Flisteridge Wood, 3-Oct-82 (KNAA)
SU17: Clout's Wood, 1965-1974 (MGM) (Morris, 1974, 1975)
SU26/27: Axford, 1869-1938 (Meyrick, 1938)
8 ST93: Stockton Wood, 1965-1974 (MGM) (Morris, 1974, 1975)
SU16: Jones's Mill, 30-May-04 (KNAA)
SU22: Blackmoor Copse, 1965-1974 (MGM) (Morris, 1974, 1975); Whiteparish Common, 1965-1974 (MGM) (Morris, 1974, 1975)
SU23: Porton Down, 1965-1974 (MGM) (Morris, 1974, 1975)
Resident. Widespread; on [a variety of broadleaved *Salix Spp.*].

***Dorytomus tortrix*** (Linnaeus, 1761)
7 SU17: Clout's Wood, 1965-1974 (MGM) (Morris, 1974, 1975)
8 SU22: Whiteparish Common, 1965-1974 (MGM) (Morris, 1974, 1975)
Resident. Very local; on poplar, *Populus tremula.*

Tribe MECININI

**CLEOPOMIARUS Pierce, 1919**
***Cleopomiarus graminis*** (Gyllenhal, 1813)
**Nb, LCGR-Fh Cl**
7 SU06: Calstone Down, 3-Jul-95 (KNAA); Cherhill Down, 1-Jun-04 (BLe)
SU16: Pewsey Downs, 1965-1974 (MGM) (Morris, 1974, 1975); Walkers Hill, 1965-1974 (MGM) (Morris, 1974, 1975)
8 ST92: Prescombe Down, 1965-1974 (MGM) (Morris, 1974, 1975)
ST94: Cotley Hill, 1965-1974 (MGM) (Morris, 1974, 1975); Ladywell (SPTAW), 3-Jun – 12-Aug-08 (Boyce & Wright, 2009)
SU01: Martin Down NNR, 1-Jan-79 (RGB); 1999 (EN, 1999)
SU03: Wylye Down NNR, 5-Jun to 10-Jul-07 (MDD) (Darby, 2009)
SU04: Shrewton, 30-Jul-82 (DRN); Blake's Firs, 1-Jan-79 (MGM) (Morris, 1979)
SU05: Black Heath (SPTAC), 3-Jun - 12-Aug-08 (Boyce & Wright, 2009)
SU14: Stonehenge Down, 5-Jul-82 (KNAA)
SU24: Bulford Ranges (SPTAE), 3-Jun - 12-Aug-08 (Boyce & Wright, 2009)
Resident. Widespread; [on bellflower, *Campanula* spp.].

***Cleopomiarus plantarum*** (Germar, 1824)
**RDBK, LCGR-Fm Cl**
7 SU16: Marlborough area, 1869-1938 (Meyrick, 1938)
Resident, possibly extinct. Very local; by sweeping. No specimen in MCC. Old records suggest that this species was formerly more widespread than now. There is only one recent record, from S. Yorkshire.

**GYMNETRON Schönherr, 1825**
***Gymnetron melanarium*** (Germar, 1821)
**Nb, LCGR-Fm Cl**
7 SU16: Marlborough area, 1869-1938 (Anon) (Morris, 1974, 1975)
SU17: Old Totterdown, nr Fyfield, 1965-1974

(MGM) (Morris, 1974, 1975)
8 ST94: Norton Down (SPTAW), 15-May-02 (PJH) (Hodge, 2002)
SU03: Grovely Wood, 11-Aug-71 (DRN)
SU04: Salisbury Plain, 16-May-02 (PJH) (Hodge, 2002)
SU14: Haxton (SPTAE), 17-May-02 (PJH) (Hodge, 2002); Wexland Hanging (SPTAC), 16-May-02 (PJH) (Hodge, 2002); 3-Jun - 12-Aug-08 (Boyce & Wright, 2009)
SU23: Porton Down, 1965-1974 (MGM) (Morris, 1974, 1975); 3-Jul-99 (PMP)
Resident. Widespread; [on germander speedwell, *Veronica chamaedrys*].

***Gymnetron rostellum*** (Herbst, 1795)
**Na**
8 SU03: Grovely Wood, 28-Jul-83 (DRN)
Resident. Very local; on vegetation near the OX Drove.

***Gymnetron veronicae*** (Germar, 1821)
**Nb**
7 SU16: Marlborough area, 1965-1974 (MGM) (Morris, 1974, 1975)
8 SU14: Netheravon, 29-Jul-83 (DRN)
Resident. Very local; on vegetation by the R. Avon [on brooklime, *Veronica beccabunga*]. No specimen in MCC.

***Gymnetron villosulum*** Gyllenhal, 1838
**Nb**
7 SU09: Clattinger Farm, 1965-1974 (MGM) (Morris, 1974, 1975)
SU26: Froxfield, 1869-1938 (Meyrick, 1938)
SU36: Stype Wood, Jun-1870 (RH) (Hislop, 1871)
8 SU13: Great Durnford, 19-Aug-69 (DRN); Idmiston, 5-Aug-82 (DRN); 1965-1974 (MGM) (Morris, 1974, 1975); Middle Woodford, 27-Jul-77 (DRN); Salisbury, 26-May-74 (DRN); Stratford sub Castle, 8-Aug-70 (DRN); Winterbourne Gunner, 1-Jan-79 (MGM) (Morris, 1979)
Resident. Local; on water-cress and other vegetation [host plants are water speedwell, *Veronica anagallisaquatica* and *V. catenata*].

**MECINUS Germar, 1821**
***Mecinus labilis*** (Herbst, 1795)
8 ST94: Bowls Barrow (SPTAW), 16-May, 22-Jun-02 (PJH) (Hodge, 2002); Ladywell (SPTAW), 15-May, 23-Jun-02 (PJH) (Hodge, 2002); Norton Down (SPTAW), 15-May-02 (PJH) (Hodge, 2002); Wadman's Coppice (SPTAW), 3-Jun - 12-Aug-08 (Boyce & Wright, 2009)
ST95: Bratton Castle, 1965-1974 (MGM) (Morris, 1974, 1975); Fore Hill, nr West Lavington, 1965-1974 (MGM) (Morris, 1974, 1975)
SU03: Grovely Wood, 1-Aug-73 (DRN)
SU04: Orcheston Down (SPTA), 18-May-02 (PJH) (Hodge, 2002); Salisbury Plain, 16-May-02 (PJH) (Hodge, 2002); Shrewton, 30-Jul, 6-Aug-82 (DRN)
SU13: Middle Woodford, 5-Aug-82 (DRN); Stratford sub Castle, 7-Aug-71 (DRN)
SU14: Haxton (SPTAE), 17-May-02 (PJH) (Hodge, 2002); Stonehenge, 22-Apr-03 (PJH) (Hodge, 2003); Wexland Hanging (SPTAC), 16-May-02, 22-Jun-02 (PJH) (Hodge, 2002); 3-Jun - 12-Aug-08 (Boyce & Wright, 2009)
SU15: Manningford, 13-Aug-87 (DRN)
SU24: Driver Training Area (SPTAE), 17-May, 27-Jun-02 (PJH) (Hodge, 2002)
Resident. Widespread; on vegetation including vetch, *Vicia cracca*, water dock and *Salix Spp.* in a wide range of habitats.

***Mecinus pascuorum*** (Gyllenhal, 1813)
**LCGR-Fl Cl**
7 ST96: Lacock, 16-Jun-82 (KNAA)
SU06: Calstone Down, 4-Jul-95 (KNAA); Morgan's Hill, 1965-1974 (MGM) (Morris, 1974, 1975)
SU16: Gopher Wood, 1965-1974 (MGM) (Morris, 1974, 1975); Pewsey Downs, 1965-1974 (MGM) (Morris, 1974, 1975); Walkers Hill, 1965-1974 (MGM) (Morris, 1974, 1975)
8 ST83: Whitesheet Down, 7-Jul-82 (KNAA)
ST91: Cranborne Chase, 14-Jun-97 (MAS)
ST92: Wardour, 5-Jun-96 (MAS)
ST93: Great Ridge Wood, 29-Oct-75 (DRN); Stockton Down, 1965-1974 (MGM) (Morris, 1974, 1975)
ST94: Bowls Barrow (SPTAW), 16-May, 22-Jun-02 (PJH) (Hodge, 2002); Cotley Hill, 1965-1974 (MGM) (Morris, 1974, 1975); Ladywell (SPTA), 23-Jun-02 (PJH) (Hodge, 2002); 3-Jun - 12-Aug-08 (Boyce & Wright, 2009); Norton Down (SPTAW), 15-May, 23-Jun-02 (PJH) (Hodge, 2002);Tinkers Track (SPTAW), 3-Jun - 12-Aug-08 (Boyce & Wright, 2009); Wadman's Coppice (SPTAW), 3-Jun - 12-Aug-08 (Boyce & Wright, 2009); Warminster Ranges (SPTAW), 3-Jun - 12-Aug-08 (Boyce & Wright, 2009)
SU01: Martin Down NNR, 21-Aug-89 (RGB)
SU03: Dinton Park, 6-Jul-00 (KNAA); Grovely Wood, 11, 13-Aug-70, 11-Aug-73 (DRN); Salisbury, 15-Aug-88 (DRN); Shrewton, 2-Aug-83 (DRN); Steeple Langford Cow Down, 1965-1974 (MGM) (Morris, 1974, 1975); Wylye Down NNR, 10-Jul-07 (MDD) (Darby, 2009)
SU04: Fox Covert (SPTAW), 3-Jun - 12-Aug-08 (Boyce & Wright, 2009); Orcheston Down (SPTA), 1965-1974 (MGM) (Morris, 1974,

1975), 18-May, 22-Jun-02 (PJH) (Hodge, 2002); Salisbury Plain, 16-May, 25-Jun-02 (PJH) (Hodge, 2002); Shrewton, 30-Jul, 6-Aug-82 (DRN); Tilshead Park (SPTAW), 14-Jul-99 (PJH) (Hodge, 1999)
SU05: Black Heath (SPTAC), 3-Jun - 12-Aug-08 (Boyce & Wright, 2009)
SU13: Camp Down, 1965-1974 (MGM) (Morris, 1974, 1975); Great Durnford, 11-Aug-69 (DRN); Salisbury, Avon Valley Reserve, 1-Apr-00 (MAS) (Darby, 2001); Winterbourne Gunner, 1974-1978 (MGM) (Morris, 1979)
SU14: Bulford Down, 1965-1974 (MGM) (Morris, 1974, 1975); Haxton (SPTAE), 17-May, 24-Jun-02 (PJH) (Hodge, 2002); Wexland Hanging (SPTAC), 16-May, 22-Jun-02 (PJH) (Hodge, 2002); 3-Jun - 12-Aug-08 (Boyce & Wright, 2009)
SU15: Everleigh (SPTAE), 18-May, 24-Jun-02 (PJH) (Hodge, 2002)
SU16: Jones's Mill, 1-Jan-97 (MAS)
SU23: Porton Down, 1965-1974 (MGM) (Morris, 1974, 1975)
SU24: Driver Training Area (SPTAE), 17-May, 24-Jun-02 (PJH) (Hodge, 2002); Dunch Hill (SPTAE), 3-Jun - 12-Aug-08 (Boyce & Wright, 2009)

Resident. Ubiquitous; [on ribwort plantain, *Plantago lanceolata*] in dry calcareous grassland. Once under bark.

***Mecinus pyraster*** (Herbst, 1795)

7 SU16: Marlborough area, 30 Mar 1879 (WFB) (Meyrick, 1880); 1869-1938 (Morris, 1974, 1975)
8 ST93: Fonthill, 19-Jul-95 (MDD)
ST94: Ladywell (SPTAW), 23-Jun-02 (PJH) (Hodge, 2002); Norton Down (SPTAW), 15-May-02 (PJH) (Hodge, 2002)
SU02: Knighton Hill, 1965-1974 (MGM) (Morris, 1974, 1975); Knighton Wood, Mar-07 (AJWA)
SU03: Grovely Wood, 11-Aug-70 (DRN); Wylye Down NNR, 4-May-07 (MDD) (Darby, 2008)
SU04: Orcheston Down (SPTA), 18-May, 22-Jun-02 (PJH) (Hodge, 2002); Salisbury Plain, 16-May-02 (PJH) (Hodge, 2002); Shrewton, 6-Aug-82 (DRN)
SU05: Redhorn Hill, 1965-1974 (MGM) (Morris, 1974, 1975)
SU13: Idmiston, 1974-1978 (MGM) (Morris, (M.G, 1979); Salisbury, 22-Aug-72 (DRN)
SU14: Haxton (SPTAE), 17-May-02 (PJH) (Hodge, 2002)
SU15: Everleigh (SPTAE), 18-May-02 (PJH) (Hodge, 2002)
SU24: Driver Training Area (SPTAE), 17-May-02 (PJH) (Hodge, 2002); North Tidworth, Devils Ditch, 30-May-04 (BLe)

Resident. Widespread; [on plantains] in calcareous grassland.

**MIARUS Schönherr, 1826**

***Miarus campanulae*** (Linnaeus, 1767)

**LCGR-Fh CI**

7 SU16: Poulton Down, Marlborough district, 1939 (det KGB) (Knight, 1940); Pewsey Downs, 24-Jun-83 (KNAA); 22-Jun-90 (APFw) (Meyrick, 1938); 1965-1974 (MGM) (Morris, 1974, 1975)
SU17: Fyfield Down, 1965-1974 (MGM) (Morris, 1974, 1975)
8 ST83: Brimsdown Hill, 1965-1974 (MGM) (Morris, 1974, 1975)
ST92: Prescombe Down, 1965-1974 (MGM) (Morris, 1974, 1975)
SU01: Martin Down NNR, 1965-1974 (MGM) (Morris, 1974, 1975); 1999 (EN, 1999)
SU02: Knapp and Barnetts Down, 1965-1974 (MGM) (Morris, 1974, 1975); Stratford Tony Down, 1965-1974 (MGM) (Morris, 1974, 1975)
SU03: Grovely Wood, 1, 4-Aug-78 (DRN); Wylye Down NNR, 10-Jul, 3-Aug-07 (MDD) (Darby, 2008)
SU13: Old Sarum, 1965-1974 (MGM) (Morris, 1974, 1975)
SU23: Porton Down, 1965-1974 (MGM) (Morris, 1974, 1975); 3-Jul-99 (PJH) (Hodge, 2002)

Resident. Widespread; [on bellflower, *Campanula* spp.].

**RHINUSA Stephens, 1829**

***Rhinusa antirrhini*** (Paykull, 1800)

7 SU26: Froxfield, 1927 (CPH) (Hurst, 1927)
8 ST83: Whitesheet Hill, 1965-1974 (MGM) (Morris, 1974, 19 75)
ST92: Prescombe Down, 1965-1974 (MGM) (Morris, 1974, 1975); Trow Down, 1965-1974 (MGM) (Morris, 1974, 1975)
ST93: Cratt Hill, 29-Jul-82 (DRN)
ST94: Bowls Barrow (SPTAW), 22-Jun-02 (PJH) (Hodge, 2002)
SU03: Grovely Wood, 1-Aug-73, 4-Aug-78, 1-Aug-88 (DRN)
SU04: Shrewton, 28-Jul, 6-Aug-82 (DRN); Salisbury Plain, 25-Jun-02 (PJH) (Hodge, 2002); Yarnbury Castle, 21-Jun-90 (APFw)
SU13: Camp Down, 1965-1974 (MGM) (Morris, 1974, 1975)
SU23: Porton Down, 3-Jul-99 (MAS)
SU24: Bulford Ranges (SPTAE), 3-Jun - 12-Aug-08 (Boyce & Wright, 2009)

Resident. Widespread; [on common toadflax, *Linaria vulgaris*].

***Rhinusa collina*** (Gyllenhal, 1813)

**Na**

8 ST94: Norton Down (SPTAW), 15-May-02 (PJH) (Hodge, 2002)
Resident. Very local; [associated with common toadflax, *Linaria vulgaris*].

***Rhinusa linariae*** (Panzer, 1795)
**Na**
8 SU15: Everleigh (SPTAE), 18-May-02 (PJH) (Hodge, 2002)
Resident. Very local; [on common toadflax, *Linaria vulgaris*].

Tribe RHAMPHINI

**ORCHESTES Illiger, 1798**
ORCHESTES Illiger, 1798
***Orchestes alni*** (Linnaeus, 1758)
7 SU08: Distillery Meadows, 9-Jun-91 (SP)
SU16: Marlborough area, 28 May 1879 (PHM) (Meyrick, 1880)
SU26: Mildenhall, 1869-1938 (Meyrick, 1938)
8 SU05: Redhorn Hill, 1965-1974 (MGM) (Morris, 1974, 1975)
SU06: All Cannings, 1869-1938 (Meyrick, 1938)
SU13: Great Durnford, 3-Aug-83, 6-Aug-87 (DRN)
Resident. Local; on elm.

***Orchestes pilosus*** (Fabricius, 1781)
7 ST86: Neston Park, 3-Jun-04 (BLe)
ST87: Corsham Park (ST8770), 3-Jun-04 (BLe)
SU16: Marlborough area, 1869-1938 (Anon) (Meyrick, 1938)
SU26: Savernake Forest, 27-May to 7-Jul-93, 28-Apr to 18-May, 1 to 19-Jul, 20-Sep to 11-Oct-94 (EN det SAW); 31-May-04 (BLe)
8 ST73: Stourhead, 4-Jul-90 (KNAA)
ST93: Pits Wood, Chilmark, 12-Jul-95 (MDD)
SU03: Grovely Wood, 11, 13-Aug-71 (DRN)
SU22: Bagfield Copse, 19-Aug-88 (DRN); Blackmoor Copse, 1974-1978 (MGM) (Morris, 1979); 17-Mar-02 (APF) (Foster, 2004); 3-Aug-03 (APF) (Foster, 2004); Landford, 12-Aug-87 (DRN); Langley Wood NNR, 27-Aug to 17-Sep-04 (MDD) (Darby, 2005)
SU23: Thorny Down Woods, 1965-1974 (MGM) (Morris, 1974, 1975)
Resident. Widespread; in woodland on oak and birch.

***Orchestes quercus*** (Linnaeus, 1758)
7 ST86: Neston Park, 3-Jun-04 (BLe)
SU16: Forest Hill, Marlborough district, 1869-1938 (Meyrick, 1938); Marlborough area, 6-Feb-1879 (TNH-S) (Meyrick, 1880); Pewsey Downs, 1965-1974 (MGM) (Morris, 1974, 1975); Walkers Hill, 1965-1974 (MGM) (Morris, 1974, 1975)
SU26: Savernake Forest, 27-May to 16-Jun, 29-Aug to 20-Sep-93; 28-Apr to 18-May, 20-Sep to 11-Oct-94 (EN det SAW); 31-May to 23 Jun, 15-Sep to 5-Oct-95 (EN det MDD)
8 ST73: Stourhead, 5-Jul-95 (KNAA)
ST91: Cranborne Chase, 29-Apr-97 (MAS)
ST93: Great Ridge Wood, 18-Apr-74 (DRN)
SU03: Grovely Wood, 11-Aug-71, 1-Aug-88 (DRN)
SU21: Burnt Ground Wood, 31-May-74 (DRN)
SU22: Bagfield Copse, 6-Aug-86 (DRN); Whiteparish Common, 1965-1974 (MGM) (Morris, 1974, 1975)
SU23: Thorny Down, Porton, 1974-1978 (MGM) (Morris, 1979)
Resident. Widespread; in woodland on oak.

***Orchestes rusci*** (Herbst, 1795)
7 SU26: Savernake Forest, 1965-1974 (MGM) (Morris, 1974, 1975)
8 ST91: Cranborne Chase, 7-May-97 (MAS)
ST93: Great Ridge Wood, 6-Aug-85 (DRN)
SU03: Grovely Wood, 1-Aug-73, 11-Aug-87, 1-Aug-88 (DRN)
SU21: Hamptworth Common, 28-Jul-77 (DRN)
SU22: Bagfield Copse, 6, 7-Aug-86 (DRN); Blackmoor Copse, 1965-1974 (MGM) (Morris, 1974, 1975); Holmere Common, 19-Aug-88 (DRN); Homan's Copse, 31-Jul-85, 19-Aug-88 (DRN); Landford, 12-Aug-87 (DRN); Whiteparish Common, 1965-1974 (MGM) (Morris, 1974, 1975)
Resident. Local; on birch.

***Orchestes signifer*** (Creutzer, 1799)
7 SU26: Savernake Forest, 28-Apr to 18-May-94 (EN det SAW); 11 to 30-May-95 (EN det MDD)
8 SU03: Grovely Wood, 11, 16-Aug-71 (DRN)
SU21: Hamptworth Estate, 25-Jul-73 (DRN)
SU22: Bagfield Copse, 7-Aug-86 (DRN)
SU23: Thorny Down Woods, 1965-1974 (MGM) (Morris, 1974, 1975)
SU25: Dod's Down, 1939 (det KGB) (Knight, 1940)
Resident. Local; in woodland on oak and hazel.

SALIUS Schrank, 1798
**Beech Leaf Miner**
***Orchestes fagi*** (Linnaeus, 1758)
7 ST86: Neston Park, 3-Jun-04 (BLe)
SU06: Morgan's Hill, 1965-1974 (MGM) (Morris, 1974, 1975)
SU26: Hen's Wood, 1869-1938 (Meyrick, 1938); Marlborough area, 12-Mar-1879 (WFB) (Meyrick, 1880)
SU17: Fyfield Down, 1-Jan-73 (MGM) (Morris, 1974, 1975)
SU26: Savernake Forest, 1869-1938 (Meyrick, 1938); 22-Aug-90 (AGD); 8 to 28-Jul-93, 28-Apr to 1-Jul-94 (EN det SAW); 31-May to 23-Jun, 4-Aug to

5-Oct-95 (EN det MDD)
8 ST73: Stourhead Woods, 11-Feb-84 (KNAA)
ST91: Cranborne Chase, 16 to 25-Jun-80 (NCC)
ST94: Cotley Hill, 1965-1974 (MGM) (Morris, 1974, 1975)
SU02: Knighton Wood, Mar-07 (AJWA)
SU03: Grovely Wood, 31-Aug-70, 11-Aug-71, 27-May-74, 3-Apr-75 (DRN); 12-Jun-00 (MDD); Little Langford Down, 14-Jul-04 (MDD)
SU05: Redhorn Hill, 1965-1974 (MGM) (Morris, 1974, 1975)
SU13: Great Durnford, 6-Apr-70 (DRN); Lower Woodford, 2-Aug-82 (DRN)
SU15: Everleigh (SPTAE), 18-May-02 (PJH) (Hodge, 2002)
SU22: Whiteparish Common, 1965-1974 (MGM) (Morris, 1974, 1975)
Resident. Widespread; in woodland on beech.

**RHAMPHUS Clairville, 1798**
***Rhamphus oxyacanthae*** (Marsham, 1802)
7 SU26: Savernake Forest, 31-May-04 (BLe)
8 SU04: Parsonage Down NNR, 2-Jul-08 (MDD); Shrewton, 1-Aug-85 (DRN)
Resident. Widespread; on hawthorn.

***Rhamphus pulicarius*** Clairville, 1798
7 SU16: Marlborough area, 1869-1938 (Meyrick, 1938)
8 ST93: Great Ridge Wood, 6-Aug-85 (DRN)
SU03: Dinton, 17-Nov-90 (SP); Grovely Wood, 21-Aug-72 (DRN)
SU21: Hamptworth Estate, 26-Jul-77 (DRN)
SU22: Glazier's Common, 16-Aug-88 (DRN)

Resident. Very local; on sallow [in damp areas].

**TACHYERGES Schönherr, 1825**
***Tachyerges pseudostigma*** (Tempère, 1982)
**RDBK**
8 SU22: Bagfield Copse, 7-Aug-86 (DRN); Holmere Common, 19-Aug-88 (DRN); Homan's Copse, 19-Aug-88 (DRN); Landford, 12-Aug-87 (DRN); Whaddon Common, 22-Jul-73 (DRN)
Resident. Very local; beaten off sallows and alder. Nationally only known from south Wiltshire. Potential local BAP species?

***Tachyerges salicis*** (Linnaeus, 1758)
7 SU16: Marlborough area, 1869-1938 (Meyrick, 1938)
8 SU12: Whaddon Common, 22-Jul-73 (DRN)
SU13: Salisbury, Avon Valley Reserve, Apr-00 (MAS) (Darby, 2001)
SU22: Bagfield Copse, 7-Aug-86 (DRN); Bentley Wood, 11-Aug-86 (DRN); Holmere Common, 19-Aug-88 (DRN); Homan's Copse, 31-Jul-85 (DRN); Landford, 12-Aug-87 (DRN)
Resident. Local; on Salix Spp. and birch.

***Tachyerges stigma*** (Germar, 1821)
8 SU26: Folly Farm, 1869-1938 (Meyrick, 1938)
SU13: Whaddon Common, 22-Jul-73 (DRN)
SU21: Landford Bog, 21-Aug-01, 22-Jun-03 (APF) (Foster, 2001, 2003)
SU22: Bagfield Copse, 6, 7-Aug-86 (DRN); Blackmoor Copse, 1965-1974 (MGM) (Morris, 1974, 1975); 31-Aug-02 (APF) (Foster, 2004); Homan's Copse, 31-Jul-85, 19-Aug-88 (DRN); Whiteparish Common, 1965-1974 (MGM) (Morris, 1974, 1975).
Resident. Local; on sallows and birches.

Tribe STOREINI

**PACHYTYCHIUS Jekel, 1861**
***Pachytychius haematocephalus*** (Gyllenhal, 1836)
**RDB1**
7 SU26: Savernake Forest, 1869-1938 (Meyrick, 1938)
Meyrick (1874) first records this weevil in Savernake Forest 'once'. Jebb repeats this in 1895 and Meyrick again in 1905. In 1938 Meyrick doesn't include 'once' and states 'on *Lotus corniculatus*'. [On exposed herb-rich grassland on steep-south facing slopes. Associated with bird's-foot trefoil, *Lotus corniculatus*. Adults at roots of grass tufts adjacent to the food plant]. Only known from one other site in South Hampshire where it has been recorded for over a hundred years. There is a single specimen without data in the MCC (conf. MGM).

Tribe STYPHLINI

**ORTHOCHAETES germar, 1824**
***Orthochaetes setiger*** (Beck, 1817)
**Nb, LCGR-Fm Ch**
7 ST06: Oldbury Castle, 1965-1974 (MGM) (Morris, 1974, 1975)
8 ST83: Brimsdown Hill, 1965-1974 (MGM) (Morris, 1974, 1975)
ST92: Prescombe Down, 1965-1974 (MGM) (Morris, 1974, 1975)
ST94: Cotley Hill, 1965-1974 (MGM) (Morris, 1974, 1975); Starveall Down, 1965-1974 (MGM) (Morris, 1974, 1975)
SU01: Martin Down NNR, 1965-1974 (MGM) (Morris, 1974, 1975)
SU02: Knapp and Barnetts Down, 1965-1974 (MGM) (Morris, 1974, 1975); Stratford Tony Down, 1965-1974 (MGM) (Morris, 1974, 1975); Woodminton Down, 1965-1974

(MGM) (Morris, 1974, 1975)
SU03: Steeple Langford Cow Down, 1965-1974 (MGM) (Morris, 1974, 1975); Wylye Down NNR, 4-May-07 (MDD) (Darby, 2008)
SU14: Bulford Down, 1965-1974 (MGM) (Morris, 1974, 1975); Stonehenge, 8-Apr-03 (PJH) (Hodge, 2003)
SU23: Porton Down, 1965-1974 (MGM) (Morris, 1974, 1975)
Resident. Widespread; [on a wide range of Compositae]. Larvae mine the leaves.

Tribe TYCHIINI

**SIBINIA Germar, 1817**
**[*Sibinia pellucens*** (Scopoli)
8 ST95/SU05: Devizes/Potterne, early June-1865 (JS) **]**
This insect was exhibited at a meeting of the Entomological Society 7-Nov-1864 by EWJ (as *Sibynes canus*) and discussed by Rye (1865) as follows: 'Two specimens... were taken by Mr Sidebotham...in a lane between Devizes and Pottern. Mr Sidebotham thinks they came off *Bryronia dioica* '. These are probably the specimens now in the NHM and the Manchester Museum.The species has not been taken in GB since and is not included in Duff (2008). It is abundant on the continent. (See p.9)

**TYCHIUS Germar, 1817**
*Tychius junceus* (Reich, 1797)
**LCGR-Fh Cm**
7 SU06: Calstone Down, 4-Jul-95 (KNAA); Oldbury Castle, 1965-1974 (MGM) (Morris, 1974, 1975); Morgan's Hill, 1965-1974 (MGM) (Morris, 1974, 1975)
SU17: Fyfield Down, 1965-1974 (MGM) (Morris, 1974, 1975)
8 ST83: Brimsdown Hill, 1965-1974 (MGM) (Morris, 1974, 1975)
ST94: Cotley Hill, 1965-1974 (MGM) (Morris, 1974, 1975); Norton Down, 23-Jun-02 (PJH) (Hodge, 2002); Wadman's Coppice (SPTAW), 3-Jun - 12-Aug-08 (Boyce & Wright, 2009)
SU02: Knapp and Barnetts Down, 1965-1974 (MGM) (Morris, 1974, 1975); Stratford Tony Down, 1965-1974 (MGM) (Morris, 1974, 1975)
SU04: Fox Covert, 25-Jun-02 (PJH) (Hodge, 2002); 3-Jun - 12-Aug-08 (Boyce & Wright, 2009); Salisbury Plain, 25-Jun-02 (PJH) (Hodge, 2002)
SU05: Black Heath (SPTAC), 3-Jun - 12-Aug-08 (Boyce & Wright, 2009); Ell Barrow (SPTAC), 3-Jun - 12-Aug-08 (Boyce & Wright, 2009)
SU14: Stonehenge, 22-Apr-03 (PJH) (Hodge, 2003)
SU23: Porton Down, 3-Jul-99 (MAS)
SU24: Beacon Hill (SPTAE), 3-Jun - 12-Aug-08 (Boyce & Wright, 2009)
Resident. Widespread; probably on [low growing Papilionaceae].

*Tychius lineatulus* Stephens, 1831
**Na, LCGR-Fl Cl**
7 SU16: Marlborough area, 1869-1938 (Anon) (Meyrick, 1938)
8 SU03: Steeple Langford Cow Down, 1965-1974 (MGM) (Morris, 1974, 1975)
SU04: Parsonage Down, 1965-1974 (MGM) (Morris, 1974, 1975)
SU14: Haxton, 24-Jun-02 (PJH) (Hodge, 2002)
Resident. Local; [probably on Papilionaceae]. Larvae unknown.

*Tychius meliloti* Stephens, 1831
**LCGR- Fl C-**
8 SU04: Tidworth Pennings, 1965-1974 (MGM) (Morris, 1974, 1975)
ST94: Ladywell (SPTAW), 23-Jun-02 (PJH) (Hodge, 2002); Tinkers Track (SPTAW), 3-Jun-12-Aug-08 (Boyce & Wright, 2009)
SU04: Shrewton, 6-Aug-82, 1-Aug-85 (DRN)
SU05: Black Heath (SPTAC), 3-Jun - 12-Aug-08 (Boyce & Wright, 2009)
SU15: Everleigh (SPTAE), 18-May-02 (PJH) (Hodge, 2002)
SU24: Driver Training Area (SPTAE), 24-Jun-02 (PJH) (Hodge, 2002)
Resident. Local; on *Melilotus* spp..

*Tychius picirostris* (Fabricius, 1787)
7 SU06: Calstone Down, 22-Aug-95 (KNAA); Morgan's Hill, 1965-1974 (MGM) (Morris, 1974, 1975)
SU16: Gopher Wood, 1965-1974 (MGM) (Morris, 1974, 1975); Marlborough area, 1869-1938 (Morris, 1974, 1975); Pewsey Downs, 1965-1974 (MGM) (Morris, 1974, 1975); Walkers Hill, 1965-1974 (MGM) (Morris, 1974, 1975)
SU17: Fyfield Down, 1-Jan-73 (MGM) (Morris, 1974, 1975); Clout's Wood, 1965-1974 (MGM) (Morris, 1974, 1975)
8 ST91: Cranborne Chase, 30-Jul-97 (MAS)
ST93: Stockton Down, 1965-1974 (MGM) (Morris, 1974, 1975)
ST94: Starveall Down, 1965-1974 (MGM) (Morris, 1974, 1975); Cotley Hill, 1965-1974 (MGM) (Morris, 1974, 1975); Norton Down, 23-Jun-02 (PJH) (Hodge, 2002)
SU01: Martin Down NNR, 21-Aug-89 (RGB)
SU02: Knapp and Barnetts Down, 1965-1974 (MGM) (Morris, 1974, 1975); Knighton Wood, Jul-08 (AJWA)

SU03: Little Langford, 3-Jul-96 (MAS); Wylye Down NNR, 5-Jun-07 (MDD) (Darby, 2008)
SU04: Fox Covert (SPTAW), 3-Jun - 12-Aug-08 (Boyce & Wright, 2009)Parsonage Down NNR, 2-Jul-08 (MDD); Shrewton, 2-Aug-83 (DRN); Tidworth Pennings, 1965-1974 (MGM) (Morris, 1974, 1975)
SU05: Black Heath (SPTAC), 3-Jun - 12-Aug-08 Boyce & Wright, 2009); Redhorn Hill, 1965-1974 (MGM) (Morris, 1974, 1975)
SU16: Woodborough, 4-Aug-83 (DRN)
SU22: Bentley Wood, 27-May-97 (MDD); Whiteparish Common, 1965-1974 (MGM) (Morris, 1974, 1975)
SU24: Dunch Hill (SPTAE), 3-Jun - 12-Aug-08 (Boyce & Wright, 2009)
Resident. Widespread; [larvae in pods of bird's-foot trefoil, *Lotus corniculatus*, and possibly clovers].

***Tychius schneideri*** (Herbst, 1795)
**Nb, LCGR-Fm Ch**
8 SU14: Stonehenge, 22-Apr-03 (PJH) (Hodge, 2003)
Resident. Very local; in a suction sample. The foodplant is kidney vetch, *Anthyllius vulneraria.*

***Tychius squamulatus*** Gyllenhal, 1836
**Nb, LCGR-Fm Ch**
8 ST94: Ladywell (SPTAW), 3-Jun - 12-Aug-08 (Boyce & Wright, 2009)
Resident. Very local; in grassland [on bird'sfoot-trefoil, Lotus corniculatus but also found at roots of kidney vetch *Anthyllis vulneraria*].

***Tychius stephensi*** Gyllenhal, 1836
**LCGR-Fm Ch**
8 ST83: Maiden Bradley, 1965-1974 (MGM) (Morris, 1974, 1975)
ST94: Bowls Barrow (SPTAW), 22-Jun-02 (PJH) (Hodge, 2002); Cotley Hill, 1965-1974 (MGM) (Morris, 1974, 1975); Norton Down, 23-Jun-02 (PJH) (Hodge, 2002)
SU02: Stratford Tony Down, 1965-1974 (MGM) (Morris, 1974, 1975); Knapp and Barnetts Down, 1965-1974 (MGM) (Morris, 1974, 1975)
SU04: Fox Covert, 25-Jun-02 (PJH) (Hodge, 2002); 3-Jun - 12-Aug-08 (Boyce & Wright, 2009); Orcheston Down (SPTA), 22-Jun-02 (PJH) (Hodge, 2002); Salisbury Plain, 16-May, 25-Jun-02 (PJH) (Hodge, 2002); Shrewton, 1-Aug-85 (DRN)
SU15: Everleigh (SPTAE), 24-Jun-02 (PJH) (Hodge, 2002)
SU22: Bentley Wood, 2-Jul-96 (MDD)
Resident. Widespread; on clovers.

***Tychius tibialis*** Boheman, 1843
**Na, LCGR-Fl Cl**
8 SU13: Salisbury, Avon Valley Reserve, 1-Apr-00 (MAS) (Darby, 2001)
Resident. Very local; in wetland. This is a remarkable record for a species that is primarily found on coastal grassland.

## Subfamily BAGOINAE

**BAGOUS Germar, 1817**
BAGOUS Germar, 1817
***Bagous tempestivus*** (Herbst, 1795)
**Nb**
8 SU13: Lower Woodford, 27-Jul-82 (DRN)
Resident. Very local; in flood refuse by a feeder to the R. Avon.

## Subfamily BARIDINAE
### Tribe BARIDINI

**AULACOBARIS Desbrochers, 1892**
***Aulacobaris lepidii*** Germar, 1824
**Na**
8 ST92: Sutton Mandeville, Jul-92 (MDD conf. MGM)
SU13: Salisbury, Avon Valley Reserve, Apr -00 (MAS) (Darby, 2001)
Resident. Very local; associated with various cresses in wetland habitats (but Sutton Mandeville example flying in garden away from wetland). Only recorded from six other vice counties in recent years.

***Aulacobaris picicornis*** (Marsham, 1802)
**Nb, LCGR-Fh Cl**
7 SU16: Marlborough area, 1869-1938 (Meyrick, 1938)
8 ST94: Bowls Barrow (SPTAW), 16-May, 22-Jun-02 (PJH) (Hodge, 2002); Ladywell (SPTAW), 3-Jun - 12-Aug-08 (Boyce & Wright, 2009)
SU04: Salisbury Plain, 25-Jun-02 (PJH) (Hodge, 2002)
SU14: Stonehenge, 8, 22-Apr-03 (PJH) (Hodge, 2003)
SU25: Sidbury Hill (SPTAE), 5-Jun-97 (MGT)
Resident. Local; [on disturbed ground] and grassland particularly on calcareous soils. Associated with wild mignonette, *Reseda lutea.*

**BARIS Germar, 1817**
**[*Baris chlorizans*** Germar
8 SU06: Devizes nd. c.1866-70 (JS)**]**
Described by Sharp 1871 as 'taken by Mr Sidebotham at Devizes'. There has been no record in GB since and the species is not included in Duff (2008). (See p.9.).

### Tribe MADOPTERINI

**LIMNOBARIS Bedel, 1885**
***Limnobaris dolorosa*** (Goeze, 1777)
7 SU16: Jones's Mill, 2-Jul-01 (APF) (Foster, 2002); 30-May-04 (KNAA)

***Limnobaris t-album*** (Linnaeus, 1758)
7 SU16: Chilton, nr Marlborough, 1869-1938 (Anon)
8 SU04: Shrewton, 2-Aug-83 (DRN)
SU12: Charlton-All-Saints, 2-Jun-95, 9-May-96 (MAS)
SU13: Great Durnford, 6-Aug-71, 27-Jul-77, (DRN); Salisbury, 26-May-74, 15-Aug-88 (DRN); Stratford sub Castle, 7-Aug-81 (DRN)
SU14: Netheravon, 29-Jul-83 (DRN)
SU16: Jones's Mill, 1-Jan-97 (MAS); 21-Jun-97 (MDD); 19-Jun, 2-Jul-01 (APF) (Foster, 2002)
Resident. Local; on vegetation by water [sedge, *Carex Spp.*].

Subfamily CEUTORHYNCHINAE
Tribe CEUTORHYNCHINI

**AMALORRHYNCHUS Reitter, 1913**
***Amalorrhynchus melanarius*** (Stephens, 1831)
7 SU17: Clout's Wood, 1965-1974 (MGM) (Morris, 1974, 1975)
8 SU12: Charlton-All-Saints, 9-May-96 (MAS)
SU13: Idmiston, 5-Aug-82, 13-Aug-87 (DRN); 1965-1974 (MGM) (Morris, 1974, 1975)
SU13: Winterbourne Gunner, 1974-1978 (MGM) (Morris, 1979)
Resident. Local; on watercress.

**CALOSIRUS Thomson, C.G, 1859**
***Calosirus terminatus*** (Herbst, 1795)
**Nb, LCGR-FI CI**
8 ST83: Whitesheet Hill, 1965-1974 (MGM) (Morris, 1974, 1975)
SU02: Middleton Down, 1965-1974 (MGM) (Morris, 1974, 1975)
SU03: Grovely Wood, 14-Apr-71 (DRN)
SU13: Camp Down, 1965-1974 (MGM) (Morris, 1974, 1975)
SU24: Driver Training Area (SPTAE), 24-Jun-02 (PJH) (Hodge, 2002)
Resident. Local; [on *Daucus carota*, but recorded from a wide variety of other Resedaceae] on calcareous soils [also by the sea].

**CEUTORHYNCHUS Germar, 1824**
***Ceutorhynchus alliariae*** Brisout, H., 1860
8 SU14: Stonehenge, 22-Apr-03 (PJH) (Hodge, 2003)
Resident. Very local; [on Aliaria petiolata at the sides of roads, tracks and paths, and on shady banks and at the margins of woods].

**Cabbage Gall Weevil**
***Ceutorhynchus assimilis*** (Paykull, 1792)
7 ST86: Neston Park, 3-Jun-04 (BLe)
SU16: Combe Farm, Marlborough district, 1873-1938 (Meyrick, 1938); Gopher Wood, 1965-1974 (MGM) (Morris, 1974, 1975); Marlborough area, 1869-1938 (Meyrick, 1938)
SU17: Clout's Wood, 1965-1974 (MGM) (Morris, 1974, 1975)
8 ST83: Brimsdown Hill, 1965-1974 (MGM) (Morris, 1974, 1975)
ST91: Cranborne Chase, 7-May-97 (MAS)
ST94: Bowls Barrow (SPTAW), 16-May-02 (PJH) (Hodge, 2002); Ladywell (SPTAW), 15-May-02 (PJH) (Hodge, 2002)
SU02: Wilton, 12-May-97 (MDD)
SU04: Salisbury Plain, 16-May-02 (PJH) (Hodge, 2002)
SU12: Charlton-All-Saints, 9-May-96 (MAS)
SU13: Salisbury, Avon Valley Reserve, 1-Apr-00 (MAS) (Darby, 2001)
SU16: Jones's Mill, 19-Jun-01 (APF) (Foster, 2002)
SU23: Thorny Down, Porton, 1974-1978 (MGM) (Morris, 1979)
SU24: Driver Training Area (SPTAE), 17-May-02 (PJH) (Hodge, 2002)
Resident. Widespread; [on a wide variety of cultivated and wild Brassicaceae].

***Ceutorhynchus chalybaeus*** Germar, 1824
8 SU12: Charlton-All-Saints, 9-May-96 (MAS)
Resident. Very local; [mainly on *Sisymbrium officiinale* at The sides of roads, tracks and paths and in open and grassy situations]. This is a variable species which has caused problems with identification.

***Ceutorhynchus cochleariae*** (Gyllenhal, 1813)
8 SU22: Whiteparish Common, 1965-1974 (MGM) (Morris, 1974, 1975)
Resident. Very local; [in wet meadows and grasslands, streamsides and the margins of lakes and ponds, and in open damp woodland, where the food plants, *Cardamine* spp., live].

***Ceutorhynchus constrictus*** (Marsham, 1802)
**Nb**
8 SU14: Stonehenge, 22-Apr-03 (PJH) (Hodge, 2003)
SU23: Porton Down, 1965-1974 (MGM) (Morris, 1974, 1975)
Resident. Very local; [on *Alliaria petiolata* at sides of roads, etc. and on shady banks and at the margins of woods].

***Ceutorhynchus contractus*** (Marsham, 1802)
7 ST06: Oldbury Castle, 1965-1974 (MGM) (Morris,

1974, 1975)
SU16: East Kennet, 1965-1974 (MGM) (Morris, 1974, 1975); Marlborough area, 4-May-1879 (WFB) (Meyrick, 1880); 1869-1938 (Meyrick, 1938); Silbury Hill, 1965-1974 (MGM) (Morris, 1974, 1975); Walkers Hill, 1965-1974 (MGM) (Morris, 1974, 1975)
SU17: Clout's Wood, 1965-1974 (MGM) (Morris, 1974, 1975); Fyfield Down NNR, 1965-1974 (MGM) (Morris, 1974, 1975); Hackpen Hill, 1965-1974 (MGM) (Morris, 1974, 1975)
SU26: Durley, 1965-1974 (MGM) (Morris, 1974, 1975)
8 ST15: Manningford, 1-Jan-02 (Meyrick, 1902)
ST83: Brimsdown Hill, 1965-1974 (MGM) (Morris, 1974, 1975)
ST92: Prescombe Down, 1965-1974 (MGM) (Morris, 1974, 1975)
ST94: Ladywell (SPTAW), 23-Jun-02 (PJH) (Hodge, 2002); Norton Down (SPTAW), 23-Jun-02 (PJH) (Hodge, 2002)
SU02: Knapp and Barnetts Down, 1965-1974 (MGM) (Morris, 1974, 1975)
SU03: Grovely Wood, 11-Aug-70, 11-Aug-71 (DRN)
SU04: Tidworth Pennings, 1965-1974 (MGM) (Morris, 1974, 1975)
SU13: Great Durnford, 16-Aug-69, 7-Aug-70 (DRN)
SU23: Thorny Down, Porton, 1974-1978 (MGM) (Morris, 1979)
SU26: Burbage, 1965-1974 (MGM) (Morris, 1974, 1975)

Resident. Widespread; [on a wide variety of Brassicaceae in waste and open places, roadside verges and field margins].

***Ceutorhynchus erysimi*** (Fabricius, 1787)

7 SU16: Gopher Wood, 1965-1974 (MGM) (Morris, 1974, 1975)
SU26: Savernake Forest, 1965-1974 (MGM) (Morris, 1974, 1975)
8 ST94: Cotley Hill, 1965-1974 (MGM) (Morris, 1974, 1975); Norton Down (SPTAW), 15-May, 23-Jun-02 (PJH) (Hodge, 2002)
SU03: Grovely Wood, 24-Aug-72, 4-Aug-78 (DRN)
SU04: Shrewton, 6-Aug-82 (DRN)
SU13: Idmiston, 13-Aug-87 (DRN); Stratford Sub Castle, 8-Aug-70 (DRN)
SU15: Manningford, 13-Aug-87 (DRN)
SU16: Jones's Mill, 1997 (MAS)
SU22: Redlynch, 4-Aug-82 (DRN)

Resident. Widespread; on mustard, watercress, weeds, etc, [on *Capsella bursa-pastoris* exclusively in the British Isles].

***Ceutorhynchus hirtulus*** Germar, 1824

**Nb**

7 SU16: Marlborough area, 6-Feb-1879 (WFB) (Meyrick, 1880)

Resident. Very local; on Cruciferae. [On whitlow grass, *Erophila verna,* in arenaceous areas, particularly maritime sand dunes, though occasionally away from such places]. No specimen in the MCC.

***Ceutorhynchus obstrictus*** (Marsham, 1802)

8 ST94: Norton Down (SPTAW), 23-Jun-02 (PJH) (Hodge, 2002)
SU02: Wilton, 6-Sep-95 (MDD det RT)
SU03: Grovely Wood, 8-Apr-74, 4-May-78, 28-Jul-83 (DRN); Wilton, 23-May-04 (AGD)
SU04: Shrewton, 2, 6-Aug-83, 1-Aug-85 (DRN)
SU13: Great Durnford, 7-Aug-70 (DRN); Idmiston, 5-Aug-82 (DRN); Salisbury, 26-May-74 (DRN); Stratford sub Castle, 8-Aug-70, 3-Aug-83 (DRN)
SU14: Stonehenge, 22-Apr-03 (PJH) (Hodge, 2003)

Resident. Widespread; [on a very wide range of Brassicaceae in arable fields, gardens, allotments and in all kinds of grassy and open places].

***Ceutorhynchus pallidactylus*** (Marsham, 1802)

7 ST96: Lacock, 16-Jun-82 (KNAA)
SU06: Calstone Down, 22-Aug-95 (KNAA)
SU16: East Kennet, 1965-1974 (MGM) (Morris, 1974, 1975); Marlborough area, 1869-1938 (Meyrick, 1938)
SU17: Clout's Wood, 1965-1974 (MGM) (Morris, 1974, 1975); Fyfield Down, 1-Jan-73 (MGM) (Morris, 1974, 1975)
8 ST94: Bowls Barrow (SPTAW), 16-May, 23-Jun-02 (PJH) (Hodge, 2002)
SU02: Wilton, 6-Sep-95 (MDD)
SU03: Grovely Wood, 16-Aug-71, 1-Aug-73, 30-May-74, 3-Apr-75, 4-Aug-78, 26-Jul-82, 1-Aug-88 (DRN)
SU04: Shrewton, 6-Aug-82 (DRN)
SU05: Chirton Gorse, 1965-1974 (MGM) (Morris, 1974, 1975)
SU13: Great Durnford, 3-Aug-83 (DRN); Idmiston, 5-Aug-82, 13-Aug-87 (DRN); Salisbury, 26-May-74 (DRN); 1-Apr-00 (MAS)
SU14: Stonehenge, 22-Apr-03 (PJH) (Hodge, 2003); Wexland Hanging (SPTAC), 22-Jun-02 (PJH) (Hodge, 2002)
SU21: Burnt Ground Wood, 31-May-74 (DRN)
SU22: Blackmoor Copse, 13-Jul-03 (APF) (Foster, 2004); Whiteparish Common, 1965-1974 (MGM) (Morris, 1974, 1975)
SU23: Porton Down, 1965-1974 (MGM) (Morris, 1974, 1975)
SU24: Driver Training Area (SPTAE), 17-May-02 (PJH) (Hodge, 2002)
SU26: Bedwyn Brail, 5-Nov-03 (MDD); Burbage, 1965-1974 (MGM) (Morris, 1974, 1975)

Resident. Widespread; [on virtually all spp of Brassicaceae at sides of raods, etc, agricultural land nd in rough an weedy places].

***Ceutorhynchus sulcicollis*** (Paykull, 1800)
7 SU16: Marlborough area, 1869-1938 (Meyrick, 1938)
Resident. Very local; on [hedge mustard]*Sisybrium officinale.*

***Ceutorhynchus typhae*** (Herbst, 1795)
7 SU16: Marlborough area, 1869-1938 (Meyrick, 1938)
SU17: Clout's Wood, 1965-1974 (MGM) (Morris, 1974, 1975)
8 ST94: Bowls Barrow (SPTAW), 16-May, 22-Jun-02 (PJH) (Hodge, 2002); Norton Down (SPTAW), 23-Jun-02 (PJH) (Hodge, 2002)
SU02: Knighton Wood, May-07 (AJWA)
SU03: Wylye, 1965-1974 (MGM) (Morris, 1974, 1975)
SU04: Shrewton, 30-Jul-82 (DRN)
SU13: Idmiston, 13-Aug-87 (DRN); Great Durnford, 22-Aug-70 (DRN); Winterbourne Gunner, 1974-1978 (MGM) (Morris, (M.G, 1979)
SU14: Stonehenge, 22-Apr-03 (PJH) (Hodge, 2003)
SU16: Jones's Mill, 19-Jun-01 (APF) (Foster, 2002)
SU22: Redlynch, 4-Aug-82 (DRN); Whiteparish Common, 1965-1974 (MGM) (Morris, 1974, 1975)
SU23: Porton Down, 3-Jul-99 (PJH) (Hodge, 2002); Thorny Down, Porton, 1974-1978 (MGM) (Morris, 1979)
Resident. Widespread; [on a wide variety of Brassicaceae in waste places, road sides, field margins, etc.]. Formerly *C.floralis.*

***Ceutorhynchus unguicularis*** Thomson, C. G, 1871
**RDB3, LCGR-Fm Cl**
7 SU16: Silbury Hill, 7-Jun-66 (MGM) (Morris, 1968)
Resident. Very local; [on hairy rock cress, *Arabis hirsuta*, in disturbed grassland and rocky places usually on calcareous soils]. Elsewhere in GB known only from W.Suffolk and Co. Clare, Ireland, but probably overlooked.

**COELIODES Schönherr, 1837**
***Coeliodes rana*** (Fabricius, 1787)
7 SU16: Pumphrey's Wood, nr Marlborough, 1873-1938 (Meyrick, 1938)
7 SU26: Puthall Wood, 1869-1938 (Meyrick, 1938); Savernake Forest, 1869-1938 (Meyrick, 1938)
8 ST91: Cranborne Chase, 4-Jul-97 (MAS); 30-Jun-99 (MDD)
SU03: Grovely Wood, 18-Jul-73 (DRN)
SU22: Blackmoor Copse, 1965-1974 (MGM) (Morris, 1974, 1975); Whiteparish, 22-Jul-73 (DRN); Whiteparish Common, 1965-1974 (MGM) (Morris, 1974, 1975)
SU23: Thorny Down, Porton, 1974-1978 (MGM) (Morris, 1979)
Resident. Widespread; on oaks. Formerly *C. dryados*

***Coeliodes ruber*** (Marsham, 1802)
**Nb**
8 SU03: Grovely Wood, 28-May-86 (DRN)
8 SU23: Thorny Down, Porton, 1869-1938 (MGM) (Morris, 1979)
Resident. Very local; on oaks.

***Coeliodes transversealbofasciatus*** (Goeze, 1777)
**Nb**
7 SU26: Savernake Forest, 29-May-76 (DHM) (Murray, 1978)
8 SU03: Grovely Wood, 1-Aug-88 (DRN)
SU22: Whiteparish Common, 1965-1974 (MGM) (Morris, 1974, 1975)
SU23: Thorny Down, Porton, 1974-1978 (MGM) (Morris, 1979)
Resident. Local; on oaks.

**COELIODINUS Dieckmann, 1972**
***Coeliodinus rubicundus*** (Herbst, 1795)
7 SU26: Mildenhall, 1-Jan-40 (Knight, 1940)
8 SU13: Idmiston, 5-Aug-82, 13-Aug-87 (DRN); 1965-1974 (MGM) (Morris, 1974, 1975); Winterbourne Gunner, 1974-1978 (MGM) (Morris, 1979)
SU21: Burnt Ground Wood, 19-Apr-74 (DRN)
SU22: Bagfield Copse, 19-Aug-88 (DRN); Blackmoor Copse, 17-May-02, 3-Aug-03 (APF) (Foster, 2004); 1965-1974 (MGM) (Morris, 1974, 1975); Whiteparish Common, 1965-1974 (MGM) (Morris, 1974, 1975)
SU25: Dod's Down, 1939 (det KGB) (Knight, 1940)
Resident. Widespread; on birch.

**DATONYCHUS Wagner, 1944**
***Datonychus melanostictus*** (Marsham, 1802)
7 SU16: Poulton, Marlborough district, 1869-1938 (Meyrick, 1938)
8 ST93: Fonthill, 19-Jul-95 (MDD)
SU13: Great Durnford, 16-Aug-69, 6-Aug-71 (DRN); Stratford sub Castle, 10-Aug-70 (DRN)
Resident. Local; on gypsy-wort and mint at the edge of water.

**DRUPENTATUS Reitter, 1913**
***Drupenatus nasturtii*** (Germar, 1824)
**Nb**
7 SU17: Clout's Wood, 1965-1974 (MGM) (Morris, 1974, 1975)
SU26: Mildenhall, 1939 (det KGB) (Knight, 1940)

SU13: Idmiston, 5-Aug-82, 13-Aug-87 (DRN)
Resident. Very local; [at sides of water on white - flowered species of water cress, *Rorippa*.

Drupentatus nasturtii

**GLOCIANUS Reitter, 1916**
***Glocianus distinctus*** (Brisout, 1870)
8
ST94: Cotley Hill, 1965-1974 (MGM) (Morris, 1974, 1975)
SU04: Salisbury Plain, 16-May-02 (PJH) (Hodge, 2002)
SU14: Wexland Hanging (SPTAC) 22-Jun-02 (PJH) (Hodge, 2002)
Resident. Very local; [in grasslands, on roadside verges and at edges of tracks, fields and paths]. Foodplant is not known with certainty.

***Glocianus moelleri*** (Thomson, C. G, 1868)
**RDBK, LCGR-Fh Cl**
8 ST94: Bowls Barrow, 22-Jun-02 (PJH) (Hodge, 2002)
Resident. Very local; in a suction sample. Not often recorded from other than chalk grassland in GB. Host plants are uncertain but probably species of *Hieracium* or *Leotodon*.

***Glocianus pilosellus*** (Gyllenhal, 1837)
**RDB2, LCGR-Fm Cl**
8 SU14: Haxton, 24-Jun-02 (PJH) (Hodge, 2002)
Resident. Very local; one male in a suction sample. [In sandy areas inland as well as on the coast, on lesser dandelion, *Taraxacum laevigatum*].

***Glocianus punctiger*** (Gyllenhal, 1837)
**Nb**
7 SU06: Cherhill Down, 1-Jun-04 (BLe)
8 SU04: Shrewton, 2-Aug-83 (DRN)
SU14: Haxton (SPTAE), 3-Jun - 12-Aug-08 (Boyce & Wright, 2009)
Resident. Very local; in a suction sample and at roots. [On species of Dandelion, *Taraxacum*].

**HADROPLONTUS Thomson, C.G., 1859**
***Hadroplontus litura*** (Fabricius, 1775)
7 SU16: Gopher Wood, 1965-1974 (MGM) (Morris, 1974, 1975); Monkton Down, 1965-1974 (MGM) (Morris, 1974, 1975); Pewsey Down, 1965-1974 (MGM) (Morris, 1974, 1975)
8 ST94: Cotley Hill, 1965-1974 (MGM) (Morris, 1974, 1975)
SU03: Grovely Wood, 24-Aug-72 (DRN)
SU04: Shrewton, 1-Aug-85 (DRN)
SU05: Redhorn Hill, 1965-1974 (MGM) (Morris, 1974, 1975)
SU21: Franchises Common, 25-Jul-73 (DRN)
SU24: Driver Training Area (SPTAE), 24-Jun-02 (PJH) (Hodge, 2002)
Resident. Local; on thistles [particularly *Cirsium arvense*].

***Hadroplontus trimaculatus*** (Fabricius, 1775)
**Nb, LCGR-Fm Cl**
8 SU04: Salisbury Plain, 25-Jun-02 (PJH) (Hodge, 2002)
SU14: Haxton (SPTAE), 3-Jun - 12-Aug-08 (Boyce & Wright, 2009)
SU15: Everleigh (SPTAE), 24-Jun-02 (PJH) (Hodge, 2002)
Resident. Very local; [mainly on thistles particularly *Carduus nutans*]

**MICRELUS Thomson, C.G, 1859**
**Small Heather Weevil**
***Micrelus ericae*** (Gyllenhal, 1813)
8 SU21: Franchises Common, 25-Jul-73 (DRN); Hamptworth Common, 28-Jul-77 (DRN); Landford Bog, 24-Jun-01, 22-Jun-03 (APF) (Foster, 2003)
SU22: Langley Wood NNR, 29-May-03 (MDD)
SU23: Porton Down, 1965-1974 (MGM) (Morris, 1974, 1975); Roche Court Down, 1965-1974 (MGM) (Morris, 1974, 1975)
Resident. Local; on heather, *Calluna vulgaris* and species of *Erica*.

*Microplontus campestris* **(Gyllenhal, 1837)**
**Nb, LCGR- Fm Cl**
7 SU06: Calstone Down, 1869-1938 (Meyrick, 1938)
Resident. Very local; [on ox eye daisy, *Leucanthemum vulgare*]. Five specimens in MCC one with label insc. *AJF* [?] *8/6/01* (conf. MGM).

***Microplontus rugulosus*** (Herbst, 1795)
7 SU17: Hackpen Hill, 1965-1974 (MGM) (Morris, 1974, 1975)
8 ST83: Chicklade, 14-Aug-87 (DRN)
ST92: Prescombe Down, 1965-1974 (MGM) (Morris, 1974, 1975)
SU16: Woodborough, 4-Aug-83 (DRN)
SU22: Redlynch, 4-Aug-82 (DRN)
Resident. Local; on [wild camomile]*Matricaria*, knotgrass, weeds, etc.

**MOGULONES Reitter, 1916**

***Mogulones asperifoliarum*** (Gyllenhal, 1813)

7 SU16: Silbury Hill, 1965-1974 (MGM) (Morris, 1974, 1975)
SU26: Savernake Forest, 1869-1938 (Meyrick, 1938)
8 SU13: Idmiston, 1869-1938 (MGM) (Morris, (M.G, 1979); Salisbury, Avon Valley Reserve, Apr-00, (MAS) (Darby, 2001)
SU14: Haxton (SPTAE), 24-Jun-02 (PJH) (Hodge, 2002)
SU24: Bulford Ranges (SPTAE), 3-Jun - 12-Aug-08 (Boyce & Wright, 2009); Dunch Hill (SPTAE), 3-Jun - 12-Aug-08 (Boyce & Wright, 2009)

Resident. Local; in [a wide range of habitats on a wide range of Boraginaceae]

***Mogulones geographicus*** (Goeze, 1777)
**Nb**

7 SU16: Marlborough area, 1869-1938 (Morris, 1974, 1975)

Resident. Very local; on [viper's bugloss]*Echium vulgare.* Two specimens in MCC, one inscribed under card *Purfleet* (conf. MGM).

Mogulones geographicus

**NEDYUS Schönherr, 1825**

***Nedyus quadrimaculatus*** (Linnaeus, 1758)

7 ST26: Flat Holm, 10-May-02 (JKG det MDD)
ST97: Chippenham, Lackham College, 29-May-04 (BLe)
SU06: Morgan's Hill, 1965-1974 (MGM) (Morris, 1974, 1975)
SU16: Marlborough area, 4-May-1879 (WFB) (Meyrick, 1880); 1869-1938 (Meyrick, 1938); Silbury Hill, 1965-1974 (MGM) (Morris, 1974, 1975)
SU17: Clout's Wood, 1965-1974 (MGM) (Morris, 1974, 1975); Fyfield Down, 1965-1974 (MGM) (Morris, 1974, 1975); Hackpen Hill, 1965-1974 (MGM) (Morris, 1974, 1975)
SU26: Savernake Forest, 1965-1974 (MGM) (Morris, 1974, 1975)
SU28: Bishopstone, 1974-1978 (MGM) (Morris, 1979)
8 ST83: Brimsdown Hill, 1965-1974 (MGM) (Morris, 1974, 1975); Whitesheet Hill, 1965-1974 (MGM) (Morris, 1974, 1975)
ST85: Green Lane Wood, 26-Jun-98 (JN)
ST93: Fonthill, 19-Jul-95 (MDD); Great Ridge Wood, 19-Apr-84 (DRN); 16-May-00 (MDD); Stockton Wood, 1965-1974 (MGM) (Morris, 1974, 1975)
ST94: Norton Down (SPTAW), 23-Jun-02 (PJH) (Hodge, 2002)
ST95: Bratton Castle, 1965-1974 (MGM) (Morris, 1974, 1975)
SU04: Yarnbury Castle, 1965-1974 (MGM) (Morris, 1974, 1975)
SU05: Redhorn Hill, 1965-1974 (MGM) (Morris, 1974, 1975)
SU13: Great Durnford, 16-Aug-69, 7-Aug-70, 6-Aug-71, 17-Jul-73, 27-Jul-77 (DRN); Salisbury, Bemerton Folly, 7-Mar - 15-Jun-02 (MDD) (Darby, 2002d)
SU14: Haxton (SPTAE), 24-Jun-02 (PJH) (Hodge, 2002); Stonehenge, 22-Apr-03 (PJH) (Hodge, 2003) : Wexland Hanging (SPTAC) 22-Jun-02 (PJH) (Hodge, 2002)
SU16: Jones's Mill, 3-May-00 (JKG det MDD)
SU22: Langley Wood NNR, 8-Jun-02 (MDD)
SU23: Porton Down, 1974-1978 (MGM) (Morris, 1979)

Resident. Widespread; on stinging nettles, *Urtica dioica.*

**PARETHELCUS Dieckmann, 1972**

***Parethelcus pollinarius*** (Forster, 1771)

7 SU16: Marlborough area, 1869-1938 (Meyrick, 1938)
SU17: Fyfield Down, 1-Jan-73 (MGM) (Morris, 1974, 1975); Hackpen Hill, 1965-1974 (MGM) (Morris, 1974, 1975)
8 ST94: Norton Down (SPTAW), 23-Jun-02 (PJH) (Hodge, 2002)
SU03: Grovely Wood, 1-Aug-88 (DRN)
SU04: Parsonage Down, 2-Jul-08 (MDD); Shrewton, 2-Aug-83 (DRN)
SU13: Great Durnford, 7-Aug-70, 17-Jul-73, 3-Aug-83 (DRN)
SU14: Haxton (SPTAE), 24-Jun-02 (PJH) (Hodge, 2002); Stonehenge, 22-Apr-03 (PJH) (Hodge, 2003)
SU22: West Grimstead, 12-Aug-87 (DRN)
SU23: Porton Down, 1974-1978 (MGM) (Morris, 1979)
SU24: Driver Training Area (SPTAE), 24-Jun-02 (PJH) (Hodge, 2002)

Resident. Widespread; on stinging nettles, *Urtica dioica.*

**POOPHAGUS Schönherr, 1837**

***Poophagus sisymbrii*** (Fabricius, 1777)

8 SU13: Idmiston, 13-Aug-87 (DRN); Stratford sub Castle, 8-Aug-70 (DRN); Winterbourne Gunner, 1869-1938 (MGM) (Morris, (M.G, 1979)
SU14: Figheldean, 28-Jul-82 (DRN)
Resident. Very local; on watercress at the sides of slow-flowing water.

**SIROCALODES Voss, 1958**
***Sirocalodes depressicollis*** (Gyllenhal, 1813)
8 ST23: Allington Farm, Porton Down, 1965-1974 (MGM) (Morris, 1974, 1975)
ST94: Norton Down (SPTAW), 23-Jun-02 (PJH) (Hodge, 2002); Grovely Wood, 16-Aug-71 (DRN)
SU03: Wylye, 1965-1974 (MGM) (Morris, 1974, 1975)
SU04: Fox Covert (SPTAW), 3-Jun - 12-Aug-08 (Boyce & Wright, 2009)
SU13: Great Durnford, 6-Aug-87 (DRN)
Resident. Local; [on species of *Fumaria* in waste places etc. on disturbed ground].

***Sirocalodes mixtus*** (Mulsant & Rey, 1858)
**Nb**
8 SU23: Thorny Down, Porton, 1965-1974 (MGM) (Morris, 1974, 1975)
Resident. Very local; [on *Fumaria* spp.].

**STENOCARUS Thomson, C.G, 1859**
***Stenocarus ruficornis*** (Stephens, 1831)
**Nb**
7 SU16: Marlborough area, 1869-1938 (Meyrick, 1938)
SU17: Fyfield Down, 1965-1974 (MGM) (Morris, 1974, 1975)
8 SU03: Wylye Down NNR, 29-Mar-07 (MDD) (Darby, 2008)
SU13: Great Durnford, 16-Aug-69 (DRN); Old Sarum, 1965-1974 (MGM) (Morris, 1974, 1975); Salisbury, 26-May-74 (DRN)
SU14: Stonehenge, 8-Apr-03 (PJH) (Hodge, 2003)
Resident. Local; on *Papaver* spp..

**THAMIOCOLUS Thomson, C.G, 1859**
***Thamiocolus viduatus*** (Gyllenhal, 1813)
**Nb**
8 SU13: Idmiston, 5-Aug-82 (DRN); Great Durnford, 6-Aug-71 (DRN)
SU14: Stonehenge, 8-Apr, 31-Aug-03 (PJH) (Hodge, 2003)
Resident. Very local; [in wet habitats on marsh woundwort,*Stachys palustris*].

**TRICHOSIROCALUS Colonnelli, 1979**
***Trichosirocalus horridus*** (Panzer, 1801)
**Na**
8 SU14: Haxton (SPTAE), 17-May-02 (PJH) (Hodge, 2002)
Resident. Very local; in a suction sample. [On thistles mainly if not exclusively *Carduus nutans* and *Cirsium vulgare* in less intensively managed grasslands].

***Trichosirocalus troglodytes*** (Fabricius, 1787)
7 ST06: Oldbury Castle, 1965-1974 (MGM) (Morris, 1974, 1975)
SU16: Great Lodge Farm, nr Marlborough, 1873-1938 (Meyrick, 1938)
SU26: Savernake Forest, 1869-1938 (Meyrick, 1938)
8 ST83: Whitesheet Hill, 1965-1974 (MGM) (Morris, 1974, 1975)
ST93: Great Ridge Wood, 29-Jul-82 (DRN)
ST94: Bowls Barrow (SPTAW), 16-May, 22-Jun-02 (PJH) (Hodge, 2002); Cotley Hill, 1965-1974 (MGM) (Morris, 1974, 1975); Norton Down (SPTA), 15-May, 23-Jun-02 (PJH) (Hodge, 2002); Starveall Down, 1965-1974 (MGM) (Morris, 1974, 1975); Tinkers Track (SPTAW), 3-Jun - 12-Aug-08 (Boyce & Wright, 2009); Warminster Ranges (SPTAW), 3-Jun - 12-Aug-08 (Boyce & Wright, 2009)
SU01: Martin Down NNR, 21-Aug-89 (RGB)
SU02: Knapp and Barnetts Down, 1965-1974 (MGM) (Morris, 1974, 1975); Knighton Hill, 1965-1974 (MGM) (Morris, 1974, 1975); Stratford Tony Down, 1965-1974 (MGM) (Morris, 1974, 1975)
SU03: Grovely Wood, 24-Aug-72, 28-Jul, 1-Aug-73, 27-May-74 (DRN); Wylye Down, 1965-1974 (MGM) (Morris, 1974, 1975); Wylye Down NNR, 20-Sep-05, 16-Oct-06 (MDD) (Darby, 2009)
SU04: Fox Covert (SPTAW), 3-Jun - 12-Aug-08 (Boyce & Wright, 2009)Orcheston Down (SPTA), 18-May, 22-Jun-02 (PJH) (Hodge, 2002); Salisbury Plain, 16-May, 25-Jun-02 (PJH) (Hodge, 2002); Shrewton, 6-Aug-82 (DRN); Tilshead Park (SPTAW), 14-Jul-99 (PJH) (Hodge, 2002)
SU05: Black Heath (SPTAC), 3-Jun - 12-Aug-08 (Boyce & Wright, 2009)
SU13: Camp Down, 1965-1974 (MGM) (Morris, 1974, 1975); Great Durnford, 7-Aug-70 (DRN); Salisbury, 12-Apr-71 (DRN)
SU14: Haxton (SPTAE), 17-May-02 (PJH) (Hodge, 2002); Stonehenge, 22-Apr-03 (PJH) (Hodge, 2003); Wexland Hanging (SPTAC), 16-May, 22-Jun-02 (PJH) (Hodge, 2002)
SU15: Everleigh (SPTAE), 18-May, 24-Jun-02 (PJH) (Hodge, 2002)
SU22: Barnridge Copse, nr Farley, 1-Jan-79 (MGM) (Morris, 1979)
SU23: Porton Down, 3-Jul-99 (BLe)
SU24: Driver Training Area (SPTAE), 24-Jun-02 (PJH) (Hodge, 2002); North Tidworth, Devils Ditch, 30-May-04 (BLe)

SU25: Tidcombe, 1869-1938 (Meyrick, 1938)
Resident. Widespread; [on ribwort, *Plantago lanceolota,* in grasslands and rough ground, on cliffs and at the sides of roads and tracks].

**ZACLADUS Reitter, 1913**
***Zacladus exiguus*** (Olivier, 1807)
**Nb**
7 SU26: Hat Gate, 1869-1938 (Meyrick, 1938)
Resident. Very local; on [small flowered] *Geranium*. No specimens in MCC.

**Meadow Cranesbill Weevil**
***Zacladus geranii*** (Paykull, 1800)
7 SU16: Gopher Wood, 1965-1974 (MGM) (Morris, 1974, 1975); Marlborough area, 1869-1938 (Meyrick, 1938); 15-Jun-60 (AAA) (Allen, 1973); Monkton Down, 1965-1974 (MGM) (Morris, 1974, 1975); Pewsey Downs, 22-Jun-90 (APFw)
SU17: Fyfield Down, 1-Jan-73 (MGM) (Morris, 1974, 1975)
SU26: Durley, 1965-1974 (MGM) (Morris, 1974, 1975)
8 ST92: Trow Down, 1965-1974 (MGM) (Morris, 1974, 1975)
ST94: Chitterne, 1965-1974 (MGM) (Morris, 1974, 1975); Ladywell (SPTAW), 3-Jun - 12-Aug-08 (Boyce & Wright, 2009)
SU15: Everleigh (SPTAE), 18-May, 24-Jun-02 (PJH) (Hodge, 2002)
SU16: Woodborough, 4-Aug-83 (DRN)
Resident. Widespread; [on large flowered *Geranium* spp.].

Tribe AMALINI

**AMALUS Schönherr, 1825**
***Amalus scortillum*** (Herbst, 1795)
7 SU09: Somerford Keynes, 4-Jun-04 (BLe)
SU16: Marlborough area, 1869-1938 (Meyrick, 1938)
8 SU22: Redlynch, 4-Aug-82 (DRN)
Resident. Very local; [on *Polygonium aviculare*].

Tribe MONONYCHINI

**MONONYCHUS Germar, 1824**
***Mononychus punctumalbum*** (Herbst, 1784)
**Na**
8 SU12: Homington, 27-Jul-77(RAJ) (Jones, 1977)
Resident. Very local; on vegetation incuding *Iris pseudacorus*, one of its foodplants, by the R.Ebble.

Tribe PHYTOBIINI

**NEOPHYTOBIUS Wagner, 1936**
***Neophytobius muricatus*** Brisout, 1867
**Na**
8 SU21: Hamptworth Estate, 5-Aug-85, 12-Aug-86 (DRN)
Resident. Very local; in moss near a flush.

***Neophytobius quadrinodosus*** (Gyllenhal, 1813)
**Na, LCGR-Fm Cl**
8 SU21: Hamptworth Estate, 5-Aug-85 (DRN)
Resident. Very local; in moss near a fishpond.

**PELENOMUS Thomson, C.G, 1859**
***Pelenomus quadrituberculatus*** (Fabricius, 1787)
7 SU09: Clattinger Farm, 1965-1974 (MGM) (Morris, 1974, 1975)
SU17: Fyfield Down, 1965-1974 (MGM) (Morris, 1974, 1975)
SU26: Froxfield, 1869-1938 (Meyrick, 1938)
8 SU21: Hamptworth Estate, 6-Apr-77 (DRN)
Resident. Local; in moss near a stream [on docks *Rumex* and *Persicaria* spp.].

***Pelenomus waltoni*** (Boheman, 1843)
**Nb**
7 SU16: Marlborough area, 1869-1938 (Anon) (Meyrick, 1938)
Resident. Very local; on *Polygonum* [In bogs, marshes, watersides, etc. where the foodplant, *Persicaria hydropiper* grows]. Two specimens without data in MCC (conf. MGM).

**RHINONCUS Schönherr, 1825**
***Rhinoncus bruchoides*** (Herbst, 1784)
7 SU07: Compton Bassett, 1965-1974 (MGM) (Morris, 1974, 1975)
8 SU22: Redlynch, 4-Aug-82 (DRN)
Resident. Very local; on weeds including *Persicaria* [spp. its foodplants].

***Rhinoncus castor*** (Fabricius, 1792)
7 ST96: Chittoe, 24-Apr-98 (EGS det MDD)
Resident. Very local; [on the dock *Rumex acetosella* and not recorded from any other host].

***Rhinoncus inconspectus*** (Herbst, 1795)
8 ST93: Fonthill, 19-Jul-95 (MDD)
SU14: Figheldean meadows, Jun-1870 (RH) (Hislop, 1871)
Resident. Very local; in wetlands [on *Persicaria amphibia* f. *terrestris*].

***Rhinoncus pericarpius*** (Linnaeus, 1758)
7 ST96: Lacock, 16-Jun-82 (KNAA)
SU16: Martinsell, 15-Sep-00 (IJG)
SU26: Folly Farm, 1869-1938 (Meyrick, 1938)
8 ST92: Wardour, 5-Jun-96 (MAS)
ST94: Warminster Ranges (SPTAW), 3-Jun - 12-

Aug-08 (Boyce & Wright, 2009)
SU03: Grovely Wood, 11-Aug-71 (DRN); Wylye, 1965-1974 (MGM) (Morris, 1974, 1975)
SU04: Shrewton, 2-Aug-83 (DRN)
SU12: Charlton-All-Saints, 9-May-96 (MAS)
SU13: Great Durnford, 3-Aug-83 (DRN); Idmiston, 1974-1978 (MGM) (Morris, 1979); Stratford sub Castle, 10-Aug-70 (DRN)
SU14: Haxton (SPTAE), 17-May, 24-Jun-02 (PJH) (Hodge, 2002); Netheravon, 29-Jul-83 (DRN)
SU22: Langley Wood NNR, 5 to 25-Jun-04 (MDD) (Darby, 2005); Redlynch, 4-Aug-82 (DRN); Whiteparish Common, 1965-1974 (MGM) (Morris, 1974, 1975)

Resident. Widespread; on a wide vaiety of docks, *Rumex spp.*

***Rhinoncus perpendicularis*** (Reich, 1797)
7 SU16: Manton, nr. Marlborough, 1869-1938 (Meyrick, 1938); Wootton R.s, 1873-1938 (Meyrick, 1938)
SU17: Clout's Wood, 1965-1974 (MGM) (Morris, 1974, 1975); Fyfield Down, 1-Jan-73 (MGM) (Morris, 1974, 1975)
8 ST83: Chicklade, 14-Aug-87 (DRN)
ST93: Fonthill, 19-Jul-95 (MDD)
SU03: Wylye, 1965-1974 (MGM) (Morris, 1974, 1975)
SU13: Salisbury, 25-Jul-78 (DRN)
SU14: Netheravon, 29-Jul-83 (DRN)
SU16: Woodborough, 4-Aug-83 (DRN)
SU22: Redlynch, 4-Aug-82 (DRN)

Resident. Widespread; at edges of water [and in wetter parts of waste ground on a wide range of *Persicaria* spp.].

## Subfamily COSSONINAE
### Tribe COSSONINI

Rhopalomesites tardyi

**RHOPALOMESITES Wollaston, 1873**
**Holly Weevil**
?***Rhopalomesites tardyi*** (Curtis, 1825)
**Nb, IEC-3, SQS-8**
7 SU05: Devizes, Roundway, 1864 (JHB) (Sidebotham, 1968)

Resident. Very local; in dead wood [of a wide range of broadleaved trees]. Not found in the county since and doubtful because of Sidebotham's other dubious species. (See p.9)

### Tribe PENTARTHRINI

**EUOPHRYUM Broun, 1909**
***Euophryum confine*** (Broun, 1881)
7 SU26: Savernake Forest, 22-Jun-90 (APFw) 16-Jun to 7-Jul-93 (EN det SAW)
8 ST73: Tucking Mill Hanging, 11-Feb-84 (KNAA)
ST85: Bullen Hill Farm, 1-Jan-82 (EGS det MDD)
ST91: Cranborne Chase, 21-May-02 (MDD)
ST92: Sutton Mandeville, 13-Jan-99, 25-Jul-07 (MDD); Wardour, 18-Jun-02 (MDD)
SU02: Knighton Wood, Apr, 6-Jul-07, May, Jul-08 (AJWA)
SU03: Grovely Wood, 26-Jul-82 (DRN); 11-May-00, 27-May-02 (MDD)
SU13: Little Durnford Down, 20-May-03 (MDD); Salisbury, Avon Valley Reserve, Apr-00 (MAS), 19-Jun-00 (Darby, 2001); Salisbury, Bemerton Folly, 7-Mar - 15-Jun-02 (MDD) (Darby, 2002d)
SU22: Bentley Wood, 11-Aug-01 (MDD); Blackmoor Copse, 24-Aug-0, 17-Mar-02 (APF) (Foster, 2004); Langley Wood NNR, 16-Jul to 6-Aug-04 (MDD) (Darby, 2005)

Resident, Widespread; in dead, generally damp rotten wood of a wide range of trees in woodland, hedgerows, etc. and sometimes indoors.

#### Subtribe PHLOEOPHAGINA

**PHLOEOPHAGUS Schönherr, 1838**
***Phloeophagus lignarius*** (Marsham, 1802)
**SQS-2**
7 SU26: Savernake Forest, 6-Dec-73 (RCS); 6-Dec-73, 22-Jun-90 (APFw); 9-May-90 (TGW)
8 ST91: Cranborne Chase, 21-May-02 (MDD)
SU03: Dinton Park, 6-Jul-01 (KNAA) (Alexander, 2004)

Resident. Very local; in dead wood of broadleaved trees [and ivy], in woodland and parkland.

#### Subtribe TYLODINA

**ACALLES Schönherr, 1825**
***Acalles misellus*** Boheman, 1844
**SQS-2**
7 SU26: Savernake Forest, 23-Oct-01 (MDD)
8 SU02: Knighton Wood, Sep, Oct-06, Apr, Sep-07 (AJWA)
SU03: Grovely Wood, 16-Apr-74, 3-Apr-75 (DRN); 3-Aug-04 (MDD); Little Langford, 14-Jul-04 (MDD)
SU22: Blackmoor Copse, 17-May-02, 13-Jul, 3-

Aug-03 (APF) (Foster, 2004); Langley Wood, 14-May to 3-Jun-04 (MDD) (Darby, 2005)
SU26: Cobham Frith, 31-Jul-80 (DRN)
Resident. Local; in woods and hedgerows. [Often found in association with dead ivy and hawthorn].

***Acalles ptinoides*** (Marsham, 1802)
**Nb**
8 ST91: Cranborne Chase, 30-Jun-99 (MDD)
SU01: Martin Down NNR, 1999 (EN, 1999)
SU03: Grovely Wood, 4-Aug-80 (DRN)
SU21: Peaked Wood, 15-Apr-74 (DRN)
SU22: Langley Wood NNR, 6 to 27-Aug-04 (MDD) (Darby, 2005)
SU26: Bedwyn Brail, 19-Jun-03 (MDD)
Resident. Local; in woodland and heathland. Ground living and not usually arboreal. Often in moss and litter.

***Acalles roboris*** Curtis, 1835
**Nb, SQS-8**
7 SU26: Savernake Forest, 1 to 19-Jul-94 (EN det SAW)
8 ST91: Cranborne Chase, 30-Jun-99 (MDD)
SU02: Knighton Wood, 6-Jul, Aug-07 (AJWA)
SU22: Langley Wood NNR, 6-Jun-02, 3-Jun to 6-Aug-04, 25-Aug-06 to 5-Jun-07 (MDD) (Darby, 2005, 2008)
Resident. Local; in deciduous woodland often on smaller dead branches of trees and also in litter.

## Subfamily ENTIMINAE
### Tribe ALOPHINI

**GRAPTUS Schönherr, 1823**
***Graptus triguttatus*** (Fabricius, 1775)
8 ST92: Sutton Mandeville, 25-May-01 (MDD)
SU04: Fox Covert, 25-Jun-02 (PJH) (Hodge, 2002); Tidworth Pennings, 1965-1974 (MGM) (Morris, 1974, 1975)
SU14: Haxton (SPTAE), 17-May, 24-Jun-02 (PJH) (Hodge, 2002)
SU15: Beach's Plantation (SPTAE), 13 - 17-Jun-97 (MGT); Everleigh (SPTAE), 24-Jun-02 (PJH) (Hodge, 2002)
Resident. Local; ground living in open and grassy places. Probably polyphagous but shows a marked preference for *Plantago lanceolota.*

### Tribe BRACHYDERINI

**NELIOCARUS Thomson, C.G, 1859**
***Neliocarus faber*** (Herbst, 1784)
**Nb, LCGR-Fm Ch**
7 SU16: Pewsey Downs, 1965-1974 (MGM) (Morris, 1974, 1975); Walkers Hill, 1965-1974 (MGM) (Morris, 1974, 1975)
8 ST83: Whitesheet Down, 7-Jul-82 (KNAA)
ST94: Tinkers Track (SPTAW), 3-Jun - 12-Aug-08 (Boyce & Wright, 2009)
SU03: Wylye Down NNR, 29-Mar to 10-Apr-07 (MDD) (Darby, 2009)
SU24: Driver Training Area (SPTAE), 1-May-05 (MGT)
Resident. Local;, in grasslands and open waste places, feeding on a wide range of plants. Wingless and ground living.

***Neliocarus nebulosus*** (Stephens, 1831)
**LCGR-Fl Cm**
8 SU22: Landford, 12-Aug-87 (DRN)
SU36: Bagshot, 1869-1938 (Meyrick, 1938)
Resident. Very local; on the ground and on vegetation in grassy and open situations. Wingless.

***Neliocarus sus*** (Stephens, 1831)
8 SU21: Franchises Common, 25-Jul-73 (DRN); Hamptworth Estate, 4-Aug-73 (DRN); Pound Bottom, 9-Apr-74 (DRN)
Resident. Very local; on heather. Ground living and wingless.

**STROPHOSOMA Billberg, 1820**
***Strophosoma capitatum*** De Geer, 1775
7 SU26: Savernake Forest, 1869-1938 (Meyrick, 1938)
8 ST91: Cranborne Chase, 16 to 25-Jun-80 (NCC)
SU03: Grovely Wood, 24-Aug-72 (DRN)
SU22: Whiteparish Common, 22-Jul-73 (DRN); 1965-1974 (MGM) (Morris, 1974, 1975)
Resident. Local; in woodland on various broadleaved trees, and in more open situations on shrubs and heathers.

***Strophosoma melanogrammum*** (Forster, 1771)
7 SU16: Marlborough area, 1869-1938 (Meyrick, 1938)
SU26: Savernake Forest, 28-Apr to 11-Oct-94 (EN det SAW); 15-Sep to 5-Oct-95 (EN det MDD)
8 ST84: 25-May-05 (DJG)
ST92: Sutton Mandeville, 29-Jan-98 (MDD); Wardour, 24-Aug-72, 1-Aug-73, 8-Apr-74, (DRN); 5-Jun-96 (MDD)
SU03: Grovely Wood, 26-Jul-82
SU13: Salisbury, Avon Valley Reserve, Apr-00 (MDD) (Darby, 2001)
SU22: Blackmoor Copse, 13-Jul-03 (APF) (Foster, 2004); Homan's Copse, 19-Aug-88 (DRN); Langley Wood NNR, 6-Jun-02, 29-May-03,, 14-May-04 to 24-Mar-05, 5-Aug-06 to 2-Jul-07 (MDD) (Darby, 2008); Whiteparish Common, 1965-1974 (MGM) (Morris, 1974, 1975)

Resident. Widespread; on various broadleaved trees in woodland and hedgerows, and often in litter on the ground underneath.

Tribe CNEORHININI

**ATTACTAGENUS Tournier, 1876**
***Attactagenus plumbeus*** (Marsham, 1802)
**Nb**
7 SU16: Marlborough area, 1869-1938 (Meyrick, 1938)
Resident. Very local; on young trees on a railway embankment. One specimen without data in MCC (conf. MGM).

Tribe GEONEMINI

**BARYNOTUS Germar, 1817**
***Barynotus moerens*** (Fabricius, 1792)
7 SU26: Savernake Forest, 15-Jun-73 (SAW); 21-Sep to 11-Oct-93 (EN det SAW); 15-Jul to 3-Aug-95 (EN det MDD)
SU27: Love's Copse, 1869-1938 (Meyrick, 1938)
8 SU03: Grovely Wood, 30-Jul-79 (DRN)
Resident. Very local; in moss at roots of dog's mercury, *Mercurialis perennis.*

***Barynotus obscurus*** (Fabricius, 1775)
**LCGR-Fl Cm**
7 SU16: Gopher Wood, 1965-1974 (MGM) (Morris, 1974, 1975); Granham Hill, Marlborough district, 1869-1938 (Meyrick, 1938); Pewsey Downs, 1965-1974 (MGM) (Morris, 1974, 1975); Walkers Hill, 1965-1974 (MGM) (Morris, 1974, 1975); West Woods, 1873-1938 (Meyrick, 1938)
8 ST83: Whitesheet Down, 7-Jul-82 (KNAA)
ST93: Stockton Down, 1965-1974 (MGM) (Morris, 1974, 1975)
SU04: Orcheston Down (SPTA), 22-Jun-02 (PJH) (Hodge, 2002)
SU05: Ell Barrow (SPTAC), 3-Jun - 12-Aug-08 (Boyce & Wright, 2009)
SU13: Great Durnford, 3-Aug-83 (DRN)
SU14: Bulford House, 3-Jun-97 (MGT)
SU21: Hamptworth Estate, 15-Apr-74 (DRN)
SU23: Porton Down, 1974-1978 (MGM) (Morris, 1979); West Winterslow, 1-Sep-01 (NB)
SU25: Sidbury Hill (SPTAE), 5-Jun-97 (MGT)
Resident. Widespread; in open grassy situations. Feeds on a wide range of plants.

***Barynotus squamosus*** Germar, 1824
7 SU16: Marlborough area, Jun-1873 (EWJ) (Meyrick, 1874, 1938)
Resident. Very local; Record requires confirmation, An almost exclusively northern species.

Tribe OTIORHYNCHINI

**OTIORHYNCHUS Germar, 1824**
**Red-Legged Weevil, Plum Weevil**
***Otiorhynchus clavipes*** (Bonsdorff, 1785)
**LCGR-Fl Cl**
7 SU06: Cherhil Down, 15-Jun-82 (KNAA); Morgan's Hill, 1965-1974 (MGM) (Morris, 1974, 1975); 1-Jun-94 (PM)
SU16: Marlborough area, 1869-1938 (Meyrick, 1938); Pewsey Downs, 1965-1974 (MGM) (Morris, 1974, 1975); Workway Drove, near Wilcot, 27-Jun-01 (IJG)
SU17: Fyfield Down, 1-Jan-73 (MGM) (Morris, 1974, 1975)
8 ST83: Whitesheet Hill, 1965-1974 (MGM) (Morris, 1974, 1975); 7-Jul-82 (KNAA)
ST91: Cranborne Chase, 16 to 25-Jun-80 (NCC); 4-Jul-97 (MAS); 22-Jun-99 (MDD)
ST92: South Down Farm, 11-Apr-82 (EGS det MDD)
ST93: Great Ridge Wood, 25-May-00 (MDD)
ST94: Ladywell (SPTAW), 15-May-02 (PJH) (Hodge, 2002); Norton Down (SPTAW), 23-Jun-02 (PJH) (Hodge, 2002)
ST95: Bratton Castle, 1965-1974 (MGM) (Morris, 1974, 1975)
SU01: Martin Down NNR, 1999 (EN, 1999)
SU02: Knighton Wood, 6-Jul-07 (AJWA); Middleton Down, 1-May-97 (BL)
SU03: Barford Down, 6, 10-Apr-72 (DRN); Grovely Wood, 13, 14, 17-Apr, 25-Oct-71, 8-Apr-75, 4-Aug-87 (DRN); 14-Jun-01 (WANHS det MDD); 3-Aug-04 (MDD); Wylye Down NNR, 4-May, 10-Jul-07 (MDD) (Darby, 2009)
SU04: Orcheston Down (SPTA), 18-May-02 (PJH) (Hodge, 2002); Shrewton, 2-Aug-83 (DRN); Yarnbury Castle, 21-Jun-90 (APFw)
SU06: Devizes, 5-Aug-83 (DRN)
SU13: Cockey Down, 14-Jun-73, 18-Jul-90 (PM)
SU14: Haxton (SPTAE), 17-May-02 (PJH) (Hodge, 2002)
SU15: Beach's Plantation (SPTAE), 16-Jun-97, 1-May-05 (MGT); Everleigh (SPTAE), 24-Jun-02 (PJH) (Hodge, 2002)
SU23: Porton Down, 11-Jul-93 (PTW det MDD); Porton, Battery Hill, 1974-1978 (MGM) (Morris, 1979); Thorny Down, Porton, 1965-1974 (MGM) (Morris, 1974, 1975)
SU24: Driver Training Area (SPTAE), 1-May-05 (MGT); Dunch Hill (SPTAE), 3-Jun - 12-Aug-08 (Boyce & Wright, 2009)
Resident. Widespread; in grassy situations on calcareous soils. Sometimes in trees as well as on

the ground and in vegetation. A minor pest of fruit, particularly raspberry and plum.

***Otiorhynchus ligneus*** (Olivier, 1807)
**LCGR-FI Cm**
8 ST94: Ladywell (SPTAW), 3-Jun - 12-Aug-08 (Boyce & Wright, 2009); Tinkers Track (SPTAW), 3-Jun - 12-Aug-08 (Boyce & Wright, 2009)
SU05: Ell Barrow (SPTAC), 3-Jun - 12-Aug-08 (Boyce & Wright, 2009)
SU13: Winterbourne Gunner, 1974-1978 (MGM) (Morris, 1979)
SU23: Porton Down, 3-Jul-99 (MAS)
SU24: Dunch Hill (SPTAE), 3-Jun - 12-Aug-08 (Boyce & Wright, 2009)
Resident. Very local; ground living under stones, debris, etc.. Primarily a coastal species.

***Otiorhynchus porcatus*** (Herbst, 1795)
8 SU03: Wylye Down NNR, 10-Apr to 10-Jul-07 (MDD) (Darby, 2008)
Resident. Very local; ground living and nocturnal, synanthropic but also found away from buildings.

**Clay-coloured Weevil, Raspberry Weevil**
***Otiorhynchus singularis*** (Linnaeus, 1767)
7 SU16: Gopher Wood, 1965-1974 (MGM) (Morris, 1974, 1975); Hen's Wood, 3-Jul-02 (JKG det MDD); Marlborough area, 1873-1938 (Meyrick, 1938)
SU17: Old Totterdown, nr Fyfield, 1965-1974 (MGM) (Morris, 1974, 1975)
SU26: Savernake Forest, 17, 31-May-90 (TGW); 29-Aug to 20-Sep-93 9 to 30-Aug, 20-Sep to 11-Oct-94 (EN det SAW); 15-Jul to 3-Aug, 15-Sep to 11-Oct-95 (EN det MDD)
8 ST84: Longleat, 25-May-05 (DJG)
ST91: Cranborne Chase, 4-Jul-97 (MAS); 22-Jun-99 (MDD)
ST92: Wardour, 5-Jun-96 (MAS)
ST93: Great Ridge Wood, 16-May-00 (MDD); Stockton Wood, 1965-1974 (MGM) (Morris, 1974, 1975); 7-Jul-96 (MAS)
SU05: Redhorn Hill, 1965-1974 (MGM) (Morris, 1974, 1975)
SU22: Langley Wood NNR, 25-May-94 (WANHS det MDD); 6-Aug-04 (MDD) (Darby, 2005); Whiteparish Common, 1965-1974 (MGM) (Morris, 1974, 1975)
SU23: Porton Down, 28-May-94 (PTW det MDD)
Resident. Widespread: on almost all species of trees and shrubs. A notorious pest.

**Vine Weevil**
***Otiorhynchus sulcatus*** (Fabricius, 1775)
7 SU16: Marlborough area, 24 Jun 1879 (RAW)
8 ST92: Sutton Mandeville, 20-Apr-92 (MDD)
SU13: Salisbury, 31-Jul, 28-Oct-73 (DRN)
SU23: Porton Down, 19-Apr-92 (PTW det MDD)
Resident. Local; a pest of house and greenhouse plants as well as vines.

Otiorhynchus sulcatus

Tribe PERITELINI

**CAENOPSIS Bach, 1854**
***Caenopsis fissirostris*** (Walton, J, 1847)
**Nb**
8 SU22: Langley Wood NNR, 17-Sep-04 (MDD) (Darby, 2005)
Resident. Very local; in a pitfall trap. Ground dwelling species associated with acid rather than calcareous soils. Probably confined to the far south of the county.

***Caenopsis waltoni*** (Boheman, 1843)
**LCGR-FI Cm**
8 ST93: Great Ridge Wood, 29-Jul-82 (DRN)
SU22: Langley Wood NNR, 14-May-04 to 24-Mar-05, 15-Sep-06 to 6-Jun-07 (MDD) (Darby, 2008)
Resident. Very local. Ground dwelling species particularly on heathland with heathers, *Calluna* and *Erica* spp..

Tribe PHYLLOBIINI

**PHYLLOBIUS Germar, 1824**
PHYLLOBIUS Germar, 1824
**Common Leaf Weevil**
***Phyllobius pyri*** (Linnaeus, 1758)
7 ST96: Chittoe, 9-May-00 (EGS det MDD)
ST97: North Draycot Park, 22-May-00 (MDD)
SU06: Cherhill Down, 31-May-04 (BLe); Morgan's Hill, 1965-1974 (MGM) (Morris, 1974, 1975)
SU09: Ashton Keynes, 7-May-01 (JKG det MDD); Blakehill Farm, 11-May-01 (JKG det MDD); Upper Waterhay, 17-May to 4-Jun-05, 5-Jun-07 (JKG det MDD)
SU16: Marlborough area, 18 Jun 1879 (AMF) (Meyrick, 1880); Clatford, May-97 (JEO);

Gopher Wood, 1965-1974 (MGM) (Morris, 1974, 1975); Lockeridge, 1994 (JEO); Marlborough area, 1869-1938 (Meyrick, 1938)
SU17: Clout's Wood, 1965-1974 (MGM) (Morris, 1974, 1975); Old Totterdown, nr Fyfield, 1965-1974 (MGM) (Morris, 1974, 1975)
SU26: Savernake Forest, 31-May-04 (BLe); 22-Jun-90 (APFw); 1965-1974 (MGM) (Morris, 1974, 1975)
8 ST83: Brimsdown Hill, 1965-1974 (MGM) (Morris, 1974, 1975)
ST85: Bullen Hill Farm, 1-Jan-82 (EGS det MDD); Green Lane Wood, 7-Jun-95 (MDD)
ST91: Cranborne Chase, 16 to 25-Jun-80 (NCC); 29-Apr, 7-May-97 (MAS)
ST93: Stockton Wood, 1965-1974 (MGM) (Morris, 1974, 1975)
ST94: Bowls Barrow (SPTAW), 16-May-02 (PJH) (Hodge, 2002); Ladywell (SPTAW), 15-May-02 (PJH) (Hodge, 2002); Norton Down SPTAW), 15-May-02 (PJH) (Hodge, 2002)
ST95: Bratton Castle, 1965-1974 (MGM) (Morris, 1974, 1975)
SU01: Martin Down NNR, 1979-1992 (RGB)
SU03: Grovely Wood, 27, 30-May-74 (DRN)
SU04: Orcheston Down (SPTA), 18-May-02 (PJH) (Hodge, 2002); Parsonage Down, 2-Jul-08 (MDD); Salisbury Plain, 16-May-02 (PJH) (Hodge, 2002)
SU13: Cockey Down, 4-May-91 (PM); Figsbury Ring, 12-May-91 (PM); Old Sarum, 31-May-97 (MAS)
SU14: Boscombe Down, 14-Jun-96 (MSt); Haxton (SPTA), 17-May, 24-Jun-02 (PJH) (Hodge, 2002); Stonehenge, 22-Apr-03 (PJH) (Hodge, 2003), Wexland Hanging (SPTAC), 16-May-02 (PJH) (Hodge, 2002)
SU15: Everleigh (SPTAE), 18-May, 24-Jun-02 (PJH) (Hodge, 2002)
SU16: Jones's Mill, 21-Jun-97 (MDD)
SU21: Hamptworth Estate, 24-Apr-84 (DRN)
SU22: Bentley Wood, 27-May-97 (MDD); Blackmoor Copse, 1965-1974 (MGM) (Morris, 1974, 1975); Langley Wood NNR, 28-May-74 (DRN); 16-Jul-04, 5-Aug-06 (MDD) (Darby, 2005, 2008); Whiteparish Common, 1965-1974 (MGM) (Morris, 1974, 1975)
SU23: Porton Down, 19-Apr-93 (PTW det MDD); Battery Hill, 1974-1978 (MGM) (Morris, 1979)
SU24: Driver Training Area (SPTAE), 17-May-02 (PJH) (Hodge, 2002)

Resident. Ubiquitous; almost anywhere where there are trees! Apparently much increased in recent years.

DIELETUS Reitter, 1916

**Silver-green Leaf Weevil**

***Phyllobius argentatus*** (Linnaeus, 1758)

7 ST96: Lacock, 16-Jun-82 (KNAA)
SU16: Marlborough area, 9-May-1879 (WFB) (Meyrick, 1880); Lockeridge, 1994 (JEO); Lockeridge Dene 14-Jun-82 (KNAA)
SU26: Savernake Forest, 1869-1938 (Anon) (Meyrick, 1938); 27-May to 16-Jun-93, 9-Jun to 19-Jul-94 (EN det SAW); 11-May to 23-Jun-95 (EN det MDD)
8 ST91: Cranborne Chase, 16 to 25-Jun-80 (NCC)
ST93: Stockton Wood, 1965-1974 (MGM) (Morris, 1974, 1975)
ST94: Bowls Barrow (SPTAW), 16-May-02 (PJH) (Hodge, 2002)
SU02: Knighton Wood, 6-Jul-07, Jul-08 (AJWA)
SU03: Grovely Wood, 27-May-74 (DRN)
SU14: Haxton (SPTAE), 17-May-02 (PJH) (Hodge, 2002)
SU15: Everleigh (SPTAE), 18-May-02 (PJH) (Hodge, 2002)
SU21: Burnt Ground Wood, 31-May-74 (DRN); Peaked Wood, 29-May-74 (DRN)
SU22: Blackmoor Copse, 4, 17-May-02 (APF) (Foster, 2004); Langley Wood NNR, 25-May-94 (WANHS det MDD)
SU23: Thorny Down, Porton, 1974-1978 (MGM) (Morris, 1979)

Resident. Widespread; on a very wide range of broadleaved trees and shrubs.

METAPHYLLOBIUS Smirnov, 1913

***Phyllobius glaucus*** (Scopoli, 1763)

7 ST98: Conygre Mead, Malmesbury, 20-Apr-02 (JKG det MDD)
SU09: Ashton Keynes, 1941 (Williamson & Cowin, 1941); Dance Common, 15-Apr-02 (JKG det MDD)
SU16: Park Copse, nr Oare, 1869-1938 (Anon) (Meyrick, 1938)
SU17: Clout's Wood, 1965-1974 (MGM) (Morris, 1974, 1975)
8 ST93: Great Ridge Wood, 10-May-00 (MDD)
SU16: Jones's Mill, 21-Jun-97 (MDD)
SU22: Whiteparish Common, 1965-1974 (MGM) (Morris, 1974, 1975)

Resident. Local; arboreal species on a wide range of broadleaved trees. Includes *P. calcaratus*.

***Phyllobius pomaceus*** Gyllenhal, 1834

7 ST96: Conigre Meadow, 20-May-01 (JKG det MDD); Lacock, 16-Jun-82 (KNAA)
SU06: Avebury, 16-Jun-82 (DBS det KNAA); Morgan's Hill, 1965-1974 (MGM) (Morris, 1974, 1975)
SU08: Red Lodge Pond, 31-May-01 (JKG det MDD)
SU09: Ashton Keynes, 1-Jan-41 (Williamson & Cowin, 1941); Upper Waterhay, 13-Jun-04, 17-May to 11-Jun-05, 5-Jun to 11-Jun-07 (JKG det MDD)
SU16: Combe Farm, Marlborough district, 1873-

1938 (Meyrick, 1938); Clatford, 1-May-97 (JEO); Gopher Wood, 1965-1974 (MGM) (Morris, 1974, 1975); Piggledene, 14-Jun-82 (KNAA); Silbury Hill, 1965-1974 (MGM) (Morris, 1974, 1975)
SU26: Savernake Forest, 1869-1938 (Anon) (Meyrick, 1938)
8 ST84: Longleat, 4-May-98, 29-Aug-99 (MB)
ST85: Green Lane Wood, 7-Jun-95 (MDD)
ST92: Wardour, 5-Jun-96 (MAS)
SU13: Cockey Down, 6-Oct-73 (PM); Salisbury, 23-May-93 (PM); Salisbury, Avon Valley Reserve, Apr-00 (MAS) (Darby, 2001);
SU14: Stonehenge, 22-Apr-03 (PJH) (Hodge, 2003)
SU15: Everleigh (SPTAE), 18-May-02 (PJH) (Hodge, 2002)
SU21: Landford Bog, 25-May-98 (MDD)
SU22: Bentley Wood, 31-May-97 (MDD); Whiteparish Common, 1965-1974 (MGM) (Morris, 1974, 1975)
SU23: Porton Down, 1974-1978 (MGM) (Morris, 1979)
SU24: Driver Training Area (SPTAE), 17-May, 24-Jun-02 (PJH) (Hodge, 2002)
SU36: Bagshot, 1869-1938 (Meyrick, 1938)

Resident. Widespread; on stinging nettles, *Urtica dioica*.

Phyllobius pomaceus

NEMOICUS Dillwyn, 1829

**Brown Leaf Weevil**

***Phyllobius oblongus*** (Linnaeus, 1758)

7 SU09: Ashton Keynes, 1941 (Williamson & Cowin, 1941); Upper Waterhay, 17-May to 4-Jun-05 (JKG det MDD)
SU16: Marlborough area, 12-Jun-1879 (AP) (Meyrick, 1880); 1869-1938 (Meyrick, 1938); Pewsey Downs, 1965-1974 (MGM) (Morris, 1974, 1975)
SU17: Clout's Wood, 1965-1974 (MGM) (Morris, 1974, 1975)
SU26: Savernake Forest, 31-May-90 (TGW)
8 ST91: Cranborne Chase, 16 to 25-Jun-80 (NCC)
ST93: Great Ridge Wood, 10, 16-May-00 (MDD)
ST94: Bowls Barrow (SPTAW), 16-May-02 (PJH) (Hodge, 2002); Norton Down (SPTAW), 23-Jun-02 (PJH) (Hodge, 2002)
SU05: Redhorn Hill, 1965-1974 (MGM) (Morris, 1974, 1975)
SU13: Salisbury, 26-May-74 (DRN)
SU14: Wexland Hanging (SPTAC), 16-May-02 (PJH) (Hodge, 2002)
SU23: Porton Down, 1965-1974 (MGM) (Morris, 1974, 1975)

Resident. Widespread; in woods and coppices, and frequent in hedges and on hedgerow trees.

PARNEMOICUS Schilsky, 1911

**Small Green Nettle Weevil**

***Phyllobius roboretanus*** Gredler, 1882

7 ST96: Lacock, 16-Jun-82 (KNAA)
SU06: Calstone Down, 3-Jul-95 (KNAA); Cherhill Down, 31-May, 1-Jun-04 (BLe); Morgan's Hill, 1965-1974 (MGM) (Morris, 1974, 1975)
SU16: Avebury, 16-Jun-82 (KNAA); Gopher Wood, 1965-1974 (MGM) (Morris, 1974, 1975); Marlborough area, 1869-1938 (Meyrick, 1938); Pewsey Downs, 1965-1974 (MGM) (Morris, 1974, 1975); 22-Jun-90 (APFw); Silbury Hill, 1965-1974 (MGM) (Morris, 1974, 1975)
SU17: Clout's Wood, 1965-1974 (MGM) (Morris, 1974, 1975)
SU26: Savernake Forest, 31-May-90 (TGW); 28-Jul to 28-Aug-93 (EN det SAW); 31-May-04 (BLe); Durley, 1965-1974 (MGM) (Morris, 1974, 1975)
8 ST83: Brimsdown Hill, 1965-1974 (MGM) (Morris, 1974, 1975); Whitesheet Hill, 1965-1974 (MGM) (Morris, 1974, 1975)
ST91: Cranborne Chase, 16 to 25-Jun-80 (NCC)
ST93: Great Ridge Wood, 6-Aug-85 (DRN); Stockton Down, 1965-1974 (MGM) (Morris, 1974, 1975)
ST94: Bowls Barrow (SPTAW), 16-May-02 (PJH) (Hodge, 2002); Ladywell (SPTAW), 15-May-02 (PJH) (Hodge, 2002); 3-Jun - 12-Aug-08 (Boyce & Wright, 2009); Norton Down (SPTA), 15-May-02 (PJH) (Hodge, 2002); Tinkers Track (SPTAW), 3-Jun - 12-Aug-08 (Boyce & Wright, 2009); Warminster Ranges (SPTAW), 3-Jun - 12-Aug-08 (Boyce & Wright, 2009)
ST95: Fore Hill, nr West Lavington, 1965-1974 (MGM) (Morris, 1974, 1975)
SU02: Knighton Wood, May-08 (AJWA)
SU03: Grovely Wood, 27, 30-May-74 (DRN); Wylye Down 1965-1974 (MGM) (Morris, 1974, 1975)
SU04: Orcheston Down (SPTA), 18-May-02 (PJH) (Hodge, 2002); Salisbury Plain, 16-May 25-Jun-02 (PJH) (Hodge, 2002);

Yarnbury Castle, 1965-1974 (MGM) (Morris, 1974, 1975); 21-Jun-90 (APFw)
SU05: Redhorn Hill, 1965-1974 (MGM) (Morris, 1974, 1975)
SU14: Stonehenge, 22-Apr-03 (PJH) (Hodge, 2003); Wexland Hanging (SPTAC), 16-May-02 (PJH) (Hodge, 2002)
SU15: Everleigh (SPTAE), 18-May-02 (PJH) (Hodge, 2002)
SU23: Porton Down, 3-Jul-99 (PMP); 1965-1974 (MGM) (Morris, 1974, 1975)
SU24: Beacon Hill (SPTAE), 3-Jun - 12-Aug-08 (Boyce & Wright, 2009); Driver Training Area (SPTAE), 17-May-02 (PJH) (Hodge, 2002); North Tidworth, Devils Ditch, 30-May-04 BLe)

Resident. Widespread; on vegetation in a wide range of habitats.

***Phyllobius viridicollis*** (Fabricius, 1792)
**LCGR-Fm Cl**
7 SU06: Cherhill Down, 31-May-04 (BLe); Morgan's Hill, 1965-1974 (MGM) (Morris, 1974, 1975)
SU16: Marlborough area, 30 May 1879 (RAW) (Meyrick, 1880); 1869-1938 (Meyrick, 1938); Pewsey Downs, 1965-1974 (MGM) (Morris, 1974, 1975); Silbury Hill, 1965-1974 (MGM) (Morris, 1974, 1975)
8 ST83: Whitesheet Hill, 1965-1974 (MGM) (Morris, 1974, 1975)
ST94: Bowls Barrow (SPTAW), 16-May, 22-Jun-02 (PJH) (Hodge, 2002); Imber Firs, 13-May-00 (EGS det MDD); Ladywell (SPTAW), 15-May-02 (PJH) (Hodge, 2002); Norton Down (SPTA), 15-May, 23-Jun-02 (PJH) (Hodge, 2002)
ST95: Bratton Castle, 1965-1974 (MGM) (Morris, 1974, 1975)
SU04: Orcheston Down (SPTA), 18-May, 22-Jun-02 (PJH) (Hodge, 2002); Salisbury Plain, 25-Jun-02 (PJH) (Hodge, 2002); Yarnbury Castle, 1965-1974 (MGM) (Morris, 1974, 1975)
SU05: Redhorn Hill, 1965-1974 (MGM) (Morris, 1974, 1975)
SU14: Haxton (SPTAE), 17-May, 24-Jun-02 (PJH) (Hodge, 2002); Stonehenge, 22-Apr-03 (PJH) (Hodge, 2003)
SU15: Everleigh (SPTAE), 18-May-02 (PJH) (Hodge, 2002)
SU23: Porton Down, 3-Jul-99 (PMP); 1965-1974 (MGM) (Morris, 1974, 1975)
SU24: Driver Training Area (SPTAE), 17-May, 24-Jun-02 (PJH) (Hodge, 2002)

Resident. Widespread; on a wide variety of plants in a wide range of habitats.

PTERYGORRHYNCHUS Pesarini, 1969
***Phyllobius maculicornis*** Germar, 1824
7 SU16: Marlborough area, 1869-1938
SU17: Fyfield Down, 1-Jan-73 (MGM) (Morris, 1974, 1975)
8 SU03: Grovely Wood, 27-May-74 (DRN)
SU15: Everleigh (SPTAE), 18-May-02 (PJH) (Hodge, 2002)
SU23: Porton Down, 1965-1974 (MGM) (Morris, 1974, 1975)

Resident. Local; on a wide variety of trees and shrubs.

SUBPHYLLOBIUS Schilsky, 1911
***Phyllobius viridiaeris*** (Laicharting, 1781)
**LCGR-Fl Cl**
7 SU09: Ashton Keynes, 7-May-01 (JKG det MDD)
SU16: Clatford, New Arbore, 1-May-97 (JEO); Marlborough area, 1869-1938 (Morris, 1974, 1975)
SU17: Fyfield Down, 1-Jan-73 (MGM) (Morris, 1974, 1975)
SU18: Rushy Platt, 20-May-01 (JKG det MDD)
8 ST91: Cranborne Chase, 7, 23-May-97 (MAS)
ST94: Bowls Barrow (SPTAW), 22-Jun-02 (PJH) (Hodge, 2002); Norton Down (SPTAW), 23-Jun-02 (PJH) (Hodge, 2002)
ST95: Bratton Castle, 1965-1974 (MGM) (Morris, 1974, 1975); Cheverell, 30-May-97 (MGT)
SU02: Middleton Down, 24-May-95 (MDD)
SU04: Fox Covert, 25-Jun-02 (PJH) (Hodge, 2002); Orcheston Down (SPTA), 22-Jun-02 (PJH) (Hodge, 2002); Salisbury Plain, 16-May, 25-Jun-02 (PJH) (Hodge, 2002)
SU05: Black Heath (SPTAC), 3-Jun - 12-Aug-08 (Boyce & Wright, 2009); Ell Barrow (SPTAC), 3-Jun - 12-Aug-08 (Boyce & Wright, 2009); Lavington Down, 5-Jul-02 (EGS det MDD)
SU13: Old Sarum, 31-May-97 (MAS)
SU14: Boscombe Down, 14-Jun-96 (MSt); Haxton (SPTA), 24-Jun-02 (PJH) (Hodge, 2002); Wexland Hanging (SPTAC), 16-May, 22-Jun-02 (PJH) (Hodge, 2002); 3-Jun - 12-Aug-08 (Boyce & Wright, 2009)
SU15: Everleigh (SPTAE), 24-Jun-02 (PJH) (Hodge, 2002)
SU16: Jones's Mill, 21-Jun-97 (MDD)
SU22: Bentley Wood, 31-May-96 (MDD)
SU24: Driver Training Area (SPTAE), 24-Jun-02 (PJH) (Hodge, 2002); Dunch Hill (SPTAE), 3-Jun - 12-Aug-08 (Boyce & Wright, 2009)

Resident. Widespread; on herbaceous vegetation in a wide range of habitats.

Tribe POLYDRUSINI

**LIOPHLOEUS Germar, 1817**
***Liophloeus tessulatus*** (Müller, O.F, 1776)

7 ST97: Chippenham, Lackham College, 29-May-04 (BLe)
SU16: Marlborough area, 1869-1938 (Morris, 1974, 1975); Pewsey Downs, 1965-1974 (MGM) (Morris, 1974, 1975)
SU17: Clout's Wood, 1965-1974 (MGM) (Morris, 1974, 1975)
8 ST84: Longleat, 1-Jun-97 (MB)
ST94: Ladywell (SPTAW), 15-May-02 (PJH) (Hodge, 2002); Norton Down (SPTAW), 15-May-02 (PJH) (Hodge, 2002)
SU15: Everleigh (SPTAE), 18-May-02 (PJH) (Hodge, 2002)
SU22: Bentley Wood, 27-May-97 (MDD); Blackmoor Copse, 17-May-02 (APF) (Foster, 2004); Langley Wood NNR, 25-Jun-04, 5-Aug-06 (MDD) (Darby, 2008); Whiteparish Common, 1965-1974 (MGM) (Morris, 1974, 1975)
SU23: Porton Down, 1965-1974 (MGM) (Morris, 1974, 1975)
Resident. Widespread; in woods and on hedges [frequently in association with ivy, *Hedera helix*].

**POLYDRUSUS Germar, 1817**

POLYDRUSUS Germar, 1817

***Polydrusus tereticollis*** (De Geer, 1775)

7 SU16: Marlborough area, 1869-1938 (Morris, 1974, 1975)
SU26: Savernake Forest, 1869-1938 (Meyrick, 1938); 31-May-90 (TGW); 25-Apr to 20-Sep-93, 28-Apr to 11-Oct-94 (EN det SAW); 11-May to 23 Jun-95 (EN det MDD); 31-May-00 (WANHS det MDD); 31-May-04 (BLe)
8 ST84: Longleat, 7-Jul-99 (MDD); 25-May-05 (DJG)
ST91: Cranborne Chase, 29-Apr-97 (MAS)
ST93: Great Ridge Wood, 2, 16-May-00 (MDD)
SU03: Grovely Wood, 16-Aug-71, 18-Jul-73, 27-May-74, 28-Jul-83 (DRN)
SU22: Bagfield Copse, 6-Aug-86 (DRN); Bentley Wood, 27-May-98 (MDD); Blackmoor Copse, 1965-1974 (MGM) (Morris, 1974, 1975); 4-May to 18-May-02 (APF) (Foster, 2004); Langley Wood NNR, 25-May-94 (WANHS det MDD); 25-Apr to 6-Jun-07 (MDD) (Darby, 2008); Whiteparish Common, 1965-1974 (MGM) (Morris, 1974, 1975)
SU25: Dod's Down, 1869-1938 (Meyrick, 1938)
Resident. Widespread; on broadleaved trees.

CHRYSOPHIS des Gozis, 1882

***Polydrusus formosus*** (Mayer, 1779)

**Na**

7 SU26: Folly Farm, 1869-1938 (Anon) (Meyrick, 1938)
8 SU22: Blackmoor Copse, 25-May-76 (MGM) (Morris, (M.G, 1978)
Resident. Very local; [on broadleaved trees, particularly hazel.] Only known from a few southern counties and Glamorgan. Formerly *P. splendidus*.

EUDIPNUS Thomson C.G, 1859

***Polydrusus mollis*** (Ström, 1768)

**Nb**

8 ST95: West Lavington, 14-May-44 (NR) (DFL, 1948)
SU23: Thorny Down, Porton, 1974-1978 (MGM) (Morris, 1979)
Resident. Very local; [on broadleaved trees, mainly oak, beech and hazel].

EUSTOLUS Thomson, C.G. 1859

***Polydrusus flavipes*** (De Geer, 1775)

**Nb**

8 SU22: Blackmoor Copse, 31-Aug-01 (APF) (Foster, 2004); Whiteparish Common, 1965-1974 (MGM) (Morris, 1974, 1975)
SU26: Bedwyn Brail, 2-Jun-03 (MDD)
Resident. Very local; on broadleaved trees.

***Polydrusus pterygomalis*** Boheman, 1840

7 SU06: Morgan's Hill, 1965-1974 (MGM) (Morris, 1974, 1975)
SU07: Yatesbury, 1869-1938 (Meyrick, 1938)
SU16: Lockeridge Dene, 14-Jun-82 (KNAA); Wooton Rivers, 8-Aug-69 (DRN)
SU17: Fyfield Down, 1-Jan-73 (MGM) (Morris, 1974, 1975); Clout's Wood, 1965-1974 (MGM) (Morris, 1974, 1975); Old Totterdown, nr Fyfield, 1965-1974 (MGM) (Morris, 1974, 1975)
SU26: Savernake Forest, 1965-1974 (MGM) (Morris, 1974, 1975); 22-Jun-90 (APFw); 8-Jul to 20-Sep-93, 9-Jun to 1-Jul, 20-Sep to 11-Oct-94 (EN det SAW); 15-Sep to 5-Oct-95 (EN det MDD)
8 ST83: Brimsdown Hill, 1965-1974 (MGM) (Morris, 1974, 1975)
ST91: Cranborne Chase, 16 to 25-Jun-80 (NCC); 22, 30-Jun-99 (MDD)
ST93: Great Ridge Wood, 6-Aug-85 (DRN); Stockton Wood, 1965-1974 (MGM) (Morris, 1974, 1975)
SU02: Knighton Wood, Jul-08 (AJWA); Middleton Down, 24-May-95 (MDD); 20-May-01 (BL)
SU03: Grovely Wood, 13-Aug-68, 13-Aug-71, 27, 30-May-74, 1-Aug-88 (DRN); 21-Aug-01 (MDD)
SU05: Redhorn Hill, 1965-1974 (MGM) (Morris, 1974, 1975)
SU13: Little Durnford, 1965-1974 (MGM) (Morris, 1974, 1975); Rox Hill, 1965-1974 (MGM) (Morris, 1974, 1975)
SU14: Stonehenge, 22-Apr-03 (PJH) (Hodge, 2003)

SU15: Everleigh (SPTAE), 18-May-02 (PJH) (Hodge, 2002)
SU22: Langley Wood NNR, 5 to 25-Jun-04 (MDD) (Darby, 2005)
SU23: Porton Down, 24-Apr-93 (PTW det MDD)
Resident. Widespread; on broadleaved trees, particularly hazel and oak. Formerly *P. undatus*.

EUSTOLUS sensu auct non Thomson, G.G, 1859
***Polydrusus cervinus*** (Linnaeus, 1758)
7 SU16: Pumphrey's Wood, nr Marlborough, 1873-1938 (Meyrick, 1938)
SU26: Savernake Forest, 1965-1974 (MGM) (Morris, 1974, 1975)
8 ST91: Cranborne Chase, 16 to 25-Jun-80 (NCC); 7- May-97 (MAS)
ST94: Bowls Barrow (SPTAW), 16-May-02 (PJH) (Hodge, 2002)
SU03: Grovely Wood, 27-May-74 (DRN)
SU05: Redhorn Hill, 1965-1974 (MGM) (Morris, 1974, 1975)
SU13: Rox Hill, 1965-1974 (MGM) (Morris, 1974, 1975)
SU14: Haxton (SPTAE), 17-May-02 (PJH) (Hodge, 2002); Stonehenge, 22-Apr-03 (PJH) (Hodge, 2003)
SU21: Peaked Wood, 29-May-74 (DRN)
SU22: Blackmoor Copse, 1974-1978 (MGM) (Morris, 1979)
SU23: Porton Down, 24-Apr-93 (PTW det MDD); Thorny Down, Porton, 1974-1978 (MGM) (Morris, 1979)
Resident. Widespread; on broadleaved trees, particularly beech and oak.

Tribe SCIAPHILINI

**BARYPEITHES Jacquelin du Val, 1855**
EXOMIAS Bedel, 1883
**Strawberry Fruit Weevil**
***Barypeithes araneiformis*** (Schrank, 1781)
7 ST06: Oldbury Castle, 1965-1974 (MGM) (Morris, 1974, 1975)
ST87: Rack Hill, 1-Jun-9630-Jun-96 (MA det MDD)
SU06: Cherhill Down, 1-Jun-04 (BLe)
SU16: Gopher Wood, 13-Jul-96 (MA det MDD)
SU27: Rabley Wood, 1869-1938 (Meyrick, 1938)
8 ST91: Cranborne Chase, 16 to 25-Jun-80 (NCC)
SU01: Martin Down NNR, 1965-1974 (MGM) (Morris, 1974, 1975); 1999 (EN, 1999)
SU02: Knighton Wood, Jul-08 (AJWA)
SU05: Ell Barrow (SPTAC), 3-Jun - 12-Aug-08 (Boyce & Wright, 2009)
SU13: Salisbury, 4, 11, 12-Apr-71 (DRN); Salisbury, Bemerton Folly, 7-Mar - 15-Jun-02 (MDD) (Darby, 2002d)
SU21: Hamptworth Estate, 29-May-74, 6-Apr-77 (DRN)
SU22: Bentley Wood, 11-Mar-78 (DRN); 27-May-97, 27-May-98 (MDD); Langley Wood NNR, 12-Apr-74 (DRN); 29-May-03, 14-May to 7-Aug-04, 25-Aug-06 to 2-Jul-07 (MDD) (Darby, 2005, 2008)
SU23: Porton Down, 1965-1974 (MGM) (Morris, 1974, 1975)
SU24: Beacon Hill (SPTAE), 3-Jun - 12-Aug-08 (Boyce & Wright, 2009)
SU26: Bedwyn Brail, 2-Jun-03 (MDD)
Resident. Widespread; in grassy places and woods. Wingless, ground living and nocturnal. Occasional pest of strawberries

***Barypeithes pellucidus*** (Boheman, 1834)
7 SU26: Savernake Forest, 22-Jun-90 (APFw); 4-Sep-90 (TGW); 8-Jul to 28 Aug-93, 9-Jun to 9-Aug-94 (EN det SAW); Jun-96 (MA det MDD)
8 ST84: Longleat, 29-Aug-99 (MB)
ST91: Cranborne Chase, 4-Jun-97 (MAS)
ST95: Bratton Castle, 1974-1978 (MGM) (Morris, 1979)
SU02: Middleton Down, 6-Nov-73 (BL)
SU04: Orcheston Down (SPTA), 18-May-02 (PJH) (Hodge, 2002)
SU22: Langley Wood NNR, 1 to 22-Apr-04 (MDD) (Darby, 2005)
SU26: Bedwyn Brail, 2-Jun-03 (MDD)
Resident. Widespread; in open grassy situations, woodland, hedgerows, etc.. Ground dwelling, wingless, nocturnal. [An occasional pest in horticultural nurseries].

**BRACHYSOMUS Schönherr, 1823**
***Brachysomus echinatus*** (Bonsdorff, 1785)
**Nb, LCGR-Fl Ch**
7 SU17: Old Totterdown, nr Fyfield, 1965-1974 (MGM) (Morris, 1974, 1975)
8 SU01: Martin Down NNR, 1999 (EN, 1999)
Resident. Very local; in grassland [usually at ground level]. Flightless.

***Brachysomus hirtus*** (Boheman, 1845)
**RDB3, LCGR-Fm Ch**
8 SU02: Knighton Wood, 6-Jul, Oct, Nov-07, (AJWA)
SU22: Blackmoor Copse, 1965-1974 (MGM) (Morris, 1974, 1975)
Resident. Very local; [usually in woods and coppices, in moss, litter, etc.]. Flightless.

**SCIAPHILUS Schönherr, 1823**
***Sciaphilus asperatus*** (Bonsdorff, 1785)
7 SU16: Marlborough area, 1869-1938 (Meyrick, 1938)
8 ST83: Maiden Bradley, 1965-1974 (MGM) (Morris, 1974, 1975)
ST84: Longleat, 29-Aug-99, 28-Aug-00 (MB)
ST91: Cranborne Chase, 16 to 25-Jun-80 (NCC)

ST94: Ladywell (SPTAW), 23-Jun-02 (PJH) (Hodge, 2002); 3-Jun - 12-Aug-08 (Boyce & Wright, 2009)
SU03: Grovely Wood, 25-Oct-71, 18-Jul-73 (DRN)
SU13: Middle Woodford, 27-Jul-77 (DRN)
SU22: Blackmoor Copse, 1965-1974 (MGM) (Morris, 1974, 1975); Whiteparish Common, 1965-1974 (MGM) (Morris, 1974, 1975)
Resident. Widespread; on a wide range of plants in various habitats,[seldom on trees].

Tribe SITONINI

**ANDRION Velázquez, 2007**
***Andrion regensteinensis*** (Herbst, 1797)
7 SU26: Bedwyn, 1869-1938 (Meyrick, 1938)
8 ST84: Longleat, 4-May-98 (MB)
SU01: Martin Down NNR, 1965-1974 (MGM) (Morris, 1974, 1975)
Resident. Very local; [on gorse, *Ulex europae*, probably other *Ulex Spp.*, broom, *Cytisus scoparius* and possibly *Genista* spp.].

**COELOSITONA González**
***Coelositona cambricus*** Stephens, 1831
7 SU16: Marlborough area, 1869-1938 (Meyrick, 1938)
8 SU21: Hamptworth Common, 28-Jul-77 (DRN); Hamptworth Estate, 12-Aug-86, 19-Aug-88 (DRN)
Resident. Very local; [chiefly on birds-foot trefoil, *Lotus corniculatus*].

**SITONA Germar, 1817**
***Sitona ambiguus*** Gyllenhal, 1834
8 ST91: Cranborne Chase, 1800-1960 (PH)
ST94: Ladywell (SPTAW), 15-May-02 (PJH) (Hodge, 2002); Norton Down (SPTAW), 15-May, 23-Jun-02 (PJH) (Hodge, 2002)
SU04: Shrewton, 30-Jul-82, 2-Aug-83, 1-Aug-85 (DRN)
SU14: Haxton (SPTAE), 17-May-02 (PJH) (Hodge, 2002)
SU22: West Grimstead, 12-Aug-82 (DRN)
Resident. Local; in grassland [on meadow vetchling,*Lathyrus pratensis,* and especially *Vicia sepium*].

***Sitona cylindricollis*** (Fahraeus, 1840)
7 SU09: Upper Waterhay, 6-Jul-07 (JKG det MDD)
8 ST94: Norton Down (SPTAW), 23-Jun-02 (PJH) (Hodge, 2002)
SU03: Grovely Wood, 27-May-74 (DRN)
SU04: Shrewton, 6-Aug-82 (DRN); Tidworth Pennings, 1965-1974 (MGM) (Morris, 1974, 1975)
SU23: Allington, 1974-1978 (MGM) (Morris, 1979)
Resident. Local; [mainly in dry places on *Melilotus* spp.].

***Sitona hispidulus*** (Fabricius, 1777)
7 ST07: Windmill Hill, nr Avebury, 1965-1974 (MGM) (Morris, 1974, 1975)
SU09: Somerford Keynes, 4-Jun-04 (BLe)
SU16: Pewsey Downs , 1965-1974 (MGM) (Morris, 1974, 1975); Poulton, Marlborough district, 1869-1938 (Meyrick, 1938); Walkers Hill, 1965-1974 (MGM) (Morris, 1974, 1975)
SU26: Savernake Forest, 20-Sep to 11-Oct-94 (EN det SAW)
8 ST84: Longleat, 29-Aug-99 (MB)
ST93: Stockton Wood, 28-Sep-95 (PJH) (Hodge, 1995)
SU01: Martin Down NNR, 21-Aug-89 (RGB)
SU03: Wylye Down NNR, 27-Jan to 29-Mar-07 (MDD) (Darby, 2009)
SU04: Shrewton, 6-Aug-82 (DRN); Tidworth Pennings, 1965-1974 (MGM) (Morris, 1974, 1975)
SU13: Idmiston, 25-Aug-72 (DRN)
SU21: Pound Bottom, 11-Aug-82 (DRN)
SU22: Bentley Wood, 3-Sep-97 (WANHS det MDD); Harestock, 26-Oct-81 (DRN)
SU23: Porton Down, 1965-1974 (MGM) (Morris, 1974, 1975)
Resident. Widespread; in grassland, waste places, etc. mainly on clovers, *Trifolium* spp.].

***Sitona lepidus*** Gyllenhal, 1834
7 SU16: Wooton Rivers, 1869-1938 (Meyrick, 1938)
8 ST93: Fonthill, 19-Jul-95 (MDD)
ST94: Cotley Hill, 1965-1974 (MGM) (Morris, 1974, 1975); Warminster Ranges (SPTAW), 3-Jun - 12-Aug-08 (Boyce & Wright, 2009)
SU03: Grovely Wood, 11-Aug-71, 21-Aug-72 (DRN)
SU13: Salisbury, 15-Aug-88 (DRN)
SU14: Stonehenge, 22-Apr-03 (PJH) (Hodge, 2003)
SU16: Woodborough, 4-Aug-83 (DRN)
SU21: Hamptworth Estate, 19-Aug-88 (DRN)
SU22: Glazier's Copse, 16-Aug-88 (DRN); Landford, 12-Aug-87 (DRN)
SU24: North Tidworth, Devils Ditch, 30-May-04 (BLe)
Resident. Widespread; in grassland, agricultural land, etc. mainly on clovers, *Trifolium* spp.].

***Sitona lineatus*** (Linnaeus, 1758)
7 SU09: Ashton Keynes, 7-May-01 (JKG det MDD); Blakehill Farm, 11-May-01 (JKG det MDD); Somerford Keynes, 4-Jun-04 (BLe)
SU16: Marlborough area, 8-May-1879 (FED-H) (Meyrick, 1880); 1869-1938 (Meyrick, 1938)
SU18: Swindon: R.mead, 7-May-01 (JKG det

MDD)
SU26: Savernake Forest, 25-Apr to 26-May, 16-Jun to 7-Jul-93; 28-Apr to 9-Jun to 11-Oct-94 (EN det SAW); 31-May to 23-Jun, 4 to 24-Aug-95 (EN det MDD)
8 SU01: Martin Down NNR, 1999 (EN, 1999)
ST84: Longleat, 4-May-98, 29-Aug-99 (MB)
ST92: Prescombe Down, 1965-1974 (MGM) (Morris, 1974, 1975)
ST93: Stockton, 28-Sep-95 (PJH) (Hodge, 1995)
ST94: Bowls Barrow (SPTAW), 16-May, 22-Jun-02 (PJH) (Hodge, 2002); Ladywell (SPTAW), 15-May-02 (PJH) (Hodge, 2002); Norton Down (SPTA), 23-Jun-02 (PJH) (Hodge, 2002); Wadman's Coppice (SPTAW), 3-Jun - 12-Aug-08 (Boyce & Wright, 2009)
SU01: Martin Down NNR, 21-Aug-89 (RGB)
SU02: Knapp and Barnetts Down, 1965-1974 (MGM) (Morris, 1974, 1975); Knighton Hill, 1965-1974 (MGM) (Morris, 1974, 1975); Middleton Down, 10-Oct-96 (BL)
SU03: Grovely Wood, 10-Apr, 29-Oct-70 (DRN) Steeple Langford Cow Down, 1965-1974 (MGM) (Morris, 1974, 1975); Wylye Down NNR, 20-Sep-05, 2-Dec-06 (MDD) (Darby, 2008)
SU04: Fox Covert (SPTAW), 3-Jun - 12-Aug-08 (Boyce & Wright, 2009)Parsonage Down, 1965-1974 (MGM) (Morris, 1974, 1975); Salisbury Plain, 16-May, 25-Jun-02 (PJH) (Hodge, 2002); Shrewton, 6-Aug-82, 1-Aug-85 (DRN)
SU05: Ell Barrow (SPTAC), 3-Jun - 12-Aug-08 (Boyce & Wright, 2009)
SU12: East Harnham, 1-Aug-82 (DRN); Little Woodbury, nr Salisbury, 1965-1974 (MGM) (Morris, 1974, 1975)
SU13: Great Durnford, 7-Aug-70 (DRN); Middle Woodford, 5-Aug-82 (DRN); Salisbury, Avon Valley Reserve, 1- Apr-00 (MAS) (Darby, 2001)
SU14: Haxton (SPTAE), 17-May-02 (PJH) (Hodge, 2002); 3-Jun - 12-Aug-08 (Boyce & Wright, 2009); Stonehenge, 22-Apr-03 (PJH) (Hodge, 2003); Wexland Hanging (SPTAC), 22-Jun-02 (PJH) (Hodge, 2002)
SU15: Everleigh (SPTAE), 18-May, 24-Jun-02 (PJH) (Hodge, 2002)
SU16: Woodborough, 4-Aug-83 (DRN)
SU21: Hamptworth Estate, 19-Aug-88 (DRN)
SU22: Blackmoor Copse, 3-Aug-03 (APF) (Foster, 2004); Langley Wood NNR, 6-Oct-06 (MDD) (Darby, 2008); Whiteparish Common, 1965-1974 (MGM) (Morris, 1974, 1975)
SU23: Porton Down, 1965-1974 (MGM) (Morris, 1974, 1975); 3-Jul-99 (PJH) (Hodge, 2002)
SU24: North Tidworth, Devils Ditch, 30-May-04 (BLe)

Resident. Widespread; in a wide variety of habitats on many plants and trees.

***Sitona lineellus*** (Bonsdorff, 1785)

7 ST98: Conygre Mead, Malmesbury, 5-Apr-02 (JKG det MDD); Rodbourne, 29-Mar-02 (JKG det MDD)
SU09: Ashton Keynes (Millenium Green), 27-Apr-00 (JKG det MDD)
8 ST94: Imber village, 13-Sep-03 (EGS det MDD)

Resident. Local; [mainly in sandt places on clovers, *Trifolium* spp.].

***Sitona macularius*** (Marsham, 1802)

**Nb, LCGR-Fl Cl**

7 SU26/27: Axford, 1869-1938 (Meyrick, 1938)
8 ST95: Bratton Castle, 1974-1978 (MGM) (Morris, 1979)
SU02: Knighton Wood, Apr-07 (AJWA)
SU23: Porton Down, 3-Jul-99 (PJH) (Hodge, 2002)

Resident. Local; on downs, grassland, etc.[on various plants but particularly sainfoin,*Onobrychis viciifolia*].

***Sitona puncticollis*** Stephens, 1831

7 SU16: Marlborough area, 28-May-1879 (EedeS) (Meyrick, 1880); 1869-1938 (Meyrick, 1938)
8 SU03: Grovely Wood, 1-Aug-73 (DRN)
SU26: Bedwyn Brail, 2-Jun-03 (MDD)

Resident. Very local; [in grassland, open places, agricultural and waste land, etc. mainly on clovers, Trifolium spp.].

***Sitona striatellus*** Gyllenhal, 1834

7 SU16: Gopher Wood, 1965-1974 (MGM) (Morris, 1974, 1975), Marlborough area, 1869-1938 (Meyrick, 1938); Silbury Hill, 1965-1974 (MGM) (Morris, 1974, 1975)
SU17: Clout's Wood, 1965-1974 (MGM) (Morris, 1974, 1975); Old Totterdown, nr Fyfield, 1965-1974 (MGM) (Morris, 1974, 1975)
8 ST83: Brimsdown Hill, 1965-1974 (MGM) (Morris, 1974, 1975)
ST93: Stockton Down, 1965-1974 (MGM) (Morris, 1974, 1975)
ST94: Bowls Barrow (SPTAW), 16-May-02 (PJH) (Hodge, 2002); Ladywell (SPTAW), 15-May-02 (PJH) (Hodge, 2002)
SU03: Little Langford, 3-Jul-96 (MAS)
SU04: Orcheston Down, 22-Jun-02 (PJH) (Hodge, 2002); Parsonage Down, 2-Jul-08 (MDD); Salisbury Plain, 16-May-02 (PJH) (Hodge, 2002)
SU05: Chirton Gorse, 1965-1974 (MGM) (Morris, 1974, 1975); Redhorn Hill, 1965-1974 (MGM) (Morris, 1974, 1975)
SU12: Charlton-All-Saints, 9-May-96 (MAS)
SU13: Salisbury, Avon Valley Reserve, 7, 13-Apr, 2-May-00 (MDD) (Darby, 2001)
SU14: Haxton (SPTAE), 17-May, 24-Jun-02 (PJH)

(Hodge, 2002); Wexland Hanging (SPTAC), 16-May, 22-Jun-02 (PJH) (Hodge, 2002)
SU21: Burnt Ground Wood, 31-May-74 (DRN); Hamptworth Estate, 18, 24-Apr-84 (DRN); Landford Bog, 24-Jun-01, 22-Jun-03 (APF) (Foster, 2001, 2003); Pound Bottom, 24-Jul-73 (DRN)
SU22: Langley Wood NNR, 27-Aug-04 (MDD) (Darby, 2005)
Resident. Widespread; on gorse and broom.

***Sitona sulcifrons*** (Thunberg, 1798)
**LCGR-Fl Cl**
7 ST06: Oldbury Castle, 1965-1974 (MGM) (Morris, 1974, 1975)
SU16: Gopher Wood, 1965-1974 (MGM) (Morris, 1974, 1975); Marlborough area, 1869-1938 (Meyrick, 1938); Pewsey Downs, 1965-1974 (MGM) (Morris, 1974, 1975)
ST83: Whitesheet Hill, 1965-1974 (MGM) (Morris, 1974, 1975)
ST94: Bowls Barrow (SPTAW), 16-May, 22-Jun-02 (PJH) (Hodge, 2002); Cotley Hill, 1965-1974 (MGM) (Morris, 1974, 1975); Norton Down, 23-Jun-02 (PJH) (Hodge, 2002); Starveall Down, 1965-1974 (MGM) (Morris, 1974, 1975)
SU02: Knapp and Barnetts Down, 1965-1974 (MGM) (Morris, 1974, 1975); Knighton Hill, 1965-1974 (MGM) (Morris, 1974, 1975); Stratford Tony Down, 1965-1974 (MGM) (Morris, 1974, 1975); Woodminton Down, 1965-1974 (MGM) (Morris, 1974, 1975)
SU03: Steeple Langford Cow Down, 1965-1974 (MGM) (Morris, 1974, 1975); Wylye Down, 1965-1974 (MGM) (Morris, 1974, 1975); Wylye Down NNR, 22-Feb to 29-Mar-07 (MDD) (Darby, 2008)
SU04: Fox Covert, 25-Jun-02 (PJH) (Hodge, 2002); 3-Jun - 12-Aug-08 (Boyce & Wright, 2009); Parsonage Down, 1965-1974 (MGM) (Morris, 1974, 1975); Tidworth Pennings, 1965-1974 (MGM) (Morris, 1974, 1975); Tilshead, 1965-1974 (MGM) (Morris, 1974, 1975)
SU05: Ell Barrow (SPTAC), 3-Jun - 12-Aug-08 (Boyce & Wright, 2009)
SU14: Wexland Hanging (SPTAC) 22-Jun-02 (PJH) (Hodge, 2002)
SU15: Everleigh (SPTAE), 24-Jun-02 (PJH) (Hodge, 2002)
Resident. Widespread; in grassland [on clovers, *Trifolium* spp. though also on Papilionaceae]

***Sitona suturalis*** Stephens, 1831
7 ST96: Bewley Common, 16-Jun-82 (KNAA)
SU06: Morgan's Hill, 1965-1974 (MGM) (Morris, 1974, 1975)
SU09: Upper Waterhay, 17-May-054-Jun-05 (JKG det MDD); 5 to 11-Jun-05 (JKG det MDD)
SU16: Gopher Wood, 1965-1974 (MGM) (Morris, 1974, 1975); Marlborough area, 24 Jun 1879 (TNH-S) (Meyrick, 1880); West Woods, 1869-1938 (Meyrick, 1938)
SU17: Clout's Wood, 1965-1974 (MGM) (Morris, 1974, 1975); Hackpen Hill, 1965-1974 (MGM) (Morris, 1974, 1975)
8 ST94: Ladywell (SPTAW), 15-May-02 (PJH) (Hodge, 2002); Norton Down (SPTAW), 23-Jun-02 (PJH) (Hodge, 2002)
SU04: Orcheston Down, 22-Jun-02 (PJH) (Hodge, 2002)
SU13: Camp Down, 1965-1974 (MGM) (Morris, 1974, 1975)
SU14: Wexland Hanging (SPTAC), 16-May-02 (PJH) (Hodge, 2002)
SU15: Everleigh (SPTAE), 18-May-02 (PJH) (Hodge, 2002)
SU16: Woodborough, 4-Aug-83 (DRN)
SU22: Barnridge Copse, nr Farley, 1974-1978 (MGM) (Morris, 1979); Langley Wood NNR, 27-Aug to 17-Sep-04 (MDD) (Darby, 2005)
Resident. Widespread; [on meadow vetchling, *Lathyrus pratensis* and vetches, *Vicia* spp.].

***Sitona waterhousei*** Walton, J, 1846
**Nb, LCGR-Fl Cl**
7 SU16: Marlborough, 26-Mar-46 (Holmes & Tennant, 1947)
Resident. Very local; [on species of *Lotus* often, but not exclusively, by the sea].

Tribe TANYMECINI

**TANYMECUS Germar, 1817**
***Tanymecus palliatus*** (Fabricius, 1787)
**Nb**
7 SU16: Marlborough area, 1869-1938 (Meyrick, 1938)
Resident. Very local; by sweeping [on various plants but showing a preference for thistles, in grassy open situations, roadside verges and grazing land].

Tribe TRACHYPHLOEINI

**TRACHYPHLOEUS Germar, 1817**
***Trachyphloeus alternans*** Gyllenhal, 1834
**Nb, LCGR-Fm Ch**
8 SU01: Martin Down NNR, 1999 (EN, 1999)
SU03: Wylye Down NNR, 1965-1974 (MGM) (Morris, 1974, 1975); 10-Oct-07 (MDD) (Darby, 2008)
SU14: Haxton, 24-Jun-02 (PJH) (Hodge, 2002)
SU24: Driver Training Area (SPTAE), 24-Jun-02 (PJH) (Hodge, 2002)

Resident. Local; in grassland. Wingless, ground living.

***Trachyphloeus angustisetulus*** Hansen, 1915
**LCGR-Fl Ch**
8 ST94: Tinkers Track (SPTAW), 3-Jun - 12-Aug-08 (Boyce & Wright, 2009)
SU05: Black Heath (SPTAC), 3-Jun - 12-Aug-08 (Boyce & Wright, 2009)
Resident. Very local; in grassland. Wingless, ground living.

***Trachyphloeus aristatus*** (Gyllenhal, 1827)
**Nb, LCGR-Fm Ch**
8 SU13: Camp Down, 1965-1974 (MGM) (Morris, 1974, 1975); Old Sarum, 1965-1974 (MGM) (Morris, 1974, 1975)
SU14: Haxton, 24-Jun-02 (PJH) (Hodge, 2002)
Resident. Local; in grassland. Wingless, ground living.

***Trachyphloeus asperatus*** Boheman, 1843
**Nb, LCGR-Fm Ch**
8 ST84: Cley Hill, 6-Jul-82 (KNAA)
ST94: Cotley Hill, 1965-1974 (MGM) (Morris, 1974, 1975)
ST95: Fore Hill, nr West Lavington, 1965-1974 (MGM) (Morris, 1974, 1975)
SU14: Stonehenge, summer-03 (PJH) (Hodge, 2003); Wexland Hanging (SPTAC) 22-Jun-02 (PJH) (Hodge, 2002)
SU24: Cross Country Driving Area, 24-Jun-02 (PJH) (Hodge, 2002)
Resident. Local; in grassland. Wingless, ground living.

***Trachyphloeus spinimanus*** Germar, 1824
**Nb, LCGR-Fh Ch**
8 SU23: Porton Down, 3-Jul-99 (MAS)
SU24: Beacon Hill (SPTAE), 3-Jun - 12-Aug-08 (Boyce & Wright, 2009)
Resident. Local; in grassland. Wingless, ground living.

### Tribe TROPIPHORINI

**TROPIPHORUS Schönherr, 1842**
***Tropiphorus elevatus*** (Herbst, 1795)
**Nb**
7 SU16: West Woods, 1869-1938 (Meyrick, 1938); Marlborough area, 1869-1938 (Meyrick, 1938)
8 SU22: Langley Wood NNR, 22-Apr to 14-May-04 (MDD) (Darby, 2005)
Resident. Very local; in litter in woodland.

***Tropiphorus terricola*** (Newman, 1838)
**Nb**
7 SU16: Marlborough area, 1869-1938 (Meyrick, 1938)
Resident. Very7 local; [in grassy places and on herbaceous vegetation, less frequently in woods].

## Subfamily HYPERINAE

**HYPERA Germar, 1817**
HYPERA Germar, 1817
***Hypera arator*** (Linnaeus, 1758)
**LCGR-Fl Cl**
7 SU16: Marlborough area, 1869-1938 (Meyrick, 1938)
8 SU03: Grovely Wood, 11-Aug-71, 1-Aug-73 (DRN)
SU21: Hamptworth Estate, 28-Oct-75 (DRN)
SU23: Thorny Down, Porton, 1974-1978 (MGM) (Morris, 1979)
Resident. Local; [in dry, open places, often on light soils, on a wide variety of Caryophyllaceae].

***Hypera nigrirostris*** (Fabricius, 1775)
**LCGR-Fl Cl**
7 SU06: Calstone Down, 3-Jul-95 (KNAA); Roundway Camp, Jul-40 (FDB) (Buck, 1956)
SU08: Somerford Common, 23-Apr-01 (JKG det MDD)
SU16: Marlborough area, 1869-1938 (Meyrick, 1938); 1965-1974 (MGM) (Morris, 1974, 1975); Wootton Rivers, 1869-1938 (Meyrick, 1938)
SU26/27: Axford, 1869-1938 (Meyrick, 1938)
8 ST92: Wardour, 5-Jun-96 (MAS)
ST93: Fonthill, 19-Jul-95 (MDD), Stockton Wood, 7-Jul-96 (MAS)
ST94: Bowls Barrow (SPTAW), 22-Jun-02 (PJH) (Hodge, 2002); Norton Down, 23-Jun-02 (PJH) (Hodge, 2002)
SU02: Knighton Hill, 1965-1974 (MGM) (Morris, 1974, 1975); Stratford Tony Down, 1965-1974 (MGM) (Morris, 1974, 1975)
SU03: Dinton Park, 6-Jul-00 (KNAA); Grovely Wood, 27-May-74 (DRN)
SU04: Shrewton, 2-Aug-83, 1-Aug-85 (DRN)
SU05: Lavington Down, 5-Jul-02 (EGS det MDD) (Meyrick, 1938)
SU13: Salisbury, 26-May-74, 15-Aug-88 (DRN)
SU16: Jones's Mill, 30-May-04 (KNAA)
SU22: Whiteparish, 1974-1978 (MGM) (Morris, 1979)
SU23: Porton Down, 3-Jul-99 (MAS) (Meyrick, 1938)
Resident. Widespread; in grassland, clover fields, waste places, etc. on clovers, *Trifolium* spp.

***Hypera plantaginis*** (De Geer, 1775)
**LCGR-Fl Cl**
7 ST87: Rack Hill, 1-Jan-95 (MA det MDD)

SU06: Cherhill Down, 1-Jun-04 (BLe)
SU16: Pewsey Down, 1965-1974 (MGM) (Morris, 1974, 1975); Walkers Hill, 1965-1974 (MGM) (Morris, 1974, 1975)
8 ST92: Prescombe Down, 1965-1974 (MGM) (Morris, 1974, 1975); Trow Down, 1965-1974 (MGM) (Morris, 1974, 1975)
ST93: Stockton Wood, 7-Jul-96 (MAS)
ST94: Ladywell (SPTAW), 3-Jun - 12-Aug-08 (Boyce & Wright, 2009)
SU01: Martin Down NNR, 1974-1978 (MGM) (Morris, 1979)
SU03: Wylye Down NNR, 22-Feb to 29-Mar-07 (MDD) (Darby, 2009)
SU14: Bulford Down, 1965-1974 (MGM) (Morris, 1974, 1975)
SU16: Jones's Mill, 30-May04 (KNAA)
Resident. Widespread; in grasslands and open places [on species of Bird's-foot trefoil, particularly *Lotus corniculatus*].

**Clover Leaf Weevil**
***Hypera postica*** (Gyllenhal, 1813)
**LCGR-Fm Cl**
8 ST94: Wadman's Coppice (SPTAW), 3-Jun - 12-Aug-08 (Boyce & Wright, 2009)
SU04: Shrewton, 1-Aug-85 (DRN)
Resident. Very local; [in grasslands, waste places and wood edges and at roadsides, most frequently on black medick, *Medicago lupulina*].

***Hypera suspiciosa*** (Herbst, 1795)
**LCGR-Fl Cl**
7 SU16: Gopher Wood, 1965-1974 (MGM) (Morris, 1974, 1975); West Woods, 9-Jul-46 (Holmes & Tennant, 1947)
8 SU05: Redhorn Hill, 1965-1974 (MGM) (Morris, 1974, 1975)
Resident. Very local; [in waste places, on roadsides, etc. in tall herb communities and in open situations with luxuriant herbaceous vegetation. On various Papilionaceae].

***Hypera venusta*** (Fabricius, 1781)
**LCGR-Fl Cl**
8 ST95: Bratton Castle, 1974-1978 (MGM) (Morris, 1979)
8 SU04: Parsonage Down, 1965-1974 (MGM) (Morris, 1974, 1975)
Resident. Very local; [on kidney vetch, *Anthyllis vulneraria* and gorse, *Ulex Spp.*].

ANTIDONUS Bedel, 1886
***Hypera zoilus*** (Scopoli, 1763)
7 SU16: Marlborough area, 30-May-1879 (PHM) (Meyrick, 1880); 1869-1938 (Meyrick, 1938)
8 SU21: Hamptworth Estate, 26-Jul-77 (DRN)
Resident. Very local; [in grasslands and croplands, on clovers, *Trifolium* spp.].

DAPALINUS Capiomont, 1868
***Hypera meles*** (Fabricius, 1792)
**Na, LCGR-Fl Cl**
7 SU09: Somerford Keynes, 4-Jun-04 (BLe)
8 SU04: Shrewton, 28, 30-Jul-82, 2-Aug-83 (DRN)
SU22: Langley Wood NNR, 22-Apr to 14-May-04 (MDD) (Darby, 2005)
Resident. Local; under cardboard and [on clovers, *Trifolium* spp. in grasslands, etc.].

ERIRINOMORPHUS Capiomont, 1868
***Hypera pollux*** (Fabricius, 1801)
7 SU16: Alton, 1869-1938 (Meyrick, 1938); Walkers Hill, 1965-1974 (MGM) (Morris, 1974, 1975)
SU06: Cherhill Down, 1-Jun-04 (BLe)
SU16: Marlborough area, 1869-1938 (Meyrick, 1938)
Pewsey Downs, 1965-1974 (MGM) (Morris, 1974, 1975)
8 ST83: Brimsdown Hill, 1965-1974 (MGM) (Morris, 1974, 1975)
ST92: Prescombe Down, 1965-1974 (MGM) (Morris, 1974, 1975)
ST93: Stockton Down, 1965-1974 (MGM) (Morris, 1974, 1975)
ST94: Cotley Hill, 1965-1974 (MGM) (Morris, 1974, 1975); Norton Down (SPTAW), 23-Jun-02 (PJH) (Hodge, 2002); 13-Jul-03 (EGS det MDD)
SU01: Martin Down NNR, 1974-1978 (MGM) (Morris, 1979)
SU02: Knighton Hill, 1965-1974 (MGM) (Morris, 1974, 1975)
SU04: Fox Covert, 25-Jun-02 (PJH) (Hodge, 2002); Salisbury Plain, 16-May-02 (PJH) (Hodge, 2002); Shrewton, 6-Aug-82, 1-Aug-85 (DRN); Tidworth Pennings, 1965-1974 (MGM) (Morris, 1974, 1975); Tilshead Park (SPTAW), 14-Jul-99 (PJH) (Hodge, 1999)
SU05: Redhorn Hill, 1965-1974 (MGM) (Morris, 1974, 1975)
SU13: Great Durnford, 17-Jul-73 (DRN)
SU14: Stonehenge, 22-Apr-03 (PJH) (Hodge, 2003)
SU16: Jones's Mill, 19-Jun, 2-Jul-01 (APF) (Foster, 2002)
SU24: Driver Training Area (SPTAE), 24-Jun-02 (PJH) (Hodge, 2002)
Resident. Widespread; [in wet and marshy places, and on the banks of ponds and water courses. On various species of Apiaceae].

***Hypera rumicis*** (Linnaeus, 1758)
7 SU09: Somerford Keynes, Swillbrook Lakes, 4-Jun-04 (BLe)
SU16: Marlborough area, 17-Feb-1879 (WFB) (Meyrick, 1880); 1869-1938 (Meyrick, 1938)
8 ST91: Cranborne Chase, 10-Jun-97 (MAS)
SU03: Grovely Wood, 12-Jun-00 (MDD)

SU04: Shrewton, 2-Aug-83 (DRN)
SU05: Redhorn Hill, 1965-1974 (MGM) (Morris, 1974, 1975)
SU13: Idmiston, 1974-1978 (MGM) (Morris, 1979)
SU14: Netheravon, 29-Jul-83 (DRN)
SU16: Woodborough, 4-Aug-83 (DRN)
SU21: Hamptworth Estate, 25-Apr-84 (DRN)
SU22: Bentley Wood, 31-May-96 (MDD)
Resident. Widespread; on docks, *Rumex spp.*.

**LIMOBIUS Schönherr, 1843**
***Limobius borealis*** (Paykull, 1792)
**Na**,
7 SU16: Gopher Wood, 1965-1974 (MGM) (Morris, 1974, 1974a, 1975); Marlborough area, 15-Jun-60 (AAA) (Allen, 1973); Pewsey Downs, 22-Jun-90 (APFw);
8 SU15: Everleigh (SPTAE), 24-Jun-02 (PJH) (Hodge, 2002)
SU16: Woodborough, 4-Aug-83 (DRN)
Resident. Local; on *Geranium pratense* [and other species of *Geranium* in various habitats].

## Subfamily LIXINAE
### Tribe LIXINI

Larinus -planus

**LARINUS Germar, 1824**
***Larinus planus*** (Fabricius, 1792)
**Nb, LCGR-Fl Cm**
7 SU06: Cherhill Down, 1-Jun-04 (BLe)
SU16: Oare, 27-Aug-01 (IJG)
SU26: Froxfield, 1869-1938 (Meyrick, 1938)
SU36: Stype Wood, Jun-1870 (RH) (Hislop, 1871)
8 ST84: Longleat, 1-Jun-97 (MB); 25-May-05 (DJG)
ST92: Sutton Mandeville, 5-Jul-95 (MDD); Wardour, 5-Jun-96 (MAS)
SU03: Wylye Down NNR, 20-Sep-05 (MDD) (Darby, 2008)
SU05: Ell Barrow (SPTAC), 3-Jun - 12-Aug-08 (Boyce & Wright, 2009)
SU14: Haxton (SPTAE), 3-Jun - 12-Aug-08 (Boyce & Wright, 2009)
SU23: Porton Down, 1974-1978 (MGM) (Morris, 1979); 3-Jul-99 (PJH) (Hodge, 2002)
SU24: Driver Training Area (SPTAE), 24-Jun-02 (PJH) (Hodge, 2002); Dunch Hill (SPTAE), 3-Jun - 12-Aug-08 (Boyce & Wright, 2009)
Resident. Widespread; on thistles in grasslands and open places.

**LIXUS Fabricius, 1801**
**[*Lixus elongatus*** (Goeze)
7 SU06: Devizes, Roundway Hill, early Jun-1864 (JS)**]**
A single specimen said to have been beaten off oak or birch by Sidebotham in a plantation on the side of Roundway Hill (Rye, 1865) and exhibited at the 7 November 1864 meeting of the Entomological Society in London (as *L. filiformis*). The British list included the species on the basis of this record up to 1977 but it has now been excluded. (Mann, et al., 2005; Morris, 2005). The whereabouts of the specimen is not known. Common species on thistles in southern Europe. (See p.9).

### Tribe RHINOCYLLINI

**RHINOCYLLUS Germar, 1817**
***Rhinocyllus conicus*** (Frölich, 1792)
**Na, LCGR-Fm Cm**
7 ST86: Neston Park, 3-Jun-04 (DJG)
8 ST91: Cranborne Chase, 24-May-97 (MAS)
Resident. Very local; [on thistles in open and disturbed places, usually on the coast].

## Subfamily MESOPTILIINAE
### Tribe MAGDALIDINI

**MAGDALIS Germar, 1817**
EDO Germar, 1819
**Plum Weevil**
***Magdalis ruficornis*** (Linnaeus, 1758)
**SQS-2**
7 SU09: Somerford Keynes, Swillbrook Lakes, 4-Jun-04 (BLe)
SU16: Marlborough area, 1869-1938 (Morris, 1974, 1975)
SU26: Savernake Forest, 6-Jun-73 (APF)
Resident. Local; [on rosaceous trees and shrubs particularly hawthorns, and plums of which it is a minor pest.The larvae in dead twigs].

ODONTOMAGDALIS Barrios, 1984
***Magdalis armigera*** (Fourcroy, 1785)
**SQS-2**
7 SU06: Morgan's Hill, 1965-1974 (MGM) (Morris, 1974, 1975)
SU16: Marlborough area, 1869-1938 (Meyrick, 1938)
SU17: Fyfield Down, 1965-1974 (MGM) (Morris, 1974, 1975)

SU26: Savernake Forest, 29-May-73(CMJ)
8 SU23: Porton Down, 1974-1978 (MGM) (Morris, 1979)
Resident. Local; [on elms in woodlands and hedgerows. Larvae in galleries under bark].

***Magdalis carbonaria*** (Linnaeus, 1758)
**Nb, SQS-4**
7 SU26: Savernake Forest, 25-Apr to 11-May-95 (EN det MDD)
ST91: Cranborne Chase, 29-Apr-97 (MAS)
SU22: Langley Wood NNR, 16-Jul to 6-Aug-04, 15-Jul to 5-Aug-06 (MDD) (Darby, 2005, 2008)
Resident. Local; in birch. Bores into dead logs, large branches and trunks. Generally regarded as a northern species.

PANUS Schönherr, 1826

**Pear Weevil**
***Magdalis barbicornis*** (Latreille, 1804)
**Na, SQS-8**
8 SU04: Parsonage Down, 2-Jul-08 (MDD)
Resident. Very local; on hawthorn [and a wide range of other rosaceous trees and shrubs].

PORROTHUS Dejean, 1821

***Magdalis cerasi*** (Linnaeus, 1758)
**Nb, SQS-4**
7 ST87: Corsham Court Park, 3-Jun-04 (KNAA)
8 SU03: Grovely Wood, 11-Aug-71 (DRN)
SU22: Whiteparish Common, 1965-1974 (MGM) (Morris, 1974, 1975)
Resident. Very local; [mainly on oak and less frequently rosaceous trees and shrubs].

## Subfamily MOLYTINAE
### Tribe MOLYTINI
#### Subtribe MOLYTINA

Liparus coronatus

**LIPARUS Olivier, 1807**
***Liparus coronatus*** (Goeze, 1777)
**Nb**
7 SU36: Shalbourne, 1869-1938 (Meyrick, 1938)
8 SU02: Bishopstone, 17-Aug-89, 2-Jun-90 (MDD)
Resident. Very local; walking in lane through cultivated fields on several occasions. [Feeds on cow parsley, *Anthriscus sylvestris*, and possibly other Apiaceae

***?Liparus germanus*** (Linnaeus, 1758)
**RDB2**
8 ST95: West Lavington (Strawberry Hill), 6-May-45 (DCL) (DFL, 1948)
This is presumably a misidentification of *L.. coronatus*. *L. germanus* is known only from E.Sussex and Kent.

#### Subtribe LEIOSOMATINA

**LEIOSOMA Stephens, 1829**
***Leiosoma deflexum*** (Panzer, 1795)
7 SU17: Clout's Wood, 1965-1974 (MGM) (Morris, 1974, 1975)
SU26/27: Axford, 1869-1938 (Meyrick, 1938)
ST84: Longleat, 4-May-98 (MB)
8 ST92: Sutton Mandeville, 2-May-92 (MDD)
ST93: Great Ridge Wood, 10-May-00 (MDD); Stockton Wood, 1965-1974 (MGM) (Morris, 1974, 1975)
SU02: Bishopstone, 28-May-91 (MDD)
SU03: Grovely Wood, 16-Aug-71, 24-Aug-72 (DRN)
SU13: Lower Woodford, 2-Aug-82 (DRN); Salisbury, Avon Valley Reserve, Apr-00 (MAS) (Darby, 2001)
SU16: Jones's Mill, 21-Jun-97 (MDD)
SU18: Hallam, 6-Aug-83 (DRN)
SU21: Landford Bog, 2003 (APF) (Foster, 2003)
SU22: Bentley Wood, 31-Mar-78 (DRN); Blackmoor Copse, 4 to 18-May-02 (APF) (Foster, 2004)
Resident. Widespread; often in damp areas, [on *Anemone nemorosa*, species of *Ranunculus* and marsh marigold, *Caltha palustris*].

***Leiosoma oblongulum*** Boheman, 1842
**Nb**
8 SU22: Langley Wood NNR, 24-Mar to 25-Apr, 6-Jun to 2-Jul-07 (MDD) (Darby, 2008)
Resident. Very local; in pitfall traps [on *Anemone nemorosa*, species of *Ranunculus* and marsh marigold,*Caltha palustris*].

### Tribe ANOPLINI

**ANOPLUS Germar, 1820**
***Anoplus plantaris*** (Naezen, 1794)
7 SU16: Marlborough area, 1869-1938 (Meyrick,

1938)
8 SU22: Blackmoor Copse, 1965-1974 (MGM) (Morris, 1974, 1975); Glazier's Copse, 16-Aug-88 (DRN); Landford, 12-Aug-87 (DRN); Whiteparish Common, 1965-1974 (MGM) (Morris, 1974, 1975)
SU23: Porton Down, 1965-1974 (MGM) (Morris, 1974, 1975)
Resident. Local; on sallows, alder, hazel [and birches] which are the foodplant, the larvae making linear mines in the leaves.

***Anoplus roboris*** Suffrian, 1840
**Nb**
8 SU22: Glazier's Copse, 16-Aug-88 (DRN); Homan's Copse, 29-May-90 (DRN) Landford, 12-Aug-87 (DRN); Langley Wood NNR, 28-May-74 (DRN)
Resident. Very local; on alder, sallow and hazel [in wet woodland]. Feeds on alder, *Alnus glutinosa*, the larvae making linear mines in the leaves].

Tribe HYLOBIINI

Hylobius abietis

**HYLOBIUS Germar, 1817**
***Hylobius abietis*** (Linnaeus, 1758)
**SQS-1**
7 SU16: Marlborough area, 1869-1938 (Meyrick, 1938)
8 ST85: Picket Wood, 14-May-82 (EGS det MDD)
ST91: Cranborne Chase, 14-Jun-97 (MAS)
ST95: West Lavington, 1938 (PLP) (DFL, 1948)
SU03: Grovely Wood, 4-Aug-80 (DRN); 21-Aug-01 (MDD)
SU21: Hamptworth Estate, 26-Jul-77, 30-Jul-85 (DRN)
SU22: Bentley Wood, 23-May-04 (AGD); Langley Wood NNR, 27-Aug-04 (MDD) (Darby, 2005)
SU23: Porton Down, 1990 (PTW det MDD)
Resident. Widespread; in coniferous woodland on pines, particularly *Pinus sylvestris* and *P. contorta*. A troublesome pest particularly of regeneration forestry.

Tribe PISSODINI

**PISSODES Germar, 1817**
**Small banded Pine Weevil**
***Pissodes castaneus*** (De Geer, 1775)
**SQS-2**
8 SU14: Haxton (SPTAE), 17-May-02 (PJH) (Hodge, 2002)
Resident. Very local; in pine woods where it is an important pest particularly of young trees. Larvae in the dead and dying wood.

## Subfamily OROBITIDINAE

**OROBITIS Germar, 1817**
**Violet Weevil**
***Orobitis cyaneus*** (Linnaeus, 1758)
7 SU16: Marlborough area, 1869-1938 (Meyrick, 1938)
8 ST91: Cranborne Chase, 16 to 25-Jun-80 (NCC)
SU02: Knighton Wood, 6-Jul-07, Jul-08 (AJWA)
SU03: Grovely Wood, 27-May-74 (DRN)
SU22: Langley Wood NNR, 14-May-04 (MDD) (Darby, 2005)
Resident. Local; on heath violet, *Viola canina*, often in moss, and on grass under trees.

## Subfamily SCOLYTINAE
Tribe SCOLYTINI
Subtribe SCOLYTINA

**SCOLYTUS Geoffroy, 1762**
***Scolytus intricatus*** (Ratzeburg, 1837)
**SQS-2**
7 ST86: Neston Park, 3-Jun-04 (KNAA)
ST87: Corsham Court Park, 3-Jun-04 (BLe, KNAA)
SU16: Marlborough area, 1869-1938 (Meyrick, 1938)
SU26: Savernake Forest, 15-Jun-73 (SAW); 25-Apr to 26-May-93 (EN det SAW); 29-May-76 (CMJ); Jan-93 (APF det KNAA)
8 ST93: Great Ridge Wood, 18-Apr-74 (DRN)
SU02: Gardiner Forest, 1-Sep-72 (DRN)
SU16: Jones's Mill, 19-Jun-01 (APF) (Foster, 2002)
SU21: Burnt Ground Wood, 19-Apr, 31-May-74 (DRN)
SU22: Blackmoor Copse, 17-Mar-02 (APF) (Foster, 2004)
Resident. Widespread; under bark of oaks.

***Scolytus multistriatus*** (Marsham, 1802)
**SQS-1**
8 SU03: Grovely Wood, 4-Aug-78 (DRN)
SU13: Middle Woodford, 27-Jul-77 (DRN)
SU22: Blackmoor Copse, 9-Jun-01 (APF) (Foster, 2004)
Resident. Very local; on hawthorn [under bark of elm].

***Scolytus rugulosus*** (Müller, P.W.J, 1818)
**SQS-2**
8 SU04: Parsonage Down, 22-Jul-08 (MDD)
SU13: Middle Woodford, 27-Jul-77 (DRN)
Resident. Very local; [on or under bark of rosaceous trees].

**Elm Bark Beetle**
***Scolytus scolytus*** (Fabricius, 1775)
**SQS-2**
7 SU16: Marlborough area, 1869-1938 (Meyrick, 1938)
8 SU23: Porton Down, 1970-2001 (Anon, 2001)
Resident. Very local; under bark of elm.

Scolytus scolytus

Subtribe CORTHYLINA

**PITYOPHTHORUS Eichhoff, 1864**
***Pityophthorus pubescens*** (Marsham, 1802)
**SQS-2**
8 SU21: Hamptworth Common, 28-Jul-77 (DRN)
SU22: Homan's Copse, 19-Aug-88 (DRN)
Resident. Very local; sweeping heather and beating alder [in thin twigs of coniferous trees].

**ERNOPERUS Thomson, C.G, 1859**
***Ernoporus fagi*** (Fabricius, 1798)
**Na, IEC-3, SQS-8**
8 SU03: Grovely Wood, 3-Apr-75 (DRN)
SU21: Peaked Wood, 1-Apr-75 (DRN)
Resident. Very local; in beech

Subtribe DRYOCOETINA

**DRYOCOETES Eichhoff, 1864**
***Dryocoetes autographus*** (Ratzeburg, 1837)
**SQS-2**
8 ST92: Wardour, 13-Nov-95 (MDD)
Resident. Very local; under bark of conifers.

***Dryocoetes villosus*** (Fabricius, 1792)
**SQS-2**
7 ST87: Corsham Court Park, 3-Jun-04 (KNAA)
SU16: Marlborough area, 1869-1938 (Meyrick, 1938)
SU26: Savernake Forest, 16-Jun to 7-Jul-93, 9-Jun to 1-Jul-94 (EN det SAW)
8 ST84: Longleat, 25-May-05 (DJG)
ST93: Great Ridge Wood, 19-Apr-84 (DRN)
SU02: Gardiner Forest, 1-Sep-72 (DRN); Knighton Wood, 6-Jul, Sep-07 (AJWA)
SU03: Grovely Wood, 18-Jul-73 (DRN)
SU22: Langley Wood NNR, 25-Jun-04 (MDD) (Darby, 2005)
SU26: Bedwyn Brail, 2-Jun-03 (MDD)
Resident. Widespread; in oak bark [and other deciduous trees].

**TAPHRORYCHUS Eichhoff, 1878**
***Taphrorychus bicolor*** (Herbst, 1793)
**Na, SQS-8**
8 SU13: Little Durnford Down, 13-Sep-99 (MDD); Old Sarum, 31-May-97 (MAS)
Resident. Very local;

**XYLOCLEPTES Ferrari, 1867**
***Xylocleptes bispinus*** (Duftschmid, 1825)
8 SU13: Salisbury, 26-May-74 (DRN)
Resident. Very local: swept near river Avon.

Subtribe IPINA

**IPS De Geer, 1775**
***Ips sexdentatus*** (Boerner, 1767)
8 SU21: Cloven Hill, 24-Apr-85 (DRN)
Resident. Very local; in bark of cut pine.

**ORTHOTOMICUS Ferrari, 1867**
***Orthotomicus laricis*** (Fabricius, 1792)
7 SU16: Marlborough area, 1869-1938 (Morris, 1974, 1975)
8 SU21: Burnt Ground Wood, 31-Oct-73 (DRN); Hamptworth Estate, 31-Oct-75 (DRN)
Resident. Very local; in pine bark.

***Orthotomicus suturalis*** (Gyllenhal, 1827)
**SQS-2**
8 SU21: Burnt Ground Wood, 31-Oct-75 (DRN); Hamptworth Estate, 1-Apr-75 (DRN)
Resident. Very local; in pine logs and bark.

**PITYOGENES Bedel, 1888**
***Pityogenes bidentatus*** (Herbst, 1783)
**SQS-2**
8 SU03: Grovely Wood, 28-Jul-83 (DRN)
SU21: Hamptworth Estate, 3-Apr-78 (DRN)
Resident. Very local; in pine bark

Subtribe XYLOTERINA

**TRYPODENDRON Stephens, 1830**
***Trypodendron domesticum*** (Linnaeus, 1758)
**IEC-3, SQS-2**
7 SU16: Marlborough area, 1869-1938 (Meyrick, 1938)
8 ST82: Oysters Coppice, 6-Oct-02 (MDD)
ST93: Great Ridge Wood, 13-Aug-80 (DRN)
SU03: Grovely Wood, 10-Apr-70, 9-Aug-71, 23-Jul-73, 16-Apr-84, 1-Aug-88 (DRN)
Resident. Local; under bark and boring in birch and beech.

***Trypodendron lineatum*** (Olivier, 1795)
8 SU03: Grovely Wood, 30-Aug-88 (DRN)
Resident, Very local; boring in beech, etc.

***Trypodendron signatum*** (Fabricius, 1792)
**Nb, IEC-3, SQS-8**
7 ST96: Spye Park, 24-Apr-87 (KNAA) (Alexander, 1988)
SU16: West Woods, May-028 (MDD)
8 ST93: Great Ridge Wood, 19-Apr-84 (DRN)

Tribe HYLESININI
Subtribe HYLESININA

**HYLESINUS Fabricius, 1801**
***Hylesinus crenatus*** (Fabricius, 1787)
**SQS-2**
7 SU16: Marlborough area, 1869-1938 (Meyrick, 1938)
SU26: Savernake Forest. NE, 22-Aug-90 (AGD)
8 SU02: Gardiner Forest, 1-Sep-72 (DRN)
Resident. Very local; under bark [of ash].

***Hylesinus orni*** (Fuchs, 1906)
**Nb, SQS-8**
8 SU03: Grovely Wood, 16-Apr-84 (DRN)
Resident. Very local: in moss on oak(?) tree

***Hylesinus toranio*** (Danthoine in Bernard, 1788)
8 SU12: Whaddon Common, 22-Jul-73 (DRN)
Resident. Very local; in cut ash bark.

**Ash Bark Beetle**
***Hylesinus varius*** (Fabricius, 1775)
**SQS-1**
7 SU26: Savernake Forest, 30-Aug to 20-Sep-94 (EN det SAW)
8 ST91: Cranborne Chase, 22-Jun-99 (MDD)
ST93: Great Ridge Wood, 29-Oct-75 (DRN)
SU02: Knighton Wood, Apr-07 (AJWA)
SU13: Salisbury, Avon Valley Reserve, Apr-00 (MAS) (Darby, 2001)
SU21: Hamptworth Common, 28-Jul-77 (DRN); Hamptworth Estate, 28-Jul-77 (DRN)
SU22: Blackmoor Copse, 3-Aug-03 (APF) (Foster, 2004)
SU25: Dod's Down, 1869-1938 (Meyrick, 1938)
Resident. Widespread; in ash and sallow.

**HYLASTINUS Bedel, 1888**
***Hylastinus obscurus*** (Marsham, 1802)
7 SU16: Marlborough area, 1869-1938 (Meyrick, 1938)
Resident. No specimens in MCC.

**PTELEOBIUS Bedel, 1888**
***Pteleobius vittatus*** (Fabricius, 1787)
**SQS-2**
8 ST92: Sutton Mandeville, 20-Apr-92 (MDD)
Resident. Very local: under bark of ash.

Subtribe HYLASTINA

**HYLASTES Erichson, 1836**
***Hylastes angustatus*** (Herbst, 1793)
8 ST92: Wardour, 13-Nov-95 (MDD)
Resident. Very local: under bark of pine.

***Hylastes ater*** (Fabricius, 1792)
**SQS-1**
8 ST92: Sutton Mandeville, 21-Mar-92 (MDD)
SU03: Grovely Wood, 8-Aug-80 (DRN)
SU21: Burnt Ground Wood, 31-Oct-75 (DRN); Hamptworth Common, 30-Mar-03 (MDD); Pound Bottom, 4-Aug-82 (DRN)
SU22: Harestock, 13-Apr-72 (DRN); Langley Wood, 14-May-04 (MDD) (Darby, 2005)
Resident. Local; in coniferous woodland.

***Hylastes attenuatus*** Erichson, 1836
8 SU03: Grovely Wood, 4-Aug-80 (DRN) 2005)
Resident. Very local: on sappy spruce.

***Hylastes brunneus*** Erichson, 1836
**SQS-2**
8 SU21: Hamptworth Estate, 3-Aug-73 (DRN)
SU22: Langley Wood NNR, 14-May to 3-Jun-04 (MDD) (Darby, 2005)
Resident. Very local; in pine bark.

***Hylastes opacus*** Erichson, 1836
**SQS-2**
8 SU21: Burnt Ground Wood, 31-Oct-76 (DRN); Hamptworth Estate, 25-Apr-84 (DRN); Pound Bottom, 4-Aug-82 (DRN)
SU22: Langley Wood NNR, 3-Jun-04 (MDD) (Darby, 2005)
Resident. Very local; in pine bark.

**HYLURGOPS LeConte, 1876**
***Hylurgops palliatus*** (Gyllenhal, 1813)SQS-1
8 ST92: Wardour, 3-Apr-96 (EN)

SU03: Grovely Wood, 26-Jul-82 (DRN)
SU21: Burnt Ground Wood, 21-Jul-73 (DRN); Cloven Hill, 25-Apr-84 (DRN); Hamptworth Common, 30-Mar-03 (MDD); Hamptworth Estate, 21-Jul-73, 31-Oct-75, 25-Apr-84 (DRN)
SU26: Bedwyn Common, 24-Jul-82 (DRN)
SU25: Tidcombe, 1869-1938 (Meyrick, 1938)
Resident. Local; in pine bark.

Subtribe PHLOEOTRIBINA

**PHLOEOTRIBUS Latreille, 1796**
***Phloeotribus rhododactylus*** (Marsham, 1802)
8 SU22: Holmere Common, 19-Aug-88 (DRN)
Resident. Very local; at rest on on sallow.

Subtribe TOMICINA

**TOMICUS Latreille, 1802**
**Pine-shoot beetle**
***Tomicus piniperda*** (Linnaeus, 1758)
**SQS-1**
7 SU16: Marlborough area, 1869-1938 (Meyrick, 1938)
8 SU03: Grovely Wood, 11-Aug-71 (DRN)
SU12: Whaddon Commmon, 22-Jul-73 (DRN)
SU21: Pound Bottom, 4-Aug-82 (DRN)
SU26: Bedwyn Common, 24-Jul-82 (DRN)
Resident. Local; on conifers and in conifer bark. One on hazel under pines.

## Family PLATYPODIDAE

Platypus cylindrus

**PLATYPUS Herbst, 1793**
**Oak Pinhole Borer**
***Platypus cylindrus*** (Fabricius, 1792)
**Nb, IEC-3, SQS-8**
7 SU26: Savernake Forest, 1970-2001 (Anon)
8 ST84: Longleat, 29-Aug-99 (MB)
SU03: Grovely Wood, 13 records 4-Aug-70 - 1-Aug-88 (DRN)
SU21: Burnt Ground Wood, 19-Apr-74 (DRN)
SU25: Ludgershall, Apr - Jul-1915 (PAH) (Harwood, 1915)
Resident. Local; boring in beech trunks and bark.

# BIBLIOGRAPHY


ALEXANDER, K.N.A. 1979. A further record of *Leistus rufomarginatus* (Dufts.) (Col, Carabidae) from Wiltshire. *Ent. Mon. Mag.* **114**: 8

____1988. Some records of *Xyloterus signatum* (F.). *Ent. Gaz.* **39**: 32

____1992. *Phytoecia cylindrica* (L.) (Col. Cerambycidae) in the Gloucestershire/Wiltshire area and Buckinghamshire. *Ent. Mon. Mag.* **128**: 258

____1996. The dead-wood beetles of Stourhead, including *Thymalus limbatus* (Fabricius) (Peltidae) new to Wiltshire. *The Coleopterist* **4**: 90

____2003. *Provisional atlas of the Cantharoidea and Buprestoidea (Coleoptera) of Britain and Ireland.* Huntingdon: Biological Records Centre

____2004. *Hypnogyra angularis* Ganglbauer from Dinton Park in South Wiltshire. *The Coleopterist* **13**: 94

____2005. Email, Beetles (British Isles), 21.X.05

____2006. *Ochrosis ventralis* (Illiger): new Cornish records and a review of its ecology in Britain. *The Coleopterist* **15**: 127-129

ALLEN, A.A. 1971. *Procraerus tibialis* Lac. (Col., Elateridae) in Wilts. and Herts. *Ent. Mon. Mag.* **107**: 12

____1973. *Limobius borealis* Payk. (Col. Curculionidae) in Wilts and Hants. *Ent. Mon. Mag.* **109**: 61

ALLEN, A.J.W. 2008. 2007 Annual Exhibition, Coleoptera.

ANONYMOUS 1882. Entomological Section-Coleoptera. *Rep.Marlboro.Coll. nat. Hist. Soc.* **30** (1881): 131-35

____1938. Obituary of Edward Meyrick. *The Marlburian* 79-82

____1999. *Martin Down NNR Pitfall Trap Survey.* Unpublished report, English Nature

BAINBRIGGE FLETCHER, T. 1938. Obituary of Edward Meyrick. *Ent. Rec.J. Var.* **50** :48-51

BOOTH, R. 1988. *Entomologist* **107**: 127-133

____2000. Annual Exhibition, Imperial College, London, SW7 – 27 November 1999. *Br. J. Ent. Nat Hist.* 13: 173-174

____2001. Annual Exhibition, Imperial College, London, SW7 – 11 November 2000. *Br. J. Ent. Nat Hist.* 14: 161-162.

BOYCE, D.C. & WRIGHT, A.S. 2009. *An assessment of the condition of invertebrate assemblages on the Salisbury Plain SSSI.* Unpublished Report to Natural England, Devizes.

BRENTNALL, H.C. 1938. Savernake Forest in the Middle Ages. *Wilts. Arch. nat. Hist. Soc. Mag.* **55**: 371-386

BUCK, F.D. 1956. Wiltshire Coleoptera. *Ent. Mon. Mag.* **92**: 360

CARR, R. 1997. Aquatic Coleoptera of Salisbury Plain. *Latissimus Newsletter of the Balfour-Browne Club*, **9**: 7-9

COOLING, D.A. 1981. Records of Aquatic Coleoptera from Rivers in Southern England. *Ent Gaz.* **32**: 103-113

DANDY, J.E. 1969. *Watsonian Vice-counties of Great Britain.* Ray Society, London

DARBY, M.D. 1961 to present. *A Biographical Dictionary of British Coleopterists.* www.coleopterist.org.uk

____1996. Coleoptera Report, *Bentley Wood Nature Notes* p.47

____1997. *Leptura fulva* in Wiltshire. *Recording Wiltshire's Biodiversity* **1** :10-11

____1998. A remarkable aggregation of Ptiliid beetles from a grass pile at Burcombe. *Recording Wiltshire's Biodiversity* **3** :9-10

____1998a. Coleoptera Report, *Bentley Wood Nature Notes* 34-35

____1999. Wiltshire Buprestidae. *Recording Wiltshire's Biodivesity* **4** : 9-11

____2000. The Stag Beetle (*Lucanus cervus* L.) in Wiltshire. *Recording Wiltshire's Biodiversity* **5**: Dec-14

____2001. The Coleoptera (Beetles) of a Salisbury Reedbed including twenty two species new to Wiltshire.*Wiltshire Studies* **94**:154-160

____2001a. The Glow-worm (*Lampyris noctiluca* Linnaeus) in Wiltshire: Records and Questions, *Wiltshire Studies* **94**: 218-19

____2001b. A first attempt to apply the Saproxylic Quality Index for Coleoptera to Wiltshire. *Recording Wiltshire's Biodiversity* **6**: 7-10

____2002. A preliminary account of the ladybirds (Coleoptera: Coccinelidae) of Wiltshire including a previously overlooked record of the five spot (*Coccinella quinquepunctata* L.). *Wiltshire Studies* **95**: 125-130

____2002a. Records of click beetles and related species (Coleoptera: Elateroidea) in Wiltshire. *Recording Wiltshire's Biodivesity* **7**: 7-11

____2002c. A Brief History of Dauntsey's School Natural History Society (fl. 1934-1960). *Wiltshire Studies* **95**: 259-268

____2002d. Beetle Survey of Bemerton Folly, Salisbury. Unpublished report for Salisbury District Council/Wiltshire Wildlife Trust

____2003. The Glow-worm (*Lampyris noctiluca* Linnaeus 1758) in Wiltshire: an update. *Wiltshire Studies* **96**: 222-224

____2003a. Some preliminary thoughts on changes to the Wiltshire Coleopterous fauna. *Recording Wiltshire's Biodiversity* **8**: 17-23

____2003b. The Wiltshire Natural History Forum 1974-2002. *Wiltshire Studies* **96**: 89-97

____2004. 'Beetles' and 'Other Notable Species' in B. Heath, The Wiltshire Wildlife Trust's Vera Jeans Nature Reserve at Jones's Mill, Pewsey. *Wiltshire Studies* **97**: 255-72

____2004a. The Rugged Oil Beetle (*Meloe rugosus* Marsham) discovered in Wiltshire. *Wiltshire Studies* **97**: 298-99

____2005. Pitfall Trap Survey of Beetles, Langley Wood NNR, Wiltshire. Unpublished report for English Nature

____2005a. *Epiphanis cornutus* Escholtz, *Hylis olexai* (Palm) and *Microrhagus pygmaeus* (Fabricius) from pitfall traps in Wiltshire. *Coleopterist* **14**: 127-128

____2005b. A previously overlooked record of the five spot ladybird, *Coccinella quinquepunctata* , in Wiltshire. *Coleopterist* **14**: 128-29

____2008. *Pitfall Trap Survey of Beetles, Outwood area of Langley Wood NNR.* Unpublished report for Natural England

____2008a. *Wylye Down NNR Beetle Survey.* Unpublished Report for English Nature

____2009.The Beetles of Wylye Down National Nature Reserve. *Wiltshire Studies* **102**:1-11

DARBY, M. and ALEXANDER, K.N.A. 2003. The use of beetles in evaluating the saproxylic quality of Savernake Forest. *Wiltshire Studies* **96**: 137-142

DRAKE, C.M. 2008. The invertebrates of two water meadows on the River Avon, Wiltshire. *Br. J. Ent. Nat. Hist.* **21**: 215-224

DUFF, A.G. 1993. *Beetles of Somerset their status and distribution.* Somerset Archaeological and Natural History Society

____(ed) 2008. *Checklist of Beetles of the British Isles*

EARP, W.J. 1965. The Marlborough College Natural History Society 1864-1964. *Rep. Marlboro. Coll. Nat. Hist. Soc.* **104**: 20-33

FOSTER, A.P.2000. *Invertebrate Survey Concentrating on Saproxylic Fauna, Devenish Reserve, Litttle Durnford Down.* Unpublished report for the Wiltshire Wildlife Trust

____2001. *An Invertebrate Survey of Landford Bog.* Unpublished report for the Wiltshire Wildlife Trust

____2002. *Jones's Mill Reserve, a Report to the Wiltshire Wildlife Trust.* Unpublished

____2003. *An Invertebrate Survey of Landford Bog.* Unpublished report for the Wiltshire Wildlife Trust

____2004. *Blackmoor Copse Reserve. An Invertebrate Survey concentrating on Saproxylic Fauna.* Unpublished report for the Wiltshire Wildlife Trust

FOSTER, G.N. 2000. *The Scarce and Threatened Coleoptera of Great Britain, Part 3 Aquatic Coleoptera.* Joint Nature Conservation Committee

FOSTER, G.N., BILTON, D.T., ROUTLEDGE, S. & EYRE, M.D., 2008. The past and present statuses of *Hydroporus rufifrons* (Muller) (Dytiscidae) in Great Britain. *The Coleopterist* **17**: 51-63

FOWLER, W.W. 1887-1891. *The Coleoptera of the British Islands* **1-5**. London, L. Reeve and Co.

FOWLER, W.W. & DONISTHORPE, H. St.J. 1913. *The Coleoptera of the British Islands* **6** London, L.Reeve & Co.

FOWLES, A.P., ALEXANDER, K.N.A. & KEY, R.S. 1999. The Saproxylic Quality Index: evaluating wooded habitats for the conservation of dead-wood Coleoptera. *The Coleopterist* **8**: 121-141

GREENSLADE, P.J.M. 1963. New County records of Carabidae (Coleoptera). *Ent. Mon. Mag.* **99**:

HAMMOND, P.M. 1974. *Changes in the British Coleopterous fauna*, Systematics Association Special Volume 6, Academic Press

HAMMOND, P.M. & HARDING, P.T. 1991. Saproxylic Invertebrate assemblages in British Woodlands: their conservation significance and its evaluation. In H.J.Read, ed. *Pollard and Veteran Tree Management.* Corporation of London, 30-37

HARDING, P. & ROSE, F. 1986. *Pasture-woodlands in Lowland Britain: a review of their importance for wildlife conservation*

HARDING, P.T. and ALEXANDER, K.N.A, The Use of Saproxylic Invertebrates in the Selection and Evaluation of Areas of Relic Forest in Pasture Woodlands. *Br. J. Ent. Nat. Hist.* **7** (Supplement): 21-26, 1994),

HARWOOD, P.H. 1915. Coleoptera, etc. in the Salisbury district. *Ent. Mon. Mag.* **51**; 265-266.

HAWKINS, R.D. 2001. *Rhizobius chrysomeloides* (Herbst) (Coleoptera: Coccinellidae) new to Britain, *Br. J. Ent. Nat. Hist.* 13: 193-195

HEATH, B. 2004. The Wiltshire Wildlife Trust's Vera Jeans Nature Reserve at Jones's Mill.*Wiltshire Studies* **97**: 255-272.

HILL, A.W. 1939. Obituary of Edward Meyrick, *Obituary Notices of the Royal Society of London* **7**: 531-548

HISLOP, R. 1871. Notes on Wiltshire Coleoptera. *Ent. Mon. Mag.* **72**: 29-30

HODGE, P.J. 1995. *An Entomological Survey for the Proposed 'A303 Improvement' between Wylye and Stockton Wood.* Unpublished report for the Department of Transport. pp.1-13

____1999. *Salisbury Plain Insect Records. Tilshead Park & Haxton Down on 14 th July 1999.* Unpublished Report for the Ministry of Defence. pp.1-4

____2002. *Salisbury Plain Coleoptera Data.* Unpublished summary of survey data

____2003. *A Survey of Insects at Stonehenge. Notes on RDB and Nationally scarce species recorded from pitfall taps and during fieldwork carried out during summer 2003.* Unpublished report.

HODGE, P.J. & JONES, R. A. 1995. *New British Beetles. Species not in Joy's practical handbook.* Reading, British Entomological and Natural History Society

HOLMES, J.R.S. AND TENNANT, P.L. 1947. Insect List. *Rep. Marlb. Coll. Nat. Hist. Soc.* **95**: 20- 21

HURST, C.P. 1927. Natural History Notes round Great Bedwyn. *Wilts. Arch. nat. Hist. Soc. Mag.* **44** : 23-29

HYMAN, P.S. & PARSONS, M. 1992, 1994. *A Review of the Scarce and Threatened Coleoptera of Great Britain*, 2 parts, Joint Nature Conservation Committee

ISR (Sav), 1994. Printout from the Invertebrate Site Register, 20 Sep. Joint Nature Conservation Committee

JEBB, A.G. 1895. Coleoptera of the Marlborough District. *Rep. Marlboro. Coll. Nat. Hist. Soc.* **44**: 55-74

JOHNSON, C. 1977. *Provisional Atlas of the Cryptophagidae-Atomariinae (Coleoptera) of Britain and Ireland,* edited for the Biological Records Centre by P.T.Harding and J.C.M.Dring. Huntingdon: Biological Records Centre

JONES, R. A. 1977. *Mononychus punctumalbum* (Herbst) (Col, Curculionidae) in Wiltshire. *Ent. Mon. Mag.* **113** : 246

JOY, N. H. 1925. Recent Captures of Coleoptera. *Ent. Mon. Mag.* **61**: 16

____1932. *A Practical Handbook of British Beetles.* vol 1 text, vol 2 plates. (Reprinted 1976)

KEEN, D. 2006. Lily Beetle in Wiltshire, *Bulletin of the Amateur Entomologists Society* **65** (466)

KIRK-SPRIGGS, A.H. 1994. *Pollen Beetles*, Handbooks for the Identification of British Insects **5 (6a),** Royal Entomological Society

KNIGHT, J.D.S. 1940. Coleoptera. *Rep. Marlboro. Coll. Nat. Hist. Soc.* **88** : 23

KRELL, F-T, JOHNSON, C., BOOTH, R. & MENDEL, H. 2005. The British *Dienerella separanda* (Reitter) is *D. clathrata* (Mannerheim): with a compilation of British records of *D. clathrata* and *D. elongata* (Curtis) (Latridiidae). *Coleopterist* **14** : 117-123

LUFF, M.L. 1998. *Provisional atlas of the ground beetles (Coleoptera, Carabidae of Britain.* Huntingdon: Biological Records Centre

____2007. *The Carabidae (ground beetles) of Britain and Ireland* (2nd edition). Handbooks for the Identification of British Insects **4 (2)**, Royal Entomological Society

MACKECHNIE-JARVIS, C. 1976. Report of a field meeting held in Savernake Forest. *Proc. Brit. Ent. Nat. Hist. Soc.* **9**: 122

MANN, D.J., HANCOCK, E.G. & MORRIS, M.G. 2005. History of the genus *Lixus* Fabricius (Curculionidae) in Britain with comments on Nomenclature. *The Coleopterist* **14**: 65-80

MCSWEENEY, P. 1991. *Botanical Survey and Assessment of Chalk Grassland in Wiltshire outside of Ministry of Defence Areas 1989-90.* Nature Conservancy Council

MENDEL, H. & CLARKE, R.E. 1996. *Provisional Atlas of the Click Beetles (Coleoptera: Elateroidea) of Britain and Ireland.* Ipswich Borough Council Museums

MEYRICK, E. 1874. Entomological Report. *Rep. Marlboro. Coll. Nat. Hist. Soc.* **9** : 41-49

____1880. Coleoptera. *Rep. Marlboro. Coll. Nat. Hist. Soc.* **29** : 131-135

____1895. Obituary of Arthur Gamul Jebb. *Rep. Marlboro. Coll. Nat. Hist. Soc.* **44** : 94

____1902. Entomological Section, *Marlboro. Coll. Nat. Hist. Soc.* **50** : 69

____1902. Entomological Section – Coleoptera. *Rep. Marlboro. Coll. Nat. Hist. Soc.* **50** : 69

____1906. List of Coleoptera of the District. *Rep. Marlboro. Coll. Nat. Hist. Soc.* **54**: 76-96

____1938. Handlist of the Coleoptera of the Marlborough District. *Rep. Marlboro. Coll. Nat. Hist. Soc.* **87**: 54-86

MOORE, B.P. 1956.Carabidae and other Coleoptera in Wiltshire. *Ent. Mon. Mag,* **92**: 134

MORRIS, M.G. 1968.*Ceuthorhynchus unguicularis* Thomson (Col. Curculionidae) in Wiltshire and Suffolk. *Ent. Mon. Mag.* **104**: 45

____1974 .A Preliminary Account of the Weevils of Wiltshire (Coleoptera, Curculionoidea), *Wilts. Archaeol. nat. Hist. Mag.* 68-6919-24, 30-38

____1974a. Wiltshire Coleoptera, with particular reference to *Limobius borealis* (Payk.) (Curculionidae), and a record from Gloucestershire. *Ent. Mon. Mag.* **110**: 222

____1978. *Polydrusus sericeus* (Schaller) (Col:Curculionidae) in South Wiltshire, *Ent. Rec. J. Var.* **90** : 22-23

____1979. Wiltshire Weevils (Coleoptera, Curculionidea) Additions and Corrections. *Wilts. Archaeol. nat. Hist. Mag.* **74**: 10-12

____1990. *Orthocerous Weevils.* Handbooks for the Identification of British Insects **5 (16)**, Royal Entomological Society

____1997, *Broad-Nosed Weevils.* Handbooks for the Identification of British Insects **5 (17a)**, Royal Entomological Society

____ 2002, *True Weevils (Part 1).* Handbooks for the Identification of British Insects **5 (17b)**, Royal Entomological Society

____ 2008, *True Weevils (Part 2).* Handbooks for the Identification of British Insects, **5 (6a)**, Royal Entomological Society

MORRIS, M.G. & JOHNSON, C. 2005. Sidebotham's Weevils (Curculionoidea). *The Coleopterist* **14**: 101-113

MURRAY, D.H. 1978. Some notable Coleoptera taken in Southern Hampshire and Savernake Forest. *Ent. Mon. Mag,* **113**: 104

NASH, D. 1975. *Silpha carinata* Herbst - a remarkable rediscovery in the British Coleoptera. *Ent. Rec. J. Var.* **87**: 285-288

____1975a. *Carpophilus marginellus* Motsch. (Col, Nitidulidae) out-of-doors in Wiltshire. *Ent. Mon. Mag.* **111**: 50

____1975b. *Silvanus bidentatus* (F.) (Col, Silvanidae) in Wiltshire. *Ent. Mon. Mag.* **110**: 20

____1976. *Nossidium pilosellum* (Marsh.) (Col. Ptiliidae) in Wilts. *Ent. Mon. Mag.* **112**: 256

____1977.*Carpophilus sexpustulatus* F. out-of-doors in Wilts. and Suffolk. *Ent. Mon. Mag.* **113**: 84

____1977a. *Laccobius atratus* (Rottenburg) in Wilts. in *Balfour-Browne Club Newsletter* **4** : 10

____1977b. *Silpha carinata* Herbst (Col: Silphidae) confirmed as a British breeding species. *Ent. Rec. J. Var.* **89**: 305-308.

____1978 Five rare Carabidae apparently new to Wiltshire. *Ent. Mon. Mag.* **114**: 81

____1978a. *Agabus brunneus* (F.) in Wilts *Ent. Mon. Mag.* **114**: 30

____1979. *Agabus brunneus* (F.) (Col, Dytiscidae) in Wiltshire. *Ent. Mon. Mag.* **114**: 30

____1979a. *Saperda populnea* (Col: Cerambycidae) and its Tachinid parasite in Wilshire. *Ent. Rec. J. Var.* **91**: 286-287

____1979b. Observations upon the occurrence of *Carabus monilis* and *Carabus nemoralis* in Suffolk, N.E.Essex, and Wilts. *Ent. Rec.* J. Var, **91**: 138

____1979c. *Selatosomus bipustulatus* L. in Wilts and Suffolk. *Ent. Rec. J. Var.* **91**: 135

____1979d. *Psylliodes weberi* Lohse (Col: Chrysomelidae) in Wiltshire. *Ent. Rec. J. Var.* **91**: 288

____1979e. *Cassida murraea* (Col. Chrysomelidae) abundant in Wilts. *Ent. Rec. J. Var.* **91** : 284

____1981. *Atomaria consanguinea* Johnson new to Wilts. *Ent. Mon. Mag.* **117**: 96

____1981b. *Hyperaspis pseudopustulata* Muls. (Col: Coccinellidae) in Wilts. *Ent. Mon. Mag.* **117**: 29

____1982. *Coccinella distincta* Fald. (Col: Coccinellidae) in Wilts. **117** : 214

____1982a *Hylastes brunneus* Er. (Col: Scolytidae) in Wilts. *Ent. Mon. Mag.* **118** : 256

____1982b. *Epierus comptus* (Erichson) (Col: Histeridae) new to Britain, *Ent Rec.J. Var.* **94** : 165-167

____1989. Notes on British *Orthotomicus* (Col: Scoltidae) including *O.suturalis* (Gyllenhal) new to Wiltshire, *Ent. Rec. J. Var.* **101**: 151-152

____1999. *Epierus comptus* (Erichson) (Histeridae) in profusion in Wiltshire, *Coleopterist* **8**: 88

____2004. A belated record of *Tachinus rufipennis* Gyllenhal from N.Wilts. *Coleopterist* **13**: 119

____2004a. *Lycoperdina bovistae* (Fabricius) (Endomychidae) in East Suffolk and South Wiltshire. *The Coleopterist* **13**: 146-147

PEARCE, E.J. 1934. A contribution towards a list of the Aquatic Coleoptera of North Wiltshire. *Ent. Mon. Mag.* **70**: 81-87

POWER, J.A. 1865. On some new and rare species of Coleoptera. *Ent. Mon. Mag.* **1**: 235-236

ROSS, A.J. and JARZEMBOWSKI, E.A. 1996. A Provisional Checklist of Fossil Insects from the Purbeck Group of Wiltshire, *Wiltshire Archaeological and Natural History Society Magazine* **89**: 106-115

RYE, E.C. 1866. Coleoptera: New British Species, Corrections of Nomenclature, etc. noticed since the Publication of the Entomologist's Annual, 1865. *Entomolgist's Annual* 47-121

____1866. Doings in Wiltshire. *Entomologists Weekly Intelligencer* **2**: *138-139*

SHARP, D. 1871. Notes on Dr Sharp's Catalogue of British Coleoptera, *Ent. Mon. Mag.* **8**: 83-84

SHAW, M.W. 1961.The Golden Buprestid *Buprestis aurulenta* L. (Col. Buprestidae) in Britain, *Ent. Mon. Mag.* **97**: 97-98

SHEPPARD, D.A. 1989. *A Preliminary Account of the Insect Fauna of Langley Wood NNR.* Unpublished typescript.

SIDEBOTHAM, I. 1868. Localities for *Mesites tardii. Ent. Mon. Mag.* **5**: 99

SMITH, D.N.1999. Analysis of beetles in smoke-blackened thatch, in J.B.Letts, *Smoke Blackened Thatch a unique source of late mediaeval plant remains from South England*', English Heritage and Reading University. pp.42-43

STEPHENS, J.F. 1828-1830. *Illustrations of British Entomology. Mandibulata* **1-4**. London, Baldwin and Cradock

____1839. *A Manual of British Coleoptera.* London: Longman, Orme Brown, Green and Longmans

STOKES, H.G. *Agrilus sinuatus* Oliv. (Col., Buprestidae) in Salisbury, Wiltshire, *Ent. Mon. Mag.* **88**: 62

TELFER, M. 2005. email, Beetles (British Isles) web site, 14 May 2005

THOMPSON, R.T. 1958. *Coleoptera, Phalacridae,* Handbooks for the Identification of British Insects **5 (5b)**, Royal Entomological Society

TWINN, P.F.G. & HARDING, P.T. 1999. *Provisional atlas of the longhorn beetles (Coleoptera, Cerambycidae) of Britain.* Huntingdon: Biological Records Centre.

UHTHOFF-KAUFMANN, R. 1995. The status of *Strangalia attenuata (*L.) (Col: Cerambycidae) in Britain. *Ent. Rec. J. Var.* **107**: 69-73

WELCH, R.C. 1979. Aphodius (Heptaulaucus villosus Gyll (Col., Scarabaeidae) in Angus. *Ent. Mon. Mag.* **115**: 209-210.

____ 1997. The British species of the genus *Aleochara* Gravenhorst (Staphylinidae). *The Coleopterist* **6**: 1-48

WILLIAMSON, K. & COWIN, W.S. 1941. Some Wiltshire Insects, *Entomologist* **74** : 174-76

# INDEX OF SCIENTIFIC NAMES

# INDEX OF ENGLISH NAMES